Bioreaction Engineering

Springer

Berlin
Heidelberg
New York
Barcelona
Hong Kong
London
Milan
Paris
Singapore
Tokyo

K. Schügerl · K.-H. Bellgardt (Eds.)

Bioreaction Engineering

Modeling and Control

With 225 Figures and 70 Tables

 Springer

Chemistry Library

Prof. em. Dr. Dr. h. c. Karl Schügerl
e-mail: schuegerl@mbox.iftc.uni-hannover.de

Prof. Dr.-Ing. Karl-Heinz Bellgardt
e-mail: bellgardt@mbox.iftc.uni-hannover.de

University of Hannover
Institute of Technical Chemistry
Callinstrasse 3
D-30167 Hannover
Germany

ISBN 3-540-66906-X Springer-Verlag Berlin Heidelberg New York

Library of Congress Cataloging-in-Publication Data

Bioreaction engineering : modeling and control / K. Schügerl, K.-H. Bellgardt (eds.)
    p. cm.
    Includes bibliographical references and index.
    ISBN 354066906X (alk. paper)
    1. Bioreactors. I. Schügerl, K. (Karl) II. Bellgardt, K.-H.
    TP 248.25.B55 B549 2000
    660'.2832–dc21   00-026598

Springer-Verlag Berlin Heidelberg New York
a member of BertelsmannSpringer Science+Business Media GmbH

© Springer-Verlag Berlin Heidelberg 2000
Printed in Germany

Cover Design: design & production, Heidelberg
Typesetting: MEDIO GmbH, Berlin

Printed on acid-free paper   SPIN 10665072   52/3020 - 5 4 3 2 1 0

Preface

The book is intended to present various examples for reactor and process modeling and control as well as for metabolic flux analysis and metabolic design at an advanced level.

In Part A, *General principles and techniques* with regard to reactor and process models, process control, and metabolic flux analysis are presented. In addition the accuracy, precision, and reliability of the measured data are discussed which are extremely important for process modeling and control. A virtual bioreactor system is presented as well, which can be used for the training of students and operators of industrial plants and for the development of advanced automation tools.

In Part B, the *General principles are applied for particular bioreactor models*. It covers the application of the computational fluiddynamic (CFD) technique to stirred tank and bubble column bioreactors. Different solution methods are presented: the Reynolds-averaging of the turbulent Navier-Stokes equations and modeling of the Reynolds stresses with an appropriate turbulence (k-ee) model, and the Euler (two fluid model), as well as the Euler-Langrange approaches.

In Part C, *Application of general principles for process models including control* are discussed with regard to the production of baker's yeast, beer, lactic acid, recombinant protein, and β-lactam antibiotics. Various models and control strategies are used for baker's yeast production with regard to the optimization of productivity, product quality and process economy. Hybrid models are applied for the state prediction, optimization, and process supervision for beer production. Kinetic models are presented for the production of lactic acid which take into account the various metabolic paths of lactic acid bacteria. A kinetic model is developed for high density cultivation of *E. coli* and a control strategy is applied for the recombinant protein production. Various models are discussed, which take the fungal morphology into account and are used for optimization of the production of β-lactam antibiotics.

In Part D, examples are given for *Metabolite flux analysis and metabolic design*. They discuss the metabolic balancing, isotopic labeling combined with NMR spectroscopy methods, and discuss their application to the flux analysis in *Saccharomyces cerevisiae*, *Zymomonas mobilis*, *Corynebacterium glutamicum*, and mammalian cells.

July 2000   Karl Schügerl and Karl-Heinz Bellgardt

Contents

Part B Application of General Principles for Reactor Models

List of Abbrevations and Symbols

Chapter 1

A	cross section area of the flow, interfacial area, m^2
a	specific interfacial area, m^{-1}
ATL	airlift tower loop
BC	bubble column
Bo	Bodenstein number of axial dispersion, 1
C	concentration, $kg\,m^{-3}$
C_{Li}	molar concentration of i in the liquid phase, $kmole\,m^{-3}$
C_{LX}	cell mass concentration in the liquid phase, $kg\,m^{-3}$
CPR	CO_2 production rate, $kg\,m^{-3}\,s^{-1}$
CSTR	continuous stirred tank reactor
D	dilution rate, s^{-1}
D_i	liquid diffusivity of substance i, $m^2\,s^{-1}$
D_L	liquid phase axial dispersion coefficient, $m^2\,s^{-1}$
D_m	molecular diffusivity, $m^2\,s^{-1}$
D_s	diameter of the column, m
E_G	gas hold up, 1
E(t)	residence time distribution
F	flow rate $m^3\,s^{-1}$
f(t)	weighting function
g	acceleration of gravity, $m\,s^{-2}$
H	Henry constant, bar^{-1}
H_j	height of the section j (j=r, d, k)
iTR	transfer rate of substance i, $kg\,m^{-3}\,s^{-1}$
K	equilibrium constant
k	fluid consistency factor, $kg\,m^{-1}\,s^{-m}$
k_H	Henry constant, $m^2\,s^{-2}$
k_L	mass transfer coefficient, $m^2\,s^{-1}$
$k_L a$	volumetric mass transfer coefficient, s^{-1}
m	mass, kg
M	molecular mass, $kg\,kmole^{-1}$
n	flow behavior index
n	particle number

OTR	oxygen transfer rate, $kg\,m^{-3}\,s^{-1}$
OUR	oxygen uptake rate, $kg\,m^{-3}\,s^{-1}$
p	pressure, bar
p	pressure, Pa
Po	power input, $kg\,m^2\,s^{-3}$
P	power input, kW
pO_2	percentage of O_2 saturation
Q	gas throughput, $m^3\,s^{-1}$
Q	reaction rate, $kg\,m^{-3}\,s^{-1}$
q_i	specific mass fraction transport rate of i
R	gas constant, $8.314\,J\,mole^{-1}\,K^{-1}$
r	model constant
RQ	respiratory quotient
RTD	distribution of residence times
r_G	gas recirculation ratio
ST	stirred tank
s	Laplace transform variable
T	temperature, K
t	time, s
t_c	circulation time, s
u_B	bubble rise velocity, $m\,s^{-1}$
u_{BT}	bubble terminal velocity, $m\,s^{-1}$
u_c	liquid circulation velocity, $m\,s^{-1}$
u_G	superficial gas velocity, $m\,s^{-1}$
u_{GD}	fraction weighted drift velocity, ms^{-1}
u_s	slip velocity of bubbles, $m\,s^{-1}$
u_L	superficial liquid velocity, $m\,s^{-1}$
V	volume, m^{-3}
w	gas throughput, s^{-1}
x	mole fraction
X	cell concentration, $kg\,m^{-3}$
$Y_{X/O}$	yield coefficient of growth with respect to oxygen consumption
z	axial coordinate
ε	volume fraction, gas holdup
η	dynamic viscosity of the medium $kg\,m^{-1}\,s^{-1}$
ρ	density of the medium, $kg\,m^{-3}$
σ	surface tension, $N\,m^{-1}$
ν	kinematic viscosity $m^2\,s^{-1}$
τ	mean residence time, s

Subscripts

C	CO_2
D	dispersed gas phase
d	down comer
E	ethanol
e	entrance
G	gas phase
H	head space

i	substance index
j	stage index
k	head section
L	liquid phase
m	measured
M	molar
N	nitrogen
O	oxygen
P	product
r	riser
S	substrate
W	water
X	cell mass

Superscripts

I	reactor inlet
N	normal (standard) conditions
O	reactor exit
Q	reaction
R	reservoir
TR	transfer

Chapter 2

A	area, m^2
c	intrinsic concentration, -
$\mathbf{c}$	vector of intrinsic concentrations, -
C	concentration, $kg\,m^{-3}$
$\mathbf{C}$	vector of concentrations
D	dilution rate, s^{-1}
D_i	diffusion coefficient of substance i, $m^2\,s^{-1}$
E	activation energy, $J\,mole^{-1}$
f	function
F	fraction of cells
g	function
J	performance index of metabolic coordinator
k	constant
$k_L a$	volumetric mass transfer coefficient, s^{-1}
K	kinetic constant
L	thickness, m
m	specific maintenance coefficient, s^{-1}
$\mathbf{m}$	vector of specific maintenance coefficients
$\mathbf{M}$	matrix of molecular weights, $kg\,mole^{-1}$
OTR	oxygen transfer rate, $kg\,m^{-3}\,s^{-1}$
p	partitioning function
$\mathbf{p}$	parameter vector
pH	pH-value
q	specific reaction rate, s^{-1}

$\mathbf{q}$	vector of specific reaction rates
Q	volumetric reaction rate, $\mathrm{kg\,m^{-3}\,s^{-1}}$
$\mathbf{Q}$	vector of volumetric reaction rates
r	radius, m
r	intrinsic reaction rate, $\mathrm{s^{-1}}$
$\mathbf{r}$	vector of intrinsic reactions
$\tilde{r}$	normalized kinetic rate expression
R	gas constant, $8.314\,\mathrm{J\,mole^{-1}\,K^{-1}}$
s	fraction of sugars
t	time, s
T	temperature, K
$\mathbf{TR}$	vector of mass transfer rates
u	cybernetic variable
x	normalized variable
$\mathbf{x}$	vector of spatial coordinates
v	volume fraction
V	volume, $\mathrm{m^3}$
X	concentration related to the biotic phase, $\mathrm{kg\,m^{-3}}$
$\mathbf{X}$	vector of concentrations of the biotic phase
Y	yield coefficient
$\mathbf{Y}$	matrix of yield coefficients
z	spatial coordinate, m
$\mathbf{z}$	vector of cellular state variables
$\mathbf{Z}$	vector of concentrations of the abiotic phase
α	fraction of viable cells, or probability for plasmid loss
β	probability for mutation
η	transcription efficiency
ν	activity controlling variable
μ	specific growth rate, $\mathrm{s^{-1}}$
ρ	density, $\mathrm{kg\,m^{-3}}$
τ	cell age, s
ξ	translation efficiency

Subscripts

B	antibiotic
crit	critical
D	death or decay
E	endogenous or enzyme
G	growth or gene
i	component index
I	inhibitor
j	component index
L	liquid phase
min	minimum value
max	maximum value
M	metabolite
O	oxygen
P	product

S   substrate
V   volumetric
X   biomass, biotic phase
Z   abiotic phase
+   plasmid bearing
–   plasmid free

Superscripts

R   reservoir
T   transpose of a matrix

Chapter 3

MFA   Metabolic Flux Analysis
$\underline{S}$ Stoichiometric matrix of metabolic network
$\underline{v}$ vector of fluxes
$\underline{r}$ vector of metabolite extracellular accumulation rates
v_i flux of reaction i
r_i extracellular accumulation rate of metabolite i
LP   Linear Programming
TCA   Tricarboxylic Acid Cycle
$\underline{c}$ vector of weight factors of fluxes in the objective function of the linear
          programming problem
c_i weight factor of flux v_i in the objective function of the linear program-
          ming problem
p_i shadow price (dual variable) of stoichiometric constraint i
z^* optimal value of the objective function of the linear programming pro-
          blem
NMR   Nuclear Magnetic Resonance spectroscopy
ATP   Adenosine triphosphate
NADH   Nicotinamide adenine dinucleotide
GC [LC]-MS Gas (or Liquid) Chromatography-Mass Spectrometry
H4D   tetrahydrodipicolinate
DAP   diaminopimelate
PEP   phospho-enol-pyruvate
G6P   glucose-6-phosphate
CHO   Chinese Hamster Ovary
MCA   Metabolic Control Analysis
FCC   Flux Control Coefficient
$C_i^{J_j}$ Flux control coefficient of reaction i (as expressed through the activity
          of enzyme i) on the flux of reaction j
E_i activity of enzyme i of the network
J_j steady-state flux through reaction j (symbol used in MCA)
gFCC   Group Flux Control Coefficient

Chapter 5

A	state matrix
A_0	matrix of the state transformation
B	input matrix
D	dilution rate (h^{-1})
f	state mapping
F	feedrate vector
F_{in}	influent flow rate (1/h)
g	input mapping
g_t	(g_0) gain matrix of the RLS algorithm (initial value)
G	carbon dioxide
G_t	gain matrix of the RLS algorithm
k_{ij}	yield coefficient (I=1–9, j=1–4)
k_m	maintenance coefficient
K	yield coefficient matrix
K_I	inhibitor constant (g/l)
K_P	controller gain
K_S	Monod constant (g/l)
L	lactate
M	number of reactions
N	number of process component
OHPA	Obligate Hydrogen Producing Acetogens
P	product concentration (g/l)
P_{sat}	saturation outflow rate vector
PI	Proportional Integral
Q	gaseous outflow rate vector
r	reaction rate vector
RLS	recursive least squares
S	substrate concentration (g/l)
S^*	desired value of S (g/l)
t	time (h)
u	input vector
x	state vector
X	biomass concentration (g/l)
y	output variable
y^*	desired value of y
Y	yield coefficient
Z	auxiliary variable
γ	forgetting factor
γ_1, γ_2	gains of the observer-based estimator
ϵ	singular perturbation variable
λ	controller again
μ	specific growth rate (h^{-1})
μ_{max}	maximum specific growth rate (h^{-1})
Π	singular perturbation variable
τ_i	integral (or reset) time
ξ	process component concentration vector

Ω steady state coefficient

Subscripts

in   influent

Superspcripts

$-$ steady-state
$\wedge$ estimate

Chapter 6

A_{JK} heat transfer area between thermostat subsystems J and K [m^2]
A_V vessel cross-sectional area [m^2]
c_{IK} concentration of component I in subsystem K [g l^{-1}]
c_K specific heat capacity in thermostat subsystem K [W h K^{-1}g^{-1}]
C_{IK} molar concentration of component I in subsystem K [mol l^{-1}]
C_{kLa} fitted k$_L$a-constant [l N^{-2}h^{-1}]
C_K volumetric heat capacity in thermostat subsystem K [W h K^{-1}]
d_2 stirrer diameter [m]
D_K reciprocal mean residence time in thermostat subsystem K [h^{-1}]
F_K flow rate from (in) subsystem K [l h^{-1}]
F_{nG} aeration rate at normalized conditions [l h^{-1}]
F_{nI} aeration rate of gas component I at normalized conditions [l h^{-1}]
H_I Henry coefficient of component I [N m^{-2}l g^{-1}]
ITR   volumetric mass transfer rate of component I from gas phase [g l^{-1} h^{-1}]
k_{Iji} inhibition constant of substrate j by substrate i [g l^{-1}]
k_J Monod limitation constant of component J [g l^{-1}]
k_La volumetric O$_2$-transfer coefficient [h^{-1}]
K_{Alvol} evaporation constant of ammonia [-]
K_{Ji} dissociation constant i of component J [mol l^{-1}]
K_W ion product of water [mol^2 l^{-2}]
K_{HSt} proportional gain of stirrer heat generation [W min^3 l^{-1}]
K_{HM} proportional gain of microbial heat generation [W h g^{-1}]
m_J exchange mass in thermostat subsystem J [g]
m_{IK} mass of component I in subsystem K [g]
$\dot{m}$ mass flow [g h^{-1}]
$\dot{m}_{OT}$ oxygen mass transfer rate from gas phase [g h^{-1}]
M_I mole mass of component I [g mol^{-1}]
N_{St} stirrer agitation speed [min^{-1}]
Ne   Newton number [-]
OUR   oxygen uptake rate [g l^{-1} h^{-1}]
p_i valency of an anion [-]
p_K total pressure in subsystem K [N m^{-2}]
p_{IK} partial pressure of component I in subsystem K [N m^{-2}]
pH   pH-value [-]
P_H electrical heating power [W]
P_{St} stirrer power input [W]

q_j	valency of a cation [-]
$q_{I/X}$	cell specific reaction rate of component I [$g\,g^{-1}\,h^{-1}$]
Q_J	supply and desupply rate of component J [$g\,l^{-1}\,h^{-1}$]
$\dot{Q}_{St}$	thermal power of the stirrer [W]
$\dot{Q}_M$	thermal power of the microorganisms [W]
r_{IL}	volumetric reaction rate of component I [$g\,l^{-1}\,h^{-1}$]
R	universal gas constant [$N\,m\,K^{-1}\,kmol^{-1}$]
R_C	cooling water entry heat resistance [$K\,W^{-1}$]
R_{JK}	overall heat transmission resistance between thermostat subsystem J and K [$K\,W^{-1}$]
R_T	thermostat circulation heat resistance [$K\,W^{-1}$]
t	time [h]
T_K	absolute temperature of subsystem K [K]
u_G	superficial gas velocity [$m\,s^{-1}$]
V_K	volume of subsystem K [l]
V_{nM}	gas mole volume [$l\,mol^{-1}$]
x_{IAIR}	mole fraction of component I in AIR [-]
x_{IG}	mole fraction of component I in the gas phase [-]
$y_{X/I}$	cell mass yield coefficient of component I [$g\,g^{-1}$]
α	stirrer power input $k_L a$-correlation parameter [-]
α_J	heat transfer coefficient in fluid J [$W\,m^{-2}\,K^{-1}$]
β	aeration rate $k_L a$-correlation parameter [-]
γ	viscosity $k_L a$-correlation parameter [-]
δ_{JK}	wall thickness between thermostat subsystems J and K [m]
$\delta_{C/O}$	ratio of CO_2/O_2-transfer coefficients [-]
η_{eff}	effective viscosity [$N\,s\,m^{-2}$]
ϑ_K	temperature in subsystem K [°C]
κ_I	environmental growth control function of parameter I [-]
λ_{JK}	thermal conductivity of the wall between thermostat subsystems J and K [$W\,m^{-1}\,K^{-1}$]
	specific growth rate [$g\,g^{-1}\,h^{-1}$]
ρ_K	density of the medium in vessel K [$g\,l^{-1}$]
τ_{JK}	thermal transport time constant between thermostat subsystems J and K [h]

Indices

Ac	Acid
AIR	air
Al	alkali (base)
C	cooling water system
C, CO2	carbon dioxide
D	subsystem double jacket
eff	effective
E	reactor environment
gr	growth fraction
G	subsystem gas phase
H	heating, harvest
H^+	hydrogen ion
I	inhibition

in   input
I/J   component I per component J
L   subsystem liquid phase
m   maintenance fraction
max   maximum value
min   minimum value
M   microbial, molar
n   normalized gas conditions
N, N2   nitrogen
opt   optimal value
out   output
O, O2   oxygen
Pi   product i
R   subsystem feed tank (reservoir)
S   sample
Si   substrate i
St   stirrer
tot   total amount of a dissociable variable
T   thermostat
T1   subsystem titration tank 1 (acid)
T2   subsystem titration tank 2 (base)
Tc   thermostat cooling system
Th   thermostat heating system
vol   volatile
V   evaporation of water, vessel
w   set point
W   water, reactor wall
X   dry biomass

Superscripts

$\star$ equilibrium with gas phase
$\cdot$ derivated over the time
$\wedge$ estimated value

Chapter 7

a specific interfacial surface area, m^{-1}
$a_{vm,i}$ array of virtual acceleration, $m\,s^{-2}$
A_b sectional area of a bubble, m^2
c_i concentration of species i, $mol\,m^{-3}$
c_D drag coefficient of a bubble, –
c_{vm}, c_{vma} coefficients of virtual mass force, –
$c_{O_2}^*$ oxygen concentration at the gas liquid interface, $mol\,m^{-3}$
d_{10} average bubble diameter, m
d_{32} Sauter mean bubble diameter, m
d_b bubble diameter, m
d_i impeller diameter, m
D_{eff} effective turbulent dispersion coefficient, $m^2\,s^{-1}$

D_{O_2}	diffusion coefficient of oxygen, $m^2\,s^{-1}$
D_T	tank diameter, m
f_k	correction factor of drag coefficient, –
F	force, N
g	gravitational acceleration, $m\,s^{-2}$
H	liquid height, m
I	inhomogeneity, –
k	turbulent kinetic energy, $m^2\,s^{-2}$
k_L	mass transfer coefficient, $m\,s^{-1}$
$k_L a$	volumetric mass transfer coefficient, s^{-1}
K_S	half saturation constant, $g\,l^{-1}$
L	Eulerian macro length scale, m
L_{res}	resultant Eulerian macro length scale, m
n	impeller speed, s^{-1}
n_i	number of impellers, –
n_t	number of tanks, –
p	pressure, Pa
p_{O_2}	partial pressure of oxygen in the gas phase, Pa
P	power input of the impeller, W
P_k	production of turbulent kinetic energy, $m^2\,s^{-3}$
q	relative number density, –
q_{O_2}	oxygen transfer rate, $mol\,m^{-3}\,s^{-1}$
Q_L	liquid pumping rate of impeller, $m^3\,s^{-1}$
r	radial coordinate, m
r_i	reaction rate of species i, $mol\,m^{-3}\,s^{-1}$
S_c	source of species c, $mol\,m^{-3}\,s^{-1}$ or $g\,m^{-3}\,s^{-1}$
S_i	interfacial coupling term, $N\,m^{-3}$
Sc	Schmidt number, –
Sc_{turb}	turbulent Schmidt number, –
t	time, s
t_c	circulation time, s
u_i	mean velocity component, $m\,s^{-1}$
u_t	rise velocity of a bubble, $m\,s^{-1}$
u_{tip}	impeller tip velocity, $m\,s^{-1}$
u_G^0	superficial gas velocity, $m\,s^{-1}$
V	volume, m^3
w	impeller blade height, m
x	coordinate, m
x_{O_2}	concentration fraction of oxygen in the liquid phase, –
y_{O_2}	molar fraction of oxygen in the gas phase, –
Y	yield coefficient, –
z	axial coordinate, m

Greek letters

δ	Kronecker symbol, –
ε	energy dissipation rate, $m^2\,s^{-3}$
ε_G	void fraction, –
φ	tangential coordinate, °

D_L	dispersion coefficient of the liquid phase, $m^2 s^{-1}$
E	gas holdup
$EtOH$	ethanol
$\underset{\sim}{f}$	population density variable
$\tilde{f}$	shape function of population density
$\mathbf{f}$	input function of regulation model
F	flow rate, $m^3 s^{-1}$
$\mathbf{F}$	system function of regulation model
FBC	fraction of budding cells
FDP	fructose-diphosphate
$F6P$	fructose-6-phosphate
G	growth or intermediate phase
GLU	glucose
$G6P$	glucose-6-phosphate
H	height, m
I	mode number of the parent cycle
J	mode number of the daughter cycle
J	profit or performance criterion
k	coefficient
$k_L a$	volumetric mass transfer coefficient, s^{-1}
K	constant
m	specific maintenance coefficient, $mole(g\,s)^{-1}$
$\mathbf{m}$	vector of specific maintenance rates
M	mitotic phase
M	molecular weight
MAC	macromolecular cell constituent
n	number of discrete cell cycle intervals
N	cell number
NAD	nicotine-amid-dinucleotide
OUR	oxygen uptake rate, $kg(m^3\,s)^{-1}$
p	specific price, kg^{-1}
$p(i)$	discrete parent cell number density in age interval i
P	parent phase
P	price
PYR	pyruvate
P/O	effectiveness of oxidative phosphorylation
$\mathbf{q}$	vector of specific reaction rates
q	specific reaction rate, $mole(g\,s)^{-1}$
r	intrinsic reaction rate, $mole(g\,s)^{-1}$
$\mathbf{r}$	vector of intrinsic reaction rates
R	volumetric reaction rate, $g(m^3\,s)^{-1}$
$R(x)$	switch function, $R(x) = \begin{cases} x \\ 0 \end{cases} for\ x \begin{matrix} > 0 \\ \leq 0 \end{matrix}$
RQ	respiratory quotient
S	synthetic phase
t	time, s
T	average cell number doubling time or length of cell cycle phase, s
T_{IJ}	oscillation period for mode IJ, s

TCC	tricarboxylic-acid cycle
u	superficial velocity, $m\,s^{-1}$
V	volume, m^3
X_N	cell number concentration, m^{-3}
$\mathbf{y}$	intrinsic state vector of regulation model
Y	yield coefficient
$\mathbf{Y}$	matrix of yield coefficients
z	normalized spatial coordinate
α	stoichiometric coefficient
α_{IJ}	decrement of oscillation for mode IJ
	specific growth rate, s^{-1}
τ	cell age, s

Subscripts

B	budded phase
C	carbon dioxide
d	downcomer
D	daughter phase
E	ethanol
ex	external
F	fermentation
G	gas phase
j	index of reactor subsystem
k	head
L	liquid phase
min	minimum value
max	maximum value
O	oxygen
P	parent phase, or interval of reactor preparation
r	riser
S	substrate, or synthetic phase
set	setpoint
tot	total
T	total
X	biomass
0	initial value

Superscripts

F	final condition
R	reservoir
0	initial condition

Chapter 12

CTR=Q_{CO2}	volumetric carbon dioxide transfer rate [g/l h]
CPR=R_{CO2}	volumetric carbon dioxide production rate [g/l h]
c_C	concentration of compound C

$c_{S,f}$	substrate concentration in feed medium [g/l]
$e_{E,C}$	mass fraction of element E in compound C [-]
f	volume flow rate [l/h]
$k_L a$	volumetric gas transfer coefficient [l/h]
m_C	mass of compound C in the reactor [g]
MW	molecular weight [g/mol]
OTR=Q_{O2}	volumetric oxygen transfer rate
OUR=R_{O2}	volumetric oxygen uptake rate [g/l h]
q	mass flow rate [g/l h]
Q	volumetric mass transfer rate [g/l h]
r	specific turnover/reaction rate [g/g h]
R	volumetric turnover/reaction rate [g/l h]
$S_{v,p}$	fully relative sensitivity of variable v to parameter p [-]
t	time [h]
t_c	cultivation time [h]
V_R	reactor volume [l]
x	gas mole fraction [-]
$y_{X,C}$	differential yield coefficient of compound X from compound C [-]
$=r_x$	specific growth rate [l/h]
$\hat{m}$	maintenance term
	estimated value
$\overset{0}{_0}$	(subscript) starting (fed-batch) conditions
	(superscript) nominal value of variable
$_*$	saturation value

Chapter 13

A	mean culture age, s
A_P	mean projected area of a pellet, m^2
a	model parameter
AAA	L-α-aminoadipic acid
ACV	L-α-aminoadipyl-L-cysteinyl-D-valine
APA	aminopenicillanic acid
AT	acyltransferase
b	model parameter
c	intrinsic concentration
c	vector of intrinsic concentrations
C	concentration, $kg\,m^{-3}$
C	vector of concentrations
CLS	corn steep liquor
CPC	Cephalosporin C
D	dilution rate, s^{-1}
D_i	diffusion coefficient of substance i, $m^2\,s^{-1}$
d	diameter, m
DAOC	deacetoxy-cephalosporin C
e	total number of hyphal elements
f	number density function
F	flow rate, $m^3\,s^{-1}$

g	spore germination frequency, s^{-1}
h	loss function in population balance, s^{-1}
IDP	Iterative Dynamic Programming
IPN	Isopenicillin N
J	performance index
k	constant
K	constant
l	length, m
m	mass, kg
m	specific maintenance coefficient, s^{-1}
$\mathbf{m}$	vector of specific maintenance coefficients
n	number of hyphal tips
N	stirrer speed, s^{-1}
OTR	oxygen transfer rate, $kg\ m^{-3}\ s^{-1}$
p	partitioning function
P	probability or price
PAA	phenylacetic acid
Pen	Penicillin
POA	phenoxyacetic acid
pO2	percentage of dissolved oxygen saturation
q	specific reaction rate, s^{-1}
$\mathbf{q}$	vector of specific reaction rates
Q	volumetric reaction rate, $kg\ m^{-3}\ s^{-1}$
r	radius, m
r	intrinsic reaction rate, s^{-1}
$\mathbf{r}$	vector of intrinsic reactions
$\tilde{r}$	normalized kinetic rate expression
t	time, s
V	volume, m^3
x	variable
X	constituent of the biotic phase
Y	yield coefficient
$\mathbf{Y}$	matrix of yield coefficients
$\mathbf{z}$	vector of process states
α	linear extension rate, ms^{-1}
δ	Dirac delta function
Φ	branching frequency, s^{-1}
ψ	specific rate of fragmentation, s^{-1}
	specific growth rate, s^{-1}
ρ	density, $kg\ m^{-3}$
τ	time, s
λ_{Shear}	shear parameter in probability distribution

Subscripts

av	average value
bra	branching
bre	breakage
C	carbon dioxide

crit	critical
cys	cysteine
Cycl	cyclase
D	death or decay
e	effective
E	enzyme
eq	equilibrium
Exp	expandase
f	final
fra	fragmentation
g	germination
gro	growth
hgu	hyphal growth unit
Hyd	hydrolase
i	component index
I	inhibitor
L	liquid phase
lag	lag-phase
ly	lysis
m	maintenance
min	minimum value
max	maximum value
O	oxygen
Oil	soy oil
P	product
p	pellet
PM	pharma medium
pr	preparation
R	averaged variable in radial layers
S	substrate
Stir	Stirrer
t	total
thr	threshold
tip	hyphal tip
val	valine
X	biomass

Superscripts

R	reservoir

Chapter 14

Metabolites

1,3-DPG	1,3-Diphosphoglycerate
3-GP	3-Glycerolphosphate
ADP	Adenine diphosphate
AMP	Adenine monophosphate

ATP	Adenine triphosphate
cAMP	cyclic AMP
F1,6bP	Fructose-1,6-bisphosphate
F2,6bP	Fructose-2,6- bisphosphate
F6P	Fructose-6-phosphate
FADH$_2$	Flavinadenine dinucleotide (reduced)
G6P	Glucose-6-phosphate
GAP	Glyceraldehyde-3-phosphate
GDP	Guanosine-diphosphate
GTP	Guanosine-triphosphate
NAD	Nicotinamide adenine-dinucleotide (oxidized)
NADH	Nicotinamide adenine-dinucleotide (reduced)
NADP	Nicotinamide adenine-dinucleotide phosphate (oxidized)
NADPH	Nicotinamide adenine-dinucleotide phosphate (reduced)
PEP	Phosphoenole-pyruvate
T6P	Trehalose-6-phosphate

Enzymes

6PG-DH	6-Phosphogluconate-dehydrogenase
C-PKA	Catalytic subunit of PKA
FBPase1	Fructose-bisphosphatase-1
G6P-DH	Glucose-6-phosphate-dehydrogenase
GAP-DH	Glyceraldehyde-3-phophate-dehydrogenase
GDH	Glutamate-dehydrogenase
HK	Hexokinase
PFK1	Phosphofructokinase-1
PFK2	Phosphofructokinase-2
PK	Pyruvate-kinase
PKA	Proteinkinase A
R-PKA	Regulatory subunit of PKA
TAL	Transaldolase
TKL2	Transketolase-2

Other abbreviations

DNA	Deoxyribonucleic acid
EMP	Embden-Mayerhof-Parnas pathway
HPLC	High performance liquid chromatography
P/O	P/O ratio
PCA	Perchloric acid
PPP	Pentose-phosphate pathway
R1	Relaxed conformation
R2	Relaxed subconformation
RNA	Ribonucleic acid
T1	Tensed conformation
T2	Tensed subconformation

Mathematical Symbols

α	Index for compartment
α_i	Non-exclusive binding coefficient for metabolite i
β	Index for compartment
c	Exclusive binding coefficient
C_i	Concentration of metabolite i
$\mathbf{C}_j$	Vector of concentration for the reaction j
$\tilde{\mathbf{C}}_j$	Vector of steady state concentration for the reaction j
$\hat{C}_i$	Approximation function for the time course of metabolite i
D	Dilution rate
$\mathbf{E}$	Elemental composition matrix
ε	Error criteria
η	Efficiency of oxidative phosphorylation
$\varepsilon_{rel,i}$	Relative error square for metabolite i
f_j	Kinetic rate expression of reaction j
Γ	Reaction matrix
k	Maintenance coefficient
K_i	Affinity with respect to metabolite i
L	General allostery coefficient
L_0	Allostery coefficient in the absence of ligands
	Growth rate
N	N function
N_i	Michaelis-Menten term for metabolite i
$\mathbf{P}_j$	Parameter vector for reaction j
Q	Effector function
$\mathbf{Q}$	Vector of net-conversion rates
$\mathbf{r}$	Vector of reaction rate
r^{max}_j	Maximal rate of reaction j
$\tilde{r}_j$	Steady state flux of reaction j
$r_{Deg,F2,6bP}$	Rate of degradation of F2,6bP
$r_{m,ATP}$	Maintenance
t	Time
$\mathbf{T}$	Transport matrix
$\mathbf{t}$	Transport vector
V	Volume
$Y_{X/S}$	Coefficient of yield
Z	Z function

Chapter 17

	null vector
ATP	adenosine triphosphate
$\alpha_{j,A}$	reaction coefficient for pathway j and metabolite A
AMM	atom mapping matrix
$\underline{\underline{A}}$	stoichiometric matrix
AT	aminotransferase
BHK	baby hamster kidney (cells)
^{13}C	carbon, isotope 13

^{14}C	carbon, isotope 14
C_i	steady state concentration of i
C_M	intracellular concentration of M
$C_{i,0}$	feed concentration of i
CO_2	carbon dioxide
CHO	Chinese hamster ovary (cells)
CS	citrate synthase
δ	tolerance
DHAP	dihydroxyacetone phosphate
$\underline{\underline{E}}$	redundancy matrix
f_j	flux through pathway j
$\underset{\approx}{f}$	column vector of all fluxes
$\hat{f}$	vector of best fit fluxes
F	volumetric flow rate
F6P	fructose 6 phosphate
$FADH_2$	flavin adenine dinucleotide, reduced
G6P	glucose 6 phosphate
GDH	glutamate dehydrogenase
GTP	guanosine triphosphate
HK	hexokinase
$\underline{\underline{I}}$	identity matrix
IDV	isoptopomer distribution vector
IMM	isotopomer mapping matrix
LDH	lactate dehydrogenase
λ	label flux
L_{Ai}	concentration of A labeled on carbon i
M_{Ai}	fractional enrichment of metabolite A, carbon i
MDH	malate dehydrogenase
ME	malic enzyme
N	cell number within a perfusion device
NADH	nicotinamide adenine dinucleotide, reduced
NADPH	nicotinamide adenine dinucleotide phosphate, reduced
NH_3	ammonia
NMR	nuclear magnetic resonance (spectroscopy)
NO_3	nitrate
OAA	oxaloacetate
OUR	oxygen uptake rate
$\underline{\underline{P}}$	permutation matrix
PC	pyruvate carboxylase
PDH	pyruvate dehydrogenase
PEPCK	phosphenolpyruvate carboxykinase
PK	pyruvate kinase
PPP	pentose phosphate pathway
PSSA	pseudo-steady-state approximation
r_A	rate of accumulation of metabolite A
$\underline{r}$	column vector of all measured fluxes
σ_i^2	variance of measurement i
TA	transaldolase
TCA	tricarboxylic acid (cycle)

TK	transketolase
V	chemstat liquid volume
V	intracellular volume
X	cell concentration
$Y_{lac/gluc}$	molar yield of lactate from glucose

Metabolites in Figs. 17.1 and 17.7

A	acetyl-CoA
AL	alanine
C	citrate
D	dihydroxyacetone phosphate
DPG	diphosphglycerate
E	erythrose
F	fructose
FU	fumarate
G	glucose
GA	glyceraldehyde
K	α-ketoglutarate
L	lactate
M	malate
N	glutamine
O	oxaloacetate
P	pyruvate
PE	phosphoenolpyruvate
PG	phosphoglycerate
R	ribose
Ru	ribulose
S	succinate
Se	sedoheptulose
T	glutamate
X	xylulose

Contributors

Karl-Heinz Bellgardt
Institut für Technische Chemie der Universität Hannover,
Callinstrasse 3, 30167 Hannover, Germany
e-mail: bellgardt@mbox.iftc.uni-hannover.de, Fax: +49 511 762 3004

Harvey W. Blanch
Department of Chemical Engineering, 201 Gilman Hall, University of California,
Berkeley, CA 94720, USA
e-mail: blanch@socrates.berkeley.edu, Fax: +1 510 643 1228

Douglas S. Clark
Department of Chemical Engineering, 201 Gilman Hall, University of California,
Berkeley, CA 94720, USA

D. Dochain
Senior Research Associate FNRS, Cesame, Université Catholique de Louvain, Bât.
Euler, 4–6 av. G. Lemaître, 1348 Louvain-La-Neuve, Belgium
e-mail: dochain@auto.ucl.ac.be, Fax: +32 10 472180

Neil S. Forbes
Department of Chemical Engineering, 201 Gilman Hall, University of California,
Berkeley, CA 94720, USA
e-mail: nforbes@socrates.berkeley.edu

K. Gollmer
University of Applied Sciences Trier, Environment Campus, 55761 Birkenfeld,
Germany
e-mail: gollmer@umwelt-campus.de

Albert A. de Graaf
Institute of Biotechnology I, Research Centre Jülich, 52425 Jülich, Germany
e-mail: a.de.graaf@fz-juelich.de, Fax: +49 2461 612710

Bernd Hitzmann
Institut für Technische Chemie, Universität Hannover, Callinstrasse 3, 30167
Hannover, Germany
e-mail: hitzmann@mbox.iftc.uni-hannover.de

Marc Jenne
Institut für Bioverfahrenstechnik, University of Stuttgart, Allmandring 31, 70569
Stuttgart, Germany

Maria I. Klapa
Department of Chemical Engineering, Massachusetts Institute of Technology,
Cambridge, MA 02139, USA
e-mail: mklapa@mit.edu

Andreas Lübbert
Martin-Luther-University, Halle-Wittenberg, 06099 Halle/Saale, Germany
e-mail: andreas.luebbert@iw.uni-halle.de, Fax: +49 345 5527 260

Reiner Luttmann
University of Applied Sciences Hamburg, Research Center of Bioprocess Engineering
and Analytical Techniques, Lohbrügger Kirchstrasse 65, 21033 Hamburg, Germany
e-mail: reiner.luttmann@rzbd.fh-hamburg.de

Klaus Mauch
Institut für Bioverfahrenstechnik, University Stuttgart, Allmandring 31, 70569
Stuttgart, Germany

M. Perrier
Département de Génie Chimique, Ecole Polytechnique de Montréal, Succursale
"Centre Ville", CP 6079, Montréal H3C 3A7, Canada

Clemens Posten
Institut für Mechanische Verfahrenstechnik und Mechanik, Universität Karlsruhe
(TH), 76128 Karlsruhe, Germany
e-mail: clemens.posten@ciw.uni-karlsruhe.de, Fax: +49 721 6086

Matthias Reuss
Institut für Bioverfahrenstechnik, University of Stuttgart, Allmandring 31, 70569
Stuttgart, Germany
e-mail: reusss@ibvt.uni-stuttgart.de, Fax: +49 711 685 5164

Ursula Rinas
GBF, National Research Center for Biotechnology Ltd., Dept. of Biochemical
Engineering, Mascheroder Weg 1, 38124 Braunschweig, Germany
e-mail: uri@gbf.de, Fax: +49 531 6181 111

Sven Schmalzriedt
Institut für Bioverfahrenstechnik, University of Stuttgart, Allmandring 31, 70569
Stuttgart, Germany

Karl Schügerl
Institut für Technische Chemie Universität Hannover, Callinstr. 3, 30167 Hannover,
Germany
e-mail: schuegerl@mbox.iftc.uni-hannover.de, Fax: +49 511 7622253

Gregory Stephanopoulos
Department of Chemical Engineering, Massachusetts Institute of Technology,
Cambridge, MA 02139, USA
e-mail: gregstep@mit.edu, Fax: +1 617 253 3122

Sam Vaseghi
Institut für Bioverfahrenstechnik, University Stuttgart, Allmandring 31, 70569
Stuttgart, Germany

John Villadsen
Department of Biotechnology, Block 223, Technical University of Denmark, 2800
Lyngby, Denmark
e-mail: kn@ibt.dtu.dk, Fax +45 4588 4148)

Introduction

Karl-Heinz Bellgardt

The Need for Modeling and Control in Biotechnical Processes

The kinetics of biotechnological processes are determined by the properties of the microorganisms, the construction of the reactor, as well as by the cultivation conditions and media. The metabolic flexibility of the cells in connection with inhomogeneities in the reactor often leads to very complex growth dynamics, which make it difficult to ensure high operational stability and reproducibility of the process, as well as constant product quality and yield. The situation is further complicated by the fact that technical substrates for industrial processes are rather undefined. The widely established on-line analytical methods are mostly not sufficient or fast enough to characterize the state of the running process. Nevertheless, before the background of increased international competition, reduced profit margins, and rigorous safety and environmental regulations, there is clearly a need for improved process control and optimization based on advanced analytical methods for substrates, products and state of the cells [1, 2].

Mathematical models can take an important part in this task although biotechnical processes are rather complex systems which still can be described only in a roughly simplifying way. The complicated structure of the metabolism and the mechanisms of its regulation are still not fully understood. Beside the intracellular processes, also the variation of the local condition in the bioreactor, caused by the fluid dynamics of the multiphase system, have to be looked at in a very simplified way. Nevertheless, the mathematical models, which naturally must be incomplete and inaccurate to a certain degree, can still be very useful and effective tools to describe those effects which are of great importance for control, optimization, or our understanding of the process [3–6]. Mathematical models provide a functional interrelation of the input variables, output variables, and inner variables of the process that can easily evaluated by computer. Thus numerical solution of the models is the fundament for the development of economic and powerful methods in the fields of process design, plant design, scale-up, optimization and automatic control [7–9] Most bioprocesses are operated under non-stationary conditions in batch or fed-batch mode. This leads to complicated optimal time profiles for the control variables, which are sometimes impossible to be determined purely experimentally. Here, mathematical or numerical optimization methods for determination of control variables and parameters can advantageously be applied to reduce the experimental effort and the required time for optimization.

The general principles of modeling of bioprocesses and related techniques are introduced in Part A. The mathematical modeling of biotechnical processes is an extremely wide field that covers all important kinds of processes with many different microorganisms or cells of plants and animals. The existing biological models are aimed at various levels of the biotic phase, beginning with a description of the complex intracellular reaction network, of the metabolic regulation on the reaction level, of the processes of replication and translation of genes, the events during the cell replication cycle, up to models for morphological processes and description of population dynamics (Chapters 2 and 3). No less manifold are the reactor models, spanning from simple homogenous single-phase systems (Chapter 1), up to complex structured models of multiphase systems and more sophisticated, detailed hydrodynamic models for description of transport and mixing processes in bioreactors (Part B).

Due to the inherent nonlinear kinetics of a bioprocess, under non-stationary operation there is an enduring change in the time constants of the system. Furthermore, most bioprocesses are essentially multi-variable systems with strong inner couplings. From these facts arise serious problems for conventional methods of automatic control, such as single-input/single-output PID-control. Here, control theory provides the advanced methods of adaptive, non-linear, and multi-variable control, which are often directly based on mathematical models. It is self-evident that the more measurements are available, the better can be the automatic control. But even with the most advanced methods of process monitoring, it will not be possible to completely determine the actual reaction rates or the intracellular state of the cells as an important subset of the state of the process. This difficulty can be overcome by the concept of model-based estimators or observers for the not measurable states and time-variable parameters. An introduction to the problems of automatic control is given in Chapter 5. A closely related field is the detection of failures or undesired process states: e.g., defects in sensors or actuators, infections by other microorganisms. The application of methods of artificial intelligence in this area is outlined in Chapter 4.

Process scheduling, supervision, automatic control, and documentation in modern biotechnical plants is done by computerized process control systems (PCS), where all the functions are implemented in software (See Fig. 1). The effort for software development can account for a great or even major percentage of the total costs for control equipment. Here the application of model simulations for testing the PCS and training of the operators can significantly reduce the time for setting to work the plant. More details of such methods are given in Chapter 6.

Part C gives examples for the modeling of selected processes of industrial importance. There, the general methods are applied to describe the kinetics of batch and fed-batch cultivations of bacteria, yeasts, and filamentous fungi. The method of metabolic flux analysis as introduced in chapter 3 is an important tool for evaluation of the intracellular reaction network, and for optimization of cultivation conditions and substrate composition. Its connection with genetic engineering for directed modification of the cellular reaction network is known as metabolic design. Some applications are given in Part D for yeast and bacteria.

Some Modeling Basics

A definition of a model can be given as follows:

A model is an image of a real system that shows analogous behavior in the important properties, and that allows within a limited region a prediction of the behavior of the original system.

The experimental study of the original system can then be replaced by the model. This can have several advantages:

Economy, cost factors: The model system is simpler, smaller, cheaper or faster then the original system

Research and development: The model is less complex than reality, certain effects can be emphasized or suppressed. This helps one to obtain a clear view of the process.

There are generally three different groups of models, physical, mathematical, and verbal models. Physical models can be realizations of the original system in a different (usually smaller) scale or with structural modifications. A second type of physical models is obtained by turning to a different physical system, e.g. from the original biotechnical process to electrical circuits. This is then called an analogous computer. Verbal models give a linguistic representation of our knowledge about the system, usually as rules (e.g. **if** this or that happens **then** the system reacts by ...) [10, 11]. They are widely applied in the area of artificial intelligence, namely expert systems. Mathematical models are not based on the real existence of a physical model system, but describe the behavior of the original system by mathematical equations. This is the highest level of abstraction. The experimental investigation is replaced by manipulation and solution of the model equations. A link between verbal models and mathematical models is established by fuzzy logic, which allows one to translate qualitative and rule-based knowledge into mathematical equations.

Mathematical models can be classified further depending on the mathematical formalism or the methods for model building. In this book, we will focus on deterministic theoretical models of biotechnical processes. Theoretical models as mechanistic models are based on physical and chemical laws and our knowledge about the inner structure and function of the system, e.g. the flows of mass and energy. They can, therefore, provide far-reaching predictions of the system behavior. In contrast, experimental or non-mechanistic models try to give – without looking into the inner of the system – a description of the observed reaction of the system in response to a certain forcing signal. These types of models are called black box models. Many "mechanistic" models in biotechnology are actually due to their over-simplification quite closer to black-box models than to mechanistic models, although mostly mechanistic interpretations are given. Very typical black-box models are Artificial Neural Networks, which have also found wide application in biotechnology [12, 13].

The universal mathematical tool of modeling are differential equations which are obtained by the balances for conservative quantities, such as mass of reactants, energy and impulse. These have the general form:

$$\underbrace{\begin{array}{c} \textit{local change in the} \\ \textit{reaction volume} \end{array}}_{\text{storage element}} = \underbrace{\begin{array}{c} \textit{local} \\ \textit{inflow} - \textit{outflow} \end{array}}_{\text{convective transport}} - \underbrace{\begin{array}{c} \textit{local} \\ \textit{inflow} - \textit{outflow} \end{array}}_{\text{diffusive transport}}$$

The models include **input variables** (e.g. Flow rate of substances), that are not determined by the model itself, **output variables** (e.g., oxygen uptake rate, OUR) and **state variables** (e.g. concentrations). The state variables represent storage elements for mass or energy of the system. The model can be seen as a calculation rule, that relates a looked for pattern of the output variables (e.g. time course) to a certain known or given pattern of input variables. Another type of model variables, the **parameters**, are fixed entities of the system. They can have a direct physical meaning. Some parameters are known or can be measured, while others have to be determined from experimental data.

The discrimination in parameters, state variables, or input variables is not an immanent property of a system, but a specific view by the developer of the model, which can be changed in relation to the mode of operation of the process or the simplifying assumptions of the model. The temperature, for example, can be controlled very tightly at a constant value, so it may be considered as a parameter in one cultivation. In a second one, the operator may set different values certain times or the control follows a given profile; this time temperature would be an input variable to the model. If the temperature varies freely it should be calculated by an energy balance, and thus became a state variable.

A **process model** as a mathematical model should provide a coherent description of the entire process on the level of plant operation, including reactor and further processing steps. The degree of complexity of the model or the possibility for a simplified modeling of some parts of the plant is determined by the intended application of the model. The model may be used in numerical simulations to answer several interesting questions: What will be the output of product per unit of time for a given input of raw material and primary energy? What are the costs of production and of treatment of waste? What is the optimum mode of operation for the reactors? Under which dynamic control of manipulating variables is the product obtained with high productivity and a certain quality? How can the profit be maximized? Very simplified balances may be sufficient as a model for the calculation of the total conversion of the process, e.g. yield of cell mass or product. Greater modeling effort and a more detailed description of the most important unit operations is generally required for a model based process optimization. One should be aware of the fact that the result of such theoretical optimizations may be significantly influenced by the model accuracy, especially for the determination of an optimal dynamic control.

To develop an accurate and complete model is not an easy task for several reasons [14]. Sometimes there are only very simplified models available for parts of the plant, e.g. cultivation. And mostly not all interdependencies between the elementary units of the process are qualitatively and quantitatively known: The cultivation is influenced in a complicated manner by medium composition, by substrate quality and preprocessing, and by the inoculum preparation. The cultivation itself may influence the downstream processing by changing rheology of the broth and varying product properties. The situation for model building is further complicated because for batch and fed-batch operation the process is never in steady state and the model must consider the process kinetics. Therefore, the modeling of the biological system, as outlined in chapters 2 and 3, is obviously an important aspect of process models for biotechnological systems.

Model building is always a combination of theoretical studies and practical experiments in a very iterative sequence. Since the problems in parameter identification and model verification increase rapidly with the model complexity, one should begin

Table 1. Step sequence of model building

Step	Action
1	Running typical experiments
2	Define the modeling goal
3	Analysis of the system and determination of structural elements
4	Simplifying assumptions (e.g. about mixing, process structure and dynamics, metabolism, kinetics)
5	Choice of important process variables: parameters, input variables, and states
6	Establishing the model by use of balances, physical laws and empirical equations
7	Simulation of the model; parameter identification to fit it to experimental data
8	Evaluation of the model quality; repeat with step 1

with as far as possible simplified assumptions and withdraw them step by step, in the case that the model quality is not sufficient. In this way, a most simple initial model grows step by step in complexity and accuracy, without becoming too complicated. Modeling includes not only the selection of correct model structure, but also the quantitative fitting to the experimental data by determining the model parameters. Unfortunately, their values are often not or only not exactly known in advance and must determined from experimental data by methods of parameter identification. The repeated steps of model building are summarized in Table 1.

Structure and Operation of Biotechnical Plant

A biotechnical production line can be roughly subdivided into three sections, the preparation and preprocessing of the input raw materials, the cultivation, and the downstream processing of the cultivation broth. An example for the general layout of a biotechnical process is shown schematically in the upper part of Fig. 1. Each of the sections is built of many elementary steps, the unit processes and unit operations, of which the cultivations of microorganisms or cells in the bioreactors are naturally the most important ones [15].

In the preparation section, the primary energy is converted into process energy in form of steam and compressed air. The substrate for the cultivations is prepared from raw substrate. Here, several steps can be involved, such as filtration, hydrolysis of higher carbon compounds, and sterilization. Usually further essential or growth supplementing compounds like mineral salts or yeast extract are added. In aerobic processes, sterile compressed air has to be provided for aeration of the bioreactors. In the preculture, the inoculation material for the first seed reactor is prepared in shaking flasks from deep-frozen cell material that is coming from the strain maintenance. The production starts with cleaning and sterilization of reactors, filling the reactors with substrate, and inoculation with the preculture.

In the cultivation section, here assumed as batch plant, the cells or microorganisms are propagated in the series of reactors. To avoid low cell concentrations and low volumetric productivity, the reactors usually have an increasing volume, beginning with the smallest seed tank in 10–100 l scale, up to the final production tank in 10–100 m^3 scale. When the cell concentration in a certain stage has reached its maximum, the fermentation broth is inoculated to the next stage. In the last production

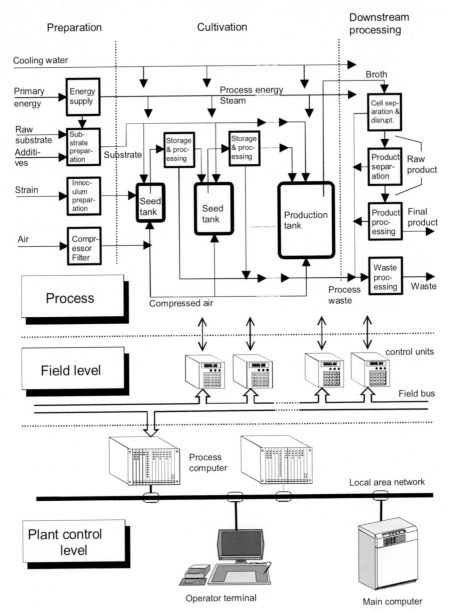

Fig. 1. The general layout of a biotechnical production process.

stage, the major amounts of cell mass and product are formed, and emphasis is put for the process control on a high product concentration, respectively productivity. During the transfer from one reactor to the next, the broth can be further treated and might be temporarily stored to meet the scheduling for the entire plant. In big production plants, several seed or production tanks are often operated in parallel, but with a certain time shift. This renders an effective and more continuous usage of the equipment for preparation and downstream processing.

In the down-stream processing section, a wide variety of processing steps can be found, depending on the properties of the product. Gaseous products can be separated in situ from the exhaust gas stream. Non-gaseous products have to be separated from the broth after the final cultivation in the production tank. For intracellular products the cells have to be disrupted at first. The cell material is then separated by filtration or centrifugation. The product purification might include filtration, extraction, adsorption, dialysis, chromatography, distillation, and other methods. Often the raw product is further modified in purely chemical steps to obtain the final product.

The treatment of waste is another important part of a biotechnological process. Waste material arises from preparation of the substrate, cleaning of reactors, and downstream processing. If it includes viable cells, the waste must be sterilized, and the load of organic material must be reduced. Furthermore, depending on the downstream processing it may be necessary to remove solvents or other reactants.

In modern production plants, the process is coupled to a computer process control system (PCS) that is shown schematically in the lower part of Fig. 1. The entire PCS can be roughly subdivided into the field level and the process control level. The main features of the different levels are summarized in Table 2. In decentralized systems, the different units are independent and spatially separated. This can improve the stability of the control system in the case of hardware failures. For smaller pilot and laboratory plants, due to economic reasons the functions of field level and process control level may be realized within only one centralized computer system.

In a decentralized system, the field level is built of small independent microcomputer control units which are directly connected to the measuring equipment and the controlling devices of the plant. Each control unit serves only for a few variables or control loops. Usually, simple signal processing functions like filtering, linearization, and normalization as well as the basic control loops, such as for temperature, flow, pressure, and pH are realized on this level. The control units also perform fixed automation sequences via Programmable Logic Control (PLC), e.g. sterilization, medium transfer. The field level provides only simple functions for displaying the values of measured and controlling variables and for direct manual operation. The control units can communicate to each other and to the process computer on the next hierarchical level via the field bus. Prefiltered data are sent to the process computer, set points and digital control commands are received from it.

The process control level covers the functions for automatic operation of the entire plant and comfortable manual operation and supervision in the control room. Graphical displays provide all information about the process and give access to the controlling variables. The entire production run is automatically controlled by recipes, which include all information on the different process phases, scheduling of the bioreactors and the other equipment, control strategies to be used, set-points during the different phases of the process, profiles for controlling variables, and composition of substrates. This is a very special feature of biotechnical processes which are mostly not operated continuously, but in batch or fed-batch mode. Another important point is that, different from chemical processes, many important variables which are essential for an optimal control can only be measured off-line. Therefore, the PCS must provide functions for handling of off-line data and for its use in the automatic control [16].

Table 2. Specific functions and features on the different levels of process and control

Level	Features	Signals and variables
Process	Mass flow Energy flow (Bio-)Chemical reactions Physical processes	Physical and chemical
Field	Measuring, analog-digital conversion Local low level control by simple control loops Programmable logic control (PLC) Signal filtering Short term data storage Linearization and normalization Local display Direct manual operation	Electrical, optical, digital data
Process control	Common features: Man-machine interface via menus and graphics Supervision, alarm handling Balancing Data processing and data reduction Long term data storage Documentation and protocolling Advanced features: Handling of off-line analysis Recipes, scheduling High level control State and parameter estimation Simulation, prediction Optimization	Digital data

On the process control level also some advanced functions can be realized that require a high computing power: Balancing of mass and energy flows, estimation of non-measurable quantities or of time-variable parameters, sophisticated control of the cultivation that considers the state of the biological system, usage of model simulations for prediction of the future course of the process, and on line optimization. Another important task is complete protocolling of all process variables and all process events, including long term data storage for future reference or further analysis of the production runs. Data reduction and calculation of economic indices provides information for the next hierarchical level of factory control or management.

Types and Structure Elements of the Bioreactor

The design of the bioreactor – as an important part of the entire production process – together with its mode of operation with respect to substrate supply, air supply and medium exchange has to ensure the optimal conditions for growth and product formation of the microorganisms, or for cells of higher organisms, such as animals and plants. As a first step of process analysis aimed at mathematical modeling and control, the main structure elements of the bioreactor and its mode of operation will be shortly summarized.

The bioreactor has to provide the proper physical and chemical environment – e.g., temperature, pH, substrate concentration – for the cells, and to ensure fast transport of substrates and products between gas phase, bulk medium and cells with as low as possible effort in material, energy, and mechanics. Different types of bioreactors have been developed to optimize transport properties, homogeneity of the liquid phase, gas dispersion, and mechanical stress. Design goals are good mixing properties resulting in low gradients of dissolved reactants and temperature, good heat exchange, and low costs of investment.

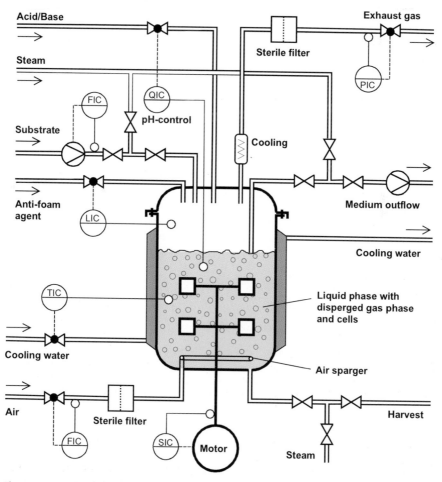

Fig. 2. Schematic diagram of a stirred tank reaktor and basic control functions: control and indication for flow rate (FIC), temperature (TIC), pressure (PIC), speed (SIC), quality variables like pH (QIC), level (LIC).

The characterization of the bioreactor as a multi-phase system means that, beside the common inner-phase transport processes of convection and diffusion, the inter-phase transport phenomena solid-liquid and liquid-gas play an important part for the over-all kinetics. Since these phenomena can be influenced by the design of the reactor, the effective minimization of inter-phase transport resistance is another central goal for reactor design. With respect to the gas-phase, the effective supply of oxygen and removal of gaseous products is desired. For a good oxygen usage the size and residence time distributions of the bubbles have to be controlled to proper values. Both can be influenced by the design of air spargers and mechanical power input. Since the oxygen transfer is often the key factor for the productivity of the process, effective methods for the dispersion of the gas stream are required [17, 18].

The bioreactor, as shown for a STR in Fig. 2, is embedded into a set of equipment to establish and maintain optimum conditions for the cells: heat exchangers and related control loops for temperature, air compressors to ensure sufficient supply of oxygen, control loops for the regulation of pH by addition of acid or base, foam control by addition of antifoam agent, and last but not least the supply of substrates and other medium additives. In several positions, controlled by a number of valves, steam can be directed into the reactor and the connecting tubes for sterilization of the entire system.

The bioreactors can be classified according to the employed methods for supply of air and mechanical energy for mixing, and are shortly introduced in the following. Beside the basic types, there exist a great number of variants.

The Stirred Tank Reactor as an Example for Reactors with Mechanical Energy Input

A typical stirred tank reactor (STR) as shown schematically in Fig. 2 has a height-to-diameter ratio of usually less than 4:1. Mechanical energy for mixing is supplied by an rotating impeller, which is also responsible for the gas dispersion. In small scale, the STRs show a high specific energy input, and in consequence, good mixing and a high possible oxygen transfer capacity. They can then usually be taken as ideally mixed. With increasing size, the energy requirement grows over-proportional. Other disadvantages are the great base area of the reactors, and the high mechanical peak-stress near to the impeller. In the outer zones of big STRs the medium can be almost stagnant. Therefore, practicable volumes of STRs range from 1 dm^3 to several 10 m^3.

Reactors with Energy Input by Compressed Air

For very big reactors up to 500 m^3, the usage of the expansion energy of compressed air as an power input is most effective. The uprising gas bubbles lead also to acceptable mixing of the liquid phase. To obtain a sufficient gas residence time, the height-to-diameter ratio must be greater than 10:1. As examples, in Fig. 3, a bubble column reactor and an airlift reactor with inner loop are shown schematically. In bubble columns, the liquid flow pattern is rather undefined, and the liquid velocity is low. This results in unsatisfactory mixing. In airlift reactors, there is a guided liquid circulation with up-flow in the gassed region (Riser) and down-flow in the degassed region (Down comer). The low gas-holdup in the down comer may lead to oxygen limitation. The liquid velocity is relatively high, resulting in good mixing of the liquid phase. Advantages of this type of reactors are the small base area, the simple design, and the low energy requirements for mixing.

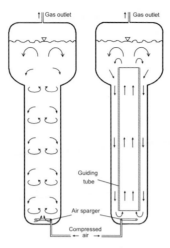

Fig. 3. Schematic diagrams of a bubble column (left) and an airlift ractor with inner loop (right).

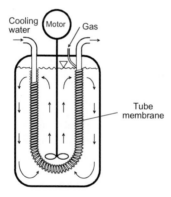

Fig. 4. Schematic diagram of membrane reactors for bubble-free aeration.

Membrane Reactors for Bubble Free Aeration

In the cultivation of animal cells the mechanical shear stress must be kept very low to avoid damage of the cells. One origin of the shear forces are small uprising bubbles, because of the high velocity gradients around the bubbles. In membrane reactors, such as shown in Fig. 4, the gas exchange with the liquid phase is via a microporous tubular membrane. In this system bubble formation can be completely avoided. The maximum oxygen transfer capacity is quite low, but sufficient for the very slowly growing animal cells. The final end of the aeration membrane can be either closed or open. In the first case, the oxygen usage is most effective, but degassification of carbon dioxide can be only via the liquid surface. Therefore, an additional gas stream through the head-space of the reactor is necessary. In the second case, the mass transfer capacity for carbon dioxide via the membrane is comparable to that for oxygen. Aeration of the head space is not necessarily required, but can be used to increase the mass transfer capacity.

A bioreaction system is typically a multi-phase system. The incorporated phases are:

Liquid-Phase

In a typical bioreactor, all substances and substrates which are required for growth or product synthesis are dissolved or suspended in the liquid medium that surrounds the cells. The medium contains organic and inorganic material that is taken up and metabolized by the cells to provide cell material and chemical energy. The main components are the sources of carbon (e.g. sugars, alcohols), nitrogen (ammonium, amino acids), phosphorus (phosphate), vitamins, and oxygen. Also the products excreted by the cells are usually dissolved in the liquid phase.

Gas-Phase

The gas phase can be subdivided into the dispersed gas phase that is contained as small gas bubbles in the liquid phase, and the gas phase in the head space of the reactor. With the exception of reactors with bubble free aeration, only the first is of much relevance for the biological reaction due to its high interfacial area. Nevertheless, for a detailed dynamic analysis of the growth kinetics also the head gas phase has to be considered, because it increases the total gas residence time, and the delay of exhaust gas measurements. Beside the inert components that might be contained in the inlet gas stream (e.g. nitrogen in supplied air), the gas-phase consists of gaseous substrates such as oxygen, gaseous products of the metabolism such as carbon dioxide, or evaporated components of the liquid phase, mainly water.

Solid-Phase

There can be several sources of solid particles in a bioprocess. Many substrates (eg. casein, peanut powder, cellulosic material) contain insoluble fractions. Often, the cells are immobilized by solid material to facilitate the retention of biomass in the reactor. The cells can be attached to the surface of solid particles (eg. sand, polymerous microcarriers), they can be immobilized within solid particles (porous beads of clay, amorphous glass, cellulosic material, or alginate), or they can be encapsulated within spherical membranes (microencapsulation, e.g. by polymers, cellulose). Even in carrier-free systems when the cells form agglomerates or pellets, it might be necessary to consider the cell-flocs as a solid phase. The presence of the solid particles can strongly influence the hydrodynamics, and the transport processes of substrates and products due to additional diffusion resistance. In the case of dense and big particles in the range of millimeters the reactor is operated as a fluidized or a fixed bed. For low particle concentrations and small particles with a density comparable to water, the solid phase might be neglected.

From the biological view-point, the gas phase, liquid phase, and solid phase may be summed up as the abiotic phase that surrounds or is in contact with the cells. The biological activity takes place in the biotic phase.

Biotic Phase

The entire population of cells in a bioreactor forms the biotic phase. Usually the corpuscular character of the cells is neglected and the biotic phase is characterized in a volume element of the liquid phase by averaged quantities such as mass of total cells or cell constituents, which are often expressed as concentrations. This is justified in the case of single cells or small cell flocs, since the cell size is only in the range of μm and the cell density is close to that of water. As biocatalysts, the cells use substrates supplied in the cultivation medium for their own survival and cell propagation, and by this synthesize a number of products among which also the desired product of the bioprocess can be found. The different functional groups of cellular metabolism for substrate degradation, energy production and biosynthesis of cell material are closely connected to each other by the network of metabolic reactions. A superseded system of control loops coordinates the reactions, connects them to the genetic level, and ensures a coordinated metabolism and cell division.

The construction of the reactor and its mode of operation have to provide a proper stream of substrates – most important are carbon sources and oxygen – to the cells to guarantee a maximum productivity. Thus the coupling of biotic phase, liquid phase and gas phase on one side, and the dynamics of inner phase transport and reaction on the other side are important aspects of any bioprocess. The dynamics of the process have time scales on rather different magnitude: quite fast phenomena with time constants below seconds are local mixing due to turbulent flow and catabolic reactions in the cells. The circulation time through the entire reactor lies in the range of seconds up to minutes, while the growth process and regulatory adaptation of the cells proceeds in the range of hours.

For the modeling and process design, the very fast phenomena are usually neglected. This does not mean that they are not important but that it is very difficult to develop proper measuring techniques for the fast mixing phenomena and metabolic reactions, and that also their simulation requires an enormous effort. When the cells pass along their way in the reactor through regions of different concentrations the fast catabolic reactions may trigger a long lasting regulatory response on the metabolism that cannot be completely understood when looking only at quantities averaged in time or space. This complex interaction between mixing and transport, fast metabolic reactions, and cellular growth and regulation can only be clarified on the basis of very detailed dynamic models of the reactors and of the cells.

Modes of Operation of a Bioreactor

Depending on the flow of medium to or from the reactor, or the supply of oxygen, the operational mode of a biotechnical processes can be classified into several groups.

Batch Cultivation

In a batch cultivation there is no exchange of liquid medium (see Fig. 5). All the substrates are contained in the medium from the beginning on, therefore, their initial concentration is quite high. After inoculation, the cells are growing uncontrolled until an essential medium component is exhausted or the accumulation of inhibiting products ceases the growth. There is a lasting change of concentrations in the reactor. The advantage of batch cultivation is the low effort for process control. It can be applied advantageously when high substrate concentrations have no negative effect on the desired biological reactions. Batch cultivation is also most suited when only relatively small amounts of the product have to be produced. In practice, a pure

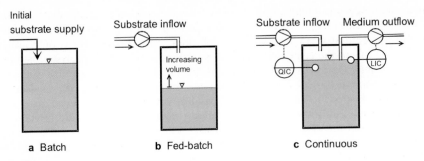

Fig. 5. The principal operational modes of bioreactors.

batch cultivation is seldom found. Often some components are fed continuously, e.g. oxygen, nitrogen source.

Fed-Batch Cultivation

In a fed-batch cultivation, medium components are continuously fed to the reactor, but no medium is taken out. This means that the liquid volume is increasing during the process. The fed-batch mode of operation is used when the substrate concentration must be kept low for optimum growth or product formation, e.g., in the case of substrate inhibition or catabolite repression of product synthesis. Since the cell mass is continuously increasing during the cultivation, the flow rate for the added medium components must be adjusted properly, either by a feedback control loop or by a predetermined flow-profile. The advantage of improved possibilities of controlling the biological reaction by the substrate flow is opposed to the greater effort for equipment.

Both, batch and fed-batch operation are preferred for most industrial processes, to avoid problems with strain stability and sterility that may arise during prolonged cultivations. The disadvantage of the previous discontinuous modes of operation is the low productivity, mainly by two reasons: after each cultivation, the reactor has to be emptied, cleaned and refilled again. This causes long unproductive intervals. Furthermore, since the reactor is inoculated with low cell densities, the maximum cell concentration and production rate is only reached close to the end of the cultivation.

Continuous Cultivation

In a continuous cultivation there is a permanent inflow of substrate to and outflow of medium including cells from the reactor, usually both with the same rate so that a steady state is reached. The cell concentration can be maintained at high values comparable to the maximum values in batch operation. Similar to the fed-batch mode, continuous operation provides good possibilities to control the biological reaction by setting a proper residence time via the flow rate. According to the criterion for controlling the flow rate, several types of steady continuous operation are obtained. The most important is the chemostat that is operated at a fixed inflow rate. In the turbidostat, the inflow rate is controlled by the measured cell concentration (e.g. via optical density of the medium). In the pH-auxostat, the respective control variable is the pH. These control methods can stabilize the operation at high flow rates close to wash-out of the reactor. In all cases, the outflow rate can be controlled in an under-lying feedback loop via the liquid level or the weight of the reactor.

The advantage a continuous process is that it usually can reach the highest pro-ductivity, since the repeated cycle of reactor preparation, cleaning, and cell propaga-tion is not required. By using reactor cascades with a proper choice of the residence times, the continuous process can be optimally adapted to many biological systems. On the other hand, failures in the equipment, infection by other microorganisms, or aging effects of the cell population limit the maximum operation time. Therefore, and because of the higher effort for control and continuous medium preparation, most processes are operated in discontinuous mode. Common exceptions where continuous operation is preferred are processes that can be operated under insterile conditions, e.g. the production of alcohols or solvents, and waste water treatment.

Cultivation with Cell Retention

In the simple continuous operation, cells are taken out from the reactor together with the medium. For products that are not synthesized in direct coupling to the cell growth, the productivity of the continuous operation can be increased by keeping the biocatalyst in the reactor (types II and III in Table 3). Three common methods for cell retention are schematically shown in Fig. 6. For large scale production, the immobilization of cells and usage of fixed- or fluidized-bed reactors is most practic-able. For suspension cultures, cell recycle reactors with centrifugation or filtering in the outflow, and membrane reactors with integrated retention can be applied. A low rate of cell harvest may be of advantage, to avoid a too high percentage of old and inactive cells.

Repeated or Cyclic Batch or Fed-Batch Cultivation

This is an extension of the discontinuous operation. Here, the cleaning-inoculation cycle is not entered after each batch or fed-batch run. Instead, only a part of the

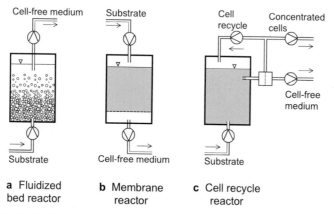

Fig. 6. Methods of cell retention in continuous operation: fixed or fluidized bed reactor with immobilized cells (a), membrane reactor (b), reactor with external sparator (c).

medium and cells is released and the cultivation then continued at once, eventually after adding new substrate. By this method, the productivity is increased because the length of the unproductive interval of the plant is reduced and the average cell concentration can be higher.

The previous modes of operation were concerned with the liquid phase and the supply of substrates. Another important substance in biotechnical processes is oxygen, and the following modes of operation with respect to it can be distinguished:

Aerobic Processes

In aerobic processes for the cultivation of strict or facultative aerobic microorganisms, the bioreactor is supplied with air that even might be supplemented by pure

Table 3. Examples for the preferred mode of operation and kinetic type of important industrial biotechnical processes

Product	Organism	Mode of operation Substrate	Oxygen	Type[a]
Ethanol	Saccharomyces cerevisiae, Zymmomonas mobilis	continuous or (repeated) batch	anaerobic/ microaerobic	I
Acetone/Butanol	Clostridium acetobutylicum	continuous	anaerobic	I
Baker's yeast	Saccharomyces cerevisiae	fed-batch	aerobic	I
Penicillin	Penicillium chrysogenum	fed-batch	aerobic	III
Cephalosporin	Acremonium chrysogenum	fed-batch	aerobic	III
Tetracycline	Streptomyces strains	fed-batch	aerobic	III
Alkaline protease	Bacillus lycheniformis	batch/fed-batch	aerobic	II
Amylases	Aspergillus oryzae	batch/fed-batch	aerobic	II
Pectinases	Aspergillus niger	batch/fed-batch	aerobic	II
Cellulases	Trichoderma resii	batch/fed-batch	aerobic	II
Invertases	Saccharomyces cerevisiae	batch/fed-batch	aerobic	
Lactic acid	Lactobacillus bulgaricus	batch	anaerobic	I
Propionic acid	Propionibacterium shermanii	batch	anaerobic/ micro-aerobic	I
Acetic acid	Gluconobacter suboxidans	batch	aerobic	I
Citric acid	Aspergillus niger	batch	aerobic	II
Polysaccharides	Xanthomonas campestris	batch	aerobic	
Amino acids	Corynebacterium glutamicum	batch	aerobic	II
Single cell protein	Candida utilis, S. cerevisiae, Methylophilus methylotrophus	batch/fed-batch	aerobic	I
Riboflavin	Eremothecium ashbyi	batch	aerobic	II
Vitamin B12	Pseudomonas denitrificans, Propionibacterium	batch aerobic	anaerobic/ microaerobic	II
Alkaloids	Claviceps paspali	batch	aerobic	III
Recombinant proteins	Saccharomyces strains Escherichia coli	batch/fed-batch	aerobic	
Immunglobulines monoclonal antibodies	Hybridoma cells	continuous	aerobic	III
Interferones	Mammalian cells	continuous	aerobic	III

[a] according to the classification of [19], see also chapter 2

oxygen. This supply normally has to cover the entire oxygen demand of the cells, since the solubility of oxygen in aqueous solutions is very low. In a typical aerobic process, e.g. the production of baker's yeast, low concentrations of oxygen have negative effects on the productivity. Therefore, if possible one tries to keep the concentration of dissolved oxygen at a relatively high percentage of the saturation value to avoid local oxygen limitation in less mixed regions of the reactor. In a large production tank, the oxygen supply is often the limiting factor for the attainable cell mass and the costs for aeration are a great part of the total production costs. Here, by economic reasons one has to accept low oxygen concentrations, or even oxygen limited conditions.

Anaerobic Processes

In anaerobic processes it is tried to keep the reactor oxygen-free. This is essential for strictly anaerobic microorganisms where oxygen is toxic. But also in the cultivation of facultative aerobic microorganisms, anaerobic operation may lead to the highest productivity or substrate turnover. An example is ethanol production by yeast. In the anaerobic case, the gas phase of the reactor consists mainly of the gaseous products of metabolism, mostly carbon dioxide.

Micro-Aerobic Processes

Some microorganisms (e.g. Propionibacteria) tolerate only very low oxygen concentrations, or the production rate is maximum under defined oxygen limited conditions (e.g. for Myxobacteria). In these cases one has to limit the oxygen transfer to the reactor by controlling the aeration rate, the stirrer speed, or the inlet oxygen concentration, for example by mixing air with nitrogen. This mode of operation is called micro-aerobic. Sometimes, typical anaerobic processes are also operated under micro-aerobic conditions. For the ethanol production, the viability of the yeast and in turn the productivity can be improved in this way.

Several examples for important industrial biotechnical processes and their modes of operation are given in Table 3. Selected processes are dealt with in detail in Parts C and D of this book.

References

1. Schügerl K (1998) Bioreaction Engineering III. Wiley, Chichester, Amsterdam
2. Rehm HJ, Reed G (1991) Biotechnology, vol 4. VCH, Weinheim
3. Bastin G, Dochain D (1990) On-line Estimation and Adaptive Control of Bioreactors, Elsevier, Amsterdam
4. Dunn IJ, Heinzle E, Ingham J, Prenosil JE (1992) Biological Reaction Engineering. VCH, Weinheim
5. Moser A (1988) Bioprocess Technology: Kinetics and Reactors. Springer, Berlin Heidelberg New York
6. Roels JA (1982) J Chem Tech Biotechnol 32:59
7. Sonnleitner B, Cheruy A (1997) J Biotechnol 53:173
8. Sonnleitner B (1997) J Biotechnol 52:175
9. Bailey JE (1998) Biotechnol Prog 14:8
10. Siimes T, Linko P, von Numers C, Nakajima M, Endo I (1995) Biotechnol Bioeng 45:135
11. Konstantinov KB, Yoshida T (1992) Biotechnol Bioeng 39:479
12. Glassey J, Ignova M, Ward AC, Montague GA, Morris AJ (1997) J Biotechnol 52:201

13. Montague G, Morris J (1994) Tibtech 12:312
14. Moser A (1993) Chem Biochem Eng Q 7(1):21
15. Bailey JE, Ollis DF (1986) Biochemical Engineering Fundamentals. McGraw-Hill, New York
16. Royce PN (1993) Crit Rev Biotechnol 13(2):179
17. Schügerl K (1991) Bioreaction Engineering I. Wiley, Chichester, Amsterdam
18. Schügerl K (1992) Bioreaction Engineering II. Wiley, Chichester, Amsterdam
19. Gaden, EL (1959) J Biochem Microbiol Technol Eng 1(4), 413

Part A
General Principles and Techniques

1 Bioreactor Models

Karl Schügerl, Karl-Heinz Bellgardt

1.1
Introduction

A common submerse bioreactor consists of a vessel provided with a mechanically moved agitator for power input, which is necessary for the intensification of the transfer processes. Various types and sizes of impellers are applied for agitator depending on the properties of the biological system, the cultivation medium, and the size of the reactor. The exact description of the fluid movement by a simple model is not possible because the main liquid flow caused by the stirrer is overlapped by turbulence fluctuations. The situation is made more complex by the presence of two or more phases. The accurate description of the biological, chemical and physical processes and their interrelation in stirred tank (ST) reactors is impossible, and therefore considerable abstraction is necessary. The same holds true for other submerse reactors, such as the bubble column (BC) and airlift tower loop (ATL) reactors, in which the power input is accomplished with the expansion of compressed gas and with a liquid pump, respectively. The prerequisite for the abstraction of the reactors is the knowledge of the influence of various parameters on the process performance. Therefore, in Sect. 1.2 the interrelation between the growth of microbial and animal cells and their physical and chemical environment is considered. Experimental investigations allow the description of the physical processes in the reactors, which are discussed in Sects. 1.3.1 and 1.4.1. The description of a simplified fluid dynamics in BC and ATL reactors is possible and considered in Sect. 1.4.2, in contrast to ST reactors, which is discussed later in Sect. 7.2. In Sects. 1.3.2 and 1.4.3 examples are given for the most common mathematical models and in Sect. 1.5 the connection is made to chapters 7 and 8 of this book.

1.2
Interrelations Between the Cells and Their Physical/Chemical Environment

The momentum, mass, and heat transfer processes have to be intensified by turbulence in the cultivation medium of high performance submersed reactors. High momentum transfer is needed for dispersion of the gas phase and increase of the interface between the gas and liquid phases in an aerated reactor. A high turbulence is necessary for high mixing intensity, i.e., quick dispersion of nutrients, dissolved oxygen, and acid/base added to the reactor, as well as for enhancing the exchange rate of gaseous and volatile components across the interface of the phases. The improved heat transfer is needed for uniform temperature in the reactor and to remove the heat produced by the living microorganisms and by the turbulent energy dissipation. Oxygen limitation of the growth is caused by low specific gas-liquid interfacial area and low exchange rate across the interfacial area. Non-uniform distributions of nutrient and dissolved oxygen in the reactor due to low mixing intensity can cause strong local concentration fluctuations, inhibition of the growth at high nutrient concentrations, and growth limitation at low nutrient and oxygen concentrations, respectively. Large pH fluctuations can impair the physiology of the cells as well. Temperature control can be a problem at low heat transfer rates. However, at high stirrer speed and excessive power input the cells can be damaged.

On the other hand, high cell density, substrate concentration, and highly viscous products, respectively, increase the viscosity of the cultivation medium and reduce the intensity of the turbulence and the rate of transfer processes. At low energy dissipation rate, large bubbles with short contact time are formed and eventually wall growth can occur. In addition, at low turbulence intensity local growth limitation and non-stirred dead water regions with exhausted nutrient and dissolved oxygen, zones appear in which cell lysis occurs.

1.3
Stirred Tank (ST) Reactors

1.3.1
Description of the Physical Processes in the Reactors

All agitator types (with axial as well as with tangential/radial main flow) cause circulation flow patterns in the reactor below and above the impeller plane, if the liquid rotation is hindered by baffles. This main flow is overlapped by turbulent fluctuations. Therefore, the flow path is chaotic and the circulation time is a stochastic quantity. In aerated liquid, the bubbles are dispersed and re-dispersed by the stirrer. Large bubbles quickly escape from the reactor, and small bubbles are dragged along with the liquid and recirculated. The higher the bubble recirculation rate, the higher is the gas hold up. At high cell concentrations, the oxygen consumption rate is high, oxygen is gradually exhausted in the bubbles along the recirculation loop. The oxygen concentrations in the gas and liquid phases depend on the local oxygen transfer rate (OTR) from the gas into the liquid phase, the age distribution of the bubbles in the reactor, and the oxygen consumption rate of the cells.

Only in the stirrer region is the oxygen transfer rate high. Some elements quickly return to the stirrer with fairly high oxygen concentration, others return later with low oxygen content, because the oxygen is consumed during the passage of the liquid elements along the different recirculation paths in the reactor, but only a low amount of oxygen is transferred from the gas phase into the liquid during this passage. The nutrient is added to the reactor close to the stirrer. It has to be quickly distributed to avoid local growth inhibitions and limitations, respectively. Its distribution in the reactor can be described by the liquid circulation as well. The nutrient concentration in the liquid depends on its consumption rate and the recirculation time distribution. Therefore, the process performance is influenced by the nutrient and oxygen consumption rates, the volumetric mass transfer coefficient, and the circulation time distributions of the phases.

   To compare the key variables, they are converted into quantities with common dimension: characteristic times [1]. This comparison indicates that the mass transfer processes and the oxygen consumption rate are the key process variables. The specific power input (Po/V), the volumetric gas flow (Q/V), and the cell concentration are the main control variables which influence these process variables.

1.3.2
Reactor Models

The ideally mixed continuous stirred tank reactor (CSTR) model assumes that in the reactor no local variation of the process variables occur. Therefore, the reactor can be described by position independent (lumped) parameters. Prerequisites are the ideally mixed gas and liquid phases, which can be tested by the determination of the residence time distributions (RTDs) of the phases in continuous reactors and by measuring-mixing time in batch reactors with global stimulus/response techniques. This model usually holds true only for small laboratory reactors with low viscosity cultivation media.

1.3.2.1
Model for the Ideal Stirred Tank Reactor

The reactor model dynamically describes the concentrations in the gas and liquid phase of the reactor in dependence on initial conditions, manipulating variables, and biological reactions of the microorganisms. For simplicity it is assumed that both phases are ideally mixed and impulse balances need not be considered. Therefore, the corresponding sub-models include only ordinary differential equations. A third sub-model describes the mass exchange between gas and liquid phase. For the ideal stirred tank reactor, the formal balance [2, 3]

$$
\underbrace{\begin{array}{c}\text{global change}\\\text{in the reactor}\end{array}}_{\text{storage element}} = \underbrace{\begin{array}{c}\text{global}\\\text{inflow} - \text{outflow}\end{array}}_{\text{convective transport}} - \underbrace{\begin{array}{c}\text{global}\\\text{inflow} - \text{outflow}\end{array}}_{\text{diffusive transport}} + \begin{array}{c}\text{global}\\\text{reaction}\end{array}
$$

yields ordinary differential equations as given below. A schematic diagram of a stirred tank reactor and the employed model variables are shown in Fig. 1.1. The reactor

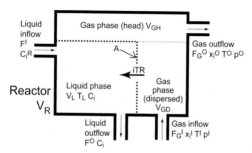

Fig. 1.1. Schematic diagram of a stirred tank reactor and related model variables

volume can be divided into three parts: liquid volume, V_L; volume of the dispersed gas phase, V_{GD}; and volume of the head space gas phase, V_{GH}:

$$V_R = V_L + V_G \tag{1.1}$$

$$V_G = V_{GD} + V_{GH} \tag{1.2}$$

which are all described by corresponding balances. By using the relative gas (ε_G) and liquid fractions (ε_L), the volume can be expressed as

$$V_L = \epsilon_L V_R \tag{1.3}$$

$$V_G = \epsilon_G V_R = (\epsilon_{GD} + \epsilon_{GH}) V_R \tag{1.4}$$

Mostly, the difference between dispersed and head gas phase is neglected, and both are considered as a sole gas volume.

Liquid-Phase Model

Here, the mass balances for a general compound i shall be considered, which are

$$\frac{dm_i(t)}{dt} = \dot{m}_i(t) = \underbrace{F^I(t) m_i^R(t)}_{\text{inflow}} - \underbrace{F^O(t) m_i(t)}_{\text{outflow}} + \underbrace{\dot{m}_i^Q(t)}_{\text{reaction}} + \underbrace{\dot{m}_i^{TR}(t)}_{\text{gas-liquid transfer}} \tag{1.5}$$

The final balances shall be given in terms of concentrations, which are related to mass by

$$m_i(t) = V_L(t) C_i(t) \tag{1.6}$$

and therefore

$$\frac{dm_i(t)}{dt} = V_L(t) \frac{dC_i(t)}{dt} + C_i(t) \frac{dV_L(t)}{dt} \tag{1.7}$$

Introducing these substitutions, together with the volumetric rate of biological reaction, Q_i (a positive value means production of compound i), and of mass exchange gas-liquid, iTR, into the above balance, Eq. (1.5), gives the model equation of the liquid phase:

$$\frac{dC_i(t)}{dt} = \frac{F^I(t)}{V_L(t)} C_i^R(t) - \frac{F^O(t)}{V_L(t)} C_i(t) - \frac{C_i(t)}{V_L(t)} \frac{dV_L(t)}{dt} + Q_i(t) + iTR(t) \tag{1.8}$$

where the change of volume is

$$\frac{dV_L(t)}{dt} = F^I(t) - F^O(t) \tag{1.9}$$

By substitution of this into Eq. (1.8), together with the definition of the dilution rate

$$D(t) = \frac{F^I(t)}{V_L(t)} \tag{1.10}$$

the concentration balance becomes for a compound i

$$\frac{dC_i(t)}{dt} = D(t) \left(C_i^R(t) - C_i(t) \right) + Q_i(t) + iTR(t) \tag{1.11}$$

Table 1.1. Modes of operation of a bioprocess with respect to the liquid phase

Mode	F^I	F^O	D	V_L
Batch	0	0	0	constant
Fed-batch	>0	0	F^I/V_L	increasing
Chemostat	>0	$=F^I$	F^I/V_L	constant
Continous	>0	>0	F^I/V_L	variable

This can be written for a number of compounds in vector notation as

$$\frac{dC(t)}{dt} = D(t)\big(C^R(t) - C(t)\big) + Q(t) + TR(t) \tag{1.12}$$

The different modes of operation of a bioreactor, depending on the rates of inflow to and outflow of the reactor, are summarized in Table 1.1. The only difference between fed-batch and continuous operation is that for the first there is no outflow of medium, $F^O=0$ [4, 5]. For multistage reactor cascades one has to set [6]

$$C_{ij}^R = C_{ij-1}, C_{i1}^R = C_i^R, \; j = \text{stage index}$$

The model equations for usual compounds are summarized in Table 1.3. In the balance of dissolved gases (d and e), the transport term by liquid exchange, $D(t)[C_i^R(t) - C_i(t)]$, is usually neglected because it is small compared to Q_i and iTR.

Gas-Phase Model

The gas-phase model is an important part of any model for a biotechnical process, if oxygen limited growth has to be considered, or measurements in the exhaust gas are used for process control. The gas phase can be further divided into dispersed gas phase and head space of the reactor, which are connected in series. The balance equations for both are identical, but with different parameters. Here, only the general form is given. The main components of the gas phase are oxygen, carbon dioxide, nitrogen, water (i=O_2, CO_2, N_2, H_2O), and additional gaseous products or substrates; they are advantageously described by their mole fractions, x_O, x_C, x_N, and x_W, respectively, which can be easily measured. To begin with, the balance for the number of particles is

$$\frac{dn_i(t)}{dt} = \underbrace{\dot{n}_i^I(t)}_{\text{inflow}} - \underbrace{\dot{n}_i^O(t)}_{\text{outflow}} - \underbrace{\dot{n}_i^{TR}(t)}_{\text{gas–liquid transfer}} \tag{1.13}$$

The volume or mole fractions are

$$x_i(t) = \frac{n_i(t)}{n_{tot}(t)} = \frac{V_i(t)}{V_G(t)} = \frac{p_i(t)}{p_G(t)} \tag{1.14}$$

By using the gas law

$$n_{tot} = \frac{pV_G}{RT} \tag{1.15}$$

the particle number becomes

$$n_i(t) = \frac{pV_G}{RT} x_i(t) \tag{1.16}$$

$$\frac{dn_i(t)}{dt} = \frac{pV_G}{RT} \frac{dx_i(t)}{dt} \tag{1.17}$$

With $\dot{V}=F$ the particle flow at gas inlet can be expressed as

$$\dot{n}_i^I(t) = \frac{p^I x_i^I(t)}{RT^I} F_G^I = Q_M^I V_L x_i^I(t) \tag{1.18}$$

where the inlet flow F_G^I is often specified under normal conditions, i.e.,

$$\dot{n}_i^I = \frac{F_G^{IN} x_i^I}{V^N}, \quad V_N = 22.5 \ \mathrm{l \, mole}^{-1} \tag{1.19}$$

The molar volumetric flow rate as used in Eq. (1.18) is defined by

$$Q_M = \frac{pF_G}{RTV_L} \tag{1.20}$$

Similarly, it follows for the outlet stream:

$$\dot{n}_i^O(t) = \frac{p^O x_i^O(t)}{RT^O} F_G^O = Q_M^O V_L x_i^O(t) \tag{1.21}$$

and the molar exchange flow with the liquid phase is

$$\dot{n}_i^{TR}(t) = \frac{iTR(t)V_L}{M_i} \tag{1.22}$$

where the positive direction of the transfer flux is defined into the liquid phase.

Using the above substitutions, Eqs. (1.18), (1.21), and (1.22) in Eq. (1.13) gives the differential equations for the mole fractions in the gas phase

$$\frac{dx_i^O(t)}{dt} = \frac{p^I T^O F_G^I(t)}{p^O T^I V_G} x_i^I(t) - \frac{F_G^O(t)}{V_G} x_i^O(t) - \frac{RT^O V_L}{M_i p^O V_G} iTR(t) \tag{1.23}$$

The resulting outlet gas stream, F_G^O, is usually not measured but it can be calculated since the summation of all mole fractions must equal one,

$$\sum_i x_i(t) \equiv 1 \qquad \sum_i \frac{dx_i(t)}{dt} \equiv 0 \tag{1.24}$$

By summing up the above differential equations, Eq. (1.23), for all compounds it follows

$$\frac{d1}{dt} = 0 = \frac{p^I T^O F_G^I(t)}{p^O T^I V_G} 1 - \frac{F_G^O(t)}{V_G} 1 - \frac{RT^O V_L}{p^O V_G} \sum_i \frac{iTR(t)}{M_i} \tag{1.25}$$

and the outlet flow can be calculated as

$$F_G^O(t) = \frac{p^I T^O F_G^I(t)}{p^O T^I} - \frac{RT^O V_L}{p^O} \sum_i \frac{iTR(t)}{M_i} \tag{1.26}$$

Table 1.2. Modes of operation of the gas-phase

Mode	Aeration	Inlet mole fractions	required balances
Aerobic, aeration by air	$F_G^I = F_{Gair}^I$	$x_O^I = 0.2096$ $x_C^I = 0.0003$ $x_N^I = 0.7901$ $x_W^I = 0$	O_2, CO_2, N_2, Water
Aerobic, aeration by oxygen	$F_G^I = F_{GO}^I$	$x_O^I = 1$ $x_C^I = 0$ $x_W^I = 0$	O_2, CO_2, Water
Anaerobic, no aeration	$F_G^I = 0$		CO_2, Water
Anaerobic, aeration by nitrogen	$F_G^I = F_{GN}^I$	$x_N^I = 1$ $x_C^I = 0$ $x_W^I = 0$	CO_2, Water
Additional gaseous products i in any mode		$x_i^I = 0$	products i (in addition)

Using this expression in Eq. (1.23) leads to the final model equations for the gas-phase

$$\frac{dx_i^O(t)}{dt} = \frac{p^I T^O F_G^I(t)}{p^O T^I V_G}\left(x_i^I(t) - x_i^O(t)\right) + \frac{RT^O V_L}{p^O V_G}\left(\sum_i \frac{iTR(t)}{M_i}x_i^O(t) - \frac{iTR(t)}{M_i}\right) \quad (1.27)$$

Some special cases for the modes of operation of the gas phase are summarized in Table 1.2. A further simplification can be obtained by considering inertial compound as being in quasi steady-state. This is justified because for such compounds there is no reaction, $Q_i=0$, which results in a very small mass transfer only when the concentrations in the gas phase are changing. The error is very small if one assumes that at any time the inflow of inertial compounds equals the outflow of the reactor, and thus neglects completely the balance equations. The mole fractions of the inertial compounds can then be eliminated from the remaining equations by using Eq. (1.24).

In many situations the gas phase can be taken as quasi-stationary, since its residence time is much lower compared to the liquid phase. This means mass storage (d/dt) can be neglected and, therefore, the reaction rates equal the transfer rates. This method can also be used for the calculation of characteristic biological parameters, such as oxygen uptake rate, OUR, carbon dioxide production rate, CPR, and respiratory quotient, RQ, by only using exhaust gas measurements. From Eq. (1.27) it follows with d/dt=0:

$$\frac{iTR}{M_i} - \sum_i \frac{iTR}{M_i}x_i^O = \frac{p^I F_G^I}{RT^I V_L}\left(x_i^I - x_i^O\right) = Q_M^I\left(x_i^I - x_i^O\right) \quad (1.28)$$

If only the compounds i=(O_2, CO_2) need to be considered, the "measured" oxygen uptake rate becomes after some calculations

$$OUR_m = -Q_{Om} \approx OTR_m \approx \underbrace{Q_M^I}_{\frac{F_G^N}{V^N V_L}} \frac{x_O^I(1 - x_C^O) - x_O^O(1 - x_C^I)}{1 - x_O^O - x_C^O} M_{O2} \qquad (1.29)$$

and the "measured" carbon dioxide production rate

$$CPR_m = Q_{Cm} \approx -CTR_m \approx -Q_M^I \frac{x_C^I(1 - x_O^O) - x_C^O(1 - x_O^I)}{1 - x_O^O - x_C^O} M_{CO2} \qquad (1.30)$$

The respiratory quotient is an important parameter for evaluation of the growth kinetics:

$$RQ = \frac{CPR}{OUR} \frac{M_{O2}}{M_{CO2}} \qquad (1.31)$$

This can be approximated by the measured respiratory quotient

$$RQ_m = \frac{x_C^I(1 - x_O^O) - x_C^O(1 - x_O^I)}{x_O^I(1 - x_C^O) - x_O^O(1 - x_C^I)} \approx RQ \qquad (1.32)$$

when substituting Eqs. (1.29) and (1.30) into Eq. (1.31).

Mass Transfer Gas-Liquid

The performance of aerobic cultivations with the need for continuous supply of oxygen is influenced significantly by the mass-transfer phenomena gas-liquid. In large production scale reactors mostly the oxygen transfer to the liquid phase is the limiting factor for the overall productivity. Many operating variables can effect the oxygen transfer in bioreactors, and the interactions between these variables are complex: bubble size distribution, medium composition, oxygen solubility, cell concentration, temperature, viscosity, power input [7, 8]. During a cultivation, all factors cannot usually be measured and analyzed separately. Therefore the complete modeling is quite difficult, and only a rough correlation with limited predictive power can be derived. The volumetric mass transfer coefficient $k_L a$ describes the nature of the mass transfer within the reactor and serves as an index of its performance. The entire transport pass, e.g., of oxygen from the gas phase to the microorganisms, involves a number of steps in series and in parallel. The overall mass transfer is determined by several factors:

1. Gas film resistance for transport between the bulk of the gas phase and the gas-liquid interface. It can be neglected compared to the resistance on the liquid side
2. Resistance of the gas-liquid interfacial area and liquid film resistance for transport to or from the bulk liquid phase. This is usually the major limiting step
3. Resistance within the bulk of the liquid phase. This is usually neglected due to eddy diffusion
4. Liquid film resistance around the cell and in the cell boundary. This can usually be neglected
5. Transport resistance through aggregates of cells. For flocculating or pellet-forming microorganisms, this can be a main limiting factor
6. The intracellular resistance with parallel reactions is very difficult to be quantified

Beside these steps in series there can be a parallel path by direct contact of cells with the gas-liquid interface. The adherence of the microorganisms at the bubble surface enhances the overall oxygen transfer.

The mass transfer rate between gas phase and liquid phase is proportional to the concentration gradient in the interfacial area and to the volumetric mass transfer coefficient. For oxygen and carbon dioxide the major mass transfer resistance is located at the liquid side and the mass transfer, either by adsorption or desorption, can be calculated by the single film model [9]:

$$iTR(t) = k_{iL}a(C_i^* - C_i) \tag{1.33}$$

where * indicates the saturation concentration in the gas-liquid interface. The transfer flux is proportional to the driving force, the concentration difference film-bulk liquid, and to the product of mass transfer coefficient, k_L, and specific interfacial area,

$$a = \frac{A}{V_L} \tag{1.34}$$

where A is the total contact area between gas and liquid phase. The saturation concentration for the dissolved gases depends on the mole fractions in the gas phase. After Henry's law,

$$x_{iL}^* = H_i p_i \tag{1.35}$$

the mole fraction in the liquid phase at saturation is proportional to the partial pressure in the gas phase. The mole fraction in the liquid phase can be approximated by

$$x_{iL} = \frac{n_i}{n_{tot}} \approx \frac{n_i}{n_W} \tag{1.36}$$

With

$$n_i = \frac{m_i}{M_i} \tag{1.37}$$

it follows

$$x_{iL} = \frac{m_i}{m_W} \frac{M_W}{M_i} \frac{V_L}{V_L} = \frac{C_i}{\rho_W} \frac{M_W}{M_i} \tag{1.38}$$

and with the mole fraction of the gas phase as

$$x_i = \frac{p_i}{p} \tag{1.39}$$

the saturation concentration is finally

$$C_i^* = \frac{H_i \rho_W M_i p}{M_W} x_i^0 = C_i'(p) x_i^0 \tag{1.40}$$

It depends on the operating pressure of the reactor and on the hydrostatic pressure of the liquid phase. For small reactors, $p=p^0$ can be taken, while for very high reactors the pressure at half of the reactor height can be used as an approximate average value. C_i' is the saturation concentration for a gas phase of a pure compound. For oxygen and carbon dioxide, these values are about $C_O' \approx 37 \ \mathrm{mg \, l^{-1}}$ and $C_C' \approx 1400 \ \mathrm{mg \, l^{-1}}$ at 37 °C and 1 bar, but these also depend on the ion concentration in the medium. In biotechnology, the concentration of dissolved oxygen is often specified as percentage of the saturation value:

$$pO_2 = \frac{C_O}{C_O' x_O^I} 100\% \tag{1.41}$$

The mass transfer resistance depends on the hydrodynamics in the reactor and the physical properties of the fermentation broth. Of special interest for further modeling is a correlation of the mass transfer coefficient to the molecular diffusion constant, which connects the parameters for different substances [10]:

$$k_{Li} \propto D_i^{r_1} \tag{1.42}$$

The exponent can be taken in the range r_1=0.5 (Higbie's model for small bubbles) to r_1=0.66 (model of Dankwerts for large bubbles). With this equation the volumetric mass transfer coefficient for a substance i can be calculated, while that for oxygen is known as

$$k_{Li} = \left(\frac{D_i}{D_O}\right)^{r_i} k_{LO} \tag{1.43}$$

The specific interfacial area, a, depends on the geometry of the system, the type of the stirrer, the power input, and the aeration rate [11–13]. The following correlation can be used [7]:

$$a\frac{V_L}{F_G^I} = r_3 \left(\frac{P}{F_G^I}\right)^{r_2} \tag{1.44}$$

For stirred tank reactors the specific power input is a function of the stirrer speed; in bubble columns it is determined by the expansion energy of the pressurized gas-inflow. This gives a final correlation, including Eqs. (1.43) and (1.44), of

$$k_{Li}a = r_4 D_i^{r_i} \frac{F_G^I}{V_L} \left(\frac{P}{F_G^I}\right)^{r_2} \tag{1.45}$$

The parameters r_1 to r_4 are difficult to predict and must be identified from experimental data. This correlation, together with Eqs. (1.11) and (1.33), gives the final model equation for dissolved oxygen. For the modeling of dissolved carbon dioxide, one has in addition to consider its dissociation according to [14–16]:

$$CO_2 + H_2O \leftrightarrow HCO_3^- + H^+ \leftrightarrow CO_3^{2-} + 2H^+ \tag{1.46}$$

Only the free, not dissociated, CO_2 contributes to the driving force for gas-liquid exchange. Bicarbonate can be mostly neglected. Nevertheless, under usual conditions for bioprocesses (pH=6.8 and T=25 °C), 75% of total carbon dioxide may be dissociated. From Eq. (1.46), the total carbon dioxide can be calculated as

$$[C_C]_{tot} = [C_C]_{free} + [HCO_3^-] + [CO_3^{2-}] \tag{1.47}$$

The dissociation reaction is very fast and always in the equilibrium given by

$$\frac{[HCO_3^-] \cdot [CO_3^{2-}]}{[C_C]_{free}} = K_C \tag{1.48}$$

Substitution of this equation into Eq. (1.47) leads to

$$[C_C]_{tot} = [C_C]_{free} + \frac{[C_C]_{free} K_C}{[H^+]} \tag{1.49}$$

and

$$[C_C]_{free} = \frac{[C_C]_{tot}[H^+]}{[H^+] + K_C} \tag{1.50}$$

where the proton concentration can be expressed as

$$[H^+] = 10^{-pH} \tag{1.51}$$

This leads to the balance of total carbon dioxide as shown in Table 1.3, Eq. (e). The lumped parameter models renounce the structure of the flow and consider only the averaged process variables. This allows to use global relationships between the process and control variables. Another typical example for the relationship of the

Table 1.3. Complete example for a CSTR model without inflow of cell mass and product

Substance	Balance equation
a) Substrate	$\dfrac{dC_S(t)}{dt} = D(t)\left(C_S^R(t) - C_S(t)\right) + Q_S(t)$
b) Cell mass	$\dfrac{dC_X(t)}{dt} = -D(t)C_X(t) + Q_X(t)$
c) Product	$\dfrac{dC_P(t)}{dt} = -D(t)C_P(t) + Q_P(t)$
d) Dissolved oxygen	$\dfrac{dC_O(t)}{dt} = Q_O(t) + \overbrace{k_{LO}a(C_{OX}O(t) - C_O(t))}^{OTR(t)}$
e) Dissolved (total) carbon dioxide	$\dfrac{dC_C(t)}{dt} = Q_C(t) + \overbrace{k_{LO}a\left(\dfrac{D_C}{D_O}\right)^{r_1}\left(C_C x_C(t) - \dfrac{10^{-pH}}{10^{-pH} + K_C} C_C(t)\right)}^{CTR(t)}$
f) Oxygen in the gas phase	$\dfrac{dx_O(t)}{dt} = \dfrac{p^I T^O F_G^I(t)}{p^O T^I V_G}\left(x_O^I(t) - x_O(t)\right)$ $+ \dfrac{RT^O V_L}{p^O V_G}\left(\dfrac{CTR(t)}{M_C} x_C(t) - \dfrac{OTR(t)}{M_O}(1 - x_O(t))\right)$
g) Carbon dioxide in the gas phase	$\dfrac{dx_C(t)}{dt} = \dfrac{p^I T^O F_G^I(t)}{p^O T^I V_G}\left(x_C^I(t) - x_C(t)\right)$ $+ \dfrac{RT^O V_L}{p^O V_G}\left(\dfrac{OTR(t)}{M_O} x_O(t) - \dfrac{CTR(t)}{M_C}(1 - x_C(t))\right)$

The superscript ^{O} for the outlet mole fraction was omitted in Eqs. d) to g) for simplicity

volumetric mass transfer coefficient as a function of the control variables in low
viscosity media is

$$k_L a \cdot \left(\frac{\nu}{g^2}\right)^{\frac{1}{3}} = r_4 \left(\frac{Po/V}{\rho(\nu g^4)^{\frac{1}{3}}}\right)^{r_1} \left(\frac{Q}{V}\left(\frac{\nu}{g^2}\right)^{\frac{1}{3}}\right)^{r_2} \tag{1.52}$$

which was developed by Schlüter et al. [17] for *Trichosporon cutaneum* cultivation
medium. Here the constants are: $r_4=7.94\times10^{-4}$, $r_1=0.62$, and $r_2=0.23$ for the flat
bladed disc turbine, and $r_4=5.89\times10^{-4}$, $r_1=0.62$, and $r_2=0.19$ for the INTERMIG impeller.

In highly viscous non-Newtonian media a more complex relationship is valid. E.g.,
Kawase and Moo-Young [18] recommended the following relationship for xanthan
production cultivation media, the validity of which was confirmed by Herbst et al.
[19]:

$$k_L a = 0.675 D_m^{0.5} \frac{\rho^{3/5}\{Po/V\rho\}^{[9+4n]/[1+n]}}{\{k/\rho\}^{[1+n]/2}\sigma^{3/4}} \left\{\frac{u_G}{u_B}\right\} \left\{\frac{\eta_{eff}}{\eta_W}\right\} \tag{1.53}$$

Here $u_G=u_B=0.265 \text{ m s}^{-1}$=bubble rise velocity, η_{eff}=effective viscosity of the cultivation medium and η_W=viscosity of water.

A large number of similar relationships have been recommended by various
authors (for review see [8]).

By means of the relationship

$$Q_X = k_L a(C_O^* - C_O)Y_{X/O} \tag{1.54}$$

the growth rate of the cells Q_X can be calculated by the measured volumetric mass
transfer coefficient $k_L a$ and the dissolved oxygen concentration C_O in the reactor if
the yield coefficient of the growth with respect to the oxygen consumption $Y_{X/O}$ is
known and the oxygen saturation at the interface is assumed.

Recirculation and Compartment Models

Recirculation time models assume the oxygen uptake in the stirrer region (micro
mixer) and the oxygen consumption along the circulation loop (macro mixer) according to Bajpai and Reuss (Fig. 1.2) [20, 21]. In batch reactors the fluid elements
enter from the micro mixer into the macro mixer and stay here for different periods

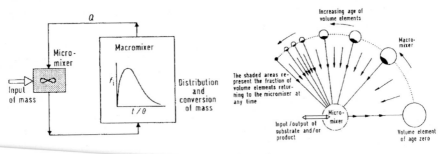

Fig. 1.2. Recirculation model of Bajpai and Reuss [20, 21]

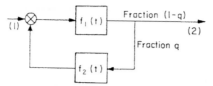

Fig. 1.3. Two compartment model of Sinclair and Brown [22]

of times before they return into the micro mixer. In continuous reactors, part of the liquid elements returns into the micro mixer, another part leaves the reactor.

In batch reactors, the recirculation time distribution can be determined by stimulus/response technique or by flow follower. In continuous reactors, the recirculation time is measured by a flow follower and the share of the fluid elements which recirculates and which leaves the reactor by stimulus response technique.

Only small bubbles recirculate in the cultivation medium many times. With increasing bubble size, the frequency of recirculation decreases and large bubbles do not recirculate at all, but leave the reactor immediately. Their residence time in the macro mixer is very short. Therefore, the distribution of residence time of the gas phase in the macro mixer is usually broad and depends on the bubble size distribution.

A two-region mixing model was recommended by Sinclair and Brown (Fig. 1.3) [22].(1–q) is the share of the fluid which does not enter into the recycle-region 2. The share q enters into the recycle-region 2; it recycles n times, before it returns into the region 1 and leaves the reactor. In the reactor with a single impeller, the weighting function of the reactor f(t) is given by the convolution ($\otimes$) integral:

$$f(t) = (1 - q) \sum_{n=0}^{n=\infty} q^n f_1(t) \otimes (f_1(t) \otimes f_2(t))^{n*} \tag{1.55}$$

where
$f_1(t)$ is the weighting function of region 1
$f_2(t)$ is the weighting function of region 2 without recycle
$f_1(t) \otimes f_2(t)$ is the weighting function of a single recycling
$(f_1(t) \otimes f_1(t))^{n*}$ is the weighting function of n-recycling through regions 1 and 2.

In reactors with m-stage impellers, each of the impeller regions can be described with a convolution integral with n-recycling in the m-regions.

Oosterhuis and Kossen [23] published a model for reactors with a two-stage impeller system and Bader [24, 25] for reactors with a multistage impeller system. These models were extended by Singh et al. [26] and by Mayr et al. [27].

Ragot and Reuss [28] developed a multiphase compartment model, which considers a recirculation in the liquid and gas including the mass transfer between the phases. This model is discussed in chapter 7 of this book.

Another model assumes three regions for the gas phase in a stirred tank reactor with a single impeller (Fig. 1.4) [29]:
1. The impeller region with gas re-dispersion (micro mixer), **region 2**
2. The recirculation region of small bubbles with long residence time (macro mixer), because the small bubbles follow the liquid flow pattern), **region 1**
3. the recirculation region of large bubbles with short residence time, **region 3**

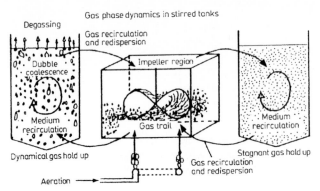

Fig. 1.4. Schematic display of the regions of gas dispersion and exchange (region 2), stagnant gas hold up of small bubbles (region 1) and dynamical gas hold up of large bubbles (region 3) of Rüffer et al. [29].

The RTD probability function is given by

$$F(t) = A/s_1(\exp(s_1t) - 1) + B/s_2(\exp(s_2t) - 1) + C/s_3(\exp(s_3t) - 1) \qquad (1.56a)$$

with

$$A = \frac{wa(s_1 + d)}{(s_1 - s_2)(s_1 - s_3)}, B = \frac{wa(s_2 + d)}{(s - s_1)(s_2 - s_3)}, C = \frac{wa(s_3 + d)}{(s_3 - s_1)(s_3 - s_2)} \qquad (1.56b)$$

Here s=Laplace transform variables, s_1 for region 1 (s_1=$-1/\tau_1$), s_2 for region 2 (s_2=$-1/\tau_2$), and s_3 for region 3 (s_3=$-1/\tau_3$) and τ_1, τ_2, τ_3 the mean residence times in regions 1, 2, and 3; a=exchange rate from region 2 to region 3; d=exchange rate from region 1 to region 2, and w=gas throughput.

In a 1 m³ stirred tank reactor with penicillin V production by *Penicillium chrysogenum*, the volume share of the regions varied with the time. The average values were: 64% (region 1), 33% (region 3), and 3% (region 2).

More detailed calculations were carried out by several authors [30–40], which are discussed in chapter 7 of this book.

1.4
Bubble Column (BC) and Airlift Tower Loop (ATL) Reactors

1.4.1
Description of the Physical Processes in the Reactors

In bubble column and airlift tower loop reactors the power input is accomplished with the expansion of compressed gas across a gas distributor or in various two-phase nozzles [8, 41, 42]. The bubbles rising in the column cause turbulence and mixing. They play a more important role in these reactors than in stirred tank ones. On account of the buoyancy forces of the bubbles, the bubble swarm moves upwards in the column. The same is valid for the riser of ATL reactors. In the down comer of ATL reactors with external loop, only the small bubbles (with buoyancy forces less than their resistance forces) are dragged along with the liquid and therefore the gas

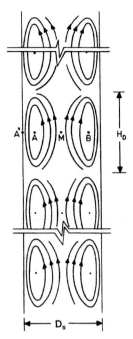

Fig. 1.5. Multiple circulation cells in a bubble column according to Joshi and Sharma [43]

bubbles follow the liquid flow with nearly plug flow character. The separate descrip-tion of the riser and the down comer of ATL reactors with internal loop is only possible for large units.

In a batch BC, the bubbles rise in the column center and transport the liquid up-wards. Close to the wall, the liquid moves downward and drags along the small bub-bles. By this means, multiple liquid circulation patterns consisting of axially superim-posed circulation cells with diameter to height ratio of unity are formed (Fig. 1.5) [43].

In ATL reactors the flow pattern in the riser depends on the transport capacity of the down comer. No back flows close to the wall are formed in the riser if the liquid transported to the top can return to the bottom through the down comer. At low transport capacity of the down comer, back flow and circulation patterns are formed similar to BC reactors.

With fed-batch operation, the liquid volume increases during the cultivation. The ATL reactor is already laid out to maintain the liquid circulation through the riser-down comer system at the beginning of the cultivation. A head part is installed to take up the increasing medium volume. Such reactors consist of a riser, down comer, and a head at the top of the column.

1.4.2
Flow Models

Three flow regimes can be distinguished in BC and ATL reactors. At low superficial gas velocity homogeneous (bubbly) flow prevails in columns of broad diameter range. The flow in down comer is always in this bubbly flow regime. By increasing superficial gas velocity above the flooding point large bubbles formed by coalescence are stabilized by the wall and fill the entire column diameter in laboratory columns, the homogeneous flow changes into slug flow. In large columns, large bubbles are formed by coalescence and bubble clusters move predominantly at the center line of the column and rise in a swarm of small bubbles with high velocity. The bubble size and velocity distribution are broad in this heterogeneous churn-turbulent flow, which is characteristic for industrial BC and in ATL reactors. The slip velocity u_s [44] and the drift flux F_{DF} [45] are only valid for bubbly flow:

$$u_s = \frac{u_G}{E_G} \pm \frac{u_L}{1 - E_G}, F_{DF} = u_s E_G (1 - E_G) = u_G (1 - E_G) \pm u_L E_G \qquad (1.57)$$

where minus and plus signs represent the co-current and counter-current flows. The model of Zuber and Findlay [46] applies for both bubble and churn-turbulent flow:

$$\frac{u_G}{E_G} = C_o(u_G + u_L) + u_{GD}, \qquad u_{GD} = u_{BT}(1 - E_G)^r \qquad (1.58)$$

where u_{GD} is the volume fraction weighted average drift velocity of the gas phase and u_{BT} is the bubble terminal velocity, which is a function of the bubble size. C_O is a constant.

Ueyama and Miyauchi [47] published a relationship for the flow pattern. Joshi and Sharma [43] developed a multiple-cell circulation model (Fig. 1.5). The average circulation velocity u_C is given by

$$u_C = 1.31 \left[gD_s \left(u_G \pm \frac{E_G u_L}{1 - E_G} - E_G u_{BT} \right) \right]^{1/3}, \qquad (1.59)$$

and for the liquid phase axial dispersion coefficient D_L:

$$D_L = 0.31 D_s u_c \qquad (1.60)$$

was obtained, which is in close agreement with the experimental results.

This relationship indicates that there is a direct relationship between the liquid circulation velocity u_c and the axial dispersion coefficient D_L in the liquid phase.

According to Eq. (1.60), the superficial gas and liquid velocities, u_G and u_L, the gas hold up E_G, and the column diameter D_s influence the process performance.

Verlaan et al. [48] used the drift flux model of Zuber and Findlay to model an ATL reactor.

1.4.3
Reactor Models

The ideally mixed BC and ATL reactor can be described by position independent (lumped) parameters. These prerequisites are fulfilled if the variation of the process parameters along the column (in BC) and riser and down comer (in ATL with internal loop) can be neglected. In continuous reactors the distribution of residence times is measured by stimulus/response technique. The ideal mixing in the liquid phase is assumed if it follows an exponential function. This model can only be applied for small laboratory reactors with internal loop and low viscosity cultivation medium. It is not suitable for describing the behavior of tall ATL reactors and reactors with external loop, because of the low intensity of the axial dispersion in liquid phase of the down comer. The flow has nearly plug flow character in the down comer.

The lumped parameter models allow the evaluation of the relationships between the process and control variables, e.g., for the volumetric mass transfer coefficient of oxygen. A typical example was developed by Akita and Yoshida [49] and confirmed by Kataoka et al. [50] in a 5.5 m diameter bubble column reactor:

$$\frac{k_L a D_s^2}{D_m} = 0.6 \left(\frac{\nu_L}{D_m}\right)^{0.5} \left(\frac{g D_s \rho_L}{\sigma}\right)^{0.62} \left(\frac{g D_s^3}{\nu_L}\right)^{0.31} E_G^{1.1} \tag{1.61}$$

The growth rate can be calculated by Eq. (1.54).

Recycling models are often used for the interpretation of the residence time distributions in ATL reactors [51–56]. However, apart from the investigations of Fröhlich et al. [54, 55], who characterized and modeled a 5-m^3 pilot plant ATL reactor with 10 m height and internal loop, in which baker's yeast was produced in continuous operation, most of the models were tested only on laboratory reactors.

In the following, a recycle model is presented, which was tested in another 5-m^3 pilot plant ATL reactor with 23 m overall height including a head section, in which baker's yeast was produced by fed-batch operation. The reactor fluid dynamics was characterized by following measurements: global gas residence time distribution (RTD) of the entire system and separate local gas RTDs in the riser, in the down comer, and in the head sections; liquid recirculation time and mixing times in the entire systems and local liquid mixing times in the riser, down comer, and head sections; radial bubble velocity profiles in the head section and radial liquid velocity profiles in the riser; radial and axial liquid dispersion by heat pulse technique and fractal analysis [56].

The residence time distribution E(t) in an ATL reactor with a head section for fed-batch operation is given by [57]:

$$E(t) = \left\{ \frac{r}{r-q} 0.5 \left(\frac{Bo_0 \tau_0}{\pi t}\right) \exp\left[-\frac{Bo_0(t-\tau_0)^2}{\pi t}\right] + \right. \tag{1.62}$$

$$\left. + \sum_{n=1}^{\infty} r^n 0.5 \left(\frac{Bo_n \tau_0}{\pi t}\right)^{0.5} \exp\left[-\frac{Bo_n(t-\tau_0)^2}{4t\tau_0}\right] \right\} \frac{(1-r)/\tau_0}{(1-r)[r/(r-q)]+r}$$

Here n=number of circulation, Bo_0=Bodenstein number in the riser-head section, Bo_c=Bodenstein number in the riser-down comer section, τ_0=mean residence time riser-head section, t_c=circulation time, r=fraction of the gas which is recirculated (small bubbles), q=fraction of the gas which does not recirculate (large bubbles),

$$t_0 = [\tau_0(1 + 2Bo_0^{-1}) + nt_c]/(1 + 2Bo_n^{-1}) \tag{1.62a}$$

$$Bo_n = X_n^{-1} - 2(X_n^{-2} + 4X_n^{-1})^{0.5} \tag{1.62b}$$

$$X_n = \left[\frac{(2Bo_c^{-1} + 8Bo_c^{-2})t_c^2}{(1 + 2Bo_c^{-1})^2} + (2Bo_0^{-1} + 8Bo_0^{-2})\tau_0^2\right] \cdot \left[\tau_0(1 + 2Bo_0^{-1}) + nt_c\right]^{-1} \tag{1.62c}$$

The identification of the model parameters was performed by a least-square fit-gradient method. The sum formation was broken off at n=5, since no higher signals could be observed.

On the other hand, microbial cultivation in ATL reactors can be simulated by axial dispersion models. Luttmann et al. [58, 59] modeled the cultivation of *Hansenula polymorpha* in an ATL reactor of 60 l volume with external loop 60 l by an axial dispersion model.

As an example, the modeling of the fed-batch cultivation of *Saccharomyces cerevisiae* on molasses in the 5-m³ pilot plant ATL reactor with a head on the top by means of an axial dispersion model is considered [60]. In addition to the fluid dynamic parameters listed above, the following cultivation variables are monitored: temperature, pH, pO_2 (in two different positions), the concentrations of cell mass, sugar, and ethanol in the cultivation medium, O_2 and CO_2 in off-gas.

The following assumptions were made:
- the reactor is isothermal
- quasi steady state prevails
- the concentrations of cell mass, sugar, and ethanol in the ATL reactor is uniform (experimentally confirmed)
- cells have the density of the cultivation medium
- the ideal gas law is valid
- mass transfer resistance in the gas film can be neglected
- for each of the three segments individual axial dispersion coefficients are valid
- the radial variation of the process variables can be neglected (uniform radial liquid velocity profile was confirmed)
- axial variation of the dissolved oxygen concentration is taken into account
- axial variations of the cross sections of the flow are taken into account
- boundary conditions between the segments were experimentally determined
- the metabolisms of the yeast is described by the Bellgardt [61] model, but the Monod model was used for the growth instead of the Monod-Blackman model

Figure 1.6 shows the scheme of the reactor system.

Regarding the one-dimensional mass transport equations for the gas phase:

$$-\frac{\partial}{\partial t}\left[A_j E_{Gj} dz \frac{P_j}{RT} x_{OGj}\right] - \frac{\partial}{\partial z}\left[A_j u_{Gj} \frac{P_j}{RT} x_{OGj}\right] dz -$$
$$- A_j(1 - E_G)dz(k_L a)_j\left(k_{Hj} P_j x_{OGj} - C_{LOj}\right) = 0 \tag{1.63}$$

Here j=r, d, k (r=riser, d=down comer and k=head).

Regarding the one-dimensional mass transport equations for the liquid phase:

$$-\frac{\partial}{\partial t}\left[A_j(1 - E_G)dz C_{Lij}\right] + \frac{\partial}{\partial z}\left[A_j\left(1 - E_{Gj}\right)D_{Lj}\frac{\partial C_{Lij}}{\partial z}\right]dz$$
$$- \frac{\partial}{\partial z}\left[A_j u_{Lj} C_{Lij}\right]dz + A_j\left(1 - E_{Gj}\right)dz(k_L a)_{ij}\left(k_H P_j x_{Gj} - C_{Lij}\right) -$$
$$- A_j\left(1 - E_{Gj}\right)dz C_{LXj} q_{ij} + Q_{LFj}\frac{dz}{H_{Fj}}C_{Lie} - Q_{LZj}\frac{dz}{H_j}C_{Lij} = 0 \tag{1.64}$$

Here C_{Li}=molar concentration of i in the liquid phase, H_j=height of the section j, H_{Fj}=height of the liquid entrance section j; i=O, S, E, X (O=oxygen, S=substrate, E=ethanol, X=cell mass). Mass transfer gas-liquid for S and X is nil. D_{Li}=axial dispersion coefficient of i in the liquid phase, C_{LX}=cell mass fraction in the liquid phase, q_i=specific mass fraction transport rate of i, Q_{LF}=volumetric rate of the feeding, C_{Lie}=concentration of i in the feed, Q_{LZ}=volumetric rate of the total liquid.

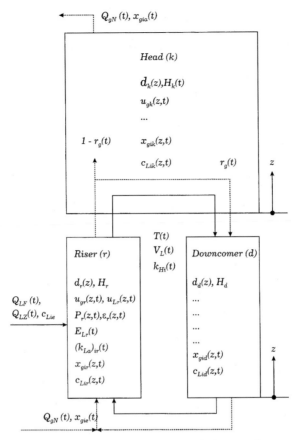

Fig. 1.6. System scheme of the pilot plant ATL reactor with the riser, down comer, head regions modeled by Strauss [60]

From the balances in the gas and liquid phase the axial dispersion model for the coupled riser, down comer, head system, a second order non-linear coupled partial differential equation system, is obtained.

For the *gas phase* ($i=O$; $j=r$, d, k):

$$-A_j E_{Gj} \frac{P_j}{RT} \frac{\partial x_{Gij}}{\partial t} - A_j u_{Gj} \frac{P_j}{RT} \frac{\partial x_{Gij}}{\partial z} - A_j \left(1 - E_{Gj}\right) (k_L a)_j \left(k_{Hi} P_j x_{Gij} - C_{Lij}\right) = 0 \quad (1.65)$$

and for the *liquid phase* ($i=O$, S, E, X; $j=r$, d, k):

$$- A_j \left(1 - E_{Gj}\right) \frac{\partial C_{Lij}}{\partial t} + D_{Lj} \left(\frac{\partial A_j}{\partial z} \left(1 - E_{Gj}\right) \frac{\partial C_{Lij}}{\partial z} - A_j \frac{\partial E_{Gj}}{\partial z} \frac{\partial C_{Lij}}{\partial z} + \right.$$

$$\left. + A_j \left(1 - E_{Gj}\right) \frac{\partial^2 C_{Lij}}{\partial z^2} \right) - A_j u_{Lj} \frac{\partial C_{Lij}}{\partial z} + A_j \left(1 - E_{Gj}\right) (k_L a)_{ij} \left(k_{Hi} P_j x_{Gij} - C_{Lij}\right) - \quad (1.66)$$

$$- A_j \left(1 - E_{Gj}\right) C_{LXj} q_{ij} + \frac{Q_{LFj} C_{Lie}}{H_{Fj}} - \frac{Q_{LZj} C_{Lij}}{H_j} = 0$$

With regard to the boundary conditions, in the gas phase the dispersion can be neglected. Therefore, differential equations are of first order, and only a single boundary condition is needed: The state of the entrance (0) corresponds to the state of the exit (H) of the section positioned before: Following boundary condition is valid for i=O; (O=oxygen):

$$x_{Gir}(0) = r_G x_{Gid}(0) + (1 - r_G)x_{Gie} \tag{1.67a}$$

$$x_{Gid}(H_d) = x_{Gir}(H_r), \; x_{Gik}(0) = x_{Gir}(H_r) \tag{1.67b}$$

Here r_G=the gas recirculation ratio, H_d=exit of the down comer, H_r=exit of the riser.

The experiments indicated that the exchange between the liquid phases of the three sections of the reactors are negligible. Therefore, the following boundary conditions were assumed at the entrance of the three sub-systems:

$$\frac{u_{Lr}(0)}{D_{Lr}(1 - E_{Gr}(0))}(C_{Lid}(0) - C_{Lir}(0)) + \frac{\partial C_{Lir}(0)}{\partial z} = 0 \tag{1.68a}$$

$$\frac{u_{Ld}(H_d)}{D_{Ld}(1 - E_{Gd}(H_d))}(C_{Lik}(0) - C_{Lid}(H_d)) + \frac{\partial C_{Lid}(H_d)}{\partial z} = 0 \tag{1.68b}$$

$$\frac{Q_{Lu}}{D_{Lk}(1 - E_{Gk}(0))A_k(0)}(C_{Lir}(H_r) - C_{Lik}(0)) + \frac{\partial C_{Lik}(0)}{\partial z} = 0 \tag{1.68c}$$

For the exit is was assumed that the dispersion coefficients are negligible, which was also proved by experiments:

$$
\begin{aligned}
&- A_r u_{Lr}\frac{\partial C_{Lir}}{\partial z} + A_r(1 - E_{Gr})(k_L a)_{ir}(k_{Hi}P_r x_{Gir} - C_{Lir}) \\
&\qquad - A_r(1 - E_{Gr})C_{LXr}q_{ir} + \frac{Q_{LFr}C_{Lie}}{H_{Fr}} = 0 \\
&- A_d u_{Ld}\frac{\partial C_{Lid}}{\partial z} + A_d(1 - E_{Gd})(k_L a)_{id}(k_{Hi}P_d x_{Gid} - C_{Lid}) \\
&\qquad - A_d(1 - E_{Gd})C_{LXd}q_{id} = 0
\end{aligned}
\tag{1.69}
$$

At the top of the head section

$$\frac{\partial C_{Lik}}{\partial z} = 0$$

is valid.

Initial conditions are

$$(j = r, d, k) : x_{GO_j}(z, 0), C_{LO_j}(z, 0) \text{ and } C_{LS_j}(z, 0) \text{ in steady state,}$$
$$C_{LE_j}(z, 0) = C_{LE0}, C_{LX_j}(z, 0) = C_{LX0} \tag{1.70}$$

The program package SLDGL, based on the self-adaptive difference method, was used [48, 49]. The computation was performed in the Regional Computer Centre of Lower Saxony (RRZN) Hannover. The agreement between the calculated and measured process variables was very good.

1.5
Conclusions

The lumped parameter models are suitable to model the performance in laboratory reactors. The liquid circulation models can be used for pilot plant reactors up to few cubic meters in volume. This holds true for the application of the axial dispersion model for BC and ATL reactors as well. However, they are unsuitable for modeling of large industrial reactors above $10\,m^3$ in volume. Only Computational Fluid Dynamics (CDF) is appropriate for modeling large stirred tank reactors [31–40, 62–66], bubble column and airlift tower loop reactors [67–73]. These methods are presented in chapter 7 for ST and chapter 8 for BC reactors in this book.

References

1. Sweere APJ, Luyben KChAM, Kossen NWF (1987) Enzyme Microb Technol 9:386
2. Dunn IJ, Heinzle E, Ingham J, Prenosil JE (1992) Biological Reaction Engineering, VCH, Weinheim
3. Schügerl K (1991) Bioreaction Engineering I. Wiley, Chichester, Amsterdam
4. Dunn IJ, Mor JR (1975) Biotechnol Bioeng 17:1805
5. Lim HC, Chen BJ, Creagan CC (1977) Biotechnol Bioeng 19:425
6. Durado A, Goma G, Albuquerque U, Sevely Y (1987) Biotechnol Bioeng 29:187
7. Mersmann A, Schneider G, Voit H, Wenzig E (1990) Chem Eng Technol 13:357
8. Schügerl K (1992) Bioreaction Engineering II. Wiley, Chichester, Amsterdam
9. Yagi H, Yoshida F (1977) Biotechnol Bioeng 19:801
10. Aiba S, Furuse H (1990) Biotechnol Bioeng 36:534
11. Kawase Y, Halard B, Moo-Young M (1992) Biotechnol Bioeng 39:1133
12. Ogut A, Hatch RT (1988) Can J Chem Eng 66:79
13. Vardar-Sukan F (1986) Process Biochemistry 4:40
14. Royce PNC, Thornhill NF (1991) AIChE Journal 37:1680
15. Pirt (1975) Principles of microbe and cell cultivation. Blackwell Scientific, Oxford
16. Noorman HJ, Luijkx GCA, Luyben K, Heinjen JJ (1992) Biotechnol Bioeng 39:1069
17. Schlüter V, Yonsel S, Deckwer WD (1992) Chem Ing Techn 64:474
18. Kawase Y, Moo-Young M (1988) Chem Eng Res Dev 66:284
19. Herbst H, Schumpe A, Deckwer WD (1992) Chem Eng Technol 15:425
20. Bajpai RK, Reuss M (1982) Can J Chem Eng 60:384
21. Reuss M, Bajpai RK (1991) Stirred tank models. In: Schügerl K (ed) Biotechnology 2nd edn, vol 4. VCH, Weinheim, p 299
22. Sinclair CG, Brown DE (1970) Biotechnol Bioeng 12:1001
23. Oosterhuis NMG, Kossen NWF (1984) Biotechnol Bioeng 26:546
24. Bader FG (1987) Improvements in multiturbine mass transfer models. In: Ho CS, Oldshue JY (eds) Biotechnology processes: scale up and mixing. AIChE, New York, p 96
25. Bader FG (1987) Biotechnol Bioeng 30:37
26. Singh V, Fuchs R, Constantinides A (1987) A new method for fermentor scale up incorporating both mixing and mass transfer effects. I. Theoretical basis. In: Ho CS, Oldshue JY (eds), Biotechnology processes: scale up and mixing. AIChE, New York, p 200
27. Mayr B, Horvat P, Nagy E, Moser A (1993) Bioprocess Eng 9:1
28. Ragot F, Reuss M (1991) A multi-phase compartment model for stirred bioreactors incorporating mass transfer and mixing. In: Reuss M, Gilles ED, Knackmuss HJ (eds) Biochemical engineering-Stuttgart. Gustav-Fischer, Stuttgart, p 184
29. Rüffer HM, Pethö A, Schügerl K, Lübbert A, Ross A, Deckwer WD (1994) Bioproc Eng 11:145
30. Spalding DB (1985) Computer simulation of two-phase flow with special reference to nuclear reactor systems. In: Lewis RW, Morgan K, Johnson IA, Smith WR (eds.) Computation techniques in heat transfer. Pineridge Press, p 1

31. Harvey PS, Greaves M (1982) Trans Inst Chem Eng 60:195
32. Placek J, Tavlarides LL (1985) AIChEJ 31:1113
33. Placek J, Tavralides LL Smith GW, Fort I (1986) AIChE J 32:1771
34. Ju SY, Mulvahill TM, Pike RW (1990) Can J Chem Eng 68:3
35. Issa RI, Gosman AD (1981) The computation of three-dimensional turbulent two-phase flows in mixer vessel. Numerical methods in laminar and turbulent flow. Preneridge Press, Swansea, p 829
36. Gosman AD, Lekakou C, Politis S, Issa RI, Looney MK (1992) AIChE J 38:1946
37. Bakker A, Akker HE (1991) A computational study on dispersion gas in a stirred reactor. Proc 7th European Congress on Mixing, p 189
38. Träghard C (1988) A hydrodynamic model for the simulation of an aerated agitated fed batch fermentor. Proc 2nd Int Conf Bioreactor Fluid Dynamics, Cambridge, p 117
39. Jenne M, Reuss M (1996) Chem Ing Techn 68:295
40. Hjertager BH (1996) Computational fluid dynamics (CFD) modelling and simulation of bioreactors. In: Larson G, Förberg C (1996) (eds) Bioreactor engineering, EFB, WP bioreactor performance and measuring and control. Saltsjöbaden
41. Deckwer WD (1991) Bubble column reactors. Wiley, Chichester
42. Chisti Y (1989) Airlift bioreactors. Elsevier Sci Publ, Amsterdam
43. Joshi JB, Sharma MM (1979) Trans I Chem E 57:244
44. Lapidus L, Elgin JC (1957) AIChE J 3:63
45. Wallis GB (1969) One-dimensional two-phase flow. McGraw Hill, New York
46. Zuber M, Findlay, JA (1965) Trans ASME J. Heat Transfer 87c:453
47. Ueyama K, Miyauchi T (1979) AIChE J 25:258
48. Verlaan P, Tramper H, van't Riet K, Luyben KChMA (1986) Chem Eng J 33:B43
49. Akita K, Yoshida F (1974) Ind. Eng, Chem Proc Des Dev 13:84
50. Kataoka H, Takeushi H, Nakao K, Yagi H, Tadahi T, Otake, T, Miauchi, T, Washini, K, Watanabe, K, Toshida, F (1979) J Chem Eng Japan 12:105
51. Fields PR, Slater NKH (1983) Chem Eng Sci 38:647
52. Warnecke HJ, Prüss J, Langemann H (1985) Chem Eng Sci 40:2321
53. Warnecke HJ, Prüss J, Leber L, Langemann H (1985) Chem Eng Sci 40:2327
54. Fröhlich S, Lotz M, Korte T, Lübbert A, Schügerl K, Seekamp M (1991) Biotechnol Bioeng 38:43
55. Fröhlich S, Lotz M, Korte T, Lübbert A, Schügerl K, Seekamp M (1991) Biotechnol Bioeng 37:910
56. Wan Liwei (1990) Detaillierte Untersuchungen des Mischverhaltens realer Mehrphasenreaktoren. VDI Verlag Reihe 8:Nr. 228
57. Rüffer HM, Wan Liwei, Lübbert A, Schügerl K (1994) Bioproc Eng 11:153
58. Luttmann R, Thoma M, Buchholz H, Schügerl K (1983) Comput and Chem Eng 7:43, 51
59. Luttmann R, Munack A, Thoma M (1985) Adv in Biochem Eng 32:95
60. Strauss G (1990) Modellbildung des Wachstums von Backhefe (Saccharomyces cerevisae) im Fed-Batch-Airliftreaktor. Diss University Hannover
61. Bellgardt KH, Kuhlmann W, Meyer HD (1982) Deterministic growth model of Saccharomyces cerevisiae. Parameter identification and simulation. In: Halme A (ed) Modelling and control of biotechnological processes, IFAC, Helsinki. Pergamon, Oxford, p 67
62. Schönauer W, Raith K (1980) Skizze der Arbeitsweise des SLDGL-Programmpaketes, Internal Report No. 18/80, Computer Centre, University of Karlsruhe
63. Schönauer W, Raith K, Glotz G (1981) The principle of the difference of the difference quotients as a key of the self-adaptive solution of non-linear partial difference equations, Comput Methods Appl Mech Eng 28:327
64. Patanakar SV (1980) Numerical heat transfer and fluid flow. McGraw Hill, New York
65. Fokema MD, Kresta SM (1994) Can J Chem Eng 72:177
66. Abid M, Xuereb C, Bertrand J (1994) Can J Chem Eng 72:184
67. Lapin A, Lübbert A (1994) Chem Eng Sci 49:3661
68. Becker S, Sokolichin A, Eigenberger G (1994) Chem Eng Sci 49:5747
69. Sokolichin A, Eigenberger G (1994) Chem Eng Sci 49:5735
70. Devanathan N, Dudukovic MP, Lapin A, Lübbert A Chem Eng Sci (1995) 50:2661

71. Lübbert A, Paaschen T, Lapin A (1996) Biotechnol Bioeng 52:248
72. Sokolichin A, Eigenberger G, Lapin A, Lübbert A (1997) Chem Eng Sci 52:611
73. Delnoij E, Kuipers JAM, van Swaaij WPM (1997) Chem Eng Sci 52:3623

2 Bioprocess Models

Karl-Heinz Bellgardt

2.1
Introduction

In this chapter, principles of modeling of biotechnical processes will be introduced. The focus is set onto models of the biological system within the entire bioprocess, as its most important steps are growth and product formation during the cultivation. The general modeling techniques will then be applied and outlined further in Parts C and D, where the methods will be applied to various processes. This chapter is structured as follows: an overview of the different types of models is given first. The models of the biotic phase are then introduced with increasing complexity, beginning with simple formal-kinetic models over structured models up to segregated population models.

For an analysis and mathematical modeling of biotechnical processes, one must look in somewhat more detail into the biological system, i.e., the microbe, plant, or animal cells that constitute the biotic phase of the bioreactor. It is well known that the cell metabolism and the resulting kinetics of growth and product formation are rather complex. Therefore, the interesting question is why this complex system often results in relatively simple growth kinetics, and can be described by only a few mathematical equations. One reason is that the functional blocks of metabolism operate together – coordinated by a network of metabolic regulation and of exchange of mass, charge and energy – to ensure the survival and reproduction of the organism. Another reason is the tremendous number of cells in the population in a bioreactor that hides individual variations in their growth and leads to a smoothed average behavior.

2.1.1
Intracellular Structure Elements

The cell of a living organism contains in a very simplified view several functional blocks and central pools of substances [1]. In the *catabolism*, the extracellular substrates, e.g., sugars, that were transported into the cell are degraded to smaller compounds, the so-called catabolites. Closely connected to the catabolism is the intermediary metabolism, in which a number of organic acids and phosphate-esters are synthesized. Well known catabolic systems that are found in many microorganisms are the glycolysis from glucose to pyruvate, and the citric-acid-cycle from pyruvate to carbon dioxide. Another important task of the catabolism is to provide chemical energy by substrate phosphorylation. This energy is required to drive energy-consuming reactions and to synthesize the macromolecular compounds of the cells. The energy yield of substrate phosphorylation is much lower compared to the respiration, but in anaerobic processes the substrate phosphorylation is often the only energy producing step of the metabolism. Many products released from microorganisms – the primary metabolites – arise directly from the catabolism, often in connection with the electron metabolism for regeneration of NADH. Examples are, beside the common product CO_2, typical "fermentation products", such alcohols, glycerol, lactate, butyrate, butanol, butane-diol, methane, and also products of aerobic "incomplete oxidation", e.g., acetic acid, gluconic acid. The formation of these products is often caused by an only partially functioning citric-acid-cycle.

In the *anabolism*, the catabolite and intermediate pool provided by the catabolism and intermediary metabolism, together with pools of chemical energy (ATP) and reduction equivalents (NADH), is used to synthesize higher-molecular cell material and structural compounds of the cell, like cell wall, membranes, organelles, and macro-molecules, such as carbohydrates, proteins, RNA, DNA, lipids. The anabolism is the main energy sink of the metabolism.

Regarding the *energy-metabolism* and *ATP-ADP-AMP-pool*, the cellular metabolism as a whole consumes energy that is produced by catabolism and respiration. The energy released in these processes is conserved in phosphorylated compounds, mainly in adenosine-tri-phosphate (ATP), the universal energy carrier of the cell. The less phosphorylated compounds are adenosine-di-phosphate (ADP) and adenosine-mono-phosphate (AMP). A second energy pool of minor importance is formed by the guanosine-phosphates (GTP, GDP, GMP).

With regard to the *electron-metabolism* and *NAD-NADH-pool*, the growth process is essentially an oxidation process with electron-transfer from the substrates to the final products. The biological oxidation is done by dehydrogenation via nicotine-amide-adenine-di-nucleotide (NAD) as a central proton acceptor. In some electron transfer reactions nicotine-amide-adenine-di-nucleotide-phosphate (NADP) and flavine-adenine-di-nucleotide (FAD) are involved. The reduction equivalents produced in the catabolism have to be continuously regenerated in parallel, either by the respiration, or by transferring hydrogen to the fermentative end-products of the metabolism.

The *respiration* is under aerobic conditions the most effective way for both regeneration of NADH to NAD and production of chemical energy as ATP by oxidative phosphorylation. The respiratory chain, located in the mitochondrial membrane, performs the biological oxy-hydrogen reaction:

$$\tfrac{1}{2}O_2 + NADH + H^+ + P/O \cdot ADP \rightarrow H_2O + NAD^+ + P/O \cdot ATP$$

The number of phosphorylations per atom oxygen (P/O-ratio) can be up to three, but is mostly found in the region from one to two.

The analysis of the flows of mass, energy, and electrons through the cell by the metabolic reactions is the basis for the mathematical modeling of the stoichiometry of the growth process. The simple fact that chemical elements, mass, energy, and charge must be conserved in any reaction, and in the metabolism as a whole, puts strong restrictions onto the microbial reaction network [2]. A few reactions can already determine the rates of many others. By using the related balances, far-reaching predictions on the yields of cell mass and products, especially primary products, can be obtained [6]. When the main substrates and final products of the growth are known, this prediction is often even possible with incomplete information on the involved intracellular reaction network.

2.1.2
Regulation of the Metabolism

In a living system as an open system, the steady flow of substances through metabolic reactions is the precondition for its lasting survival. Therefore, the cell depends greatly on the ability to stay far away from the chemical equilibrium where no reaction is possible. Most of the reactions of the metabolism are specifically catalyzed by enzymes [8]. This provides the cell with the possibility to adjust actively in a controlled manner the intracellular reaction rates, in response to the extracellular conditions (e.g., the available substrates), or depending on the intracellular state (e.g., the position in the cell propagation cycle).

There are two important mechanisms for this regulation.

Enzyme inhibition or *activation* is a very fast ad-hoc regulation on the reaction level. The regulatory substances – often substrates or products of the regulated reaction chain – form a complex with the enzyme and in this way modify immediately its catalytic activity in negative (inhibition) or positive direction (activation).

Induction or *repression* as a regulation on the genetic level influences the concentration of the enzymes by modifying the translation efficiency of genes coding for the enzyme. Compared to the previous type, this regulation is relatively slow since the synthetic capacity of the cells is rather limited. The time for full induction of enzymes is typically in the magnitude of the doubling time of the cell. The process of enzyme induction or repression often manifests itself in the so-called lag-phases, adaptation phases with time-varying growth activity of the cells. By these mechanisms the cells avoid the surplus synthesis of enzymes that are not required under the actual growth conditions. This frees energy, material, and cellular resources for synthesis of other proteins.

At the first look, with the complex regulation kept in mind, the growth kinetics seem to be too complicated for a mathematical modeling, i.e., for an exact quantitative description of the most important cellular processes. The situation becomes more difficult if one considers that structural details and the exact kinetics of many cellular regulation systems are still not known. Fortunately, as a matter of fact, both kinds of cellular regulation proceed well-coordinated in a hierarchical way to obtain a most effective metabolism. The regulation on the genetic level may be considered as the upper level of the hierarchy that is controlled rather roughly. On the lower reaction level, the fine-control with short response time is performed. For most biotechnical processes, as a very good approximation, the entire metabolism may be taken as being determined by only a few key-reactions. This may be formulated in the following rules which often can help to simplify the modeling.

2.1.2.1
Bottle-Neck Principle

The rate of the entire growth process is determined by the slowest reaction, which is called the rate-limiting step or the bottle-neck reaction.

In a linear reaction-chain, there can be only one bottle-neck at a time. In a complex branched reaction network as the cell is, there may be several bottle-necks in different branches of the network. The bottle-neck principle is sustained by the conservation laws of elements, mass, charge, and energy, by which rather independent system of the metabolism such as catabolism, anabolism, or respiration are tightly coupled. Therefore, quite often the modeling of the growth can be restricted to only a very few or even one rate-limiting reaction.

2.1.2.2
Optimality Principle

The metabolic regulation coordinates the metabolism in an optimal way. The optimality criterion is often the maximization of the growth rate.

This principle has direct consequences for the regulation on both the reaction and the genetic level. At first it means that intermediates or unnecessary end-products of the metabolism are not excessively accumulated, even if the bottle-neck is not the first reaction of the pathway. All fast reactions are down-regulated to meet the rate of the slowest reaction; they need not be considered in the model. For the modeling, the metabolism can be taken in the major parts as a quasi-stationary flow system without internal storage.

The second consequence of the optimality principle is that often, by formally solving the optimization problem for a suitable optimization criterion, an accurate and predictive model for the regulation on both hierarchical levels can be developed even without knowledge of the underlying mechanisms which are an outcome of the evolution by the competition of different species for substrates and biotopes. This property is used in the cybernetic models (Sects. 2.3.3 and 2.3.4). The main tasks of modeling are then to construct a very general model frame and to formulate the optimization criterion that mathematically represents the optimality principle.

2.1.3
Kinetics of Growth and Product Formation

The principal growth kinetics for a batch process are shown schematically in Fig. 2.1. After starting the process by inoculation, there is a maximum substrate concentration and low cell concentration. During the following cultivation with cell propagation, several different phases can be distinguished, although not all may be always present.

The *lag-phase* directly after inoculation is a period with an almost constant cell number, and only minor increase of cell mass. This can have two reason. The so-called *apparent lag-phase* is caused by a very low fraction of viable cells which first has to overgrow the inviable fraction to lead to a remarkable increase in cell number. During the *true lag-phase*, the regulatory adaptation of the cells to the growth conditions in the bioreactor take place, usually different from those in the preculture. There can also be morphological changes such as spore germination. The metabolic adaptation is performed by induction and repression of enzymes, or formation of cell organelles like ribosomes, and is accompanied by a steady increase of the growth rate.

During the following *exponential growth-phase*, cell mass and cell number increase exponentially, and the growth rate has reached its highest value for the given cultivation conditions and medium. The cellular metabolism is in a quasi-steady-

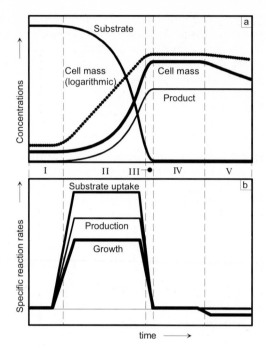

Fig. 2.1a,b. Schematic plot of a type-I bioprocess. I: Initial lag phase; II: Exponential growth phase; III: Transition phase; IV: Stationary phase; V: Declining phase: **a** kinetics of growth and production; **b** rate patterns

state, i.e., growth is balanced. The length of this phase depends mainly on the initial amounts of substrate and cell mass.

In the *transition phase*, the growth becomes limited by the low substrate concentration or accumulation of a toxic substance. In a batch process, the transition phase to complete exhaustion of the substrate is usually very short and no important part of the entire process. In a fed-batch process, this phase can be almost arbitrarily prolonged by substrate feeding.

In the *stationary phase*, there is no net growth, either due to exhaustion of substrate, or do to a balance of growth and lysis processes. The concentrations of cells and product remain constant. In a production run, the cultivation is normally stopped before this phase. The length of the stationary phase depends greatly on the microorganism strain and the cultivation condition.

In the *declining phase*, when there is no external substrate available, the cells can maintain their structural integrity by degradation of storage material, which reduces the cell mass. After exhaustion of the storage material, the cell viability decreases, followed by cell damage and autolysis. These processes not only reduce the total cell mass, but also the cell number.

Besides this pattern of the development of cell mass, the interrelation of growth rate and product synthesis is an important characteristic of any biotechnical process. Gaden [3] classified biotechnical processes according to their growth and production kinetics roughly into three different types, which all demonstrate a fairly distinctive rate pattern, as shown in Figs. 2.1–2.3.

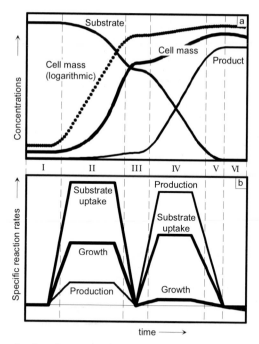

Fig. 2.2a,b. Schematic plot of a type-II bioprocess. I: Initial lag phase; II: First exponential growth phase with high growth rate; III: Lag or transition phase; IV: Production phase; V: Transition phase; VI: Stationary and declining phase: **a** kinetics of growth and production; **b** rate patterns

In processes of type I (production of primary metabolites), the main products appear directly as final products of catabolism, electron-metabolism, or energy metabolism. The production is in direct coupling with the growth. The rates of substrate uptake, cell growth, and product formation are virtually coincident, and the concentrations of cells and product show a very similar pattern. Hence these products are called growth-associated. There is basically only one key reaction or bottle-neck to be considered in a mathematical model. Examples for type-I processes are cultivations where the cells itself are the product, or anaerobic alcoholic fermentation.

To processes of type II (over-production of primary metabolites) belong such processes in which the formation of the main product is also coupled to catabolism or energy metabolism, but the product is not a final end-product of the pathway. It may be an essential intermediate of the metabolism or result from some side-reactions; therefore, the production may be partially decoupled from the growth. Under optimal growth conditions the product formation is low. Maximum product formation proceeds on the cost of growth rate, although there may still be a positive correlation of growth and product formation. Often the enhanced production is due to limitation of a certain growth factor, regulatory deficiencies of the cell, or it can be initiated by addition of a precursor. The production phase – eventually with exponential growth at slow speed – can be clearly distinguished from the first exponential phase of fast growth. The production phase might also be preceded by an additional

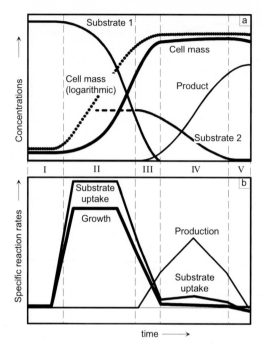

Fig. 2.3a,b. Schematic plot of a type-III bioprocess. I: Initial lag phase; II: Exponential growth phase; III: Lag or transition phase; IV: Production phase; V: Stationary and declining phase: **a** kinetics of growth and production; **b** rate patterns

transient or lag phase. The over-excretion of citric acid or amino acids belongs to this type, but also the production of some secondary metabolites. In this type of process the stoichiometry changes significantly and one has always to consider more than one rate limiting reaction in a model.

In processes of type III (non-growth associated products), the main product does not result from central growth processes – such as catabolism or energy metabolism – and it may have no obvious function in the metabolism. Antibiotic synthesis is again an example of this type. Often these products are synthesized under impaired growth conditions or in connection with morphological differentiation of the cells [4, 9]. In this type of processes there is no direct and simple coupling of growth rate and product formation; growth and production phases can be clearly distinguished. Typically, at high growth rates there is practically no product synthesis, while the maximum product formation only proceeds at very low or even without growth. In a normal batch process on a single substrate, the production phase is usually very short and the final product concentration low. The production phase can be extended by fed-batch operation, or in batch operation when the cells are grown on a mixture of two substrates. Then, in the first growth phase the substrate leading to high growth rate is used and growth is exponential. After exhaustion of the first substrate, the cells switch to the second substrate and low growth rates, where the production is induced. The production may continue into the stationary or declining phase. For this type of process, mostly structured or segregated models are appropriate (see Sects. 2.3 and 2.4).

2.1.4
General Model Structure for Biotechnical Processes

For analysis and design of biotechnical processes one must consider two aspects –
the biological reactions catalyzed by the microorganisms, and the numerous chemical and physical processes which precede, accompany, and follow them. Within the
bioreactor, the most important physical processes, which are intimately bound to the
biological reactions, are associated with the transport of material to and from the
surface of the microbial cells. The dependency of the biological reactions on the
microenvironment around the cells is called *microkinetics*. Due to the transport phenomena, the microenvironment of the cells will vary along different locations in the
bioreactor. For a continuous or fed-batch process the concentration of substrate will
be higher near to the substrate inlet, or in an aerobic process the oxygen concentration is high near to the air sparser or around the stirrer. The modeling of these
mixing effects is shown in Part B, while the modeling principles of transport effects
in films or flocs of microorganisms are presented in Sect. 2.1.5. The integral description of microkinetics in connection with the transport processes that may be
included in a reactor model, is called the *macrokinetics*. Ideally, one would always
attempt to model the transport phenomena and the microkinetics of the process
separately, because this model could be combined with proper reactor models to
predict the macrokinetics of different bioreactors. Unfortunately, at present, the effort for simulation of detailed reactor models is rather high and the methods for
analyzing the microenvironment of the cells are still rare. Therefore, what is usually
observed – and what is modeled, even for a stirred tank reactor – is always a kind of
macrokinetics. In the following, inhomogeneities in the reactor will not be explicitly
considered further.

As well as the transport phenomena not being modeled, the modeling of the biological system is also incomplete. The metabolism of a microorganism is a very complicated process, not only because the intracellular reaction network may be quite
complex, but also due to the large number of control loops on the genetic and reaction level that are overlaid to coordinate the elementary reactions. So even when the
microkinetics could be measured exactly, it would be impossible to establish a correct model in every detail because one has to introduce rigorous simplifications.
Another reason for uncertainty about the microkinetics is the inhomogeneity of the
microorganism population. The metabolism varies during the cycle of cell growth
and replication. Furthermore, there might be morphological differentiation of cells
accompanied by changes in the cell metabolisms. But what can be observed is only
an averaged behavior over a great number of cells in different states. To emphasize
all these facts and our limited knowledge about the process, the description of the
biological reaction may be called the *formalkinetic* model. This means a formal application of model equations that represent actually microkinetics to a process,
where only averaged macrokinetics can be measured. As an immediate consequence
the model parameters will change when the operating conditions of the reactor are
changed, or much more if a different reactor is used. This necessarily limits the
predictive power of models for biotechnical processes.

The general structure of a model for a bioreactor with gas phase and liquid phase,
and including the biotic phase can be drawn as in Fig. 2.4. The models for bioreactors (see Chap. 1) provide the concentrations of substrates, products, and cells as
input variables to the models of the biotic phase. To accomplish this, the reactor

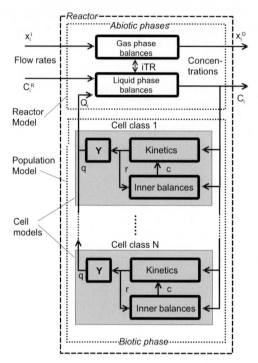

Fig. 2.4. General model structure for biotechnical processes

model must know the actual reaction rates which are the output of the biological models as described in this chapter. According to the level of the modeling and its degree of detail, different types of models of the biotic phase can be distinguished [5, 12].

Unstructured models (Sect. 2.2) have the greatest degree of simplification. The population of microorganisms or cells is viewed as homogenous biomass with their properties being time-invariant. The propagation is taken as continuous, i.e., the discontinuous processes of cell division are neglected. There are also no intracellular elements or states considered, and the model does not include inner balances. So the metabolism is assumed to be in a balanced state. The only determining variable is the cell mass or its concentration. Their properties are described by kinetics for substrate uptake, growth, and product formation that all only depend on the concentrations in the reactor [10].

Structured models (Sect. 2.3) consider the internal structural elements of the cells that may be metabolites, enzymes, or other cell constituents. By interaction of these elements, which are described by inner balances and related intrinsic reactions, the properties of the cells may become variable in time. With this approach it is possible to model the dynamics of metabolic regulation, e.g., during lag-phases of growth. But structured models also take the growth process as continuous and the population as homogenous; all cells are assumed to be identical and in the same state.

Simple segregated models (Sect. 2.4.1) discriminate inhomogeneous populations into a few classes of cells with different properties. The classes may represent different species, or within one species differences of physiological, morphological, genetic, or genealogical nature. It is assumed that the cell classes themselves are homogenous and thus can be represented by unstructured or structured models. *Segregated distribution models* (Sect. 2.4.6) include – besides – a second or further continuous independent variables. These can be, for example, mass, volume, or age of a single cell. The model equations are then partial differential equations. They are applied for very inhomogeneous populations with a continuous variation in some properties, which may be viewed as an infinite number of cell classes. Since the solution of such models requires a great numerical effort, the cell models are kept simple, e.g., unstructured models. Possible applications are descriptions of the variation of the metabolism during the cell cycle, or of the stability of plasmids in recombinant populations.

In any of the above types of models, the main task of modeling is essentially the establishment of consistent rate expressions for the main compounds of the process, cell mass, substrates, and products. Gaden [3] had already proposed in 1955 the concept of volumetric rates and specific rates. The volumetric rate Q is the rate of change of unit mass of reactant per unit volume of reactor as it appears in the balance equation, while the specific rate is the rate of change of unit mass of reactant per unit cell mass (here in vector notation for several compounds):

$$\mathbf{q}(\mathbf{C}(t), \mathbf{p}(t)) = \frac{\mathbf{Q}(\mathbf{C}(t), \mathbf{p}(t))}{C_X(t)} \tag{2.1}$$

The specific rates can be viewed as the reaction rates of a single cell. Beside the concentrations, other time varying operating parameters can also influence the microbial reactions. This is considered by the vector $\mathbf{p}(t)$. The definition of the specific reaction rates is useful since the biomass plays the role of a catalyst, i.e., the volumetric reaction rates are indeed proportional to cell mass, and usually the specific reaction rates are independent of it. The specific rates are preferred for the formulation of models for the biological system.

The net reaction rate $\mathbf{q}$ as exchange rate between cells and environment is mostly the result of several independent intracellular processes and reactions, which are called the intrinsic reaction rates, $\mathbf{r}$. Due to conservation laws and stoichiometric laws, the relation between the net reaction rates and the intrinsic specific reaction rates, catalyzed by and within the microorganisms, can be written as

$$\mathbf{q}(\mathbf{C})(t), \mathbf{p}(t)) = \mathbf{Y} \cdot \mathbf{r}(\mathbf{C}(t), \mathbf{p}(t)) \tag{2.2}$$

where $\mathbf{Y}$ is the matrix of stoichiometric or yield coefficients. When a sufficient number of kinetic expressions are specified for the intrinsic reactions, all of the net reactions can be calculated as dependent variables of the model. The stoichiometric matrix reflects the structure of the metabolic pathways within the cells. For simple models, as in the following sections, the yield coefficients are taken as formal parameters without looking at the details of the underlying reaction network. The analysis of this is done by flux analysis as shown in Chap. 3 and Part D. Another approach for determination of the yield coefficients are elemental balances [2, 10] and thermodynamic considerations [7, 11], which are not reported here.

Usual variables in growth models are as elements of the vector $\mathbf{q}$, the specific substrate uptake rate q_S, and the specific growth rate q_X. The latter is often referred to by the symbol μ. This will only be used here when the specific growth rate is specified as an intrinsic reaction.

Mostly, the kinetics of intrinsic reactions are of saturation type, i.e., for high substrate concentrations the reaction rate approaches a constant maximum value, and it vanishes if no substrate is available. The substrate uptake rate, for example, can then be written as

$$r_S = r_{S\,\max}\tilde{r}(C_S) \quad \text{where} \quad \tilde{r}(C_S) \rightarrow \begin{cases} 0 \\ 1 \end{cases} \text{for } C_S \rightarrow \begin{cases} 0 \\ \infty \end{cases} \tag{2.3}$$

Therefore, it is convenient to present the kinetics in normalized form, $\tilde{r}(C)$. The saturation characteristics of the kinetics greatly simplifies the modeling since it means that substrates or other growth factors being in excess mostly do not influence the growth and need not be considered in the model.

The maximum reaction rates, r_{max}, in the kinetics can also depend on the "physiological state" of the cells, which might be seen as a representation of the actual metabolic activity in terms of history of the growth process, concentrations of metabolites and enzymes, or the position in the cell propagation cycle. In the view of a simple unstructured model all these factors would lead to a variation of the maximum reaction rates in time, which only can be modeled by the more complex approaches.

2.1.5
Transport in Microbial Aggregates

Inhomogeneities in the liquid phase of the reactor or the near environment of the cells can have different causes:
- Insufficient mixing in the reactor, as in non-ideal STR, bubble column or airlift reactors. This leads to the development of strong spatial concentration profiles, mainly for substrates and products while the biomass mostly is more uniformly distributed in the reactor.
- Formation of relatively dense cell aggregates of more or less spherical geometry, such as clumps, flocs, or pellets. The dense biomass hinders the convective and diffusive transport within the aggregates, and causes spatial concentration gradients for substrates and products in their inner by an increased transport resistance. As a consequence, the metabolic activity of the cells, e.g., the growth rate, also shows spatial variations. This, together with the immobilization effect for the cells within the aggregates, gives rise to the development of an inhomogeneous biomass density.
- Growth of biofilms by attachment of cells to surfaces. The phenomena are similar to the previous case, but here the biomass is fixed in the reactor and the geometry of the film is flat. A simple segregated model will be given in Sect. 2.4.
- Immobilization of cells in carrier particles. This technique is used to facilitate retention of cells, the biocatalyst, in continuously operated bioreactors. This case is quite analogous to microorganism pellets although the transport resistance may be much higher.

In all of these cases an accurate modeling has to consider the transport effects and spatial variation of the concentrations. Models for transport phenomena are partial differential equations which require a much higher effort for simulation than lumped parameter models. Therefore, unstructured or simple segregated models are preferred for the description of the biological reaction in the transport models, because they do not increase the model complexity by introducing further state variables as in structured models. When applying structured models for an inhomogeneous environment, special care of its mathematical validity has to be taken as will be discussed in Sect. 2.3.

When convective flow is neglected and biomass is taken as a continuum, transport of substrates and products in aggregates of microorganisms can be described by the diffusion equation

$$\frac{\partial C_i(\mathbf{x}, t)}{\partial t} = \text{div grad}\left(D_i(\mathbf{x}, t)C_i(\mathbf{x}, t)\right) + Q_i(\mathbf{x}, t)$$
$$= \Delta(D_i(\mathbf{x}, t)C_i(\mathbf{x}, t)) + Q_i(\mathbf{x}, t) \tag{2.4}$$

where $\mathbf{x}$ is the spatial coordinate in three dimensions, and D_i the diffusion coefficient. Usually the following further simplifying assumptions are introduced for the modeling of low-density biofilms or pellets:
- The diffusion coefficient D_i is constant
- The system is highly symmetric and the concentration varies only along one spatial coordinate

A more detailed treatment can be found in [13]. Biofilms are usually taken as homogeneous along the supporting surface; thus transport is only in the perpendicular direction, z, and the balance equation simplifies to

$$\frac{\partial C_i(z, t)}{\partial t} - D_i \frac{\partial^2 C_i(z, t)}{\partial z^2} = Q_i(z, t) \tag{2.5}$$

where the reaction term Q_i is determined by the biological model. For a unique solution of this equation, boundary conditions at the attachment surface, $z{=}0$, where the transport flux vanishes

$$\frac{\partial C_i(0, t)}{\partial z} = 0 \tag{2.6}$$

and at the liquid side boundary layer, $z{=}L$, have to be specified, where L is the thickness of the biofilm:

$$C_i(L, t) = C_{iBulk}(t) \tag{2.7}$$

Here an additional transport resistance and concentration gradient within the bulk liquid phase is neglected. Furthermore, an initial condition

$$C_i(z, 0) = C_{i0}(z) \tag{2.8}$$

that must also fulfill the balance equation, Eq. (2.5), and boundary conditions, Eqs. (2.6) and (2.7), has to be given to solve a specific problem. Similarly, for a floc or pellet with spherical symmetry, the variation is only along the radial coordinate and the model becomes

$$\frac{\partial C_i(r,t)}{\partial t} - D_i\left(\frac{\partial^2 C_i(r,t)}{\partial r^2} + \frac{2}{r}\frac{\partial C_i(r,t)}{\partial r}\right) = Q_i(r,t) \tag{2.9}$$

$$\frac{\partial C_i(0,t)}{\partial r} = 0 \tag{2.10}$$

$$C_i(R,t) = C_{iBulk}(t) \tag{2.11}$$

$$C_i(r,0) = C_{i0}(r) \tag{2.12}$$

where r is the running radius and R the outer radius of the particle.

When the biomass density is high, the cells occupy a significant volume of the aggregate. This reduces the available volume for transport and increases the effective diffusion coefficient. The effect can be modeled by introducing the following dependency on the cell concentration into Eq. (2.4) [14, 15]:

$$D_{ieff} = \left(1 - \frac{C_X(\mathbf{x},t)}{C_{X\max}}\right)D_i \tag{2.13}$$

Mostly the biomass concentration is assumed as constant over time, i.e., growth of cells equals cell lysis. For a description of net biomass growth the balance equation, Eq. (2.4), can be used, where the diffusion coefficient is replaced by a factor being proportional to the growth rate of the cell [16, 17]:

$$D_{Xeff} = k_X q_X \tag{2.14}$$

The outgrowing biomass leads to a spatial extension of the cell aggregate. The balance equation then has to be solved as a moving boundary problem, or the outer boundary has to be extended to the bulk liquid.

2.2
Unstructured Models

In unstructured models, the biological reaction depends directly and only on macroscopic variables, the conditions in the bioreactor. Therefore, unstructured models are essentially combinations of elementary kinetics that mainly describe the influence of substrate and product concentrations or other variables, such as pH or temperature. The only biological state variable is the cell mass concentration C_X [10]. Nevertheless, many phenomena in biotechnological processes can be covered by this type of model. Beside the cell mass, only those other process variables that show great variations during the fermentation and have significant influence on the microbial behavior have to be considered in the model. Also some limited variations in the yield coefficients that can be incorporated by the unstructured models will be summarized here.

The most simple models, like the Logistic Law or the Cube-Root Law, do not include substrate limitation kinetics, but depend only on the cell mass [18]. They have found applications in the modeling of natural populations or filamentous microorganisms, and in such biotechnical processes where no information about substrates is available [22] or where growth is not restricted by the available substrate [19–21].

For more detailed process models this type of model is less interesting and not reported here.

2.2.1
Kinetics of Growth and Substrate Uptake

The simplest unstructured model describes the growth of a homogenous cell population, characterized by the cell concentration C_X, depending only on one limiting substrate with concentration C_S. When the specific growth rate is chosen as the only intrinsic rate, the specific rates and yield coefficients in the stoichiometric model, Eq. (2.2), become for constant parameters

$$\mathbf{q} = \begin{pmatrix} q_X \\ q_S \end{pmatrix} \quad \mathbf{Y} = \begin{pmatrix} 1 \\ -Y_{XS}^{-1} \end{pmatrix} \quad r = \mu_{max} \tilde{r}(C_S) \tag{2.15}$$

where the normalized kinetics $\tilde{r}(C_s)$ can be chosen from Table 2.1 to suit best the experimental data. The preferred application area of the different kinetics is also mentioned in the table. Especially in batch processes, the differences among them are less relevant if one keeps in mind measuring errors and remaining modeling errors. Therefore, the Monod-kinetics is mostly chosen since it is simple and has continuous derivatives with respect to C_S, which is of advantage for numerical simulation procedures.

Beside the substrate limitation, i.e., a reduction in substrate uptake rate with decreasing substrate concentration, often the reverse phenomenon is found in biotechnological processes, a reduction in growth, substrate uptake, or product formation

Table 2.1. Normalized growth kinetics for a single substrate [18]

	Name	Remark	Normalized kinetics $\tilde{r}(C_S)$
a	Monod	Only the substrate uptake step is rate limiting	$\dfrac{C_S}{K_S + C_S}$
b	Blackman	Another rate limiting step besides substrate uptake determines the maximum rate	$\min(1, K_S C_S)$
c	Teissier	Empirical equation	$1 - e^{-K_S C_S}$
d	Moser	Substrate uptake with higher order of reaction, e.g. for gaseous substrates	$\dfrac{C_S^N}{K_S^N + C_S^N}$
e	Contois	Diffusion layer around the cell	$\dfrac{C_S}{K_S C_X + C_S}$
f	Powel	Considers back diffusion of inner substrate	$\dfrac{C_S - K_1 \tilde{r}(C_S)}{K_S + C_S - K_1 \tilde{r}(C_S)}$
g	Mason and Millis	Parallel uptake by transport and diffusion	$\dfrac{C_S}{K_S + C_S} + K_1 C_S$
h	Vavilin	Extension of d) for initial inactivation by toxic substrates	$\dfrac{C_S^N}{K_S^{N-M} C_{S(t=0)}^M + C_S^N}$

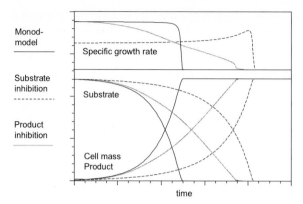

Fig. 2.5. Typical patterns of substrate and product inhibited growth in batch culture in comparison to non-inhibited growth

with increasing concentration of an effecting substance. This is usually referred to inhibition of growth or product formation by substrates or products. An example for such inhibitions is shown in Fig. 2.5 for a batch cultivation. The substrate inhibition prolongs the initial phase of the process while the product inhibition reduces the growth rate mainly in the final phase. This is due to the opposite variation in the substrate and product concentration. The latter increases during the process while the first is catabolized by the cells.

Table 2.2 contains a list of normalized pure substrate inhibition kinetics and Table 2.3 of normalized general inhibition kinetics that can be applied to products or substrates. Most of the inhibition kinetics are extensions of the Monod-equation and were derived from basic mechanisms of enzyme inhibition. Besides such true inhibitions, there may be other mechanisms leading to similar kinetics: the substrate or product might be toxic for the cell, for instance by denaturing enzymes or causing damage to the cell wall, membranes, and other cellular systems. Many of the kinetics cannot predict zero growth for a finite inhibitor concentration. Therefore, other empirical equations were proposed having such property that is mostly connected to toxic substrates (Table 2.2, h and i, and Table 2.3, g–k). The kinetics can be generalized for a joint modeling of multiple substrate limitations and inhibitions by combining substrate uptake kinetics in Table 2.1 or 2.2 and normalized inhibition terms in Table 2.3 to a product with several factors. Such generalized kinetics of different types were experimentally investigated by several authors, e.g, for anaerobic growing yeast [34, 36, 40] and for bacteria [35, 41]. The validity of different substrate inhibition kinetics for growth of yeast strains on alcohols was also successfully tested [37, 39].

As an example for combined substrate uptake/inhibition kinetics, diauxic growth on a mixture of two substrates is considered, where the uptake of one substrate is inhibited or even repressed by another, the preferred substrate. Here, the inhibition phenomenon is related to the regulation of the metabolism. A generalized kinetic expression is used for the uptake step of the second substrate (C_{S2}) that is inhibited by the preferred substrate (C_{S1}) and, furthermore, by a toxic product C_P. By choosing the Monod-kinetics (Table 2.1, a) for the uptake of the second substrate, non-compe-

Table 2.2. Substrate uptake kinetics including inhibition [18]

	Name	Normalized kinetics $\tilde{r}(C_S)$
a	Haldane (uncompetitive)[a]	$\dfrac{C_S}{K_S + C_S\left(1 + \frac{C_S}{K_I}\right)}$
b	Ierusalimsky (non-competitive)[a]	$\dfrac{C_S}{K_S + C_S}\,\dfrac{1}{\left(1 + \frac{C_S}{K_I}\right)}$
c	Aiba, Edwards	$\dfrac{C_S}{K_S + C_S}\,e^{-\frac{C_S}{K_I}}$
d	Yano (generalized uncompetitive)	$\dfrac{C_S}{K_S + C_S\left(1 + \sum\limits_{i=1}^{N}\left(\frac{C_S}{K_{Ii}}\right)^i\right)}$
e	Teissier type	$e^{-\frac{C_S}{K_I}} - e^{-K_S C_S}$
f	Webb	$\dfrac{C_S\left(1 + \frac{C_S}{K_{I1}}\right)}{K_S + C_S\left(1 + \frac{C_S}{K_I}\right)}$
g	Hill (allosteric)	$\dfrac{C_S^N}{K_S^N + C_S^N}$
h	Wayman and Tseng (toxic substrate)	$\dfrac{C_S}{K_S + C_S} - K_I \min(C_{SI} - C_S, 0)$
i	Chen (toxic substrate)	$\left(1 - \dfrac{C_S}{K_I}\right)\dfrac{C_S}{K_S + C_S - (K_{S2}C_S)^2}$
j	Tan et al. [42]	$\dfrac{\sum\limits_{i=1}^{M}\left(\frac{C_S}{K_{Ii}}\right)^i}{K_S + \sum\limits_{i=1}^{N}\left(\frac{C_S}{K_{Ii}}\right)^i}$

[a] Uncompetitive and non-competitive substrate inhibition are of the same type of function

titive inhibition by the first substrate (Table 2.3, c), and a Levenspiel-type inhibition for the product (Table 2.3, i), the resulting model becomes

$$q_{S2} = q_{S2\,\max}\,\frac{C_{S2}}{K_{S2} + C_{S2}}\,\frac{K_{I1}}{K_{I1} + C_{S1}}\left(1 - \left(\frac{C_P}{K_{I2}}\right)^N\right) \tag{2.16}$$

For a given practical problem, one should not expect that any of the kinetics gives a true mechanistic description. Instead, an equation with as few as possible parameters should be chosen that fits well to the experimental data.

A further extension of the single substrate uptake kinetics is necessary for the description of growth on mixtures of several substrates. Various attempts have been made to develop unstructured models for multiple limitations that are summarized in Table 2.4. The single substrate terms can be specified by the elementary

Table 2.3. Inhibition kinetics for single inhibitor [18]

	Name	Normalized kinetics $\tilde{r}(C_S)$
a	Competitive[a]	$\dfrac{C_S}{K_S\left(1+\frac{C_I}{K_I}\right)+C_S}$
b	Uncompetitive[a]	$\dfrac{C_S}{K_S+C_S\left(1+\frac{C_I}{K_I}\right)}$
c	Ierusalimsky (non-competitive)	$\dfrac{1}{1+\frac{C_I}{K_I}}$
d	Yano and Koya (generalized uncompetitive)	$\dfrac{1}{1+\left(\frac{C_I}{K_I}\right)^N}$
e	Aiba, Edwards	$e^{-\frac{C_I}{K_I}}$
f	Teissier type [a]	$e^{-\frac{C_I}{K_I}}-e^{-K_S C_S}$
g	Ghose and Tyagi, Dagley and Hinshelwood	$1-\dfrac{C_I}{K_I}$
h	Bazua and Wilke	$\left(1-\dfrac{C_I}{K_I}\right)^{\frac{1}{2}}$
i	Levenspiel	$\left(1-\dfrac{C_I}{K_I}\right)^N$
j	Han and Levenspiel	$\left(1-\dfrac{C_I}{K_I}\right)^N\dfrac{C_S}{K_S\left(1-\frac{C_I}{K_I}\right)^M+C_S}$
k	Luong [38]	$1-\left(\dfrac{C_I}{K_I}\right)^N$

[a] These kinetics include the substrate uptake step

Table 2.4. Kinetics for the specific growth rate by multiple limitation of N substrates [18]

	Type	Kinetics for the specific growth rate	Remarks
a	Interacting model for essential substrates	$\mu(C_{S1},C_{S2},C_{S3},\dots C_{SN})=\mu_{max}\tilde{r}_{S1}(C_{S1})$ $\tilde{r}_{S2}(C_{S2})\tilde{r}_{S3}(C_{S3})\dots\tilde{r}_{SN}(C_{SN})$	
b	Non-interacting model for essential substrates	$\mu(C_{S1},C_{S2},C_{S3},\dots C_{SN})=\min(\mu_{S1}(C_{S1}),$ $\mu_{S2}(C_{S2}),\mu_{S3}(C_{S3}),\dots,\mu_{SN}(C_{SN}))$	$\mu_{Si}(C_{Si})=\mu_{Simax}\tilde{r}_{Si}(C_{Si})$
c	Model for growth-enhancing and alternative substrates	$\mu(C_{S1},C_{S2},C_{S3},\dots C_{SN})=\mu_{S1}(C_{S1})+$ $\mu_{S2}(C_{S2})+\mu_{S3}(C_{S3})+\dots+\mu_{SN}(C_{SN})$	$\mu_{Si}(C_{Si})=\mu_{Simax}\tilde{r}_{Si}(C_{Si})$

kinetics as presented before. There are several possibilities for the interrelation of growth rate and concentrations of the substrates, depending on the kind of substrate and the microorganisms [46]. All substrates can be essential; that means there is no growth if only one of the substrates is lacking. Here, in the interactive model, all substrates together determine the growth rate. In the non-interacting model, at any time only the substrate with the strongest limitation governs the growth rate. This leads to a selection of the minimum growth rate allowed among all substrates. Equations (a) and (b) of Table 2.4 can be translated into the general vector form of the stoichiometric model, Eq. (2.2), by choosing the specific growth rate as intrinsic reaction. The vectors of the specific reaction rates, including cell mass formation and substrate uptake, and the vector of yield coefficients, are then for essential substrates

$$
\mathbf{q} = \begin{pmatrix} q_X \\ q_{S1} \\ q_{S2} \\ q_{S3} \\ \vdots \\ q_{SN} \end{pmatrix} \qquad \mathbf{Y} = \begin{pmatrix} 1 \\ -Y_{XS1}^{-1} \\ -Y_{XS2}^{-1} \\ -Y_{XS3}^{-1} \\ \vdots \\ -Y_{XSN}^{-1} \end{pmatrix} \qquad r = \mu \qquad (2.17)
$$

A typical example for the presence of two essential substrates is growth of obligate aerobic microorganisms under limitation of the carbon source and oxygen. Simulations of steady states in a chemostat after [23] are shown in Fig. 2.6 for both the interacting and non-interacting substrate uptake kinetics. The model is summarized in Table 2.5. At low dilution rates, where growth is limited by the carbon source, the cell mass decreases due to endogenous or maintenance metabolism, as explained in the following section. This can be well described by the model of Herbert, Eq. (2.20). Above a dilution rate of $D=0.16\,\mathrm{h}^{-1}$, growth switches from glycerol to oxygen limited, as can be seen by the very low oxygen concentration and increasing concentration of the carbon source. Near the critical dilution rate of about $D=0.54\,\mathrm{h}^{-1}$, both substrates are present in high concentrations. There, a second switch in the limiting compound

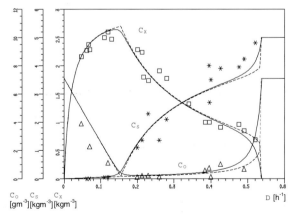

Fig. 2.6. Simulation and experimental data for carbon and oxygen limited growth in a chemostat after Sinclair and Ryder [23]

Table 2.5. Example for an unstructured model for growth on two essential substrates, glycerol and oxygen, in continuous cultivation [23]

	Description	Model equations
a	General vector notation, see Eqs. (1.12) and (2.1)	$\dfrac{d\mathbf{C}(t)}{dt} = D(\mathbf{C}^R - \mathbf{C}) + \mathbf{Q} + \mathbf{TR}$ where $\mathbf{Q} = \mathbf{q}C_X$
b	Specialized vectors, $\mathbf{r}$ considers endogeneous metabolism	$\mathbf{C} = \begin{pmatrix} C_X \\ C_S \\ C_O \end{pmatrix} \quad \mathbf{q} = \begin{pmatrix} q_X \\ q_S \\ q_O \end{pmatrix} = \mathbf{Yr}$ $\mathbf{Y} = \begin{pmatrix} 1 & -1 \\ -Y_{XS}^{-1} & 0 \\ -Y_{XO}^{-1} & 0 \end{pmatrix} \quad \mathbf{r} = \begin{pmatrix} \mu(C_S, C_O) \\ \mu_E \end{pmatrix}$
c	Intrinsic rate expression based on Monod-kinetics for carbon source and Contois-kinetics for oxygen	Interacting $\mu = \mu_{\max} \dfrac{C_S}{K_S + C_S} \dfrac{C_O}{K_O C_X + C_O}$ Non − interacting $\mu = \mu_{\max} \min\left(\dfrac{C_S}{K_S + C_S}, \dfrac{C_O}{K_O C_X + C_O} \right)$
d	Explicit model equations for the liquid phase, for the gas phase model, see Table 1.3	$\dfrac{dC_X(t)}{dt} = -DC_X(t) + \mu(C_S, C_O)C_X(t) - \mu_E C_X(t)$ $\dfrac{dC_S(t)}{dt} = D(t)\left(C_S^R(t) - C_S(t)\right) - \dfrac{\mu(C_S, C_O)C_X(t)}{Y_{XS}}$ $\dfrac{dC_O(t)}{dt} = \underbrace{k_{LO}a\left(C_O' x_O(t) - C_O(t)\right)}_{OTR} - \dfrac{\mu(C_S, C_O)C_X(t)}{Y_{XO}}$

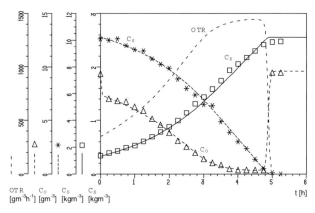

Fig. 2.7. Simulation and experimental data for carbon and oxygen limited growth of Methylomonas M15

can be found for the non-interacting model. A theoretical investigation on oxygen-limited growth by a non-interacting model was presented by Lidén et al. [43] and general comments on its applicability were given by Baltzis and Fredrickson [47]. A simulation of an oxygen limited bacterial batch cultivation with methanol as carbon source is shown in Fig. 2.7. The model resembles that of Table 2.5, but with μ given as an interactive model with double Monod-kinetics. In the oxygen limited period from $t=3.5$–4.8 h, growth proceeds only linearly in time due to the almost constant oxygen transfer (OTR) into the reactor.

Another case for growth on multiple substrates is when one of the substrates is already sufficient for growth and others are used up in parallel or sequential. For modeling of this kind of phenomena the entire kinetics can be constructed by the sum of elementary kinetics (Table 2.4, c). This model can be advantageously transformed into the general form by using each summand as an inner reaction:

$$
\mathbf{q} = \begin{pmatrix} q_X \\ q_{S1} \\ q_{S2} \\ q_{S3} \\ \vdots \\ q_{SN} \end{pmatrix} \quad \mathbf{Y} = \begin{pmatrix} 1 & 1 & 1 & \cdots & 1 \\ -Y_{XS1}^{-1} & 0 & 0 & \cdots & 0 \\ 0 & -Y_{XS2}^{-1} & 0 & \cdots & 0 \\ 0 & 0 & -Y_{XS3}^{-1} & \cdots & 0 \\ \vdots & \vdots & \vdots & \ddots & \vdots \\ 0 & 0 & 0 & \cdots & -Y_{XSN}^{-1} \end{pmatrix} \quad \mathbf{r} = \begin{pmatrix} \mu_{S1} \\ \mu_{S2} \\ \mu_{S3} \\ \vdots \\ \mu_{SN} \end{pmatrix} \quad (2.18)
$$

An example of growth-enhancing substrates is given in Fig. 2.8. In the simulation of the cultivation of *P. vulgaris* on glucose and citric acid the model developed by Tsao and Yang [24] as listed in Table 2.6 was used. The reduction of the specific growth rate can clearly be seen after exhaustion of citric acid at $t=15$ h.

In any case, compared to the simple kinetics, the multi-substrate models have to consider much more information about the structure of the metabolism, and these elementary models will only match simple cases. A number of different models for growth of yeast on the alternative, growth-enhancing substrates glucose and fructose, ranging from simple Monod-kinetics up to extended kinetics including inhibition terms, were presented by Barford et al. [44]. A satisfactory simulation of the entire cultivation could only be obtained by introducing a switch-condition for cer-

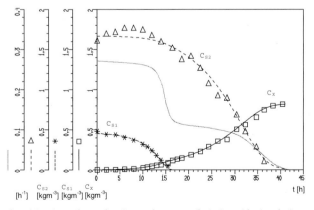

Fig. 2.8. Batch cultivation of *P. vulgaris* on glucose and citric acid, simulation and experimental data for growth enhancing substrates after Tsao and Yang [24]

Table 2.6. Example for an unstructured model for growth on two alternative substrates, glucose and citrate, in batch cultivation [24]

	Description	Model equations
a	General vector notation, see Eqs. (1.12) and (2.1)	$\dfrac{d\mathbf{C}(t)}{dt} = D(\mathbf{C}^R - \mathbf{C}) + \mathbf{Q} \quad \text{where} \quad \mathbf{Q} = \mathbf{q}C_X$
b	Specialized vectors	$\mathbf{C} = \begin{pmatrix} C_X \\ C_{S1} \\ C_{S2} \end{pmatrix} \quad \mathbf{q} = \begin{pmatrix} q_X \\ q_{S1} \\ q_{S2} \end{pmatrix} = \mathbf{Yr}$ $\mathbf{Y} = \begin{pmatrix} 1 & 1 \\ -Y_{XS1}^{-1} & 0 \\ 0 & -Y_{XS2}^{-1} \end{pmatrix} \quad \mathbf{r} = \begin{pmatrix} \mu_1(C_{S1}) \\ \mu_2(C_{S2}) \end{pmatrix}$
c	Intrinsic rate expressions based on Monod-kinetics	$\mu_1 = \mu_{\max 1} \dfrac{C_{S1}}{K_{S1} + C_{S1}} \qquad \mu_2 = \mu_{\max 2} \dfrac{C_{S2}}{K_{S2} + C_{S2}}$
d	Explicit model equations for the liquid phase	$\dfrac{dC_X(t)}{dt} = (\mu_1(C_{S1}) + \mu_2(C_{S2}))C_X(t)$ $\dfrac{dC_{S1}(t)}{dt} = -\dfrac{\mu_1(C_{S1})C_X(t)}{Y_{XS1}}$ $\dfrac{dC_{S2}(t)}{dt} = -\dfrac{\mu_2(C_{S2})C_X(t)}{Y_{XS2}}$

tain model parameters. Another case for alternative substrates is the uptake of sugar anomers, as was investigated by Benthin et al. [45] for *Lactococcus cremoris*, *Escherichia coli* and *Saccharomyces cerevisiae*. While in the yeast the uptake of glucose anomers proceeds according to competitive inhibition, in the bacteria independent uptake systems exist that are not inhibited by the other anomer. Another model for alternative substrates that uses a special type of inhibition function for every substrate component i (s_i=fraction of sugars in the feed):

$$\mu = \sum_i \frac{\mu_{\max i} s_i}{K_i + \sum_j s_j} \tag{2.19}$$

was presented by Lendenmann and Egli [50] for growth of *E. coli* on six different substrates in a chemostat.

While the models for essential substrates usually give a satisfactory description of the observed phenomena, this is not the case for alternative substrates. Here, and for more complicated situations with combinations of alternative and essential substrates, the model should consider the dynamic regulatory response of the cells and also be based on a proper description of the cellular reaction network. A sequential uptake is often found under excess of substrate. As long as the preferred substrate can support a sufficient growth rate, it suppresses and inhibits the uptake system of other substrates. After its limitation the enzymes for the substrate with the next lower preference are induced or derepressed. The result is a remarkable lag-phase for the adaptation to the next substrate. Furthermore, a purely sequential uptake is seldom found, but the substrates can be used up in parallel if the concentrations are low enough, and these alternative substrates, e.g., different sugars, can all be affected by other essential substrates like oxygen. This means a lot of regulatory phenomena have to be considered for a model of multi-substrate kinetics, which are very closely associated with the metabolic pathways and their regulation. The methodologies of structured or cybernetic models (Sect. 2.3) fit better to the modeling of such complicated growth processes.

To establish such a global model for the entire course of a cultivation that covers lag-phases, growth phases – possibly on a number of different substrates – as well as production phases, as shown in Figs. 2.2 and 2.3, is quite troublesome, and the results obtained are not always satisfactory. A much simpler task is to build several individual so-called local models for each of the phases or physiological states of a cultivation and to connect them together in an ordered time sequence to provide a full description of the process. This concept – having some analogies to the Physiological State Control proposed by Konstantinov and Yoshida [59] – was introduced by Halme [57, 58] as functional state models. To establish the correct sequence, proper switching conditions have to be found that determine the transition from one model to the next [60]. For example, in a multi-substrate cultivation, each sub-model will cover only the growth on a single limiting substrate. Naturally, the models for each phase will be relatively simple; usually unstructured ones should be sufficient, even for complicated processes. These will also have only a few parameters with great sensitivity in the respective phase. This makes functional state models well-suited for on-line application and automatic control, because at any time they contain only a minimum set of states and parameters. It also facilitates the model adaptation by on-line estimation, resulting in increased model accuracy.

Zhang et al. [61] presented a functional state model for fed-batch cultivations of baker's yeast. The possible sequence of sub-models and the related switching conditions can be represented by a state transition diagram with four states as shown in Fig. 2.9. The switching conditions – expressed in terms of respiratory quotient and dissolved oxygen concentration – worked well in the simulation of a fed-batch cultivation. But the difficulties for finding proper switching conditions that are robust against measuring and modeling errors must not be underestimated. The methods

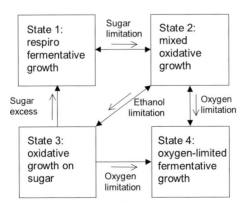

Fig. 2.9. State transition diagram of the functional state model for baker's yeast [61]

of fuzzy sets and artificial neural networks can help in this field [62]. A similar modeling concept with emphasis on the time sequence of phases during a cultivation was developed by Ruenglertpanyakul [63]. He presented models for diauxic growth on two substrates, cultivations with inhibitory products, and production of antibiotics. Local unstructured models were assigned to lag-phases, growth-phases and production-phases.

2.2.2
Endogenous and Maintenance Metabolism

In many processes, the so far presented unstructured models agree fairly well with experimental data for a wide range of specific growth rates. But notable exceptions are known, which often can be accounted for as arising from variations in the yield coefficient, i.e., the ratio of growth rate and substrate uptake rate. This variation in the efficiency of the substrate usage with growth rate can be explained as follows. The endogenous metabolism – the standing requirement for an expenditure of energy and metabolites for cell repair and maintenance that is independent of growth – reduces the yield at low growth rates. At strong limitation there can even be substrate uptake without growth. Without any external substrate being available, internal storage material can be degraded by the cells for maintaining their integrity. This can for instance be observed during the stationary/declining phase of a batch culture by a reduction in the cell mass, as shown in Fig. 2.1. Another reason for this effect can be cell lysis, when the cell repair mechanisms fail. Very satisfactory descriptions of these phenomena can be obtained by simple unstructured models.

The model of Herbert [25] took this effect into account by assuming that the observed net growth rate $q_X(C_S)$ is lower than a fictive true growth rate $\mu_G(C_S)$. The constant difference is the specific rate of endogenous metabolism μ_E:

$$q_X(C_S) = \mu_G(C_S) - \mu_E \tag{2.20}$$

An example for this model has already been presented in Fig. 2.6. At strong growth limitation, the observed growth rate can become less than zero. So, the rate μ_E also can be viewed as a specific rate of cell lysis or degradation rate of intracellular storage material. The true growth rate, which may be expressed by any of the kinetics given before, determines the rate of substrate uptake by a constant yield coefficient, Y_{XSG}:

$$q_S(C_S) = -\frac{\mu_G(C_S)}{Y_{XSG}} \tag{2.21}$$

The model can be translated into the usual vector notation by taking both, μ_G and μ_E, as inner reactions:

$$\mathbf{q} = \begin{pmatrix} q_X \\ q_S \end{pmatrix} \qquad \mathbf{Y} = \begin{pmatrix} 1 & -1 \\ \frac{-1}{Y_{XSG}} & 0 \end{pmatrix} \qquad \mathbf{r} = \begin{pmatrix} \mu_G \\ \mu_E \end{pmatrix} \tag{2.22}$$

The vector of specific reaction rates, $\mathbf{q}$, includes the specific growth rate and substrate uptake rate. The variation of yield with specific growth rate in this model results in the following hyperbolic function

$$Y_{XS}(q_X) = -\frac{q_X}{q_S} = Y_{XSG}\frac{q_X}{\mu_E + q_X} \tag{2.23}$$

Pirt [26] referred the same phenomena of the decreasing yield to an additional need for substrate consumption for maintaining the cell structure, expressed by the maintenance coefficient m_S, being a positive constant. Thus, if the uptake rate of substrate used only for growth is r_{SG}, then the total substrate uptake q_S is

$$q_s(C_S) = r_{SG}(C_S) - m_S \tag{2.24}$$

Remember that q_S, $r_{SG}<0$, since the material flux for production was defined as a positive direction. The energetic aspects of the maintenance concept were further discussed by Esener et al. [10]. By taking the specific growth rate

$$\mu(C_S) = -r_{SG}(C_S)Y_{XSG} \tag{2.25}$$

together with m_S as intrinsic reactions, Pirt's model becomes in vector notation

$$\mathbf{q} = \begin{pmatrix} q_X \\ q_S \end{pmatrix} \qquad \mathbf{Y} = \begin{pmatrix} 1 & 0 \\ \frac{-1}{Y_{XSG}} & -1 \end{pmatrix} \qquad \mathbf{r} = \begin{pmatrix} \mu \\ m_S \end{pmatrix} \tag{2.26}$$

and the variation in yield is given by

$$Y_{XS}(q_X) = -\frac{q_X}{q_S} = Y_{XSG}\frac{q_X}{m_S Y_{XSG} + q_X} \tag{2.27}$$

which is the same type of function as in Herbert's model, Eq. (2.20). A drawback of Pirt's concept is for model simulations that in batch culture the substrate concentration can become less than zero, due to the zero order maintenance reaction. But in regions of normal growth, both models are equivalent.

2.2.3
Product Formation

Modeling of product formation is, together with the modeling of cell growth, the most important part of a process model because production is just the purpose of the process. If one follows the classification of Gaden [3] (see Figs. 2.1–2.3), the typical application area of unstructured models are growth-associated products, as can be found in processes of mainly type I, but also type II with partial growth coupling. The production of primary metabolites is directly coupled with the central catabolic and anabolic reactions which proceed at a speed that is more or less proportional to the growth rate. Therefore, it can be described by a simple model, known as the Luedeking-Piret equation. For a multi-product process of N products the model can be written in vector notation for the specific rate of product formation as

$$\mathbf{q}_P = \mathbf{Y}_{PX}\mu + \mathbf{m}_P \tag{2.28}$$

where $\mathbf{Y}_{PX}$ is the vector of yield coefficients for the growth associated production and $\mathbf{m}_P$ is the vector of specific production rates due to maintenance metabolism or non-growth-associated processes.

Since in this model all production rates are linearly dependent on the specific growth rate, it cannot be applied to processes where there is a great variation in the product spectrum with growth rate as in processes of type II. Such variation means there is a switch in the fluxes of the reaction network of the cell that can, for instance, be triggered by the availability of different substrates. Then the model must contain, besides or instead of the specific growth rate, several independent intrinsic reactions that determine the product formation. This more general case is already covered by the stoichiometric model, Eq. (2.2). If one assumes that the system of equations is partitioned according to growth rate, substrate uptake rates for N compounds, and product formation rates for M compounds, it can be written for a given set of K independent intrinsic reactions as

$$
\begin{matrix}
\text{growth} \; \Big\{ \\
\text{substrate} \\
\text{uptake} \\
\\
\text{product} \\
\text{formation}
\end{matrix}
\underbrace{
\begin{pmatrix}
q_X \\ \hline
q_{S1} \\ \vdots \\ q_{SN} \\ \hline
q_{P1} \\ \vdots \\ q_{PM}
\end{pmatrix}
}_{q}
=
\underbrace{
\begin{pmatrix}
Y_{X1} & Y_{X2} & \cdots & Y_{XK} \\ \hline
-Y_{S11} & -Y_{S12} & \cdots & -Y_{S1K} \\
\vdots & \vdots & & \vdots \\
-Y_{SN1} & -Y_{SN2} & \cdots & -Y_{SNK} \\ \hline
Y_{P11} & Y_{P12} & \cdots & Y_{P1K} \\
\vdots & \vdots & & \vdots \\
Y_{PM1} & Y_{PM2} & \cdots & Y_{PMK}
\end{pmatrix}
}_{Y}
\mathbf{r}
\qquad (2.29)
$$

and after separation of the equations for growth, substrate uptake, and product formation as

$$
\begin{aligned}
q_X &= \mathbf{Y}_X \mathbf{r} \\
\mathbf{q}_S &= \mathbf{Y}_S \mathbf{r} \\
\mathbf{q}_P &= \mathbf{Y}_P \mathbf{r}
\end{aligned}
\qquad (2.30)
$$

The last line in Eq. (2.30) is a generalization of the Luedeking-Piret-equation, Eq. (2.28). As for Herbert's and Pirt's model, the endogenous or maintenance metabolism can be modeled here by a single constant element in the vector of intrinsic reactions. There is also evidence in the literature that the product formation is not always proportional to the growth rate, even for typical primary products as ethanol. This phenomenon can, for instance, be modeled by including the specific production rate into the vector $\mathbf{r}$ and using an exclusive kinetic expression for the dependency on substrate and product concentration [36].

Somewhat different notations of the model of product formation in connection with the substrate uptake rates can also be found in the literature:

$$
\mathbf{q}_P = \mathbf{Y}_{PS} \mathbf{q}_S
\qquad (2.31)
$$

$$
\mathbf{q}_S = \mathbf{Y}_{SP} \mathbf{q}_P
\qquad (2.32)
$$

These can be directly derived from Eq. (2.30) by a proper substitution of $\mathbf{r}$ and $\mathbf{Y}$. Nevertheless, Eqs. (2.30) or (2.31) should be preferred, since there the model parameters, i.e., yield coefficients, of growth and substrate uptake are completely decoupled from those of product formation [51]. This facilitates greatly the model building and parameter identification, because in the early modeling phase one can concentrate on the pure growth process without looking at the products. When the growth model is satisfactory, in a second model building phase, the product model can be added without further changes in the growth model.

The modeling of non-growth associated products is mainly concerned with the processes of type III (see Fig. 2.3) where there is by definition no direct coupling to central metabolic processes or to the growth rate. The inhibition, repression or inactivation of enzymes by catabolism of a rapidly used carbon source, e.g., glucose, does not hamper the production of primary metabolites. But the synthesis of sec-

ondary metabolites is mostly repressed under such conditions. There are two strategies to obtain maximum productivity for secondary metabolites, and to avoid the
repression: batch cultivation with slowly used carbon source, such as lactose, or
fed-batch cultivation with rapidly used carbon sources. In both cases the growth rate
of the microorganisms is controlled to such low values that production becomes
maximal. Another established fact is that the production of secondary metabolites
is often related to morphological differentiation of the cells. In this case, simple unstructured models cannot give an adequate description of the growth kinetics, because they assume homogeneous biomass. To consider the effects of morphology
one has to turn to segregated models, as described in Sect. 2.4. Since the morphological differentiation is also controlled by the available carbon source, its direct regulatory influence on the production of secondary metabolites is understandable. The
regulatory effects can also give rise to lag-phases in growth or product formation,
which should be represented by structured models.

If one nevertheless chooses to use unstructured models, it will be hard to put some
mechanistic ideas into the "formal-kinetics". There is no straightforward way for the
modeling of this more complicated type of process. One possible solution is that the
product formation is still modeled as being directly dependent on the specific
growth rate, but by a non-linear function, f_1, [2, 33]

$$q_P = f_1(\mu) \tag{2.33}$$

that may have a maximum at a certain low growth rate or be of saturation type, as
shown in Fig. 2.10. Another idea sees the production as being related to dynamic
transients of growth as a consequence of regulatory phenomena. This is expressed
by a model

$$q_P = f_2\left(\tfrac{d\mu}{dt}\right) \tag{2.34}$$

where f_2 is again some non-linear function. Finally, both could be combined to a
more general form

$$q_P = f_1(\mu) + f_2\left(\tfrac{d\mu}{dt}\right) \tag{2.35}$$

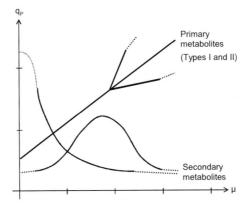

Fig. 2.10. Schematic diagram of the coupling of specific production rate and the specific
growth rate for different types of metabolites

In any case, these are pure black-box models that must be constructed for a given data set by a proper fitting procedure.

2.2.4
Other Parameters Influencing Growth

In biotechnological processes, growth and product formation depend beside the composition of the medium on many other variables and operating parameters, like temperature or pH. Temperature is an important environmental factor affecting the growth of microorganisms. In cultivations of genetically modified cells, the expression of the foreign protein is often initiated by a temperature shift. Also in the production of antibiotics the productivity can be increased by controlling the temperature to follow a certain profile. Since the kinetics of the total product proceeds according to the specific production rate multiplied by the cell mass, the temperature dependence must not only be included in the product model, but also in the growth model. The response of a culture to variation of temperature can be very complex. The composition of microorganisms with respect to DNA, RNA, protein, and lipids varies significantly with temperature [32]. The modeling of pH-influence is of special interest for cultivations without pH-control. Like temperature, growth or production is deterred when the conditions deviate from the optimum.

A convenient way is to consider the additional influencing variables, the parameter vector $\mathbf{p}(t)$ in Eq. (2.2), only in the kinetic and stoichiometric coefficients, while maintaining the form of the kinetics. When, for instance, looking at the temperature- and pH-dependence, $\mathbf{p}=(T(t), pH(t))$, a Monod-model for substrate uptake may be written as

$$q_S(C_S, T, pH) = \frac{-\mu_{\max}(T, pH)}{Y_{XS}(T, pH)} \frac{C_S}{K_S(T, pH) + C_S} \tag{2.36}$$

where the elements of $\mathbf{p}$ may be given by polynomials, e.g., for the maximum specific growth rate,

$$\mu_{\max}(T, pH) = k_0 + k_1 T + k_2 T^2 + k_3 T^3 + k_4 pH + k_5 pH^2 + k_6 pH^3 + \\ + k_7 TpH + k_8 T^2 pH + \dots \tag{2.37}$$

A polynomial degree of two is sufficient when the parameter, here μ_{max}, has a single extremum and no inflexion point. Otherwise, a higher degree has to be used. Similar polynomials can be specified for the other model parameters. All the many coefficients k of the polynomials have then to be determined from experimental data sets, which cover a sufficiently great number of combinations of different temperatures and pH-values in the region of interest. This is a tremendous experimental and numerical task. Several models of this type that are rather specific to a certain strain or process can be found in the literature. As a general finding, the yield coefficients and limitation "constants" seem to be less sensitive to variation of the growth conditions compared to the maximum reaction rates.

Eroshin et al. [27] investigated the variation of the yield with pH and temperature when *Saccharomyces cerevisiae* was cultivated on ethanol as carbon source. The following polynomial correlation was found:

$$Y_{XS} = 0.63 - 0.071x_1 - 0.012x_2 + 0.0045x_1x_2 - 0.026x_1^2 - 0.053x_2^2$$

$$x_1 = 2\frac{pH - 4.5}{pH} \quad x_2 = 2\frac{T - 29}{T} \quad , \quad \text{T in Celsius} \tag{2.38}$$

Tayeb et al. [28] gave the following linear correlation in the range of 5.5<pH<6.5 and 33 C<T<44 C for the maximum specific growth rate of *Streptococcus thermophilus*:

$$\mu_{max} = (446 - 15.9\,(T - 37) + 92\,(pH - 6.5))\,10^{-4}\,\text{min}^{-1} \tag{2.39}$$

and *Lactobacillus bulgaricus*:

$$\mu_{max} = (340 + 29.7\,(T - 44) - 115\,(pH - 5.5))\,10^{-4}\,\text{min}^{-1} \tag{2.40}$$

An analogous correlation for temperature only was presented by Kluge et al. [56] for the dependency of the specific growth and production rate of *Penicillium chrysogenum*. Chu and Constantinides [29] correlated kinetic parameters to temperature and pH for a Cephalosporin C process. For the cell yield from glucose the following function was derived:

$$Y_{XS} = 0.495 - 0.015\Delta T - 0.194\Delta pH^2 - 0.144\Delta pH^3 - 0.057\Delta T\Delta pH$$
$$+ 0.039\Delta T\Delta pH^2 + 0.134\Delta T\Delta pH^3 + 0.006\Delta T^2\Delta pH - 0.006\Delta T^2\Delta pH^2$$
$$- 0.018\Delta T^2\Delta pH^3 \tag{2.41}$$

where $\Delta T = T - 301K$ and $\Delta pH = pH - 6.8$

Reuss et al. [30] developed a model for the gluconic acid production by *Aspergilus niger*. The dependency of the maximum specific production rate on pH was approximated by

$$q_{Pmax} = q_{Pmax\,(pH = 5.5)}\,(-4.07 + 1.84pH - 0.167pH^2) \tag{2.42}$$

A similar model was given by Venkatesh et al. [52] for growth of *Lactobacillus bulgaricus*. Brown and Halsted [31] investigated the influence of pH on the growth of *Trichoderma viride* on glucose medium. In a range of 2.5<pH<4.0 a linear correlation of the maximum growth rate and the hydrogen ion concentration was found:

$$\mu_{max} = 0.1\,(1 - 309[H]^+)\,h^{-1} \tag{2.43}$$

For pH-values lower than 2.48 no growth was possible at all. This model is in agreement with the postulation that the growth rate reduction with increased hydrogen ion concentration is associated with a membrane diffusion limitation of the substrate-permease complex. Starzak et al. [34] modeled the pH-dependency of the specific growth rate of anaerobic growing yeast *Saccharomyces cerevisiae* by a variation of the half-saturation constant ($[H]_0^+$ is the initial hydrogen ion concentration):

$$K_S = \frac{1 + k_1[H]^+ + \frac{k_2}{([H]^+)^2}}{1 + k_1[H]_0^+ + \frac{k_2}{([H]_0^+)^2}} \tag{2.44}$$

A usual function to describe the variation of the specific growth rate with pH is [48, 54]

$$\mu = \frac{\mu_{max}}{1 + \frac{[H]^+}{k_1} + \frac{k_2}{[H]^+}} \tag{2.45}$$

According to this model, Zeng et al. [35] investigated the growth of *Clostridium butyricum* and *Klebsiella pneumoniae*, Åkerberg et al. [49] of *Lactococcus lactis*, and Ohara et al. [53] of *Streptococcus faecalis*. Tan et al. [55] presented a model based on a statistical thermodynamic approach that resembles the form of Table 2.2, j, when substituting C_S by $[H]^+$. This may be taken as a generalization of Eqs. (2.44) and (2.45).

More mechanistic models can be established when only looking at the temperature effects. The microorganisms can grow only over a restricted range of temperatures. The metabolism is made of a complex sequence of enzymatic reactions whose rates are individually related to temperature by activation or inactivation that should generally follow a function of the Arrhenius-type. Beside the general effect on the rates, the temperature can also exert highly selective effects on metabolic pathways, e.g., by repression or induction of particular protein synthesis (e.g., heat-shock proteins), or by irreversible denaturation of proteins and nucleic acids. According to Franks et al. [64] the general dependency of growth on temperature can be divided into several regions as shown schematic in Fig. 2.11. In the Arrhenius region, the growth rate increases exponentially with temperature up to the optimum, T_{opt}. In the superoptimal region, growth decreases by inactivation effects. Balanced growth is possible up to a maximum temperature, T_{max}. In the supermaximal region, cells can only grow temporarily before they become completely inactivated. A segregated model for this transient effect is given in Sect. 2.4. At low temperatures the growth rate falls more quickly with temperature as expected by the Arrhenius-relation.

Topiwala and Sinclair [32] analyzed the balanced growth of *Aerobacter aerogenes* in continuous culture on glucose mineral medium by means of Herbert's model. Monod-kinetics were assumed for glucose uptake. While the yield decreased only slightly with temperature, the other kinetic parameters varied strongly according to

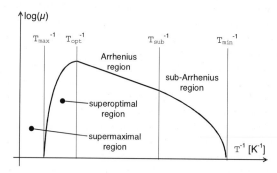

Fig. 2.11. Principal dependency of specific growth rate on temperature [64]

Arrhenius relationships. The following correlation was found in the range 293 K<T<323 K:

$$\mu_E = 2.7 \cdot 10^5 e^{\frac{-2150}{RT}}$$

$$K_S = 3.3 \cdot 10^{-11} e^{\frac{2830}{RT}} \tag{2.46}$$

$$\mu_{max} = 2.5 \cdot 10^{10} e^{\frac{-3400}{RT}} - 1.4 \cdot 10^{23} e^{\frac{-7860}{RT}}$$

The maximum specific growth rate showed a maximum at 313 K. Esener et al. [10] describes the influence of temperature on the microbial activity as superposition of activation and deactivation effects on the key-enzymes of the reaction. The resulting model for temperature dependence of the maximum specific growth rate with the parameter values from [18] is

$$\mu_{max} = \frac{6.74 \cdot 10^{14} e^{\frac{86800}{RT}}}{1 + 1.6 \cdot 10^{48} e^{\frac{287600}{RT}}} h^{-1} \tag{2.47}$$

The Arrhenius-equation can also be used to model the kinetics of temperature induced cell death: the number of viable cells decreases exponentially according to

$$\frac{dC_X}{dt} = -k_D e^{\frac{-E_D}{RT}} C_X \tag{2.48}$$

where E_D is the "activation energy" for cell death and k_D the inactivation rate constant.

2.3
Structured Models

In unstructured cell models, the biotic phase is viewed as a homogenous component with the only property, cell mass. In contrast, structured models provide information about the physiological state of the microorganisms, their composition, and regulatory adaptation to the environment. For this, the cell mass is structured into several intracellular compounds and functional groups which are connected to each other and to the environment by fluxes of material and information. The functional groups may, in one extreme, consist of detailed reaction networks considering as many as possible of known reactions or, at the other extreme, be built of roughly lumped or idealized paths of the reaction network that may represent, for instance, in only one step the entire biosynthetic system of the cell. While such a quite abstract view leads to the low-complexity compartment models, the former approach can be found in the so-called single-cell models that try to use as much as possible a-priori information on mechanisms and kinetics of the metabolic reactions. Both mark the field in which all of the structured models for the biotic phase can be found: multi-compartment models, genetically structured models, and biochemically structured models. In any case, a structured model is of higher complexity than an unstructured one because it is just the intention to give a better representation of the intracellular processes by including balances for intracellular compounds. Nevertheless, as a custom of good modeling practice, a structured model should be normally constructed of as few as possible functional groups only for the most relevant intracellular processes that provide a sufficient description for the desired application of the model. This guideline keeps the model simple and the number of state variables and parameters as small as possible. Otherwise the difficulties for the experimental model verification and parameter identification become insuperable.

2.3.1
The Constitutive Equations

To be mathematically correct, all structured models must fit into a certain framework of balance equations that is summarized in the following. A structured model is built of balances for intracellular substances, metabolites, storage material, enzymes, cell wall material, RNA, DNA, and so on. These balances are coupled together and to the outer liquid phase of the reactor by a number of reactions, as shown in Fig. 2.4. For the following, it is convenient to separate the concentrations of extracellular compounds in the liquid phase of the reactor (the abiotic phase)

$$\mathbf{Z} = \begin{pmatrix} C_S \\ C_O \\ \vdots \\ C_P \end{pmatrix} \tag{2.49}$$

from the concentration of the biotic phase, which may represent intracellular metabolites, storage material, enzymes, RNA, DNA, and so on:

$$\mathbf{X} = (X_i) = \begin{pmatrix} C_{metabolite} \\ C_{enzyme} \\ \vdots \\ C_{DNA} \end{pmatrix} \tag{2.50}$$

The entire concentration vector can be written as

$$\mathbf{C} = \begin{pmatrix} \mathbf{Z} \\ \mathbf{X} \end{pmatrix} \tag{2.51}$$

The first constitutive equation of structured models is obtained from the condition that the total cell mass C_X is just the sum of all its components:

$$C_X = \sum_i X_i = \mathbf{1}^T \mathbf{X} \tag{2.52}$$

In an unstructured model, the composition of a cell was taken as constant, $X_i =$ constant; then the biotic phase is already uniquely characterized by C_X. The intrinsic reactions and related kinetics are the main target of structured modeling. These kinetics are not only governed by the concentrations in the bulk liquid phase outside the cell but also by the intracellular concentrations. Therefore it makes sense to write down the balance equations for structured models in terms of intrinsic concentrations. When doing this, one has to take special care of the models consistency: the balance over the entire reactor must be equivalent to the balances over the biotic phase [2, 95, 96]. The volumetric intrinsic concentrations are defined in the average as mass of the substance in the cell per volume of the cell:

$$c_{iV} = \frac{X_i V_L}{C_X V_L \rho_X^{-1}} \quad \begin{matrix} \leftarrow & \text{total mass of substance i in all cells in the reactor} \\ \leftarrow & \text{total volume of all cells} \end{matrix} \tag{2.53}$$

Since the cell density ρ_X is almost constant and its remaining variations difficult to be measured, it is more convenient to use cell mass related intrinsic concentrations:

$$c = \frac{X}{C_X} \tag{2.54}$$

From Eq. (2.52) then follows the second constitutive equation for the sum of all intrinsic components,

$$\sum_i c_i = 1^T c = 1 \tag{2.55}$$

and therefore,

$$\frac{d(1^T c)}{dt} = 0 \tag{2.56}$$

The derivation of the balances for intrinsic concentrations starts with the balance equation of the reactor, Eq. (1.12), where the gas-liquid transfer here need not be considered. By partitioning the rate vectors and matrix of yield coefficients in analogy to the concentrations, Eq. (2.51),

$$Q = \begin{pmatrix} Q_Z \\ Q_X \end{pmatrix} \qquad q = \begin{pmatrix} q_Z \\ q_X \end{pmatrix} \qquad Y = \begin{pmatrix} Y_Z \\ Y_X \end{pmatrix} \tag{2.57}$$

the balances for compounds in the outer and inner of the cell can be written down separately:

$$\frac{dZ(t)}{dt} = \overbrace{\underbrace{Y_Z r(t)}_{q_Z} C_X(t)}^{Q_Z} + D(t)\left(Z^R(t) - Z(t)\right) \tag{2.58}$$

$$\frac{dX(t)}{dt} = \underbrace{\overbrace{\underbrace{Y_X r(t)}_{q_X} C_X(t) + D(t)\left(X^R(t) - X(t)\right)}}_{Q_X} \tag{2.59}$$

In the following, only the elements related to the biotic phase, Q_X and Y_X, will be further considered. By substituting Eq. (2.54) into Eq. (2.59), together with

$$X^R(t) = C_X^R c(t) \tag{2.60}$$

the balance for the cellular compounds is obtained as

$$\frac{dC_X(t)c(t)}{dt} = C_X(t)\frac{dc(t)}{dt} + c(t)\frac{dC_X(t)}{dt} \tag{2.61}$$
$$= Y_X r(t)C_X(t) + D(t)c(t)\left(C_X^R(t) - C_X(t)\right)$$

Multiplying 1^T to each summand and using the conditions of Eqs. (2.55) and (2.56) yields

$$\frac{dC_X(t)}{dt} = 1^T(Y_X r(t))C_X(t) + D(t)\left(C_X^R(t) - C_X(t)\right) \tag{2.62}$$

which is just the balance equation of the total cell mass, and the factor

$$1^T(\mathbf{Y}_x\mathbf{r}(t)) = \mu(t) \tag{2.63}$$

must equal the specific growth rate. This is the third constitutive equation of structured models. The specific growth rate is never independent of the other intrinsic reaction rates. When substituting Eqs. (2.62) and (2.63) into Eq. (2.61), the transport term due to inflow and outflow of the reactor is canceled out and the balance equation for the intrinsic concentrations, the fourth constitutive equation of structured models, follows as

$$\frac{d\mathbf{c}(t)}{dt} = \mathbf{Y}_X\mathbf{r}(t) - \overbrace{1^T(\mathbf{Y}_X\mathbf{r}(t))}^{\mu(t)}\mathbf{c}(t) \tag{2.64}$$

The second summand in this equation is due to dilution of intracellular material by the cell growth, i.e., the expansion of the cell volume. The four constitutive equations form the general frame of every structured model that may be established by different sets of $\mathbf{r}$ and $\mathbf{Y}_X$.

For biochemically structured models it is convenient to write down molar balances instead of mass balances. By the substitutions

$$\mathbf{r} = \mathbf{M}\hat{\mathbf{r}} \qquad \mathbf{Y}_X = \mathbf{M}\hat{\mathbf{Y}}_X \qquad \mathbf{c} = \mathbf{M}\hat{\mathbf{c}} \quad \text{with} \quad 1^T(\mathbf{M}\hat{\mathbf{c}}) = 1 \tag{2.65}$$

where $^\wedge$ denotes molar quantities and $\mathbf{M}$ is the diagonal matrix of molecular weights, the balance equation, Eq. (2.64), becomes

$$\frac{d\hat{\mathbf{c}}(t)}{dt} = \hat{\mathbf{Y}}_X\hat{\mathbf{r}}(t) - \mu(t)\hat{\mathbf{c}}(t) \tag{2.66}$$

and the specific growth rate is then

$$\mu(t) = 1^T\mathbf{M}\left(\hat{\mathbf{Y}}_X\mathbf{r}(\hat{\mathbf{t}})\right) \tag{2.67}$$

The consumption of metabolites for maintenance metabolism can be modeled in analogy to unstructured models by considering a zero-order reaction on the rate vector $\mathbf{r}$, as given by Eq. (2.26). If one prefers an explicit modeling of maintenance turnover, the balance equation, Eq. (2.64), becomes

$$\frac{d\mathbf{c}(t)}{dt} = \mathbf{Y}_X\mathbf{r}(t) - 1^T(\mathbf{Y}_X\mathbf{r}(t))\mathbf{c}(t) - \mathbf{m} \tag{2.68}$$

and the growth rate is then

$$\mu(t) = 1^T(\mathbf{Y}_X\mathbf{r}(t) - \mathbf{m}) = \overbrace{1^T\mathbf{Y}_X\mathbf{r}(t)}^{\mu_G} - \overbrace{1^T\mathbf{m}}^{\mu_E} \tag{2.69}$$

This equation is a unification of the models of Herbert, Eq. (2.20), and Pirt, Eq. (2.24), since application of the latter to intrinsic variables leads to an endogenous growth rate $\mu_E = 1^T\mathbf{m}$.

Some more comments on the area of validity of the above structured model and the special condition, Eq. (2.60), for the inflow from the reservoir have to be given. The cell suspension is essentially a macro-fluid. This has serious consequences for the global mass balances and reaction kinetics, and restricts the application area of structured models. Consider a continuous process with sudden cell inflow. Then, generally, the in-flowing cells will have a state **c** different from the cells already being in the reactor. Because the intrinsic substances are not exchanged between the cells ad hoc, two clearly distinguishable cell populations with different states will keep growing and the observed average reaction rate for the entire reactor will be different from the reaction rate calculated from the kinetic expressions, **r**, by using averaged concentrations for intrinsic compounds. The only exception – without practical relevance in biological systems – is the case of linear first order kinetics. Therefore, structured models are only valid for reactors without inflow of cells, or with in-flowing cells that have at any time an exactly identical state to the cells in the reactor. This is only the case for cell-recycle from the same reactor. The more general case can only be covered by distributed population balance models [120]. An equivalent problem arises when using segregated population models (Sect. 2.4) in connection with structured growth models. There, the "inflow" of cells to a population results from transitions of another population growing in the same reactor. Such models are only mathematically correct when both populations have identical intrinsic states. This is not always considered in the works published in the literature.

The situation in reactors with non-ideal mixing is similar to the ideal CSTR with inflow of cells. When the cells pass through regions with different concentrations, e.g., of substrates or oxygen, their intrinsic state is altered slightly. Since the passes of cells through the reactor follow a stochastic regime, cells of different state become neighbors in a small volume element. This means, the population is always inhomogeneous and structured models must be applied with care in connection with distributed reactor models. The best way to avoid mathematical difficulties is to assume (if justified) that the recirculation time is much lower than the time constant of intracellular processes. The biological model can then be taken as a lumped one, assuming homogeneous biomass in the entire reactor having everywhere an identical intrinsic state. Then, averaged concentrations of the abiotic phase have to be used for calculation of the intrinsic reaction rates.

2.3.2
Some Applications of Structured Models

Compartment models introduced by Williams [65] as low complexity structured models are obtained by lumping biological systems of similar function and dynamics together into a few pools. The average behavior of the i-th pool is modeled by a single variable that can be a representative concentration for a group of metabolites or enzymes. These models are mainly used to describe very general features of the biomass, such as variation of the RNA or DNA content during the cultivation or the regulation of complete functional groups of the metabolism, e.g., pathways for certain substrates, or products. Into this area fits the modeling of lag-phases of growth being observed during growth on multiple substrates. An example for diauxic growth is given below. Esener et al. [96] developed a two-compartment model for growth of *Klebsiella pneumoniae*. One compartment was assigned to genetic material, proteins, and structural compounds, the other to RNA and small metabolites. The model was able to describe lag-phases of growth on a complex substrate. Nielsen et al. [97, 98] presented a two-compartment model for the lactic acid fermentation by *Streptococcus cremoris* with similar structural pools as in the above model. The model was verified by measurements of intracellular RNA and by dynamic experiments. The model was extended to four compartments by Nikolajsen et al. [99] for cultivations on mixtures of different sugars. Mundry and Kuhn [102] published a two-compartment model for phosphate-limited growth of *Streptomyces tendae* in fed-batch culture. Korte et al. [66, 67] investigated the growth of *Escherichia coli* on complex substrates including a variety of amino acids and developed a respective multi-compartment model. A model for antibiotics production by different *Streptomyces* strains that considers, besides glucose, the uptake of ammonium and phosphate was presented by King [100, 101]. The model, including six compartments for DNA, RNA, protein, amino acids, nucleotides, and structural compounds, was successfully applied for simulation of fed-batch and shift experiments.

Compartment models have also interesting applications for genetically modified cells, since the system for foreign gene expression is usually inducible. The model of Nielsen et al. [68, 69] for run-away cultivations of *Escherichia coli* includes four compartments for RNA, host genome, plasmid genome, and foreign protein. Bentley and Compala [70] consider further in their 8-compartment model amino acids, nucleotides, lipids and native protein. By choosing proper kinetics, the models could describe the variation of total plasmid or RNA content with growth rate. Lee and Ramirez [71] modeled the expression dynamics of foreign protein as superposition of shock reaction after temperature shift and subsequent recovery of the cells. The model could mirror the effect of IPTG-inducer on production and cell growth. Coppella and Dhurjati [108] presented a model of recombinant yeast for production of human epidermal growth factor, including seven intrinsic balances. The model gave a reasonable fit to experimental data.

On the level of molecular mechanisms, the kernels of the models for regulation of substrate uptake by the operon and for foreign gene expression are quite similar. Both include the two steps of transcription of a particular gene with intrinsic concentration c_G into mRNA with intrinsic concentration c_{mRNA}, and subsequent translation of the mRNA to the key enzyme for substrate uptake, or to the foreign product (c_P). The according balance with consideration of first-order deactivation or decay – with rate constants k_{D1} and k_{D2} – can be formulated as follows:

$$\frac{dc_{mRNA}}{dt} = k_1 \eta c_G - k_{D1} c_{mRNA} - \mu c_{mRNA} \tag{2.70}$$

$$\frac{dc_P}{dt} = k_2 \xi c_{mRNA} - k_{D2} c_P - \mu c_P \tag{2.71}$$

The transcription efficiency η accounts for modulation of transcription by an operator system. It is a complex function of the concentrations of effectors, e.g., substrates, repressors, and inducers. Often the transcription is the rate-limiting step, so the mRNA balance has to be included in a model for a proper description of the induction dynamics. In more simplified models the mRNA is assumed as being in quasi-steady-state and the mRNA-concentration can be eliminated from Eq. (2.71). The translation efficiency ξ in the synthesis rate of the protein describes the affinity of the ribosome binding site to the given nucleotide sequence, and the availability of ribosomes. The kinetic expressions for both factors are crucial parts of the models and were targeted in many of the papers mentioned here. For genetically modified cells, where the foreign gene is coded on a plasmid, the concentration c_G may also vary. Then DNA synthesis and plasmid replication or loss must be modeled in a proper way. Such models are known as genetically structured models [76, 79–82, 109–115].

Table 2.7. Example for a two-compartment model for growth on two alternative substrates [18]

	Description	Model equations
a	General vector notation, see Eqs. (1.12), (2.1), and (2.64)	$\dfrac{d\mathbf{C}(t)}{dt} = \mathbf{Q} = \mathbf{q}C_X = \mathbf{Y}\mathbf{r}C_X$ $\dfrac{d\mathbf{c}(t)}{dt} = \mathbf{Y}_X\mathbf{r}(t) - \mu(t)\mathbf{c}(t)$
b	Specialized vectors and matrices	$\mathbf{C} = \begin{pmatrix} C_{S1} \\ C_{S2} \\ C_{E1} \\ C_{E2} \end{pmatrix}$ $\mathbf{c} = \begin{pmatrix} c_{E1} \\ c_{E2} \end{pmatrix}$ $\mathbf{q} = \begin{pmatrix} q_{S1} \\ q_{S2} \\ q_{E1} \\ q_{E2} \end{pmatrix}$ $\mathbf{Y} = \begin{pmatrix} -Y_{XS1}^{-1} & 0 & 0 \\ 0 & -Y_{XS2}^{-1} & 0 \\ 1 & 0 & -1 \\ 0 & 1 & 1 \end{pmatrix}$ $\mathbf{r} = \begin{pmatrix} r_{S1} \\ r_{S2} \\ r_{12} \end{pmatrix}$
c	Total cell mass	$C_X = C_{E1} + C_{E2}$ $1 = c_{E1} + c_{E2}$
d	Intrinsic rate expressions	$r_{S1} = r_{S\,max\,1}\dfrac{C_{S1}}{K_{S1} + C_{S1}}$ $r_{S2} = r_{S\,max\,2}\dfrac{C_{S2}}{K_{S2} + C_{S2}}\dfrac{K_{I1}}{K_{I1} + C_{S1}}c_{E2}$ $r_{12} = K_2 c_{E1}$
e	Specific growth rate, Eq. (2.63)	$\mu = r_{S\,max\,1}\dfrac{C_{S1}}{K_{S1} + C_{S1}} + r_{S\,max\,2}\dfrac{C_{S2}}{K_{S2} + C_{S2}}\dfrac{K_{I1}}{K_{I1} + C_{S1}}c_{E2}$
f	Explicit model equations for the liquid phase	$\dfrac{dC_{S1}}{dt} = -\dfrac{r_{S1}C_X}{Y_{XS1}}$ $\dfrac{dC_{S2}}{dt} = -\dfrac{r_{S2}C_X}{Y_{XS2}}$ $\dfrac{dC_{E1}}{dt} = (r_{S1} - r_{12})C_X \Leftrightarrow \dfrac{dc_{E1}}{dt} = r_{S1} - r_{12} - \mu c_{E1}$ $\dfrac{dC_{E2}}{dt} = (r_{S2} + r_{12})C_X \Leftrightarrow \dfrac{dc_{E2}}{dt} = r_{S2} + r_{12} - \mu c_{E2}$

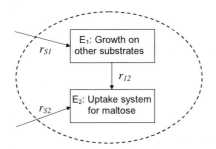

Fig. 2.12. Structure of the two-compartment model for diauxic growth of *Klebsiella terrigena* on glucose and maltose

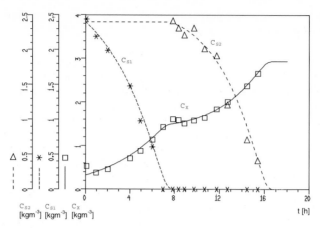

Fig. 2.13. Diauxic growth of *Klebsiella terrigena* on glucose and maltose, experimental data and simulation by a two-compartment model [18]

The most complex group of detailed mechanistic models tries to give a complete image of the biochemical processes of the cells, including replication of the genome and cell division. Although they must consider an enormous number of metabolites and reaction steps, they are in a certain aspect the most handy models: the elementary reactions have naturally the most simple kinetics, and the model parameters have direct physical or chemical meaning; both can often be examined by in vitro experiments. The single-cell growth models for *Escherichia coli* of Domach et al. [72], Ataai and Shuler [73], Shu and Shuler [74], and Peretti and Bailey [75] served as basis for model studies on genetically modified strains. One focus of model application was in the area of host-vector-interaction and mechanistic description of plasmid replication [77, 78, 116–118]. These models can provide theoretical predictions on a number of parameters that are of interest for simple segregated models, e.g., the variation of maximum growth rate, yield coefficients, or plasmid copy number with the properties of the vector or with growth conditions.

As an example of a low-complexity structured model, a two-compartment model for diauxic growth of *Klebsiella terrigena* on the substrates glucose (C_{S1}) and maltose (C_{S1}), is summarized in Table 2.7 and its structure is shown in Fig. 2.12. Glucose is the preferred substrate that inhibits and represses the uptake of the second substrate maltose. The inducible enzymes of the maltose uptake system are represented by the E_2-compartment, while E_1-compartment stands for all of the remaining metabolism. When glucose becomes limited and maltose is present in the medium, E_2 is synthesized with increased rate by derepression on r_{S2} and degradation of E_1. The diauxic lag-phase is modeled by the resulting transient on c_{E2}. One of the balances in Table 2.7 f is obsolete since, due to condition c, the concentration of either E_1 or E_2 can be calculated from the other. A simulation result for a batch cultivation is given in Fig. 2.13.

2.3.3
Cybernetic Models of the Compartment Type

A growing microbial cell breaks down high molecular weight carbon and energy sources, brings the smaller derivatives into the cell, catabolizes these further to smaller molecules, converts them to amino acids, nucleotides, carbohydrates, fatty acids, and vitamins, and finally builds cell constituents such as proteins, nucleic acids, polysaccharides, and lipids. Hundreds of enzymes have to be synthesized and must then act together in a coordinated way. Thus, regulatory mechanisms have evolved that enable a species to compete successfully with other forms of life in nature: feed-forward regulation (e.g., substrate induction and activation), feedback regulation by repression or inhibition (e.g., by catabolites or primary metabolites), or energy charge regulation of biosynthetic pathways. These control mechanisms direct the substrate and metabolite flux to proper pathways. The rate of an enzyme-catalyzed reaction is controlled by the enzyme concentration itself, and by the regulation on the reaction level by activation/inhibition to meet the requirements of metabolism and to ensure optimal growth. An ideal cell does not overproduce extremely large amounts of metabolites or enzymes, no matter what are the growth conditions. This coupling of independent reactions by the network of metabolic regulation often makes the understanding of the metabolism and its modeling so difficult. It is never enough to look at a few reactions; instead one has to adopt a more global viewpoint. In a very radical manner this is realized by the approach of cybernetic models for multi-substrate growth kinetics, introduced by Ramkrishna [83], and modified in several papers afterwards [84–88, 90, 93, 94].

Sequential and parallel uptake of more than one substrate at a time is accompanied by a complex metabolic regulation, which cannot be included satisfactorily into simple kinetics. Also usual structured models often will not give a simple and useable description in this situation, because they have to contain many variables, complicated mechanisms, and kinetics for the metabolic regulation. In the cybernetic approach, the mechanisms by which the microorganisms can coordinate their metabolic reactions to obtain optimum control are not questioned. The concept relies on the assumption that the evolution had selected the mechanisms performing just optimal control, and that it is sufficient to identify the optimality criterion. When this criterion and some general restrictions on the growth process – or more exactly, the substrate uptake steps and the biosynthetic capacity for enzymes – are known, it is possible to calculate at any time the rates of reactions by a mathematical optimization procedure without looking at the underlying mechanisms. In this way, cybernetic models can sometimes predict the sequential and parallel substrate consumption, without any assumption about coupled kinetics, by extrapolation from experiments on single substrates. The modeling proceeds in three steps:
1. Assigning an optimal control motive to the cell
2. Establishing kinetic rate expressions for uptake of a single substrate
3. Modeling of the slow dynamics of regulation by induction and repression, as determined by the synthetic capacity of the cell

The metabolic control action in cybernetic models is twofold: first, by adjusting the level of key enzymes for each substrate uptake system according to the optimality criterion, and second, by controlling the activity of the key-enzymes of reaction in analogy to a fast mechanism of activation or inhibition. Each key-enzyme forms its own compartment. In this view, the microorganisms are always able – by allocating their internal resources – to choose those substrates which fit their requirements best. For formulation of the model, the microbial growth on each substrate component S_i is represented by a reaction

$$X + S_i \xrightarrow{\text{Key–enzyme } i} (1 + Y_{XSi})X \qquad (2.72)$$

and the substrate uptake is given by a kinetic expression that depends on the concentration of the substrate, C_{Si}, and of the related key enzyme, c_{Ei}:

$$q_{Si} = \frac{-\mu_i(C_{Si}, c_{Ei}, t)}{Y_{XSi}} \qquad (2.73)$$

The enzymes are synthesized according to an optimum strategy given later. The specific growth rate on substrate i, μ_i, is a product of three factors, the intrinsic concentration of the key enzyme, c_{Ei}, a kinetic expression, $\mu_{max}\tilde{r}(C_{Si})$, chosen for instance from Table 2.1, and an activity controlling variable, v_i:

$$\mu_i(C_{Si}, c_{Ei}, t) = \mu_{\max i}\tilde{r}_i(C_{Si}(t))v_i(t)c_{Ei}(t) \qquad (2.74)$$

The total growth rate for N substrates is

$$\mu_i(C_{S1}, \cdots C_{SN}, t) = \sum_{i=1}^{N} \mu_{\max i}\tilde{r}_i(C_{Si}(t))v_i(t)c_{Ei}(t) \qquad (2.75)$$

The partition of the control action into one part for long term response by induction and repression, modeled by $c_{Ei}(t)$, and another part for immediate response by inhibition and activation, as described by $v_i(t)$, mirrors the two hierarchical levels of regulatory mechanisms. These may become visible at diauxic growth for example: the cells may switch to the preferred substrate at once when this is suddenly added to the medium, even if the less preferred substrate is present at high concentrations and the related enzymes are induced. The uptake of the less preferred substrate is then immediately inhibited, followed by a slower degradation of the enzymes. The induction-repression dynamics of key-enzymes are described by balances for their intrinsic concentrations,

$$\frac{dc_{Ei}(t)}{dt} = r_{Ei\max}\tilde{r}_{Ei}(C_{Si}(t))u_i(t) + r_{Ei0} - K_{Ei}c_{Ei}(t) - \mu(t)c_{Ei}(t) \qquad (2.76)$$

where K_{Ei} is the degradation rate constant. The synthesis rate of the key enzymes is controlled by the cybernetic variables $u_i(t)$ and by a kinetic expression depending on the substrate concentration. A small constitutive formation rate r_{Ei0} guarantees a minimum activity of the pathway. To complete the model, the control variables have to be specified. The cybernetic variables

$$\underset{\text{fully repressed}}{0} \leq u_i(t) \leq \underset{\text{fully induced}}{1} \quad \text{with} \quad \sum_i u_i(t) = 1 \qquad (2.77)$$

are determined through Herstein's matching law, which maximizes the total profit, here taken as μ, by allocating fractions of a fixed resource to alternative pathways i, according to the expected maximum relative profit for the i-th pathway, given by μ_i,

$$u_i(t) = \frac{\overbrace{\mu_{\max i}\tilde{r}_i(C_{Si}(t))c_{Ei}(t)}^{\text{expected maximum profit of pathway i}}}{\underbrace{\sum_{j=1}^{N}\mu_{\max j}\tilde{r}_j(C_{Sj}(t))c_{Ej}(t)}_{\text{total maximum profit}}} \qquad (2.78)$$

The fixed resource here is the total amount of enzymes in the cell, which cannot be synthesized in arbitrary amounts because the number of ribosomes is limited. Therefore, an increased formation of one enzyme is going on the costs of another. The activity control variables

$$\underset{\text{fully inhibited}}{0} \le v_i(t) \le \underset{\text{fully activated}}{1} \qquad (2.79)$$

for fast regulation of the key enzymes by inhibition and activation are allocated according to a heuristic strategy, in the way that pathways yielding a higher growth rate are preferred:

$$v_i(t) = \frac{\overbrace{\mu_{\max i}\tilde{r}_i(C_{Si}(t))c_{Ei}(t)}^{\text{expected maximum profit of pathway i}}}{\underbrace{\max_j\left(\mu_{\max j}\tilde{r}_j(C_{Sj}(t))c_{Ej}(t)\right)}_{\text{maximum attainable profit among all pathways}}} \qquad (2.80)$$

Obviously, the above optimization strategy is a local one. It only considers the actual growth conditions without trying to optimize the response over a certain interval in the future.

A formalism similar to the above can be used to consider maintenance metabolism. Another extension of the concept was given by Straight and Ramkrishna [89, 91, 92] for the modeling of connected metabolic pathways, and Yoo and Kim [103] have proposed a variant for non-growth-associated product synthesis (poly-β-hydroxybutyric acid). Doshi et al. [105] and Venkatesh et al. [106] suggested a simplified procedure for determination of the cybernetic variables for sequential substrate utilization. Chetan et al. [119] presented a cybernetic model for growth of baker's yeast on glucose, galactose, and melibiose.

2.3.4
Cybernetic Models of the Metabolic Regulator Type

Another modeling method with similarities to the above cybernetic models is the metabolic regulator approach [18]. The ideas behind cybernetic modeling were outlined in the previous section. Again, the microorganism is modeled as an optimal strategist, which tries to use the different metabolic pathways optimally. This model concept is not restricted to substrate uptake steps, but includes the general model of the reaction network as given by Eq. (2.2), including further intracellular reactions and product synthesis. Here, the role of the net reaction rates $\mathbf{q}$ as dependent variables of the intrinsic reaction rates $\mathbf{r}$ is not fixed for any growth condition, but elements of $\mathbf{q}$ and $\mathbf{r}$ can change their role in different growth situations. Furthermore, the intrinsic reactions are not solely determined by kinetic expressions, but also by the optimization strategy behind the metabolic regulation. Therefore, it is convenient to rewrite the stoichiometric model in the form

$$\mathbf{Yr}(t) = \mathbf{m}(t) \tag{2.81}$$

without explicitly distinguishing between net and intrinsic reactions. Here, $\mathbf{r}$ is an extended reaction vector also containing elements of $\mathbf{q}$, and $\mathbf{m}$ is a vector of M growth independent specific reaction rates, like maintenance terms or rates of constitutive reactions. But mostly the elements of $\mathbf{m}$ will be zero. In the above stoichiometric model the number of reactions, N, will usually be larger than the number of stoichiometric equations M, otherwise the microorganisms would have no freedom to regulate the relative activity of their pathways. The way to derive more conditions and finally to solve Eq. (2.81) uniquely is to introduce a cybernetic criterion, the metabolic coordinator J, for determination of the optimum metabolic regulation:

$$J(\mathbf{r}(t)) \xrightarrow{\mathbf{r}(t)} \text{optimum} \tag{2.82}$$

under consideration of boundary conditions given by Eq. (2.84). The optimization strategy is again a local one, since J depends only on the actual reaction rates. Similar to the cybernetic models of the compartment type, the optimization strategy followed by the microorganisms is often successfully modeled by maximization of the specific growth rate,

$$J = \mu(t) \xrightarrow{\mathbf{r}(t)} \text{maximum} \tag{2.83}$$

Since the stoichiometric model, Eq. (2.81), is underdetermined, additional conditions have to be considered to obtain a meaningful solution of the optimization problem. These are possible rate limiting steps of the metabolism, which can be formulated as inequalities,

$$\mathbf{r}_{\min}(t) \le \mathbf{r}(t) \le \mathbf{r}_{\max}(t) \tag{2.84}$$

as further explained by Table 2.8. Remember that $r_i<0$ means uptake and $r_i>0$ production. Such possible rate-limiting steps might be an inherent maximum biosynthetic capacity for macro-molecules and cell material (a) or kinetics of substrate

Table 2.8. Possible boundary conditions in Eq. (2.84)

	Kinetics	Expression for $\mathbf{r}_{\min}$ or $\mathbf{r}_{\max}$	Meaning
a	Zero order	r_{min} = constant or r_{max} = constant	Saturation
b	Single substrate uptake	$r_{min}\tilde{r}(C_S)$ or $r_{max}\tilde{r}(C_S)$	Substrate limitation
c	Time function, as given by Eq. (2.85)	$r_{min}(t) = r_{Ej}(t)$ or $r_{max}(t) = r_{Ej}(t)$	Regulation of the key-enzyme of the pathway by induction or repression
d	Infinite	$\pm \infty$	Never limiting
e	Combinations of (a) to (c)	$\max(r_{min})$, $\min(r_{max})$	Switch between different mechanisms

uptake (b). By using case (a) with $r_0=0$ on one side of Eq. (2.84), the rates can also be restricted to either uptake or production. Of special importance for the modeling of microbial adaptation are rate-limiting steps due to regulation of key enzymes by induction and repression (c). Beside this slow regulation, implicit in the model there is a fast regulation on the level of enzyme activity, since, as an outcome of the optimization, the turnover of some reactions may be lower than the maximum turnover determined by the boundary conditions. The vectors of possible rate-limiting steps must contain constraints (finite rates) on at least N-M reactions; by the optimization, at least N-M elements of $\mathbf{r}$ will always equal a constraint, i.e., the corresponding element of $\mathbf{r}_{min}$ or $\mathbf{r}_{max}$. Such reactions are said to be actually rate limiting. Recently, Bonarius et al. [104] have discussed the problem of finding the inherent constraints of metabolic networks that could be presented by a model as given above. They also suggested an experimental procedure for their determination.

The regulation on reaction level by inhibition/activation is completely covered by the above model without further assumptions. When condition (c) in Table 2.8 is not rate-limiting or when such conditions are not used, the model describes the fully adapted, optimum state of growth, in which all inducible enzymes are present in sufficient amount. Then the model is essentially an unstructured one that uses kinetics of the non-interacting type and a refined reaction network. For consideration of the dynamics of adaptation, balance equations for key-enzymes have to be specified which determine $r_{Ej}(t)$. In this approach, a strategy is used which is local to the regulated pathway. Their properties are as follows: as long as the coordinator J uses fully the capacity of the pathway up to the actual maximum rate $r_{Ej}(t)$, the concentration of the key enzyme has to be increased by further induction. If the actual rate allocated to the pathway is not rate-limiting, then the concentration of the enzyme should be reduced by stronger repression to manage the resources economically. These conditions can be automatically met by a tracking controller, a metabolic regulator, for the concentration of the key enzyme, which is regulated according to the actual requirements $r_j(t)$:

$$\frac{dr_{Ej}(t)}{dt} = \left(K_{1j} - \mu(t)\right)r_{Ej}(t) + K_{2j}\left(r_j(t) + r_{0j}\right) \tag{2.85}$$

where r_{Ej} is proportional to a fictive intrinsic concentration of key enzyme for the pathway j and r_{0j} is a small constitutive level of enzyme synthesis. The regulation has sufficient gain and the required tracking properties when

$$K_{1j}<0 \quad \text{and} \quad K_{2j}>\mu_{max}-K_{1j} \tag{2.86}$$

The balance at Eq. (2.85) for synthesis of a key enzyme is under growth limitation of an autocatalytic nature, since then the actual rate r_j is proportional to r_{Ej} and to the enzyme concentration. Thus, during limitation, the regulator is unstable and therefore increases the enzyme concentration until the regulated pathway has sufficient capacity. The importance of autocatalytic enzyme synthesis for modeling of growth dynamics on mixtures of substrates was recently discussed by Narang et al. [107].

An example of this type of model is given in Chap. 9. By this concept, growth on multiple substrates with sequential and parallel uptake, including essential compounds, as well as shifts in the product spectrum can be successfully described.

2.4
Segregated Models

Unstructured and structured models presumed a homogenous population of cells and only one species in the bioreactor. Many phenomena cannot be described by these unsegregated models:

- Irreversible alteration or disturbances in physiology and cell metabolism
- Morphological differentiation of the cells
- Mutations in the genome with related variations in the metabolism
- Spatial segregation of growth regions, attachment of cells to surfaces, or aggregation of cells
- Growth of more than one species in the bioreactor by intentional mixed cultures or infections

Simple segregated models for these phenomena can be built on the basis of ordinary differential equations by discriminating several classes of independent cells. Each cell class is mostly described by unstructured models, but with great mathematical care structured models can also be used. Here, only models with a finite number of cell classes will be considered, while segregated models with a continuous variation in cell properties that have to be based on partial differential equations are referred to in Sect. 2.4.6. The applications of segregated models are quite widespread and manifold so that it is difficult to construct a general and useful framework as for structured models. So, this chapter will rely mainly on some selected examples. Modeling of growth of multiple species that interact only indirectly, for instance by competing for substrate, is beyond the scope of this chapter. A review on this topic is given in [1].

2.4.1
Simple Segregated Models

For the general formulation of simple segregated models, which are also schematically represented by Fig. 2.4, the vector of concentrations of the liquid phase is similarly to structured models (see Eq. 2.51) split into two parts: the first, $\mathbf{Z}$, contains the concentrations of the abiotic phase such as substrates and cell-external products, and the second, $\mathbf{X}$, the concentrations of the biotic phase. When restricting the models for the N cell classes of the biotic phase to the unstructured type, the vector X includes only the biomass concentrations for each class:

$$\mathbf{X} = \begin{pmatrix} c_{X1} \\ c_{X2} \\ \vdots \\ c_{XN} \end{pmatrix} \tag{2.87}$$

Now every cell class or species is described by a stoichiometric model, as given by Eq. (2.2):

$$\mathbf{q}_i(\mathbf{C}(t)) = \mathbf{Y}_i \mathbf{r}_i(\mathbf{C}(t)) \tag{2.88}$$

where i denotes the index of the cell class. The vector $\mathbf{r}$ of intrinsic reactions now not only includes kinetic expressions as shown in Sect. 2.2, but also describes the transition of cells from class i to the other classes by appropriate rate equations. In the balance equation of the liquid phase, the contributions of all classes have to be considered. Equation (1.12) for a CSTR is then modified to

$$\frac{d\mathbf{C}(t)}{dt} = \underbrace{\sum_{i=1}^{N} \mathbf{q}_i(t) C_{Xi}(t)}_{\mathbf{Q}_i(t)} + D(t)\big(\mathbf{C}^R(t) - \mathbf{C}(t)\big) + \mathbf{TR}(t) \tag{2.89}$$

When all net reactions $\mathbf{q}_i$ of one cell class are concatenated together to a matrix of net reactions for all cell classes

$$\hat{\mathbf{q}} = (\mathbf{q}_1, \mathbf{q}_2, \cdots, \mathbf{q}_N) \tag{2.90}$$

the notation of the previous balance is simplified to

$$\frac{d\mathbf{C}(t)}{dt} = \hat{\mathbf{q}}(t)\mathbf{X}(t) + D(t)\big(\mathbf{C}^R(t) - \mathbf{C}(t)\big) + \mathbf{TR}(t) \tag{2.91}$$

The possible transition of cells between classes causes some problems when structured models are employed for the cell classes. For the target class, the transition means just the same as an inflow of cells from a reservoir in a continuous cultivation. Therefore the same restrictions are valid as discussed for structured models in Sect. 2.3: the application of such models is only correct when both the source and target cell classes have identical state at any time. Therefore, structured segregated models with cell transition from one class to another are only valid when cells of both classes have been grown from a single initial population, and when their different properties do not lead to different intrinsic states afterwards. These restrictions are not always observed in structured-segregated models in the literature.

2.4.2
Segregated Models for Physiological Properties

Many factors can harm microorganisms or cells and change irreversibly their physiology: heat, radiation, extreme pH-values, long periods of starvation, and so on. The damages caused under those circumstances often alter the structure of cell constituents such as membranes, cell wall, proteins, or DNA, but the cell as a whole may survive without being completely destroyed. Such extreme treatment of the cell population may prevent some of its parts from further regular cell growth, the population becomes inhomogeneous with respect to its growth and propagation activity. Tempest et al. [121] found, during continuous culture of *Aerobacter aerogenes*, only 40% of the cells were active at strong limitation due to low dilution rates ($D=0.004\,h^{-1}$), while under better substrate supply at high dilution rates ($D>0.5\,h^{-1}$) the percentage increased to more than 90%. This variation has great influence on the yield and productivity of the process. Another related phenomenon is a reduced but clearly non-zero cell concentration at very low dilution rates. Under such conditions the observed cell yield can be much higher than that predicted by the simple maintenance models, Eqs. (2.20) and (2.24). Sinclair and Topiwala [123] presented a simple segregated model (see Table 2.9) that discriminates for the entire population two cell classes: viable (or active) cells, X_1, and non-viable (or dormant, dead, inactive) cells without metabolic activity, X_2:

$$C_X = C_{X1} + C_{X2} \tag{2.92}$$

Table 2.9. Segregated model including viable and non-viable cells [18]

	Description	Model equations
a	Specialized vectors, r considers endogenous metabolism and maintenance	$\mathbf{C} = \begin{pmatrix} C_{X1} \\ C_{X2} \\ C_S \end{pmatrix} \quad \mathbf{Y}_1 = \begin{pmatrix} 1 & 0 & -1 & -1 \\ 0 & 0 & 1 & 0 \\ \frac{-1}{Y_{XS1}} & -1 & 0 & 0 \end{pmatrix}$
		$\mathbf{r}_1 = \begin{pmatrix} \mu_1 \\ m_S \\ K_{12} \\ \mu_E \end{pmatrix} \quad \mathbf{Y}_2 = \mathbf{r}_2 = 0$
b	Explicit model equations for the liquid phase	$\dfrac{dC_{X1}}{dt} = (\mu_1 - \mu_E - K_{12} - D)C_{X1}$
		$\dfrac{dC_{X2}}{dt} = K_{12}C_{X1} - DC_{X2}$
		$\dfrac{dC_S}{dt} = \dfrac{-\mu_1}{Y_{XS1}}C_{X1} - m_S C_{X1} + D(C_S^R - C_S)$

By the fraction of viable cells,

$$\alpha = \frac{C_{X1}}{C_X} = \frac{C_{X1}}{C_{X1} + C_{X2}} \leq 1 \tag{2.93}$$

which is under steady-state conditions a function of the specific growth rate, respective dilution rate in a chemostat,

$$\alpha = \frac{D}{K_D + D} \tag{2.94}$$

the cell concentrations of both classes can be expressed as

$$C_{X1} = (1 - \alpha)C_X \quad C_{X2} = \alpha C_X \tag{2.95}$$

In steady-state, the true specific growth rate μ_1 of the viable cell population is then increased compared to the observed growth rate q_X of the entire population:

$$\mu_1 = \frac{q_X}{\alpha(q_X)} + \mu_E \quad \mu_2 = 0 \tag{2.96}$$

where μ_E is the rate of endogenous metabolism. In the model, $\mu_1(C_S)$ can be chosen as Monod-Kinetics; K_{12} is the specific formation rate of inviable cells. Veilleux [124] applied such a model for simulation of a two-stage continuous cultivation. In both stages the concentrations of viable and non-viable cells could be nicely fitted. In addition to the original model [123], maintenance requirement of the viable cell fraction was introduced for the substrate uptake by the specific rate m_S [18]. This extended model also covers Pirt's concept of dormant cells [122]. For steady-states in a chemostat, the observed yield can then be calculated as

$$Y_{XS} = Y_{XS1} \frac{K_D + D}{K_D + D + m_S Y_{XS1}} \tag{2.97}$$

At high dilution rates, observed yield and the fraction of viable cells attain their maximum values. For D→0 the viable cells disappear completely, while the observed yield – and the observed biomass – clearly remains at non-zero values.

The influence of temperature on the growth can often be described by unstructured models by means of a functional variation in the model parameters, as discussed in Sect. 2.2. But the temperature influence can be more complex. For growth of *Streptococcus cremoris* it was found that anabolism already breaks down at lower

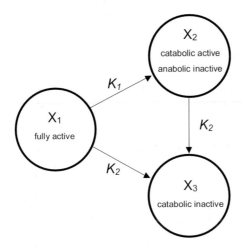

Fig. 2.14. Structure of the segregated model for growth at superoptimal temperatures [64]

Table 2.10. Segregated model for temperature inactivation [64]

	Description	Model equations
a	Specialized vectors	$\mathbf{C} = \begin{pmatrix} C_{X1} \\ C_{X2} \\ C_{X3} \\ C_S \end{pmatrix}$ $\mathbf{Y}_1 = \begin{pmatrix} 1 & -1 & -1 \\ 0 & 1 & 0 \\ 0 & 0 & 1 \\ -Y_{XS1}^{-1} & 0 & 0 \end{pmatrix}$ $\mathbf{r}_1 = \begin{pmatrix} \mu_1 \\ K_1 \\ K_2 \end{pmatrix}$
		$\mathbf{Y}_2 = \begin{pmatrix} 0 & 0 \\ -1 & 0 \\ 1 & 0 \\ 0 & -1 \end{pmatrix}$ $\mathbf{r}_2 = \begin{pmatrix} K_2 \\ m_S \end{pmatrix}$ $\mathbf{Y}_3 = \mathbf{r}_3 = \mathbf{0}$
b	Explicit model equations	$\dfrac{dC_{X1}}{dt} = (\mu_1 - K_1 - K_2)C_{X1}$
		$\dfrac{dC_{X2}}{dt} = K_1 C_{X1} - K_2 C_{X2}$
		$\dfrac{dC_{X3}}{dt} = K_2 C_{X1} + K_2 C_{X2}$
		$\dfrac{dC_S}{dt} = \dfrac{-\mu_1}{Y_{XS1}} C_{X1} - m_S C_{X2}$

temperatures than catabolism, the latter being decoupled from the first. The anabolic inactive cells are still able to catabolize substrate and secret metabolites, but are unable to grow. Furthermore, after temperature-shifts cells can grow for a while at temperatures that lead, after a prolonged time, to a complete inactivation of all cells, thus preventing balanced growth. Franks et al. [64] developed a segregated model for this phenomenon of temperature inactivation. The cell population can be represented by three classes – fully active cells, X_1, catabolic active/anabolic inactive cells, X_2, and catabolic inactive cells, X_3. The structure of the model is shown in Fig. 2.14 and the model equations are summarized in Table 2.10. The rate constants were assumed to have Arrhenius-form:

$$\mu, K_1, K_2, m_S \propto e^{\frac{-E}{RT}} \tag{2.98}$$

The model could successfully describe the experimental results for balanced growth at various temperatures, as well as transient growth at supermaximal temperatures according to Fig. 2.11.

2.4.3
A Model for Spatial Segregation by Wall Attachment

Attachment of cells to surfaces is an often observed phenomenon in biotechnology. In continuously operated bioreactors it leads to the effect that cells do not wash out even when the dilution rate is above the critical value. The reason is a continuous inoculation of the suspended population by cells detaching from the wall. Kreikenbohm and Stephan [125] published a segregated model for anaerobic growth of *Pelobacter acidigallici* on gallic acid, containing two cell classes, the first, X_1, is homogeneously suspended, the second, X_2, forms a thin homogenous biofilm on the reactor wall with small constant thickness as shown in Fig. 2.15. The biofilm is assumed as being well supplied with substrate. Such a model can help to estimate the rate constants of cell attachment and detachment. After introducing the volume fraction v of cells attached to the wall,

$$v = \frac{AL}{V_L} \tag{2.99}$$

where V_L is the reactor liquid volume, A the surface of the wall and L the thickness of the biofilm, the model can be noted as in Table 2.11. There, K_{12} and K_{21} are the rate constant of adsorption and desorption, respectively, K_{1L} and K_{2L} are specific rates of cell lysis.

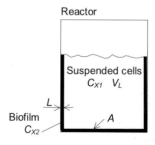

Fig. 2.15. Schematic diagram for a suspension culture that exhibits wall growth

Table 2.11. Segregated model for continuous cultivations with cell attachment to the reactor wall [125]

	Description	Model equations
a	Specialized vectors	$\mathbf{C} = \begin{pmatrix} C_{X1} \\ C_{X2} \\ C_S \end{pmatrix} \quad \mathbf{Y}_1 = \begin{pmatrix} 1 & -1 & -1 \\ 0 & 0 & \nu^{-1} \\ \frac{-1}{Y_{XS1}} & 0 & 0 \end{pmatrix} \quad \mathbf{r}_1 = \begin{pmatrix} \mu_1 \\ K_{1L} \\ K_{12} \end{pmatrix}$
		$\mathbf{Y}_2 = \begin{pmatrix} 0 & 0 & \nu \\ 1 & -1 & -1 \\ \frac{-\nu}{Y_{XS2}} & 0 & 0 \end{pmatrix} \quad \mathbf{r}_2 = \begin{pmatrix} \mu_2 \\ K_{2L} \\ K_{21} \end{pmatrix}$
b	Explicit model equations	$\dfrac{dC_{X1}}{dt} = (\mu_1 - K_{12} - K_{1L} - D)C_{X1} + K_{21}C_{X2}\nu$
		$\dfrac{dC_{X2}}{dt} = (\mu_2 - K_{21} - K_{2L})C_{X2} + \dfrac{K_{12}C_{X1}}{\nu}$
		$\dfrac{dC_S}{dt} = -\dfrac{\mu_1}{Y_{XS1}}C_{X1} - \dfrac{\mu_2}{Y_{XS2}}\nu C_{X2} + D(C_S^R - C_S)$

By a steady-state analysis of the model, the following approximate condition for the critical dilution rate can be derived for high substrate concentrations in the medium:

$$D_{crit} \approx \underbrace{\mu_{1\,max} - K_{12} - K_{1L}}_{\substack{D_{crit1} \\ \text{Critical dilution rate of suspended cells}}} + \underbrace{\frac{K_{12}K_{21}}{K_{21} + K_{2L} - \mu_{2\,max}}}_{\substack{\Delta D_{crit} \\ \text{Increase by wall growth}}} \qquad (2.100)$$

It is increased by ΔD_{crit} over the value D_{crit1} where a culture without wall attachment would wash out totally. Here, above D_{crit1} only the concentration of suspended cells begins to fall, but the steady states remain stable up to D_{crit}, where the final wash-out occurs. This is a major difference to the model of Topiwala and Hamer [126]. There the biofilm was assumed to form a constant monolayer, C_{X2}=constant, which results for high dilution rates in a hyperbolic relation

$$C_{X1} = \frac{\mu_{2\,max}C_{X2}}{D - \mu_{1\,max}} \qquad (2.101)$$

without wash-out.

2.4.4
Segregated Models for Morphological Differentiation, Morphologically Structured Models

Growth of filamentous microorganisms proceeds quite differently from symmetrically dividing bacteria or yeast. The hyphae, long tubular cells, only grow at free tips. Cell propagation is due to branching, i.e., formation of new tips, and segmentation of the hyphae, usually without separation of the cells. Therefore, big filamentous flocs or even dense pellets of more or less spherical form may be formed that contain a great number of cells with a wide distribution of age and morphological state. Cells without growing tips within the mycelia can have a metabolic activity that is different from those having free tips. Growth conditions, substrate supply and cell age as well have great influence on morphology of filamentous organisms, and in turn the morphological form is an important factor for product synthesis. The hyphal form is preferred at optimal growth while, at impaired conditions or stress, morphological differentiation to more resistant forms or (arthro)spores can be observed. Therefore, the growth kinetics of filamentous organisms are rather complex and it is difficult to identify the key factors to control growth and production. In processes for secondary metabolites the optimum conditions for growth differ substantially from those for biosynthesis of the desired product. This is not only true for the substrate supply but also temperature and pH. Mathematical models can help to clarify the view of the process and offer means for its optimization. The methodology of simple segregated models, when describing morphological differentiation, also called morphologically structured models, provide the appropriate tools for modeling: the clearly distinguishable forms within the population, e.g., growing and non-growing, hyphae and spores, young and old, may be lumped into respective cell classes [1, 146].

In the production of Cephalosporin C by *Cephalosporium acremonium* (*Acremonium chrysogenum*) three clearly distinguishable forms can be found: slim hyphae, thick-walled or swollen hyphal fragments, and arthrospores. Under optimal growth conditions in the early phase of the cultivation, mainly slim hyphae can be found which represent the growing fraction of the population. With time hyphae differentiate into highly septated thick-walled cells which differentiate further into arthrospores under impaired growth conditions or limitation. Maximum production of the antibiotic is correlated with a high percentage of swollen hyphal fragments.

The model of Chu and Constantinides [144] describes growth of *Cephalosporium acremonium* on glucose and sucrose (Table 2.12), as well as its pH and temperature dependency (not shown). Product synthesis is repressed by glucose, but not during growth at slowly catabolizible substrates such as sucrose, maltose, or lactose. This property can be used to optimize productivity in batch culture: the process is started on a mixture of glucose and sucrose. The growth is then diauxic with a first phase on glucose where at high growth rates mainly slim hyphae are formed. In a second phase on sucrose with low growth rates, the morphology changes to the high-producing thick-walled cells. Two cell types are discriminated in the model: slim hyphae, X_1, and producing thick-walled hyphal fragments, X_2, as schematically drawn in Fig. 2.16. Slim hyphae can develop from both forms. The ring expansion enzyme in the product synthesis chain is synthesized by the slim hyphae during metamorphosis. Product synthesis by thick-walled cells proceeds proportional to the enzyme concentration that is modeled by a respective balance. The model was successfully applied for optimization of pH and temperature profiles during batch cultivations, leading to a doubled product concentration.

Matsumura et al. [145] published an extended model of Cephalosporin C production with additional consideration of metabolic inactive arthrospores, X_3, that are formed at limitation (see Table 2.13). The structure of the model is shown in Fig. 2.17. Furthermore, balances for intracellular methionine are included into the model. Methionine stimulates product formation and differentiation of slim hyphae to thick-walled forms. Synthesis of ring-expansion enzyme and product is assumed to take place only in thick-walled cells. In the model, the accumulation of intracellular methionine is calculated for all cell types. The sum is compared to experimental

Table 2.12. Excerpts of the segregated model for growth of *Cephalosporium acremonium* including morphological differentiation of thin hyphae to swollen hyphae [144]

	Description	Model equations
a	Slim hyphae, lysis at glucose limitation	$\dfrac{dC_{X1}}{dt} = \mu_1(C_{X1} + C_{X2}) - \left(\mu_{12} + \dfrac{K_{D1\,max}K_{I1}}{K_{I1} + C_{S1}}\right)C_{X1}$
b	Thick-walled hyphae, lysis by first-order reaction	$\dfrac{dC_{X2}}{dt} = \mu_{12}C_{X1} - K_{D2}C_{X2}$
c	Glucose	$\dfrac{dC_{S1}}{dt} = \dfrac{-\mu_{S1}}{Y_{XS1}}(C_{X1} + C_{X2})$
d	Succrose, used for maintenance when glucose is limiting	$\dfrac{dC_{S2}}{dt} = -\left(\dfrac{\mu_{S2}}{Y_{XS2}} + \dfrac{m_{S2}C_{S2}}{K_{S2} + C_{S2} + \frac{C_{S1}}{K_{I2}}}\right)(C_{X1} + C_{X2})$
e	Product Cephalosporin C (c_E: intrinsic concentration of ring-expansion enzyme)	$\dfrac{dC_P}{dt} = K_P c_E C_{X2} - K_{DP}C_P$
f	Specific growth rate	$\mu_1 = \mu_{S1} + \mu_{S2}$
		$\mu_{S1} = \dfrac{\mu_{1\,max}C_{S1}}{K_{S1} + C_{S1}} \quad \mu_{S2} = \dfrac{\mu_{2\,max}(t)C_{S2}}{K_{S2} + C_{S2}}$
g	Metamorphosis reaction for slim to thick-walled cells	$\mu_{12} = \dfrac{\mu_{12\,max}K_{I1}}{K_{I1} + C_{S1}}$

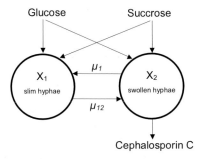

Fig. 2.16. Structure of the segregated model with slim and swollen hyphae for growth of *Cephalosporium acremonium* [144]

data. An intrinsic balance for ring-expansion enzyme is established under the assumption that enzyme synthesis is induced after a lag-phase by methionine and immediately repressed by glucose. Product synthesis rate in swollen hyphae is proportional to the intracellular enzyme concentration. The model was used for evaluation of fed-batch strategies for production of the antibiotic, by which the product concentration could be increased almost twice.

Simulations of the Penicillin production by *Penicillium chrysogenum* in fed-batch and repeated fed-batch cultivations were carried out by Zangirolami et al. [147]. The model, summarized in Table 2.14, discriminates three morphological forms, begin-

Table 2.13. Excerpts of the segregated model for growth of *Cephalosporium acremonium* including arthrospores [145]

	Description	Model equations
a	Slim hyphae	$\dfrac{dC_{X1}}{dt} = (\mu_1 - \mu_{12} - K_{D1})C_{X1}$
b	Thick-walled cells (swollen hyphea)	$\dfrac{dC_{X2}}{dt} = \mu_{12}C_{X1} - \mu_{23}C_{X2} - K_{D2}C_{X2}$
c	Inactive arthrospores	$\dfrac{dC_{X3}}{dt} = \mu_{23}C_{X2} - K_{D3}C_{X3}$
d	Glucose, used for growth of slim hyphae and maintenance metabolism of swollen hyphae	$\dfrac{dC_{S1}}{dt} = \dfrac{-\mu_1}{Y_{XS1}}C_{X1} - \dfrac{m_{S1}C_{S1}}{K_{S1} + C_{S1}}C_{X2}$
e	Methionine, uptake by the active forms	$\dfrac{dC_{S2}}{dt} = \dfrac{q_{S21\,max}C_{S2}}{K_{S2} + C_{S2}}C_{X1} + \dfrac{q_{S22\,max}C_{S2}}{K_{S2} + C_{S2}}C_{X2}$
f	Product formation (c_E: intrinsic enzyme concentration)	$\dfrac{dC_P}{dt} = K_P c_E C_{X2} - K_{DP}C_P$
g	Specific growth rate on glucose	$\mu_1 = \dfrac{\mu_{1\,max}C_{S1}}{K_{S1} + C_{S1}}$
h	Metamorphosis rate to swollen hyphae	$\mu_{12} = \left(K_{11} + \dfrac{K_{12}C_{S2}}{K_{S2} + C_{S2}}\right)\dfrac{C_{S1}}{K_{S1} + C_{S1}}$
i	Metamorphosis rate to arthrospores	$\mu_{23} = K_{21} + \dfrac{K_{22}K_{S1}}{K_{S1} + C_{S1}}$

Table 2.14. Excerpts of the model for *Penicillium chrysogenum* including three morphological forms [147]

	Description	Model equations
a	Apical cells	$\dfrac{dC_{X1}}{dt} = (\mu_1 - K_{12})C_{X1} + K_{21}C_{X2} - DC_{X1}$
b	Subapical cells	$\dfrac{dC_{X2}}{dt} = (\mu_2 - K_{21} - \mu_{23})C_{X2} + K_{12}C_{X1} - DC_{X2}$
c	Hyphal cells, (K_3: active fraction of the cell)	$\dfrac{dC_{X3}}{dt} = \mu_3 K_3 C_{X3} + \mu_{23}C_{X2} - DC_{X3}$
d	Penicillin formation	$\dfrac{dC_P}{dt} = K_P \dfrac{C_{S1}}{C_{S1} + K_{S1} + \frac{C_{S1}^2}{K_{I1}}}(C_{X2} + K_3 C_{X3}) - DC_P$
e	Specific growth rates	$\mu_1 = \dfrac{\mu_{1\,max}C_S}{K_S + C_S} \qquad \mu_2 = \dfrac{\mu_{2\,max}C_S}{K_S + C_S} \qquad \mu_3 = \dfrac{\mu_{3\,max}C_S}{K_S + C_S}$
f	Metamorphosis rate to hyphal cells	$\mu_{23} = \dfrac{K_{23}}{C_S K_{IS} + 1}$

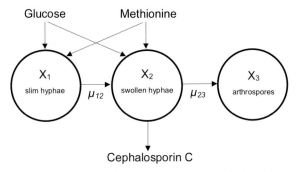

Fig. 2.17. Structure of the segregated model for growth of *Cephalosporium acremonium* including arthrospores [145]

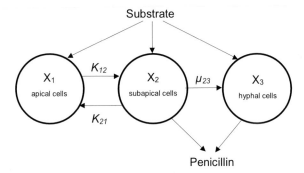

Fig. 2.18. Structure of the segregated model for hyphal growth of *Penicillium chrysogenum* [147]

ning from the tip: apical cells of the hyphae, subapical cells, and hyphal cells. The metamorphosis reactions are shown schematically in Fig. 2.18. During their life cycle, cells pass through the sequence of the three forms, but differentiation to hyphal cells runs only under limitation. By branching, new tips with apical cells can grow out of subapical cells. Product synthesis, inhibited by glucose, is assumed to take place in subapical and hyphal cells. A similar model that was also fitted to experimental data from cultivations of *Getrichun candium* and *Streptomyces hygroscopicus* was published by Nielsen [148]. There, the average properties of the hyphal elements were derived from a distributed population model. Another example of segregated modeling of Penicillin production will be given in Chap. 13.

2.4.5
Segregated Models for Recombinant Organisms

In recombinant organisms, the ability to produce foreign proteins is often mediated by introducing a relatively short circular DNA-sequence, the plasmid, into the host or wild-strain. The number of plasmids per cell is usually in the range of a few dozen. The plasmid-bearing, producing cells are denoted in the following as X_+. During cell division, the plasmids from the mother cell are randomly distributed to the daughter cells. Therefore, there is a certain probability the case will happen that one daughter cell receives all the plasmids and the other none (segregational instability). The further fate of the plasmid-free non-producing cells, i.e., the host strain, here denoted as X_-, depends on the applied selection mechanism. In the case without selection force, the plasmid-free cells have a growth-advantage because they do not carry the burden of propagating the plasmids and synthesizing the foreign protein. This results in a higher growth rate compared to the recombinant cells and the percentage of the X_--population will increase with time. This effect lowers greatly the yield and productivity for the foreign protein. In continuous culture, the X_--population eventually overgrows the X_+-population [127].

By applying a selection mechanism that prefers the plasmid-bearing cells, the stability of the plasmids can be increased to some degree. In the case where the plasmid is coding for a resistance against some antibiotic, the plasmid-free cells cannot grow at all when the medium contains the antibiotic. Nevertheless, it is possible that non-producing cells appear which still have resistance against the antibiotics. The mechanism can be an incorporation of the resistance-marker of the plasmid into the genome, or a mutation in the product-coding region of the plasmid (mutational instability). Again these cells have a growth-advantage compared to the producing cells. Another selection mechanism is based on host strains that are unable to synthesize an essential growth factor. The plasmid then contains a complementing gene that is coding for an enzyme by which the essential component can be synthesized. If the medium lacks the essential component, plasmid-free cells are again unable to propagate normally. The causes of instability in this case are identical to the antibiotics-resistance.

Therefore, in cultivations of genetically modified cells, a genetically inhomogeneous population is usually found. Segregated models can describe the stability of the plasmid-coded foreign DNA and the interference of growth and plasmid synthesis by introducing at least two cell classes, plasmid-free and plasmid-bearing [128]. The kinetics of the entire mixed population are then determined by the transition or mutation probabilities between the classes, and by the differences in their growth parameters such as maximum specific growth rate or yield. These can be estimated from experimental data [129, 130]. The plasmid content together with the percentage of plasmid-bearing cells determines the foreign protein productivity of the culture.

A minimal model for growth of recombinant cells on non-selective media considering only segregational instability of the plasmid includes the two cell classes X_+ and X_-. With the probability α for loss of plasmid the balances for the cell concentrations in a continuous cultivation are

$$\frac{dC_{X+}}{dt} = \mu_+(1-\alpha)C_{X+} - DC_{X+} \tag{2.102}$$

$$\frac{dC_{X-}}{dt} = \mu_+\alpha C_{X+} + \mu_- C_{X-} - DC_{X-} \tag{2.103}$$

Ollis and Chang [131] studied batch-growth kinetics based on this segregated model using Monod-expressions for the specific growth rates. The Luedeking-Piret-equation described product formation by the X_+-population. Such a model was used by Park et al. [132] to investigate optimal control policies for batch, fed-batch, and continuous cultivation of *Escherichia coli* with cloned *trp* operon. Satyagal and Agrawal [136] introduced an inhibition by the plasmid concentration into the kinetics for μ_+ to consider the additional burden for plasmid synthesis. Mosrati et al. [137] used this model with slight modification for an accurate analysis of experimental data. They conclude that α increases with growth rate.

Parulekar et al. [134] published the results of theoretical studies for continuous cultures of recombinant methylotrophs. Methanol is an inhibitory substrate. Plasmid stability can therefore be improved when it contains a gene that removes this inhibition. For this case, the following kinetics were used for the specific growth rates:

$$\mu_+ = \mu_{+\,max} \frac{C_S}{K_{S+} + C_S} \tag{2.104}$$

$$\mu_- = \mu_{-\,max} \frac{C_S}{K_{S-} + C_S + \frac{C_S^2}{K_I}} \tag{2.105}$$

The model simulations showed regions of coexistence of both cell types at various dilution rates during steady states. Cycling of the feeding could improve the stability. This was also found by Ryder and DiBiasio [135].

A study on continuous cultivation of recombinant yeast in air-lift reactors was carried out by Zhang et al. [133]. Based on the model the maximum specific growth rates and plasmid loss probability were estimated. Coppella and Dhurjati [108] applied the segregated model, Eqs. (2.102) and (2.103), to recombinant yeast for production of human epidermal growth factor in connection with a structured growth model. Cell growth, plasmid segregation, and gene product synthesis were successfully predicted for three reactor configurations: batch, fed-batch, and a hollow-fiber bioreactor. For production of Hepatitis B virus antigen by recombinant yeast a model was presented by Shi et al. [138]. Their unstructured growth model also took the effects of ethanol and leucine into account. The model was used to construct a Kalman-filter for state estimation.

Chang and Lim [139] published a model for continuous cultivation of antibiotic-resistant recombinant strains of *Escherichia coli* and *Proteus mirabilis* carrying a resistance marker. For modeling of antibiotic-induced death, the balance equation of plasmid-free cells was extended by a death rate μ_D,

$$\frac{dC_{X-}}{dt} = \mu_+ \alpha C_{X+} + (\mu_- - \mu_D)C_{X-} - DC_{X-} \tag{2.106}$$

with the following kinetic expression:

$$\mu_D = \mu_{D\,max} \frac{C_B}{K_B + C_B} \tag{2.107}$$

where C_B is the antibiotic concentration. Model studies were carried out for optimization of the antibiotic level in the medium.

Scrienc et al. [140] developed two models for growth of a recombinant yeast mutant with an impaired URA3 gene on selective media, an age distribution model and a simple segregated model. The employed yeast strain is unable to form an enzyme that catalyzes the synthesis of the essential growth factor uracil. The complementing gene for this enzyme, which is called the complementing gene product, is also contained on the plasmid. The authors stress that an important and general aspect for all selection systems is that it is not the selection gene itself that endows the host cell with the selective growth phenotype but the product of that gene. This fact has major significance for the growth of unstable recombinant cultures. Since the number of enzyme-molecules is relatively high, one can assume that the enzymes are always regularly partitioned between the daughter cells. After loss of the plasmid, the daughter cells can grow further for a while, because they still contain a certain amount of the enzyme coding for the essential growth factor. The age distribution model is compared with the simple segregated model, Eqs. (2.102) and (2.103), assuming constant average growth rate of plasmid-free cells, $\mu_- = \bar{\mu}_-$. This is related to the fraction of the plasmid-bearing cells, F_+, obtained by the distribution model under stationary conditions:

$$\bar{\mu}_- = \mu_+ \left(1 - \frac{\alpha}{1 - F_+(\infty)} \right) \tag{2.108}$$

where

$$F_+(t) = \frac{C_{X+}(t)}{C_{X+}(t) + C_{X-}(t)} \tag{2.109}$$

In case the X_--population is not overgrown, the frequency of the segregational instability can be estimated for stationary exponential growth from the observed over-all growth rate, $\mu(\infty)$, as

$$\alpha = 1 - \frac{\mu(\infty)}{\bar{\mu}_-} \tag{2.110}$$

The model of Sardonini and DiBiasio [141] for growth of recombinant *Saccharomyces cerevisiae* on selective media was aimed to explain the unexpectedly high number of plasmid-free cells in the population. The employed selection mechanism is also based upon the incapability of the host to synthesize an essential metabolite (M). Ideally, plasmid-free cells should be unable to propagate. Plasmid-bearing cells have the complementing gene and synthesize the essential metabolite for their own growth. But since the metabolite is also released into the media, it can support growth of plasmid-free cells to some extent, which increases their percentage in the entire population. The explanation of the effect that the X_--cells not only originate from segregational instability but also from independent growth here is slightly different from the previous model. An extended Monod-kinetics describes the dependency of the specific growth rate of plasmid-free cells on the essential metabolite:

$$\mu_- = \mu_{-\,\mathrm{max}} \frac{C_S}{K_S + C_S} \frac{C_M}{K_M + C_M} \tag{2.111}$$

The simulation showed that 67% of the plasmid-free cells originated in independent growth and only 33% form the unequal plasmid partitioning.

An extension of the models for growth on non-selective media by another cell class "plasmid-carrying with mutation in the product gene" for consideration of the mutational/structural stability causes formally no problems [79, 142], although the determination of the model parameters raises some difficulties. The model for both segregational and mutational instability of recombinant yeast is based on three cell classes as shown in Fig. 2.19: the non-producing host, X_-, the producing plasmid bearing cells, X_+, and plasmid-bearing but non-producing cells, X_-^*. When the

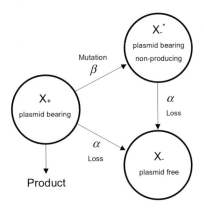

Fig. 2.19. Structure of the model for segregational and mutational/structural instability of re-combinant yeast [79]

probability for segregational loss of plasmid is denoted as α, and for mutational loss of only the product-coding gene as β, the model becomes

$$\frac{dC_{X+}}{dt} = \mu_+(1 - \alpha - \beta)C_{X+} \tag{2.112}$$

$$\frac{dC_{X-}}{dt} = \mu_+\alpha C_{X+} + \mu_{*+}\alpha C_{X*+} + \mu_- C_{X-} \tag{2.113}$$

$$\frac{dC_{X*+}}{dt} = \mu_+\beta C_{X+} + \mu_{*+}(1 - \alpha)C_{X*+} \tag{2.114}$$

For the modeling of the interference of plasmid copy number, cloned gene product synthesis, and growth, they are assumed as being separable into independent factors. A Monod-kinetics for substrate uptake and inhibition kinetics by the plasmid con-centration, c_G, and product concentration, c_P, are chosen:

$$\mu_+ = \mu_{max}\left(1 - \frac{c_G}{c_{G\,max}}\right)^{r_1}\left(1 - \frac{c_P}{c_{P\,max}}\right)^{r_2}\frac{C_S}{K_S + C_S} \tag{2.115}$$

$$\mu_- = \mu_{max}\frac{C_S}{K_S + C_S} \tag{2.116}$$

$$\mu_{*+} = \mu_{max}\left(1 - \frac{c_G}{c_{G\,max}}\right)^{r_3}\frac{C_S}{K_S + C_S} \tag{2.117}$$

The formation of the product that is accumulated within the cells is modeled ac-cording to Eqs. (2.70) and (2.71) with specialized expressions for η and ξ.

Schwartz et al. [143] developed a segregated model for plasmid instability of yeast grown in selective and non-selective media. The plasmid contains the gene for anti-biotic resistance. In the model, the effects of plasmid loss and mutation are consid-ered. By mutation (or crossover) the yeast cells can acquire antibiotic resistance. Four cell classes are considered: plasmid bearing cells, X_+ that produce foreign pro-tein and do have antibiotics resistance; non-producing plasmid free cells without antibiotics resistance, X_-; plasmid bearing cells that have acquired antibiotics resistance by mutation, X^*_+; and plasmid-free cells with acquired resistance, X^*_-. The model balances are

$$\frac{dC_{X+}}{dt} = \mu_+(1 - \alpha - \beta)C_{X+} \tag{2.118}$$

$$\frac{dC_{X-}}{dt} = \mu_+\alpha C_{X+} + \mu_-C_{X-} \tag{2.119}$$

$$\frac{dC_{X*+}}{dt} = \mu_+\beta C_{X+} + \mu_{*+}(1 - \alpha)C_{X*+} \tag{2.120}$$

$$\frac{dC_{X*-}}{dt} = \mu_{*+}\alpha C_{X*+} + \mu_{*-}C_{X*-} \tag{2.121}$$

The influence of the antibiotic, C_B in $mg\,l^{-1}$, given by the following inhibition kinetics,

$$\mu_+ = \mu_{max}\left(1 - \frac{0.41C_B}{159 + C_B}\right) \qquad \mu_- = \mu_{max}(1 - 0.023C_B) \tag{2.122}$$

is analyzed and compared to experimental data for the case of constant maximum growth rate. The experimental findings could be explained that in non-selective media after 50 generations, 80% of the cells still exhibited the desired phenotype, and that only marginal improvement of plasmid stability by addition of antibiotics could be observed due to mutations.

2.4.6
Population Balance Models

The unstructured or structured models took the cell population in the bioreactor as a homogenous biomass. Segregated models divided an inhomogeneous population into a few cell classes that again were each assumed as being homogeneous. The advantage of this simplified view is that the models are based on ordinary differential equations which are easy to handle and solve numerically. On the other hand, these models may be inadequate or inaccurate in several situations as it was already discussed in the related sections. The simple segregated modeling approach may be inaccurate when there is a great continuous variation of cell properties that cannot be matched to few discrete classes. This case can, for instance, be found for plasmid copy number and product synthesis in cultivations of unstable recombinant organisms [152–154].

The situation can also be met that the unstructured, structured, or segregated models usually work fine but fail only under some circumstances, for instance, when the inoculation procedure or the process control scheme was changed. The reason can be a synchronization of the growth of a majority of cells in the division cycle. Usually, growth and division of a single cell is independent of the others in the population. This, together with a slight variation in the lengths of the cell cycles, results in a uniform distribution of the division events over time. On average, the population seems then to be homogeneous and the cell number increases steadily. But growth may become synchronized by a sudden change in conditions. A typical case is a preculture that was running into substrate limitation. Under starvation most of the cells rest at a similar point in the cell cycle, e.g., shortly before initiation of division. When a bioreactor is inoculated with this preculture, all cells resume growing at the same time, and subsequently also enter division synchronously. This may last for several generations until the synchrony is lost by individual variations in the cycle lengths. Under synchronous growth the modulation of the metabolism over the cell cycle becomes visible on the measured variables of the bioreactor, e.g., as oscillations on the concentrations or a step-like increase in cell number. When there exists some feed-back mechanism from such an oscillating variable in the reactor to the cell metabolism, e.g., by a regulatory influence, this can even stabilize synchronous growth for a long time.

The above-mentioned cases can be described to some extent by population balance models. These models, based on partial differential equations, not only look at the time-dependence of the system but consider its development as governed by further independent variables z representing the state of the cell. These can be the age of a cell, its position in the cell cycle, its total mass or volume, the mass of cell constituents, or other properties. The related balance equation is for symmetric cell division [149]

$$\frac{\partial f(t,\mathbf{z})}{\partial t} + \frac{\partial(\dot{\mathbf{z}}f(t,\mathbf{z}))}{\partial \mathbf{z}} = 2\int p(\mathbf{z},\mathbf{z}')g(\mathbf{z}')f(t,\mathbf{z}')d\mathbf{z}' - g(\mathbf{z})f(t,\mathbf{z}) \tag{2.123}$$

with initial condition

$$f(0,\mathbf{z}) = f_0(\mathbf{z}) \tag{2.124}$$

where $\mathbf{z}$ is the vector of cellular state variables, $\dot{\mathbf{z}}$ is the mean growth rate vector of $\mathbf{z}$, f is the density function of the distribution in the population state, g is the division rate or probability of division of a cell, and p is the partitioning function, i.e., the conditional probability that a mother cell in state $\mathbf{z}'$ will divide and form two new cells in state $\mathbf{z}$.

The partitioning function must have the following properties:

$$p(\mathbf{z},\mathbf{z}') = 0 \qquad\qquad for\ \mathbf{z} > \mathbf{z}'$$
$$\int p(\mathbf{z},\mathbf{z}')d\mathbf{z} = 1 \tag{2.125}$$
$$p(\mathbf{z},\mathbf{z}') = p(\mathbf{z}' - \mathbf{z},\mathbf{z}') \qquad for\ \mathbf{z} < \mathbf{z}'$$

The solution of the balance equation is, in general, rather difficult, but it can be significantly simplified by an iterative procedure based on successive generations of cells [150]. In age distribution models, the state of the population is characterized by the time since birth of the cell, i.e., the cell age τ. The balance equation then becomes

$$\frac{\partial f(t,\tau)}{\partial t} + \frac{\partial f(t,\tau)}{\partial \tau} = -g(\tau)f(t,\tau) \tag{2.126}$$

with initial condition

$$f(0,\tau) = f_0(\tau) \tag{2.127}$$

and boundary condition

$$f(t,0) = 2 \int\limits_0^\infty g(\tau)f(t,\tau)d\tau' \tag{2.128}$$

Equation (2.126) is the so-called M'Kendrick-von Foerster equation for the development of cell number density [151]. An application of age distribution models for simulation of Baker's yeast production is given in Chap. 9.

References

1.   Bailey JE, Ollis DF (1986) Biochemical engineering fundamentals, 2nd edn. McGraw-Hill, New York
2.   Roels JA (1983) Energetics and kinetics in biotechnology. Elsevier Biomedical Press, Amsterdam
3.   Gaden EL (1959) J Biochem Microbiol Technol Eng 1:413
4.   Aharonowitz Y, Demain AL (1980) Biotechnol Bioeng 22, Suppl 1:5
5.   Bailey JE (1998) Biotechnol Prog 14:8
6.   Heijnen JJ (1994) Tibtech, 12:483
7.   van Gulik WM, Heijnen JJ (1995) Biotechnol Bioeng 48:681
8.   Toda K (1981) J Chem Tech Biotechnol 31:775
9.   Calam CT (1979) Folia Microbiol 24:276
10. Esener AA, Roels JA, Kossen NWF (1983) Biotechnol Bioeng 25:2803
11. Noorman HJ, Heijnen JJ, Luyben KCAM (1991) Biotechnol Bioeng 38:603
12. Nielsen J, Villadsen J (1992) Chem Eng Sci 47:4225
13. Wood BD, Whitaker S (1998) Chem Eng Sci 53:397
14. van Suijdam JC, Hols H, Kossen NWF (1982) Biotechnol Bioeng 24:177
15. de Gooijer CD, RH Wijffels, Tramper J (1991) Biotechnol Bioeng 38:1991
16. Buschulte TK, Gilles ED (1990) Modelling and simulation of hyphal growth, metabolism and mass transfer in pellets of streptomyces. In: Christiansen C, Munck L, Villadsen J (eds) Proceedings of the 5th European Congress on Biotechnology, p 279
17. Buschulte TK (1992) PhD thesis, Universität Stuttgart
18. Bellgardt KH (1991) Cell models. In: Rehm HJ, Reed G (eds) Biotechnology, vol 4. VCH, Weinheim
19. Cui QH, Lawson L, Gao D (1984) Biotechnol Bioeng 26:682
20. Frame KK, Hu WS (1988) Biotechnol Bioeng 32:1061
21. Monbouquette HG (1992) Biotechnol Bioeng 39:498
22. Weiss RM, Ollis RF (1980) Biotechnol Bioeng 22:859
23. Sinclair CG, Ryder DH (1975) Biotechnol Bioeng 17:375
24. Tsao GT, Yang CM (1976) Biotechnol Bioeng 18:827
25. Herbert D (1959) In: Tenvall D (ed) Recent progress in microbiology. Alquist and Wiksell, Stockholm
26. Pirt SJ (1965) Proc R Soc Ser B 163:224
27. Eroshin VK, Utkin IS, Ladynichev SV, Samoylov VV, Kuvshinnikov VD, Skryabin GK (1976) Biotechnol Bioeng 18:289
28. Tayeb YJ, Bouillanne C, Desmazeaud MJ (1984) J Ferment Technol 62:461
29. Chu WB, Constantinides A (1988) Biotechnol Bioeng 32:277
30. Reuss M, Fröhlich S, Kramer B, Messerschmidt K, Pommerening G (1986) Bioproc Eng 1:79
31. Brown DE, Halsted DJ (1975) Biotechnol Bioeng 17:1199
32. Topiwala H, Sinclair CG (1971) Biotechnol Bioeng 13:785
33. Bellgardt KH (1998) Adv Biochem Eng Biotechnol 60:153
34. Starzak M, Krzystek L, Nowicki L, Michalski H (1994) Chem Eng J 54:221

35. Zeng AP, Ross A, Biebl H, Tag C, Günzel B, Deckwer WD (1994) Biotechnol Bioeng 44:902
36. Dourado A, Goma G, Albuquerque U, Sevely Y (1987) Biotechnol Bioeng 29:187
37. Kowda M, Wasungu M, Simard RE (1982) Biotechnol Bioeng 24:1125
38. Luong JHT (1885) Biotechnol Bioeng 27:280
39. Luong JHT (1987) Biotechnol Bioeng 29:242
40. Han K, Levenspiel O (1988) Biotechnol Bioeng 32:430
41. Kosaric N, Ong SL, Duvnjak Z, Moser A (1984) Acta Biotechnol 4:153
42. Tan Y, Wang ZX, Marshal KC (1996) Biotechnol Bioeng 52:602
43. Lidén G, Franzén CJ, Niklasson C (1994) Biotechnol Bioeng 44:419
44. Barford JP, Phillips PJ, Orlowski JH (1992) Bioproc Eng 7:297
45. Benthin S, Nielsen J, Villadsen J (1992) Biotechnol Bioeng 40:137
46. Pavlou S, Fredrickson AG (1989) Biotechnol Bioeng 34:971
47. Baltzis BC, Fredrickson AG (1988) Biotechnol Bioeng 31:75
48. Sinclair CG (1989) Microbial process kinetics. In: Bu'Lock J, Kristiansen B (eds) Basic biotechnology. Academic Press, London, p 102
49. Åkerberg C, Hofvendahl K, Zacchi G, Hahn-Hägerdal (1998) Appl Microbiol Biotechnol 49:682
50. Lendenmann U, Egli T (1998) Biotechnol Bioeng 59:99
51. Lam JC, Ollis DF (1981) Biotechnol Bioeng 23:1517
52. Venkatesh KV, Okos MR, Wankat PC (1993) Process Biochemistry 28:231
53. Ohara H, Hiyama K, Yoshida T (1992) Appl Microbiol Biotechnol 38:403
54. Jackson JV, Edwards VH (1975) Biotechnol Bioeng 17:943
55. Tan Y, Wang ZX, Marshal KC (1998) Biotechnol Bioeng 59:724
56. Kluge M, Siegmund D, Diekmann H, Thoma M (1992) Appl Microbiol Biotechnol 36:446
57. Halme A (1989) Expert system approach to recognize the state of fermentation and to diagnose faults in bioreactors. In: Fish NM, Fox RI, Thornhill NF (eds) Computer application in fermentation technology. Elsevier Applied Science Publishers, London, p 159
58. Halme A, Visala A, Kankare J (1992) Functional state concept in modelling of biotechnical processes. In: Stephanopoulos G, Karim MN (eds) IFAC Symposium on Modelling and Control of Biotechnical Processes, Pergamon Press, New York, p 153
59. Konstantinov K, Yoshida T (1989) Biotechnol Bioeng 33:1145
60. Zhang XC, Visala A, Halme A, Linko P (1994) Process Control 4:1
61. Zhang XC, Visala A, Halme A, Linko P (1994) J Biotechnol 27:1
62. Linko P, Siimes T, Kosola A, Zhu YH, Eerikäinen T (1994) Hybrid fuzzy knowledge-based and neural systems for fed-batch baker's yeast bioprocess control. In: Proceedings of the 1st Asian Control Conference, 27–30 July 1994, Tokyo:491
63. Ruenglertpanyakul W (1996) PhD Thesis, Universität Hannover
64. Franks PA, Hall RJ, Linklater PM (1980) Biotechnol Bioeng 22:1465
65. Williams FM (1967) J Theor Biol 15:190
66. Korte G, Rinas U, Kracke-Helm HA, Schügerl K (1991) Appl Microbiol Biotechnol 35:185
67. Korte G, Rinas U, Kracke-Helm HA, Schügerl K (1991) Appl Microbiol Biotechnol 35:189
68. Nielsen J, Emborg C, Halberg K, Villadsen J (1989) Biotechnol Bioeng 34:478
69. Nielsen J, Pedersen AG, Strudsholm K, Villadsen J (1991) Biotechnol Bioeng 37:802
70. Bentley WE, Kompala DS (1989) Biotechnol Bioeng 33:49
71. Lee J, Ramirez WF (1992) Biotechnol Bioeng 39:635
72. Domach MM, Leung SK, Cahn RE, Cocks GG, Shuler ML (1984) Biotechnol Bioeng 26:203
73. Ataai MM, Shuler ML (1985) Biotechnol Bioeng 27:1027
74. Shu J, Shuler ML (1989) Biotechnol Bioeng 33:1117
75. Peretti SW, Bailey JE (1986) Biotechnol Bioeng 28:1672
76. Lee SB, Bailey JE (1984) Biotechnol Bioeng 26:66
77. Peretti, SW, Bailey JE (1987) Biotechnol Bioeng 29:316
78. Ataai MM, Shuler ML (1987) Biotechnol Bioeng 30:389
79. Lee SB, Seressiotis A, Bailey JE (1985) Biotechnol Bioeng 27:1699
80. Lee SB, Bailey JE (1984) Biotechnol Bioeng 26:1372

81. Lee SB, Bailey JE (1984) Biotechnol Bioeng 26:1383
82. Betenbaugh MJ, Dhurjati PA (1990) Biotechnol Bioeng 36:124
83. Ramkrishna D (1982) A cybernetic perspective of microbial growth. In: Papoutsakis E, Stephanopoulos GN, Blanch HW (eds) Foundations of biochemical engineering kinetics and thermodynamics in biological systems. American Chemical Society, Washington DC
84. Kompala DS, Ramkrishna D, Tsao GT (1984) Biotechnol Bioeng 26:1272
85. Dhurjati P, Ramkrishna D, Flickinger MC, Tsao GT (1985) Biotechnol Bioeng 27:1
86. Kompala DS, Ramkrishna D, Jansen NB, Tsao GT (1986) Biotechnol Bioeng 28:1044
87. Turner BG, Ramkrishna D, Jansen NB (1988) Biotechnol Bioeng 32:46
88. Turner BG, Ramkrishna D (1988) Biotechnol Bioeng 31:41
89. Straight JV, Ramkrishna D (1990) Biotechnol Bioeng 37:895
90. Baloo S, Ramkrishna D (1990) Biotechnol Bioeng 38:1337
91. Straight JV, Ramkrishna D (1994) Biotechnol Prog 10:574
92. Straight JV, Ramkrishna D (1994) Biotechnol Prog 10:588
93. Ramakrishna R, Ramkrishna D, Konopka AE (1997) Biotechnol Bioeng 54:77
94. Narang A, Konopka A, Ramkrishna D (1997) Chem Eng Sci 52:2567
95. Fredrickson AG (1976) Biotechnol Bioeng 18:1481
96. Esener AA, Veerman T, Roels JA, Kossen NWF (1982) Biotechnol Bioeng 24:1749
97. Nielsen J, Nikolajsen K, Villadsen J (1991) Biotechnol Bioeng 38:1
98. Nielsen J, Nikolajsen K, Villadsen J (1991) Biotechnol Bioeng 38:11
99. Nikolajsen K, Nielsen J, Villadsen J (1991) Biotechnol Bioeng 38:24
100. King R (1997) J Biotechnol 52:219
101. King R, Büdenbender C (1997) J Biotechnol 52:235
102. Mundry C, Kuhn KP (1991) Appl Microbiol Biotechnol 35:306
103. Yoo S, Kim WS (1994) Biotechnol Bioeng 43:1043
104. Bonarius HPJ, Schmid G, Tramper J (1997) Tibtech 15:308
105. Doshi P, Rengaswamy R, Venkatesch KV (1997) Process Biochem 32:643
106. Venkatesch KV, Doshi P, Rengaswamy R (1997) Biotechnol Bioeng 56:635
107. Narang A, Konopka A, Ramkrishna D (1997) J Theor Biol 184:301
108. Coppella SJ, Dhurjati P (1990) Biotechnol Bioeng 36:356
109. Ray NG, Vieth WR, Venkatasubramanian K (1987) Biotechnol Bioeng 29:1003
110. Lee SB, Bailey JE (1984) Plasmid 11:151
111. Lee SB, Bailey JE (1984) Plasmid 11:166
112. Ryu DY, Park SH (1987) Ann NY Acad Sci 506:396
113. Chen W, Bailey JE, Lee SB (1991) Biotechnol Bioeng 38:679
114. Koh BT, Yap MGS (1993) Biotechnol Bioeng 41:707
115. Axe DD, Bailey JE (1994) Biotechnol Bioeng 43:242
116. Kim BG, Shuler ML (1991) Biotechnol Bioeng 1076
117. Laffend L, Shuler ML (1994) Biotechnol Bioeng 43:388
118. Laffend L, Shuler ML (1994) Biotechnol Bioeng 43:399
119. Chetan J, Gadgil P, Bhat PJ, Venkatesh RV (1996) Biotechnol Prog 12:744
120. Fredrickson AG (1992) AIChE Journal 38:835
121. Tempest DW, Herbert D, Phipps PJ (1967) Microbial physiology and continuous culture. HMSO, London, p 240
122. Pirt SJ (1987) J Ferment Technol 65:173
123. Sinclair CG, Topiwala HH (1970) Biotechnol Bioeng 12:1069
124. Veilleux BG (1980) European J Appl Microbiol Biotechnol 9:165
125. Kreikenbohm R, Stephan W (1985) Biotechnol Bioeng 27:296
126. Topiwala HH, Hamer G (1971) Biotechnol Bioeng 13:919
127. Summers DK (1991) Tibtech 9:273
128. Imanaka T, Aiba S (1981) Ann NY Acad Sci 396:1
129. Park SH, Ryu DDY, Lee SB (1991) Biotechnol Bioeng 37:404
130. Shoham Y, Demain AL (1991) Biotechnol Bioeng 37:927
131. Ollis DF, Chang HT (1982) Biotechnol Bioeng 24:2581
132. Park TH, Seo JH, Lim HC (1989) Biotechnol Bioeng 34:1167

133. Zhang Z, Scharer JM, Moo-Young M (1997) J Biotechnol 55:31
134. Parulekar SJ, Chang YK, Lim HC (1987) Biotechnol Bioeng 29:911
135. Ryder DF, DiBiasio D (1984) Biotechnol Bioeng 26:942
136. Satyagal VN, Agrawal P (1989) Biotechnol Bioeng 34:265
137. Mosrati R, Nancib N, Boudrant J (1993) Biotechnol Bioeng 41:395
138. Shi Y, Ryu DY, Yuan WK (1993) Biotechnol Bioeng 41:55
139. Chang YK, Lim HC (1987) Biotechnol Bioeng 29:950
140. Scrienc F, Campbell JL, Bailey JE (1986) Biotechnol Bioeng 18:996
141. Sardonini CA, DiBiasio D (1987) Biotechnol Bioeng 29:469
142. Hjortso MA, Bailey JE (1984) Biotechnol Bioeng 26:528
143. Schwartz LS, Jansen NB, Ho NWY, Tsao GT (1988) Biotechnol Bioeng 32:733
144. Chu WB, Constantinides A (1988) Biotechnol Bioeng 32:277
145. Matsumura M, Imanaka T, Yoshida T, Taguchi H (1981) J Ferment Technol 59:115
146. Nielsen J (1992) Adv Biochem Eng Biotechnol 46:187
147. Zangirolami TC, Johansen CL, Nielsen J, Jørgensen SB (1997) Biotechnol Bioeng 56:595
148. Nielsen J (1993) Biotechnol Bioeng 41:715
149. Fredricksen AG, Ramkrishna D, Tsuchiya HM (1967) Math Biosci 1:327
150. Liou JJ, Scrienc F, Fredrickson AG (1997) Chem Eng Sci 52:1529
151. Webb GF (1986) J Math Biology 23:269
152. Seo JH, Bailey JE (1985) Biotechnol Bioeng 27:156
153. Wittrup KD, Bailey JE (1988) Biotechnol Bioeng 31:304
154. Srienc F, Campbell JL, Bailey JE (1986) Biotechnol Bioeng 18:996

3 Metabolic Flux Analysis

Maria I. Klapa, Gregory Stephanopoulos

3.1
Introduction

Metabolic pathways are sequences of enzyme-catalyzed reaction steps, converting substrates to a variety of products to meet the needs of the cell. Metabolic pathways can interact to create complex networks. Manipulation of metabolic pathways to improve the cellular properties is an old concept in biological sciences. This approach initially relied on random mutagenesis and creative selection techniques to identify superior strains with respect to a certain objective. Despite impressive successes in many biological areas, the mutations that played important role in achieving the desired cellular properties were poorly characterized and the strain development process remained random, combining science with elements of art [1]. Recombinant DNA technology introduced a new dimension to pathway manipulation, because it allowed precise modifications of enzymatic reactions in metabolic pathways. Metabolic engineering emerged then as the scientific field aiming at the directed modification of the enzymatic, regulatory, or transport activities of the cell to improve cellular properties, with the use of recombinant DNA technology [2–4].

One could argue that metabolic engineering is just the technological manifestation of applied molecular biology and there is no justification for the introduction of a new scientific field. However, metabolic engineering is indeed a new scientific field [2] based on the rational, as opposed to the *ad hoc*, process to identify the targets for pathway manipulation [1]. To determine the required biological changes, an enhanced understanding of metabolism and cellular functions must be obtained. Thus, the analysis and correct interpretation of the structure and control of metabolic networks constitute the first critical task of metabolic engineering towards the goal of rational pathway manipulation.

One important novel aspect of metabolic engineering is the emphasis it places on the concept of metabolic networks, as opposed to individual reactions. By considering integrated networks rather than reactions or parts of networks in isolation from one another, an enhanced understanding of metabolism can be obtained [4]. The complexity though of cellular metabolic networks, involving large numbers of metabolites and enzymes under the control of intricate regulatory mechanisms, and the difficulty of assessing enzyme kinetic properties in vivo pose two key obstacles in the analysis of metabolic networks. Metabolic fluxes emerge then as the preferred parameter to determine the physiological state of the cell. Perhaps the most significant contribution of metabolic engineering is the emphasis it places on in vivo metabolic fluxes and their control [5] for the investigation of metabolic networks.

Flux is defined as the rate at which material is processed through a metabolic pathway. Consequently, the flux of a linear pathway at steady state is equal to the steady-state rates of all the individual reactions. The value of the flux does not carry information about the activity of the corresponding pathway enzymes, but it defines the extent to which these enzymes participate in a conversion process. Thus, metabolic fluxes provide a measure of the degree of engagement of the various pathways in the overall cellular functions and metabolic processes at specific genetic and environmental conditions, constituting a fundamental determinant of cell physiology [1]. Fluxes are also useful for the determination of maximum theoretical yields [6–10] and the observation of the function of metabolic pathways in vivo [11]. Flux determination can play an important role in elucidating metabolic pathways to a finer detail [1, 12]. By comparing and analyzing flux distributions throughout a metabolic network at various biological conditions, information concerning the *control* of flux and the regulatory structure of metabolic networks can be extracted [3, 12]. Therefore, the accurate quantification and analysis of metabolic fluxes is an essential step in the study of the structure and control of metabolic networks. The quantification of metabolic fluxes is termed *Metabolic Flux Analysis* (MFA) and is a major objective of metabolic engineering.

The accurate quantification of flux distribution throughout a metabolic network is not a trivial task, considering the complexity of metabolic networks and the scarcity of useful experimental data to describe the intracellular reactions. As flux determination is coupled with reaction stoichiometries, as shown in Sect. 3.2, the accuracy of flux quantification depends on the degree to which the assumed intracellular biochemistry describes the actual in vivo structure of metabolic networks. Unfortunately, many reaction steps are unknown, even qualitatively. It is critical then that the available experimental measurements and other biochemical constraints provide sufficient redundancy to validate the accuracy of measurements used [13–15], as well as to test the consistency of the assumed intracellular biochemistry with the overall data [6, 16]. Barring experimental error, any identified inconsistency of the biochemistry with the existing data indicates the presence of errors in the assumed structure of the metabolic network. In this regard, issues such as pathway modification (e.g., considering reaction reversibility [17, 18], or alternate conversion routes [8, 19]) and discovery of new pathways [19–21] are an important part of the metabolic flux determination process. The complexity of the metabolic flux quantification problem should not discourage researchers from flux analysis, as it is an essential part of the analysis of intracellular metabolism, which can greatly improve the efficacy of future genetic manipulations.

In the next section, we present methods that have been employed to date for flux quantification. Their advantages and weaknesses, the latter leading to the introduction of new methods or to combinations of several existing techniques, will be discussed in an effort to provide a systematic way to flux quantification. The last section of this chapter will include examples of the use of fluxes for the elucidation of metabolic pathways, emphasizing the importance of Metabolic Flux Analysis as a major task of metabolic engineering.

3.2
Flux Quantification Methods

The starting point of MFA is the reaction network stoichiometry for the conversion of substrates to metabolic products and biomass constituents. The extent to which intracellular flux estimates can be accepted as reliable measures of the actual in vivo fluxes depends on the validity of the assumed intracellular biochemistry. Therefore, flux determination methods require a high degree of redundancy to check the accuracy of experimental measurements and the consistency of the assumed biochemistry. This is a critical point that must be considered before the selection of a particular flux determination method to be applied on a metabolic network.

3.2.1
Metabolite Balancing

Metabolite balancing is a very powerful methodology and usually is the first step in the determination of fluxes. Intracellular fluxes are determined as functions of the measurable extracellular fluxes, using a stoichiometric model for the major intracellular reactions and applying a mass balance around each intracellular metabolite. The method is readily applied and does not require enzyme kinetics information. The only assumption is the pseudo-steady state approximation for the intracellular metabolites, which is justified in most cases, as there is a very high turnover rate of the pools of most metabolites [1]. Experimentally a metabolic steady-state can be achieved in a continuous flow reactor. The resulting set of equations can be expressed in a matrix form as:

$$S \cdot \underline{v} = \underline{r} \tag{3.1}$$

where, S is the stoichiometric matrix of the metabolic network. The number of rows is equal to the number of metabolites in the network and the number of columns is equal to the number of unknown fluxes at steady state. $\underline{v}$ is the vector of unknown fluxes and $\underline{r}$ is the vector of metabolite extracelluar accumulation rates. The elements of $\underline{r}$ corresponding to intracellular metabolites are zero, whereas those corresponding to extracellular metabolites are equal to the measured accumulation rates.

Since the number of reactions is, in general, greater than the number of metabolites, the system of Eq. (3.1) is in general underdetermined. Additionally, there are some structural singularities that increase the degrees of freedom of the system, like pathways that branch apart and rejoin later in the network, reversible reactions, or metabolic cycles, that cannot be resolved by metabolite balancing alone [1]. In mathematical terms, these structures introduce linearly dependent equations in the stoichiometric matrix, decreasing its rank (and consequently increasing the degrees of freedom). Figure 3.1 illustrates this point for the case of an artificial network containing two competing pathways and one reversible reaction; the forward and the backward directions correspond to linearly dependent columns decreasing the rank of the matrix by 1. Using metabolite balancing, one may be able to determine the net flux of the reversible reaction, but the forward and the backward fluxes cannot be differentiated.

In these cases, where metabolic networks cannot be resolved by metabolite balancing constraints alone, additional constraints – at least as many as the degrees of

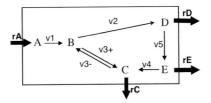

System of Mass Balances:

$$
\begin{array}{ccccc}
 & & \overbrace{}^{v3net} & & \\
v1 & v2 & v3+ & v3- & v4 \quad v5
\end{array}
$$

$$
\begin{bmatrix}
-1 & 0 & 0 & 0 & 0 & 0 \\
1 & -1 & -1 & 1 & 0 & 0 \\
0 & 0 & 1 & -1 & 1 & 0 \\
0 & 1 & 0 & 0 & 0 & -1 \\
0 & 0 & 0 & 0 & -1 & 1
\end{bmatrix}
\begin{bmatrix}
v1 \\ v2 \\ v3+ \\ v3- \\ v4 \\ v5
\end{bmatrix}
=
\begin{bmatrix}
rA \\ 0 \\ rC \\ rD \\ rE
\end{bmatrix}
\rightarrow
\left\{
\begin{array}{l}
v1 = -rA \\
v2 = v1 - v3net = v1 - \alpha \\
v4 = rC - v3net = rC - \alpha \\
v5 = v1 - rD - \alpha \\
v5 = rE + rC - \alpha \\
\alpha = [(v3+)-(v3-)]
\end{array}
\right.
$$

Fig. 3.1. Using only extracellular accumulation rate measurements, structural singularities like reversible reactions cannot be resolved. The forward and backward directions of a reversible reaction correspond to linearly dependent columns of the stoichiometric matrix, as it is shown for reversible reaction 3 in the network above. Note that, in the depicted network, another source of singularity are the two competing pathways B→C and B→D→E→C. The flux split ratio at B cannot be determined from metabolite balancing (rank(S) = 4, 2 degrees of freedom and one redundant equation).

freedom of the system of Eq. (3.1) – are needed before the system can be solved for the unknown fluxes. If more constraints than the degrees of freedom are introduced, then the system becomes overdetermined and the redundant equations can be used to test the consistency of the overall balances and the validity of the pseudo-steady state assumption [1]. These additional constraints can be obtained from the following sources:

– Theoretical assumptions concerning the intracellular biochemistry [7, 8, 16, 21, 22]. Some reactions introducing singularities could be lumped together, considered irreversible, or even neglected, if there is no experimental evidence of the presence or high activity of the corresponding enzyme [6, 7, 23, 24]. Information about the relative flux ratio of two pathways could be obtained by in vitro enzyme activity measurements [21, 24, 25]. This information should be considered with caution, because in vitro enzyme activity measurements are found to bear little relationship to actual in vivo flux distributions. It should also be taken into consideration that most of the theoretical assumptions are usually based on observations of the wild-type microorganisms in their native environment. Thus, they might not hold true for highly engineered organisms and/or under extreme environmental conditions [26].

As the reliability of flux estimates is based strongly on the assumed biochemistry, the validity of the theoretical assumptions must be tested through satisfaction of redundancies. Any conclusions about the physiological state of the cell based on the determined flux distribution must be evaluated in the context of the assumptions that have been made as part of the analysis. When comparing results from various analyses, they should be based on similar set of assumptions.

– Linear programming (LP) [27–29]. The use of linear programming for the analysis of metabolic networks dates back to the mid-1980s. It has been applied to determine the maximum yields of fermentation products from glucose in butyric and propionic acid bacteria [30, 31], to adipocyte metabolism [32], and to analysis of tricarboxylic acid (TCA) cycle and acetate overflow in *Escherichia coli* [33]. With this approach, metabolic fluxes are determined such as to be stoichiometrically feasible (subject to the metabolite balance constraints) and optimize a certain objective function. In mathematical terms this is equivalent to the solution of the following LP problem [27]:

$$\text{minimize } \underline{c}\,\underline{v} = \sum c_i v_i$$
$$\text{subject to } \mathbf{S} \cdot \underline{v} = \underline{r} \tag{3.2}$$

where $\underline{c}$ is the vector of the weight factors (or *cost* vector [34]) of fluxes in the objective function. The weights c_i are selected in such a way as to allow the formulation of physiologically meaningful objective functions [27]. For example, if the objective function is to maximize the production rate of a metabolite, then the weight factor corresponding to the flux of the reaction(s) producing this metabolite should be –1 and the weight factors of the other fluxes should be 0. Objective functions that have been used for flux estimation, as well as their interpretation, are listed in Table 3.1, as it has been adopted from [35]. By imposing an objective function, one tests the limits of a cell in carrying out a certain task (such as growing, utilizing ATP, etc.) subject only to the constraints imposed by the assumed biochemistry under specific genetic and environmental conditions. The LP problem of Eq. (3.2) is usually solved using the simplex method [34, 36].

Linear optimization solutions tend, by definition, to occur at extreme points ('corners') of the feasible domain [34]. When LP is applied to a metabolic network, the obtained flux distribution is an end point of the stoichiometrically feasible domain [24, 28, 35], termed *metabolic genotype* [29, 35]. In general, the feasible domain is a

Table 3.1. Objective functions that can be imposed on the flux distribution under specific genetic and environmental conditions to study the limits of cellular function in carrying out a certain task (adopted from [35])

Question	Objective	Reference
What are the biochemical production capabilities?	Maximize metabolite production rate	Varma et al. [41]
What is the maximal growth rate and biomass yield?	Maximize growth rate	Varma and Palsson [39]; Varma and Palsson [94]
How efficiently can metabolism channel metabolites through the network?	Minimize the Euclidean norm	Bonarius et al. [42]
How energetically efficient can metabolism operate?	Minimize ATP production rate or minimize nutrient uptake rate	Majewski and Domach [33]; Savinell and Palsson [27]; Fell and Small [32]
What is the tradeoff between biomass production and metabolite overproduction?	Maximize biomass production rate for a given metabolite production rate	Varma et al. [41]

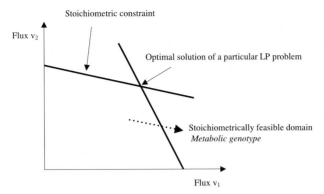

Fig. 3.2. In two-dimensional space, the stoichiometric constraints correspond to lines and the feasible domain is a plane region bounded by them. The linear optimization solutions tend to occur at extreme points ('corners') of the feasible region

polyhedron [34] formed by the intersection of hyperplanes [34] corresponding to the stoichiometric constraints imposed on the metabolic fluxes as these are expressed by the rows of the system of Eq. (3.1). Figure 3.2 illustrates the notion of the stoichiometrically feasible domain in the case of a network with two unknown fluxes, v_1 and v_2. In two-dimensional space, the stoichiometric constraints correspond to lines and the feasible domain is a plane region bounded by them.

Application of LP to flux determination is a very useful technique, since it provides boundaries for the in vivo achievable flux distributions. The technique shares all the advantages of metabolite balancing: it is simple and does not require any kinetic information. It can be used to test the *metabolic flexibility* of a microorganism [35, 37], i.e., how the cell readjusts its metabolism to meet a specific objective after mutations of its genotype. In addition, through duality theory [34], it is possible to determine the effect of the presence, absence, or modification of a stoichiometric constraint on the objective function [27, 28, 38–41]. For example, if the objective function is to maximize the production rate of metabolite j, then the dual variables ('shadow prices')

$$p_i = \frac{\partial z^*}{\partial r_i} \tag{3.3}$$

reflect the effects that changes in the extracelluar accumulation rate of metabolite i have on the theoretical yield of metabolite j. z^* is the optimal value of the objective function, which, in this example, is equal to the theoretical yield of metabolite j. Savinell and Palsson [27, 38] used LP to examine mammalian cell growth in culture. They were able to compare the value of different nutrients with respect to their overall growth supporting ability and also their ability to satisfy energy or other requirements of the cell.

The drawbacks of the application of LP to flux quantification derive from the fact that (1) it is not certain that the intracellular biochemistry remains unchanged under various cellular "objectives" or that the rest of the metabolism, besides the mutation site, is not affected after a mutagenesis, and (2) that the obtained flux distribution may not be the one achievable in vivo, because of "optimization" criteria used by the cells in vivo are not the same as the considered ones. Bonarius et al. [42] calculated the flux distribution in hybridoma cells imposing the minimum-norm constraint, i.e., efficient channeling of metabolites through the metabolic pathways, considering it as the situation most likely to exist in vivo. However, using data from ^{13}C-nuclear magnetic resonance (NMR) spectroscopy experiments [43], the in vivo flux distribution was actually found to be relatively similar to that obtained if one of the objective functions 'maximize ATP' or 'maximize NADH' was imposed to the system instead of the 'minimum Euclidean norm' constraint, that they had initially considered.

- Inclusion of additional measurements, such as those obtained from isotopic-tracer techniques. Such methods provide additional information about the in vivo metabolic fluxes, decreasing the number of required assumptions and increasing the observability of the system. Combination of these techniques with metabolite balancing can elucidate the in vivo flux distribution in finer detail and provide enough redundancy to check the validity of estimated fluxes and the consistency of the assumed intracellular biochemistry. They will be discussed in more detail in the next session.

Aiba and Matsuoka [19] could be considered the first to present the use of extracellular metabolite measurements in combination with intracellular reaction stoichiometry for the determination of metabolic fluxes. Their work, as well as the work of Papoutsakis [44] on solvent production, was focused on the validation of the bioreaction network responsible for product formation through the estimation of intracellular fluxes. Vallino and Stephanopoulos [3, 6, 16, 45, 46] were the first to use the extracellular fluxes and metabolite balancing technique for the elucidation of intracellular regulatory mechanisms. Considering changes in the intracellular fluxes and flux split ratios at key branch points of *Corynebacterium glutamicum* under different genetic and environmental conditions, they were able to derive useful conclusions about the control of flux. Computer programs have been written to facilitate flux determination using metabolite balancing in a user friendly environment [6, 47, 48]. We discuss other applications of metabolite balancing in elucidating metabolic networks in Sect. 3.3.

3.2.2
Isotopic-Tracer Techniques

Isotopic tracer techniques have been used extensively in biochemistry for the elucidation of metabolic pathways [49–57]. In isotopic-tracer techniques, cells are provided with a substrate labeled with a detectable isotope at (a) specific atom(s). It is assumed that the metabolic system is not disturbed by the introduction of the isotope. One class of isotopic-tracer methods is the use of substrates labeled with ^{13}C or ^{14}C at specific carbon locations. ^{13}C isotopes have the advantage of being stable, non-radioactive, and detectable by Nuclear Magnetic Resonance (NMR).

The isotope distribution among the metabolites of the network for specific labeled substrate(s) and known biochemistry is a function of the in vivo metabolic fluxes. Information on the isotope distribution of various extracellular and/or intracellular metabolites can be obtained by measuring the label enrichment distribution by NMR, studying the fine structure of NMR spectra or measuring the mass isotopomer distribution by Gas (or Liquid) Chromatography-Mass Spectrometry (GC [LC]-MS). Analysis of these experimental measurements can be used to obtain additional constraints on in vivo metabolic fluxes beyond those of metabolite balancing. We show how data from isotopic-tracer experiments can be used systematically to resolve parts of the network that cannot be elucidated by metabolite balancing alone.

Flux quantification through isotopic tracer techniques requires either transient or steady-state isotope intensity measurements. Information about the transient approach, which uses radioactive isotopes and can be used to small networks, can be found in [1]. It should be noted that the flux determination analysis presented here is based on the second approach and it is valid only if the experimental data are obtained at metabolitic and isotopic steady-state. The chemostat is the preferred system to conduct metabolitic steady-state experiments. In addition the labeled substrate has to be introduced to the system in such a way (1) that the metabolitic steady-state is not disturbed and (2) to allow isotopic steady state to be reached. It is possible to assume metabolitic and isotopic steady-state for a limited duration during a transient experiment in a batch reactor. It is important though, to seek for internal checks providing evidence for the validity of the metabolitic and isotopic steady-state assumption, as inappropriate application of metabolite balancing- and isotopic-tracer analysis will produce incorrect and misleading results.

Whether the measured isotopic distribution of metabolites in the network contains information to elucidate parts of networks unresolved by metabolite balancing only depends on (1) the assumed biochemistry, (2) the type and labeling of substrate used, and (3) the extracellular metabolite(s), whose degree of label enrichment and isotopomer distribution is measured [1]. The reason for this is that, based on the assumed structure of the network, some metabolites are affected differently to others from the introduction of a particular labeled substrate. Figure 3.3 shows the splitting pathway at the tetrahydrodipicolinate (H4D) branch point, which is part of lysine biosynthesis, a metabolic pathway that cannot be elucidated using material balances as the only constraints. These pathways, that branch apart and rejoin later, can be differentiated through the use of isotopic labels, provided that they introduce asymmetries in the distribution of carbon atoms of intermediate metabolites, generating different labeling patterns in the common secreted metabolite [1]. The measurement of the isotopic distribution of the common metabolite provides information to determine the competing pathways. In the case of Fig. 3.3, the asymmetry is introduced by the epimerase reaction in the last step of the four-step pathway, so that the flux ratio can be determined from the difference in the ^{13}C label recovered on the 3rd and 5th carbon atom of lysine, if [1-^{13}C] glucose or [3-^{13}C] pyruvate is used as substrate [58, 25].

The redundant information obtained from the isotopic-tracer techniques can be used to confirm the flux estimates obtained by material balancing and test the accuracy of the assumptions made about intracellular biochemistry. In the first use of ^{13}C-labeled glucose with hybridoma cultures [59–62] it was possible to confirm the flux estimates that have been independently obtained through extracellular metabolite measurements and material balancing. In this method, the labeling pattern of secreted lactate was predicted from the fluxes estimated by material balancing and found to agree well with the isotope enrichments measured by NMR spectroscopy. This method uses isotope enrichment measurements (henceforth referred to as 'carbon enrichment analysis') to determine metabolic fluxes by applying isotope mass balance around every carbon atom of the metabolic network. Because of isotopic steady-state assumption, the label flux into a carbon atom should be equal to the label flux out. The isotope mass balances provide additional constraints to the metabolite balancing ones of Eq. (3.1) and could enhance the observability of the system.

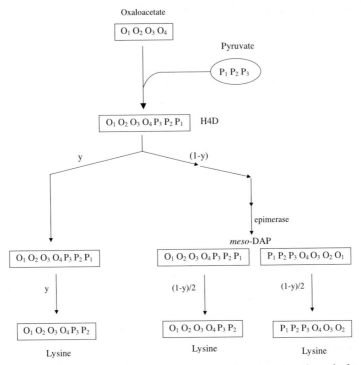

Fig. 3.3. Template for lysine labeling from pyruvate and oxaloacetate through the splitting pathway at the H4D branch point, which is part of lysine biosynthesis. The four-step succiny-lase pathway (right) contains the enzymatic sequence N-succinyl-2,6-ketopimelate synthase, N-succinylaminoketopimelate:glutamate aminotransferase, N-succinyl-diaminopimelate desucci-nylase and diaminopimelate epimerase. The one-step pathway (left) involves the action of $meso$-diaminopimelate ($meso$-DAP) dehydrogenase. $Subscripts$ on O and P indicate the original carbon positions in oxaloacetate and pyruvate. y denotes the fraction of the flux out of the H4D node that goes via the one-step pathway

In Fig. 3.1, we showed that metabolite balancing cannot differentiate between for-ward and backward flux of a reversible reaction. In Fig. 3.4 we show how this singu-larity can be resolved in the particular network of Fig. 3.1 using carbon enrichment analysis. We need to note that the examples in this section refer to carbon enrich-ment analysis, but the same analysis can be applied if isotope of another element (such as phosphorus or oxygen) is used to resolve the structural singularities of the metabolic network.

In order to get a matrix representation for carbon enrichment balancing, as in Eq. (3.1), and facilitate the use of the method for the analysis of complex networks, Zupke and Stephanopoulos [63] introduced the concept of $atom$ $mapping$ $matrices$. Marx et al. [64, 65] used the carbon enrichment analysis to elucidate the overall net-work of $C.$ $glutamicum$. The method is simple and, in combination with metabolite balancing, very powerful, but its complexity increases with the size of the network, since the carbon enrichments are non-linear function of metabolic fluxes. Addition-ally, the resolution of ^{13}C NMR prevents the accurate determination of label enrich-ment at many carbon atoms of intracellular metabolites, because of low intracellular

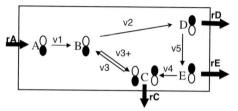

Isotope Mass Balance around carbon:

D_1: $\quad v_2 B_2 = (r_D + v_5) D_1 \quad$ => $B_2 = D_1 \quad$ (because of the mass balance around D-see Fig. 3.1)

D_2: $\quad v_2 B_1 = (r_D + v_5) D_2 \quad$ => $B_1 = D_2 \quad$ (because of the mass balance around D)

E_1: $\quad v_5 D_1 = (r_E + v_4) E_1 \quad$ => $D_1 = E_1 \quad$ (because of the mass balance around E)

E_2: $\quad v_5 D_2 = (r_E + v_4) E_2 \quad$ => $D_2 = E_2 \quad$ (because of the mass balance around E)

$$B_2 = D_1 = E_1$$
$$B_1 = D_2 = E_2$$

Isotope Mass Balance around carbon:

B_1: $\quad v_2 A_1 + v_3^- C_1 = (v_2 + v_3^+) B_1$

B_2: $\quad v_2 A_2 + v_3^- C_2 = (v_2 + v_3^+) B_2$

C_1: $\quad v_3^+ B_1 + v_4 E_1 = (v_3^- + rC) C_1 \quad$ => $v_3^+ B_1 + v_4 B_2 = (v_3^- + rC) C_1$

C_2: $\quad v_3^+ B_2 + v_4 E_2 = (v_3^- + rC) C_2 \quad$ => $v_3^+ B_2 + v_4 B_1 = (v_3^- + rC) C_2$

} Solve for v_3^+ and v_3^-

Net flux of reactions 2 and 4 (see Fig. 3.1)

$$v_2 = -rA - (v_3^+ - v_3^-)$$
$$v_4 = rC - (v_3^+ - v_3^-)$$

Fig. 3.4. Differentiation of the forward and backward fluxes of a reversible reaction using carbon enrichment measurements. Assuming that all network metabolites have 2 carbon atoms and the transfer of label through the network above is done as shown with the black and white carbon atoms (i.e. $A_1 \rightarrow B_1$, $A_2 \rightarrow B_2$, $B_1 \rightarrow D_2$, $B_2 \rightarrow D_1$, $B_1 \rightarrow C_1$, $B_2 \rightarrow C_2$, $D_1 \rightarrow E_1$, $D_2 \rightarrow E_2$, $E_1 \rightarrow C_1$, $E_2 \rightarrow C_2$), then measuring the label enrichment of D1 and D2 (or E1, E2 or D1, E2 or D2, E1) will allow the determination of the forward and backward flux of reaction 3 of the network. The flux split ratio at B (see caption of Figure 3.1) is simultaneously determined. Since the system is now fully resolved, any additional label enrichment measurement provides redundant information that can be used in consistency analysis

concentrations, especially in microorganisms. This reduces the amount of information obtained about the system and there are parts of networks that cannot be observed with carbon enrichment analysis because of lack of data [64]. In addition, carbon enrichment analysis cannot elucidate to a detailed extent networks containing metabolic cycles [1, 51].

To maximize the amount of information that can be extracted from experiments with ^{13}C labeled substrates, one needs to account for all isotopomers [51, 55, 56, 66] that arise in a pathway for a certain labeled substrate. Isotopomers are the molecules of the same metabolite with different labeling patterns. Isotopomer distribution analysis is based on the formation of steady-state isotopomer balances [66, 67] around every metabolite of the network. The isotopomer balances allow the determination of the isotopomer population as a function of metabolic fluxes.

Isotopomer distribution analysis can use information from carbon enrichment measurements, fine structure of NMR spectra, and mass isotopomer distribution measurements from GC-MS. The degree of label enrichment at any carbon atom of the network can be derived from the isotopomer distribution of the corresponding metabolite as the sum of the relative populations of metabolite isotopomers labeled at this particular carbon atom [66, 68]. Therefore, the carbon enrichment analysis is an inherent component of isotopomer distribution analysis. The observed multiplet pattern of NMR spectra at various carbon atoms is the result of the superposition of all metabolite isotopomers that are labeled at those carbon atoms [66–67, 69–70]. The line splitting is due to ^{13}C-^{13}C coupling between adjacent carbon atoms. Therefore, the ratio of lines in a particular multiplet is equal to the ratio of isotopomers contributing to the formation of these lines (for more details see [51, 55, 56, 66]). Mass isotopomer distribution measurements obtained from GC [LC]-MS can provide further information on the population of isotopomers with different molecular weight [66, 67]. Thus, all three kinds of measurements can, through the isotopomer balances, provide additional constraints on the metabolic fluxes, besides those imposed from metabolite balances.

The general flow of calculations for the determination of metabolic flux distribution is illustrated in Fig. 3.5. Although the flux quantification methods are in principle rather simple, their implementation on integrated complex networks can be involved and computationally intensive. It requires the solution of steady state balances for all metabolites and their isotopomers in the network, which is a demanding task for realistic networks, since the isotopomer balances are non-linear functions of fluxes. They have to be solved in an iterative way to allow estimation of the flux distribution that is consistent with the imposed constraints. The number of unknowns increases with the complexity of the system and the reaction reversibility. Whereas reaction reversibility is not important for the application of metabolite balancing technique, it has to be considered in the isotopic tracer techniques, since it affects labeling scrambling through the network. Neglecting it will lead to wrong estimates of metabolic flux distribution and wrong conclusions about the intracellular biochemistry [17, 18, 65, 71].

A great degree of redundancy is built into the calculations if all the available measurements presented above are considered. This allows one to test the accuracy of measurements used and the consistency of the assumed intracellular biochemistry with the overall data through satisfaction of redundancies. Identification of inconsistencies between the theoretical predictions and the overall data will be an indication of the presence of errors either in the measurements or in the assumed biochemical pathways. If some measurements are suspected of containing gross errors, either they are eliminated and the flux estimation process is repeated with the rest of the measurements or a new set of measurements is used (for more details see [13]). If none of the measurements is identified as containing gross errors and the inconsistency remains, this is an indication of errors in the assumed biochemistry. Then the flux quantification process can be repeated by altering the biochemical pathways, until a satisfactory fit is obtained (see schematic diagram in Fig. 3.5). However, because of the nonlinearity of the isotopomer balancing equations with respect to the metabolic fluxes, classical statistical methods for the analysis of the quality of the experimental measurements [13] and the estimated fluxes [6, 14, 16] cannot be readily applied. The development and application of more sophisticated statistical techniques to include the nonlinear case of carbon label systems is required (see [70,

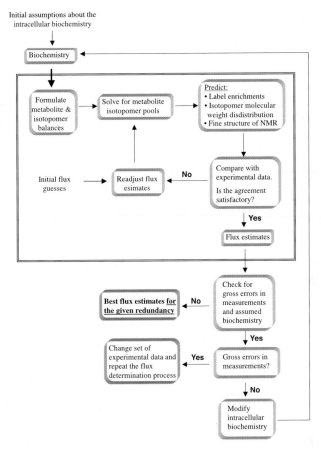

Fig. 3.5. Schematic diagram of the flux determination process based on information obtained from isotope enrichments, fine structure of NMR spectra, and molecular weight distribution by GC [LC]-MS. It consists of two parts:1) the iterative process leading to the determination of flux estimates that satisfy the non-linear system of constraints for an assumed biochemistry (it is illustrated by the *part of the diagram in the box*) and 2) consistency analysis to identify gross errors in the measurements or the assumed biochemistry (see text) (it is illustrated by the *part of the diagram outside the box*)

71]). In addition, suitable computer programs that facilitate the analysis of label transfer and calculation of isotopomer distribution through the metabolic network (e.g., [72, 73]) can be very useful for the selection of the type of labeled substrate(s) that facilitates the elucidation of a particular metabolic network. In the beginning of this section we noted that, in order to delineate the flux distribution in a part of the network using isotopic-tracer techniques, it is important to select carefully the labeling of the substrate and to be able to measure the isotope distribution of this (or those) metabolite(s) that contain information about this part of the network. Still we lack a systematic way of experimental design. To date it is not clear, before the actual experiment is carried out, if and to what extent the selected labeling of substrate, or if and which experimental measurements, can enhance the observability of the metabolic system and delineate structural singularities that cannot be resolved using only extracellular accumulation rate measurements.

In addition to the development of sophisticated and systematic computational methods for accurate analysis of the experimental data, the interest should also be focused in the enhancement of the resolution of the experimental analytical techniques. The low resolution of ^{13}C NMR for low intracellular metabolite concentrations, which is mostly the case in microorganisms, may be the source of gross errors in the experimental measurements. Recently, the isolation and hydrolysis of intracellular proteins and nucleic acids [66, 74] enabled the collection of high precision carbon enrichment measurements for intracellular metabolites that were previously unattainable [18, 65, 66]. These analytical methods require that the system is at metabolite and isotopic steady state, so that the measured enrichment of the macromolecules reflects the actual in vivo enrichment of their precursors. Recently, two-dimensional NMR, using small amounts of uniformly labeled substrates, was successful as an additional method for the determination of metabolic fluxes [26, 70, 75, 76].

3.3
Applications of Metabolic Flux Analysis in the Elucidation of Metabolic Networks

A large number of publications on the applications of metabolic flux determination to the analysis of metabolic networks can be found in the recent literature. In the short period of time since fluxes became a central focus of metabolic engineering as fundamental determinants of cell physiology, their quantification at various genetic or environmental conditions has provided valuable information about many biological systems. The examples below were selected to illustrate how metabolic flux analysis can upgrade the information contained in experimental measurements to provide further insights into the cellular metabolism.

Metabolic flux distribution at specific genetic and environmental conditions represents a 'snapshot' [76] of metabolic function. It shows the relative activity of the various pathways under these conditions. It is, though, through the comparison of many 'snapshots' that information about the control of flux and a better understanding of the intracellular metabolism can be obtained. Utilizing different carbon sources, mutants and other perturbations, and the corresponding flux distributions, Vallino and Stephanopoulos [45, 46] analyzed the rigidity of the two 'principal nodes' of lysine production network of *C. glutamicum*. The branch points of the network that have the most direct impact on product yield are defined as 'principal' or critical [3, 4]. Rigidity is the 'inherent resistance to flux alterations' [3]. Thus, if a node is found to be rigid, the flux distribution around this node is expected to remain relatively insensitive to perturbations of the enzymes surrounding it [3]. Analysis of the rigidity of principal nodes can provide insights about the control of the network. In the lysine production network, the phosphoenolpyruvate (PEP)/pyruvate node was found to be rigid, as opposed to the glucose-6-phosphate (G6P) node, which was identified as flexible. These results suggest that genetic modifications to enzymes around the PEP/pyruvate node are more promising with respect to the impact on product yield than those around the G6P node.

Another example of MFA in providing insights about network structure is the work of Park et al. [20]. Through the use of metabolite balancing combined with isotopomer distribution analysis in selected mutants, they obtained direct evidence of the presence of pyruvate carboxylating activity in *C. glutamicum*. The pyruvate carboxylase gene was indeed cloned later in *C. glutamicum* [77]. This is an important finding as the PEP/pyruvate node is one of the principal nodes of the *C. glutamicum* network responsible for the production of glutamate and lysine – two main products of this microorganism. In another analysis, Nissen et al. [21] studied the anaerobic growth of *Saccharomyces cerevisiae* under various growth conditions, considering a compartmentalized intracellular network. By adding or ignoring reactions corresponding to the function of a particular isoenzyme and testing the feasibility of the flux distribution for each network configuration, they were able to analyze the possible role of various isoenzymes in the network. Nyberg et al. [78, 79] investigated potential causal factors of glycosylation variability in cultures of Chinese Hamster Ovary (CHO) cells producing recombinant human interferon-γ (IFN-γ) using metabolite balancing analysis. A correlation between glycosylation site occupancy and the TCA cycle activity in glucose limited chemostats was identified, suggesting the possible limitation of glycosylation by the availability of activated sugar precursors, verified experimentally later [78, 79]. Metabolic flux analysis was also applied to the analysis of *Bacillus subtilis* [26, 80], *E. coli* [81], and *Penicillium chrysogenum* [8, 82] among other studies. The pentose phosphate pathway and the pyruvate shunt were identified as major pathways of glucose catabolism in riboflavin-producing *B. subtilis* [26]. Analyzing the intracellular flux distribution of two phenylalanine-producing recombinant *E.coli* strains, Chen et al. [83] studied the effects of using different glucose uptake systems in phenylalanine production. Analyzing the penicillin production by *P. chrysogenum*, Jørgensen et al. [8] and Henriksen et al. [82] found a correlation between the flux split ratio at the G6P node and the yield of penicillin on glucose.

Varma and Palsson [28, 39] studied the metabolic capabilities of *E. coli* using linear optimization analysis. This allowed them to identify the limiting factors in the production of the biosynthetic precursors and the relative importance of the biosynthetic precursors to biomass generation. Sensitivity analysis enabled the characterization of the effect of factors such as changes of the active pathways, metabolic demands, P/O ratio or maintenance requirement on the maximum biomass yield. Sauer et al. [10] used linear programming to investigate the metabolic capacity of *B. subtilis* for the production of purine pathway related biochemicals. This analysis is important because it provides valuable insights of the stoichiometric limitations of bacilli bacteria, a strain commonly used in industry. Their results showed that the maximum theoretical yields of purine pathway related biochemicals are limited by pathway stoichiometry, providing the host generates sufficient energy. PEP was identified as the bottleneck for the maximum yield when glucose is the substrate. Knowledge of this limitation allowed the determination of potential ways to overpass it. Linear programming was also used for the analysis of growth of *S. cerevisiae* on various glucose ethanol/mixtures [9].

It was mentioned before that an important goal of metabolic engineering is to elucidate the parameters responsible for the control of flux. We also presented some examples of how metabolic fluxes, and in particular their variations at different conditions, can be used to study the interaction between different pathways, as well as to identify the principal nodes in a metabolic network and examine their rigidity. However, the analysis of flux distribution at different genetic and environmental conditions can provide qualitative measures of the control of flux. Only in the proper framework can flux analysis be used to quantify the regulation of a metabolic network. Metabolic Control Analysis (MCA) [84, 85], developed in 1970s for the quantification of the degree of control exercised by specific enzymes, metabolites, and effectors in a network upon each network flux, provides such a framework. MCA relates system variables, such as the flux, to the system parameters, i.e., enzyme activities, through a set of control coefficients. The most important of these coefficients is the 'flux control coefficient' (FCC) defined as

$$C_i^{J_j} = \frac{E_i dJ_j}{J_j dE_i} = \frac{d\ln J_j}{d\ln E_i} \qquad (3.4)$$

where J_j is the steady-state flux through reaction j and E_i is the activity of the i-th enzyme of the network [1]. FCCs quantify the effect of each enzyme of the network on each of the fluxes through the different reactions. It should be noted that experimental determination of individual FCCs requires the measurement of flux changes through a reaction j following a known change in the activity of the enzyme i. Hence, accurate determination of flux control coefficients, which is essential for the reliability of the MCA analysis, depends on the accuracy of flux determination process and the validity of the measured flux distribution.

An extension of control analysis through individual FCCs, suitable for the analysis of complex metabolic networks, is to collect individual reactions around various branch points in *groups* and then describe the degree of control exercised by each group of reactions through the introduction of *group control coefficients* [86]. Methods have been developed for the determination of group control coefficients from the measurement of fluxes and flux changes caused by the introduction of specific genetic and environmental perturbations to the system [87]. Figure 3.6 shows the calculation of such control coefficients for the three groups of reactions formed around the PEP/pyruvate branchpoint of *C. glutamicum*, along with the list of the various perturbations that were applied for this purpose [88]. The magnitude of these coefficients suggests that the main control in the production of lysine lies within the group of reactions in the lysine biosynthetic pathway, while a moderate effect is expected from reactions in the TCA cycle.

MCA is one of several means to analyze metabolism. Kinetic models have also been used to quantify the control of flux. The kinetic models describe reaction rates in the network as a function of the parameters that affects them [12], such as metabolite concentrations, enzyme activities, effector concentrations, etc. (e.g., [88, 89]). Schlosser et al. [91] were able to determine sensitivity parameters characterizing metabolic control based on the formulation of a linear kinetic model. A more sophisticated (log)-linear kinetic model [92], which takes into consideration the nonlinearity of enzyme kinetics, has been used to identify the optimal regulatory structure to achieve a certain objective in a metabolic network [93]. It should be noted that both MCA and metabolic optimization through kinetic models rely on the knowledge of flux. The degree to which both approaches can describe the regulation of a metabolic network depends on the amount of information that can be extracted from the metabolic system based on the analysis of fluxes.

It is thus seen that metabolic fluxes are indeed fundamental determinants of cell physiology. Knowledge of intracellular fluxes is essential in resolving the metabolic state of the cells to a finer detail. Consequently, flux quantification supports the ana-

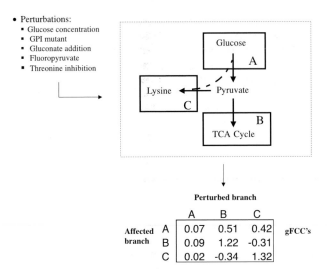

- Perturbations:
  - Glucose concentration
  - GPI mutant
  - Gluconate addition
  - Fluoropyruvate
  - Threonine inhibition

		A	B	C	
Affected	A	0.07	0.51	0.42	gFCC's
branch	B	0.09	1.22	-0.31	
	C	0.02	-0.34	1.32	

Perturbed branch

Fig. 3.6. Determination of group control coefficients (gFCCs) for the groups of reactions formed around the PEP/pyruvate branch point in the lysine biosynthesis pathway of *C. gluta-micum*. gFCCs were determined from flux experimental data obtained after the indicated genetic and environmental perturbations. The gFCCs of the table represent the consensus estimates from all these perturbations. Rows of the gFCC table refer to the affected branch of reactions, and columns to the perturbed branch of reactions. For example, the element of the first column and third row (0.09) is the gFCC of branch A on branch C and measures the effect of branch A on the flux of branch C

lysis of the regulatory mechanisms of metabolic networks [5, 12, 26], a major objective of metabolic engineering towards the goal of rational pathway manipulation.

3.4
Conclusions

Metabolic fluxes are the fundamental determinants of cell function and carry valuable information about the metabolism, its structure, and its control. It has been shown how experimental information can be upgraded through metabolic flux analysis to provide insight about the metabolic state of the cells, about their stoichiometric limitations and theoretical yields. They can be used to identify intracellular regulatory mechanisms and calculate control coefficients or equivalent sensitivity parameters. The accurate quantification of metabolic fluxes is a major part of the analysis of metabolic networks. Such knowledge forms the basis for directed modifications of metabolic networks to achieve a certain objective. It has to be emphasized that the accuracy of any metabolic control models, which are used to unravel the intracellular regulatory structure and examine the metabolic flexibility of the cells, is strongly connected with the accuracy of the flux estimates. Therefore, the flux becomes a focal point of metabolic engineering and justifies further research into the improvement of methods for flux quantification, even though the latter is not a trivial task and it requires extensive instrumentation along with expensive media, as well as sophisticated analytical and computational techniques.

Although the emphasis has been on metabolic fluxes and metabolic pathways, the concepts and methods of metabolic engineering can be used, in general, for the analysis of any bioreaction network. Defining the flux as the rate at which material is processed through a metabolic pathway, all the flux determination methods described above are based on the law of mass conservation, expressed either as metabolite or isotope balances. We can introduce the same concept for pathways transducing fluxes of energy and their determination will be based on the law of energy conservation. In a similar way, the field of metabolic engineering can be broadened and its principles can be applied in the analysis of signal transduction or other pathways. When what today is called "signal" is identified in terms of physicochemical laws, the metabolic flux analysis methods could be readily applied for the elucidation of signal transduction pathways or any corresponding networks. The only difference will be the use of 'signal' balances instead of mass or energy ones.

References

1. Stephanopoulos G, Nielsen J, Aristidou A (1998) Metabolic engineering. Academic Press, San Diego
2. Bailey JE (1991) Science 252:1668
3. Stephanopoulos G, Vallino JJ (1991) Science 252:1675
4. Stephanopoulos G, Sinskey AJ (1993) TIBTECH 11:392
5. Stephanopoulos G (1998) Metabolic Engineering 1:1
6. Vallino JJ (1990) PhD Thesis, Massachusetts Institute of Technology, Cambridge, MA, USA
7. Vallino JJ, Stephanopoulos G (1993) Biotechnol Bioeng 41:633
8. Jørgensen H, Nielsen J, Villadsen J, Møllgaard H (1995) Biotechnol Bioeng 46:117
9. van Gulik WM, Heijnen JJ (1995) Biotechnol Bioeng 48:681
10. Sauer U, Cameron DC, Bailey JE (1998) Biotechnol Bioeng 59:227
11. Takiguchi N, Shimizu H, Shioya S (1997) Biotechnol Bioeng 55:170
12. Bailey JE (1998) Biotechnol Prog 14:8
13. Wang NS, Stephanopoulos G (1983) Biotechnol Bioeng 25:2177
14. van der Heijden RTJM, Heijnen JJ, Hellinga C, Romein B, Luyben KCAM (1994) Biotechnol Bioeng 43:3
15. van der Heijden RTJM, Heijnen JJ, Hellinga C, Romein B, Luyben KCAM (1994) Biotechnol Bioeng 43:11
16. Vallino JJ, Stephanopoulos G (1989) Flux determination in cellular bioreaction networks. In: Sikdar SK, Bier M, Todd P (eds) Frontiers in bioprocessing. CRC Press, Boca Raton, p 205
17. Wiechert W, de Graaf AA (1997) Biotechnol Bioeng 55:101
18. Follstad BD, Stephanopoulos G (1998) Eur J Biochem 252:360
19. Aiba S, Matsuoka M (1979) Biotechnol Bioeng 21:1373
20. Park SM, ShawReid C, Sinskey AJ, Stephanopoulos G (1997) Appl Microbiol Biot 47:430
21. Nissen TL, Schulze U, Nielsen J, Villadsen J (1997) Microbiology 143:203
22. Park SM, Sinskey AJ, Stephanopoulos G (1997) Biotechnol Bioeng 55:864
23. Papoutsakis ET (1984) Biotechnol Bioeng 26:174
24. Bonarius HPJ, Schmid G, Tramper J (1997) TIBTECH 15:308
25. Shaw-Reid CA (1997) PhD Thesis, Massachusetts Institute of Technology, Cambridge, MA, USA
26. Sauer U, Hatzimanikatis V, Bailey JE, Hochuli M, Szyperski T, Wuthrich K (1997) Nat Biotechnol 15:448
27. Savinell JM, Palsson BO (1992) J Theor Biol 154:421
28. Varma A, Palsson BO (1993) J Theor Biol 165:477
29. Varma A, Palsson BO (1994) Bio-Technol 12:994
30. Papoutsakis ET, Meyer CL (1985) Biotechnol Bioeng 27:50

31. Papoutsakis ET, Meyer CL (1985) Biotechnol Bioeng 27:67
32. Fell DA, Small JR (1986) Biochem J 238:781
33. Majewski R, Domach M (1990) Biotechnol Bioeng 35:732
34. Bertsimas D, Tsitsiklis JN (1997) Introduction to linear optimization. Athena Scientific, Belmont, Massachusetts
35. Edwards JS, Palsson BO (1998) Biotechnol Bioeng 58:162
36. Dantzig GB (1963) Linear programming and extensions. Princeton University Press, Princeton
37. Pramanik J, Keasling JD (1997) Biotechnol Bioeng 50:398
38. Savinell JM, Palsson BO (1992) J Theor Biol 154:455
39. Varma A, Palsson BO (1993) J Theor Biol 165:503
40. Varma A, Boesch BW, Palsson BO (1993) Appl Environ Microbiol 59:2465
41. Varma A, Boesch BW, Palsson BO (1993) Biotechnol Bioeng 42:59
42. Bonarius HPJ, Hatzimanikatis V, Meesters KPH, de Gooijer CD, Schmid G, Tramper J (1996) Biotechnol Bioeng 50:299
43. Bonarius HPJ, Timmerarends B, de Gooijer CD, Tramper J (1998) Biotechnol Bioeng 58:258
44. Papoutsakis ET (1984) Biotechnol Bioeng 26:1741
45. Vallino JJ, Stephanopoulos G (1994) Biotechnol Prog 10:320
46. Vallino JJ, Stephanopoulos G (1994) Biotechnol Prog 10:327
47. Pissarra PN (1996) PhD Thesis, University of London, UK
48. Pissarra PN, Henriksen CM (1998) FLUXMAP. A visual environment for metabolic flux analysis of biochemical pathways. In:Yoshida T, Shioya S. (eds) Preprints of the 7th International Conference on Computer Applications in Biotechnology, May 31-June 4, 1998, Osaka, Japan, p 339
49. Shulman RG, Brown TR, Ugurbil K, Ogawa S, Cohen SM, den Hollander JA (1979) Science 205:160
50. Dickinson JR, Dawes IW, Boyd ASF, Baxter RL (1983) Proc Natl Acad Sci USA 80:5847
51. Chance EM, Seeholzer SH, Kobayashi K, Williamson JR (1983) J Biol Chem 258:13,785
52. Walsh K, Koshland DE (1984) J Biol Chem 259:9646
53. den Hollander JA, Ugurbil K, Brown TR, Bednar M, Redfield C, Shulman RG (1986) Biochemistry 25:203
54. Crawford A, Hunter BK, Wood JM (1987) App Environ Microbiol 53:2445
55. Malloy CR, Sherry AD, Jeffrey FMH (1988) J Biol Chem 263:6964
56. Malloy CR, Sherry AD, Jeffrey FMH (1990) Am J Physiol 259:H987
57. Di Donato L, Des Rosiers C, Montgomery JA, David F, Garneau M, Brunengraber H (1993) J Biol Chem 268:4170
58. Sonntag K, Eggeling L, de Graaf AA, Sahm H (1993) Eur J Biochem 213:1325
59. Mancuso A, Sharfstein ST, Tucker SN, Clark DS, Blanch HW (1994) Biotechnol Bioeng 44:563
60. Sharfstein ST, Tucker SN, Mancuso A, Blanch HW, Clark DS (1994) Biotechnol Bioeng 43:1059
61. Zupke C, Sinskey AJ, Stephanopoulos G (1995) Appl Microbiol Biotechnol 44:27
62. Zupke C, Stephanopoulos G (1995) Biotechnol Bioeng 45:292
63. Zupke C, Stephanopoulos G (1994) Biotechnol Prog 10:489
64. Marx A, de Graaf AA, Wiechert W, Eggeling L, Sahm H (1996) Biotechnol Bioeng 49:111
65. Marx A, Striegel K, de Graaf AA, Sahm H, Eggeling L (1997) Biotechnol Bioeng 56:168
66. Klapa MI, Park SM, Sinskey AJ, Stephanopoulos GN (1999) Biotechnol Bioeng 62:375
67. Park SM (1996) PhD Thesis, Massachusetts Institute of Technology, Cambridge, MA, USA
68. Wiechert W, de Graaf AA (1996) In vivo stationary flux analysis by ^{13}C labeling experiments. In: Sahm H, Wandrey C (eds) Advances in biochemical engineering biotechnology. Springer, Berlin Heidelberg New York, p 109
69. Park SM, Klapa MI, Sinskey AJ, Stephanopoulos GN (1999) Biotechnol Bioeng 62:392
70. Schmidt K, Nielsen J, Villadsen J (1999) J Biotech 71:175
71. Wiechert W, Siefke C, de Graaf AA, Marx A (1997) Biotechnol Bioeng 55:118
72. Schmidt K, Carlsen M, Nielsen J, Villadsen J (1997) Biotechnol Bioeng 55:831

73. Zupke C, Tompkins R, Yarmush M (1997) Anal Biochem 247:287
74. Szyperski T (1995) Eur J Biochem 232:433
75. Szyperski T, Bailey JE, Wuthrich K (1996) TIBTECH 14:453
76. Nielsen J (1998) Biotechnol Bioeng 58:125
77. Koffas MAG, Ramamoorthi R, Pine WA, Sinskey AJ, Stephanopoulos G (1998) Appl Microbiol Biotechnol 50:346
78. Nyberg GB, Balcarcel R, Follstad BD, Stephanopoulos G, Wang DIC (1999) Biotechnol Bioeng 62:336
79. Nyberg GB (1998) PhD Thesis, Massachusetts Institute of Technology, Cambridge, MA, USA
80. Sauer U, Hatzimanikatis V, Hohmann HP, Manneberg M, vanLoon APGM, Bailey JE (1996) Appl Environ Microb 62:3687
81. Chen R, Bailey JE (1994) Biotechnol Progr 10:360
82. Henriksen CM, Christensen LH, Nielsen J, Villadsen J (1996) J Biotechnol 45:149
83. Chen RZ, Hatzimanikatis V, Yap WMGJ, Postma PW, Bailey JE (1997) Biotechnol Progr 13:768
84. Kacser H, Burns JA (1973) Symp Soc Exp Biol 27:65
85. Heinrich R, Rapoport TA (1974) Eur J Biochem 42:39
86. Brown GC, Hafner RP, Brand MD (1990) Eur J Biochem 188:321
87. Stephanopoulos G, Simpson TW (1997) Chem Eng Sci 52:2607
88. Simpson TW, Shimizu H, Stephanopoulos G (1998) Biotechnol Bioeng 58:149
89. Savageau MA (1969) J Theor Biology 25:365
90. Savageau MA (1969) J Theor Biology 25:370
91. Schlosser OM, Holcomb T, Bailey JE (1993) Biotechnol Bioeng 41:1027
92. Hatzimanikatis V, Floudas C, Bailey JE (1996) AIChE Journal 42:1277
93. Hatzimanikatis V, Emmerling M, Sauer U, Bailey JE (1998) Biotechnol Bioeng 58:154
94. Varma A, Palsson BO (1994) Appl Environ Microbiol 60: 3724

4 Accuracy and Reliability of Measured Data

Bernd Hitzmann

4.1
Accuracy and Reliability of Measured Data

All measurements have the ultimate goal of creating information. Measurements are also the basis for new bioprocess development. The study of cell metabolism and regulation would be impossible without reliable analytical methods. Furthermore, process optimization and control are also based on reliable measurements. Bioprocess analyzers have the fundamental task to provide the information for optimizing microbial growth, yields, and product quality. Therefore, measurements are not only a key issue in modern process development, but they also open the door to process supervision and control, and avoid restricted views when analyzing processes just through a keyhole.

Increasingly sophisticated equipment has been developed due to the growing demand on high quality measurements. Almost no measurement performed in bioreaction engineering is direct, i.e., none is obtained by immediate comparison with a reference quantity. Most measurements are achieved by means of some specific physical or chemical property of the analyte, often hidden in the very complex sample matrix common to bioprocesses. This very complex matrix, which consist of the numerous substances surrounding the analyte and their mutual influences, makes the task of measuring difficult and leads to the development of complicated measurement systems. Validation of these systems under real process conditions requires even more ingenuity than their development. However, the greater the measuring system complexity, the more difficult is it to ensure the accuracy and the reliability of the collected data. The need for accuracy and reliability increases even more if the data of a process analyzer is to be used as input for a controller. If unusual measurements are made the system operator is required to decide if the fault is from the analyzer or from the bioprocess itself.

An additional typical requirement is that the operation of the analyzer and its service must be as simple as possible. The aim of using such a measurement system is to get more information, not to increase the chances of malfunction of the whole process. Due to the complex nature of many process analyzers this requirement can only be met if the degree of automation of such a system is very high. To achieve this goal, the automation system must not only collect and evaluate the measured data and generate direct actions if necessary (i.e., control of valves) but also make decisions based on sophisticated data analysis. Making such complex decisions requires an additional module. This decision-making module must be able to supervise the analyzer in a similar way to the operator. Causes of faults must be identified and fixed, if possible. The operator (and potentially other entities) must be informed using a high level of communication. The satisfactory operation of such a module demands knowledge-based systems.

The accuracy and reliability of measured data will be discussed in this contribution. Recommendations are given to guarantee the accuracy and reliability of measured data in the complex environment that characterizes the field of bioprocess engineering.

4.1.1
Accuracy and Precision of Measurements

When the terms accuracy and precision are applied to measurements in a colloquial way, they are frequently used as synonyms. In a more scientific way, they have totally different meanings. In this work, the words accuracy and precision are used as suggested in the literature (i.e., Guide for the use of terms in reporting data in analytical chemistry, Anal Chim 54(1982):157). The different meaning of these two terms can be expressed by a cause-effect relationship. The accuracy of a measurement is influenced by the method bias or systematic errors, which are caused by malfunctioning parts of the measurement system, inhibitors in the sample, unstable reagents, or an inappropriate sampling point or system and will be recognized as a shift of the measurement results. On the other hand, the precision of a measurement will be affected by random errors, which will cause a distribution of measurements and can be taken into account by means of statistical methods. Therefore, measurements can be any combination of precise/imprecise and accurate/inaccurate.

4.1.2
Accuracy

A new analyzer must be validated, before use, by means of an alternative method that guarantees the required accuracy. Otherwise, the complex sample matrix can influence the results significantly without the user's knowledge. Recovery rates can be used as a measurement of the accuracy of the measurement system. Measurements of samples of known analyte concentration can also be obtained. The measurement system should work under realistic conditions, as similar as possible to the ones prevailing on the actual system. A common fault is not to use a routinely used sampling device and procedure during accuracy tests. However, sampling devices and procedures are a tremendous source of systematic errors and must be calibrated in the same manner as the process analyzer [1]. Schügerl [2] presents examples of errors that are produced by the sampling system during a cultivation process. The sample used during validation of a data collection system typically should have a matrix which is representative of the real sample. However, this requirement is not directly applicable to bioprocess engineering, due to the almost unavoidable time variability of sample composition. This intrinsic complexity of living systems should not be used as an excuse to use unrealistic samples consisting of pure solutions of the analyte when validating a measurement system. A more realistic sample matrix that mimics a real bioprocess environment is, for instance, a yeast extract solution. Such a solution has been used by Weigel et al. [3] to compare different FIA systems for sucrose measurement. For validation purposes the absolute accuracy of an individual measurement has to be calculated. This parameter is given by

$$x_i - \hat{x} \tag{4.1}$$

In this expression, x_i is the measured value and $\hat{x}$ is the estimate of the true value of the measured quantity (i.e., the concentration of a standard solution). For replicate measurements, the mean value $\bar{x}$ is used which yields (for the measure of accuracy)

$$\bar{x} - \hat{x} \tag{4.2}$$

However, it is unlikely that the mean value will be exactly equal to the true value, even if no systematic error is involved in the corresponding measurement. Of course, the reason for this is the influence of small, unavoidable random errors. Therefore, a statistical test is necessary to decide if the observed difference is significant. In that case, it could be caused by a systematic error, or by random errors. In fact, the null hypothesis tested is that – apart from random variation – there is no significant difference between observed and known values. As a threshold, the probability of 5% is often used. If the probability that the difference is produced by chance is small, the null hypothesis is rejected. For example, if the probability is less than 5%, the null hypothesis is rejected and the difference between the observed value and its estimate is said to be significant at the 5% (or 0.05) level. Sometimes the number of measured values is small. In that case, the t-distribution has to be applied instead of the normal (or Gaussian) distribution. Even when the assumption of approximating a set of data to a normal distribution is not strictly valid, it is still considered an acceptable approximation. A testing value TV is calculated and compared with a threshold value of t from a table, considering the significance level as well as the degrees of freedom:

$$TV = \sqrt{n}\frac{|\bar{x} - \hat{x}|}{s} \tag{4.3}$$

In this equation n is the number of measurements and s the standard deviation, calculated as follows:

$$s = \sqrt{\frac{\sum_{i=1}^{n}(x_i - \bar{x})^2}{n - 1}} \tag{4.4}$$

If TV is lower than the tabulated critical value of t with n–1 degrees of freedom and a significance level of 0.05 (which can be found in most statistical textbooks, for example Miller and Miller [4], Massart et al. [5], Funk et al. [6]), then no significant difference can be demonstrated, and the null hypothesis cannot be rejected.

As an example, 15 glucose measurements are presented in Table 4.1, by using a standard solution of $0.25\,\mathrm{g\,l^{-1}}$ and a Yellow Springs analyzer. The mean value is $0.255\,\mathrm{g\,l^{-1}}$ and the standard deviation is $0.010\,\mathrm{g\,l^{-1}}$.

Table 4.1. Glucose measurements performed on a standard solution of $0.25\,\mathrm{g\,l^{-1}}$

Glucose measurements $[\mathrm{g\,l^{-1}}]$				
0.278	0.235	0.261	0.256	0.262
0.247	0.267	0.253	0.255	0.252
0.258	0.253	0.252	0.244	0.249

The calculated TV=1.85 is smaller than the tabulated value of 2.14. No differences between the expected and measured value can be proved and therefore the null hypothesis cannot be rejected. If, however, the null hypothesis would have been rejected due to a higher TV value, then there is a risk of 5% that the hypothesis has been rejected, even though it is true (this is called type I error). This risk can be reduced by reducing the significance level to 1% or even to 0.1%. It is also possible to retain the null hypothesis even if it is false (what consists of a type II error). To avoid this error, an alternative hypothesis has to be postulated, for which a new value – the true value, related to the tested mean value – must be defined.

4.1.3
Precision

To determine how precise a measurement is, a statistic evaluation of a measurement set of repeated measurements x_i (i=1,...n) has to be performed and the standard deviation s of this measurement set is calculated. The mean measured value $\bar{x}$ and its standard deviation s are considered estimates of the true values of these quantities. The relative standard deviation S_{rel} or the variance $var(x)$ are also used to characterize the precision,

$$S_{rel} = \frac{s}{\bar{x}} \tag{4.5}$$

$$var(x) = s^2 \tag{4.6}$$

For a set of data that consists of a few measurements, then the mean deviation d or its relative value d_{rel} can be calculated as a measurement of precision,

$$d = \frac{\sum_{i=1}^{n} |x_i - \bar{x}|}{n} \tag{4.7}$$

$$d_{rel} = \frac{d}{\bar{x}} \tag{4.8}$$

While these formulas apply to a set of data, there is no universal convention to estimate the precision of a single measurement. A common practice in analytical chemistry is to present estimates in the form

$$\bar{x} \pm \Delta x = \bar{x} \pm s \tag{4.9}$$

or

$$\bar{x} \pm \Delta x = \bar{x} \pm d \tag{4.10}$$

The standard error of the mean is calculated less often than the standard deviation

$$\bar{x} \pm \Delta x = \bar{x} \pm \frac{s}{\sqrt{n}} \tag{4.11}$$

Also the 95% confidence limits are frequently used:

$$\bar{x} \pm \Delta x = \bar{x} \pm t_{n-1}^{0.95} \frac{s}{\sqrt{n}} \tag{4.12}$$

In this equation, the value $t_{n-1}^{0.95}$ represents the corresponding value of the t-distribution with a confidence level of 95% and n–1 degrees of freedom. This confidence level provides a high probability of inclusion of the true value in the interval. Such values can be found in most statistical textbooks as mentioned above. All these expressions can be easily transformed, given that the number of measurements is known.

If a calibration model (regression line) is used to calculate the concentration of a sample solution, then the precision depends not only on the experimental error, as discussed above, but also on the confidence limits of the calibration model y_{conf}, which are given by the expression

$$y_{conf} = b_1 x + b_0 \pm t_{n-2}^{p} s_{Rsd} \sqrt{\frac{1}{n} + \frac{(x - \bar{x})^2}{\sum\limits_{i=1}^{n} (x_i - \bar{x})^2}} \tag{4.13}$$

Here b_0 and b_1 are the regression coefficients, x is the concentration at which the confidence limit is calculated, t_{n-2}^{p} is the t value corresponding to a given significance level p and n–2 degrees of freedom, and x_1 and $\bar{x}$ are the concentrations of the calibration solution and the mean value, respectively. The term s_{Rsd} is the residual standard deviation with respect to the regression line, which is calculated as

$$s_{Rsd} = \sqrt{\frac{\sum\limits_{i=1}^{n} (y_{i,mea} - y_{i,lin.mod})^2}{n-2}} \tag{4.14}$$

Here $y_{i,mea}$ is the i-th measurement datum, and $y_{i,lin.mod}$ is its corresponding predicted value using the linear model. It is apparent from the equation to calculate the confidence limit, as well as from Fig. 4.1, that the larger the difference between x and $\bar{x}$, the larger the term containing the square root becomes. This widens the confidence interval at the edges of the calibration line. The most precise value is obtained in the center of the measurement range. Another aspect which can be derived from the equation to calculate the confidence interval is the fact that the term of the

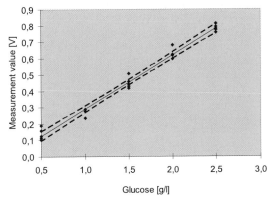

Fig. 4.1. A glucose calibration function (*solid line*) and its 95% confidence interval (*dashed lines*)

square root is reduced if the number of measurements n is large or if $\sum_{i=1}^{n}(x_i - \bar{x})^2$ is large. For a constant n, the summation term is directly proportional to the difference between x_i and $\bar{x}$. This means that the values x_i must be far away from the edges of the measurement range in order to increase the sum and to decrease the error of the prediction. In practice, however, this can only be done if the appropriateness of a linear calibration model is assured. Otherwise, the calibration points should be evenly spaced over the measurement range of interest and a test should be performed to assure that a linear model is really more appropriate than a quadratic one. The way that such a test is performed will be described below.

Since the regression line cannot be calculated exactly, it is obvious that the predicted concentration is subject to error too. The confidence limits for the value of the analyte concentration can be estimated by the following formula:

$$x_{conf} = \frac{y - b_0}{b_1} \pm t_{n-2}^p \frac{s_{Rsd}}{b_1} \sqrt{\frac{1}{m} + \frac{1}{n} + \frac{(y_m - \bar{y})^2}{b_1^2 \sum_{i=1}^{n}(x_i - \bar{x})^2}} \tag{4.15}$$

where y_m is the mean value of the measured values for which the confidence limits will be determined, and m is the number of replicates. If just one measurement has been performed, it will be y_m and m=1. $\bar{y}$ is the mean value of the measurements of the calibration set. The discussion of the formula is analogous to the one for the confidence limits of the calibration model. For the latter, a narrow confidence band requires large values for n, m, b, and the difference between $\bar{x}$ and x_i. As mentioned before, this condition can only be achieved if the adequate fit of a linear model is guaranteed.

As has been mentioned before in this work, the complex sample matrix characteristic of bioprocess systems will often influence the precision as well as the accuracy of the measurement. The confidence limit of the predicted concentration will not be very helpful under this condition. The matrix of the calibration solution should be as similar as possible to the matrix of the sample solution. However, this condition cannot be guaranteed in bioprocess techniques because of the time-changing matrix. Substances that interfere, either inhibiting or increasing the detection reaction, can experience changes in concentration during a bioprocess run and will therefore influence the determination of analyte concentration. A procedure which can account for this consideration is the standard addition method. Using this method, the original sample solution is divided into several samples of the same volume (aliquots). Variable known amounts of the analyte are added to these samples and all are diluted to the same volume. The measured signals of all samples are determined and a regression line is calculated, for which as usual the independent variable (x-axis) is the amount of analyte added and the dependent variable (y-axis) is the signal obtained with the corresponding sample. The absolute value of the root of the regression line is the concentration of the original sample.

The disadvantage of this procedure is that a large amount of sample solution is required and several measurements have to be performed. As described, the method uses an extrapolation model, which is less precise than an interpolation technique. Furthermore, it is difficult to automate.

To compensate this disadvantage, the technique can be transferred to FIA [7–9]. A multiple injection FIA system is required, where three different solutions are injected in a fast sequence. The solutions are a standard solution, the sample solution, and a

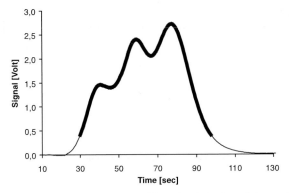

Fig. 4.2. Typical measurement signal obtained by a fast injection of three solutions containing 0.35, 0.55, and 0.6 g l^{-1} glucose, respectively. The values marked by the *bold line* style were used for network evaluation

second standard solution. Due to dispersion inherent in FIA a mixing of the three solutions will occur. A typical measurement signal of such a triplicate injection ($\Delta t_{\text{Injection time}}$=20 s) is presented in Fig. 4.2.

Partial least squares regression [10] as well as neural networks can be used for the evaluation of such signals. The later technique will be discussed in this work in more detail. The neural network was trained to predict the concentration of the three injected solutions, using the individual measurements as input values. As an example, if the enzyme activity is changing due to effects hidden in the sample matrix, the signal intensity of the whole superimposed signal will change. The known analyte concentration of the first and last injected standard solutions allow the reliable calculation of the concentration of the injected sample by linear interpolation. Solving for the sample concentration yields

$$C_2 = \frac{\hat{C}_2 - \hat{C}_1}{\hat{C}_3 - \hat{C}_1}(C_3 - C_1) + C_1 \qquad (4.16)$$

In this equation $\hat{C}_1$, $\hat{C}_2$, and $\hat{C}_3$ are the predicted analyte concentrations of samples 1, 2, and 3 respectively. C_1 and C_3 are the known analyte concentrations of the two standard solutions (which have been injected first and last). C_2 is the corrected analyte concentration of the solution injected in second place. Due to the mixture of the three solutions, the matrix effects of the bioprocess sample will also affect the signals obtained from the standards. Therefore, these effects will be compensated due to the application of linear interpolation. The analyte concentration is reliably determined. Results of the evaluation of multiple injection signals can be seen in Table 4.2. In the first three columns the concentrations of the injected solutions (standard, sample, and second standard) are presented. Signals from some of these measurements have been used for the training of a feed forward back-propagation neural network (consisting of 20 input neurons, 4 hidden neurons and 3 output neurons). The other signals were applied as a test set to evaluate the training. The relative mean deviation (called error) is used to express the precision of the evaluation as mentioned above. The best results obtained can be seen in the fourth and fifth column, where the mean error (out of four repeated measurements) of the uncorrected prediction for the

Table 4.2. Mean % errors (based on the relative mean deviation) of glucose prediction from FIA signals of a threefold injection system (for more information see the text)

			Mean error of prediction of sample concentration [%]				
Standard 1 Sample		Standard 2	Predicted directly		Applying a linear interpolation correction		
Glucose [g l⁻¹]			Training set	Test set	Training set	Test set	Validation set
0.10	0.15	0.35	6.0		4.7		8.9
0.10	0.20	0.35		2.4		4.6	4.2
0.10	0.25	0.35	2.8		2.5		6.7
0.10	0.30	0.35		1.2		1.3	6.7
0.35	0.40	0.60	1.9		2.2		0.9
0.35	0.45	0.60		0.8		0.8	2.4
0.35	0.50	0.60	1.8		1.4		1.8
0.35	0.55	0.60		6.1		0.9	0.3
0.60	0.65	0.85		3.2		3.7	1.5
0.60	0.70	0.85	1.0		1.0		2.7
0.60	0.75	0.85		1.1		1.1	1.4
0.60	0.80	0.85	1.0		1.1		2.7
Mean values			2.4	2.5	2.2	2.1	3.3

sample concentration (given by the neural network) is shown. The overall mean error of the training set and the test set are 2.4% and 2.5%, respectively. Similar results have been obtained for the prediction of the concentration of the two standards. The correction using linear interpolation has been applied to the corresponding values, although it is not necessary. As can be seen from the results, presented in column six and seven of Table 4.2, the errors of prediction (2.2% for the training, 2.1% for the test set) are slightly lower.

For the validation of the matrix effect on the compensation technique, the temperature of the enzyme reactor thermostat was reduced from 30 °C to 26.5 °C, and all samples were measured again. The signal intensity was reduced by more than 20% due to the temperature shift. This value also represent a rough estimation of the measurement error if no calibration was performed. Using the proposed compensation technique, a correction is inherent in the evaluation. The results of an evaluation can be seen in the last column of Table 4.2. Although, the mean overall error of 3.3% is slightly larger than the ones presented before, it is still a good result and demonstrates the usefulness of the compensation technique.

4.2
Measurement Reliability

The reliability of measured data is especially of utmost importance if the data are to be used for process control. An operator must use his experience and judgment when supervising a process analyzer. For example, he can make an assessment of the order of magnitude of the data collected during a special process state and how different measurements can be correlated to each other. Furthermore, during a (re)-calibration procedure the operator knows the way the raw measured data should look. He can check if the calculated calibration model is only changing marginally and, moreover, if this change is continuous rather than abrupt. The process operator is able to distinguish between unusual process measurements due to a faulty analyzer or due to a fault in the production process itself. Also, a fault can happen so that the operator has to follow the whole process as well as the analyzers very closely. This, however, is unrealistic for industrial applications because it will be excessively time-consuming, resulting in additional expense.

The calibration model will not be calculated correctly if a fault occurs during the measurement of calibration data. Therefore, such a fault will have a tremendous influence on all further measurements. Because of this essential significance of the calibration, a system is required to validate the calibration measurements as well as the calibration model on-line. Such a system must contain the knowledge about measurement interpretation, as mentioned above. Therefore a knowledge-based system that performs self-testing and self-correcting procedures is required.

The significance of knowledge-based systems in analytical chemistry has been pointed out by Wünsch and Gansen [11]. These researchers built various expert systems employing commercial available shells, and demonstrated that analytical decisions can be automated in this way. Schoenmakers [12] presented an overview of knowledge-based systems developed for analytical chemistry. So far, only a few applications have been published in which knowledge-based systems were used for the analysis of on-line process analyzers. Wolters et al. [13] presented a generally applicable validation program called VALID. This system was based on an expert system and evaluates the calibration procedure, the drift of the analytical system, and the effect of the sample matrix in an interactive way with the user. Peris et al. [14] reported using an expert system developed in Modula-2 for the specification and intelligent control of an FIA system for bioprocess monitoring. This expert system runs in real time, in contrast to VALID. The tasks of this knowledge-based system are to control the analysis procedure, to select the measurement that will be processed next, and to identify faults in the operation. Brandt and Hitzmann [15] reported a knowledge-based system that has been applied to the fast detection and diagnosis of general faults in FIA systems. This system also runs under real-time condition without a dialog with a user, in the same way as the system described by Peris et al. [14]. The system presented by Brandt and Hitzmann [15] has been programmed by using Vax-OPS5, a tool for the development of production systems. It runs on a Vax computer and obtains the FIA measurements from another computer, which has been used to automate fully the FIA system.

4.2.1
Assessment of Measured Data Reliability by Means of a Knowledge-Based System

The knowledge-based system was applied to supervise the FIA system, as a process analyzer. This FIA system has been used for glucose measurements during various bioprocesses, as described in detail by Dullau and Schügerl [16]. The glucose concentration was converted to glucono lactone using an enzyme reactor with immobilized glucose oxidase. Oxygen is utilized during this reaction. The oxygen uptake was measured by means of an oxygen probe. The typical peak shaped signal was evaluated using the peak height. In the following paragraph the peak height is just called "measurement" and can be correlated with the glucose concentration. Eighteen samples per hour have been processed with this FIA system.

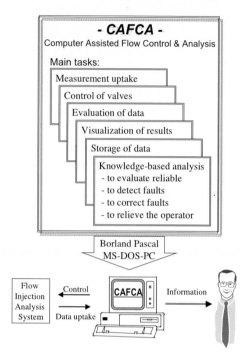

Fig. 4.3. The tasks of the automation system CAFCA containing the knowledge-based analysis of measurements

An automated data evaluation and management system called CAFCA (Computer Assisted Flow Control & Analysis) was developed [17] for the control of the FIA system. This system is already commercially available as a professional version (ANASYSCON, Hannover). CAFCA was programmed by using Borland Pascal 7.0 and two toolboxes. One toolbox has been used to guarantee the real time requirements (RTKernel Version 4.0, On Time Informatik, Hamburg) and the other for the development of a user-friendly graphic output. CAFCA runs on standard PC computers using the operating system MS-DOS, and processes all the tasks in a multi-task manner. As can be seen in Fig. 4.3, one of the tasks is the knowledge-based procedure called XP-CAFCA. All the implemented numerical procedures implemented in XP-CAFCA have been validated previously [18] and complemented with heuristic rules. Various statistical tests are executed by XP-CAFCA, which can be obtained from any statistical textbook (for example Miller and Miller [4], Massart et al. [5], Funk et al. [6]).

Information required for the development of the knowledge-based system [19] was obtained by observing the FIA system during several cultivation processes. Depending on the running time of these bioprocess (which lasted from a few days up to several weeks), a (re)calibration was performed almost every day. As described elsewhere [17] CAFCA can perform a user-activated re-calibration whenever it is required, or after a user-specified number of measurements. Since a fault can occur any time during calibration and data collection, supervision of the system would be very time-consuming. A systematic error analysis of FIA system is presented by

Chen and Zeng [20]. The knowledge-based system was developed to relieve the operator of this task. For this, all the faults collected during a run of the analyzer had to be collected. Examples of typical faults were plugging of injector, selector valve, or tubes; standard solutions of incorrect concentration; microbial contamination of enzyme cartridges; empty vial; fluctuating flow rate; air bubbles; disconnection of tube; malfunction of valves; failure of the pump; rupture of a tube fitting; changing of temperature; reduced sensitivity; damaged detector system; electronic disturbance; baseline drift; wrong A/D-range; hardware breakdown; and wrong measurement range.

An experienced operator is able to identify these faults, using his judgment. The knowledge-based system XP-CAFCA was developed following this same principle. It consists of a numerical-statistical and a knowledge-based module. The numerical-statistical module is composed of several functions for calculating the characteristics of the calibration. These characteristics are employed for the judgment of the calibration, using the knowledge-based module. Faulty measurements are selected and eliminated for modeling purposes. After XP-CAFCA has identified a fault, it reports the decisions and actions taken to the operator. The main tests and analysis of the numerical-statistical and knowledge-based module are described in detail in the following section.

4.2.2
Numerical and Statistical Tests Performed by the Knowledge-Based System

The numerical-statistical module uses validated algorithms for time series and statistical processing, and for data management. Every algorithm of the numerical module calculates parameters which are used by the knowledge-based module for decision-making purposes. A model-based approach (called Projective Reference Evaluation, or PRE) was in order to increase the reliability of raw measurement evaluation. Another model-based approach was given by Wu and Bellgardt [21], which uses a fault detection algorithm that relies on a recursive parameter identification. Other evaluation methods presented for a reliable evaluation of FIA measurements are digital filters [22, 23], Fourier transformation [24], wavelet transform [25], as well as neural networks [26–28]. However, PRE allows not only unreliable measurements to be recognized, but even distorted measurements can be evaluated reliably. This new evaluation technique will be described in the following section. A more detailed presentation is available [29].

In FIA, each element of the dispersed sample zone has a different concentration ratio of sample and carrier solution. Usually the shape of all signals is equal if signals obtained from different concentrations are normalized so that the peak maximum has a value of 1.0. However, the information relevant to the way that the analysis system responds to a measurement is hidden in that peak shape. The theory underneath this response is too complex to provide simple modeling alternatives. Therefore, additional information resulting from the evaluation of disturbed signals is not readily available, but can be exploited in the following way.

A typical FIA-measurement of an unknown sample can be represented as a vector **m.** For this purpose, each individual measurement of the peak shaped signal is considered a component of such a vector. This vector can be related to a reliable reference measurement **r** (which can also be obtained by averaging many signals) as follows:

$$\mathbf{m} = \lambda \mathbf{r} + \mathbf{e} \tag{4.17}$$

For this calculation the measurement signals are transformed so that the baseline is equal to zero. The residual vector $\mathbf{e}$ represents those contributions to $\mathbf{m}$ which cannot be explained by $\lambda \mathbf{r}$. From this equation λ can be estimated (using the least squares estimator) by the formula

$$\lambda = \mathbf{mr}/\mathbf{rr} \tag{4.18}$$

and can be used as a latent variable – instead of the peak height or area – in a (linear) regression model to calculate the concentration of the unknown sample in a very reliable fashion. It is apparent from the above equation that the higher parts of a peak, such as the part around the maximum, contribute more to λ than the lower parts (i.e., the peak tail). Because the components of $\mathbf{r}$ that belong to the baseline are equal to zero, they do not contribute to λ.

Using this procedure, it is possible to validate the quality of λ by inspecting $\mathbf{e}$. If some components of $\mathbf{e}$ are higher than a specific threshold – such as, for example, more than twice the standard deviation – it is most likely that the corresponding part of the measurement signal is faulty. Therefore, it should not be used for the estimation of λ. So, λ can be estimated again (neglecting those components of $\mathbf{m}$ and $\mathbf{r}$ which correspond to the components of $\mathbf{e}$ whose values are too high) and $\mathbf{e}$ can be inspected iteratively, until all the components of the residual vector $\mathbf{e}$ are small. The following paragraph provides an example of the usefulness of this technique, even if the original measurement signal is faulty. In this example, inherent information contained on $\mathbf{r}$ is used to perform the correction mentioned above.

Various patterns of distortions to raw measurement signals of a sucrose FIA system can be added to demonstrate the ability of PRE as a reliable signal evaluation technique. One original measurement (a), as well as three different distortions (b, c, and d) are presented in Fig. 4.4.

Figure 4.4b represents a distortion by which the measurements between 50 s and 60 s are set to 3.0 units. In Fig. 4.4c, each second measured value between 30 s and 85 s was set to a random number from the interval $(-0.5, 0.5)$. Figure 4.4d represents a distortion obtained by adding a minus function to the measured values between 45 s and 65 s. All three distortions represent typical observed faults and have been applied to signals of sucrose measurements. The precision of this evaluation is expressed as the relative deviation to the evaluation of the signals without the distortion. The mean value of the five individual measurements is called average error and is presented in Table 4.3. The measured signal shown in Fig. 4.4 was obtained from a standard concentration of $2.6\,\mathrm{g\,l^{-1}}$. As a result, the relative percentage of the distortions is much bigger for lower concentrations. It can also be observed that the error is lower than 6.8%, even for severely distorted signals. The average error of all distortions is 1.7%, which demonstrates the usefulness of the PRE technique.

The knowledge-based system XP-CAFCA uses this evaluation technique to determine reliably the concentration of each measurement, as well as the calibration data used for the regression model. Furthermore, it uses the number of rejected raw measurements and the sum of the remaining error (calculated by means of PRE) as a measure of the quality of the analyte determination. This can also be used as a measure of the actual performance of the analyzer itself.

Further tests were carried out in order to obtain more evidence of reliable analyzer performance. XP-CAFCA does a simple monotony-test as a simple method to

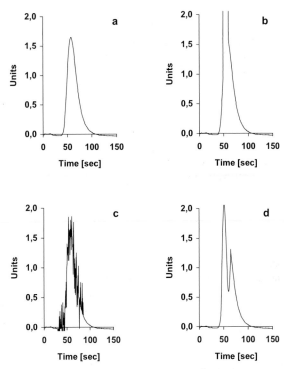

Fig. 4.4.a Measured signal of a sample containing 2.6 g l^{-1} sucrose. **b–d** Distortions applied to A

Table 4.3. Average error of distorted signals

Sucrose Conc. [g l^{-1}]	Distortion b Error [%]	Distortion c Error [%]	Distortion d Error [%]
0.6	6.8	0.6	6.1
1.0	4.0	0.4	3.1
1.4	1.5	0.3	1.2
1.8	0.7	0.1	0.5
2.2	1.1	0.1	0.9
2.6	1.6	0.1	0.8
3.0	3.4	0.1	2.4
Ave. error	2.7	0.2	2.1

show if a fault occurred during the calibration. The test counts the local maximum and minimum points contained in the actual calibration data, which are sorted by standard solution concentrations. If no fault occurred, the calibration data should include the minimum measurement of the lowest concentration standards, as well as the maximum measurement of the highest concentration standards. It is expected that the calibration measurements are monotonically climbing or falling. The characteristic value returned is zero if the measurements of the calibration are monotonic without any local maximum or minimum in the mid-range. Otherwise the number of local extremes is returned. If the value is not equal to zero, then the knowledge-based module uses this information to select the tests that will be further performed. The advantage of this simple test is that no additional information about the method or previous calibrations is necessary to identify a fault.

XP-CAFCA stores the results of each numerical procedure of all calibrations in a file for further testing. The knowledge-based system checks the deviation of the actual calibration parameters from the previous ones during the validation of a new calibration. It also performs a linear regression with all previous calibration parameters vs time to make predictions of the actual values. The next step is to inspect, if the measured values for each standard fit into an interval set by experience, using the preceding calibration model parameters. The measurements are marked as faulty if some values are out of the chosen range. These marked values will be further processed by the knowledge-based system. For example, it decides if a test on swapped standards has to be performed.

The detection of outliers in a calibration data set is very important. For instance, outliers can be caused by plugged tubes and valves, as well as by standard solutions of incorrect concentration. XP-CAFCA uses three different methods (outlier tests A, B, and C) to detect outliers included in the calibration data set.

If the number of multiple determinations is equal to or higher than three, then the Outlier-A-Test is used. It calculates the standard deviation and the average standard deviation of all measurements of a standard solution. A measurement is marked as an outlier if its deviation is more than twice the average standard deviation. This results in a 5% error probability, considering a normal distribution. An example of an outlier included on a multiple determination is presented in Fig. 4.5a.

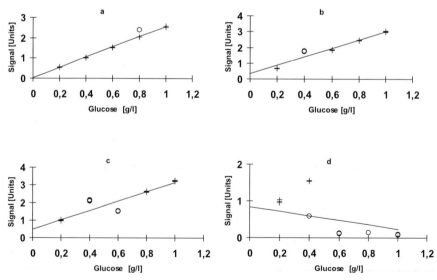

Fig. 4.5a–d. Examples of disturbed calibration data: **a** outlier of a multiple determination; **b** outlier of multiple measurements of a standard; **c** swapped standard solutions; **d** calibration invalid. (+=reliable measurements; O=faulty measurements)

Outlier-B-Test is based on a cross-validation procedure. First, the calibration model with the standard deviation is calculated using all data. A single measurement is selected and eliminated out of the data set, and a new calibration model is calculated using the remaining data. The residual standard deviations of both models are F-tested for a significant difference. For this, a test-value (TV) is calculated:

$$TV = \frac{(n_1 - 2)s_{Rsd1} - (n_2 - 2)s_{Rsd2}}{s_{Rsd2}^2} \qquad (4.19)$$

TV is the test-value, n_1, n_2 are the numbers of measurements, and s_{Rsd1}, s_{Rsd2} are the residual standard deviations of the calibration model using all measurements and all measurements except one, respectively. The calculated TV is compared with the corresponding value obtained from the tabulated F-Test values

$$F(f_1 = 1, f_2 = n_2 - 2, P = 99\,\%) \qquad (4.20)$$

which represents the F value calculated by the Fisher-Test with f_1 and f_2 degrees of freedom and a significance level of 99%. TV is considered an outlier if its value is greater than the corresponding tabulated value. If no outliers are present, no significant difference will be detected. This procedure is repeated with all measurements. This test can be performed even when single determinations of each standard are carried out.

The Outlier-B-Test and Outlier-C-Test are similar. The main difference is that on the Outlier-B-Test not only one measurement is eliminated but so are all the values measured for a special standard. This test is used to detect outliers caused by faulty standard solutions, for example. Figure 4.5b shows a calibration data that includes one outlier of this type.

XP-CAFCA calculates the confidence intervals for all standard concentrations applied to the actual calibration model as a further test for reliability, as described above. The knowledge-based system checks if the measurements of a standard solution marked by the above-mentioned trend-analysis fits into the confidence interval of another marked standard measurement. The results are used in the knowledge-based module to decide if standard solutions have been swapped by the operator. Figure 4.5c shows a calibration data with swapped standards.

XP-CAFCA performs weighted least-squares fits on first-order, second-order, and third-order regression model, depending on the number of standards. A lack-of-fit test is applied to all models. The residual standard deviations of all three models are F-tested for a significant difference for this purpose. The lowest order regression model is used, provided that the next higher order model does not provide a significantly improved fit. This criterion guarantees an optimal calibration model.

4.2.3
Knowledge-Based Module

The calculated characteristics of the numerical-statistical module are judged using the knowledge-based module. The knowledge implemented in this module is represented in logical rules of the form IF AND THEN ELSE . The conditional part of these rules compares the characteristics returned by the numerical-statistical module with experience-based limits. These rules are used to decide whether a fault occurred or not, and also to decide which rule will be performed next. The simplified

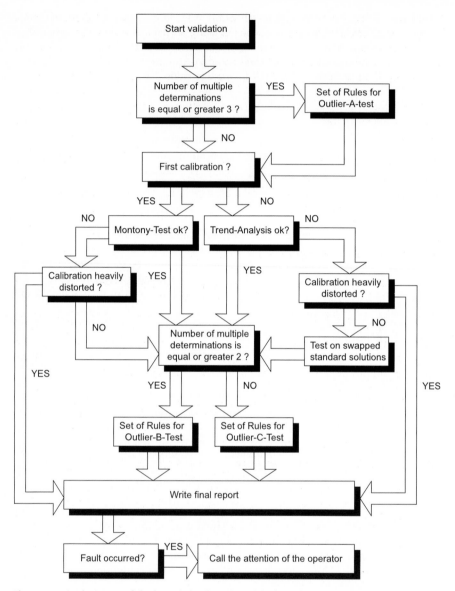

Fig. 4.6. Main decisions of the knowledge-based module for calibration analysis

flow diagram of the overall process is shown in Fig. 4.6. Every node of the network represents a set of rules. Every set of rules is partitioned into subsets. Beginning with the knowledge-based analysis the characteristics returned by the numerical-statistical procedures are inspected. The subset rules are activated to validate or to reject this lack of consistency in case an unusual deviation between calculated and expected values occurs. The marked values as well as the calculated trends of the trend-analysis are used for this and other considerations.

If a fault is detected, XP-CAFCA generates a list of actions that depends on the magnitude of the fault. The actions are different for the first and subsequent calibrations because the first calibration is used as a reference to analyze new calibrations. XP-CAFCA stores all faults found and the corresponding required actions during a validation. Then it returns the highest priority action to CAFCA. This action can only be generated during the first calibration and will cause CAFCA to repeat automatically the calibration and inform the operator by means of a light signal. XP-CAFCA decides to use the previous validated calibration for the evaluation of the new measurements if a subsequent calibration is validated and a serious fault is found. When faults susceptible to correction by the knowledge-based system are detected, such as outliers or swapped standard solutions, XP-CAFCA performs these corrections and informs the operator by a text message and a light signal. The actual values are confirmed if they are not found faulty during the calibration performed by the knowledge-based system.

4.2.4
Methodology of the Knowledge-Based System

The methodology on which the XP-CAFCA is based can be explained by an example. Table 4.4 contains the calibration data obtained from the FIA system. The same data is presented in Fig. 4.7, in which two faults are apparent. One of these is caused by the operator, by mistakenly swapping two standard solutions. The other is caused by a plugged valve and results in the outlier that corresponds to a glucose concentration of $0.8\,g\,l^{-1}$. The system knows that a linear calibration model is required due to the user selection of the method, and also due to previous calibrations obtained with the FIA system and validated by XP-CAFCA.

The first step during the investigation is the Outlier-A-Test, which eliminates the outlier that corresponds to $0.8\,g\,l^{-1}$ standard solution. The trend-analysis finds that the measurements of two standards ($0.2\,g\,l^{-1}$ and $0.4\,g\,l^{-1}$) deviate significantly from the previous calibrations. Therefore, XP-CAFCA decides to perform a test on swapped standards. By these considerations, the knowledge-based system recognizes that the measurements of both marked standards have been swapped. The last step of the validation is the test for the optimum calibration model. This is performed because a loss in enzyme activity can cause a shift of the model from first to second order. XP-CAFCA determines that the fit provided by second or third order regressions is not improved, finishes the investigation, and generates the action list saying that the validated calibration is chosen for evaluation of further measurements to

Table 4.4. Measured values and calibration model parameters

Standard solution $[g\,l^{-1}]$	Multiple determinations [units]		
0.2	1.022	1.032	1.048
0.4	0.534	0.571	0.558
0.6	1.556	1.541	1.503
0.8	2.059	2.603	2.042
1.0	2.511	2.577	2.579

Function: y=0.1653+2.3618x
Rest.Std.Dev=0.4088

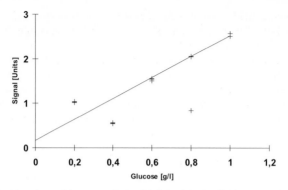

Fig. 4.7. Calibration data with swapped standards and an outlier

Table 4.5. Validated values and calibration model parameters

Standard solution [g l⁻¹]	Multiple determinations [units]		
0.2	0.534	0.571	0.558
0.4	1.022	1.032	1.048
0.6	1.556	1.541	1.503
0.8	2.059	2.042	outlier
1.0	2.511	2.577	2.579

Function: $y = 0.0398 + 2.5096x$
Rest.Std.Dev = 0.0122

CAFCA. The calibration data and the calibration model parameters obtained after the validation are presented in Table 4.5. The final report of the validation is saved to disk.

This example demonstrates that the reliability of the measurements of the process analyzer has been significantly increased. The advantages of the knowledge-based system are that it is as helpful as the human expert, except that it will always be ready, attentive, and will perform at the same level. Using XP-CAFCA during several measurements was very helpful. Some faults have been corrected by the system, but still the operator must be called when some other faults occur. Although the building of a real-time knowledge-based system requires much effort, the benefits of the automated supervision provide a positive balance.

4.3
Conclusions

The accuracy and reliability of measurements in bioprocess technique are discussed in this contribution. The terms accuracy and precision are explained and common examples in analytical chemistry are provided. The application of these concepts to bioprocess engineering is not a simple task, due to the higher variability of living systems. A method that enables one to compensate the complex sample matrix typical to bioprocesses is presented. However, not only the effects due to sample composition have to be considered, but also the effects of the sampling itself. The latter are related to positioning of the sample module and sample preparation as well, and can have tremendous influence on the reliability of measured data. The concentration of the analyte should be determined by means of two independent measurement methods if a new analyzer system has to be validated. The results from this validation have to be analyzed using statistical tests.

The way that the reliability of measurements can be guaranteed is shown by the application of a knowledge-based system to the supervision and validation of a complex process analyzer for glucose measurements performed during bioprocess monitoring. The application of the knowledge-based system demonstrated its usefulness to reduce problems presented during process automation. The description of the entire analytical process in terms of models and heuristic rules allowed the application of the knowledge-based system. In this manner, the reliability of the analyzer system as well as the whole bioprocess can be significantly enhanced. The analyzer operating under this system requires reduced supervision, and the operator is relieved from repetitive work. The reliability of measurements can be improved not only by the application of analyzer-related knowledge but also by the application of knowledge related to the bioprocess itself. Balance equations can be tested and even approximately applicable models should be used to test the consistency of the data. Data are the ultimate basis for process analysis, supervision, and control. Therefore, their reliability is of utmost importance. The supervision of other analyzers (such as HPLC) and whole processes can be applied in the same manner as presented for the FIA system. The expense of such a system can be justified by the improved performance of the analyzer or the process.

References

1. Mattiasson B, Håkanson, H. (1993) Tibtech 11:136
2. Schügerl K (1993) J Biotechnol 31:241
3. Weigel B, Hitzmann B, Kretzmer G, Schügerl K, Huwig A, Gifforn F (1996) J Biotechnol 50:93
4. Miller JC, Miller JN (1993) Statistics for analytical chemistry. Ellis Horwood, Chichester
5. Massart DL, Vandeginste BGM, Deming SN, Michotte Y, Kaufman L (1988) Chemometrics: a textbook. Elsevier, Amsterdam
6. Funk W, Dammann V, Donnevert G (1995) Quality assurance in analytical chemistry. VCH Verlagsgesellschaft mbH, Weinheim
7. Ruzicka J, Hansen EH (1988) Flow injection analysis, 2nd edn. Wiley, New York
8. Valcárcel M, Luque de Castro MD (1987) Flow-injection analysis, principles and applications. Ellis Horwood, Chichester
9. Scheller F, Schmid RD (eds) (1992) Biosensors: fundamentals, technologies and applications. GBF Monographs, vol 17. VCH, Weinheim
10. Schöngarth K, Hitzmann B (1998) Anal Chim Acta 363:183
11. Wünsch G, Gansen M (1989) Fresenius Z Anal Chem 333:607
12. Schoenmakers P (1993) Knowledge-based systems in chemical analysis. In: Buydens LMC, Schoenmakers PJ (eds) Intelligent software for chemical analysis. Elsevier, Amsterdam, p 13
13. Wolters R, van Opstal MAJ, Kateman G (1990) Anal Chim Acta 133:65
14. Peris M, Maquieira A, Puchades R, Chirivella V, Ors R, Serrano J, Bonastre J (1993) Chemometrics Intell Lab Syst 21:243
15. Brandt J, Hitzmann B (1994) Anal Chim Acta 291:29

16. Dullau T, Schügerl K (1991) Process analysis and control by enzyme-FIA-systems. In: Schmid RD (ed) Flow injection analysis (FIA) based on enzyme or antibodies. GBF Monographs, vol 14. VCH, Weinheim, p 27
17. Hitzmann B, Löhn A, Reinecke M, Schulze B, Scheper T (1995) Anal Chim Acta 313:55
18. Press WH, Flannery BP, Teutolsky SA, Vetterling WT (1987) Numerical recipes – the art of scientific computing. Cambridge University Press, Cambridge
19. Van Leeuwen H (1993) Developing expert systems. In: Buydens LMC, Schoenmakers PJ (eds) Intelligent software for chemical analysis. Elsevier, Amsterdam, p 79
20. Chen D, Zeng Y (1990) Anal Chim Acta 235:337
21. Wu X, Bellgardt KH (1995) Anal Chim Acta 313:161
22. Szostek B, Trojanowicz M (1992) Anal Chim Acta 261:509
23. Trojanowicz M, Szostek B (1992) Anal Chim Acta 261:521
24. Szostek B, Trojanowicz M (1994) Anal Chim Acta 22:221
25. Bos M, Hoogendam E (1992) Anal Chim Acta 267:73
26. Hartnett M, Diamond D, Barker PG (1993) Analyst 118:347
27. Hitzmann B, Kullick T (1994) Anal Chim Acta 294:243
28. Hitzmann B, Ritzka A, Ulber R, Schöngarth K, Broxtermann O (1998) J Biotechnology 65:15
29. Hitzmann B, Löhn A, Arndt M, Ulber R, Müller C (1997) Anal Chim Acta 348:161

5 Bioprocess Control

D. Dochain, M. Perrier

5.1
Introduction

Industrial-scale biotechnological processes have progressed vigorously over the last decades. Generally speaking, the problems arising from the implementation of these processes are similar to those of more classical industrial processes and the need for monitoring systems and automatic control in order to optimize production efficiency, improve products quality, or detect disturbances in process operation is obvious. Nevertheless, automatic control of industrial biotechnological processes is clearly developing very slowly. There are two main reasons for this:

1 The internal working and dynamics of these processes are as yet badly grasped and many problems of methodology in modeling remain to be solved. It is difficult to develop models taking into account the numerous factors which can influence the specific bacterial growth rate. The modeling effort is often tedious and requires a great number of experiments before producing a reliable model. Reproducibility of experiments is often uncertain due to the difficulty in obtaining the same environmental conditions. Moreover, as these processes involve living organisms, their dynamic behavior is strongly nonlinear and non-stationary. Model parameters cannot remain constant over a long period – they will vary, e.g., due to metabolic variations of biomass or to random and unobservable physiological or genetic modifications. It should also be noted that the lack of accuracy of the measurements often leads to identifiability problems.

2 Another essential difficulty lies in the absence, in most cases, of cheap and reliable instrumentation suited to real-time monitoring. To date, the market offers very few sensors capable of providing reliable on-line measurements of the biological and biochemical parameters required to implement high performance automatic control strategies. The main variables, i.e., biomass, substrate, and synthesis product concentrations, generally need determining through laboratory analyses. The cost and duration of the analyses obviously limit the frequency of the measurements.

Figure 5.1 shows a schematic view of a computer controlled bioreactor. In the illustrated situation, the influent flow rate is modulated via a control algorithm implemented in the computer. As will be explained below, the control algorithm combines different informations about the process provided by on-line measurements and some knowledge about the process dynamics. The on-line measurements can sometimes undertake some mathematical treatment by software sensors in order to provide ex-

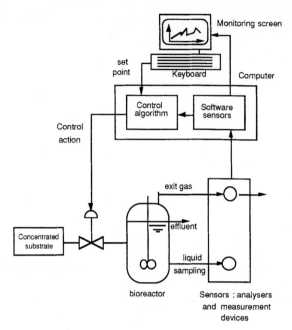

Fig. 5.1. Schematic view of a computer controlled bioreactor

tra information about some key process parameters (like specific growth rates) and unmeasured variables.

The chapter is organized as follows. Section 5.2 is dedicated to the basic concepts of automatic control of bioprocesses with a particular emphasis on the notions of stability and disturbances. Section 5.3 introduces the basic ingredients of automatic control (dynamical model, feedback, proportional, integral, and feedforward actions), and briefly compares linear control vs nonlinear control, and adaptive control vs non adaptive control. Finally Sect. 5.4 is concerned with the design of adaptive linearizing control schemes.

5.2
Bioprocess Control: Basic Concepts

The objective of automatic control when applied to any process (including bioprocesses) is to run and maintain the process in stable optimal operating conditions in spite of disturbances.

Let us first give some more insights about the concepts that are covered by the words *stable*, *optimal*, and *disturbances*. And let us consider the following simple example to illustrate these concepts: a simple microbial growth reaction in a stirred tank reactor. The microbial growth reaction is characterized by the following reaction scheme:

$$S \to X \tag{5.1}$$

with S the limiting substrate and X the biomass. Mass balances for both components S and X leads to the following set of dynamical equations:

$$\frac{dS}{dt} = -DS - \frac{1}{Y_{X/S}}\mu X + DS_{in} \qquad (5.2)$$

$$\frac{dX}{dt} = -DX + \mu X \qquad (5.3)$$

where D is the dilution rate (h^{-1}) (i.e., the ratio of the influent flow rate F_{in} (l/h) over the volume V (l) occupied by the reacting medium in the reactor), $Y_{X/S}$ is the yield coefficient, μ is the specific growth rate (h^{-1}), and S_{in} is the influent substrate concentration (g/l). Note that:

1. The above model is a *dynamical* model in order to capture the time evolution of the process: this is an essential ingredient of automatic control (we shall go back to this in Sect. 5.3).
2. The model is valid whatever the operation mode is: continuous (then the reactor is sometimes called a *chemostat* or a CSTR (Continuous Stirred Tank Reactor)), batch (then D is equal to zero), fed-batch (then the volume is not constant and its time variation is given by the equation $\frac{dV}{dt} = F_{in}$.
3. We use the same notation for the components (in the reaction scheme at Eq. 5.1) and their concentration (in Eqs. 5.2 and 5.3).

Let us now consider that the specific growth rate here is of the Haldane type:

$$\mu = \frac{\mu^* S}{K_s + S + S^2/K_I} \qquad (5.4)$$

with K_s the Monod constant, K_I the inhibitor constant, and μ^* a constant, which is indeed connected to the maximum specific growth rate μ_{max} as follows:

$$\mu = \mu_{max}(1 + 2\sqrt{\frac{K_s}{K_I}}) \qquad (5.5)$$

Let us also consider that the objective is to control the process by using the dilution rate D (or more precisely, the influent flow rate F_{in}) or the influent substrate concentration S_{in} as the control action.

5.2.1
Disturbances

Let us illustrate the notion of disturbances with the above example. If the dynamical model (Eqs. 5.2–5.4) is assumed to be perfect, and if the dilution rate D is the only manipulated input, then the influent substrate concentration S_{in} is a disturbance: variations of S_{in} will *disturb* the process by inducing variations in the time evolution of the variables S and X.

Assume now that the dynamical model is not perfect, e.g., that some of the parameters μ^*, K_s, and K_I of the Haldane model (Eq. 5.4) are changing with time (e.g., due to micro-organisms adaptation): these variations will also be disturbances with respect to the considered process model. Another similar disturbance may come from the presence of a side-reaction (e.g., maintenance or micro-organisms death) that has been neglected in the above dynamical model but which may not be not negligible under some operating conditions. Then, in case of maintenance for instance, the *true* dynamical model of the process is described by Eq. (5.3) and the following modified mass balance equation for S:

$$\frac{dS}{dt} = -DS - \frac{1}{Y_{X/S}}\mu X - k_m X + DS_{in} \tag{5.6}$$

where k_m is the maintenance coefficient.

Finally, typical disturbances may be due to process failures (e.g., actuator, sensor, or even stirring system). Even if those disturbances are inherently different, it is important to consider all the disturbances that are likely to take place during the process operation when analyzing and testing the performances of a designed controller that has to be implemented on a real bioprocess.

5.2.2
Stability

The notion of stability has been largely considered and studied in dynamical systems and automatic control, and has led to different mathematical definitions and tests (see, e.g., [17]). Here we shall concentrate on Lyapunov stability. Qualitatively the notion of stability can be roughly explained as follows. Let us consider that the process is at some equilibrium point (or *steady state*). The equilibrium point will be stable if the process does not go far from this state except for small deviations. It will be unstable if the variables of the process moves away from the equilibrium point and if their variations becomes larger and larger as time increases. Stability is obviously of vital concern in automatic control: it is essential to design controllers that keep processes in stable conditions on the one hand, and that are capable of stabilizing unstable processes on the other.

Let us illustrate the concept of stability on the above microbial growth example, and how to test it.

The Haldane model is represented in Fig. 5.2. What we shall show now is that values of the substrate concentration on the left hand side of the maximum specific growth rate $\mu^*/(1 + 2\sqrt{\frac{K_S}{K_I}})$ correspond to stable equilibrium points, while those on the right hand side correspond to unstable ones.

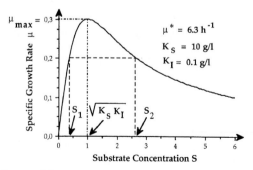

Fig. 5.2. The Haldane model

5.2.2.1
Equilibrium Points

An equilibrium point is, by definition, a *constant* state (denoted here $[\bar{S}, \bar{X}]$) which satisfies the equation of the dynamical model, i.e., the following algebraic equations in our example:

$$\frac{dS}{dt} = 0 \Rightarrow -\bar{D}\bar{S} - \frac{1}{Y_{X/S}}\mu(\bar{S})\bar{X} + \bar{D}\bar{S}_{in} = 0 \tag{5.7}$$

$$\frac{dX}{dt} = 0 \Rightarrow -\bar{D}\bar{X} + \mu(\bar{S})\bar{X} = 0 \tag{5.8}$$

for given constant values of $\bar{D}$ and $\bar{S}_{in}$. If we consider the Haldane model, it is straightforward to check that there are three possible equilibrium points:
1. Operational equilibrium point number 1: $\bar{S}_1 \leq \sqrt{K_2 K_1}, \bar{X}_1 = Y_{X/S}(\bar{S}_{in} - \bar{S}_1)$
2. Operational equilibrium point number 2: $\bar{S}_2 > \sqrt{K_s K_I}, \bar{X}_2 = Y_{X/S}(\bar{S}_{in} - \bar{S}_2)$
3. Wash-out equilibrium point: $\bar{X}_3 = 0, \bar{S}_3 = \bar{S}_{in}$.

The wash-out may occur for any value of $\bar{D}$ and $\bar{S}_{in}$. It is also the only possible equilibrium point for $\bar{D} > \mu_{max}$. It is obviously an undesirable steady-state. The values of $\bar{S}$ of the first two equilibrium points corresponds indeed to the two solutions of the second order algebraic equation obtained by introducing the Haldane equation (Eq. 5.4) into the equilibrium point equation of X (Eq. 5.8):

$$\frac{\bar{D}}{K_I}\bar{S}^2 + (\bar{D} - \mu^*)\bar{S} + \bar{D}K_s = 0 \tag{5.9}$$

5.2.2.2
Stability Analysis

For the stability analysis, let us consider the linear approximation of the model (Eqs. 5.2–5.4) around the equilibrium points (i.e., the *linearized tangent model*). Whatever the input u (D, S_{in}, or both), the linearized tangent model around each operational equilibrium point is written as follows:

$$\frac{dx}{dt} = Ax + Bu \tag{5.10}$$

where x is the vector of the deviations of S and X from the equilibrium values:

$$x = \begin{bmatrix} S - \bar{S} \\ X - \bar{X} \end{bmatrix} \tag{5.11}$$

and the matrix A is given by the following expression:

$$A = \begin{bmatrix} 0 & \Omega \\ -\frac{1}{Y_{X/S}}\bar{D} & -\frac{1}{Y_{X/S}}\Omega - \bar{D} \end{bmatrix} \tag{5.12}$$

with

$$\Omega = \frac{\mu^* \bar{X}(K_s - \frac{\bar{S}^2}{K_I})}{(K_s + \bar{S} + \frac{\bar{S}^2}{K_I})^2} \tag{5.13}$$

The stability of the equilibrium points is determined by the eigenvalues of the matrix A: the equilibrium point will be stable if the real parts of all the eigenvalues are negative; it will be unstable if one eigenvalue has a positive real part. In our example, the eigenvalues are equal to $(-D)$ and $(-\frac{1}{Y_{X/S}}\Omega)$. If the first eigenvalue is always negative, the second one is negative for the first equilibrium point $(\bar{S}_1, \bar{X}_1)$ and is positive for the second one $(\bar{S}_2, \bar{X}_2)$. This means that $(\bar{S}_1, \bar{X}_1)$ is a stable equilibrium point, and that $(\bar{S}_2, \bar{X}_2)$ is an unstable one.

5.2.3
Regulation vs Tracking

Two typical situations are usually considered in automatic control: constant set point control (regulation) and trajectory tracking. Regulation is typical of the control of continuous reactors, where the objective is usually to maintain the process in an a priori chosen steady state despite the disturbances. But in many instances like in batch and fedbatch reactors but also for start-up of reactors and in case of production grade changes, regulation is not appropriate: the objective is then to drive the process from some initial state to some final desired state by following some predetermined trajectory. In the above simple microbial growth example, if we consider a fedbatch reactor and if the objective is to optimize the production of biomass, then there is an optimal trajectory which is an "exponential" profile for X: the controller has then to be designed so as to maintain the process as close as possible to the optimal profile.

5.3
Bioprocess Control: Basic Ingredients

5.3.1
Dynamical Model

The first essential basic ingredient for the design of automatic control algorithms is a dynamical model of the process. The need for dynamical models in the control design is motivated by the intrinsic need to handle transients, either for transferring the process from one state to another, from one set point to another, for following a desired trajectory, or for rejecting or reducing the effect of disturbances. In the latter, disturbances influence the dynamical behavior of the process: good knowledge of the process dynamics, and in particular of the dynamical effect of the disturbance, will be helpful to select the appropriate strategy to reduce and possibly reject their negative influence on the process operation.

The dynamical models to be considered for control design can take different forms. The preceding sections suggest that the most obvious model forms are the mass balance models. But many other forms are possible. These include linear models deduced, e.g., by linearization of the mass balance models, or by identification from input-output data (e.g., [8]) (such models are also called *black-box* models).

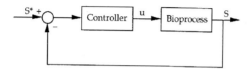

Fig. 5.3. Feedback loop for the control of S in a bioprocess

The input-output models may be nonlinear; neural network models are presently a quite popular version of nonlinear black box models (e.g., [9]). A possibly attractive form of dynamical models of bioprocesses are hybrid models, which are an intermediate form between mass balance and neural network models (e.g., [6]). In the following, mainly for reasons of conciseness and pedagogical reasons, we concentrate on mass balance models.

5.3.2
Feedback

Another essential basic ingredient of automatic control is the introduction of a *feedback* loop: on-line (real-time) information about the process (typically on-line measurement of the variable to be controlled) is provided to the controller which then provides the control action to the process. This concept of feedback loop is illustrated in Fig. 5.3.

5.3.3
Proportional Action

Most feedback control algorithms (except very simple controllers like on-off controllers) modulate the amplitude of the control action proportionally to the *control error*, i.e., the difference between the desired value of the controlled variable and its measured value: this is known as the proportional action.

5.3.4
Integral Action

The proportional action is usually not sufficient to guarantee a control error equal to zero in steady-state: this motivates the introduction in the control scheme of an integral action, i.e., a term proportional to the integral of the control error.

5.3.5
Feedforward Action

If the effects of a disturbance on the bioprocess dynamics are known, the controller may also be designed by incorporating a *feedforward* action in order to *anticipate* the possibly negative effects of the disturbance. We shall illustrate below how to incorporate feedforward action when the influent substrate concentration S_{in} is a measured disturbance.

5.3.6
Linear Control vs Nonlinear Control

The controller can be designed on the basis either of a linear model, or of a non-linear model of the process. As above, let us illustrate the design with a simple example: the control of the substrate concentration S by acting on the dilution rate D (or more precisely the influent flow rate F_{in}) in a stirred tank reactor with a simple microbial growth reaction.

Let us start with the first option: linear control. The control design is based on a linear model of the process, e.g., the linearized tangent model (Eqs. 5.10 and 5.11) around some steady state $(\bar{S}, \bar{X}, (\bar{D}, \bar{S}_{in})$. If we consider the control of S at a prescribed value S^* (e.g., equal to $\bar{S}$) a linear PI (*Proportional Integral*) regulator will then have typically the following form (see e.g., [15]):

$$D = \bar{D} + K_p[(S^* - S) + \frac{1}{\tau_i} \int_0^t (S^* - S(\tau))d\tau], K_P > 0, \tau_i > 0 \tag{5.14}$$

where K_P is the controller gain, and τ_i is the integral time or reset time.

Indeed a similar PI regulator can be computed on the basis of a nonlinear model of the process, e.g., the mass balance equation of the substrate concentration S (Eq. 5.2). The control algorithm will then be written as follows:

$$D = \frac{\lambda_1(S^* - S) + \lambda_2 \int_0^t (S^* - S(\tau))d\tau - \mu X/Y_{X/S}}{S_{in} - S}, \lambda_1 > 0, \lambda_2 > 0 \tag{5.15}$$

The above controller equation could have been obtained by combining the mass balance equation (Eq. 5.2) with the following desired closed-loop second-order dynamical equation:

$$\frac{d(S^* - S)}{dt} = -\lambda_1(S^* - S) - \lambda_2 \int_0^t (S^* - S(\tau))d\tau \tag{5.16}$$

for S^* constant[1]. In simple terms, Eq. (5.16) imposes that, if at some time instant the substrate concentration S is different from its desired value S^*, the controller will force S to converge to S^* with a rate determined by the design parameters λ_1 and λ_2.

Let us present at this point the advantages and drawbacks of the approaches introduced above. Before doing so, it is important to note that the second algorithm (Eq. 5.15) is more sophisticated than the first one (Eq. 5.14). In particular the second controller requires the knowledge (on-line measurement) of the biomass concentration X and of the influent substrate concentration S_{in}, and the knowledge of the specific growth rate model. When comparing both approaches, it is important to have these differences in mind.

5.3.6.1
Linear Control

The linear controller, if it is correctly tuned (see, e.g., [15] for tuning rules), will most probably do a good job if it is used for regulation around a steady-state. It also

---

1 Indeed by differentiating Eq. (5.16) once with respect to time, one obtains the usual second order linear equation $\frac{d^2x}{dt^2} + \lambda_1 \frac{dx}{dt} + \lambda_2 x = 0$ with $x=S^*-S$

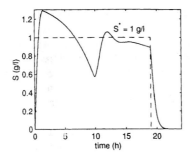

Fig. 5.4. Numerical simulation of a linear PI control of a fedbatch reactor

has the advantage of being easily implemented via standard industrial PI regulators. However, its performance (in particular its transient performance) might be degrading if the process is moving away from the steady-state for which it has been calibrated and tuned: this happens typically for start-up and grade changes in CSTRs, and this might even be more critical for (fed-)batch reactors in which the process state typically follows large variations (just think of the "exponential" growth of the biomass in this type of reactor!). This is illustrated in Fig. 5.4 where the PI regulation of S has been numerically simulated with the following model parameter values for Eqs. (5.2)–(5.4) and under the following conditions:

$\mu^* = 5\,h^{-1}$, $K_s = 10\,g/l$, $K_I = 0.1\,g/l$, $Y_{X/S} = 1$

$S(0) = 0\,g/l$, $(X(0) = 0.1\,g/l$, $V(0) = 10\,l$, $V(t_f) = 20\,l$, $S_{in} = 10\,g/l$

$S^* = 1\,g/l$, $K_P = 2$, $\tau_I = 0.5$

The values of the design parameters of the PI have been chosen so as to correspond to a closed loop dynamics close to the mean open loop dynamics. We note that the controller gain is too high at the beginning (and it induces large variations of the controlled output S), and then too small (the controller is not able to maintain the controlled output S to its desired value during the second part of the "exponential" growth phase).

5.3.6.2
Nonlinear Control

The main advantage of the nonlinear controllers is directly related to the above-mentioned drawback of the linear ones: if correctly designed and tuned, it will be able to handle a larger spectrum of the operating domain of the bioprocess. In the above example, the main source of nonlinearity is the term $\mu X/Y_{X/S}$. The presence of this term in the controller will be particularly important in the control of fedbatch reactors: as has been pointed out in [1], the integration of a term which characterizes the "exponential" biomass growth is a key factor for the efficiency of the control action. The other additional term $S_{in}-S$ also plays an important role: if S_{in} is a disturbance in the context of the control of the bioprocess, then this term is indeed a *feedforward* term that allows one to anticipate the effect of variations of S_{in} on the performance of the closed-loop system (process+controller).

However, first the implementation of the control algorithm will require a computer instead of a standard industrial PI (but can this still be considered as a drawback today?). Secondly (as for any controller) the performance will depend greatly on the quality and reliability of the model on which the design has been based. In other words, in the above example this means that we assume that the hydrodynamics in the reactor are fairly represented by completely mixed conditions, and that the simple growth reaction is largely dominant with regard to other possible side reactions. Besides, the implementation as such of the controller (Eq. 5.15) requires the values of X and also of S_{in}. In many cases one may expect that if S is measured on-line it should not be too difficult to measure S_{in} too (in many instances you do not even need to measure it since the influent substrate concentration is known by user's choice). But the situation is usually different for the biomass concentration: very often X is not measured on-line, or let us say it differently, S and X are often not available at the same time for on-line measurement. Finally, the nonlinear controller (Eq. 5.15) relies on the knowledge of (the model of) the specific growth rate μ: as suggested previously, this is a major source of uncertainty in bioprocess models.

5.3.7
Adaptive Control vs Non-Adaptive Control

One possible way to handle the above difficulties (lack of on-line measurements, uncertainties on the kinetics models) is to use adaptive (linearizing) control (see [2] and Sect. 5.4). In the above example this means that the control law will be implemented by considering on-line estimates of X and μ. An on-line estimate of the biomass concentration can be provided by an *asymptotic observer* (for a systematic design, see [2] and Sect. 5.4). In our example, it will take the following form. We first define an auxiliary variable Z:

$$Z = S + \frac{S}{Y_{X/S}} \tag{5.17}$$

The dynamics of Z are readily derived from the mass balance equations of X and S:

$$\frac{dZ}{dt} = DS_{in} - DZ \tag{5.18}$$

The on-line estimate $\hat{X}$ of X is then given by computing Z from Eq. (5.18) and by rewriting Eq. (5.17):

$$\hat{X} = Y_{X/S}(Z - S) \tag{5.19}$$

or equivalently, one can replace the term $X/Y_{X/S}$ in the controller equation by $Z-S$.

Let us now consider the on-line estimation of μ. Different methods can be used to give an on-line estimate $\hat{\mu}$ of μ. One possible estimator is the observer-based estimator (see [2]), which specializes in our example as follows:

$$\frac{d\hat{S}}{dt} = -\hat{\mu}(Z - S) + DS_{in} - DS + \gamma_1(S - \hat{S}), \gamma_1 > 0 \tag{5.20}$$

$$\frac{d\hat{\mu}}{dt} = -\gamma_2(S - \hat{S}), \gamma_2 > 0 \tag{5.21}$$

$\hat{S}$ is an "estimate" of S whose role is to drive the estimation of the unknown value of the specific growth rate μ: this is achieved by comparing $\hat{S}$ to the measured value S (this is the role of the correction terms $S\text{-}\hat{S}$ in Eqs. (5.20) and (5.21). γ_1 and γ_2 are design parameters that have to be calibrated to obtain the best on-line estimates (see [2]).

Since the parameter estimation introduces an integral action in the control loop ([2, 11]), the term $\gamma_2 \int_0^t (S^* - S(\tau))d\tau$ in the control law (Eq. 5.15) is not necessary any more. Then in our example, the adaptive linearizing controller is given by the following equations:

$$D = \frac{\lambda_1(S^* - S) - \hat{\mu}(Z - S)}{S_{in} - S}, \lambda_1 > 0 \qquad (5.22)$$

with Z and $\hat{\mu}$ given by Eqs. (5.18), (5.20), and (5.21), respectively.

Different applications of adaptive linearizing control to bioprocesses have been described, e.g., in [3, 12–14, 16]. One of the key feature of the adaptive linearizing can be illustrated on the basis of the above example: it allows one to incorporate the well-known characteristics of the process dynamics, while keeping the usual features of classical controllers:

– proportional action: via the term $\lambda_1(S^*\text{-}S)$
– integral action: via the adaptation mechanism (Eqs. 5.20 and 5.21)
– feedforward action: via the presence of S_{in}. The controller is capable of anticipating the effect of a variation of the influent substrate concentration: for instance, an increase of S_{in} results in a decrease of the control input D, inversely proportional to this increase.

Besides, as for the non-adaptive nonlinear PI, the controller contains a state "estimate" via the term $\mu X/Y_{X/S}$ (or more precisely $\hat{\mu}(Z\text{-}S)$). This is particularly important in fedbatch fermentation since it gives the exponential biomass growth term, and also in continuous operation since it provides a means to modulate the control action in connection with the level of biomass (e.g., if the process is in good working conditions, in particular if the biomass concentration is high, it is possible to treat high amounts of substrate, since the control action D is proportional to X).

Finally, note that, beside the integral action, the estimation of "physical" parameters (here μ) has the further advantage of giving useful information that can be used for monitoring the process, and possibly also for analyzing the internal working of the process. Figure 5.5 illustrates the performance of the adaptive linearizing controller in numerical simulation. The same model parameters as in Fig. 5.4 have been used. The following design parameters and initial values have been considered:

$$S^* = 1g/l, C_1 = 2, \gamma_1 = 5, \gamma_2 = \frac{\gamma_1^2}{4(Z - S)} \qquad (5.23)$$

$$\hat{\mu}(0) = 0.1\,h^{-1}, \hat{S}(0) = S(0), Z(0) = S(0) + 0.5X(0), S(0) = 0.5\,g/l \qquad (5.24)$$

The initial value of S has been set to 0.5 g/l: this gives the opportunity to see that the controller has no problem to force S to converge to its desired value (see Fig. 5.5a). Figure 5.5b,c exhibits the exponential profile of the biomass X, and the control input F_{in}. Finally Fig. 5.5d gives the on-line estimation of the specific growth rate μ.

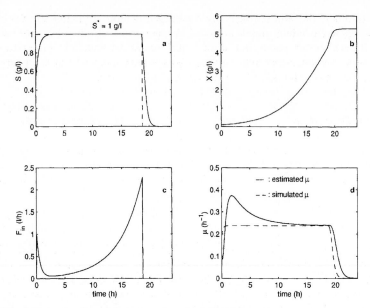

Fig. 5.5a–d. Adaptive linearizing controller of a fedbatch reactor

5.3.8
Other Approaches

The (adaptive) nonlinear controller presented in the preceding section belongs indeed to a large class of control strategies, often labelled *model-based* control. This includes control approaches like *model predictive control* in which constraints (like input constraints) may be handled explicitly in the design (see, e.g., [15]). We have restricted the presentation to mass balance models, but obviously any dynamical model of the bioprocess (e.g., hybrid model), as long as it is reliable, can be considered in the control design.

5.4
Adaptive Linearizing Control of Bioprocesses

In this section we shall concentrate on the design of *adaptive linearizing controllers* on a systematic basis (via the General Dynamical Model introduced below). The motivation for this choice of control approach has already been suggested above, and can be further justified. As has already been said, the dynamics of bioprocesses are usually described by material balance equations. If the hydrodynamics and the underlying reaction mechanisms of the bioprocess are often rather well characterized, the kinetics of these reactions are usually badly characterized or defined. Symptomatically there exist many (heuristic) models available in the literature for the specific growth rate (more than 60, see, e.g., [2]). Beside the difficulty of choosing an appropriate model, the parameters of the selected model are often practically unidentifiable, i.e., they cannot be given unique values from the available experimental data (see, e.g., [2, 7]). The adaptive linearizing control approach has the basic following features: it incorporates the well-known features about the process dynamics (basically, the reaction network and the material balances) in the control algorithms which are moreover capable of dealing with the process uncertainty (in particular of the reaction kinetics) via an adaptation scheme.

5.4.1
General Dynamical Model

A biotechnological process can be defined as a set of M biochemical reactions involving N components. The dynamical model of a bioprocess in a stirred tank reactor can be deduced from mass balance considerations and written in the following compact form:

$$\frac{d\xi}{dt} = -D\xi + Kr + F - Q \tag{5.25}$$

where ξ is the vector of the bioprocess component (g/L) ($\dim(\xi)$=N), D is the dilution rate (h^{-1}), K is the yield coefficient matrix ($\dim(K)$=N×M), r is the reaction rate vector ($g/(L.h)$) ($\dim(r)$=M), F ($g/(L.h)$) is the feed rate vector, and Q ($g/(L.h)$) the gaseous outflow rate vector ($\dim(F)$=$\dim(Q)$=N). The model equation (Eq. 5.25) has been called the *General Dynamical Model* for stirred tank bioreactors (see [2]). As already mentioned in Sect. 5.2, note that the model covers the different types of operating conditions: batch, fed-batch (as in example 2 below) or continuous reactors (as in example 1 below).

5.4.1.1
Example 1: Anaerobic Digestion

Four metabolic paths can be identified in anaerobic digestion [10]: two for acidogenesis and two for methanization. In the first acidogenic path (Path 1), glucose is decomposed into fatty volatile acids (acetate, propionate), hydrogen, and inorganic carbon by acidogenic bacteria. In the second acidogenic path (Path 2), OHPA (Obligate Hydrogen Producing Acetogens) decompose propionate into acetate, hydrogen, and inorganic carbon. In a first methanization path (Path 3), acetate is transformed into methane and inorganic carbon by acetoclastic methanogenic bacteria, while in the second methanization path (Path 4), hydrogen combines with inorganic carbon to produce methane under the action of hydrogenophilic methanogenic bacteria. The process can then be described by the following reaction network:

$$S_1 \rightarrow X_1 + S_2 + S_3 + S_4 + S_5 \tag{5.26}$$

$$S_2 \rightarrow X_2 + S_3 + S_4 + S_5 \tag{5.27}$$

$$S_3 \rightarrow X_3 + S_5 + P_1 \tag{5.28}$$

$$S_4 + S_5 \rightarrow X_4 + P_1 \tag{5.29}$$

where S_1, S_2, S_3, S_4, S_5, X_1, X_2, X_3, X_4, and P_1 are respectively glucose, propionate, acetate, hydrogen, inorganic carbon, acidogenic bacteria, OHPA, acetoclastic methanogenic bacteria, hydrogenophilic methanogenic bacteria, and methane. The dynamical model of the anaerobic digestion process (N=10, M=4) in a stirred tank reactor can be described within the above formalism (Eq. 5.25) by using the following definitions:

$$
\xi = \begin{bmatrix} X_1 \\ S_1 \\ X_2 \\ S_2 \\ X_3 \\ S_3 \\ X_4 \\ S_4 \\ S_5 \\ P_1 \end{bmatrix}, K = \begin{bmatrix} 1 & 0 & 0 & 0 \\ -k_{21} & 0 & 0 & 0 \\ 0 & 1 & 0 & 0 \\ k_{41} & -k_{42} & 0 & 0 \\ 0 & 0 & 0 & 1 \\ k_{61} & k_{62} & -k_{63} & 0 \\ 0 & 0 & 0 & 1 \\ k_{81} & k_{82} & 0 & -k_{84} \\ k_{91} & k_{92} & k_{93} & -k_{94} \\ 0 & 0 & k_{03} & k_{04} \end{bmatrix}, F = \begin{bmatrix} 0 \\ DS_{in} \\ 0 \\ 0 \\ 0 \\ 0 \\ 0 \\ 0 \\ 0 \\ 0 \end{bmatrix}, Q = \begin{bmatrix} 0 \\ 0 \\ 0 \\ 0 \\ 0 \\ 0 \\ 0 \\ Q_1 \\ Q_2 \\ Q_3 \end{bmatrix} \quad (5.30)
$$

$$
r = \begin{bmatrix} r_1 \\ r_2 \\ r_3 \\ r_4 \end{bmatrix} = \begin{bmatrix} \mu_1 X_1 \\ \mu_2 X_2 \\ \mu_3 X_3 \\ \mu_4 X_4 \end{bmatrix} \quad (5.31)
$$

where μ_1, μ_2, μ_3, μ_4 are the specific growth rates (h^{-1}) of the reactions at Eqs. (5.26)–(5.29), respectively, and S_{in}, Q_1, Q_2, and Q_3 represent respectively the influent glucose concentration and the gaseous outflow rates of H_2, CO_2, and CH_4.

5.4.1.2
Example 2: Animal Cell Culture

Let us consider one animal cell culture process: a human embryo kidney (HEK-293) cell culture [16]. It is characterized by the following reaction network:

$$
S + C \rightarrow X + G \quad (5.32)
$$

$$
S \rightarrow X + L \quad (5.33)
$$

where S, C, X, G, and L represent the glucose, dissolved oxygen, cells, carbon dioxide and lactate, respectively. Reactions at Eqs. (5.32) and (5.33) are an oxidation (respiration) reaction on glucose, and a glycolysis (fermentation) on glucose. The dynamics of the process in a stirred tank reactor are given in the matrix format (Eq. 5.25) by the following vectors and matrices:

$$
\xi = \begin{bmatrix} S \\ C \\ X \\ G \\ L \end{bmatrix}, K = \begin{bmatrix} -k_1 & -k_4 \\ -k_2 & 0 \\ 1 & 1 \\ k_3 & 0 \\ 0 & k_5 \end{bmatrix}, r = \begin{bmatrix} \mu_R X \\ \mu_F X \end{bmatrix}, F = \begin{bmatrix} DS_{in} \\ Q_{in} \\ 0 \\ 0 \\ 0 \end{bmatrix}, Q = \begin{bmatrix} 0 \\ 0 \\ 0 \\ Q_1 \\ 0 \end{bmatrix} \quad (5.34)
$$

k_j (j=1–9) are yield coefficients, S_{in} is the influent glucose concentration (g/l), Q_{in} is the oxygen feedrate (g/l/h), Q_1 is the CO_2 outflow rate (g/l/h), and μ_i (i=R and F) are the specific growth rates (1/h) associated to each growth reaction.

5.4.2
Model Reduction

The above anaerobic digestion example illustrates that the dynamical model may be fairly complex with a large number of differential equations. But there are many practical applications where a simplified reduced order model is sufficient from an engineering viewpoint. Model simplification can be achieved by using singular perturbation, which is a technique for transforming a set of $n+m$ differential equations into a set of n differential equations and a set of m algebraic equations. It is suitable when neglecting the dynamics of substrates and of products with low solubility in the liquid phase. The method will be illustrated with one specific example (low solubility product) before stating the general rule for order reduction.

5.4.2.1
Singular Perturbation Technique for Low Solubility Products

Let us consider a biochemical reaction with a volatile product P which gives off in gaseous form and has low solubility in the liquid phase. The dynamical equation of P is as follows:

$$\frac{dP}{dt} = kr - DP - Q \tag{5.35}$$

The consistency of this model requires that the product concentration P be proportional to a saturation concentration representative of the product solubility, which is expressed as $P=\Pi P_{sat}$ where P_{sat} is the saturation concentration which is constant in a stable physico-chemical environment. The model (Eq. 5.35) is rewritten in the standard singular perturbation form, with $\varepsilon=P_{sat}$:

$$\varepsilon\frac{d\Pi}{dt} = kr - \varepsilon D\Pi - Q \tag{5.36}$$

If the solubility is very low, we obtain a reduced order model by setting $\varepsilon=0$ and replacing the differential equation (Eq. 5.36) by the algebraic one:

$$Q = kr \tag{5.37}$$

5.4.2.2
A General Rule for Order Reduction

The above example shows that the rule for model simplification is actually very simple and that an explicit singular perturbation analysis is not really needed. Consider that, for some i, the dynamics of the component ξ_i are to be neglected. The dynamics of ξ_i are described by Eq. (5.25):

$$\frac{d\xi_i}{dt} = -D\xi_i + K_i r + F_i - Q_i \tag{5.38}$$

where K_i is the row of K corresponding to the component ξ_i. The simplification is achieved by setting ξ_i and $d\xi_i/dt$ to zero, i.e., by replacing the differential Eq. (5.38) by the following algebraic equation:

$$K_i r = -F_i + Q_i \tag{5.39}$$

It has been shown that the above model order reduction rule is not only valid for low solubility products but also for bioprocesses with fast and slow reactions. Then the above order reduction rule (Eq. 5.39) applies to substrates of fast reactions (as long as they intervene only in fast reactions).

5.4.2.3
Example 1: Anaerobic Digestion

Let us see how to apply the above model order reduction rule to the anaerobic digestion. First of all, it is well-known that methane is a low solubility product. Therefore the above procedure applies. Furthermore, assume that for instance the first acidogenic path (reaction at Eq. 5.26) is limiting, i.e., that the last three reactions (Eqs. 5.27–5.29) are fast and the reaction at Eq. (5.26) is slow. We can then apply the model order reduction rule to the propionate concentration S_2, the acetate concentration S_3, the hydrogen concentration S_4, and the dissolved methane concentration P_1. By setting their values and their time derivatives to zero, we reduce their differential equations to a set of algebraic equations. Let us further consider that the gaseous hydrogen outflow rate Q_1 is negligible. If one considers the expressions of the reaction rate r_1 which can be drawn from these algebraic equations, and introduces them in the dynamical equation of the glucose concentration S_1, this one can be rewritten as follows:

$$\frac{dS_1}{dt} = -DS_1 - k_1 Q_3 + DS_{in} \tag{5.40}$$

where k_1 is defined as follows:

$$k_1 = \frac{k_{21} k_{42} k_{63} k_{84}}{k_{03} k_{84}(k_{61} k_{42} + k_{62} k_{41}) + k_{04} k_{63}(k_{81} k_{42} + k_{82} k_{41})} \tag{5.41}$$

Note that the coefficient k_1 is a nonlinear combination of the yield coefficients k_{ij}.

5.4.3
Control Design

Let us now concentrate on the design of adaptive linearizing controllers. It is based on the dynamical material balances of the process (Eq. 5.25). More specifically, the control design will be based on the dynamical equation that relates the controlled variable and the control variable. If we denote the controlled variable (typically, some process component concentration(s)) by y and the control variable (typically flow rate(s) or influent substrate concentration(s)) by u, then the dynamical equation between y and u is given from Eq. (5.25) (possibly after model reduction) by the following expresion:

$$\frac{dy}{dt} = f(\xi) + g(\xi)u \tag{5.42}$$

where f and g are functions of the process component concentration. In the following, we shall illustrate the control design with the animal cell culture. In this example, if the control objective is to control the glucose concentration S ($y=S$) by acting on the influent flow rate F_{in} ($u=F_{in}$), then Eq. (5.42) before model reduction specializes as follows:

$$\frac{dS}{dt} = -k_1 \mu_R X - k_4 \mu_F X - \frac{F_{in}}{V} S + \frac{F_{in}}{V} S_{in} \qquad (5.43)$$

In our example, we consider that CO_2 is a low solubility product. Then the dynamical equation (Eq. 5.43) becomes:

$$\frac{dS}{dt} = -\frac{k_1}{k_3} Q_1 - k_4 \mu_F X - \frac{F_{in}}{V} S + \frac{F_{in}}{V} S_{in} \qquad (5.44)$$

i.e.,

$$f = -\frac{k_1}{k_3} Q_1 - k_4 \mu_F X, g = \frac{S_{in} - S}{V} \qquad (5.45)$$

An interesting feature of the above dynamical equation is its independence with respect to the specific growth rate μ_R: the largely uncertain or unknown specific growth rate μ_R has been replaced by a variable usually easy to measure: the gaseous outflow rate of CO_2, Q_1.

The design of the adaptive linearizing controllers is also based on the following (realistic) assumptions:

A1. Some of the (combinations of) the yield coefficients are known
A2. The kinetics are unknown
A3. The influent flow rates and substrate concentration, and the gaseous outflow rates are known (either by measurement or by user's choice)

5.4.3.1
The Monitoring Tool 1: An Asymptotic Observer

The key result of this section is the use of a state transformation by which part of the dynamical model (Eq. 5.25) becomes independent of the reaction kinetics r (see [2]). This transformation will play a very important role since it will allow us in our example to implement a software sensor called *asymptotic observer* for the (unmeasured) biomass based on the general dynamical model independently of the unmeasured variables. In other instances (see [4, 5]), it allows to implement a controller independent of unmeasured variables (see also Sect. 5.3.7).

The transformation is defined as follows. Let us denote rank(K)=ρ, dim(ξ)=n (number of process components), dim(r)=m (number of reactions), and consider a state partition:

$$\xi = \begin{bmatrix} \xi_a \\ \xi_b \end{bmatrix} \qquad (5.46)$$

where ξ_b contains ρ (arbitrarily chosen) process variables and ξ_a the others, but such that the corresponding submatrix K_a is full rank (rank(K_a)=r). Let us define the state transformation Z (dim(z)=n-r):

$$Z = A_0 \xi_a + \xi_b \qquad (5.47)$$

where A_0 is solution of the matrix equation:

$$A_0 K_a + K_b = 0 \tag{5.48}$$

The dynamical equation of Z can be readily derived from the material balance equations (Eq. 5.25):

$$\frac{dZ}{dt} = -DZ + A_0(F_a - Q_a) + F_b - Q_b \tag{5.49}$$

where the indices F_a-Q_a and F_b-Q_b correspond to the feed rates and gaseous outflow rates of ξ_a and ξ_b, respectively. Equation (5.49) is independent of the reaction kinetics r. An important special case of the above state transformation is μ=m (independent irreversible reactions), A_0 is equal to:

$$A_0 = -K_b K_{Ra}^{-1} \tag{5.50}$$

Equation (5.49) is a system linear-in-the state Z; we can check that it is asymptotically stable if D is positive (or more precisely, as long as D does not remain equal to zero for too long. The stability condition is indeed written as follows: if there exist positive constants β and δ such that $\int_t^{t+\delta} D(\tau)d\tau > \beta$ for all t). Therefore Eq. (5.49) can be used to compute Z on-line from the knowledge of the feed rates F, the gaseous flow rates Q, and the knowledge of some combinations of the yield coefficients. The variables Z can then be used to estimate on-line the unmeasured variables from the measured variables by implementing an asymptotic observer which is independent of the reaction kinetics (see [2]):

$$\frac{d\hat{Z}}{dt} = -D\hat{Z} + A_0(F_a - Q_a) + F_b - Q_b \tag{5.51}$$

$$\hat{\xi}_b = \hat{Z} + K_b K_a^{-1} \xi_a \tag{5.52}$$

where $\hat{Z}$ and $\hat{\xi}_b$ are the on-line estimates of Z and ξ_b, respectively. Let us apply the above transformation to our example. Here, n=5 and r=m=2. A particular form of the asymptotic observer is when ξ_a contains the measured process components and ξ_b the unmeasured ones. However, as it is illustrated here, we have chosen another state partition. Let us choose, e.g., $\xi_b=[S, C, L]^T$ and $\xi_a=[K, G]^T$ and define z as follows:

$$z = \begin{bmatrix} Z_1 \\ Z_2 \\ Z_3 \end{bmatrix} = \begin{bmatrix} S - \frac{k_4-k_1}{k_3}G + k_4 X \\ \frac{k_2}{k_3}G + C \\ L - k_5 X + \frac{k_5}{k_3}G \end{bmatrix} \tag{5.53}$$

The dynamical equations of Z are readily derived from Eqs. (5.49), (5.50), and (5.53):

$$\frac{dZ_1}{dt} = -DZ_1 + DS_{in} + \frac{k_4 - k_1}{k_3}Q_1 \tag{5.54}$$

$$\frac{dZ_2}{dt} = -DZ_2 + Q_{in} - \frac{k_2}{k_3}Q_1 \tag{5.55}$$

$$\frac{dZ_3}{dt} = -DZ_3 - \frac{k_5}{k_3}Q_1 \tag{5.56}$$

Let us see how it specializes in the animal cell culture example in the particular situation where we are only interested to have an on-line software measurement of the biomass X. The asymptotic observer is then written as follows by using Eqs. (5.53) and (5.56); Eqs. (5.57) and (5.58) represent *Adaptive Linearizing Control (Part I): On-line Estimation of X*:

$$\frac{d\hat{Z}_1}{dt} = -D\hat{Z}_1 + DS_{in} + \frac{k_4 - k_1}{k_3} Q_1 \tag{5.57}$$

$$\hat{X} = \frac{Z_1 - S}{k_4} \tag{5.58}$$

where $\hat{X}$ represents the on-line estimate of X provided by the aymptotic observer. Note that in the second equation of the above software sensor (Eqs. 5.57 and 5.58), we have considered the low solubility assumption for the CO_2 ($G=0$). Note that the asymptotic observer requires the knowledge of the *ratio* $(k_4-k_1)/k_3$ and the yield coefficient k_4.

5.4.3.2
The Monitoring Tool 2: The Parameter Estimation

The parameter estimation has a double objective here. The first one is to give on-line estimates of uncertain parameters (the reaction kinetics in our example, see assumption A3). As already mentioned in Sect. 5.3.7, a second important objective of the parameter estimation is to incorporate integral action in the control scheme (to eliminate offsets). In adaptive control, there are typically two different approaches corresponding to the way the parameter adaptation is introduced in the controller: the direct and the indirect methods. Here we have considered the latter (this choice is indeed somewhat arbitrary). In Sect. 5.3.7, we have considered an observer-based estimator; here we consider an alternative approach: the unknown parameter(s) will be estimated on-line via a recursive least-square (RLS) estimation algorithm with a forgetting factor. The usual form of the RLS form with forgetting factor for a system described in discrete-time by the equation

$$x_t = \phi_t^T \theta \tag{5.59}$$

(where θ is the vector of parameters to be estimated, and t the time index) is written as follows:

$$\hat{\theta}_{t+1} = \hat{\theta}_t + G_t \phi_t (x_t - \phi_t^T \hat{\theta}_t) \tag{5.60}$$

$$G_t = \frac{G_{t-1}}{\gamma}\left(I - \frac{\phi_t \phi_t^T G_t}{\gamma + \phi_t^T G_t \phi_t}\right), \; 0 < \gamma \leq 1, \; G_0 > 0 \tag{5.61}$$

where G_t is the RLS gain matrix and γ the forgetting factor. Let us derive the RLS estimation algorithm in our animal cell culture example. Here we assume that k_1/k_3 is (sufficiently well) known and to report all the uncertainty on the kinetics parameter μ_F. We further assume (in line with basic kinetics laws) that there is no growth in absence of the limiting substrate S (μ_F ($S=0$)=0): the simplest way to express it is to write μ_F as follows:

$$\mu_F = \theta S \tag{5.62}$$

where θ is an unknown, possibly time varying, parameter. The next step in the derivation of the estimation algorithm is to rewrite the glucose equation (Eq. 5.44) in the format at Eq. (5.59) by discretizing the time derivative dS/dt via a first order Euler approximation ($dS/dt \cong S_{t+1}-S_t)/T$ with T the sampling period). This gives:

$$S_{t+1} - S_t + T\frac{k_1}{k_3}Q_{1,t} - T\frac{F_{in,t}}{V_t}(S_{in,t} - S_t) = -Tk_4\theta_t S_t X_t \tag{5.63}$$

i.e., $x_t = S_{t+1} - S_t + T\frac{k_1}{k_3}Q_{1,t} - T\frac{F_{in,t}}{V_t}(S_{in,t} - S_t)$ and $\phi_t = -Tk_4 S_t X_t$. The RLS estimation algorithm with a forgetting factor specializes as follows (with $\hat{X}_t$ given by the asymptotic observer): Eqs. 5.64 and 5.65 represent *Adaptive Linearizing Control (Part II): On-line Estimation of θ*:

$$\hat{\theta}_{t+1} - \hat{\theta}_t + g_t Tk_4 S_t \hat{X}_t(S_{t+1} - S_t + T\frac{k_1}{k_3}Q_{1,t} - T\frac{F_{in,t}}{V_t}(S_{in,t} - S_t) + Tk_4\hat{\theta}_t S_t \hat{X}_t \tag{5.64}$$

$$g_t = \frac{g_t - 1}{\gamma + g_t T^2 k_4^2 S_t^2 \hat{X}_t^2}, 0 < \gamma \le 1, g_0 > 0 \tag{5.65}$$

5.4.3.3
The Control Tool: The Adaptive Linearizing Controller

Recall that the control problem under study here is the control of some process component concentration(s) (denoted by y) by acting on the flow rate(s) or influent substrate concentration(s) (denoted by u). Assume that we wish to have the following linear stable closed-loop dynamics for y:

$$\frac{dy}{dt} = \Lambda(y^* - y) \tag{5.66}$$

with y^* the desired values for y, and Λ an (arbitrary) stable matrix (see Sect. 5.2.2.2). By combining Eqs. (5.42) and (5.66) and eliminating dy/dt, we obtain the following linearizing control law:

$$u = g(\xi,\theta)^{-1}[\Lambda(y^* - y) - f(\xi,\theta)] \tag{5.67}$$

By considering that some process parameters are estimated on-line, the above control algorithm (Eq. 5.67) is written in its discrete-time version as follows:

$$u = g(\xi,\hat{\theta})^{-1}[\Lambda(y^* - y) - f(\xi,\hat{\theta})] \tag{5.68}$$

In the animal cell culture example, if we consider the following closed-loop desired dynamics (with S^* the desired ethanol concentration):

$$\frac{dS}{dt} = \lambda(S^* - S), \lambda > 0 \tag{5.69}$$

the adaptive linearizing control, Eq. (5.68), specializes as follows: Eq. (70) represents *Adaptive Linearizing Control (Part III): The Controller*:

$$F_{in,t} = \frac{V_t}{S_{in,t} - S_t}[\lambda(S^* - S_t) + \frac{k_1}{k_3}Q_{1,t} + k_4\hat{\theta}_t S_t \hat{X}_t] \tag{5.70}$$

The above equation (Eq. 5.70), together with Eqs. (5.57), (5.58), (5.64), and (5.65) constitutes the adaptive linearizing controller of animal cell culture process. Note

that the controller structure is similar to the one presented in the preceding section
(Eq. 5.22) (with the addition of the CO_2 gaseous outflow rate term), and the comments mentioned then obviously readily apply here also.

5.4.4
Experimental Results

The adaptive linearizing controller (Eqs. 5.57, 5.58, 5.64, 5.65 and 5.70) has been
implemented on a 22-l pilot-scale bioreactor (see [16]). Figure 5.6 presents one set
of experimental results. The glucose has been measured via an FIA biosensor device.
The temperature and pH were maintained at constant values of 37 °C and 7.2, respectively. The initial conditions were equal to

$$V_0 = 19L, \ X_0 = 0.18 \times 10^6 \ cells/ml, \ S_0 = 1.3 \ mmol/l, \ S_{in} = 3.3 \ mol/l$$

In this experiment, CO_2 was not available for on-line measurement. Therefore the
adaptive linearizing controller without the carbon dioxide term has been implemented. The values of the initial conditions and design parameters have been chosen via
numerical simulation and simple computation:

$$\hat{\theta}_0 = 0.25 \ h^{-1} \ mM^{-1}, \hat{X} = X_0, \ \lambda = 0.4 \ h^{-1}, \ \gamma = 0.9, \ S^* = 1 \ mmol/l$$

The initial value of $\hat{\theta}$ is given by $\mu_{F,0}/S_0$ where μ_F is given by the simulation model
validated on batch experimental data. The controller gain has been chosen so as to
have closed time constant shorter than in open loop: the highest open loop time
constant was around 5 h, the chosen λ corresponds to 2.5 h time constant. The value
of the forgetting factor λ (0.9) allows one to handle time variations of the parameter
θ. Figure 5.6 presents the controlled variable, i.e., the glucose concentration S

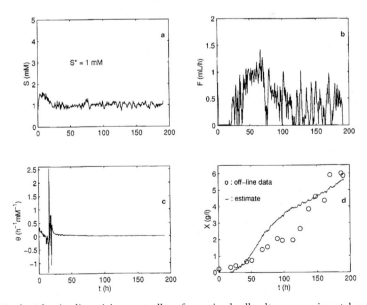

Fig. 5.6a–d. Adaptive linearizing controller of an animal cell culture: experimental results

(Fig. 5.6a), the control variable, i.e., F_{in} (Fig. 5.6b), and the estimated values of θ (Fig. 5.6c), and X compared to off-line measurements (Fig. 5.6d), respectively.

Acknowledgments. This paper presents research results of the Belgian Programme on Inter-University Poles of Attraction initiated by the Belgian State, Prime Minister's Office for Science, Technology and Culture. The scientific responsibility rests with its authors.

References

1. Axelsson JP (1989) Modelling and control of fermentation processes. PhD thesis, Lund Inst Technology
2. Bastin G, Dochain D (1990) On-line estimation and adaptive control of bioreactors. Elsevier, Amsterdam
3. Chen L, Bastin G, Van Breusegem V (1995) A case study of adaptive nonlinear regulation of fed-batch biological reactors. Automatica 31:55–65
4. Dochain D, Perrier M (1993) Control design for nonlinear wastewater treatment processes. Water Sci Technol 28:283–293
5. Dochain D, Perrier M (1997) Dynamical modelling, analysis, monitoring and control design for nonlinear bioprocesses. Advances in Biochemical Engineering/Biotechnology 56:147–197
6. Feyo De Azevedo F, Dahm B, Oliviera FR (1997) Hybrid modelling of biochemical processes: a comparison with the conventional approach. Computers Chem Eng 21[Suppl]S751–S756
7. Holmberg A (1982) On the practical identifiability of microbial growth models incorporating Michaelis-Menten type nonlinearities. Math Biosci 62:23–43
8. Ljung L (1987) System identification – theory for the user. Prentice-Hall, Englewood Cliffs, NJ
9. Montague G, Morris J (1994) Neural network contribution in biotechnology. TIBTECH 12:312–323
10. Mosey FE (1983) Mathematical modelling of the anaerobic digestion process: regulatory mechanisms for the formation of short-chain volatile acids from glucose. Water Sci Technol 15:209–232
11. Perrier M, Dochain D (1993) Evaluation of control strategies for anaerobic digestion processes. Int J Adaptive Cont Signal Proc 7:309–321
12. Pomerleau Y, Perrier M, Bourque D (1995) Dynamics and control of the fed-batch production of poly-β-hydroxybutyrique by *Methylobacterium extorquens*. Proc 6th Int Conf Computer Appl in Biotechnol, pp 107–112
13. Pomerleau Y, Viel G (1992) Industrial application of adaptive nonlinear control for baker's yeast production. In: Karim NM, Stephanopoulos G (eds) Proc 5th Int Conf Computer Appl in Biotechnol, Pergamon, Oxford, pp 315–318
14. Renard P, Dochain D, Bastin G, Naveau H, Nyns EJ (1988) Adaptive control of anaerobic digestion processes. A pilot-scale application. Biotechnol Bioeng 31:287–294
15. Seborg DE, Edgar TF, Mellichamp DA (1989) Process dynamics and control. Wiley, New York
16. Siegwart P, Male K, Côté J, Luong JHT, Perrier M, Kamen A (1998) Adaptive control at low-level glucose concentration to study HEK-293 cell metabolism in serum pure cultures. Biotechnol Progress 28:927–937
17. Willems JL (1970) Stability theory of dynamical systems. Nelson, London

6 On-Line Simulation Techniques for Bioreactor Control Development

Reiner Luttmann, Klaus-Uwe Gollmer

6.1
Introduction

During the last decade, bioprocess development automation tasks have become more significant for overall process performance. Quality requirements on the one hand and growing process knowledge on the other have resulted in applied control strategies of increasing complexity. At the same time modern information technology facilitates higher flexibility of the technical systems used for practical realization of control tasks [1, 2]. Nevertheless, one of the most important goals in bioprocess automation is the manipulation of the process to meet desired performance criteria. Examples are the control of pO_2 to a prescribed limit or the realization of an exponential substrate feeding strategy. In practice, the process and the controller form a closed loop, often realized in a feedback fashion (Fig. 6.1a). For design and stability investigations of these control-loops both process and controller require treatment using a theoretical mathematical model (Fig. 6.1b). Obviously the time-dependent changes of the relevant process variables are of major interest for these applications. Therefore the underlying model is highly dynamic in nature and typically a set of differential equations based on mass and energy balances is used for simulation of the process behavior.

The main advantage of this model-based design strategy is that the optimization procedure is almost independent of application. Therefore well established methods from control theory can be applied to bioprocess automation tasks. Finally, the theo-

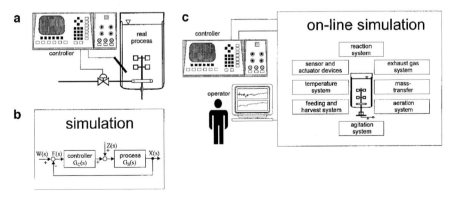

Fig. 6.1a–c. Different types of process simulation

retically optimized strategy has to be transferred to the real-world application by setting the parameters of commercial PID-controllers or using the high level language of a process control system.

Because of the inherent complexity of the underlying software program a number of test fermentations has to be performed to ensure proper functionality and interaction of all control tasks. Due to economic reasons, the verification of a new strategy by on-line simulation is gaining influence in the design cycle of bioprocess automation methods.

The essential idea behind on-line simulation is to use a model (virtual bioreactor) to emulate the process behavior and connect this artificial system to the real-world automation hardware (Fig. 6.1c). Therefore, the performance of a new implementation, for example, can be validated without running a real-world process. In contrast to the above-mentioned model-based design methods, on-line simulation has to be done in real-time interacting with the existing automation hardware. Nevertheless, the computational power of today's personal computers enables on-line simulation even for complex models.

On-line simulations have been applied to a wide field of technological systems and engineering problems. One of the most popular examples is the flight simulator, which runs on almost any personal computer. In this case, a mathematical model of the airplane is manipulated in real-time by the human operator to ensure a proper flight. "Virtual reality" is one of the strongest arguments for using on-line simulation tools for training and instruction of operators.

6.2
Application

The fields of application may be classified into two groups, industrial and educational. Whereas industry is predominantly interested in maximizing profit, academics are looking for efficient teaching tools.

6.2.1
Application in the Biochemical Industry

Recently the chemical industry has been directing its attention to the development and application of on-line simulation methods. Remarkable benefits are expected in the following areas.

6.2.1.1
Plant Set Up

Today, modern software packages allow us to simulate the combination of discrete and continuous units that are involved in the production and purification of biochemical products [3]. A typical plant model for batch processes consists of unit operations such as product dilution, filter press, solvent extraction chromatography operation, ultrafiltration, and pasteurization. Furthermore, cleaning in place (CIP) equipment is required for various processing areas. Plant-wide simulation challenges for computer aided process design and the identification of bottle-necks in the whole production process [4]. From the automation point of view, checking a new control system in a simulated environment is a typical procedure for the chemical and petrochemical industries. The results of preliminary tests have significant influence on the design of production plants. Operators are able to familiarize themselves with the behavior of the new units months before the real plant is ready. Based on simulation, the reliability of the planning process and the time for putting the plant into operation could be reduced significantly.

6.2.1.2
Economy

Yield improvement caused by efficient control strategies has an evident influence on the overall production costs. Application of model-based control algorithms (e.g., OLFO or adaptive control) guarantees an optimal performance even under small disturbances of actual process conditions. On-line prediction gives deeper insight into future process behavior and enables, for example, the determination of an optimal scheduling strategy for harvest or transfer operations. Deviations from the expected optimal trajectory could be detected and faults could be identified easily based on such predictions. As a consequence the operators are able to reduce the economic loss due to interruption.

6.2.1.3
Quality

Reproducibility is a prerequisite for success of biotechnology in research and industrial production. Producing a product of constant high quality is one of the main goals in process automation. As a consequence of biological uncertainties, all process conditions as far as they can be influenced must be brought under control. As already mentioned, on-line simulation is a powerful tool to optimize and verify the characteristics of a particular control algorithm even in extracurricular situations. With it, there is considerable evidence that the outcoming product has the expected properties.

6.2.1.4
Validation

Reproducibility of product quality is strongly coupled with the question of validation. Validation plays an important role, especially in pharmaceutical industries, where the production process has to be in compliance with the requirements of GMP rules [5]. Integration of on-line simulation into the life cycle for automated equipment provides the chance to detect failures (e.g., electrical cabling, software bugs, calibration errors) at a very early stage. As a result, the number of validation runs could be reduced significantly.

6.2.1.5
Complexity

In the past, a lot of supervision and control tasks in industry has been done by the human operators. Today, increasing quality requirements and economic reasons have led to more complex automation structures. The current situation is characterized by a high degree of automation and the application of sophisticated control [6]. That applies to the whole automation hierarchy, beginning from the sensor-actuator system with built-in fault detection capacity [7] to the plant control system using complex fuzzy rules. The present state of efficient biotechnology processes, for example high cell density cultivation techniques, requires fully instrumented bioreactors and complex fermentation strategies. Today's plants for production of pharmaceuticals are more or less fully automated. A good example is the production of Erythropoietin using robots to assist the human workers [8]. Particularly in research and development applications with its highly flexible experimental set up, there is a need for rapid realization and verification of efficient automation methods. To get the novel structure under control, on-line simulation is inevitable. After running a "virtual fermentation" the probability of correct functional behavior is greatly increased.

6.2.1.6
Training

A fundamental field of application is in professional training of industrial staff. Due to the fact that nearly every process situation could be simulated, operators are able to grasp the behavior and influence of control actions by the technique of learning by doing. In contrast to real plants, fault conditions and accidental situations could be simulated without danger. This experience and the challenge to evolve and check new ideas for process improvement leads to a higher motivation. Moreover, the financial expenditure for development and maintenance of simulation equipment is considerably lower than for a real training plant.

6.2.2
Application in Education

The restricted financial situation is one of the strongest arguments for using on-line simulation at schools and universities. From the educational point of view, teaching modern methods using "virtual reality" is much more meaningful than obsolete hardware. Combining real-time and speed-up simulations allows for the skipping of uninteresting phases, whereas important situations could be replayed in slow-motion. Obviously, in educational concepts the emphasis is more on models for unit operations than for simulation of complete plants. Building their own automation program and verifying the strategy with on-line simulation increase students' knowledge about interaction in complex technological systems and helps to bring the theoretical solution into practice.

6.3
General Architecture of On-Line Simulation Systems

The automation infrastructure on the one hand and the user-interface on the other has formative influence on the general architecture of a typical on-line simulation system. Maximum reality could be achieved by using an independent device for bioreactor imitation and the existing automation infrastructure for process visualization and control. As shown in Fig. 6.2a, the on-line simulation is coupled to the control system by using A/D- and D/A-converters claiming the electrical plugs in the field.

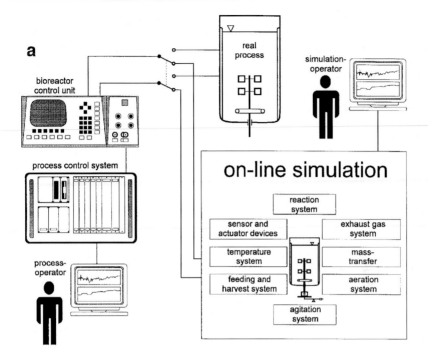

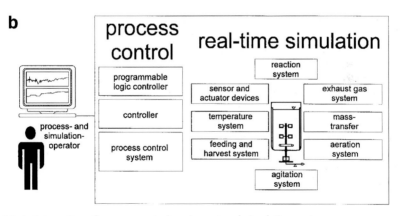

Fig. 6.2a,b. Integration of process control equipment and simulation systems

One advantage of such an architecture is the complete review of all automation re-levant devices including the established cabling and the existing user-interface for operation and visualization. Although quite expensive, such strategy is well suited for validation purposes in the pharmaceutical industries. After plant set-up, a sec-ond set of automation devices is needed for running a training-simulation and the real-world plant in parallel.

However, an integrated approach shown in Fig. 6.2b emulates the control facilities and simulate the process simultaneously using only one computer. The main drawback of this very economic solution is the lower performance in terms of reality due to software-imitation of control and visualization devices. It is up to the user to adjust the expenditure for reproducing the exact behavior of the emulated process control components. Nevertheless, a high functionality combined with low-cost hardware makes this architecture convenient for educational or small plant applications. Some of the commercial simulation systems incorporate Microsoft OLE automation technology or DDE support and offer the capabilities to build powerful hybrid programs by integrating the visualization facilities of SCADA systems like LabView or InTouch.

6.3.1
Components of Simulation Systems

In spite of conceptional differences, three components are common to all on-line simulation systems. Namely the type of model, the build in mathematical methods, and the user interface characterize a particular solution.

6.3.1.1
Models

Depending on the intended use, the mathematical description could be more or less complicated. Precise models for almost all chemical unit operations are available from the literature or application of mass- and energy conservation laws. After adjusting the parameter of such mechanistic formalism, the models coincide with the real-world application in a wide range of experimental conditions. In contrast to modeling and simulation for process optimization, the range of validity is one of the key issues in on-line applications. Especially in training and educational use the model has to be able to describe process behavior under normal and exceptional conditions. Moreover the complete process including all instrumentation and auxiliary facilities (full scope model) has to be covered by the set of equations. Fortunately, it is often sufficient to describe the dynamic behavior of peripherical devices in tendency rather than to be in exact quantitative coincidence with the reality. That is the reason for simple approximation of typical sensor devices by using linear differential equations (low-pass filter). Most often, the biological activity is included by using formal kinetic models.

6.3.1.2
Numerical Methods

From the mathematical point of view, the simulation of typical biochemical engineering models is equivalent to the problem of finding a solution for a set of ordinary differential equations (odes). Generally there is no analytic expression available, so it is necessary to approximate the solution by numerical means. Beginning at initial time and with initial conditions, the well known solver (e.g. Runge-Kutta formula) steps through the time interval, computing a solution at each time step. If the solution for a time step satisfies the solver's error tolerance criteria, it is a successful step. Otherwise, it is a failed attempt; the solver shrinks the step size and tries again

[9]. To meet the real-time requirements in on-line applications, first make sure that there is enough computing power to finish this iterative process in a defined time interval. By observation of process- and simulation-time a synchronization task is responsible for real-time, slow-motion or speed-up simulation mode. At the end of each predefined time step there has to be an interaction with the external interfaces. This could be the export of output variables (measurements, controlled variables) and the import of manipulated variables from the automation devices or operator action. In addition to ode-solver and synchronization, some optimization procedures should be implemented for identification of unknown parameters or on-line optimization application.

6.3.1.3
User Interface

To achieve maximum reality, the operation and visualization of the process under investigation should be done by using the user interface of the existing process control system. By means of this system, the user is able to redefine set points and change parameters for the control-loops. At the same time, the simulation itself has to be operated. Mainly the mode (real-time, speed-up, slow-motion, pause) or model events (sensor/actuator fault, calibration errors etc.) should be manipulated on-line. In any case, the user has be able to change or extend the build-in model by using a high level modeling language or a block oriented diagram.

6.4
Full Scope Model of the Fermentation Process

The principle of virtual processing requires an extended model going far beyond the classical description of the gas and fluid phase reaction space of the reactor.

Figure 6.3 shows a photograph of a 7.5-l bioreactor BIOSTAT ED5, for which a complete dynamic model is to be developed later.

The model, in the form of a virtual process in conjunction with the real fermentor process control system, is a useful tool in education and research.

This principle, termed on-line simulation, is used in the research field to develop and test complex processing strategies, which can then be implemented in the real reactor.

The methodology to be developed requires a description of the complete technical and biological bioreactor system. This includes the final control elements and local controllers, the processing, physicochemical and biological behavior, and the individual in-line and on-line measurement systems.

Figure 6.4 shows the required subsystems of a model for on-line simulation.

The model covers not only the engineering aspects (temperature control, mixing, aeration and gas removal, and the mass transport behavior), but also the biological reaction process (growth, substrate and oxygen uptake, CO_2- and product formation) and their interactions (pH, viscosity, foam formation, etc.).

When modeling the actual reaction process the major problems are found in the exact description of mass transport as a function of agitation speed, aeration rate and rheology, notation of the interaction of the relevant cell-specific reaction rates, and the determination of the parameters influencing pH.

Fig. 6.3. Highly instrumented 7.5-l bioreactor with integrated on-line simulation. Photo: E. Stagat, University of Applied Sciences Hamburg

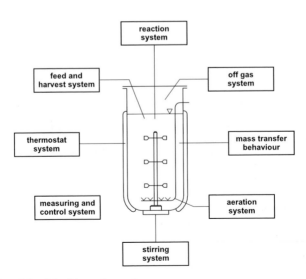

Fig. 6.4. Submodels of the bioreactor system

In addition, feed, harvest and pH titration systems, and the reactors measurement and control systems must be described.

The individual submodels are introduced in detail below and then integrated to form the aggregate model.

In contrast to classical modeling of biotechnological processes, a complete description of the process behavior is required in this case.

Thus, the general function is assured and may be checked when the real process control system is connected to the simulator.

Furthermore, the real-time and control behavior of the modeled parameters, e.g., temperature, agitation speed, aeration rate, or pH may be studied on-line.

6.5
Submodels of the Bioreactor Process

6.5.1
Engineering Components

6.5.1.1
Temperature Control System

The bioreactor temperature control system comprises the reactor liquid phase, the surrounding jacket, and the supply unit connected by piping. This in turn includes the heat exchangers for cooling (water) and heating (steam or electrical).

Figure 6.5 shows a schematic diagram of the temperature control system.

Exact modeling requires partial differential equations as both space- and time-dependent heat transfer processes take place.

The complete model will be treated with a package for the solution of odes within the scope of this contribution. Thus, a number of simplifications are assumed: ide-

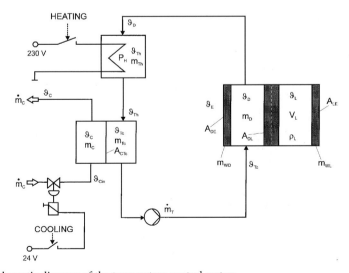

Fig. 6.5. Schematic diagram of the temperature control system

ally mixed subsystems in the temperature control circulation and no heat loss from the pipework.

The temperature control system thus comprises the submodel's thermostat heating (Th), cooling water (C), thermostat cooling (Tc), double jacket (D), and liquid phase reaction system (L). The heat capacities of the reactor walls (W) are distributed proportionately to the systems D and L.

Temperature control takes place in a pressurized closed circuit in which water is pumped with a mass flow $\dot{m}_T$ through the subsystem K, K=Th, Tc, and D. The reciprocal mean residence time D_K in the exchanger system K,

$$D_K = \frac{\dot{m}_T}{m_K} \tag{6.1}$$

where
$\dot{m}_T$ = thermostat circulation mass flow rate
m_K = exchange mass in subsystem K,

and the equivalent for the cooling water system D_C,

$$D_C(t) = \frac{\dot{m}_C(t)}{m_C} \tag{6.2}$$

where
$\dot{m}_C$ = temperature control manipulated cooling water flow rate
m_C = exchange mass in primary cooling circuit,

are defined here.

The heat transmission coefficient k_{JK},

$$k_{JK} = \frac{1}{\frac{1}{\alpha_J} + \frac{\delta_{JK}}{\lambda_{JK}} + \frac{1}{\alpha_K}} \tag{6.3}$$

where
α_I = heat transfer coefficient in subsystem I; I=J,K
δ_{JK} = wall thickness between J and K; J,K $\in$ {C,Tc,D,L,E}
λ_{JK} = thermal conductivity of the wall between J and K,

is used to describe classical heat transfer through a plane wall, which in simplification may be assumed for the reactor geometry in this case.

Thus both, the overall heat transmission resistance R_{JK},

$$R_{JK} = \frac{1}{k_{JK} \cdot A_{JK}} \tag{6.4}$$

where
A_{JK} = heat transfer area between J and K,

and the thermal transport time constant τ_{JK},

$$\tau_{JK} = R_{JK} \cdot C_J \tag{6.5}$$

where
C_J = volumetric heat capacity in system J,

are defined.

In subsystems C, Th, and Tc the volumetric thermal capacity C_J,

$$C_J = m_J \cdot c_{H2O} \tag{6.6}$$

where
c_{H2O} = specific heat capacity of water

is determined by the mass of water present therein.

The heat capacity of the reactor wall is taken into consideration by fractions allotted to the double jacket D and the liquid phase L of the reactor

$$C_D = m_D \cdot c_{H2O} + m_{WD} \cdot c_W \tag{6.7}$$

and

$$C_L(t) = \rho_L(t) \cdot V_L(t) \cdot c_{H2O} + m_{WL} \cdot c_W \tag{6.8}$$

where
m_{WI} = mass of reactor wall fraction in subsystem I; I=D,L
c_W = specific heat capacity of the wall
ρ_L = reactors liquid phase density
V_L = liquid volume in reactor.

Thus, the odes of the temperatures in the thermostat system are as follows.

The thermostat heating system Th,

$$\dot{\vartheta}_{Th}(t) = -D_{Th} \cdot \vartheta_{Th}(t) + D_{Th} \cdot \vartheta_D(t) + \frac{P_H(t)}{C_{Th}} \tag{6.9}$$

where
ϑ_{Th} = temperature in thermostat heating system
P_H = required electrical heating power
ϑ_D = temperature in double jacket,

contains only convective heat flux and has the pulse-modulated heating power P_H of the temperature controller as manipulating variable.

The cooling heat exchanger has on the primary side the cooling water system C,

$$\dot{\vartheta}_C(t) = -\left[D_C(t) + \frac{1}{\tau_{CTc}}\right] \cdot \vartheta_C(t) + D_C(t) \cdot \vartheta_{Cin}(t) + \frac{\vartheta_{Tc}(t)}{\tau_{CTc}} \tag{6.10}$$

where
ϑ_C = temperature of the cooling water system
ϑ_{Cin} = temperature of the cooling water feed,

and on the secondary side the thermostat cooling system Tc,

$$\dot{\vartheta}_{Tc}(t) = -\left[D_{Tc} + \frac{1}{\tau_{TcC}}\right] \cdot \vartheta_{Tc}(t) + D_{Tc} \cdot \vartheta_{Th}(t) + \frac{\vartheta_C(t)}{\tau_{TcC}} \tag{6.11}$$

where
ϑ_{Tc} = temperature of thermostat cooling system,

which is connected convectively to the heating system.

The double jacket system D,

$$\dot{\vartheta}_D(t) = -\left[D_D + \frac{1}{\tau_{DL}} + \frac{1}{\tau_{DE}}\right] \cdot \vartheta_D(t) + D_D \cdot \vartheta_{Tc}(t) + \frac{\vartheta_L(t)}{\tau_{DL}} + \frac{\vartheta_E(t)}{\tau_{DE}} \tag{6.12}$$

where
ϑ_L = temperature of liquid phase
ϑ_E = temperature of reactor environment,

is connected convectively to the cooling system and is involved in heat exchange with the liquid phase L and the reactor environment E.

The liquid phase of the reactor, the reaction volume L,

$$\dot{\vartheta}_L(t) = -\left[\frac{1}{\tau_{LD}(t)} + \frac{1}{\tau_{LE}(t)}\right] \cdot \vartheta_L(t) + \frac{\vartheta_L(t)}{\tau_{LD}(t)} + \frac{\vartheta_E(t)}{\tau_{LE}(t)} + \frac{\dot{Q}_{St}(t) + \dot{Q}_M(t)}{C_L(t)} \qquad (6.13)$$

where
$\dot{Q}_{St}$ thermal power of the stirrer
$\dot{Q}_M$ thermal power of the microorganisms,

is involved in heat exchange with the double jacket D and surroundings E via the reactor lid and base.

The liquid phase receives heat generated by the stirrer $\dot{Q}_{St}$,

$$\dot{Q}_{St}(t) = K_{HSt} \cdot V_L(t) \cdot N_{St}^3(t) \qquad (6.14)$$

where
K_{HSt} = proportional gain of stirrer heat generation
N_{St} = agitation speed,

and heat produced by the microorganisms $\dot{Q}_M$,

$$\dot{Q}_M(t) = K_{HM} \cdot V_L(t) \cdot OUR(t) \qquad (6.15)$$

where
K_{HM} = proportional gain of microbial heat generation
OUR = microbial oxygen uptake rate.

The electrical circuit representation in Fig. 6.6 shows the connection of the reactor thermostat model to the control hardware of the (laboratory) system of Fig. 6.3.

The cooling water entry heat resistance R_C,

$$R_C(t) = \frac{1}{\dot{m}_C(t) \cdot c_{H2O}}, \qquad (6.16)$$

describes the convective primary cooling water flow, whereas the thermostat circulation heat resistance R_T,

$$R_T = \frac{1}{\dot{m}_T \cdot c_{H2O}}, \qquad (6.17)$$

accounts for the convective heat flux in the thermostat system.

There is no feedback from the convective streams which are thus depicted as a trap amplifier.

The jacket entry temperature ϑ_{Tc} and the liquid phase temperature ϑ_L are fed as actual values to a cascade control system. The slave controller output pulses the signal COOLING to the cooling water valve, described by the connection of resistance R_C, or the signal HEATING to switch on the electric heating power.

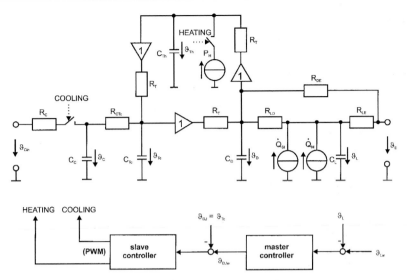

Fig. 6.6. Electrical circuit representing the temperature control system

6.5.1.2
Pressure Behavior

The pressure in the liquid phase p_L and gas phase p_G are equal for the reactor described here.

During cultivation ($\vartheta_L < 100\,^\circ$C) they correspond to the set point p_{Gw} of an ideal pressure controller according to

$$p_L(t) = p_G(t) = p_{Gw}(t).$$ (6.18)

During sterilization ($\vartheta_L > 100\,^\circ$C) both air inlet and outlet are closed.
In this case, fermentor steam pressure, given by

$$p_G(t) = p_L(t) = K_{pL1} \cdot 10^{\left[K_{pL2} - \frac{K_{pL3}}{T_L(t)} - K_{pL4} \cdot \log\left(\frac{T_L(t)}{T_{nG}}\right) \right]}$$ (6.19)

with interpolation constants according to Duprès-Rankine,
$K_{pL1} = 9.807\ \text{N/m}^2$
$K_{pL2} = 10.9$
$K_{pL3} = 2461\ \text{K}$
$K_{pL4} = 2.065,$

is controlled using the liquid phase temperature T_L,

$$T_L(t) = \vartheta_L(t) + T_{nG}$$ (6.20)

where
$T_{nG} = 273.15\ \text{K}$, (normalized conditions).

6.5.1.3
Aeration Behavior

A well-equipped bioreactor has a gas-mixing station with four massflow controllers for air, O_2, N_2 and CO_2.

These are controlled via the corresponding set points F_{nIw}. Assuming ideal control the corresponding gas throughput at normalized conditions is obtained with

$$F_{nI}(t) = F_{nIw}(t) \tag{6.21}$$

for I = AIR, O2, N2, and CO2.

Hence, total aeration rate F_{nG} is obtained,

$$F_{nG}(t) = F_{nAIR}(t) + F_{nO2}(t) + F_{nN2}(t) + F_{nCO2}(t), \tag{6.22}$$

again referring to normalized conditions.

Together with the O_2 mass fraction x_{OGin},

$$x_{OGin}(t) = \frac{F_{nO2}(t) + x_{OAIR} \cdot F_{nAIR}(t)}{F_{nG}(t)} \tag{6.23}$$

where
$x_{OAIR} = 0.2094$ O_2 mass fraction of air,

and the CO_2 mass fraction x_{CGin},

$$x_{CGin}(t) = \frac{F_{nCO2}(t) + x_{CAIR} \cdot F_{nAIR}(t)}{F_{nG}(t)} \tag{6.24}$$

where
$x_{CAIR} = 0.0003$, CO_2 mass fraction of air,

all conditions being defined at the gas entry point.

6.6
Mass Balances of the Complete Aerobic Growth Process

The model to be developed refers to the description of a growth process of *Escherichia coli* (cell mass X) on the substrates glucose (S1) and glycerol (S2). The cells consume oxygen (O) and generate carbon dioxide (C).

The cell mass is initially viewed as a dissolved component.

In addition, the production of acetate (P1) and its reuse (S3) is taken into account.

In order to simulate the whole process the pH system with the components buffer acid (B1), buffer base (B2), added acid (Ac), and added base (Al) is required. This requires not only mass balancing of the liquid phase reaction space (L) but also of the acid (T1) and base (T2) feed vessels. The substrate feed vessel (R) must also be included in the model, as substrate is fed during the process.

Since O_2 supply and CO_2 removal is realized via a flowing gas phase, both these components must be balanced over this phase.

The general mass balances of the process form the starting point of modeling [10]. This is done over the macroscopic volume element of an aerated biphasic system shown in Fig. 6.7.

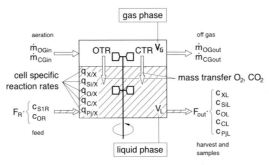

Fig. 6.7. Macroscopic volume element used to develop the mathematical model of an aerobic growth process

It will be assumed that the reactor may be described as an ideally mixed vessel, i.e., exit stream component fractions correspond to the values in the related subsystems.

6.6.1
Gas Phase Balances

The oxygen and carbon dioxide mass balances are the most common on-line tools for the observation of cell activity of aerobic processes.

By introducing an inert balance for N_2 and assuming ideal gas cooling we have the basis for a global gas phase mass balance.

The oxygen supply rate Q_{O2},

$$Q_{O2}(t) = \frac{\dot{m}_{OGin}(t) - \dot{m}_{OGout}(t)}{V_L(t)} =$$
$$\frac{F_{nG}(t) \cdot M_{O2}}{V_{nM} \cdot V_L(t)} \cdot \frac{x_{OGin}(t) \cdot [1 - x_{CGout}(t)] - x_{OGout}(t) \cdot [1 - x_{CGin}(t)]}{1 - x_{OGout}(t) - x_{CGout}(t)} \qquad (6.25)$$

where
$\dot{m}_{OGin/out}$ = O_2 mass flow at gas phase entry and exit, respectively
F_{nG} = aeration rate at normalized conditions
M_{O2} = mole mass of O_2
V_{nM} = gas mole volume
V_L = liquid volume
$x_{IGin/out}$ = mass fraction of component I; I=O(O_2),C(CO_2),

is calculated with respect to the reaction volume V_L, as is the CO_2 removal rate Q_{CO2},

$$Q_{CO2}(t) = \frac{\dot{m}_{CGout}(t) - \dot{m}_{CGin}(t)}{V_L(t)} =$$
$$\frac{F_{nG}(t) \cdot M_{CO2}}{V_{nM} \cdot V_L(t)} \cdot \frac{x_{CGout}(t) \cdot [1 - x_{OGin}(t)] - x_{CGin}(t) \cdot [1 - x_{OGout}(t)]}{1 - x_{OGout}(t) - x_{CGout}(t)} \qquad (6.26)$$

where
$\dot{m}_{CGin/out}$ = CO_2 mass flow at gas phase entry and exit, respectively
M_{CO2} = mole mass of CO_2.

By introducing the molar respiration quotient RQ,

$$RQ(t) = \frac{Q_{CO2}(t) \cdot M_{O2}}{M_{CO2} \cdot Q_{O2}(t)}, \tag{6.27}$$

and a rating for the gas supply with the maximum oxygen supply rate Q_{O2max} which may be achieved at equal aeration rate with pure oxygen,

$$Q_{O2max}(t) = \frac{F_{nG}(t) \cdot M_{O2}}{V_{nM} \cdot V_L(t)}, \tag{6.28}$$

one obtains, in conjunction with Eq. (6.25), a clear representation of the O_2 supply rate Q_{O2},

$$Q_{O2}(t) = Q_{O2max}(t) \cdot \frac{x_{OGin}(t) - x_{OGout}(t)}{1 - [1 - RQ(t)] \cdot x_{OGout}(t)}. \tag{6.29}$$

In the special case of RQ$\equiv$1 the O_2 and CO_2 balances become decoupled with

$$Q_{O2}(t) = Q_{O2max}(t) \cdot [x_{OGin}(t) - x_{OGout}(t)]. \tag{6.30}$$

In this case the difference in mass fractions,

$$x_{CGout}(t) - x_{CGin}(t) = x_{OGin}(t) - x_{OGout}(t), \tag{6.31}$$

is the same for O_2 and CO_2, and both Eq. (6.25) and Eq. (6.26) decoupled.

Balancing the gas volume in contact with the liquid phase shown in Fig. 6.7 leads to the gas phase O_2 mass balance,

$$\frac{d}{dt}(m_{OG}(t)) = V_L(t) \cdot [Q_{O2}(t) - OTR(t)] \tag{6.32}$$

where
m_{OG} = O_2 mass in the reactor gas phase
OTR = O_2 transfer rate between gas and liquid phases,

and to the gas phase CO_2 mass balance,

$$\frac{d}{dt}(m_{CG}(t)) = V_L(t) \cdot [-Q_{CO2}(t) - CTR(t)] \tag{6.33}$$

where
m_{CG} = CO_2 mass in the reactor gas phase
CTR = CO_2 transfer rate between gas and liquid phases.

An exact solution of the odes (Eqs. 6.32 and 6.33) requires knowledge of the gas volume V_G. Since the accumulation terms may generally be neglected, only the quasi-steady state terms,

$$OTR(t) = Q_{O2}(t) \tag{6.34}$$

and

$$CTR(t) = -Q_{CO2}(t) \tag{6.35}$$

are used.

Thus, both transfer rates may be measured via the exit gas balances.

6.6.2
The O_2- and CO_2-Transfer Equations

The descriptions of the O_2 supply to and CO_2 removal from the liquid phase are significantly different.

The oxygen transfer rate OTR,

$$OTR(t) = k_La(t) \cdot [c_{OL}^*(t) - c_{OL}(t)] \tag{6.36}$$

where

k_La = volumetric O_2 transfer coefficient
c_{OL}^* = O_2 equilibrium concentration with the gas phase
c_{OL} = O_2 concentration in liquid bulk

is coupled to the gas phase via the equilibrium concentration c_{OL}^*,

$$c_{OL}^*(t) = \frac{p_{OG}(t)}{H_{O2}(t)} = \frac{p_G(t)}{H_{O2}(t)} \cdot x_{OG}(t) \tag{6.37}$$

where

p_{OG} = oxygen gas phase partial pressure
H_{O2} = O_2 Henry coefficient
p_G = total gas phase pressure
x_{OG} = O_2 mass fraction in the gas phase.

By defining two O_2 transfer parameters, the maximum O_2 saturation concentration,

$$c_{OLmax}(t) = \frac{p_G(t)}{H_{O2}(t)}, \tag{6.38}$$

and the theoretical maximum O_2 transfer rate,

$$OTR_{max}(t) = k_La(t) \cdot c_{OLmax}(t), \tag{6.39}$$

and substituting Eq. (6.29) and Eq. (6.36) in Eq. (6.34) with $c_{OL} \to 0$ the only useful O_2 transfer scale-up criterion is obtained, the oxygen transfer capacity OTC,

$$OTC(t) = \frac{Q_{O2max}(t) \cdot OTR_{max}(t) \cdot x_{OGin}(t)}{Q_{O2max}(t) + OTR_{max}(t)} \cdot \frac{2}{1 + \sqrt{1 - \frac{4 \cdot [1 - RQ(t)] \cdot Q_{O2max}(t) \cdot OTR_{max}(t) \cdot x_{OGin}(t)}{(Q_{O2max}(t) + OTR_{max}(t))^2}}} \cdot \tag{6.40}$$

The same k_La value is used to describe the carbon dioxide transfer rate CTR,

$$CTR(t) = \delta_{c/o} \cdot k_La(t) \cdot [c_{CL}^*(t) - c_{CL}(t)] \tag{6.41}$$

where

$\delta_{C/O}$ = ratio of CO_2/O_2 transfer coefficients
c_{CL}^* = CO_2 equilibrium concentration with the gas phase
c_{CL} = non-dissociated fraction of the CO_2 concentration in the liquid phase,

as used for O_2 transfer (Eq. 6.36).

The gas phase driving force c_{CL}^*,

$$c_{CL}^*(t) = \frac{p_{CG}(t)}{H_{CO2}(t)} = \frac{p_G(t)}{H_{CO2}(t)} \cdot x_{CG}(t) \qquad (6.42)$$

where
H_{CO2} = Henry coefficient for CO_2
x_{CG} = mass fraction of CO_2 in the gas phase,

is derived from Henry's law.
    The driving CO_2 concentration in the liquid phase c_{CL},

$$c_{CL}(t) = \frac{(C_{HL}^+(t))^2 \cdot M_{CO2}}{(C_{HL}^+(t))^2 + K_{C1} \cdot C_{HL}^+(t) + K_{C1} \cdot K_{C2}} \cdot C_{CLtot}(t) \qquad (6.43)$$

where
C_{HL}^+ = molar H^+ concentration in the liquid phase
C_{CLtot} = molar CO_2 total concentration in the liquid phase
K_{Cj} = dissociation constant of CO_2; j=1,2,

is obtained solely from the non-dissociated fraction of the molar CO_2 total concentration C_{CLtot} which is composed of CO_2, carbonate, and bicarbonate.
    The molar hydrogen ion concentration C_{HL}^+ is calculated from the pH model described later.

6.6.3
The k_La Correlation

The literature reports numerous k_La correlations based on aeration rate, power input from the stirrer, and the viscosity of the fermentation broth [11],

$$k_La(t) = C_{kLa} \cdot \left[\frac{P_{St}(t)}{V_L(t)}\right]^\alpha \cdot (u_G(t))^\beta \cdot (\eta_{eff}(t))^\gamma \qquad (6.44)$$

where
C_{kLa} = fitted constant
P_{St} = stirrer power input
u_G = superficial gas velocity
η_{eff} = effective viscosity
α,β,γ = correlation parameters.

For the on-line simulation the k_La value must in some way be linked to the agitation speed, aeration rate, and viscosity.
    In the following the stirrer power input P_{St},

$$P_{St}(t) = Ne(t) \cdot \rho_L(t) \cdot d_2^5 \cdot N_{St}^3(t) \qquad (6.45)$$

where
Ne = Newton number
d_2 = stirrer diameter
N_{St} = stirrer speed,

will be substituted by the stirrer speed N_{St}, and the gas superficial velocity u_G,

$$u_G(t) = \frac{p_{nG} \cdot T_{Gin}}{p_{Gin} \cdot T_{nG} \cdot A_V} \cdot F_{nG}(t) \qquad (6.46)$$

where
T_{nG}, p_{nG} = state of the gas phase at normalized conditions
T_{Gin}, p_{Gin} = state of the gas phase at reactor entry
A_V = vessel cross-sectional area,

by the aeration rate F_{nG}.

Assuming the simplification of constant Ne, ρ_L, η_{eff}, a simple $k_L a$ correlation may be derived,

$$k_L a(t) = k_L a_{max} \cdot \left[\frac{N_{St}(t)}{N_{Stmax}}\right]^{3\alpha} \cdot \left[\frac{F_{nG}(t)}{F_{nGmax}}\right]^{\beta} \cdot \left[\frac{V_{Lmin}}{V_L(t)}\right]^{-\alpha} \qquad (6.47)$$

where
N_{Stmax} = maximum stirrer speed
F_{nGmax} = maximum aeration rate
V_{Lmin} = minimum working volume
$k_L a_{max}$ = maximum $k_L a$ value.

The minimum working volume V_{Lmin} is that volume which, under maximum aeration rate F_{nGmax} and maximum stirrer speed N_{Stmax}, does not result in stirrer flooding. In this case the maximum $k_L a$ value is obtained.

6.6.4
The Liquid Phase Balances

In order to describe the reaction process with the three modes of operation – batch, fed batch, and continuous – a mass balance on the liquid volume V_L is required,

$$\dot{V}_L(t) = F_{in}(t) - F_{out}(t), \qquad (6.48)$$

in which it is presumed that the material properties in the entry streams and in the reactor are the same.

The volumetric flow into the reactor F_{in},

$$F_{in}(t) = F_R(t) + F_{T1}(t) + F_{T2}(t) - F_V(t) \qquad (6.49)$$

where
F_R = substrate feed rate
F_{T1} = acid titration rate during pH control
F_{T2} = base titration rate during pH control
F_V = water evaporation rate due to aeration,

takes into account medium components such as feed and the loss due to evaporation.

The volumetric flow from the reactor F_{out},

$$F_{out}(t) = F_H(t) + F_S(t) \qquad (6.50)$$

where
F_H = harvest or transfer rate
F_S = sampling rate,

describes the loss of medium from the reactor.

Hence, the general mass balance of component I in the liquid phase L may be given with respect to the component's (mass) concentration c_{IL},

$$\dot{c}_{IL}(t) = \frac{F_R(t)}{V_L(t)} \cdot c_{IR}(t) - \frac{F_{in}(t)}{V_L(t)} \cdot c_{IL}(t) + r_{IL}(t) + ITR(t) \tag{6.51}$$

where

c_{IR} = concentration of I in the substrate vessel
r_{IL} = volumetric reaction rate of component I
ITR = transfer rate of component I from the gas phase.

The volumetric reaction rates r_{IL},

$$r_{IL}(t) = \pm q_{I/X}(t) \cdot c_{XL}(t) \tag{6.52}$$

where

$q_{I/X}$ = cell-specific reaction rate of component I
c_{XL} = dry biomass concentration in the liquid phase,

will be replaced in the following by the cell-specific activities and the cell concentration, whereby the cell-specific reaction rate μ is termed $q_{X/X}$.

The following balance equations are stated for dissociable components with molar concentrations C_I.

The cell mass balance (dry biomass X),

$$\dot{c}_{XL}(t) + \frac{F_{in}(t)}{V_L(t)} \cdot c_{XL}(t) = q_{X/X}(t) \cdot c_{XL}(t), \tag{6.53}$$

the glucose balance (substrate S1),

$$\dot{c}_{S1L}(t) + \frac{F_{in}(t)}{V_L(t)} \cdot c_{S1L}(t) = \frac{F_R(t)}{V_L(t)} \cdot c_{S1R}(t) - q_{S1/X}(t) \cdot c_{XL}(t), \tag{6.54}$$

the glycerol balance (substrate S2),

$$\dot{c}_{S2L}(t) + \frac{F_{in}(t)}{V_L(t)} \cdot c_{S2L}(t) = -q_{S2/X}(t) \cdot c_{XL}(t), \tag{6.55}$$

the acetate balance (product P1 and substrate S3),

$$\dot{C}_{P1Ltot}(t) + \frac{F_{in}(t)}{V_L(t)} \cdot C_{P1Ltot}(t) = \left[\frac{q_{P1/X}(t) - q_{S3/X}(t)}{M_{P1}}\right] \cdot c_{XL}(t), \tag{6.56}$$

the dissolved oxygen balance (O),

$$\dot{c}_{OL}(t) + \frac{F_{in}(t)}{V_L(t)} \cdot c_{OL}(t) = \frac{F_R(t) \cdot c_{OR} + F_{T1}(t) \cdot c_{OT1} + F_{T2}(t) \cdot c_{OT2}}{V_L(t)} +$$
$$+ OTR(t) - q_{O/X}(t) \cdot c_{XL}(t), \tag{6.57}$$

the dissolved carbon dioxide balance (C),

$$\dot{C}_{CLtot}(t) + \frac{F_{in}(t)}{V_L(t)} \cdot C_{CLtot}(t) = \frac{F_R(t) \cdot C_{ORtot} + F_{T1}(t) \cdot C_{CT1tot} + F_{T2}(t) \cdot C_{CT2tot}}{V_L(t)} +$$
$$+ \frac{CTR(t) + q_{C/X}(t) \cdot c_{XL}(t)}{M_{CO2}}, \tag{6.58}$$

and the ammonia balance (pH base Al and nitrogen source),

$$\dot{C}_{AlLtot}(t) + \frac{F_{in}(t)}{V_L(t)} \cdot C_{AlLtot}(t) = \frac{F_{T2}(t)}{V_L(t)} \cdot C_{AlT2tot} + \frac{AlTR(t) - q_{Al/X}(t) \cdot c_{XL}(t)}{M_{Al}}, \quad (6.59)$$

comprise the reaction model.

The pH is controlled by addition of alkali component ammonia at the titration rate F_{T2}, fed from vessel T2.

The loss of ammonia by evaporation,

$$AlTR(t) = -\frac{K_{Alvol} \cdot F_{nG}(t)}{V_L(t)} \cdot \frac{K_W \cdot C_{AlLtot}(t) \cdot M_{Al}}{K_{Al} \cdot C_{HL}^+(t) + K_W} \quad (6.60)$$

where

K_{Alvol} = evaporation constant of ammonia
K_W = ion product of water
K_{Al} = dissociation constant of ammonia

is given as analogue to Raoult's Law for the non-dissociated fraction.

A model of the pH behavior requires the description of the pH buffer and added acid in addition to the correction by the alkali component ammonia.

In the case example the pH will be corrected by phosphoric acid H_2PO_4.

The phosphoric acid balance (pH acid Ac),

$$\dot{C}_{AcLtot}(t) + \frac{F_{in}(t)}{V_L(t)} \cdot C_{AcLtot}(t) = \frac{F_{T1}(t)}{V_L(t)} \cdot C_{AcT1tot}, \quad (6.61)$$

contains only the pH controller correction with the acid titration rate F_{T1}.

The buffer chosen comprised potassium dihydrogen phosphate KH_2PO_4 (B1) and potassium hydrogen phosphate K_2HPO_4 (B2).

Both components are diluted during feeding.

Thus, two purely hydrodynamic balances for the buffer acid (j=1) and base (j=2) are required,

$$\dot{C}_{BjLtot}(t) + \frac{F_{in}(t)}{V_L(t)} \cdot C_{BjLtot}(t) = 0. \quad (6.62)$$

6.6.5
The Feed and Titration Vessels System

Measurement of the weights of the substrate, acid, and base feed vessels can be used to obtain information on the integral activity of the cells.

A description of the whole bioreactor thus requires a mass balance of the substrate feed vessel,

$$\dot{m}_R(t) = -\rho_R \cdot F_R(t) \quad (6.63)$$

the acid feed vessel,

$$\dot{m}_{T1}(t) = -\rho_{T1} \cdot F_{T1}(t) \quad (6.64)$$

and the base feed vessel,

$$\dot{m}_{T2}(t) = -\rho_{T2} \cdot F_{T2}(t) \quad (6.65)$$

where in each case
ρ_K = density of the medium in vessel K.

6.7
The pH Model

The mass balances mentioned form the prerequisite basis for modeling the pH value which is influenced by the solution of Eqs. (6.56), (6.58), (6.59), (6.61), and (6.62).

By definition the pH value is a dimensionless measure of the number of hydroxonium hydrogen ions (H_3O^+) present in the electrolyte solution,

$$pH(t) = -\log_{10} C_{HL}^+(t) \qquad (6.66)$$

which may be found in various hydration forms and are given as a mass concentration C_{HL}^+ in $mol\,l^{-1}$.

The starting point of a pH model is defined by the ionic product of water,

$$C_{HL}^+(t) \cdot C_{OHL}^-(t) = K_W = 10^{-14}\ mol^2\,l^{-2}, \qquad (6.67)$$

and the electroneutrality condition which states that the sum of the product of concentration and ionic valency must be equal for positive and negative ions [12, 13].

From this one derives the basic pH equation,

$$C_{OHL}^-(t) + \sum_i p_i \cdot C_{iL}^{p_i-}(t) = C_{HL}^+(t) + \sum_j q_j \cdot C_{jL}^{q_j+}(t) \qquad (6.68)$$

in which
i   = an arbitrary anion (except OH^-)
j   = an arbitrary cation (except H_3O^+)
p_i, q_j = valency of the anion i or cation j
C_{iL}, C_{jL} = molar concentration of anion i or cation j.

Substituting Eq. (6.66) in Eq. (6.68) and setting

$$x(t) = C_{HL}^+(t) \qquad (6.69)$$

one obtains the solution polynomial,

$$f(x(t)) = x^2(t) + \left[\sum_j q_j \cdot C_{jL}^{q_j+}(t) - \sum_i p_i \cdot C_{iL}^{p_i-}(t) \right] \cdot x(t) - K_W = 0, \qquad (6.70)$$

for the unknown hydrogen ion concentration C_{HL}^+.

The modeling objective is to find all ions in the feed medium, to determine the degree of dissociation for them, and to describe the changes to processes due to cell activity or actions taken externally.

In the present example the pH is influenced by the following:
1. By the pH buffer mixture of potassium dihydrogen phosphate (KH_2PO_4), C_{B1Ltot}, and potassium hydrogen phosphate (K_2HPO_4), C_{B2Ltot}.
   These two salts of phosphoric acid dissociate completely, resulting in three positive potassium ions,

$$C_{BL}^+(t) = C_{B1Ltot}(t) + 2 \cdot C_{B2Ltot}(t). \qquad (6.71)$$

The liberated phosphoric acid,

$$C_{BLtot}(t) = C_{B1Ltot}(t) + C_{B2Ltot}(t) \tag{6.72}$$

dissociates in three stages with constants K_{B1}, K_{B2}, and K_{B3}, resulting in three ionic fractions.

2. By acetate (CH_3COOH), a metabolic product, C_{P1Ltot}.
   Acetic acid protolyses to a single ionic charge with a dissociation constant K_{P1}.
3. By carbon dioxide (CO_2), a metabolic product, which may enter the system via the aeration system or exit via the exhaust, C_{CLtot}.
   Carbon dioxide can give up two protons in aqueous solution as carbonic acid.
4. By the ammonia (NH_3) added for pH control and as a nitrogen source, C_{AlLtot}.
   Ammonia forms ammonium cations (NH_4^+).
5. By the phosphoric acid (H_2PO_4) added for pH control, C_{AcLtot}.
   In the same way as the buffer, three ionic fractions are obtained.

After substituting the variables involved for all dissociation stages in Eq. (6.70) a solution is obtained for the protons C_{HL}^+ determining pH,

$$
\begin{aligned}
f(x(t)) = x^2(t) + \Bigg\{ &\left[C_{B1Ltot}(t) + 2 \cdot C_{B2Ltot}(t) + \frac{K_{Al} \cdot x(t)}{K_{Al} \cdot x(t) + K_W} \cdot C_{AlLtot}(t) \right] \\
&- \left[\frac{K_{P1}}{x(t) + K_{P1}} \cdot C_{P1Ltot}(t) + \frac{K_{C1} \cdot x(t) + 2 \cdot K_{C1} \cdot K_{C2}}{x^2(t) + K_{C1} \cdot x(t) + K_{C1} \cdot K_{C2}} \cdot C_{CLtot}(t) \right. \\
&+ \frac{K_{B1} \cdot x^2(t) + 2 \cdot K_{B1} \cdot K_{B2} \cdot x(t) + 3 \cdot K_{B1} \cdot K_{B2} \cdot K_{B3}}{x^3(t) + K_{B1} \cdot x^2(t) + K_{B1} \cdot K_{B2} \cdot x(t) + K_{B1} \cdot K_{B2} \cdot K_{B3}} \cdot \\
&\left. \cdot (C_{B1Ltot}(t) + C_{B2Ltot}(t) + C_{AcLtot}(t)) \right] \Bigg\} \cdot x(t) - K_W = 0. \tag{6.73}
\end{aligned}
$$

6.8
The Reaction Model

The reaction model chosen is based on the bottle-neck principle in which a single enzymatic catalysis determines the reaction [14].

The cell-specific growth rate, $q_{X/X}$,

$$q_{X/X}(t) = \mu(t) = \min\{\mu_{Sgr}(t), \mu_{Ogr}(t)\} - q_{X/Xm} \le \mu_{1max} \tag{6.74}$$

where

μ	= cell-specific growth rate observed
μ_{1max}	= maximum possible cell-specific growth rate on glucose
μ_{Sgr}	= substrate-determined cell-specific growth rate
μ_{Ogr}	= O_2-determined cell-specific growth rate
$q_{X/Xm}$	= cell-specific (death) maintenance rate

is controlled by the rate-determining step in the substrate or O_2 supply to the cells and is limited by the maximum specific growth rate on glucose, μ_{1max}.

In order to avoid zero order reactions in the substrate mass balances, the uptake for maintenance purposes is substituted by $q_{X/Xm}$ in the cell mass balance equation. Thus, the cell obtains its maintenance energy from storage substances.

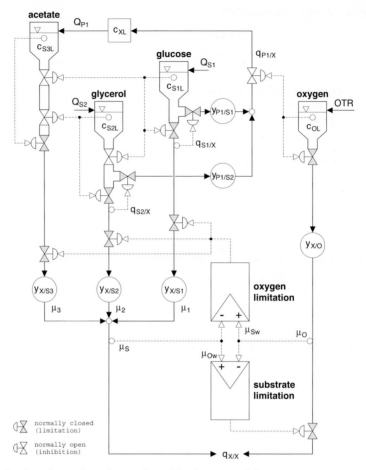

Fig. 6.8. Bottle-neck reaction scheme of a multi-substrate model

Figure 6.8 shows the multisubstrate reaction scheme described in a simplified form.

With adequate O_2 supply the cell-specific uptake rate of the preferred substrate 1, glucose,

$$q_{S1/Xopt}(t) = q_{S1/Xmax}(t) \cdot \frac{c_{S1L}(t)}{k_{S1} + c_{S1L}(t)} \tag{6.75}$$

where

$q_{S1/Xopt}$ = optimum cell-specific glucose uptake rate
$q_{S1/Xmax}$ = maximum cell-specific glucose uptake rate
k_{S1} = glucose limitation constant

is only limited by the associated concentration, c_{S1L}. This is depicted in the bottle-neck as a spring-loaded valve (normally closed).

Glucose inhibits glycerol uptake,

$$q_{S2/Xopt}(t) = q_{S2/Xmax}(t) \cdot \frac{c_{S2L}(t)}{k_{S2} + c_{S2L}(t)} \cdot \frac{k_{I21}}{k_{I21} + c_{S1L}(t)} \qquad (6.76)$$

where

$q_{S2/Xopt}$ = optimum cell-specific glycerol uptake rate
$q_{S2/Xmax}$ = maximum cell-specific glycerol uptake rate
k_{S2} = glycerol limitation constant
k_{I21} = inhibition constant for glycerol uptake (S2) due to glucose (S1),

and both these inhibit the uptake of acetate,

$$q_{S3/Xopt}(t) = q_{S3/Xmax}(t) \cdot \frac{c_{S3L}(t)}{k_{S3} + c_{S3L}(t)} \cdot \prod_{j=1}^{2} \frac{k_{I3j}}{k_{I3j} + c_{SjL}(t)} \qquad (6.77)$$

where

$q_{S3/Xopt}$ = optimum acetate uptake rate
$q_{S3/Xmax}$ = maximum cell-specific acetate uptake rate
k_{S3} = acetate limitation constant
k_{I3j} = inhibition constant for acetate uptake due to glucose (j=1) and glycerol (j=2).

The inhibition mechanism is symbolized by the characteristic curve of a spring-loaded (normally open) valve in Fig. 6.8.

The respective maximum cell-specific uptake rate, $q_{Si/Xmax}$,

$$q_{Si/Xmax}(t) = \frac{\mu_{imax}(t)}{y_{X/Sigr}} + q_{Si/Xm} \qquad (6.78)$$

where

$y_{X/Sigr}$ = cell mass growth yield for substrate i
$q_{Si/Xm}$ = cell-specific maintenance rate for substrate i,

contains the pH and temperature dependent maximum cell-specific growth rate on substrate i [15],

$$\mu_{imax}(t) = \mu_{iopt} \cdot \kappa_T(t) \cdot \kappa_{pH}(t) \qquad (6.79)$$

where

μ_{iopt} = optimum specific growth rate on substrate i
κ_l = environmental growth control function of parameter l, l=ϑ_L, pH

In the temperature range $\vartheta_L \in [\vartheta_{Lmin}, \vartheta_{Lmax}]$ the growth rate is controlled by

$$\kappa_T(t) = \frac{(\vartheta_L(t) - \vartheta_{Lmin})^2 \cdot (\vartheta_L(t) - \vartheta_{Lmax})}{(\vartheta_{Lopt} - \vartheta_{Lmin}) \cdot [(\vartheta_{Lopt} - \vartheta_{Lmin}) \cdot (\vartheta_L(t) - \vartheta_{Lopt}) - (\vartheta_{Lopt} - \vartheta_{Lmax}) \cdot (\vartheta_{Lopt} + \vartheta_{Lmin} - 2 \cdot \vartheta_L(t))]}$$

$$(6.80)$$

where

ϑ_{Lmin} = minimum temperature boundary of growth
ϑ_{Lmax} = maximum temperature boundary of growth
ϑ_{Lopt} = temperature of optimal growth.

Similar behavior of growth is observed in a pH-range of $[pH_{min}, pH_{max}]$,

$$\kappa_{pH}(t) = \frac{(pH(t) - pH_{min}) \cdot (pH(t) - pH_{max})}{(pH(t) - pH_{min}) \cdot (pH(t) - pH_{max}) - (pH(t) - pH_{opt})^2} \tag{6.81}$$

where
pH_{min} = minimum pH boundary of growth
pH_{max} = maximum pH boundary of growth
pH_{opt} = pH-value of optimal growth.

Both the temperature control function κ_T and the pH control function κ_{pH} are zero outside their defined ranges.

Due to considerations of consistency, the cell-specific maintenance rate, $q_{X/Xm}$, must be of equal magnitude for all substrates,

$$q_{X/Xm} = y_{X/Sigr} \cdot q_{Si/Xm}, i = 1, 2, 3. \tag{6.82}$$

In order to maximize its metabolic efficiency, the cell will take up all three carbon sources simultaneously (mathematically) when there is adequate O_2 supply,

$$\mu_{Sgr}(t) = \sum_{i=1}^{3} y_{X/Sigr} \cdot q_{Si/Xopt}(t) \leq \mu_{1max}(t) + q_{X/Xm}. \tag{6.83}$$

In reality, due to catabolite repression, the substrates are taken up in sequence. This is brought about by the inhibition terms.

The supply of oxygen is the second bottle-neck in Eq. (6.74).

The cell-specific O_2 uptake rate, $q_{O/X}$,

$$q_{O/X}(t) = q_{O/Xgr}(t) + q_{O/Xm}, \tag{6.84}$$

consists of a growth fraction, $q_{O/Xgr}$, i.e., the energy for substrate uptake, and a fraction for conversion of maintenance storage substances, $q_{O/Xm}$.

In the case of an O_2 bottle-neck the possible growth fraction, μ_{Ogr},

$$\begin{aligned} \mu_{Ogr}(t) &= y_{X/Ogr} \cdot q_{O/Xgr}(t) \\ &= (\mu_{1max}(t) + q_{X/Xm}) \cdot \frac{c_{OL}(t)}{k_O + c_{OL}(t)} \end{aligned} \tag{6.85}$$

where
$y_{X/Ogr}$ = O_2 growth yield coefficient
k_O = O_2 limitation constant,

is controlled by the dissolved oxygen concentration, c_{OL}.

After evaluation of the rate-determining step in Eq. (6.74),

$$q_{X/Xgr}(t) = \min\{\mu_{Sgr}(t), \mu_{Ogr}(t)\} \tag{6.86}$$

where
$q_{X/Xgr}$ = resulting cell-specific growth fraction,

the associated cell-specific reaction rates may be calculated.

The cell-specific growth rate, $q_{X/X}$,

$$q_{X/X}(t) = q_{X/Xgr}(t) - q_{X/Xm}, \tag{6.87}$$

the cell-specific O_2 uptake rate, $q_{O/X}$,

$$q_{O/X}(t) = \frac{q_{X/Xgr}(t)}{y_{X/Ogr}} + q_{O/Xm},$$
(6.88)

the cell-specific glucose uptake rate, $q_{S1/X}$,

$$q_{S1/X}(t) = \frac{q_{X/Xgr}(t)}{y_{X/S1gr}},$$
(6.89)

and the cell-specific ammonia (N_2) uptake rate, $q_{Al/X}$,

$$q_{Al/X}(t) = \frac{q_{X/Xgr}(t)}{y_{X/Algr}}$$
(6.90)

where
$y_{X/Algr}$ = cell mass yield for ammonia,

may be calculated directly from the growth fraction, $q_{X/Xgr}$.
    The cell-specific glycerol uptake rate, $q_{S2/X}$,

$$q_{S2/X}(t) = \frac{q_{X/Xgr}(t) - y_{X/S1gr} \cdot q_{S1/X}(t)}{y_{X/S2gr}},$$
(6.91)

and the cell-specific acetate uptake rate, $q_{S3/X}$,

$$q_{S3/X}(t) = \frac{q_{X/Xgr}(t) - \sum_{i=1}^{2} y_{X/Sigr} \cdot q_{Si/X}(t)}{y_{X/S3gr}}$$
(6.92)

are calculated recursively, resulting in the residual fraction for the optimum growth rate.
    Finally, the cell-specific acetate production rate, $q_{P1/X}$,

$$q_{P1/X} = \left[\sum_{i=1}^{2} y_{P1/Si} \cdot q_{Si/X}(t) \right] \cdot \frac{k_{IP1O}}{k_{IP1O} + c_{OL}(t)}$$
(6.93)

where
$y_{P1/Si}$ = acetate yield from substrate i, i=1, 2
k_{IP1O} = inhibition constant for acetate production due to oxygen,

completes the reaction model.
    Acetate is formed by uptake of both glucose and glycerol. Product formation is inhibited by the oxygen content of the medium.
    The simulation of a complete process in a bioreactor is possible using the model developed in Sects 6.3–6.6.
    For a discussion of further details of the model described, e.g., foam development and control, the reader is referred to [16].
    In the next section the model is used for the development of processing strategies.

6.9
Application Examples of On-Line Simulation Techniques

Figure 6.9 shows the Digital Control Unit DCU of the bioreactor BIOSTAT ED5, shown in Fig. 6.3. The DCU is linked to the on-line simulation system BIOSIM and the bioreactor host process control system UBICON.

The UBICON system on the right is the master control system of the actual fermentation process [17]. It is developed for biotechnological research and based on a VMEbus system. This master computer is connected to its peripheral I/O units via a CAN-bus. The I/O station shown, CTERM, is responsible among other things for communication with the DCU and other serial peripherals by translation of serial protocols to the CAN data transfer.

The VMEbus system on the left of Fig. 6.9 is the on-line simulator BIOSIM which includes the I/O structure of the basic bioreactor, the reactor periphery, and the measurement amplifier outputs. The simulator is part of another UBICON system and forms the virtual bioreactor process.

6.9.1
Training with Virtual Reaction Processes

It is frequently impossible to carry out complex cultivations in the course of student practicals. The extensive preparation and the slow course of a fermentation stand in stark contrast to the tight scheduling of such practicals. Furthermore, most students lack the practical abilities necessary to guarantee reproducible and thus useful results. On-line simulation is a possible remedy. The students run the virtual process with the same automation technology which they use for other real but simpler experiments in the bioreactor. The advantage is that the students simultaneously accustom themselves to the system but can make mistakes without grave consequences.

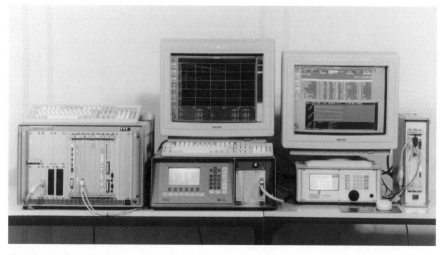

Fig. 6.9. On-line simulation setup. Photo: E. Stagat, University of Applied Sciences Hamburg

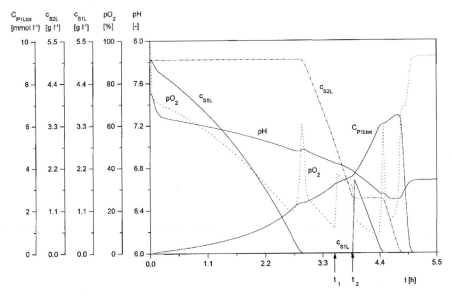

Fig. 6.10. Course of experiment during students' training

By alteration of the reaction equations and definition of the associated parameters it is thus possible for the supervisor to specify well-defined processes. In addition, checking the experimental evaluation is facilitated.

Figure 6.10 shows an example of a practical experiment. The process corresponds to the reaction behavior of *Escherichia coli* K12 introduced in Sect. 6.6, with an inoculation concentration of $1 \, g \, l^{-1}$. The cells grow on the two substrates, glucose and glycerol, initially present at concentrations of $5 \, g \, l^{-1}$. The experiment shows the shift from oxygen to substrate limitations, the diauxic behavior, the formation and reuse of acetate, and its influence on the pH. Addition of glucose during the glycerol growth phase at $t=t_2$ clearly shows the preference for glucose uptake. By alteration of the settings for the various process phases, e.g., the agitation speed at $t=t_1$, the students can run virtual experiments to determine unknown reaction parameters. This is done either off-line after the experiment is finished, e.g., with Excel, or on-line using a MATLAB subsystem running simultaneously.

6.9.2
Development of a High Cell Density Cultivation

6.9.2.1
The μ-Stat Problem

Figure 6.11 shows in principle the reaction behavior commonly encountered with recombinant production.

The cell-specific product formation rate $q_{P2/X}$ exhibits a marked dependency on the specific growth rate μ. At low and high growth rates the cell ceases to produce.

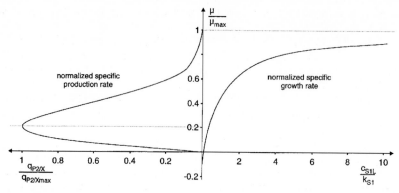

Fig. 6.11. Dependency of cell-specific production rate on substrate concentration and growth rate

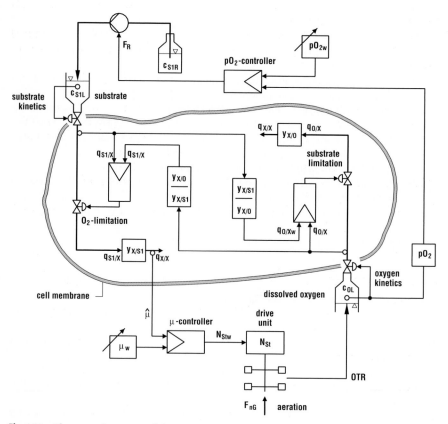

Fig. 6.12. The control concept of the μ-stat

Between these extremes a production maximum is observed under strong substrate limitation.

The processing objective is to hold the cell-specific growth rate at an optimum μ_{opt} using a dynamic fed batch scheme. In order to obtain a high product yield the process is to be executed as a high cell density cultivation with a cell dry weight up to $120\,\mathrm{g\,l^{-1}}$ [18].

It is not possible to control the specific growth rate using the substrate concentration as reliable on-line measurements of glucose in the k_{S1} range of $E.\ coli$ ($<5\,\mathrm{mg\,l^{-1}}$) are not available [19].

The objective of the strategy which will be developed below is to allow a freely definable cell-specific growth rate $q_{X/X}$ during production of recombinant proteins [20]. This mechanism, termed a μ-stat, can be forced by running under substrate or oxygen limitation. The latter, however, leads to the formation of undesirable byproducts. It is especially important to suppress acetate production as it leads to disturbing growth and production inhibition. Hence, the measured relative dissolved oxygen partial pressure pO_2 is kept by addition of substrate in an uncritical O_2 range, as shown in Fig. 6.12.

This control mechanism simultaneously fulfills the requirement of substrate limitation. However, the process is highly sensitive to falsely set control parameters. There is a resultant over-supply of substrate for a short period and a shift in the limitation type.

Constant stirrer speed N_{St} and aeration rate F_{nG} result in an almost constant feed rate F_R. The cell mass in the reactor, m_{XL}, grows linearly in this case and the specific growth rate μ falls hyperbolically.

A second control mechanism is therefore required to increase the feed rate in order to hold μ constant.

This is done with the μ/agitation controller. μ is determined using reactor mass balances or a Kalman filter and kept constant by a cascaded stirrer speed controller.

A reduction in μ results in a stirrer speed increase which in turn leads to a higher pO_2. The pO_2/feed controller increases the feed rate F_R and thus μ.

From a control viewpoint a number of problems must be solved:

1. The two controllers for pO_2 and μ are coupled within the cell by unknown regulation mechanisms.
2. The processing objective requires that for a given set point μ_w exponential growth of cell mass m_{XL},

$$m_{XL}(t) = m_{XL1} \cdot e^{\mu_w \cdot (t-t_1)}, \tag{6.94}$$

takes place after feeding has started at t_1. This is only possible by coupling the two controllers.
3. The control variable μ is not directly measurable. Hence, observers must be implemented in the automation scheme.
4. Badly set controller parameters lead to a shift in the reaction-limiting component and thus to a structural switch in the controlled process.
5. The variability in the dynamical process behavior due to the exponential increase in cell mass requires adjustment of the controller parameters.

These problems will be solved below using appropriate process control technology and with the aid of on-line simulation.

6.9.2.2
Observation of Cell-Specific Growth Rate

The observation of the cell-specific growth rate is based, in the case of the μ-stat, on the cell mass balance of a substrate limited cultivation,

$$\dot{m}_{XL}(t) = q_{X/X}(t) \cdot m_{XL}(t) = y_{X/O}(t) \cdot \dot{m}_{OT}(t) = y_{X/S1}(t) \cdot F_R(t) \cdot c_{S1R} \tag{6.95}$$

where
$y_{X/I}$ = effective cell yield of component I.

The O_2 transfer mass flux, $\dot{m}_{OT}$,

$$\dot{m}_{OT}(t) = V_L(t) \cdot Q_{O2}(t), \tag{6.96}$$

is obtained form the gas balance Eq. (6.25) without knowledge of the reaction volume V_L.

Formal integration of Eq. (6.95) results in the solution for $q_{X/X}$,

$$
\begin{aligned}
q_{X/X}(t) = \mu(t) &= \frac{y_{X/O}(t) \cdot \dot{m}_{OT}(t)}{m_{XL1} + \int\limits_{t_1}^{t} y_{X/O}(\tau) \cdot \dot{m}_{OT}(\tau)d\tau} \\
&= \frac{y_{X/S1}(t) \cdot F_R(t) \cdot c_{S1R}}{m_{XL1} + \int\limits_{t_1}^{t} y_{X/S1}(\tau) \cdot F_R(\tau) \cdot c_{S1R}d\tau}
\end{aligned}
\tag{6.97}
$$

where
t_1 = time-point begin of μ estimation
m_{XL1} = cell mass at time t_1.

Equation (6.97) can unfortunately not be used on-line because both yield coefficients are themselves functions of the unknown specific growth rate,

$$y_{X/O}(t) = \frac{\mu(t) \cdot y_{X/Ogr}}{\mu(t) + q_{O/Xm} \cdot y_{X/Ogr} + q_{S1/Xm} \cdot y_{X/S1gr}} \tag{6.98}$$

and

$$y_{X/S1}(t) = \frac{\mu(t) \cdot y_{X/S1gr}}{\mu(t) + q_{S1/Xm} \cdot y_{X/S1gr}}. \tag{6.99}$$

However, for the application in question μ is controlled to the set point μ_w.

Assuming perfect control, both yield coefficients in Eqs. (6.98) and (6.99) are constant.

As a result, the μ-observer is derived from the O_2 mass balance (Eq. 6.96) and its integration,

$$\hat{\mu}_O(t) = \frac{\dot{m}_{OT}(t)}{K_{Ow} + \int\limits_{t_1}^{t} \dot{m}_{OT}(\tau)d\tau} \tag{6.100}$$

where
$\hat{\mu}_O$ = estimated μ from O_2 mass balance.

The integration constant K_{Ow} at time t_1,

$$K_{Ow} = \frac{m_{XL1} \cdot (\mu_w + q_{O/Xm} \cdot Y_{X/Ogr} + q_{S1/Xm} \cdot Y_{X/S1gr})}{\mu_w \cdot Y_{X/Ogr}} \qquad (6.101)$$

must be estimated before the μ observation is activated. Equation (6.100) is convergent even in the case of incorrect estimation of K_{Ow}, since the integral in the denominator increases.

Equations (6.100) and (6.101) may also be applied for non-limited batch phase growth on glucose ($\mu_w = \mu_{1max}$).

The second equation for determining μ,

$$\hat{\mu}_S(t) = \frac{F_R(t)}{K_{S1w} + \int_{t_1}^{t} F_R(\tau)d\tau} \qquad (6.102)$$

where

$\hat{\mu}_S$ = estimated μ from the glucose feed mass balance,

is applicable only for the substrate limited fed batch phase and is dependent on the glucose integration constant K_{S1w},

$$K_{S1w} = \frac{m_{XL1} \cdot (\mu_w + q_{S1/Xm} \cdot Y_{X/S1gr})}{\mu_w \cdot Y_{X/S1gr} \cdot c_{S1R}}, \qquad (6.103)$$

which must itself be estimated at time t_1.

The integration of the feed rate F_R in Eq. (6.102),

$$\int_{t_1}^{t} F_R(\tau)d\tau = \frac{m_{R1} - m_R(t)}{\rho_R}, \qquad (6.104)$$

may be substituted by on-line measurement of the feed vessel mass.

6.9.2.3
Course and Testing of Processing Strategies

Figure 6.13 shows an on-line simulation of a high cell density cultivation (HCDC) for the production of recombinant DNA products with *Escherichia coli* in a 75-l bioreactor. One aim of the process was to cultivate the cells with air aeration and without oxygen transfer limitation up to $100\,\mathrm{g\,l^{-1}}$ dry weight [21].

The process starts in a batch mode, where cells grow on the main substrate glucose. In order to avoid O_2-transfer limited growth conditions, at t=4.9 h, a classical pO_2/agitation-control is switched on. During this process stage the pO_2-controller is not allowed to reduce the agitation speed by more than 25% of the previous maximum value. Therefore, the end of the batch phase is observable by a steep increase in the pO_2-level. This is followed by a double diauxic growth phase. First the cells change to a by-substrate, glycerol, and then to their own product, acetate.

At t=12.85 h all substrates are consumed and the μ-stat procedure is switched on automatically, where the process changes into a fed batch-cultivation. During this stage the cells grow limited on glucose. A set point μ_w=0.11 h^{-1} avoids any oxygen-transfer limitation. Figure 6.13 demonstrates the powerful control method. At the

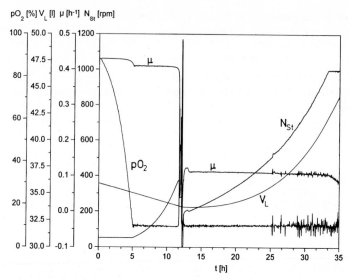

Fig. 6.13. Course of a simulated high cell density cultivation with μ-stat strategy; pO_2=dissolved oxygen tension, V_L=liquid volume, μ=specific growth rate, N_{St}=agitation speed

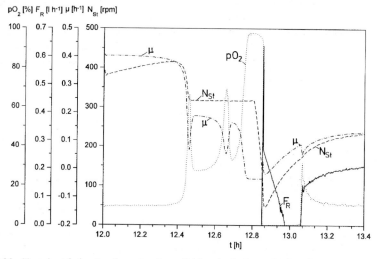

Fig. 6.14. Transient behavior from batch to fed batch during a high cell density cultivation; pO_2=dissolved oxygen tension, F_R=substrate feed rate, μ=specific growth rate, N_{St}=stirrer speed

end of batch, the cells grow up to $8\,\mathrm{g\,l^{-1}}$ and at the end of the process, $110\,\mathrm{g}$ dry mass per l can be achieved.

The problem of HCDC is an automatic switch from batch to fed batch processing and the tuning of control parameters during the μ-stat procedure. The advantage of on-line simulation lies in a simple reproduction of complex process behavior without any preparation of bioreactor and biological material.

The process discussed in Fig. 6.13 was simulated in real time conditions [20]. As an example of on-line control, interactions between the process control system UBI-CON and the simulator BIOSIM is shown in Fig. 6.14. The transient region between batch and fed batch starts with a step decrease in the specific growth rate, μ, due to the loss of glucose. The pO_2/agitation control is switched off after the agitation speed, N_{St}, is reduced by about 25% of the previous maximum value.

The change of growth from substrate 1, glucose, to substrate 2, glycerol, is now observable with the first pO_2 peak. After consumption of glycerol, pO_2 increases once more. The second pO_2 peak indicates the change to substrate 3, the product acetate. After consumption of the last carbon source, pO_2 increases again.

The specific growth rate, μ, is less than zero, due to the substrate maintenance (death) rate (Eq. 6.87), whereas a small amount of oxygen is consumed further on, due to the O_2-maintenance rate (Eq. 6.88).

The last pO_2-peak is identified by the control system in order to start the fed batch process. The agitation speed, N_{St}, is reduced to a low level and the simultaneous control of pO_2 by substrate addition and μ by agitation speed alteration is switched on.

A high pO_2 control deviation leads to a strong substrate addition after the beginning of control. The pO_2 decreases drastically and the substrate-limited process changes to an oxygen-transfer-limited one. Now the pO_2 controller reduces the substrate feed rate, F_R, slowly to zero. The last steep increase of pO_2 is due to the glucose consumption and the next change to substrate limitation. The restart of substrate feeding ends in a stable pO_2-control at the set point of 10% with substrate-limited conditions and a slowly increase of feed. The time course of μ indicates the discussed process behavior. The μ-control transient behavior is more or less smooth and ends at the set point of $0.11\,h^{-1}$ with a slightly increase of agitation speed.

Figure 6.15 shows an application of high cell density cultivation based on the developed μ-stat strategy [22].

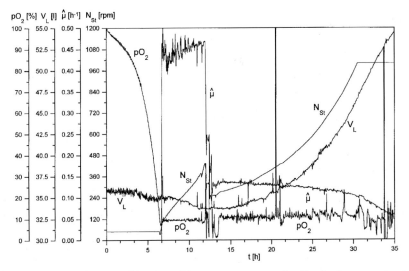

Fig. 6.15. High cell density cultivation with μ-stat strategy; pO_2=dissolved oxygen tension, V_L=liquid volume $\hat{\mu}$=observed specific growth rate, N_{St}=agitation speed

Similar control hardware is used as in the previous on-line simulation in Fig. 6.14. During batch phase pO_2/agitation control and $\hat{\mu}$-calculation are switched on at t=6.6 h.

The end of batch is observed by the double diauxic growth phase.

The consumption of all carbon sources is followed by the μ-stat procedure.

Experiment in Fig. 6.15 and simulation in Fig. 6.13 show similar behavior.

At t=30 h, the end of the controlled range, the cells grow up to $95\,g\,l^{-1}$ and at the end of the process, $110\,g\,l^{-1}$ dry biomass can be achieved.

6.10
Summary

The integration of on-line simulation methods in the process control system of bioreactors is an attractive addition both for training of students as well as for operators of industrial plants and for the development of advanced automation tools.

The method of virtual processing introduced here relies on complete modeling of the bioreactor system and its replacement by the real-time simulator with duplicate I/O connections.

Furthermore, the real process control system of the bioreactor serves to control the simulation.

This offers an inexpensive method for validating the automation hardware and software and enables the development of complex processing strategies on the basis of advanced functions.

The fully automated high cell density cultivation with controlled specific growth rate (μ-stat procedure) described demonstrates the power of the methods.

Even beginners in fermentation can successfully carry out real high cell density cultivations with concentrations of over $100\,g\,l^{-1}$ dry cell weight after training on the simulator.

References

1. Bailey JE (1998) Mathematical modeling and analysis in biochemical engineering: past accomplishments and future opportunities. Biotechnol Prog 14:8–20
2. Gollmer K, Posten C (1995) Fieldbus application in the hierachical automation structure of a biotechnological pilot plant. J Biotechnol 40:99–109
3. Zhou YH, Holwill ILJ, Titchener-Hooker NJ (1997) A study of the use of computer simulation for the design of integrated downstream processes. Bioprocess Engineering 16:367–374
4. Hwang F (1997) Batch pharmaceutical process design via simulation. Pharmaceutical Engineering, Jan/Feb, 26–43
5. GAMP, Good Automated Manufacturing Practice. International Society for Pharmaceutical Engineering (ISPE). http://www.activa.co.uk/gamp/
6. Sonnleitner B (1997) Bioprocess automation and bioprocess design. J Biotechnol 52:175–179
7. Clarke DW, Fraher PMA (1995) Model-based validation of DOx Sensor. Preprint of IFAC Workshop on on-line fault detection and supervision in chemical process industries, pp 216–223
8. Kunitake R, Suzuki A, Ichihashi H, Matsuda S, Hirai O, Morimoto K (1997) Fully automated roller bottle handling system for large scale culture of mammalian cells. J Biotechnol 52:289–294
9. Mathworks (1997) Using Matlab. The Mathworks Inc

10. Dunn IJ, Heinzle E, Ingham J, Prenosil JE (1992) Biological reaction engineering. VCH, Weinheim
11. Reuss M (1993) Oxygen transfer and mixing: Scale up implications. In: Rehm HJ, Reed G (eds) Biotechnology vol 3. VCH, Weinheim, pp 188–217
12. Fomferra N (1993) Mathematical modelling and real-time simulation of biotechnological processes. Diploma thesis, University of Applied Sciences, Hamburg
13. Gustafsson T (1982) Calculation of the pH value of a mixture of solutions. Chem Eng Sci 37:1419–1421
14. Bellgardt KH (1991) Cell models. In: Rehm HJ, Reed G (eds) Biotechnology, vol 4. VCH, Weinheim, pp 267–298
15. Rosso L, Lobry JR, Bajard S, Flandrois JP (1995) Convenient model to describe the combined effect of temperature and pH on microbial growth. Am Society Microbiol 61:610–616
16. Luttmann R, Gollmer K (1998) BIOSIM, an on-line simulation system for education and research. In UBICON – universal bioprocess control system, user's manual
17. Gollmer K, Gäbel T, Nothnagel J, Posten C (1992) UBICON – an universal bioprocess control system. DECHEMA biotechnology conferences 5. VCH, Weinheim
18. Luttmann R, Hartkopf J, Roß A (1994) Development of control strategies for high cell density cultivations. Mathematics and computers in simulation 37:153–164
19. Luttmann R, Slamal H, Evtimova V, Berens M, Scheffler U, Elzholz O (1997) On-line determination of reaction kinetics. Proceedings of the 8th European congress on biotechnology, Budapest, Hungary
20. Luttmann R, Bitzer G, Müller D, Scheffler U, Friedriszik U (1995) Development of a mystat with online simulation methods. In: Schügerl K, Munack A (eds) Postprint volume of CAB 6 – International conference on computer applications in biotechnology, Garmisch-Partenkirchen, Germany
21. Lee SY (1996) High cell density culture of *Escherichia coli*. TIBTECH 14:98–104
22. Riesenberg D, Schulz V, Knorre WA, Pohl HD, Korz D, Sanders EA, Roß A, Deckwer W-D (1991) High cell density cultivation of *Escherichia coli* at controlled specific growth rate. J Biotechnol 20:17–28

Part B
Application of General Principles for Reactor Models

7 Application of Computational Fluiddynamics (CFD) to Modeling Stirred Tank Bioreactors

Matthias Reuss, Sven Schmalzriedt, Marc Jenne

7.1
Introduction

The stirred and aerated tank fermentor is still the most important type of bioreactor for industrial production processes. The agitator or agitators are required to perform a wide range of functions: adequate momentum, heat and mass transfer, and mixing as well as gas-dispersion and homogenization of suspensions. As these various operations make different demands, optimization of the individual tasks would result in different design of the impellers. Therefore, agitators used in practice always reflect compromises. Conventional impellers used in fermentation are typically classified into axial and radial flow impellers. Examples of the two groups are illustrated in Fig. 7.1. Of the many impeller geometries, the six-blade disk impeller (Rushton turbine) with gas sparging below the impeller is most often used in standard configurations. However, an interest in using high-flow, low-power-number agitators such as Intermig (Fig. 7.2a), Lightnin A315 (Fig. 7.2b), Prochem Maxflow T (Fig. 7.2c), and Scaba 6SRGT (Fig. 7.2d) has been developed and improved performance of fermentation processes has been reported with such alternative designs [1–3].

The physiological state of microorganisms and their related behavior (growth and product formation) is determined by the immediate environment. During design of a bioreactor, and in particularly during scale-up, an attempt is made to recreate this environment in the large scale reactor as similar as possible to that established in bench scale and/or pilot plant vessels. Engineering solutions to this problem include maintaining geometric similarities whenever possible, and also criteria such as constant power per unit volume, volumetric mass transfer coefficient, circulation time, terminal mixing time, shear rate, tip speed, etc. (Table 7.1). The logic behind the different criteria is that preserving each of these singly represents maintaining the constancy of the corresponding characteristic of the extracellular environment at different scales. And, if this environment property is the one that most critically influences the desired microbial productivity, a successful scale-up might result. Indeed, a number of microbial systems are in accord with these techniques and permit a production scale operation reasonably in agreement with that established in the laboratory. However, it is easy to see that these criteria are mutually exclusive and, therefore, do not allow an exact replication of environmental similarity at any two different scales. Under such circumstances, the behavior of microorganisms in different fermentors remains uncertain. As a matter of fact, the use of volumetric properties as scale-up criteria demands a guarantee of uniformity of properties throughout the system, something that may be impossible even in a small vessel.

Radial Flow Impellers

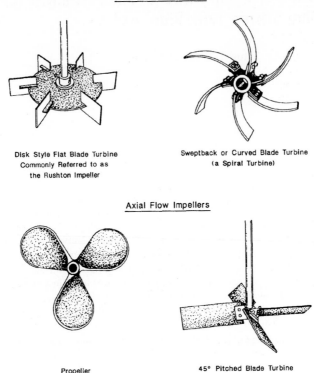

Disk Style Flat Blade Turbine
Commonly Referred to as
the Rushton Impeller

Sweptback or Curved Blade Turbine
(a Spiral Turbine)

Axial Flow Impellers

Propeller

45° Pitched Blade Turbine

Fig. 7.1. Impellers used in stirred tank bioreactors [58]

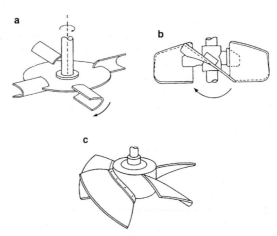

a

b

c

Fig. 7.2a–c. Low-power number agitators: **a** Scaba; **b** Lightning A 315; **c** Prochem Maxflow T [3]

Table 7.1. Common criteria for scale-up of stirred tank bioreactors

Volumetric oxygen transfer coefficient	$k_L a$
Volumetric power input	P/V
Volumetric gas flow	Q_G/V
Impeller tip speed	nd_i
Agitation speed	n
Terminal mixing time	θ_∞

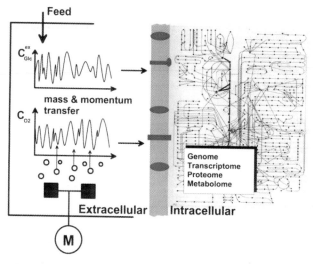

Fig. 7.3. Illustration of interactions between extracellular environment and metabolism

Complex interactions between transport phenomena and reaction kinetics characterize bioreactors and determine their performance. A quantitative description of these phenomena should consequently rest upon the two interwoven aspects of structured modeling (Fig. 7.3). The first aspect concerns the complex interaction of the functional units of the cells, including the mathematical formulation of reaction rates and the key regulation of these networks in response to changes in the environment. The second aspect has to do with the structure of the abiotic phases of the bioreactor in order to analyze the quality of mixing and other transport phenomena such as mass transfer between the phases causing gradients in the concentrations of various substrates and products.

These problems are particularly important for those processes in which nutrients are continuously introduced into the broth. For specific nutrients such as oxygen and sometimes other nutrients such as carbon source, the time constant for their distribution (mixing-time) may be of the same magnitude as those of their consumption in any reasonable sized reactor beyond the bench-scale. If we accept that spatial variations exist, we are faced with the problem of dynamically changing environment conditions. This in turn may result in drastic changes in metabolism and final outcome of the process. The long-term mathematical description of these phenomena requires flexible tools to be adapted to different systems and able to integrate the process and reactor.

Despite the strong interdependence, this chapter will concentrate on the detailed modeling of the abiotic phases of the reactor. Various tools are available to tackle problems of incomplete mixing. The models suggested for stirred tank reactors may be classified into two groups – reactor flow models and computational fluiddynamics (CFD) including turbulence models.

The most important tools for modeling situations of incomplete mixing based on reactor flow models are compartment models and recirculation models.

The application of the various approaches to bioreactor modeling has been extensively described by Reuss and Bajpai [4] and Reuss [5]. As a result of these critical reviews, several serious limitations of these modeling strategies have been outlined. One of these limitations is related to the omission of backmixing of the gas phase and, therefore, the disregard of the material balance equation for oxygen in the gas phase. Second, in many of the models the number of compartments is connected to the intensity of mixing and, thus, the structures are virtual in space. In other words, the compartments are without identity because of missing coordinates. It is easy to see that the two mentioned problems are closely related if attempts are made to couple mixing of gas and liquid phase including local mass transfer across the interface.

In contrast to these classical approaches of reactor flow models, this chapter summarizes the state of the art in the application of computational fluiddynamics (CFD), making use of the numerical solution of the state equations for mass, momentum, kinetic energy, and energy dissipation. The application of this approach is illustrated for single-phase and two-phase flow. Single as well as multiple impellers will be treated.

With the aid of some examples of industrial importance the use of the fluiddynamic simulations will be demonstrated. The first examples deal with the three-dimensional transient concentration field in mixing experiments at different scales of operation. The process examples chosen are treating the problem of substrate distribution during fed batch operation as well as oxygen distribution during the oxygen sensitive microaerobic production of acetoin and butanediol with *Bacillus subtilis*.

7.2
Modeling and Simulation of Gas/Liquid Flow in Stirred Tank Reactors

It is generally accepted now that Reynolds-averaging the turbulent Navier-Stokes equations and modeling the Reynolds-stresses with an appropriate turbulence model is a promising way of modeling the flow behavior. Ongoing development of commercial computational fluiddynamics software (CFD) and increasing computer power are continuously improving the conditions for the simulation of the three-dimensional and turbulent flow structure in stirred tanks. Among the variety of impellers the Rushton turbine is well established for many tasks, mainly due to good gas dispersion and mixing of liquids with low viscosities. The Rushton turbine generates a flow leaving the impeller in radial and tangential direction. This radial-tangential jet flow divides at the vessel wall, and the flow then recirculates back into the impeller region. Besides turbulent dispersion, recirculation of the flow is the main reason for the mixing capability of stirred tanks.

In spite of the improved hardware and software, which have greatly expanded the tools available for simulating fluid flow in stirred tank reactors, a number of unsolved problems and open questions still exist.

A critical analysis of the many publications concerning the simulation of the liquid flow in baffled stirred tank reactors equipped with a Rushton turbine shows several discrepancies. The most important differences between the simulations concern the dimensionality of the simulations (three-dimensional or axisymmetric), turbulence modeling, the modeling approaches for the Rushton turbine as well as the accuracy of the numerical predictions, which depends on the grid size.

In what follows, different modeling approaches for the Rushton turbine are examined and critically reviewed. For more than a decade (see, e.g., [6–8]) it has been an established method to specify boundary conditions for the impeller with the aid of experimental data. One advantage of this method is the reliable description of the outflow region of the impeller, which can essentially be considered as a circumferentially and time averaged radial-tangential jet. A resulting additional advantage is the reduced computational expense of stationary simulations compared to transient simulations. The resolution of the vortex system behind the stirrer blades (see, e.g., [9]) in applying this method, however, is not possible. In order to specify boundary conditions for other types of impellers, one has to perform time-consuming experiments in advance. To remove the two last disadvantages mentioned, recent attempts have been made to simulate the unsteady flow within and outside the impeller swept region in applying the so-called sliding-mesh technique (see, e.g., [10, 11]). A critical comparison of the results from the sliding mesh technique and simulations with measured data in the impeller region has been presented by Brucato et al. [12]. However, the sliding mesh technique requires excessive computational resources and for most engineering applications knowledge of the full time varying and periodic flow field may not be necessary. Another possibility to simulate flow details between the impeller blades is the so-called snapshot approach (see, e.g., [26]). This is often also called a multiple reference frame method (see [13]). Experimental data to specify boundary conditions are not necessary. An advantage compared with the sliding-mesh technique is that the full time dependent transport equations need not be solved. This offers an interesting and promising approach. However, the essential comparisons with experimental observations are lacking.

Besides the modeling approaches for the Rushton turbine, the dimensionality of the simulation is another point of discussion. Table 7.2 shows an overview of publications concerning simulations of the liquid flow in stirred tank reactors using experimental data as impeller boundary condition for the Rushton turbine. The fourth column of Table 7.2 informs about the dimensionality of the simulations. In axisymmetric simulations the stationary baffles fixed at the vessel wall are not resolved as geometrical bodies, but modeled with the drag resistance force of the baffles to match the experimentally observed velocity fields (see, e.g., [15]). The coefficient required in the formulation of the drag resistance force lacks physical meaning and needs to be adjusted. Axisymmetric simulations are then able to describe the experimentally observed bulk circulation in radial-axial direction as the most important flow characteristic in the stirred tank (see, e.g., [15, 17]). Increasing computer power has continuously improved the conditions to resolve the flow field in three dimensions. Three-dimensional simulations show flow details in the vessel (e.g., vortex formation behind the baffles), which cannot be observed in axisymmetric simulations.

Table 7.2. Summary of fluiddynamic simulations of stirred tank vessels

Author	Year	turbulence model	dimensionality
Single-phase flow			
Platzer [14]	1981	standard k-ε	axisymmetric
Harvey and Greaves [15]	1982	standard k-ε	axisymmetric
Placek et al. [16, 17]	1986	three-equation k_p-k_{Te}	axisymmetric
Middleton et al. [18]	1986	standard k-ε	3D
Ju [6]	1987	modified k-ε	3D
Joshi and Ranade [7]	1990	standard k-ε	3D
Kresta and Wood [8]	1991	standard k-ε	3D
Bakker and Van den Akker [19]	1994	algebraic stress	3D
Togatorop et al. [20]	1994	standard k-ε	axisymmetric
Brucato et al. [12]	1998	standard k-ε	3D
Two-phase flow			
Issa and Gosman [21]	1981	standard k-ε	axisymmetric
Trägardh [22]	1988	standard k-ε	axisymmetric
Politis et al. [23]	1992	standard k-ε	3D
Morud and Hjertager [24]	1993	standard k-ε	axisymmetric
Bakker and Van den Akker [25]	1994	algebraic stress	3D
Ranade and Van den Akker [26]	1994	standard k-ε	3D

The third column of Table 7.2 concerns the choice of the turbulence model. Most of the authors use the standard form of the so-called k-ε model, although it is well known that this approach of turbulence modeling may fail in flow regimes with strong streamline curvature and vortex generation. Some authors use modifications of the k-ε model or other turbulence models like the three-equation k_P-k_T-ε model or the algebraic stress model.

An important feature in modeling the two-phase flow is to distinguish between Eulerian and Langrangian approaches. In the Langrangian approach, the continuous phase is treated as a continuum while the dispersed gas bubbles are modeled as single particles. In the Eulerian approach the dispersed phase is also considered as a continuum resulting in the so-called two fluid model. Only the Eulerian approach has been considered for aerated stirred tank reactors so far. If only gravitation, pressure, and drag force are taken into account in the momentum equation for the gas phase, the relative velocities of the gas phase are calculated from algebraic equations. This is the so-called algebraic slip model. The disadvantage of this simple approach is the fact, that additional interface forces are neglected. Issa and Gosman [21] calculated the flow in a gassed and stirred vessel equipped with a Rushton turbine using the algebraic slip model. Furthermore, they used very coarse grids because of limited computing power. Experimental verification of their simulations was not shown. Trägardh [22] reported about 2D-simulations with the algebraic slip model for a stirred vessel equipped with three impellers. Politis et al. [23] performed three dimensional simulations with the two fluid and k-ε model. They considered different interfacial forces and critically examined their influence. These authors were able to show that, in addition to the drag force, the virtual mass force particularly needs to be considered. For boundary conditions in the impeller region, values for averaged tangential velocities as well as k and ε from measured data were used.

Morud and Hjertager [24] followed an axisymmetrical approach based on the two fluid and k-ε model. The virtual mass force was neglected. These authors observed a considerable deviation between measured and simulated data.

Bakker and Van den Akker [25] proposed a single phase model in which a reduction of pumping capacity due to aeration was taken into account. The turbulence was modeled with the aid of an algebraic stress model. A more detailed model for the flow in the impeller region was presented by Ranade and Van den Akker [26]. These authors used the two-fluid approach and modeled the two-phase flow in the impeller region with the aid of a snapshot method.

7.3
Single Phase Flow

7.3.1
Transport Equations

The transport equations describing the instantaneous behavior of turbulent liquid flow are three Navier-Stokes equations (transport of momentum corresponding to the three spatial coordinates r, z, φ in a cylindrical polar coordinate system) and a continuity equation. The instantaneous velocity components and the pressure can be replaced by the sum of a time-averaged mean component and a root-mean-square fluctuation component according to Reynolds. The resulting Reynolds equations and the continuity equation are summarized below:

$$\frac{\partial(\rho u_i)}{\partial t} + \frac{\partial(\rho u_i u_j)}{\partial x_i} = -\frac{\partial}{\partial x_i}\left(\tau_{ij} + \rho \overline{u_i' u_j'}\right) - \frac{\partial p}{\partial x_i} + \rho g_i \tag{7.1}$$

$$\frac{\partial \rho}{\partial t} + \frac{\partial}{\partial x_i}(\rho u_i) = 0 \tag{7.2}$$

A reasonable compromise for model accuracy and computational expense are eddy viscosity models relating the individual Reynolds stresses to mean flow gradients:

$$\rho \overline{u_i' u_j'} = -\rho \nu_{turb}\left(\frac{\partial u_i}{\partial x_j} + \frac{\partial u_j}{\partial x_i}\right) + \frac{2}{3}\rho \delta_{ij} k \tag{7.3}$$

where ν_{turb} is the turbulent eddy viscosity. The transport of momentum, which is related to turbulence, is thought of as turbulent eddies, which, like molecules, collide and exchange momentum.

The family of two-equation k-ε models is the most widely used of the eddy viscosity models. A k-ε model consists of two transport equations, one for the turbulent kinetic energy k and one for the energy dissipation rate ε. The turbulent eddy viscosity is calculated from

$$\nu_{turb} = c_\mu \frac{k^2}{\varepsilon} \tag{7.4}$$

where C_μ is a parameter which depends on the specific k-ε model.

The standard k-ε model, as presented by Launder and Spalding [27], is by far the most widely-used two-equation eddy viscosity model, also for modeling turbulence

Table 7.3. Parameter values in the k-ε model

Parameter	Value
$C_\mu m_{,s}$	0.09
$C_{1,s}$	1.44
$C_{2,S}$	1.92
$\sigma_{k,s}$	1.00
$\sigma_{\varepsilon,s}$	1.314

in stirred tank reactors (see Table 7.2). The popularity of the model and its wide use and testing has thrown light on both its capabilities and its shortcomings, which are well-documented in the literature [27–33]. For high turbulent Reynolds numbers, the model may be summarized as follows:

$$\frac{\partial(\rho k)}{\partial t} + \frac{\partial}{\partial x_i}(\rho u_i k) = \frac{\partial}{\partial x_i}\left(\rho \frac{v_{eff}}{\sigma_{k,S}} \frac{\partial k}{\partial x_i}\right) + \rho(P_k - \varepsilon) \tag{7.5}$$

$$\frac{\partial(\rho \varepsilon)}{\partial t} + \frac{\partial}{\partial x_i}(\rho u_i \varepsilon) = \frac{\partial}{\partial x_i}\left(\rho \frac{v_{eff}}{\sigma_{\varepsilon,S}} \frac{\partial \varepsilon}{\partial x_i}\right) + \rho\left(c_{1,S}\frac{\varepsilon}{k}P_k - c_{2,S}\frac{\varepsilon}{k}\varepsilon\right) \tag{7.6}$$

The model parameters of the standard k-ε model are listed in Table 7.3.

The production of turbulent kinetic energy P_k is modeled with the aid of the eddy viscosity hypothesis:

$$P_k = v_{turb}\left(\frac{\partial u_i}{\partial x_i} + \frac{\partial u_j}{\partial x_i}\right)\frac{\partial u_i}{\partial x_j} \tag{7.7}$$

The dissipation rate ε can be regarded as the rate at which energy is being transferred across the energy spectrum from large to small eddies. The standard k-ε model assumes spectral equilibrium, which implies that once turbulent kinetic energy is generated at the low-wave-number end of the spectrum (large eddies), it is dissipated immediately at the same location at the high-wave-number end (small eddies). In other words, the standard k-ε model assumes that P_k is near to ε. As far as the stirred vessel is concerned, this is a very restrictive assumption, because there is a vast size disparity between those eddies, in which turbulence production takes place (mainly at the stirrer), and the eddies, in which turbulence dissipation occurs.

The standard k-ε model employs a single time scale $\tau_d = k/\varepsilon$ called dissipation range time scale in the equation to characterize the dynamic processes occurring in the energy spectrum. Thus, Eq. (7.6) can be rewritten as:

$$\frac{\partial(\rho \varepsilon)}{\partial t} + \frac{\partial}{\partial x_i}(\rho u_i \varepsilon) = \frac{\partial}{\partial x_i}\left(\rho \frac{v_{eff}}{\sigma_{\varepsilon,S}} \frac{\partial \varepsilon}{\partial x_i}\right) + \rho\left(c_{1,S}\frac{P_k}{\tau_d} - c_{2,S}\frac{\varepsilon}{\tau_d}\right) \tag{7.8}$$

The energy spectrum, however, comprises fluctuating motions with a spectrum of time scales, and a single time scale approach is unlikely to be adequate under all circumstances. Consequently, the model has been found to perform less satisfactorily in a number of flow situations, including separated flows, streamline curvature, swirl, rotation, compressibility, axisymmetrical jets etc.

Because of its wide use, variants and ad-hoc modifications aimed at improving its performance abound in the literature. The most well-known modifications are the Chen-Kim and the RNG variant of the k-ε model.

In order to ameliorate the previously mentioned deficiencies in the standard k-ε model, Chen and Kim [34] proposed a modification which improves the dynamic response of the ε equation by introducing an additional time scale

$$\tau_P = \frac{k}{P_k} \tag{7.9}$$

which is called the production-range time scale. The final expression of the transport equation for the dissipation rate is given as

$$\frac{\partial(\rho\varepsilon)}{\partial t} + \frac{\partial}{\partial x_i}(\rho u_i \varepsilon) = \frac{\partial}{\partial x_i}\left(\rho\frac{\nu_{eff}}{\sigma_{\varepsilon,CK}}\frac{\partial\varepsilon}{\partial x_i}\right) + \rho\left(\overbrace{c_{1,CK}\frac{P_k}{\tau_d}}^{\text{1st part}} + \overbrace{c_{3,CK}\frac{P_k}{\tau_p}}^{\text{2nd part}} - c_{2,CK}\frac{\varepsilon}{\tau_d}\right) \tag{7.10}$$

production term

The parameters of the Chen-Kim model are summarized in Table 7.4.

The first part of the production term corresponds with the production term of the standard k-ε model. Notice that the second production term is related to the time scale τ_p. The introduction of this additional term enables the energy transfer to respond more efficiently to the mean strain than does the standard k-ε model. Thus, τ_p enables the development of a field of ε suppressing the well-known overshoot phenomenon of the turbulent kinetic energy k. This overshoot appears when the standard k-ε model is applied to flow conditions with large values of mean strain (see [29, 32, 33]).

The modification may be summarized as follows. Production of ε appears in two energy fluxes divided by two different time scales τ_d and τ_p. The multiplying coefficients might be seen as weighting factors for these two energy fluxes. One may expect that this feature offers advantages in separated flows and also in other flows where turbulence is far from local equilibrium (P_k is far from ε). If P_k is near to ε (local equilibrium) the Chen-Kim modified k-ε model is almost identical to the standard k-ε model. Then, τ_p equals τ_d and summing up the two ε production terms leads to the ε production term of the standard k-ε model. The resulting coefficient $c_{1,CK} + c_{3,CK} = 1.4$ is only slightly lower than $c_{1,S}$. This is the reason why for simple boundary type flows the Chen-Kim modified k-ε model gives results similar to those predicted by the standard k-ε model. However, for complex elliptic turbulent flow problems (internal turbulent recirculating flows) involving rapid changes of turbu-

Table 7.4. Parameter values in the Chen-Kim model

Parameter	Value
$C_{\mu,CK}$	0.09
$C_{1,CK}$	1.15
$C_{2,CK}$	1.90
$C_{3,CK}$	0.25
$\sigma_{k,CK}$	0.75
$\sigma_{\varepsilon,CK}$	1.15

lent kinetic energy production and dissipation rates the Chen-Kim modified $k\text{-}\varepsilon$ model has been shown to give much better results than the standard $k\text{-}\varepsilon$ model [34].

To improve further the agreement between simulations and experimental observations the ratio of two parameter $c_{1,CK}/c_{3,CK}$ in the Chen-Kim model was slightly modified. For modification of these parameters the ratio of the Eulerian macro length scale L_i to the impeller blade height w have been employed. The property can be compared in geometric similar vessels.

According to Batchelor's [35] energy cascade theory the energy dissipation rate ε is given by

$$\varepsilon = \frac{3}{2}A\frac{(u'^2)^{3/2}}{L_{res}} \tag{7.11}$$

Introducing the kinetic energy k, Eq. (7.11) reads

$$\varepsilon = A\frac{k^{3/2}}{L_{res}} \tag{7.12}$$

Wu and Patterson [36] normalized their own observations as well as results from other observations with the blade height w:

$$\frac{L_{res}}{w} = \frac{A}{w}\frac{k^{3/2}}{\varepsilon} \tag{7.13}$$

If the macro length scale for anisotropic turbulence $L_{res} = \sqrt{\sum L_i^2}$ is replaced by the length scale for isotropic turbulence the following equation holds:

$$\frac{L}{w} = \frac{A}{\sqrt{3}}\frac{k^{3/2}}{\varepsilon} \tag{7.14}$$

For the estimation of the improved parameters in the turbulence model, the simulated values of k and ε are used to predict the local values of Lw^{-1}, which are then compared with the experimental observations reported by Wu and Patterson [36] in the impeller region. For more details, the interested reader is referred to the original paper of Jenne and Reuss [37]. At conditions of spectral equilibrium the dynamic response of the optimized ε transport equation should always be identical to that of the standard $k\text{-}\varepsilon$ model. Thus, the sum of the two parameters of the Chen-Kim model is constrained by

$$C_{1,CK} + C_{3,CK} = 1.4 \tag{7.15}$$

Only one parameter has been varied. The values obtained for the two parameters at minimal deviations between computed and measured length scale in the impeller region are

$$(C_{1,CK})\text{opt} = 1.36$$

and

$$(C_{3,CK})\text{opt} = 0.04$$

7.3.2
Simulations and Comparison with Experimental Observations

The Reynolds equations, the continuity equation, which is turned into an equation for pressure correction (see [38]), and the transport equations for the turbulence quantities k and ε are integrated over the respective finite volume elements resulting from the discretization of the stirred tank domain. The convection and diffusion terms in the transport equations are approximated using the hybrid-scheme of Patankar [38]. The resulting algebraic equations are then solved with the aid of the commercial CFD software Phoenics (Version 2.1). So-called false-time-step relaxation is used in order to achieve stationarity. The semi-implicit method, which considers the pressure-link of the pressure correction equation and the Reynolds equations, is the Simplest algorithm. The sets of algebraic equations for each variable are solved iteratively by means of the ADI technique.

With the aid of a systematic comparison of simulations and experimental data obtained by Costes and Couderc [39] the performance of the following turbulence models has been analyzed:
- standard k-ε model
- optimized Chen-Kim k-ε model
- Chen-Kim k-ε model
- RNG k-ε model (renormalized group method, Yakhot and Smith [40])

Figure 7.4 demonstrates that in some positions within the tank it is easy to discriminate between the various k-ε models on the basis of a comparison with measured mean velocities. Thus, the profiles of axial mean velocities belonging to the baffle plane ($\varphi=45°$) in Fig 7.4, show significant deviations for the velocities obtained with the aid of the Chen-Kim and the RNG k-ε model. The standard and the optimized Chen-Kim k-ε model fit the experimental data well.

However, the differences in performance of the four k-ε models are even more pronounced by inspecting Figs. 7.5 and 7.6. Figure 7.5 concentrates on the outflow of the impeller (radial-tangential jet flow) exactly between two baffles ($\varphi=0°$). Figure 7.5a shows radial profiles of radial mean velocities at the impeller centerline and at the height of the lower impeller blade edge. The optimized Chen-Kin k-ε model fits the experimental data well, whereas the other k-ε models deviate significantly. The values obtained with the Chen-Kim and RNG k-ε model at the impeller centerline are higher than the measured data. The standard k-ε model overestimates the energy loss in the outflow of the impeller. Consequently, the values of the predicted radial mean velocities are too low. The values of radial mean velocities obtained with the Chen-Kim and the RNG k-ε model at the height of the lower impeller blade edge are lower than the measured ones, because the predicted turbulent shear stress does not drag the entrainment flow sufficiently. The standard k-ε model overestimates the turbulent shear stress and consequently, the values of the radial mean velocities are too high. Figure 7.5b shows radial profiles of tangential mean velocities; again at the impeller centerline and at the height of the radial velocity components equally apply to the tangential components.

Figure 7.6 shows radial profiles of tangential mean velocities at different heights in the bulk of the vessel. Again, the optimized Chen-Kim k-ε model fits the experimental data well, whereas the other k-ε models deviate significantly. Figure 7.7a illustrates the flow field between two baffles at $0°$. One can see the flow leaving the im-

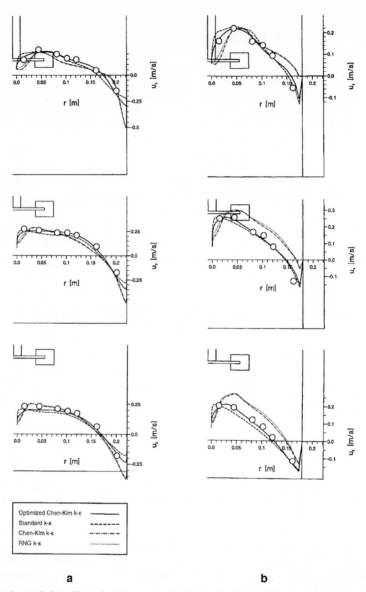

Fig. 7.4a,b. Radial profiles of axial mean velocities in the bulk of the vessel. Simulated profiles obtained with four different k-ε models. Measurements from [39]: **a** between two baffles ($\varphi=0°$); **b** baffle plane ($\varphi=45°$)

peller in a radial direction and the subsequent entrainment flow into the outflow of the impeller. Thus, the flow divides at the vessel wall and recirculates back either directly into the impeller or again as entrainment flow into the outflow of the impeller. The aim of Fig.7.8 is to demonstrate that the flow in the r-φ direction is in fact three-dimensional. Figure 7.8a shows the flow field at a short distance below the

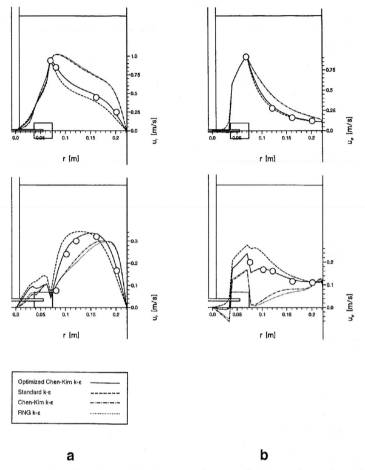

Fig. 7.5a,b. Radial profiles of: **a** radial; **b** tangential mean velocities in the outflow of the impeller between two baffles (φ=0°). Simulated profiles obtained with four different k-ε models. Measurements from [39]

impeller. The flow is directed to the vessel wall and has a positive tangential velocity component like the radial-tangential jet originating from the impeller. The radial-tangential jet is able to drag the flow in its vicinity due to turbulent shear stress. Again, the impeller tip velocity u_{tip} is shown as reference vector and indicates high velocities. Behind the baffle, vortex formation can be observed.

Figure 7.8b illustrates the flow field at the height of the lower bulk circulation center. The reference vector indicates lower values of velocity. The flow, originally directed to the vessel wall above the lower bulk circulation center (see Fig. 7.7a), is turned to the bottom of the vessel. This is the effect of the baffles. And the flow, originally directed to the vessel axis below the lower bulk circulation center, is turned to the impeller region. Consequently, the larger amount of momentum is associated to the flow directed in negative and positive z direction, respectively, and not in r-φ direction. The vortex behind the baffle still exists. The vortex occupying the vessel axis

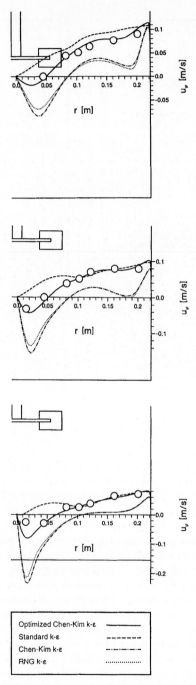

Fig. 7.6. Radial profiles of tangential mean velocities in the bulk of the vessel between two baffles ($\varphi=0°$). Simulated profiles obtained with four different k-ε models. Measurements from [39]

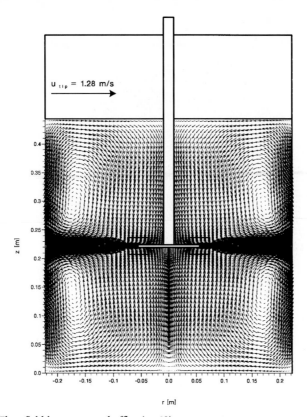

Fig. 7.7. Flow field between two baffles ($\varphi=0°$)

with a slight angular motion in opposite direction to the impeller rotation (see [39, 41] can also be observed.

Figure 7.8c shows the flow field near the bottom of the vessel. The flow accumulates in the high pressure region in front of the baffle. The main part of the flow has to leave the baffle with a negative tangential velocity component. A smaller part of the flow evades in a low pressure region with a positive tangential velocity component. The smaller part meets the main part of the flow leaving the baffle, which is shifted 90° in positive φ direction, in a sharp angle (see [76]). The two parts of the flow join each other and form the basis of the vortex occupying the vessel axis. Due to the prevailing negative tangential momentum in the united flow the vortex obtains its slight angular motion in the opposite direction to the impeller rotation. For more details (distribution of pressure and turbulence properties) the interested reader is referred to the original publication of Jenne and Reuss [37] as well as Jenne [42].

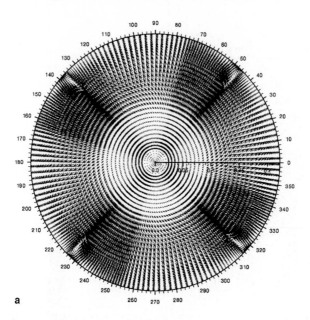

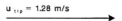

a

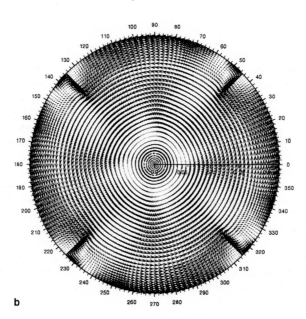

b

u$_{tip}$ = 1.28 m/s

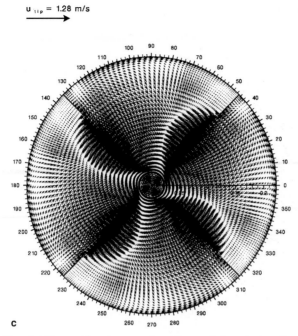

Fig. 7.8a–c. Flow field in r-φ direction: **a** at short distance below the impeller; **b** at the height of the lower bulk circulation center; **c** near the bottom of the vessel

7.4
Multiple Impellers

In industry, reactors are usually equipped with two or more impellers. Very little data are found concerning the details of flow patterns, particularly quantitative information about velocity fields and distribution of turbulence intensities. However, many workers have investigated the effect of different impeller types and configurations on mass transfer gas liquid and mixing. Important and quite useful results have been summarized by Bouaifi et al. [43] and John et al. [44]. Improved reactor performance has been observed when incorporating mixed flow systems (e.g., with lower impeller acting radially and the upper impeller axially) in a baffled system [45–47]. In particular for large scale fed batch fermentations these configurations should offer advantages because of improved axial mixing. A few CFD simulations together with simulations of mixing behavior presented in Sect. 7.6.2 will serve to elucidate these phenomena.

To reduce complexity and computation time, two-dimensional simulations have been performed for this comparison. In these simulations the baffles are modeled as a momentum sink in the Reynolds equation for tangential direction [15]. The assumption that these simplified simulations are able to reasonably approximate the radial-axial flow behavior seems to be justified because of the large height-to-diameter ratio.

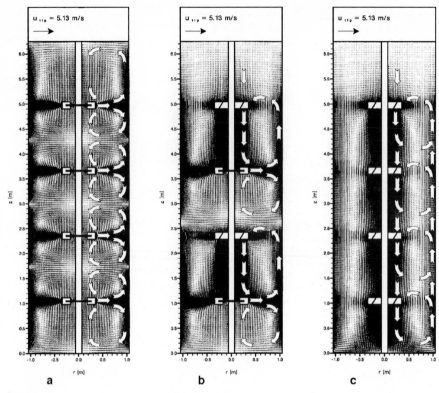

Fig. 7.9a–c. Flow field for different multiple impeller systems: **a** 4 Rushton turbines; **b** 2 Rushton turbines and 2 pitched blade impellers; **c** 4 pitched blade impellers

Fig. 7.10. Pitched blade axial-flow impeller

In Fig. 7.9a velocity fields are shown for a system of four Rushton turbines. In addition to the velocity vector field, large arrows are used to illustrate the flow behavior. Each impeller creates a more or less independent symmetrical flow field. The multiple impeller system therefore shows very poor axial convection. The transport between the individual cells is performed mainly with the aid of axial turbulent dispersion.

The results from similar simulations with two Rushton and two pitched blade impellers as well as four pitched blade impeller are shown in Fig. 7.9b and c, respectively. The pitched blade axial-flow impeller is shown in Fig. 7.10. In both cases an improved convection in axial direction can be observed. The results of simulated mixing experiments presented in Sect. 7.6.2 will confirm more rapid mixing for both systems.

7.5
Gas-Liquid Flow

The simulations of the gas-liquid flow are based on the Eulerian two-fluid model originally derived by Ishii [50]. In this approach, each phase is treated as a continuum. After averaging the general transport equations, we get the following set of multi-phase conservation equations [23, 26]:
CONTINUITY:

$$\frac{\partial}{\partial t}(\rho_k \varepsilon_k) + \frac{\partial}{\partial x_i}\left(\rho_k \varepsilon_k u_{k,i} - \rho_k \frac{\nu_{turb}}{Sc_{turb}}\frac{\partial \varepsilon_k}{\partial x_i}\right) = 0 \qquad k = L, G \qquad (7.16)$$

A dispersive transport of gas bubbles and liquid has been considered in both continuity equations. Sc_{turb} is the Schmidt number for turbulent transport which is assumed to be 1 [26]. The global mass conservation is given by:

$$\varepsilon_G + \varepsilon_L = 1 \qquad (7.17)$$

MOMENTUM:
Liquid phase:

$$\frac{\partial(\rho_L \varepsilon_L u_{L,i})}{\partial t} + \frac{\partial(\rho_L \varepsilon_L u_{L,i} u_{L,j})}{\partial x_j} = -\varepsilon_L \frac{\partial}{\partial x_j}\left(\tau_{L,ij} + \rho_L u'_{L,i} u'_{L,j}\right) - $$
$$- \varepsilon_L \frac{\partial p}{\partial x_i} + \rho_L \varepsilon_L g_i + S_i \qquad (7.18)$$

with the laminar shear stress $\tau_{L,ij}$ and turbulent Reynolds-stresses given by the Boussinesq approximation:

$$-\rho_L u'_{L,i} u'_{L,j} = \rho_L \nu_{turb}\left(\frac{\partial u_{L,i}}{\partial x_j} + \frac{\partial u_{L,j}}{\partial x_i}\right) - \frac{2}{3}\rho_L \delta_{ij} k \qquad (7.19)$$

Gas phase:

$$\frac{\partial(\rho_G \varepsilon_G u_{G,i})}{\partial t} + \frac{\partial(\rho_G \varepsilon_G u_{G,i} u_{G,j})}{\partial x_j} = -\varepsilon_G \frac{\partial p}{\partial x_i} + \rho_G \varepsilon_G g_i - S_i \qquad (7.20)$$

Reynolds-stresses in the gas phase can be neglected.

7.5.1
Interfacial Forces

The interfacial coupling term S_i in Eqs. (7.18) and (7.20) is a linear combination of several forces. Politis et al. [23] have compared the order of magnitude of the various forces and concluded that only the drag force and the virtual mass force need to be considered.

7.5.1.1
Drag Force

From the definition of the drag coefficient c_D of a single bubble the following expression for the drag force can be derived:

$$F_{D,i} = \frac{\rho_L}{2} c_d A_b |\Delta u| \Delta u_i \qquad (7.21)$$

A_b is the sectional area of the bubble $= (0.25\pi) d_b^2$, Δu_i is the relative velocity between the bubble and the liquid in the direction i, $|\Delta u|$ is the absolute value of the relative velocity vector. The momentum equation (Eq. 7.18) is related to the total volume dV which contains gas and liquid. The volumetric force S_i is therefore

$$S_i = \frac{F_i}{dV} = \varepsilon_G \frac{F_i}{dV_G} \qquad (7.22)$$

and with Eq. (7.21):

$$S_{D,i} = \varepsilon_G \frac{3}{4} c_D \frac{\rho_L}{d_b} |\Delta u| \Delta u_i \qquad (7.23)$$

The following correlations for air bubbles rising in distilled and tap water have been proposed by Kuo and Wallis [49]. For distilled water the equations for the drag coefficent read:

$Re < 0.49$	$C_D = 24\ Re^{-1}$
$0.49 < Re < 33$	$C_D = 20.68\ Re^{-0.643}$
$33 < Re < 661$	$C_D = 72\ Re^{-1}$
$661 < Re < 1237$ and $We \approx 4$	$C_D = 0.02083\ Re^4\ Mo$
$Re > 1237$ and $We < 8$	$C_D = 0.125\ We$

For tap water Kuo and Wallis [49] proposed the following equations:

$Re < 0.49$	$C_D = 24\ Re^{-1}$
$0.49 < Re < 100$	$C_D = 20.68\ Re^{-0.643}$
$100 < Re < 717$	$C_D = 6.3\ Re^{-0.385}$
$Re \gg 717$ and $We < 8$	$C_D = We/3$

The dimensionless numbers in these equations are defined by:

Reynolds number $Re = \frac{\rho_L |\Delta u| d_b}{\mu_L}$

Weber number $We = \frac{\rho_L |\Delta u|^2 d_b}{\sigma}$

Morton number $Mo = \frac{(\rho_L - \rho_G) g \mu_L^4}{\rho_L^2 \sigma^3}$

Making use of the correlation proposed by Ishii and Zuber [48] for correction of the drag force in a bubble swarm, the drag coefficient C_D is multiplied by the correction factor f_k which is given by:

$$f_k = \left(\frac{1 + 17.67 f^{\frac{6}{7}}}{18.67 f} \right)^2 \qquad (7.24)$$

with

$$f = \sqrt{1 - \varepsilon_G} \frac{\mu_L}{\mu_m} \qquad (7.25)$$

and

$$\frac{\mu_L}{\mu_m} = (1 - \varepsilon_G)^{2.5 \frac{\mu_G + 0.4\mu_L}{\mu_G + \mu_L}} \tag{7.26}$$

for $d_b > 1.8$ mm. For bubble diameter smaller than 1.8 mm the Reynolds number is calculated with μ_m from Eq. (7.26).

7.5.1.2
Virtual Mass Force

The virtual mass force represents the force required to accelerate the apparent mass of the surrounding continuous phase in the immediate vicinity of the gas bubble. Drew and Lahey [51] have proposed the following formulation:

$$S_{vm,i} = \rho_L C_{vm} \varepsilon_G a_{vm,i} \tag{7.27}$$

$$a_{vm,i} = \frac{\partial(u_{G,i} - u_{L,i})}{\partial t} - u_{G,i} \frac{\partial u_{G,i}}{\partial x_i} - u_{L,i} \frac{\partial u_{L,i}}{\partial x_i} \tag{7.28}$$

The virtual volume coefficient C_{vm} for potential flow around a sphere is 0.5. For ellipsoidal bubbles with a ratio of semiaxes 1:2 C_{vm} is 1.12. For ellipsoidal bubbles with random wobbling motions Lopez de Bertodano et al. [52] calculated C_{vm} to be about 2. In addition, C_{vm} is a function of the specific gas hold up [77–79]:

$$C_{vm} = C_{vma}(1 - \varepsilon_G) \tag{7.29}$$

with

$$C_{vm} = C_{vma}(1 - 2.78 \min [0.20, \varepsilon_G]) \tag{7.30}$$

and $0.5 < C_{vma} \leq 2.0$.

In context with the discussion on the effects of the virtual mass force it is worthwhile to insert a short comment on the so-called algebraic slip model. The momentum equation of the gas phase can be simplified if the inertial forces are neglected and only drag is considered as interfacial force. Equation (7.20) then becomes

$$0 = -\varepsilon_G \frac{\partial p}{\partial x_i} + \rho_G \varepsilon_G g_i - S_{D,i} \tag{7.31}$$

The relative velocities between gas and liquid phases are then calculated from simple algebraic equations. The computing effort is therefore similar to one phase simulations which makes the algebraic slip model very attractive. However, it is not possible to include the virtual mass force, because the algebraic characteristic disappears if derivatives for gas and liquid velocities, Eq. (7.28), are introduced.

7.5.2
Turbulence Model

For applications in two-phase flow the k-ε models have been modified in different ways. One possibility is to insert additional sources into the transport equations for k and ε [53–56]. An alternative is to consider an increase of the turbulent viscosity in the liquid phase caused by the bubbles. According to Sato et al. [57] and Lopez de Bertodano et al. [52] this effect can be described by

$$\nu_{turb} = c_\mu \frac{k^2}{\varepsilon} + C_{\mu,b} d_b \varepsilon_G |\Delta u| \tag{7.32}$$

For the parameter $C_{\mu,b}$ Lopez de Bertodano et al. [52] suggested a value of 0.6. Assuming that the optimized version of the Chen-Kim model is still valid, the transport equations for the turbulence quantities k and ε for the two phase system are given by

$$\frac{\partial(\rho_L \varepsilon_L k)}{\partial t} + \frac{\partial}{\partial x_i}(\rho_L \varepsilon_L u_{L,i} k) = \frac{\partial}{\partial x_i}\left(\rho_L \varepsilon_L \frac{\nu_{eff}}{\sigma_{k,s}} \frac{\partial k}{\partial x_i}\right) + \\ + \frac{\partial}{\partial x_i}\left(\rho_L k \frac{\nu_{turb}}{Sc_{turb}} \frac{\partial \varepsilon_L}{\partial x_i}\right) + \rho_L \varepsilon_L (P_k - \varepsilon) \tag{7.33}$$

and

$$\frac{\partial(\rho_L \varepsilon_L \varepsilon)}{\partial t} + \frac{\partial}{\partial x_i}(\rho_L \varepsilon_L u_{L,i} \varepsilon) = \frac{\partial}{\partial x_i}\left(\rho_L \varepsilon_L \frac{\nu_{eff}}{\sigma_{\varepsilon,CK}} \frac{\partial \varepsilon}{\partial x_i}\right) + \frac{\partial}{\partial x_i}\left(\rho_L \varepsilon \frac{\nu_{turb}}{Sc_{turb}} \frac{\partial \varepsilon_L}{\partial x_i}\right) \\ + \rho_L \varepsilon_L (C_{l,CK} \frac{P_k}{\tau_d} + C_{3,CK} \frac{P_K}{\tau_d} - C_{2,CK} \frac{\varepsilon}{\tau_d}) \tag{7.34}$$

with

$$P_k = \nu_{turb}\left(\frac{\partial u_{L,i}}{\partial x_j} + \frac{\partial u_{L,j}}{\partial x_i}\right)\frac{\partial u_{L,i}}{\partial x_j} \tag{7.35}$$

7.5.3
Impeller Model

An important effect of the aeration is the reduction of the pumping capacity (volumetric rate of liquid leaving the impeller in radial direction). As already discussed in context with the single phase simulations, again the boundary conditions at the impeller are predicted from measured data of the averaged velocities. The first step towards estimation of these velocities is a reasonable estimate of the pumping capacity at non-aerated conditions. Taking into account an averaged value from the large variations in pumping capacities for Rushton turbines and the strong influence of the impeller off-bottom clearance [58], a pumping capacity of $Q_{L,0}=1.53nd^3_i$ was estimated for the impeller configuration used by Bombac et al. [59, 60]. The data from these investigations will be used for the following comparison between measured and simulated values of the specific gas hold up (see Sect. 7.5.4).

The quantitative estimation of the effect of aeration is even more difficult. Rousar and Van den Akker [61] reported the following correlation between power consumption and pumping capacity for aerated impellers:

$$\left(\frac{Q_{L,G}}{Q_{L,0}}\right)_1 = \left(\frac{P_G}{P_o}\right)^{0.341} \tag{7.36}$$

which is close to the effect on the circulation time for low values of the aeration number reported by Reuss and Bajpai [4]. On the other side, Joshi et al. [62] recommended

$$\left(\frac{Q_{L,G}}{Q_{L,0}}\right)_2 = \left(\frac{P_G}{P_o}\right)^{1.0} \tag{7.37}$$

Due to these uncertainties, it was decided to fit this parameter from the comparison between measured and simulated values of the specific gas hold up with the constraint

$$\left(\frac{Q_{L,G}}{Q_{L,0}}\right)_1 < \left(\frac{Q_{L,G}}{Q_{L,0}}\right)_{estimated} < \left(\frac{Q_{L,G}}{Q_{L,0}}\right)_2 \tag{7.38}$$

7.5.4
Simulation Results

The simulations discussed in the following have been performed for the vessel and impeller geometries used by Bombac et al. [59, 60] in their systematic investigations of the distribution of specific gas hold-up at different speed of agitation. These measurements were performed by using conductivity sensors. Table 7.5 summarizes the geometrical properties of the system as well as operation conditions (aeration and agitation).

For the prediction of the interfacial forces (Sect. 7.5.1) it is necessary to estimate a representative bubble diameter. From measurements of bubble size distributions in stirred vessels reported by Barigou [63, 64] for coalescing and non-coalescing systems, an average value of the bubble diameter was calculated with

$$d_{10} = \sum_{i=1}^{n_b} q_i d_{b,i} \tag{7.39}$$

where q_i=relative number fraction and $d_{b,i}$=bubble diameter in class i. Additionally, the Sauter mean diameter was predicted according to

$$d_{32} = \sum_{i=1}^{n_b} q_i d_{b,i}^3 / \sum_{i=1}^{n_b} q_i d_{b,i}^2 \tag{7.40}$$

For the bubble size distribution measured by Barigou [63, 64] a value of

$$\frac{d_{10}}{d_{32}} = 0.67 \tag{7.41}$$

Table 7.5. Geometrical and operational parameters in the investigations of Bombac [59, 60]

Nr.	N min^{-1}	Q m^3 s^{-1}	P_G/Po	$Q_{L,G}/Q_{L,0}$ Eq. (7.36)	Eq. (7.37)	Estimated
1	376	5.56×10^{-4}	0.77	0.91	0.77	0.91
2	376	1.67×10^{-3}	0.43	0.75	0.43	0.64
3	266	1.67×10^{-3}	0.48	0.78	0.48	0.62

D_T=0.44 m, $H\,D_T^{-1}$=1, $d_i\,D_T^{-1}$=0.33, off-bottom clearance/liquid height $C\,H^{-1}$=0.25

Table 7.6. Predicted bubble diameters for the three operation conditions of Bombac [59, 60]

Nr.	N $[\text{min}^{-1}]$	Q $[\text{m}^3\,\text{s}^{-1}]$	P_O $[\text{W}]$	P_G/P_O $[-]$	P_G $[\text{W}]$	P_G/V_L $[\text{W}\,\text{m}^{-3}]$	$\varepsilon_{G,Ca}$ $[\%]$	d_{32} mm	d_{10} mm
1	367	5.56×10^{-3}	99	0.77	76	1058	2.7	3.1	2.1
2	376	1.67×10^{-3}	99	0.43	42	591	6.0	5.0	3.4
3	266	1.67×10^{-3}	35	0.48	17	235	5.4	6.5	4.4

was calculated. To estimate bubble diameters for the operation conditions in the work of Bombac [59, 60], in the first step the Sauter mean diameter (in cm) was predicted from the well known Calderbank equation [65]:

$$d_{32} = 4.15 \left(\frac{\sigma^{0.6}}{(P_G/V_L)^{0.4}\rho_L^{0.2}} \right) \overline{\varepsilon_{G,Ca}}^{-0.5} + 0.09\,cm \tag{7.42}$$

with

$$\overline{\varepsilon_{G,Ca}} = \left(\frac{u_G^0\,\overline{\varepsilon}_{G,Ca}}{u_t} \right)^{0.5} + 0.0216 \left(\frac{\sigma^{0.6}}{(P_G/V_L)^{0.4}\rho_L^{0.2}} \right)^{-1} \left(\frac{u_G^0}{u_t} \right)^{0.5} \tag{7.43}$$

For the rise velocity Calderbank assumed a constant value of $0.26\,\text{m s}^{-1}$.

The information necessary to apply Eq. (7.42) for the three operation conditions from the work of Bombac [59, 60] are summarized in Table 7.6.

Results from the simulations performed with the software package Phoenics are shown in Figs. 7.11–7.14 [42]. Figure 7.11a,b shows exemplary simulated flow velo-

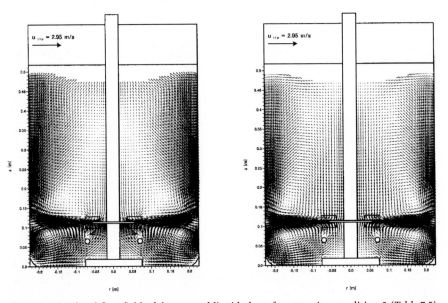

Fig. 7.11. Simulated flow field of the gas and liquid phase for operation condition 2 (Table 7.5)

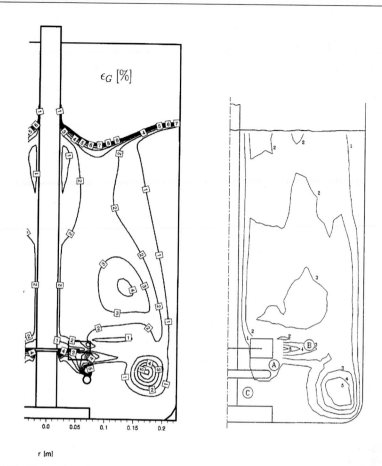

Fig. 7.12. Comparison between simulated (*left*) and measured (*right*) local value of the gas-hold up at operation condition 1 of Bombac [59, 60]

Table 7.7. Comparison of integral values for the specific gas hold up

Operation condition No.	$\varepsilon_{G,Ca}$ Eq. (7.43) [%]	ε_G Bombac [59, 60] experimentally observed [%]	ε_G, Simulations [%]
1	2.7	2.2	1.8
2	6.0	4.2	4.0
3	5.4	3.3	4.3

cities for the gas and liquid phase. Figures. 7.12–7.14 summarize comparisons between simulated and predicted local values of the specific gas hold up. Figure 7.15 shows contour plots of volume fractions at three different angles between the baffles as well as at two different levels of the tank. In total, the simulated values of the local gas hold up show a satisfactorily agreement with experimental data for the three different operation conditions. Table 7.7 shows the comparison of the

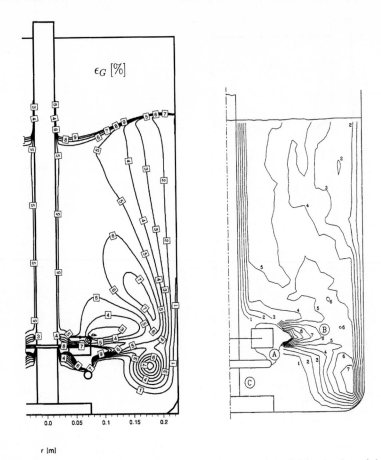

Fig. 7.13. Comparison between simulated (*left*) and measured (*right*) local value of the gas-hold up at operation condition 2 of Bombac [59, 60]

integrated value of the specific gas hold up. With the exception of operation condition 3 again, a reasonable agreement is observed. A careful analysis of these deviations illustrates that the intensity of gas recirculation is slightly overpredicted in the simulations. This is probably caused by the uncertainties in the prediction of the average bubble diameter.

7.6
Application of CFD to Simulations of Mixing and Biotechnical Processes

7.6.1
Methodology

The approach used for the application of CFD is illustrated in Fig. 7.16. It is based on the assumption that the stationary flow field is not affected by mass transfer and

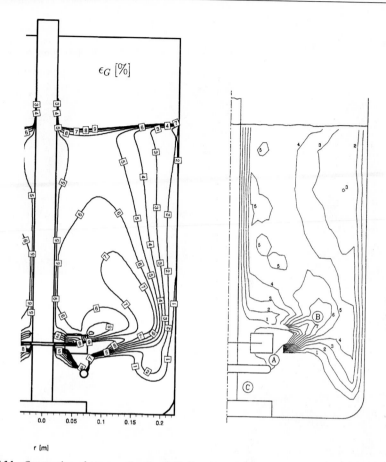

Fig. 7.14. Comparison between simulated (*left*) and measured (*right*) local value of the gas-hold up at operation condition 3 of Bombac [59, 60]

reactions. This is a reasonable assumption for many biotechnical processes with Newtonian flow behavior. Also the depletion of oxygen from the gas phase is rather low and usually compensated by the desorption of carbon dioxide. The methodology is attractive because it permits a separation of fluiddynamics (momentum balances, continuity equations, and turbulence model) from material balance equations for the state variables of interest. Figure 7.16 illustrates how results from the fluiddynamic simulations (mean velocities $u_{r,z,\varphi}^{G,L}(r, z, \varphi)$, turbulent dispersion coefficient D_{eff} (r,z,φ), and local gas hold up ε_G (r,z,φ)) can be used as parameters in the material balance equations.

The main advantage of the separation is the reduction of the computational effort. Another aspect is the fact, that the two sets of equations can be solved with different numerical methods and on different numerical grids. Due to the nature of the nonlinearities in material and momentum equations they usually require different grid refinements in different areas of the computational domain. Additionally, if the assumption of a stationary flow field is valid, the simulation

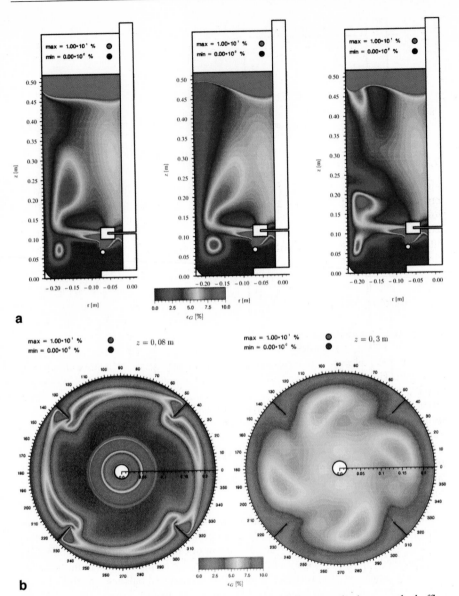

Fig. 7.15. a Contour plots of specific gas hold up at three different angles between the baffles. b Contour plots of specific gas hold up at two different levels of the tank

of the coupled set of equations would be unnecessarily slowed down by solving the momentum equations. The material balance to be solved for each of the reacting components reads

$$\frac{\partial c}{\partial t} + u_i \frac{\partial c}{\partial x_i} = -\frac{\partial}{\partial x_i}\left(D_{eff}\frac{\partial c}{\partial x_j}\right) + S_c \qquad (7.44)$$

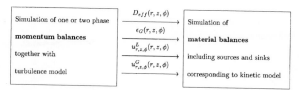

Fig. 7.16. Separation of momentum and material balance equations

S_c = reaction+mass transfer gas/liquid+inlets/outlets

$D_{eff}(r,z,\varphi)$ = turbulent dispersion coefficient=$v_{eff}(r,z,\varphi)/Sc_{turb}$

with Sc_{turb}=turbulent Schmidt number, assumed to be 1, and $v_{eff}(r,z,\varphi)$ from turbulence modeling. Discretization of material balance equations is made using finite volume elements. They are solved either with the differential-algebra solver Limex (two-dimensional simulations) or ug (unstructured grids, development of the Institute for Computer Applications, University of Stuttgart, Prof. G. Wittum) for three-dimensional simulations.

7.6.2
Simulation of Tracer Experiments

Tracer experiments are used to determine mixing characteristics. The most important tracer substances are acids or bases, electrolytes, colored agents and heated liquid. Pulse experiments can be simulated by solving the material balance equation of the tracer. Figure 7.17 illustrates an example of the simulated dynamics of a tracer distribution for a single, symmetrically positioned Rushton turbine. The terminal mixing time θ, used as a quantitative measure, can be easily predicted from the simulated response within the finite volume element which corresponds with the sensor position.

From systematic simulations of tracer experiments in a stirred vessel equipped with a Rushton turbine, a height/tank diameter ratio $HD_T^{-1}=1$, an impeller/tank diameter ratio=0.3125, and an impeller clearance/height of liquid ratio=0.31, the following correlation was obtained:

$$n\theta_{95} = 27.5 \tag{7.45}$$

with θ_{95}=time for 95% homogeneity. This result is in reasonable agreement with numbers reported in the literature ($n\theta=35$ [66] and =32 [67]).

Despite this good agreement, the terminal mixing time itself does not provide a very informative measure of the mixing process. Because transient concentrations in all volume elements are available from the simulations, it is also possible to calculate the time course of inhomogeneity defined by Landau and Prochazka [68]:

$$I(t) = \frac{1}{V(c^\infty - c^0)} \sum_{i=1}^{n} V_i |(c_i(t) - c^\infty)| \tag{7.46}$$

Figure 7.18 summarizes the transient inhomogeneity for three simulated tracer experiments differing in the position of the tracer input.

The following example serves to illustrate that tracer experiments also help to discriminate between different turbulence models. For this purpose, the classical tracer

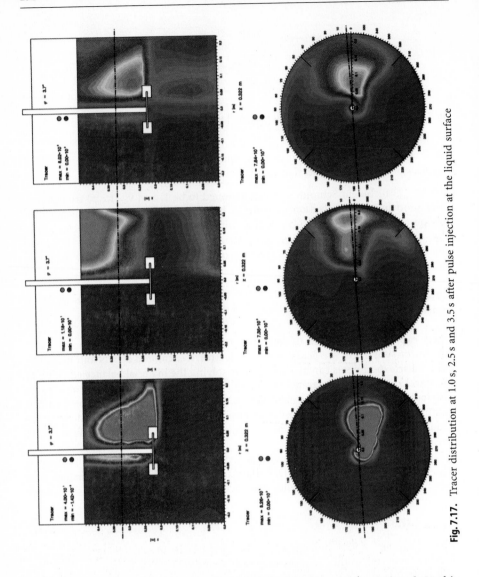

Fig. 7.17. Tracer distribution at 1.0 s, 2.5 s and 3.5 s after pulse injection at the liquid surface

experiments suggested by Khang and Levenspiel [69] have been employed. In this experiment (also described by Tatterson [58] and Reuss and Bajpai [4]), the pulse injection of the tracer is made with the aid of a concentric ring near the stirrer tips and the response measured with a concentric ring electrode nearby. As a simple representation of the recirculation flow for an impeller symmetrically placed in a tank with $HD_T^{-1}=1$ Khang and Levenspiel [69] suggested a tank-in-series model with recirculation loop. The approximated pulse response for such a recycle system takes the form

$$y(t) = 1 + 2 \, \exp\left(-\frac{2\pi^2}{n_t t_c}t\right) \, \cos\left(\frac{2\pi}{t_c} + \frac{2\pi}{n_t}\right) \tag{7.47}$$

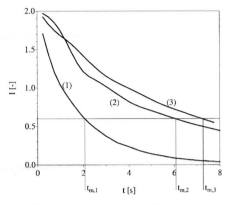

Fig. 7.18. Time course of inhomogenity for three different tracer pulses, 1: circular pulse into stirrer zone, 2: pulse onto liquid surface, 3: pulse into stirrer zone

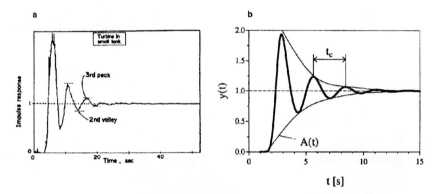

Fig. 7.19a,b. Response to a circular pulse into the stirrer zone: **a** measured (Khang and Levenspiel [69]); **b** simulated

The two required parameters – circulation time t_c and number of tanks in the cascade n_t – can be predicted from frequency and decrease of the amplitude of the measured or simulated response, respectively. Figure 7.19 shows a measured [69] and a simulated (material balance Eq. 7.44 and CFD) response. Simulated and measured responses agree qualitatively well. The more interesting results of this exercise are illustrated in Fig. 7.20. The simulated response, obviously, is very sensitive to the turbulence model used in the simulations. As a consequence of the overestimation of the turbulent viscosity by the standard k-ε model, the corresponding simulations show a response in which, in contrast to the experimental observations, the oscillations are completely damped down.

Large scale fermentation equipment usually contains more impellers. Based on the CFD simulations shown in Sect. 7.4, it is then also possible to investigate the mixing behavior of these systems. Figure 7.21 depicts the flow field and distribution of a tracer 60 s after introduction of the pulse from the top in a 22 m^3 fermentor (geometrical properties: see [70]). Figure 7.22 shows a comparison of measured [70] and simulated responses for the same reactor.

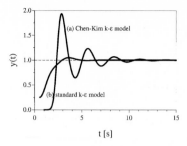

Fig. 7.20. System response to a pulse into the stirrer zone: graph (a) flow field calculated with the optimized Chen-Kim model, graph (b) flow field calculated with standard k-ε model

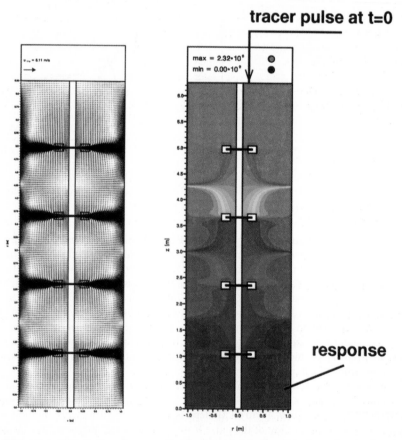

Fig. 7.21. Flow field (*left*) and tracer distribution (*right*) 60 s after the pulse (tank volume $22\,\mathrm{m}^3$)

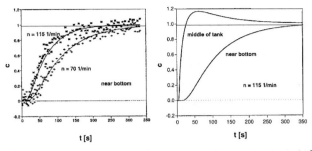

Fig. 7.22. Comparison of measured [70] and simulated responses for the 22-m^3 reactor

A careful analysis of these comparisons illustrated that the measured signals always show a slightly faster response. With the aid of further investigations it was possible to indicate that these differences are caused by the insufficiency of the two-dimensional approximations. The two-dimensional approximations show a very poor axial mixing between the impeller regions of the individual Rushton turbines resulting in a more or less independent symmetrical flow field. The three-dimensional simulations of the multiple impeller systems (not shown) indicate an additional tangential mixing effect primarily caused by the flow around the baffles. Despite these insufficiencies of the two-dimensional simulations, a reasonable agreement between simulated and measured terminal mixing times could be observed. The mixing time calculated from the equation for multiple impeller systems proposed by Groen [71] which was experimentally verified in vessels between 0.015 m^3 and 130 m^3

$$\frac{\theta_{95}}{D_T^{2/3}} \varepsilon^{1/3} = 0.11 \left(\frac{w}{D_T}\right)^{-4/3} \left(\frac{H + n_i D_T}{H}\right)^2 \left(\frac{H}{D_T}\right)^2 \tag{7.48}$$

where n_i=number of impellers, H=liquid height, w=height of impeller blade, and D_T=tank diameter, is $(\theta_{95})_{measured}$=257s. This value is in reasonable agreement with the mixing time estimated from the simulations, $(\theta_{95})_{simulated}$=288s.

7.6.3
Simulations of Substrate Distribution in Fed Batch Fermentations

The following example shows how simulations can be applied for optimization of fed batch processes which are very common for a variety of biotechnical processes to avoid oxygen limitations, heat transfer problems, over-flow metabolism, or catabolite regulation. The concentration of the carbon and energy source in the feed are as high as possible (in the range of 500 kg m^{-3}). The concentration inside the tank is very often in the range of the saturation constant, e.g., in the concentration range of 1–100 mg l^{-1}. The challenge in the scale up of these processes is then to prevent concentration gradients resulting in further limitations ($c_S < c_{S,crit}$) as well as unwanted byproduct formation or inhibition of production rates ($c_S > c_{S,crit}$). Examples of over-flow metabolism are the growth of *Saccharomyces cerevisiae* (production of ethanol) and *Escherischia coli* (production of acetate).

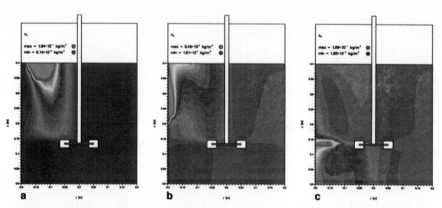

Fig. 7.23a–c. Substrate distribution for different positions of the inlet of the concentrated feed solution: **a** liquid surface; **b** wall; **c** stirrer zone

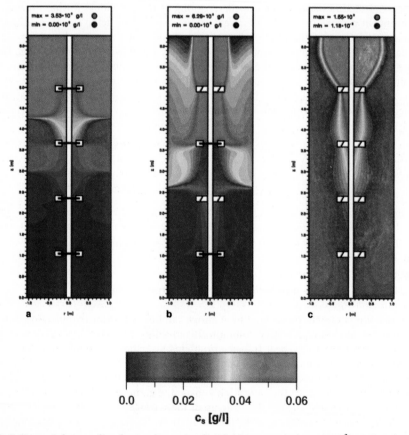

Fig. 7.24a–c. Substrate distribution in a stirred tank bioreactor (volume 22 m^3) equipped with different impellers: **a** 4 Rushton turbines; **b** 2 Rushton turbines and 2 pitched blade impellers; **c** 4 pitched blade impellers

Assuming that oxygen supply is sufficient to avoid local oxygen limitations, the kinetic model required for the simulation includes only the material balance equation for the substrate. As suggested in earlier simulations based on recirculation models (micro-macromixer) by Bajpai and Reuss [80] the uptake kinetics is only considered in the vicinity of the so-called critical sugar concentration. Thus, a rather simple unstructured empirical model was chosen for the purpose of this study. It involves a Monod type of kinetics for substrate uptake

$$r_S = r_S^{max} \frac{c_S}{K_S + c_S} c_X = -S_C \tag{7.49}$$

which holds at a certain time of the process for the corresponding biomass concentration c_X. If substrate concentration c_S locally exceeds the critical concentration $c_{S,crit}$, an ethanol production rate is superimposed which is given by

$$r_P = r_P^{max} \frac{c_S - c_{S,crit}}{K_S + (c_{S,crit} - c_S)} c_X \tag{7.50}$$

Equations (7.49) and (7.50) are then used as source terms in the material balance equation, Eq. (7.44). Additionally, the feeding rate is considered as a source term in the volume element corresponding to the feeding point. Figure 7.23 shows results of simulations of the substrate distribution at three different positions for the substrate inlet for a vessel with a volume of 68 l [81]. As expected, feeding of the concentrated sugar solution into the impeller region leads to the best equidistribution of substrate.

Again simulations have been performed for large scale vessels with multiple impellers. Figure 7.24 summarizes the distribution of sugar in a vessel of $22\,m^3$ equipped with different combinations of Rushton turbines and axial impellers. As expected, the axial impellers generally lead to a better distribution of the substrate.

7.6.4
Production of Acetoin/Butanediol with *Bacillus subtilis*

Bacillus subtilis has been used several times as a model for an oxygen sensitive culture for the characterization of the effects of inhomogeneities on the intensity of oxygen transfer gas liquid in stirred tank reactors [73]. At dissolved oxygen concentrations below 1% saturation, the ratio of the two production rates of acetoin and butanediol strongly depends on the dissolved oxygen concentration. In other words, the selectivity of the process is very sensitive to changes in dissolved oxygen under microaerobic conditions. A detailed simulation of these effects requires a model for the oxygen gradients as well as local production rates for the two products as a function of dissolved oxygen concentration.

For the simulations presented in the following, a kinetic model proposed by Moes et al. [73, 74] has been used. This model takes into account the formation of biomass and the two products acetoin and butanediol as well as substrate and oxygen consumption. The source terms in the material balance equations (Eq. 7.44) for the six state variables then read

$$S_X = r_{ATP \to X} Y_{X/ATP} c_X$$

$$S_S = -\frac{r_{S \to ATP}}{Y_{P/S}} c_X$$

$$S_{Ac} = (r_{S \to Ac} - r_{Ac \to Bu} + r_{Bu \to Ac}) c_X$$

$$S_{Bu} = (r_{Ac \to Bu} - r_{Bu \to Ac}) c_X \qquad (7.51)$$

$$S_{O_2^L} = -r_{NADH} Y_{o_2/NAD} c_X + q_{O_2}$$

$$S_{O_2^G} = q_{O_2} \frac{1 - \varepsilon_G}{\varepsilon_G}$$

Mass transfer gas/liquid is modeled as local mass transfer rate:

$$q_{O_2}(r, z, \varphi) = k_L(r, z, \varphi) a(r, z, \varphi) \left(c_{O_2}^*(r, z, \varphi) - c_{O_2}^L(r, z, \varphi) \right) \qquad (7.52)$$

The kinetic expressions are given by:

$$r_{ATP \to X} = \left[\left(Y_{ATP,aer} \left(\frac{1}{Y_{P/S}} - \frac{1}{Y_{Ac/S}} \right) + Y_{ATP,anaer} \right) r_{S \to AC} + \right.$$

$$\left. + Y_{ATP/Bu}(r_{Ac \to Bu} - r_{Bu \to Ac}) R_S \right] / \left(1 + \frac{Y_{ATP,aer} Y_{X/ATP}}{Y_{X/S}} \right)$$

$$r_{S \to E} = \left(\frac{1}{Y_{P/S}} - \frac{1}{Y_{Ac/S}} \right) r_{S \to Ac} - \frac{Y_{X/ATP}}{Y_{X/S}} r_{ATP \to X}$$

$$r_{NADH} = Y_{NAD,resp} r_{S \to E} + Y_{NAD/Ac} r_{S \to Ac} + r_{Bu \to Ac} - r_{Ac \to Bu} + r_{ATP \to X} Y_{X/ATP} Y_{NADH/X}$$

$$r_{S \to Ac} = k_1 R_S$$

$$r_{Ac \to Bu} = k_2 R_{Ac} R_S + k_{2,eq} R_{Ac}(1 - R_S)$$

$$r_{Bu \to Ac} = k_3 R_{Bu} R_S + k_{3,eq} R_{Bu}(1 - R_S)$$

$$R_S = \frac{c_S}{K_S + c_S} \qquad R_{Ac} = \frac{c_{Ac}}{K_{Ac} + c_{Ac}} \qquad R_{Bu} = \frac{c_{Bu}}{K_{Bu} + c_{Bu}} \qquad R_{O_2} = \frac{c_{O_2}}{K_{O_2} + c_{O_2}}$$

$$k_1 = k_{1,0} + m_1 R_{O_2} \qquad k_2 = k_{2,0} + m_2 R_{O_2} \qquad k_3 = k_{3,0} + m_3 R_{O_2}$$

$$Y_{P/S} = Y_{P/S,0} + m_4 R_{O_2} \qquad Y_{ATP/NADH_2} = Y_{ATP,0} + ATP_{max} R_{O_2}$$

$$Y_{ATP,aer} = Y_{ATP/PYR} + Y_{NAD,resp} Y_{ATP/NADH_2}$$

$$Y_{ATP,anaer} = Y_{ATP/PYR} + Y_{NAD,Ac} Y_{ATP/NADH_2}$$

$$Y_{ATP,Bu} = Y_{NAD,Bu} Y_{ATP/NADH_2} \qquad (7.53)$$

The values of the model parameters are given in the original work of Moes [73]. The concentration at the gas liquid interface in Eq. (7.52) is calculated from Henry's law,

$$H = \frac{p_{O_2}(r, z\varphi)}{x_{O_2}(r, z, \varphi)} \qquad (7.54)$$

with $x_{O_2}(r, z\varphi) = c_{O_2}(r, z, \varphi) / \sum_i c_i$ and $p_{O_2}(r, z, \varphi) = y_{O_2}(r, z, \varphi) P(r, z, \varphi)$

Additionally, it is possible to estimate local values for the volumetric mass transfer coefficient. For this purpose the mass transfer coefficient is estimated with the aid of the equation suggested by Kawase and Moo Young [75]

$$k_L(r,z,\varphi) = 0.301(\varepsilon(r,z,\varphi)\upsilon_L)^{0.25}Sc^{-0.5} \tag{7.55}$$

and the local value of the specific surface area from

$$a(r,z,\varphi) = \frac{6\varepsilon_G(r,z,\varphi)}{d_{32}} \tag{7.56}$$

Due to the reduced coalescence in the fermentation broth, a constant Sauter mean diameter has been assumed in these simulations.

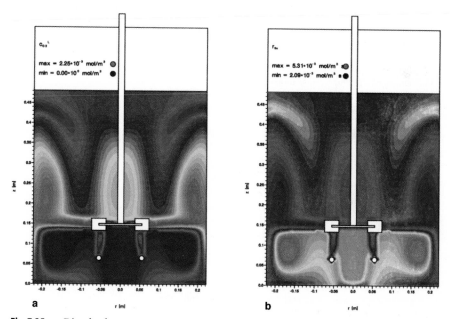

Fig. 7.25. **a** Dissolved oxygen concentration after 8 h of batch fermentation of *Bacillus subtilis*. **b** Production rate of butanediol after 8 h of batch fermentation of *Bacillus subtilis*

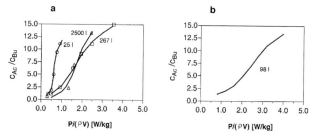

Fig. 7.26. Ratio of acetoin to butanediol as a function of specific power input: **a** measured (data from [72]); **b** simulated

Figure 7.25 shows typical results from the simulation of the system of material balance equations which are parameterized from the CFD simulations as described before. In this figure, distributions of dissolved oxygen concentration and the production rate of butanediol after 8 h of simulated batch fermentation are shown. Figure 7.26 illustrates a comparison of measured [72] and simulated ratios of the two products acetoin and butanediol at the end of the fermentation as a function of specific power input. The simulations show a behavior qualitatively similar to that observed by Griot [72]. No attempt was made to obtain a quantitative fit due to the diversity of sources of parameter values.

References

1. Balmer GJ, Moore IPT, Nienow, AW (1987) Aerated and anaerated power and mass transfer characteristics. In: Ho ES, Oldshue JY (eds) Biotechnology, scale-up and mixing. AIChE, New York, p 116
2. Gbewongyo K, Dimasi D, Buckland BC (1987) Characterization of oxygen transfer and power absorption of hydrofoil impellers in viscous mycelial fermentations. In: Ho ES, Oldshue JY (eds) Biotechology, scale-up and mixing. AIChE, New York, p 128
3. Nienow AW (1990) Trends Biotechnol 8:224
4. Reuss M, Bajpai R (1991) Stirred tank models. In: Schügerl K (ed) Biotechnology, A multi-volume comprehensive treatise, vol. 4: Measuring, modelling and control. VCH, Weinheim, New York, Basel, Cambridge, p 299
5. Reuss M (1995) Stirred tank bioreactors. In: Asenjo JA, Merchuk JC (eds) Bioreactor system design. Marcel Dekker, New York Basel Hong Kong, p 207
6. Ju SY (1987) Three-dimensional turbulent flow-field in a turbine stirred tank. Ph. D. Thesis, Louisiana State University
7. Joshi, JB, Ranade VV (1990) Trans I Chem Eng 68:19
8. Kresta SM, Wood PE (1991) AIChE J 37:448
9. Van't Riet K, Smith JM (1975) Chem Eng Sci 30:1093
10. Perng CY, Murthy JY (1993) AIChE Symp Ser 89:37
11. Takeda H, Narasaki K, Kitajima H, Sudoh S, Onofusa M, Iguchi S (1993) Computers and Fluids 22:223
12. Brucato A, Ciofalo M, Grisafi F, Micale G (1998) Chem Eng Sci 53:3653
13. Gosman AD (1998) Trans I Chem Eng 76:153
14. Platzer B (1981) Chem Tech 33:16
15. Harvey PS, Greaves M (1982) Trans I Chem Eng 60:201
16. Placek J, Tavlarides LL (1985) AIChE J 31:1113
17. Placek J, Tavlarides LL, Smith GW, Fort I (1986) AIChE J 32:1771
18. Middleton JC, Pierce F, Lynch PM (1986) Chem Eng Res Des 64:18
19. Bakker A, Van den Akker HEA (1994) Trans I Chem Eng 72:583
20. Togatorop A, Mann R, Schofield DF (1994) AIChE Symp Ser 299:19
21. Issa RI, Gosman AD (1981) The computation of three-dimensional turbulent two phase flows in mixer vessels. In: Numerial methods in laminar and turbulent flow. Pineridge Press, Swansea, p 829
22. Trägardh C (1988) A hydrodynamic model for the simulation of an aerated agitated fed-batch fermentation. In: Bioreactor fluid dynamics. Elsevier Applied Science, p 117
23. Politis S, Issa RI, Gosman AD, Lekakon C, Looney MK (1992) AIChE J 38:1946
24. Morud K, Hjertager BH (1993) Computational fluid danymics simulations of bioreactors. In: Mortensen U, Noorman H (eds) Bioreactor performance. IDEON, Lund, Sweden, p 47
25. Bakker A, Van den Akker HEA (1994) Trans I Chem Eng 72:594
26. Ranade VV, Van den Akker HEA (1994) Chem Eng Sci 49:5175
27. Launder BE, Spalding DB (1972) Mathematical models of tubulence. Academic Press
28. Launder BE, Spalding DB (1974) Comp Meth Appl Mech Eng 3:269

29. Kim JJ (1978) Three dimensional turbulent flow-field in a turbine stirred tank. PhD thesis, Louisiana State University
30. Pope SB (1978) AIAA J 16:279
31. Hanjalick K, Launder BE (1980) Trans ASME 102:34
32. Kline SJ, Cantwell BJ, Lilley GM (1981) The 1980–1981 HFOSR-HTMM-Stanford Conference on Complex Turbulent Flow, Standfort University, I, II, III
33. Roback R, Johnson BV (1983) NASA CR-168,252
34. Chen YS, Kim SW (1987) NASA CR-179,204
35. Batchelor GK (1953) The theory of homogeneous turbulence. Cambridge University Press, Cambridge
36. Wu H, Patterson G (1989) Chem Eng Sci 44:2207
37. Jenne M, Reuss M (1999) Chem Eng Sci (in press)
38. Patankar SV (1980) Numerical heat transfer and fluid flow. Mc Graw-Hill
39. Costes J, Couderc JP (1988) Chem Eng Sci 43:2765
40. Yakhot V, Smith LM (1992) J Sci Comput 7(1)
41. Nagata S (1975) Mixing: principles and application. Wiley
42. Jenne, M (1999) Modellierung und Simulation der Strömungsverhältnisse in begasten Rührkesselreaktoren. PhD thesis, Universität Stuttgart
43. Bouafi M, Roustan M, Djebbar R (1997) Mixing IX, multiphase systems. Récents progrès en génie des procédés 11(52):137
44. John AH, Bujalski W, Nienov AW (1997) Mixing IX, multiphase systems. Récents progrès en génie des procédés 11:169
45. Nienow AW, Elson TP (1988) Chem Eng Res Des 66:5
46. Cooke M, Middleton JC, Bush JR (1988) Proc 2nd Int Conf Bioreactor Fluid Dynamics. BHRA/Elsevier, p 37
47. Abradi V, Rovera G, Baldi G, Sicardi S, Conti R (1990) Trans I Chem E 68:516
48. Ishii M, Zuber N (1979) AIChE J 25:843
49. Kuo JT, Wallis GB (1988) Int J Multiphase Flow 14:547
50. Ishii M (1975) Thermo-fluid dynamic theory of two-phase-flow. Eyrolles
51. Drew DA, Lahey TJ (1987) Int J Multiphase Flow 13:113
52. Lopez de Bertodano M, Lahey TJ, Jones OCC (1994) Trans SME 116:128
53. Lopez de Bertodano M, Lee SJ, Lahey RT, Drew DA (1990) ASME J Fluids Enging 112:107
54. Svendsen HF, Jakobsen HA, Torvik R (1992) Chem Eng Sci 47:3297
55. Johansen ST, Boysan F (1988) Metall Trans B 19B:755
56. Lahey RT, Lopez de Bertodano M, Jones OC (1993) Nuclear Enging Des 141:177
57. Sato Y, Adatomi M, Sekoguchi K (1981) Int J Multiphase Flow 7:167
58. Tatterson GB (1991) Fluid mixing and gas dispersion in agitated tanks. McGraw Hill
59. Bombac A (1994) PhD thesis, University of Ljubljana
60. Bombac A, Zun I, Filipic B, Zumer M (1997) AIChE J 43:2921
61. Rousar I, Van den Akker HEA (1994) Proc 8th Europ Conf on Mixing, Cambridge, UK, p 89
62. Joshi JB, Pandit AB, Sharma MM (1982) Chem Eng Sci 37:813
63. Barigou M (1987) PhD thesis, University of Bath
64. Barigou M, Greaves M (1992) Chem Eng Sci 47:2009
65. Calderbank PH (1958) Trans I Chem Eng 36:443
66. Voncken RM (1966) Circumlatie stromingen en menjing in geroerde vaten. PhD thesis, Delft University of Technology
67. Hoogendoorn CJ, Hartog AP (1967) Chem Eng Sci 22:1689
68. Landau J, Prochazka J (1961) Coll Czechoslov Chem Commun 26:1976
69. Khang SJ, Levenspiel O (1976) Chem Eng Sci 31:569
70. Cui YQ, van der Lans RGJM, Noorman HJ, Luyben KCAM (1966) TransI ChemE 74 (A):261
71. Groen DJ (1994) Macromixing in bioreactors. PhD thesis, Delft University of Technology
72. Griot M (1987) Maßstabsvergrößerung von Bioreaktoren mit einer sauerstoffempfindlichen Testkultur. PhD thesis, ETH Zürich

73. Moes J (1985) Untersuchung von Mischphänomenen mit Hilfe von *Bacillus subtilis*. PhD thesis, ETH Zürich
74. Moes J, Griot M, Keller J, Heinzle E, Dunn LJ, Bourne JR (1985) Biotech Bioeng 27:482
75. Kawase Y, Moo-Young M (1990) Chem Eng J 43:B19
76. Yianneskis M, Popiolek Z, Whitelaw JH (1987) Fluid Mech 175:537
77. Watanabe T, Hirano M, Tanabe F, Kamo H (1990) Nuclear Engng Design 120:181
78. Huang B (1989) Modelisation numérique d'écoulements disphasiques à bulles dans des réacteurs chimiques. PhD thesis, Lyon
79. Kowe R, Hunt JCR, Hunt A, Couet B, Bradbury LJS (1988) Int J Multiphase Flow 14:587
80. Bajpai R, Reuss M (1982) Can J Chem Eng 60:384
81. Schmalzriedt S, Reuss M (1997) Mixing IX, Recent advances in mixing, Récent Progrès en Génie des Procédés 11(51):171

8 Bubble Column Bioreactors

Andreas Lübbert

8.1
Introduction

From the various types of bioreactors discussed in the literature, only a few are actually used in industry. By far the majority of processes are performed in stirred tank bioreactors. Besides these standard reactors, only bubble columns or, more generally, airlift reactors reached a noticeable number of applications.

In bubble column bioreactors, agitation is performed by density driven fluid motions induced by sparging air into the continuous liquid phase at the bottom of the reactor. Air sparging leads to density inversion in the culture medium and, thus, to buoyancy motions in the dispersion. At sufficiently high gas flow rates, these convective flows become turbulent, which has the practical advantage that the fluid becomes intensively mixed. Since in bioreactors for aerobic cultures, air dispersion is necessary anyway in order to provide sufficient oxygen mass transfer rates, gas sparging serves two purposes in these reactors: agitation and mass transfer. In many practical systems, mixing by these buoyancy-induced flow effects makes additional mechanical agitation needless. This is the domain where airlift reactors are applied in industrial practice.

Bubble column bioreactors are usually constructed as cylindrical vessels with an aspect ratio (reactor height/diameter) greater than two. The air can be dispersed at the bottom of the vessel by any gas sparger. For example, one can use the conventional ring sparger most often used in stirred tank bioreactors. Thus, these simple bubble column reactors do not contain any moving part and are thus not only inexpensive in terms of investment cost, but also easy to maintain, e.g., easy to clean and sterilize.

Often, these simple bubble column reactors are divided into two segments by means of additional wall elements. This may most easily be accomplished by inserting a tube concentrically into the cylindrical reactor. When then the liquid is aerated in one segment only, e.g., in the annulus, a substantial density difference is generated between both segments. Then a strong circulatory fluid motion results. The net effect is that dispersion rises or is lifted in the aerated part, named the riser section, and, by continuity, the liquid flows back in the other section referred to as the downcomer. Because of this circulatory motion, we then speak about an airlift loop reactor. There are many different ways to construct such segmented airlift loop reactors.

Since a bioreactor, as we will understand it in this chapter, is a container in which a production process is being performed, there are some obvious requirements to be met. The criteria are almost the same for all of them. One most obvious primary task of the bioreactor is mixing. The central objective in most bubble column bioreactors is to obtain a high degree of homogeneity with respect to the species dissolved in the continuous liquid phase, in order to provide nearly the same chemical and physical environmental conditions for all cells, if possible close to the optimum with respect to a maximal rate of the desired biochemical conversion. Additionally, the reactor must be able to provide sufficiently high oxygen mass transfer rates in order to supply the cells in aerobic cultures with sufficient oxygen, and, finally, it must be able to keep the culture temperature close to its optimal value.

When obtaining a high mass transfer rate is the main objective in a particular process, a stirred tank reactor might be preferable as compared to a bubble column since these reactors allow one to apply higher agitation power inputs and thus higher mass transfer rates. However, when bubble columns are able to provide sufficiently high oxygen transfer rates, they do this at a considerably higher energetic efficiency. This is often a decisive argument for using bubble column reactors, e.g., in the yeast manufacturing industry. Another essential advantage as compared to stirred tank reactors is that there is virtually no serious restriction with respect to the size of bubble column bioreactors. Thus, bioreactors larger than 300 m^3 are nearly exclusively constructed as airlift reactors.

In order to design a bubble column, it is necessary to understand the multiphase fluid motion in the reactor. Intensive bubble column research over many decades has shown that simple assumptions about the flow patterns do not allow one to make the quantitative predictions necessary for serious reactor design. Hence, physically based detailed fluid mechanical models are required for such a purpose. From the design point of view, models represent that part of our current knowledge about these reactors that can be exploited quantitatively. This chapter deals with bubble column models. Only recently has it become possible to perform detailed computational fluid dynamical simulation of the complex flow in these reactors. The focus in this chapter is on mechanistic insight into the complex two-phase gas-liquid flows in bubble column reactors.

8.2
Phenomenology

As the driving mechanism for the fluid flow in bubble column reactors, the density differences of the fluid elements in this system were identified. These density differences are mainly caused by the dispersed gas not being homogeneously distributed in the column. The gas hold-up ε, defined as the fraction of the gas-phase of the total volume V_D of the dispersion considered, is thus of main importance for airlift reactors:

$$\varepsilon = \frac{V_G}{V_D} \tag{8.1}$$

V_G is the total gas volume within V_D. The gas hold-up is thus a volume-related quantity.

When the entire culture within a bioreactor is considered, we more precisely refer to the corresponding hold-up as the total or global gas hold-up. We may, however, particularly in systems where the gas phase is not quasi-homogeneously distributed, also speak about spatial distributions of the gas phase. Then we prefer to speak about local gas holdups ε_l, which are defined by the differential quotient

$$\varepsilon_l = \frac{dV_G}{dV_D} \tag{8.2}$$

In this case we assume that we are allowed to speak about the dispersion as a quasi homogeneous fluid. This is a reasonable assumption when we speak about technical reactor sizes where the bubbles are usually very small as compared to the reactor dimensions. However, when we discuss effects on small scales, we should keep in mind that in gas-liquid dispersions the gas-phase is composed of individual bubbles of volume V_j. Then this assumption might be violated, e.g., it does not make sense to speak about continuous gas hold-ups when we are looking for effects on scales of the order of the mean bubble size. On such small scales, the gas-liquid two-phase flow system must be treated in a discontinuous way.

When we consider the individual bubbles of volume V_j, we may rewrite the total gas hold-up as

$$\varepsilon = \frac{V_G}{V_D} = \frac{\sum V_j}{V_D} \tag{8.3}$$

where the sum is drawn over all bubbles in the entire volume V_D of the dispersion. The volume V_j characterizes the bubble size. However, it is often more convenient to use its diameter instead, since the diameter can most easily be estimated by visual inspection of the dispersion. Unfortunately, the bubble diameter is not well defined since different bubbles of the same volume might have different forms. This problem has been circumvented by defining the so-called *equivalent diameter*, d_j, which is the diameter of a sphere with the bubble's volume V_j. Then we can write the gas volume V_G as

$$V_G = \sum V_j = \frac{\pi}{6} \sum d_j^3 \tag{8.4}$$

Bubbles have several functions in bioreactors, their most important one being the transport of a gaseous reactant from the dispersed gas phase across the bubble surfaces into the liquid phase. In this respect, the transport cross section A available for this mass transfer is of primary importance. A is the total interfacial area between the gas and the liquid phase. This is, apart from the top surface of the dispersion, the sum of all bubble surfaces

$$A = \sum A_j \approx \pi \sum d_j^2 \tag{8.5}$$

where A_j is the surface area of the j-th bubble. In biochemical reaction engineering, one prefers to discuss the transport cross section as a specific, in this particular case a volume related quantity. This specific interfacial area a is defined as

$$a = \frac{A}{V_D} = \frac{\sum A_j}{V_D} \approx \frac{\pi \sum d_j^2}{V_D} \tag{8.6}$$

With these definitions we are able to relate the gas hold-up with the quantities of primary interest in biochemical reaction engineering. Starting from the definition at Eq. (8.1), we get

$$\varepsilon = \frac{V_G}{V_D} = \frac{AV_G}{V_D A} = \frac{A \frac{\pi}{6} \sum d_j^3}{V_D \pi \sum d_j^2} = \frac{\frac{\pi}{6} \sum d_j^3}{\pi \sum d_j^2} = a \frac{\frac{\pi}{6} \sum d_j^3}{\pi \sum d_j^2} \qquad (8.7)$$

Since the bubbles in biotechnologically relevant dispersions are not of uniform size, such expressions are of little practical use. One is more interested in characterizing the bubbles by some mean bubble diameter. In the light of the preceding discussion it is convenient, as proposed by Sauter, to take the quantity

$$d_s = \frac{\sum d_j^3}{\sum d_j^2} \qquad (8.8)$$

as such a mean bubble diameter. The sums are drawn over all bubbles j. In the literature, d_s is referred to as the Sauter diameter.

With the Sauter diameter d_s, we obtain a simple, practically very important relation between the relevant gas-phase characteristics of a bubble dispersion:

$$\varepsilon = a \frac{d_s}{6} \quad \text{or} \quad a = \frac{6\varepsilon}{d_s} \qquad (8.9)$$

which essentially says that the important specific interfacial area a is proportional to the gas hold-up ε, the proportionality constant being dependent in a most simple way from the Sauter mean bubble diameter d_s. In other words, we can increase the specific interfacial area a by increasing the gas hold-up ε and we are more effective when we are able to keep the mean bubble size smaller.

Assuming that the volume V_L of the liquid-phase of a culture within a bioreactor is given, then the higher the gas hold-up ε, the larger the total volume V_D of the dispersion will be. Usually, in discussions about the gas-phase behavior within bioreactors, the solid phase, in particular the cells, can be attributed to the liquid phase. In reactors where the reactor horizontal cross sectional area A_{RCS} is uniform with respect to the height H_D of the dispersion, we can write the gas hold-up as a function of the height H_D and the height H_L the liquid will assume within the reactor when there are no gas bubbles would be present.

$$\varepsilon = \frac{V_G}{V_D} = \frac{V_D - V_L}{V_D} = \frac{(H_D - H_L)A_{RCS}}{H_D A_{RCS}} = \frac{(H_D - H_L)}{H_D} \qquad (8.10)$$

This relative enhancement of the culture level in the reactor by aeration or simply the presence of the gas bubbles is of practical importance, since one must take care of this level during the operation of the bioreactor.

In a global view, a higher gas hold-up of the dispersion leads to a lower mean density ρ_D of the dispersion. When the gas phases can be assumed to be quasi-homogeneously distributed, the density ρ_D can be represented as a linear function of the gas hold-up ε:

$$\rho_D = \varepsilon \rho_G + (1-\varepsilon) r_L \approx (1-\varepsilon) r_L \qquad (8.11)$$

since the gas density ρ_G is very much smaller than the density ρ_L of the liquid phase. This has an immediate consequence for the static pressure at a particular height H within the reactor. Since the static pressure drop Δp over a height interval ΔH is dependent on the density of the dispersion, it is possible to relate directly the static pressure drop and the gas hold-up

$$\Delta p = \rho_D g \Delta H \tag{8.12}$$

$$\varepsilon = 1 - \frac{\Delta p}{\rho_L g \Delta H} \tag{8.13}$$

a relationship that is often used to measure the gas hold-up in multiphase reactors. g is the gravitational acceleration.

The gas hold-up can also be viewed from the point of view of the aeration rate. The aeration rate is often represented by the gas throughput Q_G relative to the cross sectional area A_{RCS} of the reactor. This quantity is referred to as the superficial gas velocity w_{sg}, since it formally has the dimension of a velocity:

$$W_{sg} = \frac{Q_G}{A_{RCS}} \tag{8.14}$$

When the mean bubble velocity in the laboratory coordinate system is w_{bl}, then with

$$W_{bl} = \frac{H_D}{\tau_B} \tag{8.15}$$

where τ_B is the mean residence time of the bubbles in the reactor, and

$$Q_G = \frac{V_G}{\tau_B} \tag{8.16}$$

we get

$$\varepsilon = \frac{V_G}{D_D} = \frac{V_G}{A_{RCS} H_D} = \frac{V_G \tau_B}{\tau_B A_{RCS} H_D} = \frac{W_{sg} \tau_B}{V_D} = \frac{W_{sg}}{W_{bl}} \tag{8.17}$$

In other words, the gas hold-up is proportional to the superficial gas velocity, the proportionality constant being the reciprocal value of the mean bubble rise velocity in the laboratory coordinate system. The superficial gas velocity is the most important manipulatable variable. The effective bubble velocity, w_{bl}, however, depends on the fluid-dynamic properties within the reactor as well of the broth rheology.

8.3
Basic Equations of Motion

8.3.1
Fundamental Laws of Fluid Motion

The state of a fluid flow system is basically determined by the quantitative description of dynamics of the flow field, i.e., the spatial (r) and time (t) dependent fluid velocity $u(r,t)$. The basic equations of motion for fluid dynamical systems are derived from the elementary conservation laws for mass and momentum (e.g., [1]).

8.3.1.1
Mass Conservation

If we consider some finite fluid volume, the mass conservation law says that the amount of mass m within this volume can only be changed by transport of mass across its boundary surface. If we consider the density ρ, i.e., the mass per volume instead of the mass, the mathematical formulation of this statement is

$$\frac{\partial \rho}{\partial t} = -\text{div}\ (\rho \mathbf{u}) \tag{8.18}$$

8.3.1.2
Conservation of Momentum

The momentum equation is Newton's second law applied to fluid motion, saying that a substantial change of the momentum mu requires the action of a force F. In a volume related representation, the substantial derivative of $\rho \mathbf{u}$ is thus equal to the sum f of all volume related forces acting on the fluid element under consideration:

$$\frac{D(\rho \mathbf{u})}{DT} = \mathbf{f} \tag{8.19}$$

The volume-related forces acting on a fluid element are usually distinguished by body forces $\mathbf{f}_b$, pressure forces $\mathbf{f}_p$, and surface forces $\mathbf{f}_s$.

From the forces acting on the fluid volume, the field or body forces, the gravitation force is the most important in our case as long as we are not dealing with special, e.g., magnetic particles which would be influenced by magnetic fields. The gravitation leads to

$$\mathbf{f}_b = \rho \mathbf{g} \tag{8.20}$$

where $\mathbf{g}$ is the gravitational acceleration.

The pressure term is the pressure gradient across the fluid element considered:

$$\mathbf{f}_p = \nabla p \tag{8.21}$$

It can also be interpreted as a force resulting from normal stresses on the fluid element. Finally, there are forces acting tangentially on the surface of the fluid element. These are the viscous shear forces, which can be described by

$$\mathbf{f}_s = \nabla T \tag{8.22}$$

where T is the viscous stress tensor. In a viscous fluid flow, the two terms $\mathbf{f}_p$ and $\mathbf{f}_s$ describe the immediate momentum transfer between adjacent fluid elements in a viscous fluid flow.

8.3.1.3
Navier-Stokes Equation System

In fluid dynamics, the fluid elements considered as control elements are usually represented by the cells of the numerical grid used during the solution of the equations of motion. The flow velocities $\mathbf{u}$ are thus considered at discrete, spatially fixed points in the numerical grid only. This representation is referred to the Eulerian representation. In Cartesian coordinates, the substantial (or total or convectional) derivative operator becomes

$$\frac{D}{Dt} = \frac{\partial}{\partial t} + \frac{\partial}{\partial x_1}\frac{\partial x_1}{\partial t} + \frac{\partial}{\partial x_2}\frac{\partial x_2}{\partial t} + \frac{\partial}{\partial x_3}\frac{\partial x_3}{\partial t} = \frac{\partial}{\partial t} + \nabla \mathbf{u} \tag{8.23}$$

where x_i denote the components of the position vector $\mathbf{r}$. The stress tensor $\mathbf{T}$ is then defined by its components in the following form:

$$T_{ij} = \mu_{\text{eff}}\left(\frac{\partial u_i}{\partial x_j} + \frac{\partial u_j}{\partial x_i} - \frac{2}{3}\delta_{ij}\frac{\partial u_n}{\partial x_n}\right) \tag{8.24}$$

and μ_{eff} the effective dynamic viscosity of the dispersion. δ_{ij} is the Kronecker symbol.

Hence, the momentum balance reads

$$\frac{\partial \rho \mathbf{u}}{\partial t} + \nabla(\rho \mathbf{u}\mathbf{u}) = -\nabla p + \nabla \mathbf{T} \tag{8.25}$$

and the equation of mass conservation reads in a corresponding representation

$$\frac{\partial \rho}{\partial t} + \nabla(\rho \mathbf{u}) = 0 \tag{8.26}$$

It is known as the continuity equation.

The combination of both equations is usually referred to as the Navier-Stokes equation system. The Navier-Stokes equation system, known for about 150 years, is a most general equation system and encompasses all effects of fluid motions.

When we are dealing with two-phase gas-liquid flows, as we are forced to do in bioreactors, the fluid density ρ becomes a further quantity that dynamically changes with space and time. Hence, it is necessary to provide independent information about the density variations. The appropriate approaches will be discussed at a later point. First it is necessary to recognize the main practical problems that appear during current attempts to solve the Navier-Stokes equation system, since this will shed some light on our possibilities to extend this equation system.

8.3.1.4
Problems with Solving the Equations of Motion

The Navier-Stokes equation system is a quite general differential equation system. Its adjustment to the flow field in a particular technical apparatus, for example to the flow in a cylindrical vessel used as a biochemical reactor, is made by definition of the boundary conditions, which must include the precise shape of the reactor. In the case that we are dealing with, dynamic flow situations, we additionally need initial conditions to the flow.

The initial conditions are usually very simple. We can, for example, assume that initially there is no flow within the reactor. The boundary conditions thus define the technical problem under consideration. In the particular case of a fluid flow in a cylindrical vessel, the boundary conditions are as follows: the fluid velocity at the walls of the container are zero. At the top surface (excluding the effects of the vessel wall) only the axial velocity is restricted to zero because the horizontal components are free.

To a novice, the Navier-Stokes equation system looks quite simple and the boundary conditions for the fluid motions in the cylindrical vessel are very simple as well. And from this point of view, one would expect that the flow problem is easy to solve. Unfortunately, however, things are not as simple as they would seem.

The first problem is the spatial resolution which would be necessary to cover all relevant fluid motions in these reactors. Turbulent fluid motions are of primary importance, as they are induced in order to obtain good mixing properties. According to Kolmogoroff's turbulence theory, the order of magnitude of the diameter of the smallest eddies, which are of importance, is roughly 10^{-4} m. It was estimated that one needs at least 10 grid points to resolve the motion of an individual eddy [2]. If we consider a biochemical reactor of a characteristic length scale of 1 m, a simple estimation shows that we would need 10^5 elements in each direction of the numerical grid, or 10^{15} elements in a three-dimensional flow. With our current computers we are able to handle numbers of grid points of the order of magnitude of 10^6 only. This immediately shows that a straightforward approach covering all scales, which might be of importance, will be impossible in the near future.

The second severe problem is that the pressure p cannot be determined separately. Both the continuity and the momentum equation influence the pressure. Hence, solutions of both must be determined in such a way that both lead to the same pressure distribution p(**r**) at the same time t. In practice this must be obtained by a time-consuming iterative approach.

The third major problem is that there are significant nonlinear terms in the Navier-Stokes equation system. The solution of the differential equation system thus requires implicit strategies, since explicit ones are known to lead to considerable instabilities. Implicit schemes, however, require time-consuming iterations as well.

These three problems significantly reduce the number of computational grid elements that can be tackled within an acceptable computation time. Unfortunately, one must deal with a rather low spatial resolution and, consequently, only large scale motions can be resolved. Thus, these calculations of fluid motions in technical scale reactors are often called Large-Eddy-Simulations. Flow effects expected to result from motions on smaller length scales must then be lumped together in the material parameters of the equation of motion. The main material parameters in the equations are the fluid density and the fluid viscosity. In the two-phase flow of bubble column reactors both are dependent on the resolution of the numerical grid used in the calculations. The effective viscosity depends on the fluid motions with characteristic length scales significantly smaller than the fluid elements that can be resolved in the calculations. The density depends on the local positions of the bubbles. Since the bubbles are not homogeneously distributed, the density of the dispersion in a particular computational block is also dependent on its size. We need so-called subgrid-models to account for these effects. We will go into more detail later on.

Generally the physical flow is independent of its representation. The representation must be chosen for convenience reasons. Instead of using cylindrical coordinates to solve the equations of motion discussed, as would be the natural first choice, it proved to be computationally more efficient to use Cartesian coordinates in order to reach a predefined accuracy. The reason is that the matrices which result from these representations during the solution algorithm get a form that can be solved much faster. The disadvantage is that the spherical cross section of the column is more difficult to describe in this representation. However, this is overcompensated by the computational advantage.

One might argue that the problem can be treated in a two-dimensional way when dealing with cylindrical column reactors, since this would allow one to increase considerably the numerical resolution. Such a representation, however, is not possible, since the crucial ingredients of turbulent flows are vortex interactions [3], and interaction between eddies are three-dimensional effects. Comparisons of computations in so-called waver columns, vessels of rectangular cross section with one dimension of the cross section being much smaller than the other, showed a significant difference between both results from two- and three-dimensional calculations, although it was assumed that in this geometry the two dimensional approach would be possible without problems, as the degree of freedom in the third direction is much restricted in these devices.

One often made further approaches to reduce the computational load such as time averaging of the Navier-Stokes equation system. In such a way the random fluctuations of higher frequency in the fluid flow, associated with the turbulent motion, are smoothed out, while the slow variations of the flow field remain. The idea behind this averaging/smoothing is to reduce drastically the number of grid points necessary to simulate the flow in this way. The most often performed average procedure is based on the decomposition, referred to as the Reynolds decomposition, of the flow velocity **u** into two parts:

$$\mathbf{u} = \hat{\mathbf{u}} + \mathbf{u}' \tag{8.27}$$

where $\hat{\mathbf{u}}$ is a slow varying average velocity and $\mathbf{u}'$ is the fast and randomly fluctuating component that is averaged away by this particular averaging procedure, called Reynolds averaging. The averaged equations are called Reynolds equations (e.g. [2]).

Evidently a physical problem cannot be solved by averaging alone [3]. The advantage of the smaller number of grid elements is obtained at the cost of additional assumptions about the influences of the essential physics averaged or separated away, namely the turbulent fluctuations. In other words, the Reynolds-averaged equations need additional information about the influence of the turbulence in the form of apparent turbulent stresses on the fluid motion. The problem of modeling these relationships is referred to as the closure problem. Thus in order to solve the Reynolds equations, an additional turbulence model is required that is not necessary when the original Navier-Stokes equation system is solved directly. The best known of such models is the so-called k-ε model.

The k-ε model was often discussed in turbulence theory (e.g., [[2]). According Boussinsq's approximation, the Reynolds' stress, i.e., the additional term appearing with Reynolds' averaging, is proportional to the changes of the mean velocity components $\hat{u}_i$ in a perpendicular direction, the proportionality constant being the turbulent viscosity μ_T:

$$-\rho \overline{u_i' u_j'} = \mu_T \left(\frac{\partial \hat{u}_i}{\partial x_j} + \frac{\partial \hat{u}_j}{\partial x_i} \right) \tag{8.28}$$

Thus, the problem is to estimate the turbulent viscosity μ_T. General kinetic theory shows that

$$\mu_T = \rho v l \tag{8.29}$$

where v and l are the velocity scale and the corresponding length scale of the turbulent motion. Thus, these quantities have to be estimated. Kolmogoroff and Prandtl showed in the 1940s that v can be estimated by the root mean value of the mass related kinetic energy k [m^2/s^2] of the turbulent fluid motions (k=u'$_i$u'$_i$/2). It was also shown that the corresponding length scale l can be estimated to be proportional to k$^{3/2}$/ε, where ε [m^2/s^3] is the mass related energy dissipated per unit time. Then we get

$$\mu_T = C\rho k^{1/2}l = c\rho k^2/\varepsilon \tag{8.30}$$

where C and c are constants. Thus the turbulent viscosity is essentially estimated by the kinetic energy k in the turbulent fluid oscillations and the energy dissipation density ε, i.e., the power loss of the turbulently fluctuating fluid. This relationship is often called the k-ε model of turbulent viscosity (e.g., [2]). Sometimes people try to use this expression to estimate the effective viscosity in the large eddy simulation procedure as well.

In this chapter we do not consider averaged equations for the entire flow system but concentrate on the Large-Eddy-solution of the Navier-Stokes equation system.

8.3.1.5
Numerical Aspects

There are well established solution procedures that can be used to solve the Navier-Stokes equations system (Eqs. 8.25 and 8.26) provided sufficient computation power is available, e.g., the SIMPLER technique [4].

Here only a few main problems will be mentioned qualitatively in order to show the practical problems which may appear in practice. As shown before, only a very limited number of computational grid elements is possible, and the accuracy in the solution, which can be obtained at a fixed number of grid elements, very much depends on the discretization procedure of the equations of motion, i.e., the particular algorithm used to solve the equations numerically.

There are two types of errors due to the discretization method, which are of considerable importance to bubble column simulation. The first is that sharp gradients in a process variable, e.g. a concentration of a species dissolved in a fluid, are smoothed out by the numerical procedure. Since this effect is similar to a physical diffusion process or a process with an increased momentum transport due to an enhanced viscosity, this error is referred to as numerical dispersion or artificial viscosity. Sokolichin et al. [5] showed that the most often used UPWIND discretization may lead to numerical diffusion effects which are much larger than the real physical ones so that the final results are qualitatively different (damped fluctuations) to the observed patterns of the quantity investigated. The consequence is that one is forced to use so-called higher order solving procedures that can maintain the gradient. However, these higher order methods lead to another error.

This second typical error also appears at gradients and leads to overshooting effects on one or both sides of the position of the gradient. This error is referred to a the dispersion error, since it significantly influences the phase relationships of wave propagation in the fluid. One of the most problematic results of this error to bioreactor simulations is that it might result in negative concentrations.

The numerical schemes are intensively being developed in order to avoid these errors as much as possible. One currently much investigated method is the implicit total variation diminishing (TVD) method that considerably reduces the artificial dispersion effects [5]. This method tries to reduce both errors at the cost of additional iteration steps which are, of course, time consuming.

8.3.2
Two-Fluid Model

In two-phase gas-liquid flows, the fluid density varies with time and spatial position. Hence we need an additional model describing the density variations. One can relate the density $\rho = \rho_D$ of the dispersion to the hold-up or volume fraction ε of the gas phase by Eq. (8.11).

The volume-related mass $\varepsilon \rho_G$ of gas in the control volume element may be balanced [5]. Assuming that the gas-holdup can be regarded as a continuous quantity, we can then formulate a continuity equation for $\varepsilon \rho_G$, which essentially means that we make a balance across the mass of the gas content per unit volume. Under the assumption that there is no significant mass transfer or chemical reaction, we obtain

$$\frac{\partial \rho_G \varepsilon}{\partial t} + \nabla(\rho_G \varepsilon \, \mathbf{u}_G) = \nabla[D\nabla(\rho_G \varepsilon)] \qquad (8.31)$$

where the term on the right hand slide of accounts for some generalized effective gas phase losses due to random fluctuations of the bubble paths, which is described by a diffusion term. When gas absorption, desorption, or chemical reaction effects are significant, a further source term must be added on the right hand side of the equation. Thus, since ρ_G can be considered constant and known, we obtained a dynamic equation for the local gas holdup ε.

A problem might be to determine the velocity $\mathbf{u'}_G$ of the gas phase. However, this can be estimated from the liquid phase velocity $\mathbf{u}$ by assuming a slip velocity $\mathbf{u'}_{slip}$ between the gas phase and the liquid phase:

$$\mathbf{u}_G = \mathbf{u} + \mathbf{u'}_{slip} \qquad (8.32)$$

There are several proposals for the slip force in the literature, which account for the slip between individual bubbles of different sizes and a stagnant continuous liquid phase or a bubble swarm and the liquid in which it rises (e.g., [6]). As experiments showed, a fix slip velocity of 20 cm/s in a vertical direction is not too bad an approximation for $\mathbf{u}_G$. The reason for this roughly constant bubble velocity as a function of their size is that larger bubbles more easily change their form and they do so in such a way that their flow resistance coefficient c_w is roughly constant for the relevant bubble Reynolds numbers range between 500 and 5000. In water we obtain for such a range

$$c_w = 50 \left[\frac{g}{cm^3 s} \right] \qquad (8.33)$$

The assumption of the local gas-holdup ε to be a continuous quantity is equivalent to the assumption that the gas phase is considered a continuous, i.e., a second space-filling fluid. The corresponding equation of motion can then also be formulated in a Eulerian representation. Hence, the simulation of the two-phase gas-liquid flow requires a simultaneous solving of two fluid flow systems – that for the dispersion and that for the gas phase. This approach is thus termed the Euler-Euler representation or simply the two-fluid-model.

8.3.3
Euler-Lagrange Approach

The number of grid points that can currently be handled in simulation software for two-phase gas-liquid flows on the available workstations is about $33 \times 33 \times 200$. When we are dealing with a reactor diameter of 1 m, the computational grid elements have a volume of roughly 30 cm^3, while a bubble with an equivalent diameter of 4 mm has a volume of only 0.03 cm^3. Hence, the typical bubble is small as compared to the grid element. This fact that the bubble motion cannot be resolved on such grids seems to justify the assumption of considering the gas phase in numerical simulation as a continuous phase. However, small dispersed particles are often considered part of the continuous liquid phase, just as the microorganisms dissolved in the continuous liquid phase must be considered part of the liquid.

Landau and Lifschitz [7] showed that the Navier-Stokes equation system can be applied to two-phase flows when the dispersed phase elements are small, do not drastically change the overall fluid density, and if the momentum of the particles or bubbles can be neglected. Then, obviously, the corresponding material properties appearing in the Navier-Stokes equation system must be properly corrected: The density must be chosen as the effective density of the dispersion and, similarly, the usual viscosity μ must be replaced by an effective viscosity μ_{eff}.

In bioreactors, these conditions are widely fulfilled. The sizes of the dispersed bubbles are small as compared with the characteristic reactor scale, and hence we are dealing with small particles. The density changes of the dispersion within the computational grid elements due to the appearance of bubbles, however, is not negligible. It is inhomogeneously distributed and changing with time, since the bubbles move relative to the continuous liquid phase.

The natural way of considering the gas-phase motion, and hence to estimate the fluid density, is to follow the self-evident physical picture and regard the bubbles as individuals that rise in their turbulently agitated liquid surrounding. As already mentioned, this cannot be done to a sufficient accuracy on the same grid as the fluid flow simulation. The bubble motion must be determined with a much higher spatial resolution, within the numerical grid of the Navier-Stokes calculation. At each time step of the solution of the Navier-Stokes equations, the positions and the sizes of the individual bubbles then allow one to determine the average density within the individual grid elements simply by averaging. This determination of the apparent density in the numerical grid elements can thus be termed a subgrid-model for the density.

When the dispersion is treated in the Eulerian representation and the gas phase motion in a Lagrangean one, the entire two-phase flow representation is referred to as a Euler-Lagrange representation.

8.3.3.1
Dynamics of the Dispersed Gas-Phase

The Lagrangean treatment of the motion of individual gas bubbles has the advantage that many effects observed in gas-liquid flow investigations can be treated in a direct way. However, a large number of bubbles must be followed on their path through the bubble column in order to get accurate results. Lapin and Lübbert [8] and Lapin et al. [12] were able to follow simultaneously up to roughly 200,000 bubbles. Hence, one needs simply to apply equations of motion to determine the bubble trajectories.

The slip velocities $\mathbf{u}_{slip}$ of the bubbles relative to the velocity $\mathbf{u}_l$ of the surrounding liquid-phase of the dispersion is dominated by pressure gradients as well as their flow resistance. These effects can be taken into account by the force law:

$$c_{vm}\rho_L \frac{d\mathbf{u}_{slip}}{dt} = -\nabla p - c\mathbf{u}_{slip} \tag{8.34}$$

The bubble velocity $\mathbf{u}_b$ in the laboratory coordinate system, which is needed to determine the bubble position, is simply related to these two velocity values by

$$\mathbf{u}_b = \mathbf{u}_L + \mathbf{u}_{slip} \tag{8.35}$$

The positions $\mathbf{s}_b$ of the bubbles after one time step of length Δt, can then directly be determined by simply applying the equation

$$\mathbf{s}_b(t) = \mathbf{u}_b(t)t + \mathbf{s}_o \tag{8.36}$$

assuming that their positions at the previous time step in the larger-scale calculation was $\mathbf{s}_o$.

8.3.3.2
Effective Viscosity

For the effective viscosity appearing in the Navier-Stokes equation one also needs a model. Here, it is necessary to remember that the effective viscosity in a turbulent flow is essentially determined by the energy dissipation in the eddy motion. As is well known, the random fluid motions on the smaller scale do dissipate more kinetic energy than the motions on the larger scale.

In the large-eddy simulation of bubble column reactors, the mesh width is, as already estimated, of the order of magnitude of some centimeters. With respect to the statistical turbulence theory this scale falls into the range of inertia. This is the range of scales in which inertia and buoyancy forces are significantly larger than the viscous shear forces. Consequently, the influence of the viscosity on the fluid flow patterns is expected to be small. This can be confirmed by simple numerical experiments, which show that a change in the numerical value of the viscosity by 100% leads to an integral change of the velocity patterns in the range lower than 5%. Hence, the influence of the viscosity is rather low for the large scale fluid motion. The effective viscosity can thus be taken as constant for simulations of the fluid flow. The concrete value can either be taken from experiments or from turbulence models. As such models, use of the $k\varepsilon$ model (Eq. 8.30) was tried. Experimental information is expected to provide more reliable results.

8.3.3.3
Mass Transfer and Chemical Reaction

In order to consider correctly the concentrations of the gaseous component, for example oxygen, dissolved in the liquid phase, the simulation must also contain the gas-liquid mass transfer. As usual in chemical and biochemical engineering, the mass transfer resistance can be assumed to be the liquid-side boundary layer around the bubbles. In this picture, the mass m of gas transferred per unit time into the continuous liquid phase is

$$\frac{dm}{dt} = k_L A(O^* - O) \tag{8.37}$$

where k_L is the mass transfer constant, A the interfacial area, and (O^*-O) the driving concentration difference, i.e., the difference between the saturation concentration O^* of the dissolved oxygen assumed at the physical interface and the local concentration O of oxygen dissolve in the liquid bulk. Once the gas has been transferred into the liquid phase, it is mixed across the liquid phase by fluid motions. This mixing process is described by a combination of convection and diffusion:

$$\frac{\partial O}{\partial t} + (\mathbf{u}\nabla)O = D_{eff}\Delta O - R_c + R_{mt} \tag{8.38}$$

Besides the diffusive and convectional transport of the dissolved gas within the liquid, a source (R_{mt}) and a consumption term (R_{Cg}) must be considered in this rate equation to account for gas-liquid mass transfer and consumption by chemical reaction.

The gas consumption by means of chemical reaction can be described by a kinetic expression. A typical example is a Monod-type rate expression of the form

$$R_C \sim \frac{O}{K_o + O} f(T) \tag{8.39}$$

where K_o is a Monod constant. The conversion rate is often a function f of the temperature T as well, which may be approximated by a set of two Arrhenius-terms as shown in standard text books (e.g., [10]). The source term corresponds to the mass transfer rate already mentioned in Eq. (8.37):

$$R_{mt} \sim k_L a \, \Delta O \tag{8.40}$$

The biochemical conversion process usually produces heat which must be removed via the heat exchangers, which are most often welted as jackets onto the reactor wall. A heat balance can than be used to model the temperature behavior of the dispersion:

$$\frac{\partial T}{\partial t} + (\mathbf{u}\nabla)T = K_{eff}\Delta T + R_T \tag{8.41}$$

where K_{eff} is the effective temperature conductivity. The source term, the heat production rate, may be considered proportional to the chemical reaction rate:

$$R_T \sim R_C \tag{8.42}$$

This heat transfer as well as the subsequent temperature increase within the cooling jackets can easily be considered in such an extended bubble column reactor model.

8.3.3.4
Mixing Due to the Bubble Rise

Bubbles are known to carry liquid in their wakes relative to the continuous bulk liquid phase. They continuously pick up liquid and, on average they release this liquid after some residence time from their wakes. Thus, the bubbles actively mix liquid [11] at least on a scale of some bubble diameters. This is a scale which is still smaller than the typical resolution of the large eddy simulation described so far. The question is whether or not this wake mixing effects have some influence on the global mixing in the bubble column. This question was addressed by Lapin et al. [12].

These authors developed a model that starts by describing the exchange of liquid elements between wake and bulk. For φ, denoting the local wake volume per unit volume (wake holdup), and Φ the local volumetric flow rate of the liquid through the wakes, the following equations were established:

$$(1 - \varphi)\frac{\partial c_b}{\partial t} = \Phi(c_w - c_b) \tag{8.43}$$

$$\varphi)\left(\frac{\partial c_w}{\partial t} + u_b \frac{\partial c_w}{\partial x}\right) = -\Phi(c_w - c_b) \tag{8.44}$$

u_b is the bubble or wake velocity relative to the bulk. Index w refers to wake properties while index b refers to bubble properties.

The analytical solution of these equations for a typical tracer dispersion by the bubble wakes with the initial conditions:

$$c_b(x, t = 0) = \delta(x), \; c_w(x, t = 0) = 0 \tag{8.45}$$

looks a little bit simpler when we take $\kappa = (1 - \varphi)/\varphi$, i.e., the relative volume of bulk and wake liquid.

$$c_w(x, t) = \kappa \eta(x) e^{-\kappa x - t + x} I_0\left(2\sqrt{\kappa x(t - x)}\right) \tag{8.46}$$

$$c_b(x, t) = \delta(x)e^{-t} + \eta(x)\sqrt{\frac{\kappa}{x}}(t - x)e^{-\kappa x - t + x} I_1\left(2\sqrt{\kappa x(t - x)}\right) \tag{8.47}$$

where $\eta(x)$ is a unit step function where I_0 and I_1 are modified Bessel functions of the first kind. Both expressions are only defined for $0 < x < t$. Elsewhere, the solutions are identically equal to zero. Physically, these solutions show how a tracer spot becomes dispersed with time over the coordinate x. The property that most clearly characterizes the dispersion effect is the variance σ^2 of such a tracer distribution, which is obviously a function of time. The expression, Lapin et al. [12] found, is rather complex, however, for with large times a rather simple expression was obtained for the variance:

$$\sigma^2 = 2u_b^2(1 - \varphi)^2\varphi\,\tau_r t \tag{8.48}$$

This variance is proportional to time t and $\tau_r = \varphi/\Phi$, the mean residence time of the liquid in the bubbles' wake. For the often used axial dispersion model, the standard deviation is

$$\sigma^2 = 2D_{ax}t \tag{8.49}$$

where D_{ax} is the axial dispersion coefficient. By comparison of the coefficients, we thus obtain for the axial dispersion coefficient with respect to the bubble wake dispersion

$$D_{axw} = \frac{u_b^2(1 - \varphi)^2\varphi^2}{\Phi} \tag{8.50}$$

The wake additionally leads to a convective transport of liquid with a mean velocity U, which immediately turns out to be

$$U = u_b \varphi \tag{8.51}$$

The velocity U of the center of mass of the tracer cloud can easily be interpreted in physical terms. It is equal to the velocity u_b of the bubble or wake relative to the bulk liquid times a factor between 0 and 1, namely the wake hold-up. The physical explanation is simple: a given fluid element is either transported within or outsides the wake. If it is within the wake, it is being convected with the wake's velocity for a while. The mean time it is transported with the wake is equal to the mean residence time of the liquid in the wake. The probability of a liquid fluid element of being within the wake is equal to the wake hold-up φ.

An order of magnitude estimations shows that this bubble wake contribution to axial mixing accounts for roughly 20% of the total mixing in terms of the axial dispersion coefficient.

8.3.3.5
Problem of Bubble Coalescence and Redispersion

Coalescence and redispersion of bubbles change the bubble number and size distributions. The influence to be expected on the large-scale motion of the dispersion is not dramatic as long as one is dealing with low viscous liquids where the mean bubble size is not to be expected to become very large. Small bubbles easily follow the larger-scale liquid motions. Preliminary calculations showed that the characteristic quantity of the bubble size distribution.– which is by far the most important to the flow pattern of the dispersion – is its first moment, i.e., the mean bubble diameter. The second, i.e., the width, is of much less importance to the bubble column fluid dynamics. The main influence of the bubble size distribution on the performance of a bubble column bioreactor is to be expected to be due to its influence on the mass transfer and the micro-mixing properties of the reactor.

Both effects, coalescence and redispersion, can easily be incorporated into the Euler-Lagrange simulation software, as the paths of individual bubbles are followed up in this approach. The problem is to provide probability distributions for the coalescence for the case that two or more bubbles approach each other to a distance where coalescence may occur. Well accepted physically based theories are not yet available. For redispersion, a corresponding probability distribution must be provided, which basically depicts the probability of a bubble decay as a function of the energy dissipation density in the fluid flow around the bubble. Once again, expressions that reliably map such processes are still not available.

8.3.3.6
Rating of the Euler-Lagrange Representation

Since all the effects, the fluid flow, the mass, the heat transfer, as well as the chemical reaction appear simultaneously, all these equations must be solved at the same time. Lapin and Lübbert [13] performed such dynamical simulation directly in three dimensions on a Cartesian frame of computational nodes. The grid size used was $33 \times 33 \times 200$. A typical example of a velocity pattern in a yeast fermenter is shown in Fig. 8.1. The reactor diameter in this test case was 1 m.

Fig. 8.1. Typical instantaneous flow structure of the dispersion in the bubble column. *The scale at the lower left* marks the velocity of 10 cm/s. Since the arrow length is not easy to recognize, the velocity is additionally represented by the size of the triangle which marks the direction of the arrow. *Blue arrow color* indicates that the vectors have a component coming out of the plane, while *red arrows* are directed into the plane, *black ones* fairly remain within the plane

In simulations of the flow in such large bubble columns we are not able to increase the number of elements in the numerical grid, and thus the sizes of the grid elements become larger. At the same gas holdup we thus have more bubbles in the individual computational grid elements. At the same time the total number of bubbles in the reactor is larger as well, and it may easily become larger than the number that can be dealt with individually. In this case it is of advantage to remember that bubbles are moving in swarms or bubble clusters through the reactor. The motion of these swarms can be treated in roughly the same way as the motion of single bubbles. In this case the center of a bubbles swarm can be followed up on its way through the bubble column. For more details see [5].

Finally, as also shown in [5], both representations, the Euler-Euler and the Euler-Lagrange techniques, lead to practically the same results where they could be applied to the same fluid flow simulations. Hence, in many cases one can choose one of these two alternatives. However, some aspects can be represented more directly with the Euler-Lagrange representation. Some applications of the Euler-Lagrange approach will be discussed in the next section.

8.4
Modeling of Particular Aspects of Bubble Column Reactors

8.4.1
Velocity Patterns in Bubble Column Reactors

The most general characteristics of the flow in a bubble column bioreactor is the pattern of the velocity vectors across the volume of the dispersion. A three-dimensional velocity field is difficult to represent graphically. Figure 8.1 depicts a typical pattern of velocity vectors computed for points on a plane through the column axis. Only the projections of the velocities on this surface can be shown. However, in order to provide a little more information about the three-dimensional velocity vectors, the arrows were colored. The red arrows indicate that the corresponding vector has another component showing in the projection area, the blue ones represent components coming out of the area. Those which do not have a significant component perpendicular to the projections are shown in black. The velocity scale is shown as an arrow at the lower left side of the left graph.

The velocity pattern depicted in Fig. 8.1 represents the flow computed for a particular instant in time. It shows a typical situation which appears after the system had sufficient time to reach a quasi stationary state. The structure of the flow pattern changes continuously with time in a chaotic way, although the assumed superficial gas velocity, $w_{sg}=2$ cm/s, is not too high. About 200,000 bubbles clusters were assumed to rise in the column in that situation, leading to a gas holdup of about 10%.

The outer appearance of this transient fluid flow shows the pattern expected for turbulent fluid flow in a biochemical reactor. It is dominated by an eddy motion with eddies of various sizes that are being generated and are decaying. From the videos, which were constructed from many successive such velocity patterns, one can observe that the eddies are generated randomly, they move for some distance within the reactor, and then they decay to form smaller eddies.

Since the eddy motion is chaotic, it is practically impossible to compare images calculated for different times. Quantitative comparisons of the flow patterns are only possible on a statistical basis. The easiest-to-obtain of such representations is time averaging of the transient flow patterns. They must be obtained by averaging over times spans that are long compared to the characteristic time-constants of the motion of the characteristic randomly appearing flow structures, i.e., over periods which are long enough to average out all random flow structures.

Such an averaged velocity pattern is shown in Fig. 8.2 in the same representation as before. A fairly symmetric flow profile appears after averaging. This profile is the type of flow patterns usually discussed in bubble column literature. It was obtained after averaging over a period of 30 min.

As can be seen from the figure, long-time averaging of the transient flow velocities leads to the well known gulf stream patterns (a term proposed by Freedman and Davidson [14]). Obviously all the random motions, which are induced for mixing purposes, are averaged away. The long-time-averaged velocity profile shows the profiles that have been measured many times over the past 30 years. Comparisons of the results of the simulation software and such velocity profiles have been performed on several different scales (e.g., [15 9 or 12] to validate the computational code.

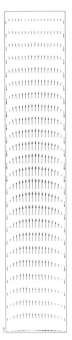

Fig. 8.2. Long-time averaged version of the velocity pattern depicted in Fig. 8.1. The averaging time is 30 min. The scale, 10 cm/s, is shown in form of the *arrow at the lower left side of the graph*

8.4.2
Fate of Individual Cells in the Bubble Column Bioreactor

Cells submersed in the liquid phase of a culture medium respond in their metabolism to various concentrations in their environment. In larger reactors we expect, as we will see later on in some detail, inhomogeneities in the concentrations of substrate components. The flow patterns in bioreactors show that the cells move through the reactor and will thus experience different concentrations in their environment. The transient eddy motion suggests that the environment of an individual cell changes randomly, while the averaged profiles suggest a periodical change on the average. Hence, the question appears as to what concentrations an individual cell really sees.

For questions like these, the Euler-Lagrange representation of the fluid-mechanical bubble column reactors is optimal. A single cell can be simulated by a fluid particle of neutral buoyancy just as a small bubble with the gas density equal to the cell's density, which is practically equal to the liquid density, and followed up on its path through the column. Figure 8.3 [16] presents a typical trajectory calculated for a bubble column of 21 cm diameter operated at a superficial gas velocity of 2 cm/s. The path appears to be nearly totally random. There is no similarity to the well-ordered flow field that appears after long-time averaging of the local velocities. The particular model reactor, for which the computations were performed, was chosen

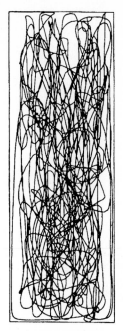

Fig. 8.3. Simulated trajectory of a single fluid particle in the bubble column reactor. The trajectory appears to be a random path

since measurement values of the path of radioactive flow-follower particles were available for a column of this size, so that the simulations could be validated experimentally [17]. The radioactive particle could be tracked for hours on its path through this bubble column, which was operated under the conditions assumed in the numerical simulations.

The question as to whether or not the dispersed particle will see some periodicity in the flow was answered by the question how long an intelligent cell would need to recognize a ordered average gulf stream flow when it would be able to record and analyze all the velocities it saw on its trip through the reactor. This problem could also be investigated by the particle tracking method, since the velocities of the particle are available for the entire path. The answer to the question is that the particle would need times of the order of hours to recognize the gulf stream. This time is well above any time constant relevant to the biochemical conversion processes. Hence, in practice the particle does not recognize periodic flows. From the point of view of a single particle, the flow is entirely chaotic.

As the measurement of the particle trajectory and its numerical counterpart show, the individual particles do not notice the deterministic orderly motion which appeared after averaging the local velocities over a sufficiently long time. Thus, from the point of view of the cells suspended in the liquid, the averaged gulf-stream motion can be regarded as an artifact.

That means that model calculations assuming the particles to pass regions of high and low oxygen concentration periodically do not seem to be very realistic. Scale-down experiments, which try to mimic periodic changes in the environment conditions, may provide information on the dynamics by which the cells react to changes, but do not approximate true conditions within bubble column bioreactors. For stirred tank reactors the situation is expected to be much similar.

8.4.3
Influence of Tilted Columns

The Lagrangean representation of the gas-phase motion is perfectly suited to determining gas residence time distributions, simply by recording the time for each bubble from its entrance into the reactor at the sparger to the time at which the bubbles leave the dispersion at its top surface. Figure 8.4 depicts two typical examples calculated for a bubble column operated at different vertical alignments [15].

In the figure, the behavior of a truly vertical column is compared with a column slightly tilted by only 0.5° away from the vertical. As already found experimentally, the influence of a column inclination is considerable [18–20].

In Fig. 8.5, the corresponding velocity profiles are recorded.

The considerable changes in the mean residence time effect cannot immediately be recognized in the corresponding transient flow patterns. These remain practically unchanged. However, the long-time averaged patterns disclose the fundamental difference between both flow situations: While the gulf stream pattern characterizes the averaged flow in the vertical column, the mean flow pattern in the tilted column changed to a different flow mode. This is characterized by a single large asymmetric circulation mode (Fig. 8.4). Such a mode depicts larger mean flow velocities, and hence the mean residence times become smaller.

8.4.4
Oxygen Distribution in a Yeast Fermenter

Sufficient oxygen supply for the culture is one of the key issues in bioreactor engineering. A main objective is to keep the dissolved oxygen concentration in the medium above some critical value throughout the entire reactor. From this point of view the dissolved oxygen concentration pattern in bubble column fermenters during a cultivation process is of primary interest. Here we discuss the results of such a pat-

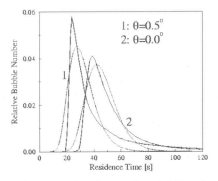

Fig. 8.4. Typical gas residence time distributions for a bubble column, perfectly aligned to the vertical and one tilted by 0.5° away from the vertical

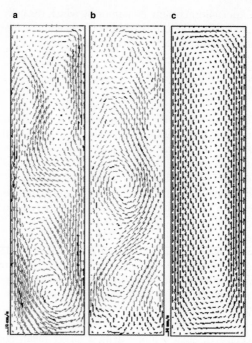

Fig. 8.5. *Left and middle:* transient velocity patterns in a bubble column tilted by 0.5° from the vertical. Both represents velocity patterns on planes through the axis of the reactor. They are perpendicular to each other. The *right graph* shows the long-time-average of transient patterns

tern simulated with the Euler-Lagrange representation. The bioreactor considered is a cylindrical pilot-scale bubble column bioreactor with 1 m diameter, aerated homogeneously across its entire flat bottom. Since the time scale at which the yeast grows is much larger than the time scale at which the fluid flow structures in the reactor are changing, we assume that the yeast concentration is not changing with time during the situation simulated.

The dissolved oxygen concentration pattern $O(\mathbf{r})$, compiled for one cross sectional area through the axis of the fermenter, is shown on the left hand side of Fig. 8.6. The corresponding fluid flow situation is the chaotic velocity pattern shown in Fig. 8.1; i.e., both these plots depict the same momentary situation for different variables. The most striking result which can be notices at first glance is that, due to the oxygen consumption by the cells, the dissolved oxygen concentration does not reach a state of saturation in the reactor and the oxygen does not become homogeneously distributed across the reactor. The patterns remain chaotic with respect to space and time. Nevertheless, globally, a quasi-equilibrium is obtained in the sense that the distribution density $p(O)$ of all the dissolved oxygen concentration O values at a given time in the reactor does not change with time, as long as the yeast concentration can be considered constant.

The long-time averaged dissolved oxygen concentration pattern, however, which is depicted on the right side of Fig. 8.6, exhibits a clear structure, indicating that the concentrations near the walls are considerably smaller than in the center of the column.

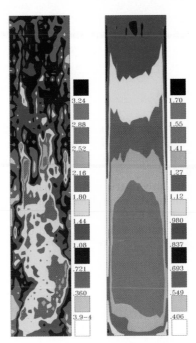

Fig. 8.6. Patterns of the dissolved oxygen concentration on a plane through the axis of a bubble column bioreactor during a yeast cultivation: On the *left*, a transient situation is shown, while on the *right*, a profile is shown that was obtained by averaging the transient results over 30 min real time

This has an immediate consequence for dissolved oxygen concentration measurements, which are most often performed by means of electrochemical electrodes. In practice, such electrodes reach only a few centimeters into the reactor and are usually mounted near the bottom of the vessel, where they are easily accessible to the personnel. Each point in the graph of the right hand side of Fig. 8.6 can be interpreted as a time-averaged dissolved oxygen concentration signal. It is immediately clear from the figure that the values to be expected at typical measurement positions are far away from the true representative mean concentration for the entire reactor. In other words the true mean oxygen concentration in a yeast production fermenter is extremely difficult to measure.

In indirect measurements of key parameters of the fermentation process, e.g., the specific growth rate μ, one tries to make use of values of simple-to-measure variables. Usually only a few measurements are performed at large pilot or production reactors that could be used for that purpose. Oxygen partial pressure in the vent line of the reactor and the dissolved oxygen concentration O are proper candidates, which are closely related to growth. In cases where the growth or product development rate is oxygen limited, we need kinetic data about the oxygen dependency of these parameters. Assume that the oxygen limitation effect can be described by a Monod-type kinetic limitation term. In such cases, an estimation of the growth rate would require knowledge of the corresponding Monod constant K. This value must be determined from process data. For that reason, dissolved oxygen concentration measurements must be performed. In oxygen-limited process states one can assume that the oxygen transferred into the reactor is immediately consumed by the cells, so that no oxygen accumulation occurs. That means that the oxygen consumption rate OCR is equal to the oxygen transfer rate OTR:

$$\text{const.} \frac{O}{K+O} = k_L a(O^* - O).\qquad(8.52)$$

Let us assume that the $k_L a$ value can be estimated from off-gas measurements and O^*, the solubility, is available from the literature. Then, when the constant const. is also available, K can be determined with a measurement value of O, provided O is homogeneously distributed.

When O is not homogeneously distributed, and that is why this problem has been addressed at this point, one needs the probability density function p(O) in order to determine the correct average of OCR. Since its oxygen dependency is nonlinear, one cannot simply take the expectation value $\hat{O}$ of O and use Eq. (8.52). For the oxygen consumption rate the correct averaging is

$$\text{OCR} \propto \int_0^{O_{max}} p(O) \frac{O}{K_o + O} dO \neq \frac{\hat{O}}{K_o + \hat{O}}\qquad(8.53)$$

Since it is linear, the averaging procedure is not necessary for the OTR relationship:

$$\text{OTR} = k_L a(O^* - \hat{O}) = E\{k_L a(O^* - O)\}\qquad(8.54)$$

From Fig. 8.6 we just learned that the homogeneity assumption cannot be justified. Hence, the average of the expressions in Eq. (8.53) must be determined using the dissolved oxygen distribution p(O). Since this probability density cannot be properly provided in practice, one must estimate the mean $\hat{O}$ and determine a rough estimate K_o from Eq. (8.52).

Qualitatively, there is obviously an error when one uses the average value $\hat{O}$ in Eq. (8.52). The question, however, is how big this effect is, quantitatively in typical biotechnical applications, and whether or not it worth considering. This answer can be given by numerical simulation, since in this case we have access to p(O).

As numerical experiments have shown, the answer largely depends on the actual dissolved oxygen concentration probability density p(O). Different situations concerning the oxygen supply in the reactor must be distinguished. When the oxygen concentrations in the reactor are rather low and the corresponding probability distribution p(O) is sufficiently narrow, only the nearly linear "small-O-interval" in the Monod-like oxygen uptake kinetics is affected. Hence, the influence of the nonlinearity is small. The same applies for very high oxygen concentrations. The latter case, however, does not often appear in industrial practice, since the corresponding oxygen transfer rates would be too costly. In between these two extreme cases, the effects of the nonlinearities are expected to be significantly higher.

An easy way to characterize the error is to determine the ratio K_o by the K obtained via a correct averaging procedure, which can be determined without specifying the constant const.. When we consider the effects in the turbulent flows in bubble column fermenters as depicted in Figs. 8.1–8.5, then the quotient is shown in Fig. 8.7.

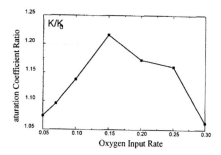

Fig. 8.7. Ratio of K obtained with a correct averaging by K_o obtained using an estimate for the average dissolved oxygen concentration and the value as a function of the aeration rates

As expected from the preceding discussion, the deviations are maximal for a mediate oxygen concentration interval and smaller at small and high oxygen concentrations. At its maximum, at mediate aeration rates, the effect exceeds 20%.

At this point it is interesting to recall that the mean oxygen concentration Ô used in the preceding discussion, which is the first moment of the distribution density $p(O)$, is nearly impossible to measure as the long-time averaged results in Fig. 8.6 indicate. With measured dissolved oxygen concentration values the effect is expected to be significant larger. As the long-term average shows, there are considerable oxygen concentration profiles to be expected in bioreactors, a fact that was measured several times by different groups even in stirred tank bioreactors (e.g., [21].

This discussion essentially gives some idea of the errors in determining kinetic parameters in inhomogeneous reactors under the wrong assumption that the reactor would be well mixed.

This study clearly shows that the turbulent gas-liquid two-phase flow in biochemical bubble column production reactors lead to chaotic patterns of the dissolved oxygen concentration. An equilibrium in the sense of a homogeneous distribution of the dissolved oxygen concentration cannot be obtained. In order to characterize the dissolved oxygen concentration pattern, two statistical measures can be used. The first is the time averaged oxygen pattern as shown in Fig. 8.2. This pattern does not have much relevance to the production process as such, but serves as a medium for comparison with measurements, which are usually time averages as well. The second statistical measure is the dissolved oxygen concentration probability density, characterizing the probability by which a yeast cell traveling through the reactor will see an oxygen concentration O in its environment.

The probability density $p(O)$ is highly dependent on the flow conditions, the local mass transfer rate, as well as the oxygen consumption rate by the biomass. The effects of the oxygen concentration probability density $p(O)$ on the integral oxygen uptake by the culture is considerable at medium oxygen concentration where the kinetics depicts its nonlinearities. Interestingly the influence of the bioreactor construction shows that more regular global flow structures increase the effect. This was demonstrated at flat bubble columns with rectangular cross section. Similar large-scale global flow structures are to be expected in stirred tank reactors. Hence, it is recommended to investigate this effect at stirred tanks as well.

What can be learned from the aspects considered in this section is that measurements cannot provide accurate mean oxygen concentrations in larger bioreactors. The determination of kinetic parameters from measurements made in large reactors is usually biased. Since the error is due to the fluid flow properties of the particular scales, this bias is obviously scale-dependent. This means that values of kinetic parameters, which are generally based on measured values in finite reactors, are in most cases scale-dependent values and cannot be used to an arbitrary accuracy for scale-up studies.

8.5
Conclusions

Modeling is understood as a quantitatively exploitable formulation of our current knowledge about the processes under consideration. It takes a lot of effort and is thus justified only when it does help to solve open questions of significant importance. Different objectives require different models, since the models must consider those mechanisms which are relevant to the problem to be solved.

The current problem with two-phase gas-liquid flows in bubble column bioreactors is that we do have the model but cannot solve it to an arbitrarily chosen spatial resolution. That means the problem lies in the possibility of us exploiting the model. The decisive limitation is, and will be in the near future, the available computing power to solve the Navier-Stokes equation system. Currently we can tackle detailed fluid motions on a relatively rough scale only. However, as was shown in this chapter, several basic questions about fluid flow could be answered with the currently available software.

The random nature of the flow can be shown quite directly. It has immediate consequences for all effects which are usually discussed on the single particle basis. The long-time averaged flow patterns do not have the importance that might be assumed from their frequent appearance in fluid dynamical discussions in literature. It is of importance only with respect to those questions which directly consider the long-time averaged behavior of the fluid flow. An example of practical importance is the gas residence time distribution. Another more technical aspect is that comparisons – may they be between computational results or between computational results and measurement data – can only be performed on averages. All other problems in particular mass transfer and mixing, which are the key objectives of inducing flows in bioreactors, are governed by the chaotic flow components.

The main activity to cope with coarse computational grids is compensating for the fluid flow effects originating from motions on the small scale by means of appropriate subgrid models. Since we cannot resolve the details of these fluid flow effects, we must lump them together and project them into the material properties of the fluid, namely the effective viscosity and the fluid density.

Although the current CFD results are qualitatively good enough to serve as basic reactor models, they unfortunately cannot yet be incorporated into process optimization studies, simply because their evaluation takes too much time. With the computers available, full three-dimensional dynamical simulations take several days and thus cannot be used in the iterative procedures by which we optimize the reactor performance. Such optimization procedures would require us to recompute the flow patterns many times during a single optimization study. However, if the computing power of our workstations continues to rise at the rates we have seen in the last decade, the time at which we will be able to incorporate such models will not be too far ahead.

Acknowledgement. Deutsche Forschungsgemeinschaft DFG generously supported my work on fluid dynamics with several different projects over the previous year.

References

1. Feynman RP, Leighton RB, Sands M (1963) The Feynman lectures on physics. Addison Wesley, Reading, Ma
2. Anderson DA, Tannehill JC, Pletcher RH (1984) Computational fluid mechanics and heat transfer. Hemisphere/McGraw-Hill, Washington New-York
3. Liepmann HW (1979) The rise and fall of ideas in turbulence. Am Scientist 67:221–228
4. Patankar SV (1980) Numerical heat transfer and fluid flow. McGraw-Hill, New York
5. Sokolichin A, Eigenberger G, Lapin A, Lübbert A (1997) Dynamic numerical simulation of gas-liquid two-phase flows: Euler-Euler versus Euler-Lagrange. Chem Eng Sci 52:611–626
6. Abou-El-Hassan ME (1986) Correlations for bubble rise in gas-liquid systems. In: Cheremisinoff NP (ed) Encyclopedia of fluid mechanics, vol 3: Gas-liquid flows, chap 6. Gulf, Houston, pp 110–120
7. Landau LD, Lifschitz EM (1971) Lehrbuch der Theoretischen Physik, vol VI, Hydrodynamik, 2. Aufl., Akademie, Berlin
8. Lapin A, Lübbert A (1994) Numerical simulation of the dynamics of two-phase gas-liquid flow in bubble columns. Chem Eng Sci 49:3661–3674
9. Lapin A, Maul C, Junghans K, Lübbert A (2000) Industrial-scale bubble column reactors, gas-liquid flow and chemical reaction (to be published)
10. Nielsen J, Villadsen J (1994) Bioreaction engineering principles. Plenum, New York
11. Fan LS, Tsuchiya T (1990) Bubble wake dynamics in liquids and liquid-solid suspensions, Butterworth-Heinemann, Boston
12. Lapin A, Paaschen T, Lübbert A (2000) Local mixing in bubble column reactors: Bubbles as local mixing elements (to be published)
13. Lapin A, Lübbert A (1999) Problems with measurements of kinetic constants in bioreactors (to be published)
14. Freedman W, Davidson JF (1969) Hold-up and liquid circulation in bubble columns. Trans Inst Chem Eng 47:T251–T262
15. Lübbert A, Lapin A (1996) Dynamic numerical experiments with 3D-bubble columns. Revue de l'Institut Francais du Pétrole 51:269–277
16. Lübbert A, Paaschen T, Lapin A (1996) Fluid dynamics in bubble column bioreactors: experiments and numerical simulations. Biotechnol Bioeng 52:248–258
17. Devanathan N, Dudukovic M, Lapin A, Lübbert A (1995) Chaotic flow in bubble column reactors. Chem Eng Sci 50:2661–2667
18. König B, Buchholz R, Lücke J, Schügerl K (1978) Longitudinal mixing of the liquid phase in bubble columns. Ger Chem Eng 1:199–205
19. Rice RG, Littlefield MA (1987) Dispersion Coefficients for ideal bubbly flow in truly vertical columns. Chem Eng Sci 42:2043–2051
20. Rice RG, Barbe DT, Geary NW (1990) Correlation of nonvertically and entrance effects in bubble columns, AIChE J 36:1421–1424
21. Oosterhuis NMG, Kossen NWF (1984) Dissolved oxygen concentration profiles in a production scale bioreactor. Biotechnol Bioeng 26:546

Part C
Application of General Principles for
Process Models Including Control

9 Baker's Yeast Production

Karl-Heinz Bellgardt

9.1
Introduction

Yeast production is an important industry in several fields: besides baker's yeast, these are animal feed yeast, yeast for hydrolyzates used in biotechnology and the food industry, and in biotechnology as host organism for foreign genes. Thus there is a wide application area of models as presented in this chapter, which is structured as follows. An introduction to important phenomena of yeast growth is given first. Only the important subjects for modeling and control will be summarized. A detailed survey of yeast metabolism and the baker's yeast process can be found, e.g., in [1].

The next two sections are concerned with modeling of growth kinetics in ideally mixed stirred-tank reactors (Sect. 9.2) and airlift tower-loop reactors (Sect. 9.3). The baker's yeast production is an example for a process that cannot be described in all its important aspects by simple growth models. Yeast metabolism is very flexible with respect to usage of different substrates and oxygen, and several regulatory phenomena have to be taken into account for modeling of growth. The proper approach are structured models based on a representation of the metabolic reaction network and a cybernetic criterion for its regulation.

The product quality of the final yeast product is influenced by the state of the cells within their propagation cycle. This requires the extension of the growth models by a population balance model to cover the distribution of cells in the cell cycle. Population balances are introduced in Sect. 9.4. At first both models, the growth and population models, are combined to provide information on the development of the fraction of budding cells during fed-batch cultivations. Then emphasis is put on the analysis of self-synchronized growth in chemostat cultivations.

The fraction of budding cells is then used in Sect. 9.5 as a product quality index to evaluate optimal fed-batch control policies of the yeast production process with respect to product quality and profit. The economic optimization requires a global view of interconnected processing steps of the whole process and a detailed look at the fermentation steps. Possibilities for automatic feed-back control of the substrate supply by means of measurements in the liquid phase and exhaust gas are then discussed in Sect. 9.6.

9.1.1
Metabolic Types of Yeast Growth and Regulatory Effects

Yeast can react on the availability of substrates such as carbon and nitrogen sources or oxygen by a flexible choice of different metabolic pathways. A simplified map of the metabolic network is shown in Fig. 9.1. Oxygen supply is one of the main determining factors of yeast metabolism and the limiting parameter for the production process. In excess of oxygen, at low sugar concentrations, yeast can achieve an oxidative metabolism. For energy production, the carbon source is completely oxidized to carbon dioxide and water via glycolysis and tricarboxylic-acid cycle (TCC), with almost no by-products. The reduction equivalents ($NADH_2$) are transferred mostly into the respiratory chain. These pathways can produce up to 36 moles of ATP per mole glucose, resulting in a very high yield for cell mass of up to 0.5 g dry cell mass per gram of sugar. Therefore, oxidative growth is the preferred metabolic type for baker's yeast production, although the growth rate is limited to about $\mu < 0.25\,h^{-1}$. Ethanol can also serve as a sole substrate for oxidative metabolism. The related enzymes of gluconeogenesis are inducible, which usually results in a lag phase of growth after a switch from growth on sugars to ethanol. Although glucose is the preferred substrate, parallel ethanol uptake is also possible up to a maximum rate that is determined by the availability of glucose and oxygen.

On sugar substrates, a second metabolic type is fermentative metabolism. This can be due to complete absence of oxygen under anaerobic conditions, due to partial oxygen limitation under aerobic conditions, or by a saturation in the rates of oxidative pathways (oxidoreductive or respiro-fermentative metabolism). Then, ethanol is produced as a final proton acceptor for $NADH_2$. Glycerol, acetaldehyde, and organic acids can be found in small amounts as by-products. The chemical energy then comes mainly from glycolysis with a yield of 2 moles ATP per mole of sugar, plus 2 moles of ethanol and CO_2 as by-products. By this low efficiency compared to oxidative growth and due to the loss of much carbon in the products ethanol and CO_2, the cell yield is also very low. Under anaerobic conditions, the yield reaches only

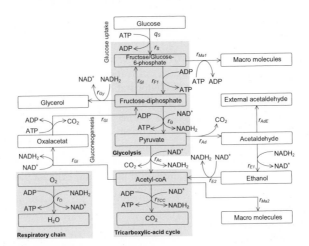

Fig. 9.1. Simplified map of yeast metabolism

$Y_{XS}=0.1$ and the growth rate stays below $\mu=0.3\,h^{-1}$, compared to $\mu<0.5\,h^{-1}$ for oxidoreductive growth. The production of a large volume of carbon dioxide is used in baking to leaven the dough. The fermentative activity – the ability to produce a certain amount of CO_2 in a given time – is, beside storage properties and resulting flavor of the dough, an important quality index for the yeast product.

The respiratory quotient can be used to estimate the metabolic type of yeast growth. For $RQ>1$ growth is fermentative, $RQ=1$ indicates respiratory growth on sugar substrate, and $RQ<1$ can be found during ethanol consumption.

For the directed coordination of its metabolism in dependence on substrate and oxygen supply the yeast disposes of a complex regulatory system on the epigenetic and reaction levels. This regulation affects uptake of carbon sources, glycolysis, gluconeogenesis, respiration, and product synthesis.

The Pasteur-effect – suppression of fermentative activity by respiration – is the earliest regulatory effect discovered in yeast. It is connected to the repression of enzymes of the respiratory chain and TCC under fermentative growth. Since the induction level under aerobic conditions depends on the available oxygen, a sudden increase in oxygen supply usually leads to short lag phase until the enzyme level is adjusted to the new condition and the metabolism is directed towards more effective energy production.

The Crabtree-effect is referred to as oxidoreductive metabolism under excess of oxygen and sugar, caused by a saturated capacity of the respiratory system of the cell. Metabolic fluxes are directed towards fermentative pathways with ethanol production. This is very important for the control of baker's yeast production because an over-supply of sugar in relation to a limited capacity of oxidative metabolism – either by internal saturation or by external oxygen supply – leads to fermentative type of growth with low yield of cell mass.

The Glucose-effect is the repression of uptake systems or pathways needed for other substrates at high concentrations of glucose, the substrate that can most easily be used by the cells. As a result of this repression, glucose is preferred and catabolized first, even when other substrates are present at high concentrations. For baker's yeast production, the glucose-effect with respect to ethanol – as a possible substrate for oxidative growth – is most important, because ethanol can be either produced or metabolized during the cultivation. Ethanol and sugar can even be taken up in parallel, then both compete for the oxidative pathways.

9.1.2
The Asymmetric Propagation of Yeast

The baker's yeast *Saccharomyces cerevisiae* is a single-cell micro-fungus. Cells are spherical to ellipsoid with a diameter ranging from 2.5 μm to 10.5 μm and a length from 4.5 μm to 21 μm. The vegetative multiplication by budding is most important for its propagation in technical processes. The budding cycle of yeast is a dynamic process of a series of ordered steps as schematically shown in Fig. 9.2: growth in mass and/or volume of single cells in the intermediate phase G_1, DNA-synthesis phase (S), a phase of bud formation and bud growth including mitotic phase (G_2, M), and a further intermediate phase G_1^* before separation of parent cell and bud, which becomes the new daughter cell. The period for the completion of the cycle is asymmetric for parent and daughter cells. These are smaller than the parent cell and must increase in size before initiation of chromosome duplication [2, 3]. For this reason, a newly

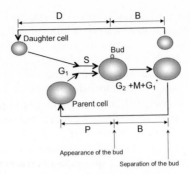

Fig. 9.2. Schematic diagram of the asymmetric budding cycle of yeast with single cell phases of daughter cells (D) and parent cells (P), and double cell budded phase (B)

formed daughter cell needs more time in the G_1-phase for the formation of its first bud than for the subsequent buds. Therefore, yeast populations are heterogeneous mixed populations with a certain distribution of distinguishable daughter cells, parent cells, and budding cells. The duration of the several cycling phases determines the genealogical age distribution of the yeast population. The genealogical age of the cell can be easily determined by counting the number of bud scars on the cell surface. For exponentially growing cultures these data was used to calculate the age distribution of a population and the duration of the cell cycle phases [4, 5]. The duration of the G_1-phases of parent and daughter cells is positively correlated to the cell number doubling time. In contrast, it is found that the division process – the duration of the budded phase (G_2+M+G_1^*) – is almost constant and independent of the external growth conditions. Only for very slow growth does the length of this phase increase slightly. Hartwell and Unger [3] found that the length of the budding interval and the generation time of parent cells do not vary with the genealogical age, at least not for a few generations. In addition, the budding intervals of daughters and parents are practically identical in length and undergo only minor changes for variations in the growth rate. The control mechanism for the cell cycle of *Saccharomyces cerevisiae* in batch culture can be smoothly explained by the concept of critical mass for the initiation of division. In this concept, the generation time of daughter cells is longer than that of parent cells, because they must first accumulate more cell mass. Since the size of the bud at separation becomes smaller at low growth rates, the enforced increase in daughter generation time with average doubling time is evident.

Based on these findings, the whole cycling process of yeast cells can be divided by a simplified view into three cycling phases (see Fig. 9.2): unbudded daughter cell phase D with length T_D, unbudded parent cell phase P (length T_P), and budding phase B (T_B), which has equal length for both budding daughter cells and budding parent cells. The experimental results demonstrated [3, 4] that T_D, T_P and T_B can be linearly correlated with the average cell number doubling time T, even for a variety of strains, different temperatures, and substrates. The lengths of the cell cycle phases can be described by the following simple correlations, the cycling phase equations:

$$T_D = K_{D1} T + K_{D2}$$
$$T_P = K_{P1} T + K_{P2} \tag{9.1}$$
$$T_B = K_{B1} T + K_{B2}$$

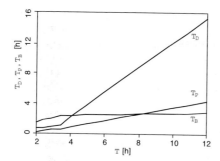

Fig. 9.3. Graphical representation of the cycling phase equations [5]: lengths of the cell cycle phases (T_D, T_P, T_B) over averaged doubling time T

Table 9.1. Experimentally determined parameters in the cycling phase equations for *Saccharomyces cerevisiae* S288c/1

K_{D1}	K_{D2}	K_{P1}	K_{P2}	K_{B1}	K_{B2}	Remark
1.302	−1.30 h	0.440	−0.367 h	0.180	0.767 h	batch culture [4]
0.330	0.05 h	0.250	−0.233 h	0.710	0.100 h	T<3.17 h
1.600	−3.93 h	0.390	−0.433 h	0.110	1.800 h	T>3.17 h, continuous culture [5]

Model parameters from the literature are given in Table 9.1. For an extended range of growth rates in chemostat cultivation, Thompson and Wheals [5] found biphasic correlations for the generation times of parent and daughter cells and the length of the budding phase as shown in Fig. 9.3: for $T<3.2$ h, growth is practically symmetric with $T_D = T_P < 1$ h, and $T_B < 2.5$ h, the length of the cycling phases are nearly proportional to T. For $T>3.2$ h, growth becomes asymmetric and T_D is more and more prolonged at the cost of T_P while T_B is almost constant at about 2.5 h. The parameters for both regions of doubling times are also given in Table 9.1.

There were a lot of investigations to clarify the interconnection of metabolism and cell cycle, which are summarized in [6]. It was found that most of the proteins, cell wall components, and RNA are synthesized with a constant rate during the cell cycle. The activity of most enzymes is also not strongly modulated or their concentrations are much higher than needed for the catalytic activity. So, while most cellular processes may be considered as continuous, the periodic events originate from the replication and partitioning of the DNA and the separation of the bud. The rate-limiting step of the cell cycle is obviously the growth process, i.e., the rates for cell mass or volume increase, which are directly connected to the continuous processes of metabolism. The cells need to reach a certain mass or volume to initiate the budding. This also explains why the G_1-phase of the first generation is longer than that of parent cells. The genetic program for cell division can normally be executed faster than the accumulation step. Growth and division are coordinated by monitoring the level of key metabolites, which then trigger the start of the next cell cycle phase. Alberghina et al. [68] and Mariani et al. [69] investigated the protein distribution in the cell cycle by means of extended population balance models. They proposed that the protein content of the cells controls bud initiation and the critical protein level increases at each generation of parent cells. The dominant role of the mass accumulation facilitates modeling since the continuous increase of mass can be described by structured growth models averaging over the population. The population dynamics are then influenced mainly by the average growth rate.

A very special feature of baker's yeast is self-synchronized growth in continuous culture, where oscillations appear on population variables such as the fraction of budding cells, or on process variables like the rates of oxygen uptake and carbon dioxide evolution, even when the cultivation conditions are kept constant. It has been subject to several investigations that were reviewed in [32, 33]. Self-sustained oscillations that are directly related to the asymmetric budding cycle of yeast are mostly found in the region of oxidative growth. These can appear spontaneously or be induced by shifts in the dilution rate or by small pulses of substrates such as glucose, ethanol, acetic acid, and pyruvic acid. The oscillation period varies in general with growth rate, but eventually multiple oscillation frequencies at a single growth rate could be observed. During synchronized growth, the variation of the metabolism during the cell cycle becomes visible on macroscopic variables. This facilitates experimental investigations of the related intracellular biochemical processes and physiology [7]. At the beginning of the budding phase, the metabolism is oxidoreductive with ethanol production. The ethanol is reconsumed in the later stages of the cell cycle. Also a periodic built up and degradation of storage material and proteins is found during the cycle. For sustained oscillations, a feed-back mechanism must be assumed, which controls the cell cycle by variations in the concentrations of substrates or other growth factors. Many works lead to the conclusion that ethanol together with glucose is responsible for the stabilization of synchronous

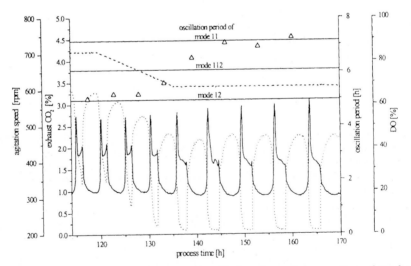

Fig. 9.4. Experimental data for self-sustained oscillations in continuous culture of *Saccharomyces cerevisiae* at $D=0.097h^{-1}$ during mode change from 12 to 11: (- - - - - -) agitation speed, (·····) dissolved oxygen, (——) CO_2 in exhaust gas, (Δ) experimental period length in comparison to the expected values. Reprinted by permission of Wiley-Liss. Inc., a subsidiary of John Wiley & Sons, Inc. from [34] "Oxygen, pH and carbon source induced changes of the mode of oscillation in synchronous continuous culture of *Saccharomyces cerevisiae*". Copyright © 1999.

oscillations. The oscillations can only exist when a periodic switch from ethanol uptake to ethanol formation is possible during the cell cycle. Under reduced dissolved oxygen concentration, the oscillations disappeared or oscillation periods are greatly prolonged and rather irregular, because ethanol uptake is hindered. An example for self sustained oscillation during continuous cultivation of yeast is given in Fig. 9.4.

The quality of the yeast product is influenced by fermentative activity, storage material, and protein content. Since all these vary over the cell cycle, controlled synchronization and harvest at a certain point in the cell cycle may lead to increased product quality. This property can be used for determination of optimum fed-batch control policies as shown in Sect. 9.5.

9.2
Growth Modeling

A model of growth kinetics in bioprocesses includes the two sub-models of the reactor, i.e., of the abiotic phase, and of the population of microorganisms, the biotic phase. This section is concerned with growth of baker's yeast in stirred tank reactors under batch, fed-batch, and continuous operation. The general reactor model for stirred tank reactors is given in Chap. 1 in Table 1.3. For the yeast process, it includes the main components of the liquid phase: cell mass, substrate (e.g., glucose or molasses), ethanol, dissolved oxygen, and dissolved carbon dioxide. The vector of concentrations and specific reaction rates are then

$$
C = \begin{pmatrix} C_X \\ C_S \\ C_E \\ C_O \\ C_C \end{pmatrix} \qquad q = \begin{pmatrix} q_X \\ q_S \\ q_E \\ q_O \\ q_C \end{pmatrix} \tag{9.2}
$$

The substrate flow rate F is the main manipulating variable of the reactor during fed-batch operation. It determines the increase of the volume of the liquid phase, Eq. (1.9), and the dilution of dissolved substances. For an accurate description of oxygen limited growth, the model of the gas phase (Table 1.3, f and g) must also be included.

A review of models for baker's yeast is given by Nielsen and Villadsen [8]. The models range from very simple compartment models up to complex structured ones [9, 10]. Cybernetic models (see Sects. 2.3.3 and 2.3.4 in Chap. 2) are well suited for baker's yeast because they simplify greatly the modeling of the complex growth phenomena and related regulatory effects. The cybernetic model of the biotic phase presented here [45, 47, 85] is built of three parts: the stoichiometric model for a representation of the metabolic reaction network, a cybernetic criterion for determination of the optimal flux distribution in the network under consideration of the rate limiting steps of metabolism, and a dynamic model for slow regulation by induction/repression of key enzymes for some of the rate-limiting steps.

9.2.1
Stoichiometric Model

The stoichiometric model is based on the simplified map of the metabolic network of yeast as shown in Fig. 9.1. It includes the following building blocks: glycolysis, tricarboxylic-acid cycle – approximated by a linear chain of net-throughput of pyruvate to CO_2, respiratory chain, gluconeogenesis, and product formation for ethanol, CO_2, glycerol, and acetaldehyde. Synthesis of macromolecular cell constituents is assumed to start only from fructose-6-phosphate and TCC, by this representing quite a number of reactions that were analyzed in more detail in [11, 12]. The very simplified modeling of the TCC also requires a simplified view of the metabolism during growth on ethanol. As accurate investigations have shown [13, 14] a substrate shift from 100% sugar to 100% ethanol proceeds via at least four different metabolic flux regimes that also influence several reactions in the TCC. Nevertheless, the simplified models are also quite well in agreement with experimental data. The pentose-phosphate way was neglected for the model since it is of greater importance for other substrates such as glucose. These simplifications are compensated for during parameter identification by slight deviations in some model parameters. The metabolic map can be represented by a system of reactions as given in Table 9.2. From that scheme of reactions the stoichiometric model can be derived by establishing balances for each considered metabolite. According to the principle of optimality (see Sect. 2.1.2.2 in Chap. 2), no surplus intracellular metabolites are accumulated during the growth process, so that for a modeling of cultivation processes with a relatively long time horizon the reaction model can be introduced as quasi-stationary. For investigation of short-term effects and intracellular kinetics, dynamic modeling of cellular reactions may be of interest [15–17].

Before establishing the stoichiometric model, further simplifications are introduced. The biomass is assumed to have constant composition and consist only of structural compounds. Then the rates of both biosynthetic pathways must be proportional to the specific growth rate,

$$r_{Ma1} = K_{Ma1}\mu \tag{9.3}$$
$$r_{Ma2} = K_{Ma2}\mu$$

Ideally, the parameters K_{Ma1} and K_{Ma2} are not independent of each other but should satisfy the condition

$$K_{Ma1}M_{F6P} + K_{Ma2}M_{ACY} = 1 \tag{9.4}$$

Since the model reflects a simplified view of the metabolism, the fitting of experimental data can be improved by taking both as independent parameters. The energy consumption for growth and maintenance requirements, r_{ATP} is modeled in analogy to Pirt's concept, Eq. (2.24) in Chap. 2, as

$$r_{ATP} = \frac{\mu}{Y_{ATP}} + m_{ATP} \tag{9.5}$$

The rates of glycolysis, r_{F1}, and gluconeogenesis, r_{Gy}, are joined to only one rate, $r_F = r_{F1} - r_{Gy}$, and the differences in ATP-yield are considered by a switch function $R(r_F)$ in the ATP-balance.

Table 9.2. Metabolic reactions included in the stoichiometric model

	Step	Reaction	Remarks
a	Glucose uptake, 1st phosphorylation and isomerization	$GLU + ATP \xrightarrow{r_S} F6P + ADP$	F6P and G6P are assumed to be in equilibrium
b	2nd phosphorylation	$F6P + ATP \xrightarrow{r_{F1}} FDP + ADP$	
c	Biosynthetic pathway 1	$F6P + \alpha_{ATP} \cdot ATP$ $\xrightarrow{r_{Ma1}} \alpha_{Ma1} \cdot MAC_1 + \alpha_{ATP} \cdot ADP$	representative reaction
d	Glycerol synthesis	$\frac{1}{2}FDP + NADH_2 \xrightarrow{r_{Gy}} GLY + NAD^+$	r_{Gy} can be modeled as starting from F6P with respective modification in the ATP balance
e	Glycolysis	$FDP + 4ADP + 2NAD^+$ $\xrightarrow{r_G} 2PYR + 4ATP + 2NADH_2$	
f	Acetaldehyde synthesis	$PYR \xrightarrow{r_{Ad}} ACD + CO_2$	
g	Acetaldehyde excretion	$ACD \xrightarrow{r_{AdE}} ACD_{ex}$	
h	Ethanol synthesis	$ACD + NADH_2 \xrightarrow{r_{E1}} EtOH + NAD^+$	
i	Ethanol uptake	$EtOH + NAD^+ \xrightarrow{r_{E2}} ACY + NADH_2$	simplified to only one step
j	Acetyl-coA synthesis	$PYR + NAD^+ \xrightarrow{r_{Ac}} ACY + NADH_2 + CO_2$	
k	Biosynthetic pathway 2	$ACY \xrightarrow{r_{Ma2}} \alpha_{Ma2} MAC_2$	representative reaction
l	Tricarboxylic-acid cycle	$ACY + ADP + 4NAD^+$ $\xrightarrow{r_{TCC}} 2CO_2 + ATP + 4NADH_2$	TCC is approximated by linear net-reaction, FAD is considered as NAD
m	Respiratory chain	$O_2 + 2P/O \cdot ADP + 2NADH_2$ $\xrightarrow{r_O} 2H_2O + 2P/O \cdot ATP + 2NAD^+$	FAD is considered as NAD
n	Gluconeogenesis	$4ACY + 4ATP$ $\xrightarrow{r_{Gl}} F6P + 4ADP + 2CO_2 + P_i$	via malate, oxalacetic acid, phosphoenol-pyruvate, and FDP

The formation of by-products acetaldehyde and glycerol is according to experimental findings correlated to the ethanol production,

$$r_{AdE} = K_{Ad} r_{E1}$$
$$r_{Gy} = 2K_{EG} r_{E1}$$

$$(9.6)$$

Both rates were not independent if the simplified map of the metabolism would be exact. Then, since NADH must be balanced, the following relation can be derived for anaerobic growth

$$K_{Ad} = 2K_{EG}$$

$$(9.7)$$

One reason for deviations from Eq. (9.7) is a possible excretion of other by-products.

For modeling of technical cultivations, the knowledge of the intracellular rates r_G, r_{Ad} and r_{TCC} is of less interest. So they can be eliminated from the system of equations of the stoichiometric model for simplicity. By further use of the conditions at Eqs. (9.3), (9.5), and (9.6), the stoichiometric model can be written with molar specific reaction rates in the notation of Eq. (2.81) in Chap. 2 as

$$\mathbf{Y} \mathbf{r} = \mathbf{m}$$

$$(9.8)$$

including the following matrix and vectors:

$$\mathbf{Y} = \begin{pmatrix} 1 & K_{Ma1} & 0 & 0 & 0 & K_{EG} \\ 2 & -4K_{Ma2} & 5 & -2 & 1 & -1 - 2K_{EG} - K_{Ma3} \\ 2 + R(-r_F) & -K_{Ma2} - \frac{1}{Y_{ATP}} & 1 & 2P/O & 0 & -2K_{EG} \\ 2 & 0 & -1 & 0 & 1 & -1 - K_{AD} \end{pmatrix}$$

$$\mathbf{r} = \begin{pmatrix} r_F \\ \mu \\ r_{Ac} \\ r_O \\ r_{E2} \\ r_{E1} \end{pmatrix} \qquad \mathbf{m} = \begin{pmatrix} r_S \\ 0 \\ m_{ATP} \\ 0 \end{pmatrix}$$

$$(9.9)$$

The net reaction rates $\mathbf{q}$, defined by Eq. (2.1) together with Eq. (9.2), appearing in the reactor model, are then given by

$$\mathbf{q} = \begin{pmatrix} q_X \\ q_S \\ q_E \\ q_O \\ q_C \end{pmatrix} = \begin{pmatrix} \mu \\ -r_S \\ r_{E1} - r_{E2} \\ -r_O \\ 2(r_S + r_{Ac} - (K_{Ma1} + K_{Ma1})\mu - K_{EG}R(r_F)r_{E1}) \end{pmatrix}$$

$$(9.10)$$

Although the stoichiometric model may cover most of the complex growth phenomena of yeast, it contains – when considering the conditions of Eqs. (9.4) and (9.7) – only five independent stoichiometric parameters, which all have a direct biological interpretation.

The sugar uptake system is assumed to be constitutive. Therefore, the sugar uptake rate r_S is included into the vector $\mathbf{m}$ of growth independent reaction rates and modeled according to Monod-kinetics as

$$r_S = r_{S\,max} \frac{C_S}{K_S + C_S} \tag{9.11}$$

although an accurate analysis shows that it is also controlled by ATP [15, 18], especially at high glucose concentrations. There is also evidence in the literature of dynamic regulation of sugar uptake in medium and long time scales above 1 h that may be relevant for cultivation processes under excess of substrate [19, 20]. This is also neglected here for the sake of simplicity. For growth on substrates of natural sources such as molasses, the uptake of other sugars beside glucose is important. A respective model including fructose and sucrose was published by Barford et al. [21, 22]. In batch cultivation the uptake of these sugars proceeded in parallel at different rates, while for maltose a sequential uptake was found [23]. A cybernetic model for adaptation during yeast growth on galactose and melibiose was published by Gadgil et al. [24]. Under sugar limited growth in fed-batch cultivation all sugar components are normally used immediately and it is then not necessary to discriminate between them.

9.2.2
Cybernetic Modeling of Metabolic Regulation

In the cybernetic modeling, the growth kinetics are determined by an optimum strategy for the metabolic regulation. When applying the metabolic regulator approach (see Sect. 2.3.4 of Chap. 2) for the modeling of yeast growth kinetics, the optimum strategy of maximization of the specific growth rate, Eq. (2.83), was found to work quite well:

$$\mu(t) \xrightarrow[\text{Eq.}(2.84)]{r(t)} \text{maximum} \tag{9.12}$$

The solution of Eq. (9.8) for $\mathbf{r}(t)$ by the above optimization procedure, also called the metabolic coordinator, has to consider constraints by additional restrictions for possible rate limiting reactions as given by Eq. (2.84). The inherently rate limiting steps in the model are on one hand the uptake steps for ethanol r_{E2}, and oxygen r_O. For simplicity, they are introduced as first-order kinetics. In conjunction with the other constraints this leads to Blackman-kinetics for these uptake steps. For high ethanol concentration inhibition effects should be considered in the uptake kinetics [25]. A careful investigation of oxygen-limited growth also reveals a more complicated nature of the oxygen uptake being dependent on the intracellular ATP levels [26]. This can even cause hysteresis effects with respect to variations in the dissolved oxygen concentration. The intracellular rate-limiting steps on the other hand are the limited capacity in the oxidative pathway, r_{Acmax}, the dynamic regulation of respiration $r_{Omax}(t)$, and of gluconeogenesis, $r_{Fmax}(t)$. For each of the latter two reactions, metabolic regulators are introduced into the model as given below. Beside the TCC-start-reaction, r_{Acmax}, as assumed here, alternative mechanisms for causing oxidoreductive growth are also possible, e.g., saturation of the respiration. This slightly different view of the Crabtree-effect is adopted in the model of Sonnleitner and Käppeli [28]. Such a mechanism could also easily be used instead in the constraints given below. But rate-limiting steps may also be located within the TCC [27]. With these assumptions, the minimum and maximum reaction rates in Eq. (2.84) are as follows:

$$\mathbf{r}_{min} = \begin{pmatrix} -r_{F\,max}(t) \\ -\infty \\ 0 \\ 0 \\ 0 \\ 0 \end{pmatrix} \qquad \begin{matrix} \text{regulation of gluconeogenesis} \\ \text{specific growth rate, not restricted} \\ \text{entry of TCC, irreversible} \\ \text{oxygen uptake only} \\ \text{ethanol uptake only} \\ \text{ethanol production only} \end{matrix} \qquad (9.13)$$

$$\mathbf{r}_{max} = \begin{pmatrix} \infty \\ \infty \\ r_{Acmax} \\ \min(K_O C_O, r_{O\,max}(t)) \\ K_E C_E \\ \infty \end{pmatrix} \qquad \begin{matrix} \text{glycolysis, not restricted} \\ \text{specific growth rate, not restricted} \\ \text{Crabtree effect} \\ 1^{\text{st}} \text{ order oxygen uptake kinetics and regulation} \\ 1^{\text{st}} \text{ order ethanol uptake kinetics} \\ \text{ethanol production, not restricted} \end{matrix}$$

$$(9.14)$$

Since the oxidative metabolism is most effective with respect to energy yield (ATP), the coordination strategy for the metabolism, Eq. (9.12), directs as much as possible of the substrate fluxes into the tricarboxylic-acid cycle, as long as this is below saturation (r_{Acmax}) or the respiratory chain can regenerate all of the produced $NADH_2$. This also gives an explanation of the Pasteur-effect. The rate of the respiratory chain is restricted by the oxygen supply or by the actual induction state of its key enzymes. When the sugar flux cannot saturate the capacity of these pathways, ethanol is used as substrate, when available, up to that limit. Here, the glucose effect is explained by the lower energy yield from ethanol compared to hexoses. In contrast, when the sugar flux exceeds the capacity of oxidative metabolism, the surplus amount is directed to ethanol production and growth becomes fermentative. Then energy production becomes less effective, but nevertheless the growth rate can be increased further by using the additional ATP produced during glycolysis. The switch from ethanol production to uptake into the TCC proceeds instantaneously; a lag-phase is only observed when gluconeogenesis is being involved [29].

The above metabolic model is a static one for the instantaneous adaptation of the cells to the actual supply of substrate and oxygen. By the action of the metabolic coordination, Eq. (9.12), it covers the fast metabolic regulation on the reaction level by activation/inhibition. The model can already be used for simulation of steady states in chemostat cultivations, as presented in the next section. But the respiratory activity and the pathways for gluconeogenesis are subject to long-term regulation by enzyme induction and repression. The resulting lag-phases of growth during phases of regulatory adaptation must be described by corresponding dynamic models: two metabolic regulators for the oxidative capacity and induction of gluconeogenesis during growth on ethanol. The latter one is modeled according to Eq. (2.85) as

$$\frac{dr_{F\,max}(t)}{dt} = (K_{1F} - \mu)r_{F\,max}(t) + K_{2F}(R(-r_F(t)) + r_{F\,min}) \qquad (9.15)$$

In extension of the presentation in Sect. 2.3.4 of Chap. 2, here the metabolic regulator of respiratory activity is established as a dynamic system of third order, which

Table 9.3. Model parameters used for simulations in Figs. 9.5–9.7

Parameter	Fig. 9.5	Fig. 9.6	Fig. 9.7	Units
r_{Smax}	0.019	0.016	0.018	mole $(gh)^{-1}$
K_S	0.8	0.45	0.45	g l^{-1}
K_E	0.5	1.1	1.1	l $(gh)^{-1}$
K_O	–	12.8	–	l $(mgh)^{-1}$
K_{Ma1}	0.0075	0.0053	0.002	mole $(gh)^{-1}$
K_{Ma2}	0.0	0.0044	0.007	mole $(gh)^{-1}$
K_{Ma3}	0.0002	0.39	0.016	–
K_{EG}	0.036	0.05	0.05	–
K_{Ad}	0.033	0.08	0.07	–
P/O	0.82–1.5	0.93–2.7	1–2.5	–
Y_{ATP}	10.2	12	12.9	g mole^{-1}
m_{ATP}	0.002	0.0025	0.002	mole $(gh)^{-1}$
r_{Acmax}	0.0028	0.0034	0.016	mole $(gh)^{-1}$

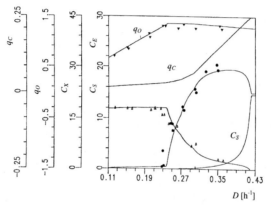

Fig. 9.5. Simulation (*lines*) of steady-states of chemostat cultivation of *Saccharomyces cerevisiae* H1022 on a mixture of 1.5% glucose and 1.5% ethanol in comparison to experimental data (*symbols*) of Rieger et al. [48]: concentrations in kg m^{-3} of glucose C_S, ethanol C_E (●), cell mass C_X (△), specific rates in g(g h)$^{-1}$ of oxygen uptake q_O (▼) and carbon dioxide production q_C. Reprinted from [45] with permission by VCH Weinheim

shows a damped oscillatory behavior, as can be seen in Fig. 9.6. The model is in state space representation

$$\frac{d\mathbf{y}(t)}{dt} = \mathbf{F}(\mu)\mathbf{y}(t) + \mathbf{f}(\mu)(r_O(t) + r_{O\min}) \tag{9.16}$$

where the elements of the state vector

$$\mathbf{y} = \begin{pmatrix} r_{O\max} \\ c_{E1} \\ c_{E2} \end{pmatrix} \tag{9.17}$$

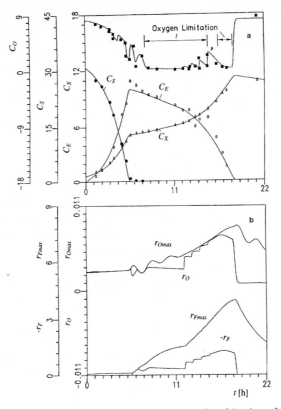

Fig. 9.6a,b. Simulation (*lines*) of diauxic growth in batch cultivation of *Saccharomyces cerevisiae* H1022 on glucose in comparison to experimental data (*symbols*): **a** concentrations in kg m⁻³ of glucose C_S (●), ethanol C_E (○), cell mass C_X (Δ), and dissolved oxygen C_O [g m⁻³] (■); **b** specific oxygen uptake rate r_O and output of the metabolic regulator r_{Omax} in mole(g h)⁻¹, specific rate of gluconeogenesis $-r_F$ and output of the metabolic regulator r_{Fmax} in mmole (g h)⁻¹. Reprinted from [45] with permission of VCH Weinheim

are the maximum specific oxygen uptake rate, r_{Omax}, and intrinsic concentrations of two fictitious enzymes, c_{E1} and c_{E2}. Due to the μ-proportional dilution terms in intrinsic balances, Eq. (2.64), the elements of **F** and **f** are non-linear functions of the specific growth rate μ [85], and can be written as follows:

$$\mathbf{F} = \begin{pmatrix} -3\mu - 0.63 & 1 & 0 \\ -3\mu^2 + 2.7\mu - 7.9 & 0 & 1 \\ -\mu^3 + 13.6\mu^2 - 9.6\mu + 1.2 & 0 & 0 \end{pmatrix} \quad \mathbf{f} = 0.27 \begin{pmatrix} 1 \\ 2\mu - 0.36 \\ \mu^2 + 10.9 \end{pmatrix} \qquad (9.18)$$

where μ is in h⁻¹ and r_{Omax}, r_O in mmole(gh)⁻¹. The output variables of the regulation models, Eqs. (9.15) and (9.16) are as possible rate-limiting steps of the metabolism included into the constraints, Eqs. (9.13) and (9.14).

The assumption of a minimum respiratory capacity as described by the parameter r_{Omin} in Eq. (9.16) is supported by the investigation of Sonnleitner and Hahnemann

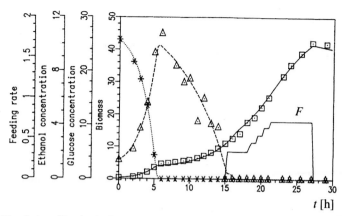

Fig. 9.7. Simulation (*lines*) of a batch/fed-batch cultivation of *Saccharomyces cerevisiae* H620 on molasses and experimental data (*symbols*): concentrations in kg m^{-3} of molasses C_S (✱), ethanol C_E (Δ), cell mass C_X (❑), and substrate flow rate F [l h^{-1}]. Reprinted from [49]

[30]. It was also found that ethanol can have a very long lasting negative effect over several days on the maximum oxygen uptake rate. This was not included in the above model. After prolonged adaptation the respiratory capacity may even be increased over the value corresponding to saturation of the respiratory system [31]. The above model is much simpler than the more detailed structured models including a number of intrinsic balances presented by Steinmeyer and Shuler [9] and Coppella and Dhurjati [10]. But an accurate description of a wide range of process conditions that are important for baker's yeast production is provided.

9.2.3
Application of the Model for Simulation of Batch, Fed-Batch, and Continuous Cultivations

In this section, simulation results with the above model for continuous, batch, and fed-batch cultivations of baker's yeast in laboratory-scale stirred-tank reactors are presented. The model parameters used in the simulations are summarized in Table 9.3. The simulation results for growth of yeast in chemostat culture on a mixed substrate of 1.5% glucose and 1.5% ethanol and experimental data of [48] are given in Fig. 9.5. At low growth rates, controlled by low dilution rates, both substrates are used up completely for growth by oxidative metabolism. For $D>0.23$ h^{-1} some of the ethanol is left in the medium because of the limited capacity of oxidative metabolism by saturation of the TCC (r_{Acmax}). In this range of dilution rates, glucose is the preferred substrate due to the higher efficiency in energy production. The increase in ethanol is accompanied by a decrease in cell concentration. For $D>0.3$ h^{-1} even ethanol is produced since the flux from glucose alone already exceeds the oxidative capacity of the cells. All these phenomena are predicted by the metabolic coordinator that maximizes the growth rate with respect to the actual inherently rate-limiting steps. With usual structured models for growth on multiple sugars the effect cannot be described so smoothly [46].

Figure 9.6 shows simulation results from [45] for an aerobic batch cultivation in glucose medium in comparison to experimental data of Kuhlmann (cited in [45]). In

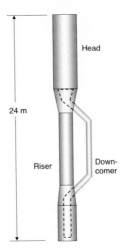

24 m

Head

Riser

Down-
comer

Fig. 9.8. Schematic drawing of the pilot-scale 5-m^2 airlift-tower-loop reactor for fed-batch cultivation of baker's yeast [52]

the first growth phase up to 5.5 h, the sugar concentration is high and ethanol is produced due to the Crabtree-effect. The yeast is growing on the preferred substrate glucose and the uptake of ethanol is suppressed by the glucose-effect until all sugar is used up. This behavior leads to the well-known growth-diauxy of yeast. There are two mechanisms for suppression of ethanol uptake: inhibition by the metabolic coordinator and repression of the key-enzymes for gluconeogenesis. By the latter effect, r_{Fmax} stays at low values. In the transient phase when the metabolism is switching from glucose to ethanol growth, both regulatory systems for r_F and r_O become rate limiting for a short period, and therefore their activities are induced. But growth is mainly limited by low oxygen concentrations during the second growth phase with few exceptions between $t=14$ h and 16 h, where the actual capacity of the respiratory system is at the limit. At $t=12$, 13, 14, and 15 h the agitation speed was increased to follow the higher oxygen demand by growth of cell mass. During growth, the energy requirements for maintenance are covered by the substrate catabolism. In the starvation period, beginning from $t=18$ h, the cell mass concentration decreases. Here the model predicts a degradation of cell material to produce energy as required for maintenance by the rate m_{ATP}.

Figure 9.7 presents a combined batch and fed-batch cultivation of *Saccharomyces cerevisiae* strain H620. After the end of the diauxic batch phase at $t=15$ h, substrate flow is switched on and later increased further according to the profile shown. The aim of a fed-batch control is to ensure purely oxidative growth without ethanol production by limited feeding of the sugar. During oxidative growth, yeast has high requirements for oxygen of about 1 g O_2 per g of dry cell mass. The profile of the molasses feed in these stages has to be adjusted to meet the oxidative capacity of the cells or the maximum oxygen transfer of the reactor. If one tries to keep the growth rate as high as possible, it follows more or less a two-phase feeding scheme: in the first exponential growth phase under excess of oxygen a constant specific growth rate of about 0.25 h^{-1} – just at the critical value for the Crabtree-effect – is set by an exponentially increasing substrate flow. In the second phase the growth rate has to

Table 9.4, Correlations for fluid-dynamic parameters in the reactor model, Eqs. (9.63)–(9.70)

Parameter	Correlation	Section	Reference
Dispersion coefficient of the liquid phase	$D_{Lj} = 1.23\ cm^2\ s^{-1} \left(\dfrac{d_j}{cm}\right)^{1.5} \left(\dfrac{u_{Gj}}{cm\ s^{-1}}\right)^{0.5}$	$j=r,d,k$ all sections	Towel and Ackerman[a]
Volumetric mass transfer coefficient	$(k_L a)_r = \dfrac{2 u_{Lr}}{H_d} \left(\dfrac{u_{Gr}}{u_{Lr}}\right)^{0.87} \left(1 + \dfrac{A_d}{A_r}\right)^{-1}$	riser	Bello et al. [58]
	$(k_L a)_d = 0.89\,(k_L a)_r$	downcomer	Strauß [52]
	$(k_L a)_k = 0.011\ s^{-1} \left(\dfrac{u_{Gk}}{cm\ s^{-1}}\right)^{0.82}$	head	Kastanek[a]
Gas holdup	$E_{Gr}(z) = \dfrac{u_{Gr}(z)}{0.24 + 1.35\frac{u_{Gr}(z)}{m\,s^{-1}} + \left(\frac{u_{Lr}(z)}{m\,s^{-1}}\right)^{0.93}}$	riser	Hills [59]
	$E_{Gd}(z) = 0.89\,E_{Gr}(z)$	downcomer	Bello et al. [58]
	$E_{Gk}(z) = 0.63 \left(\dfrac{u_{Gk}(z)}{m\,s^{-1}}\right)^{0.775}$	head	Weiland [60]

[a] Recommended in [57]

be limited by an almost constant feed-rate according to the maximum oxygen transfer capacity into the reactor to avoid oxygen limitation and ethanol production. An optimum production with maximum yield and good productivity can only be achieved by automatically controlling the substrate feed just at the critical value for the onset of the Crabtree-effect. This will be further discussed in Sect. 9.6.

9.3
Growth in Airlift Tower-Loop Reactors

In this section the model is applied for baker's yeast fed-batch production in a pilot-scale airlift tower-loop reactor. In this type of reactor aeration is very energy efficient, but inhomogeneities must be expected in the liquid phase, especially for the concentrations of substrate and oxygen, that can influence the growth and be very critical for baker's yeast production. The reactor is schematically shown in Fig. 9.8. Among the three sub-systems – riser, downcomer, and head – only the head space has a variable working volume due to the substrate feed. To cover extended fed-batch periods, its height and diameter are unusually large for airlift reactors. Substrate feeding is distributed along the height of the riser. The model is used to analyze the process and study possible directions for its optimization.

With respect to high productivity, the process has to be operated at the critical substrate feed rate for switch of metabolism to oxidoreductive growth with ethanol

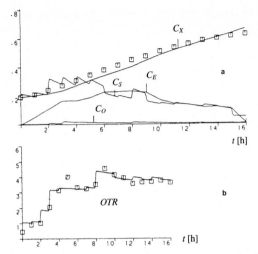

Fig. 9.9a,b. Simulation (*lines*) and experimental data (*symbols*) of a fed-batch cultivation of *Saccharomyces cerevisiae* in the airlift tower-loop reactor [52]: **a** spatially averaged concentrations of glucose, ethanol, cell mass ($\square$), and dissolved oxygen; scaling: $C_O = 0.5\,\mathrm{mmole\,l^{-1}}$, $C_S = 2\,\mathrm{mmole\,l^{-1}}$, $C_E = 0.5\,\mathrm{mmole\,l^{-1}}$, $C_X = 100\,\mathrm{g\,l^{-1}}$; **b** total oxygen transfer rate *OTR* $[\mathrm{kg\,m^{-3}\,h^{-1}}]$ ($\square$)

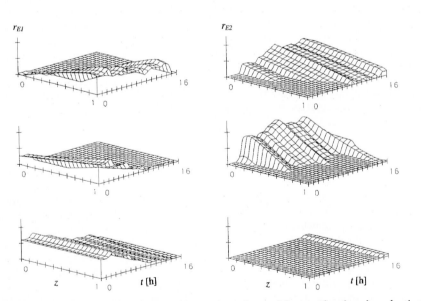

Fig. 9.10. Simulation results for spatial and time dependence of the specific ethanol production rate r_{E1} (*left side*) and uptake rate r_{E2} (*right side*) for the fed-batch cultivation in Fig. 9.9 [52]; *top*: riser; *middle*: downcomer; *bottom*: head; The normalized spatial coordinate in all three sections is $z = 0$ for bottom and $z = 1$ for top, the scaling unit for r_{E1} and r_{E2} is $5 \times 10^{-7}\,\mathrm{mmole(gs)^{-1}}$

production. To avoid loss of productivity in an inhomogeneous environment, the flow rate must not be decreased so much that in every location of the reactor the metabolism is purely oxidative. Therefore, the cells will pass through various regions where by different substrate concentrations the substrate uptake rate is either above or below the critical value. This value may also vary due to different levels of dissolved oxygen. Abel et al. [53] concluded from experiments with enforced periodic O_2-variation that as long as the residence time in the oxygen-limited region is below 1 min and the volume of the downcomer less than one-third of the riser, the yield should not be negatively influenced. The yeast can switch very fast between oxidative growth with ethanol uptake and oxidoreductive growth with ethanol production, as long as ethanol is used in parallel to sugar and only catabolized in the TCC for energy production. On the other hand, comparative investigations by George et al. [51] on a 215-m^3 bubble column production reactor and a scaled-down reactor point to a more critical role of inhomogeneous substrate distribution by localized feeding. There, with a recirculation interval of 60 s, the yield decreased by about 6%, while at the same time the gassing power of the yeast product increased. But in ATL reactors with a higher recirculation flow the inhomogeneities are less pronounced.

For an investigation of the influence of concentration profiles on the yeast growth, the model of the previous section was extended by a distributed reactor model to describe the spatial dependencies of process variables [52]. The reactor model, as presented in Chap. 1, Eqs. (1.63)–(1.70), has to describe the time and space-dependent concentrations in the gas and liquid phase as a function of initial conditions, manipulating variables, and biological reactions of the yeast cells. Beside reaction and mass transfer, convection and dispersion are considered for gas and liquid phase. In airlift reactors, the fluid-dynamics and flow characteristics are controlled by the aeration rate which also influences the mass transfer and mixing properties. The correlations given in Table 9.4 were used for the dispersion coefficients, transport parameters, and gas holdup. The parameters were estimated from experimental data based on measurements by Lübbert et al. [54, 55]. Since it is ensured in normal fed-batch operation that sufficient amounts of sugar are always available, gluconeogenesis is avoided and the respiratory system is fully induced. Therefore, the dynamic regulation models as part of the metabolic model can be omitted for this study. This also avoids the problem with structured segregated models as was discussed in Sect. 2.3 in Chap. 2.

After verification of the model with experimental data, simulation studies for the pilot plant were carried out to investigate the influence of non-ideal mixing and concentration gradients. Some results are shown in Fig. 9.9 for averaged concentrations of the liquid phase, and in Fig. 9.10 for specific ethanol uptake and production rates. The spatial concentration profiles (not shown) are relatively flat and the cell mass could be considered as a lumped variable. Dissolved oxygen is limiting during the entire cultivation. Therefore, the cell mass increases almost linearly. To meet the increased oxygen need of the culture, the aeration rate was increased stepwise several times as can be seen from the oxygen transfer rate. In the average, there is ethanol accumulation in the first half of the cultivation which is later on reconsumed.

As expected, the yeast metabolism switches from fermentative with ethanol production to oxidative with ethanol uptake during each circulation. Under the given operating conditions in the head space of the reactor, only ethanol production is found. After an initial phase of about 3 h with overfeed of sugar, there are both ethanol production and uptake in downcomer and riser (Fig. 9.10); ethanol is produced in the upper region of the two sections, where the oxygen transfer is lowest. At these locations there is also a slightly higher sugar concentration due to distributed feeding over the height of the reactor. In the bottom regions of the system ethanol uptake can be found due to lower sugar and higher dissolved oxygen concentration. Productivity and yield are not greatly influenced as expected from [53], since ethanol serves as a buffer that equalizes the differences in growth conditions for the growth-determining intracellular energy production. When sufficient sugar is available and ethanol flux is only directed to the TCC, there is practically no loss in energy efficiency.

The analysis reveals that for the given system substrate, feeding in the riser is not optimal because then the substrate concentration is highest just in those parts of the reactor where the oxygen supply is lowest. This limits the productivity because the substrate flow rate has to be kept low to avoid too much ethanol production and loss of yield. It can be concluded from the simulation studies that in this special reactor a distributed substrate feeding in the downcomer would be favorable in order to flatten the profiles for oxygen in the entire reactor. The reason is the very high recirculation flow that transports enough gas bubbles and oxygen into the downcomer down to its bottom, where oxygen transfer is increased by pressure effects. This at first unexpected effect was also found by Pollard et al. [56].

9.4
Population Balance Models for the Asymmetric Cell Cycle of Yeast

In this section, age distribution models for growing yeast populations are presented, aimed at two aspects: first, to provide a quality parameter for the yeast production that can be used for optimal dynamic control of fed-batch processes, and second, analysis of stable synchronous oscillations in continuous processes. There were different attempts to develop such models for balanced and non-balanced growth, mostly on the basis of the general population models presented in Chap. 2, Sect. 2.4.6, Eq. (2.123), or the age distribution model, Eq. (2.126). The variables used to characterize the population depend on the desired application of the model and the kind of analysis that should be done. A model for studying biochemical processes during the cell cycle will naturally use other quantities than a model that is aimed at a description of some population parameters during industrial cultivations. Of course one would always like to establish a model that is as close as possible to the biochemical mechanisms of cell growth and proliferation. But this is presently far out of reach. Therefore, usually some representative variables without direct correspondence in the metabolism are used, such as total mass, age, or volume of a cell, or mass of a representative intrinsic constituent, e.g., protein content or storage material. The next question for population models is whether the length of the cell cycle phases is assumed as stochastic with a continuous variation around an average value – so individual cells may behave differently – or assumed as deterministic, where all cells finish their phase at the same discrete value of the characteristic variable. The latter type of model has a higher degree of simplification, but has the advantage of a facilitated solution, either numerically or even analytically for some special cases. This property makes it much easier to check the model against a great number of experimental data and to apply it for data interpretation, e.g., for balanced asynchronous growth of yeast [3]. As any model, unstructured or structured, a population model also adopts a very simplified view of reality, and one must always be aware of it [62]; but one should not succumb to the illusion that less simplified models must give a more exact representation. They generally have a higher degree of freedom and include more parameters and mathematical functions, so the experimental verification becomes much more difficult – often due to the limited availability of experimental data, or difficulties in the model solution – or even impossible when the model can be arbitrarily fitted to almost any data set. So the general guideline is also imperative for population models: a model is never reality, but a model should be as simple as possible as long as it is in agreement with experimental data. Conceptual models that are compared to no or limited experimental data are of less use for practical application [63–66].

9.4.1
Age Distribution Model of Yeast for Batch and Fed-Batch Processes

Many microorganisms such as bacteria propagate by symmetrical division. The parent cell "dies" as it is reborn in two identical daughter cells. This is reflected in the model, Eqs. (2.123) and (2.128) in Chap. 2, by the factor 2 that is also called the birth rate. The propagation of yeast by budding follows an asymmetric mechanism as shown schematically in Fig. 9.2: the bud grows out of the parent cell. After bud separation, the remaining bud scar is closed and the parent cell keeps growing and can give birth to further buds. A new-born bud behaves differently from its parent cell since it is usually much smaller. One of its characteristics is a longer generation time compared to parent cells. Only after its first budding does the former daughter cell, now in the first generation of parents, behave quite similar to the elder generations of parent cells. Therefore, a population model for yeast has at least to consider two clearly distinguishable sub-populations of parent and daughter cells. A short review of population models for baker's yeast is given in [32, 33].

The early investigations of the population dynamics [3–5], where the cycling phase equations, Eq. (9.1), were established and verified, can be represented by an age distribution model, as given by Eq. (2.126) in Chap. 2, with discrete deterministic division ages and a loss function g that equals the dilution rate D of the reactor. But the clearly distinguishable sub-populations of daughter and parent cells have to be described by at least two coupled population balances. The respective cells number densities in dependence of age τ and time t, $f_D(t, \tau)$ and $f_P(t, \tau)$, are used as characterizing variables of the population. The M'Kendrick-von Foerster equations according to the simplified scheme of the asymmetric cell cycle as shown in Fig. 9.2 then become

$$\frac{\partial f_P(t,\tau)}{\partial t} + \frac{\partial f_P(t,\tau)}{\partial \tau} = -D \cdot f_P(t,\tau) \quad \text{with} \quad 0 \leq \tau \leq T_P + T_B \tag{9.19}$$

$$\frac{\partial f_D(t,\tau)}{\partial t} + \frac{\partial f_D(t,\tau)}{\partial \tau} = -D \cdot f_D(t,\tau) \quad \text{with} \quad 0 \leq \tau \leq T_D + T_B \tag{9.20}$$

The processes of bud separation and formation of new unbudded daughter and parent cells, each with a birth rate equal to one, is then introduced into the model by the boundary conditions

$$f_D(t,0) = f_P(t,0) = f_P(t, T_P + T_B) + f_D(t, T_D + T_B) \tag{9.21}$$

For an the initial condition at $t=0$ with exponential age distribution

$$f_D(0,\tau) = f_P(0,\tau) = f_0 e^{-\mu\tau} \tag{9.22}$$

growth is balanced and the solutions of the population balances for the number densities become

$$f_D(t,\tau) = f_P(t,\tau) = f_0 e^{-Dt} e^{\mu(t-\tau)} \tag{9.23}$$

where $D=0$ for batch cultivation. The number of cell in the cell cycle intervals and in total are

$$N_D(t) = \int_0^{T_D} f_D(t,\tau)d\tau \quad N_P(t) = \int_0^{T_P} f_P(t,\tau)d\tau$$

$$N_B(t) = \int_{T_D}^{T_D+T_B} f_D(t,\tau)d\tau + \int_{T_P}^{T_P+T_B} f_P(t,\tau)d\tau \tag{9.24}$$

$$N_{tot}(t) = N_D(t) + N_P(t) + N_B(t) \tag{9.25}$$

and the fraction of budding cells is then

$$FBC(t) = \frac{N_B(t)}{N_{tot}(t)} \tag{9.26}$$

Hartwell and Unger [3] found from the above solutions the following condition – which can be called stationary condition – for the length of the cell cycle intervals of an exponential asynchronously growing yeast population,

$$e^{-\mu(T_B+T_P)} + e^{-\mu(T_B+T_D)} = 1 \tag{9.27}$$

where the specific growth rate and average doubling time are connected by

$$T = \frac{\ln(2)}{\mu} \tag{9.28}$$

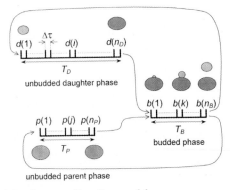

Fig. 9.11. Structure of the discrete cell cycling model

Tyson et al. [106] derived from the previous model the relation between the length of the budding phase and the fraction of budding cells as

$$FBC = e^{\mu T_B} - 1 \tag{9.29}$$

that is in good agreement with experimental data. Adams et al. [61] presented a population model for asymmetric division with discrete age classes. The transitions from one age class to the next or to the next cycle are determined by probabilities. For steady state the model is equivalent to that of [3]. The simulation showed strongly damped oscillations during transients of growth. The dynamics of the cell cycle are of interest for fed-batch cultivations when it is desired to predict the product quality which depends on the fermentative activity and storage properties. Takamatsu et al. [71] have found that the fermentative activity is correlated to the fraction of budding cells (FBC) in the yeast population. Since the storage properties are also related to FBC [50], this single parameter can be used in a process model to describe several quality aspects. A low FBC at the end of the final fermentation stage gives a good product quality. In an industrial-scale baker's yeast production process, the cell growth rate generally changes with time. Moreover, the whole fermentation period is only about 10 h. A true steady exponential growth state is not obtained under these operating conditions.

A time-discrete cell cycling model (CCM) – combined with a growth model for *Saccharomyces cerevisiae* – to describe this asymmetric budding process under dynamic conditions was developed and verified with experimental data by Yuan et al. [47, 49, 50]. The growth model, given in Sect. 9.2, represents the continuous reactions of growth. The oxidative pathways and the gluconeogenesis are subject to long-term regulation by enzyme induction and repression. The respective dynamic regulation models were omitted here because in the final stage of fed-batch production the respiratory system is fully adapted and gluconeogenesis is avoided. The age distribution model covers the discontinuous processes of the asymmetric cell multiplication process and the age distribution of the population. As was discussed in Sect. 9.1, the cell metabolism at large does not vary greatly during the cycling process and the increase of cell mass is the rate limiting step of cell multiplication, i.e., the cell division cycle is controlled to a great extent by the specific growth rate. Since in the range of interest the mean cell volume and mean cell density vary only slightly but in opposite direction, the cell number doubling time was assumed to be equal to the cell mass doubling time, and Eq. (9.28) was applied under dynamic conditions to connect the growth model with the age distribution model. This makes it possible to simulate the control of the cell cycling process by manipulating the common controlling variables, e.g., substrate feeding rate. For a limited degree of synchrony, a feed back from the cell cycling model to the metabolic model need not be considered. For simplification, parent cells and daughter cells were assumed to have the same metabolic activity as described by a unique specific growth rate.

To facilitate simulation, the continuous balances of the population model, Eqs. (9.19) and (9.20), are approximated by discrete shift-registers with n_D, n_P and n_B storage elements, as shown in Fig. 9.11. The cycling phases D, P, and B (see

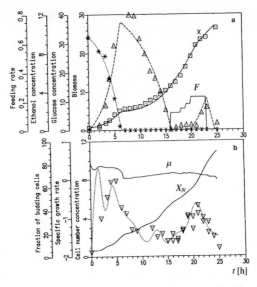

Fig. 9.12a,b. Experimental data (*symbols*) and simulation results (*lines*) by the process model for a combined batch and fed-batch cultivation of baker's yeast *Saccharomyces cerevisiae* H1022 on glucose: **a** sugar feed rate F [$l\,h^{-1}$] and concentrations in $kg\,m^{-3}$ of cell mass C_X (☐), substrate C_S (✳) and ethanol C_E (Δ); **b** specific growth rate μ [h^{-1}], cell number concentration X_N [$10^{-5}\,l^{-1}$] and fraction of budding cells *FBC* [%] (∇). Reprinted from [49]

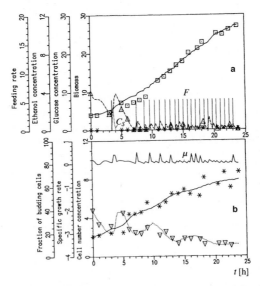

Fig. 9.13a,b. Experimental data (*symbols*) and simulation results (*lines*) for a pulsed-feeding fed-batch cultivation of baker's yeast *Saccharomyces cerevisiae* H1022 on molasses: **a** sugar feed rate F [$l\,h^{-1}$] and concentrations in $kg\,m^{-3}$ of cell mass C_X (☐), substrate C_S (✳), and ethanol C_E (Δ); **b** specific growth rate μ [h^{-1}], cell number concentration X_N [$10^{-5}\,l^{-1}$] (✳) and fraction of budding cells *FBC* [%] (∇). Reprinted from [49]

Table 9.5. Identified parameters in the growth model for *Saccharomyces cerevisiae* H1022

Parameter	Fig. 9.12	Fig. 9.13	Units
r_{Smax}	0.014	0.019	mole $(gh)^{-1}$
K_{Ma1}	0.003	0.007	mole $(gh)^{-1}$
K_{Ma2}	0.012	0.005	mole $(gh)^{-1}$
K_{Ma3}	0.006	0.0001	–
K_{EG}	0.05	0.05	–
K_{Ad}	0.02	0.031	–
P/O	1.2–3.9	1.5–2.4	–
Y_{ATP}	10.3	10.5	g mole^{-1}
m_{ATP}	0.001	0.003	mole $(gh)^{-1}$
r_{Acmax}	0.019	0.003	mole $(gh)^{-1}$
C_S^R	351	325	kg m^{-3}

Table 9.6. Identified parameters in the cycling phase equations for *Saccharomyces cerevisiae* H1022

Experiment	K_{D1}	K_{D2}	K_{P1}	K_{P2}	K_{B1}	K_{B2}	Substrate	Temperature
Fig. 9.12	1.84	−1.09 h	0.46	−0.1 h	0.29	1.39 h	glucose ethanol	303 K
	3.32	−0.26 h	0.36	−0.1 h	0.35	0.08 h		
Fig. 9.13	1.88	−1.11 h	0.59	−0.1 h	0.26	1.2 h	molasses	307 K

Fig. 9.2) are divided into small cycling age intervals $\Delta\tau$. The number of these intervals in each cycling phase are given by

$$n_D = \frac{T_D}{\Delta\tau} \quad n_P = \frac{T_P}{\Delta\tau} \quad n_B = \frac{T_B}{\Delta\tau} \tag{9.30}$$

where the lengths of the cell cycle phases for single daughter, single parent and budded cells, T_D, T_P and T_B, are determined by Eq. (9.1). The continuous cell number densities are replaced by discrete ones in the following manner:

$$\begin{aligned} f_D(1\ldots T_D) &\to d(1\ldots n_D) \\ f_P(1\ldots T_P) &\to p(1\ldots n_P) \\ f_D(T_D+1\ldots T_D+T_B) + f_P(T_P+1\ldots T_P+T_B) &\to b(1\ldots n_B) \end{aligned} \tag{9.31}$$

In each simulation time step with a length $\Delta\tau$, the following substitutions are performed

$$\begin{aligned} d(i-1) &\Rightarrow d(i) & b(n_B) &\Rightarrow d(1) \\ p(i-1) &\Rightarrow p(i) & b(n_B) &\Rightarrow p(1) \\ b(i-1) &\Rightarrow b(i) & d(n_D)+p(n_P) &\Rightarrow b(1) \end{aligned} \tag{9.32}$$

For varying T, the cells are redistributed over more or less cycling intervals by keeping the old distribution pattern and the relative positions of the cells in the cycling phases. When T decreases, only cells in D are redistributed by an exponential non-uniform pattern, so as to adapt to the dynamic dividing behavior during the cycling process [49]. The averaged population variables can be calculated by simple summations as for the total number of cells, the cell number concentration and the fraction of budding cells,

$$N_{tot} = \sum_{i=1}^{n_D} d(i) + \sum_{j=1}^{n_P} p(j) + \sum_{k=1}^{n_B} b(k) \tag{9.33}$$

$$X_N = \frac{N_{tot}}{V_L} \tag{9.34}$$

$$FBC = \sum_{k=1}^{n_B} \frac{b(k)}{N_{tot}} \tag{9.35}$$

The model was tested with a number of experiments. The results of two fed-batch experiments are shown in Figs. 9.12 and 9.13. The identified model parameters are given in Tables 9.5 and 9.6. Although the latter parameter values show some variation, the calculated lengths of the cycling intervals are close to the literature values for balanced growth. In the first experiment, the time series of the feeding rate was eventually equivalent to an industrial incremental feeding pattern. At the beginning, the model simulation for FBC differs somewhat from the measurement. This is due to an initial lag-phase that was not taken into account in the model: there the simulated specific growth rate is too high, resulting in the sharp peak in the simulated FBC, as illustrated in Fig. 9.12. There were actually three different growth phases in the experiment according to the available substrates: oxidoreductive on glucose only, oxidative on ethanol only, and oxidative on glucose during the fed-batch. In Fig. 9.12. the cycling model is compared with experimental data for FBC. Besides the tendency of increased FBC with higher growth rates, some oscillations in FBC can be observed. The final value of FBC is about 20%. In the ethanol growth phase, the duration of the daughter cell phase was much longer than on sugar at comparable μ. This is probably caused by the changes of cell morphology due to poor nutrition [67]. Therefore, the parameters of the cycling phase equations were identified again for ethanol growth as given in Table 9.6.

In the second experiment shown in Fig. 9.13, a pulsed feeding pattern was used, based on the CO_2 content in the exhaust gas. If CO_2 was below a preset lower threshold, the flow was turned on, whereas when it had reached a preset upper limit, the pump was shut down. The feeding pulse sequence is shown in Fig. 9.13a. The width of the pulses was about 1–1.5 min and the pulse period 0.6–0.9 h. At the beginning, cells grow on ethanol that remains from a previous batch culture and FBC decreases. The first feeding pulse at about $t=3$ h leads to a short period of oxidoreductive growth with ethanol production and high growth rate so that FBC increases again. During the remaining part of the cultivation the average growth rate decreases moderately, accompanied by a similar trend on FBC with slight oscillations.

In the experiments, the simulated cell number concentration, X_N, was in good agreement with the measurements, though the specific growth rate varied over a wide range. Also the FBC simulated by the CCM follows the measurements well. From these results, one may conclude that the CCM with its simple structure is able to describe the oscillatory transients by partial synchronization of the cell population of baker's yeast under dynamic conditions, and can be applied in process optimization, for example minimization of FBC. For minimization of FBC at the harvesting time in fed-batch culture, the feeding strategies should be in accordance with such an intrinsic oscillation and reinforce it in a way as to make the FBC approach its minimum at a desired harvesting time. This will be further outlined in Sect. 9.5

9.4.2
Age Distribution Model for Data Analysis of Stable Synchronous Oscillations in a Chemostat

Population synchrony is an important factor for autonomous and forced oscillations in continuous culture of yeast. For a rather mechanistic modeling of self-synchronization and sustained oscillations, the population balances must be extended by a structured growth model that introduces a non-linear feed-back mechanism for the stabilization of the oscillation due to differences in the metabolism of budded and unbudded cells. In the model of Strässle [35–37] a bottleneck in the respiratory pathway and description of the metabolism of storage carbohydrates was included. The range of dilution rates and the frequency of sustained oscillations could roughly be reproduced. Cazzador et al. [38] established a model for a size control mechanism that is modulated by the substrate concentration. The model was in qualitative agreement with observations of frequency and amplitude of oscillations at different dilution rates. Unfortunately, there are no analytical solutions of the balance equations of such models, and one must turn to numerical methods, which require an enormous expenditure in computer time. This limits the simulation studies to only a few points of the parameter space and makes it difficult to draw some general conclusions. For a broad evaluation of experimental data these models are too complicated.

Sustained synchronous oscillations of baker's yeast were also analyzed by means of an age distribution model without consideration of cell metabolism [32, 33]. Many characteristics of the oscillations – with the exception of the actual shape of the age distribution – are already determined to a great extent by the asymmetric cell cycling behavior of yeast in conjunction with the mathematical properties of the solution of the population balance equations. The assumption of discrete division ages without variation in the lengths of the cell cycle intervals of individual cells of the population seems to be a very natural approximation, because such variations should be reduced by the feed-back, which is responsible for self-stabilization of the oscillation. The accompanying oscillations on the concentrations of substrates actively force the cells into a narrow distribution and reduce the dispersion in cy-

Table 9.7. Oscillation periods for a few mode numbers

Mode		$T_{IJ}T^{-1}$	T_{IJ} ($T = 2.5$ h)	α_{IJ}	Remark
I	J				
1	1	1	2.5	0.5	Symmetric division
1	2	0.694	1.736	0.6180	Golden mean
1	3	0.551	1.379	0.6823	
2	2	0.5	1.25	0.7071	Symmetric division
2	3	0.406	1.014	0.7549	
2	4	0.347	0.868	0.7861	

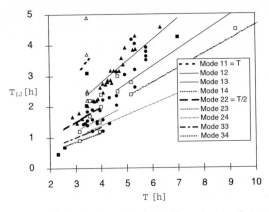

Fig. 9.14. Calculated oscillation periods (*lines*) of self-synchronized growth over doubling time in comparison to experimental data (*symbols*) from the literature: ▲ [35], ● [39], ❑ [40], ■ [43], △ [44]

cling events. The model considers incomplete synchronization and an age distribution of cells within the cell cycle, but only due to a phase shift in the growth of individual cells. The assumption of discrete division age is actually not a great restriction for analyzing stable synchronous oscillations, because under such conditions also in a model with division age distribution, the statistical fluctuation must be exactly leveled out during the oscillations with an average value identical to the discrete division age. For a distribution being conserved between subsequent oscillation periods it makes no difference whether all cells exactly keep their order during a cycle or some of them exchange their positions by stochastic processes. Furthermore, when the cell metabolism during one oscillation period is determined by the average doubling time, analytical solutions for the model can be obtained. That precondition is supported by the results on yeast physiology reported in Sect. 9.1 and also by the work of Münch et al. [39] who showed that very slight disturbances are sufficient to establish or alter the oscillations.

For extending the above model to stable synchronous oscillations the initial condition, Eq. (9.22), can be rewritten as

$$f_D(0, \tau) = f_P(0, \tau) = f_0 e^{-\mu\tau} \tilde{f}(-\tau) \tag{9.36}$$

and the solutions of the population balances become for chemostat processes, with $D \equiv \mu$,

$$f_D(t, \tau) = f_P(t, \tau) = n_0 e^{-\mu\tau} \tilde{f}(t - \tau) \tag{9.37}$$

Here, an additional shape function, $\tilde{f}$, of the age distribution appears, which is responsible for the oscillatory behavior of the solution,

$$\tilde{f}(t - \tau) = \tilde{f}(t - \tau + T_{IJ}) \tag{9.38}$$

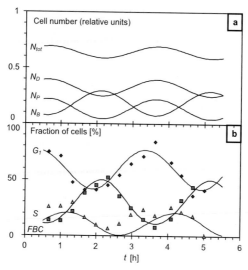

Fig. 9.15a,b. Simulation (*lines*) of synchronous growth in comparison to experimental data (*symbols*) from Strässle [35]: Mode 12, T=4.4 h, T_B=1.6 h, T_S=0.5 h: **a** relative numbers of total (N_{tot}), unbudded daughter (N_D) and unbudded parent cells (N_P); **b** fraction of cells in the budding (■), S- (▲), and G_1-phase (◆)

Equation (9.37) describes a sustained oscillation with constant amplitude over time, but exponentially decreasing amplitude over cell age due to the dilution. It also matches the stationary condition at Eq. (9.27). The possible oscillation periods are

$$T_{IJ} = \frac{T_B + T_P}{I} = \frac{T_B + T_D}{J} \quad \text{for} \quad I = 1, 2, 3..., \quad J = 1, 2, 3..., I \tag{9.39}$$

During sustained oscillations in population variables the lengths of the parent and daughter cycles must be multiples of the oscillation period as defined by the mode numbers I and J. By introducing the decrement of the oscillation,

$$\alpha = e^{-\mu T_{IJ}} \tag{9.40}$$

Eq. (9.27) changes to

$$\alpha^I + \alpha^J = 1 \tag{9.41}$$

which can be solved for α when the mode numbers are given, yielding a value α_{IJ}. The oscillation period of a certain mode then becomes

$$T_{IJ} = -\frac{\ln(\alpha_{IJ})}{\mu} = -T\frac{\ln(\alpha_{IJ})}{\ln(2)} \tag{9.42}$$

Some values of α_{IJ} and T_{IJ} for low mode numbers are given in Table 9.7. The solution for sustained oscillations in chemostat processes, Eqs. (9.37)–(9.42), has some very interesting properties due to the asymmetric population dynamics. Although the main determining factor for the frequency of the oscillation is the average doubling time T, the oscillation frequency is not unique for a certain T, but may change with the mode of oscillation; fixed discrete but multiple oscillation frequencies are possible. The actual mode present in the culture depends on initial conditions, cultivation conditions, and growth kinetics. Therefore, it can be possible that under identical cultivation conditions different oscillation frequencies are observed as shown in Fig. 9.4. Oscillation periods from the experimental work of several authors and calculated from the model for different modes of oscillation are presented in Fig. 9.14 over doubling time. The predicted oscillation frequencies agree well with experimental data. The observed variability of the frequency at a fixed doubling time can be explained to a great extent by the multiple modes of oscillations. The prevailing number of measurements close to mode 12 fall into the region $T<7$ h, and for higher modes into the region $T<4.5$ h. This coincides with situations where the daughter cycle is shortened by the oscillation and the parent cycle is prolonged compared to asynchronous growth. For $T<3.2$ h, the asynchronous growth follows a nearly symmetric division pattern, which is very close to the oscillation modes 11, 22, and 33. It can be expected that under such conditions one of these modes is preferred. In a very wide region for medium doubling times, the modes 12, 24, and 36 – where the length of the daughter cycle is twice that of the parent cycle – come very close to the asynchronous case. Hence, these modes may be called the natural modes of oscillation of a synchronous growing yeast population.

Further evaluation of experimental data by the model revealed that sustained oscillations can only exist in a region where the length of the parent cycle is increased, and the length of the daughter cycle is decreased, compared to asynchronous growth [33]. The preferred region of oscillations of a certain mode seems to be determined by the length of the budding phase which also controls the oscillation amplitude. The predicted tendency of the oscillation amplitude in dependency on the oscillation frequency was also in good agreement with experimental data for the CO_2 evolution. The theoretical analysis showed that two different types of synchronous oscillations exist under excess of oxygen and oxygen limitation. The latter one, with oscillation periods greater than the doubling time, could not be explained by the model.

While the oscillation period of a certain mode is exactly known from the above theory, the direct determination of the actual mode of oscillation in the experiments still remains difficult since exact age distribution data is mostly not available. One method is the careful analysis and simulation of the time course of the oscillation on average population parameters, such as the total number of cells in a certain phase of the cell cycle. In Fig. 9.15 a sinusoidal shape of the age distribution

$$f(t,\tau) = 1 + \sin\left(\frac{2\pi}{T_{II}}(t-\tau)\right) \tag{9.43}$$

was used to compare the model to experimental data of Strässle [35]. The culture is growing at $T=4.4$ h under the natural mode 12. The budding phase was estimated at $T_B=1.6$ h and the length of the S-phase for the DNA-synthesis to $T_S=0.5$ h. These estimates are in good agreement with the data of Münch [41] and Beuse et al. [42]. The number of cells in the P- and B-phases does not return to zero in this experiment, although the age distribution does. The simulated phase difference between the oscillating cell number in the S-phase and the other phases of the cell cycle is consistent with the assumption that the S-phase is located at the end of the single cell phase just before the bud appears.

The important property of the model is that the general characteristics of the oscillation are already fixed by the mode numbers and the doubling time, while cultivation conditions, growth kinetics, and cell metabolism mainly influence the shape of the age distribution. For all other parameters of the oscillating population variables, such as relative amplitude, average value, or phase shift, there is no remaining degree of freedom in the model; they are all uniquely determined by the mode numbers. By a different mode the model calculations as given in Fig. 9.15 would clearly disagree with the experimental data. The quite good agreement with the experimental data for a wide range of experimental conditions supports the simplifying assumption of the model. Nevertheless, synchronous oscillations may be more complicated in nature than can be reflected by the simple pure age distribution model. Beuse et al. [42] found additional modes of oscillations that can be explained by a further daughter cell class and extended the model accordingly. There were also hints that the length of the budding phase may be different for parent and daughter cells. Hjortso and Nielsen [70] extended a population model by a substrate balance and could show that in such a model multiple modes of oscillations could also exist. But for a full understanding of all the details there is still a long way to go.

9.5
Considerations for Process Optimization

In this section the dynamic process model is used to investigate optimum control strategies in the baker's yeast process and to determine optimal operating parameters.

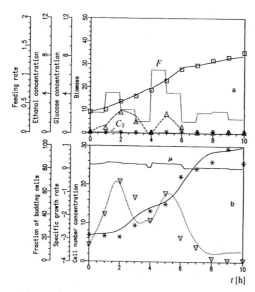

Fig. 9.16a,b. Comparison of simulations (*lines*) and experimental data (*symbols*) for a fed-batch optimum quality control strategy: **a** sugar feed rate F [$l h^{-1}$] and concentrations in kg m^{-3} of cell mass C_X (□), substrate C_S (✱), and ethanol C_E (Δ); **b** specific growth rate μ [h^{-1}], cell number concentration X_N [$10^{-5} l^{-1}$] (✱) and fraction of budding cells FBC [%] (∇). Reprinted from [49]

Table 9.8. Identified parameters in the model for *Saccharomyces cerevisiae* H1022 used in the simulation of Fig. 9.16

Parameter	Value	Units
r_{Smax}	0.018	mole $(gh)^{-1}$
K_{Ma1}	0.003	mole $(gh)^{-1}$
K_{Ma2}	0.01	mole $(gh)^{-1}$
K_{Ma3}	0.034	–
K_{EG}	0.05	–
K_{Ad}	0.024	–
P/O	1.8–2.1	–
Y_{ATP}	11	g mole^{-1}
m_{ATP}	0.001	mole $(gh)^{-1}$
r_{Acmax}	0.018	mole $(gh)^{-1}$
K_{D1}	1.23	–
K_{D2}	−2.04	h
K_{P1}	0.435	–
K_{P2}	−0.43	h
K_{B1}	0.023	–
K_{B2}	1.49	h

The baker's yeast production process is mostly optimized with respect to yield because the substrate costs account for the greatest cost factor of production. By fed-batch operation, the substrate concentration and uptake rate are kept relatively low at the critical level just before oxygen limitation or Crabtree-effect. This strategy avoids ethanol production with loss of yield, but also limits productivity. An optimization should also take into account, in addition to yield, productivity and economic profit, as well as the quality of the yeast product. The latter aspect is often neglected in model-based optimization because it is difficult to relate quality to the usual model variables. In practice the proper control of the relevant manipulating variables with the aim of getting the end-product with desired quality, i.e., good storage stability and leavening power for baking, is still more an art than science [72, 73].

9.5.1
Optimization of Product Quality

For baker's yeast, several aspects of product quality are connected to the cell cycling process and the fraction of budding cells, as was mentioned above. The FBC at harvest is the result of the entire dynamic development of the cell cycling process during the cultivation, which is difficult to optimize without a mathematical model. Therefore, a combined growth and age-distribution model should be applied for such an extended optimization. For simplicity, only the final production tank is considered here in detail, because it has a major influence on yield, productivity, and product quality, and it uses a great percentage of the resources. A complete model for the entire series of cultivations in the multi-stage production process could be built by multiplication of the elementary model. Quality parameters for yeast that can be influenced by cultivation conditions are the fermentative activity in terms of the ability to produce CO_2 for leavening the dough, and the storage stability. For active dry yeast the loss of activity during the dehydration-rehydration cycle is also of importance. The storage stability of compressed baker's yeast depends on several factors besides storage temperature [74, 75]: trehalose content, protein content, and maturity. It has been shown that storage carbohydrates such as trehalose and the average protein content [68] of yeast cells are strongly influenced by the fraction of budding cells and the age distribution of the population. Dairaku et al. [82] and Takamatsu et al. [71] pointed out that a minimized FBC at harvesting time of the cultivation corresponded to a high leavening activity. Some investigations revealed a dual effect of FBC on the storage properties [50]: if the yeast is used within a period of a few days, samples with lower FBC values have less leavening activity than those with higher FBC values. However, if the compressed yeast cells are stored for more than ten days before being used, the samples with higher FBC values lose their leavening activity very rapidly. Therefore, minimization of the FBC is favorable for enhancing the storage stability of compressed baker's yeast. Control of product quality may also be of interest for the production of active dry yeast (ADY). It was found that the leavening activity of ADY samples decreases, and the loss of leavening activity during the dehydration-rehydration cycle increases strongly with FBC. This suggests that minimization of the final FBC is also important during the ADY-production.

A symmetric division model for optimization of the μ-profile was applied in [71, 82] for fed-batch cultivation. But there the cycling pattern of the budding yeast will be far from symmetric division [4, 5]. The optimal control of the fraction of budding cells, or equivalently, the maturity of baker' yeast, by dynamic optimization of the substrate feeding rate for fed-batch cultivation was therefore studied by means of the asymmetric CCM [47], as described in Sect. 9.4.1. The growth model takes the real control variables, e.g., substrate feeding rate, as the input and delivers an accurate estimation of μ to the cell-cycling model, where it is translated into the proper lengths of the cycling phases in the asymmetric age distribution model. An optimal feeding strategy was designed to minimize the FBC at harvesting time, while keeping yield and productivity at a relatively high level [49, 50]. A stepwise flow rate, $F(t_i)$, as shown in Fig. 9.16, was determined by minimizing the following objective function:

$$J(T_F) = \left[C_S^2 + C_E^2 + FBC^2 + (45 - C_X)^2 \right]_{t=T_F} \xrightarrow[F_{min}<F(t_i)<F_{max}]{F(t_i),i=1,2,..9} \min \qquad (9.44)$$

where T_F is the cultivation period. The flow rate is restricted by the installed equipment in the range F_{min} to F_{max}. The sugar concentration in the feed was $250\,\mathrm{g\,l^{-1}}$. The model parameters used for the optimization were obtained from the experiment given in Fig. 9.12 including the ethanol growth phase. The optimal control policy was applied in an open loop without feedback from process variables. Figure 9.16 shows the simulation results of the experiment in comparison with measured data. The model parameters were slightly readjusted after the experiment was carried out to eliminate the influence of ethanol-growth. Table 9.8 gives the results of the parameter identification.

The principal strategy for quality optimization by minimization of FBC as quality index is as follows. At constant conditions, low FBC could be achieved only with long doubling times and, therefore, low productivity. But for unsteady operation and step changes in the growth rate the cell cycling process of the yeast population tends to synchronization, and oscillations in FBC can be observed. This property can be used for maximization of fermentative activity of the final product under high productivity by suitable dynamic control of the specific growth rate. The cells are then har-

Table 9.9. Consideration of the upstream and downstream processing steps in the model

Type of cost	Model	Remark
Fixed	$P_F=4.72T_T$	operation, capital of investment and staff
Cultivation medium	$P_S=P_{Me}+P_{Sa}+P_{Sp}$	Total
	$P_{Me}=(V_L^0C_S^0+K_{S1}(V_L^F-V_L^0)C_S^R)p_{Me}$	Substrate
	$P_{Sa}=(1-K_{S1})(V_L^F-V_L^0)p_{Sa}$	Salts and other media components
	$P_{Sp}=K_{S2}V_L^F$	Preparation and storage, including waste water treatment
Aeration	$P_{Ae}=F_GT_Fp_{Ae}$	Aeration rate is assumed constant
Downstream processing	$P_D=P_{DC}+P_{DD}$	Total
	$P_{DC}=K_{D1}V_L^F$	Centrifugation
	$P_{DD}=V_L^FC_X^Fp_{DD}$	Drying
Proceeds for the sold product minus pitching yeast	$P_E=V_L^F(C_X^F-C_X^0)(1-K_QFBC^F)p_E$	Considering the quality of the yeast product by the fraction of budding cells with an empirical correlation

vested in the dynamic minimum of FBC at the end of the fermentation, being much lower than the stationary value for the averaged specific growth rate. In this way the productivity and quality can be decoupled to some extent to optimize both of them. The minimum FBC value in the experiment is as low as 1.7%, which is very close to complete synchronization. This is normally unattainable with the conventional feeding strategies. Compared with conventional incremental feeding, the flow profile is characterized by pulsing. During the periods of high flow rate, ethanol is formed due to the Crabtree-effect at high growth rates. However, since ethanol is reconsumed completely afterwards – but in parallel to sugar – the yield is not lowered significantly in comparison to conventional control strategies, where usually also some ethanol is accumulated. In the period of ethanol uptake the growth rate is reduced. By this alternation in the specific growth rate, the pulsed feeding course enforces to some extent synchronization of the population growth to prepare a trough of the FBC at the desired harvesting point, and final FBC is minimized. The steeply decreased feeding during the last portion of the cultivation (about 3 h) supports the down-swing of the FBC further and allows the uptake of ethanol. In contrast to some industrial practice, the feeding is not stopped completely before harvest. This is beneficial to the cells for maintaining a higher trehalose content and, hence, a better storage stability. The studies in a number of experiments demonstrated that the quality index of the storage stability of compressed baker's yeast can be controlled and, indeed, be improved by the model-based open-loop feeding strategy in fedbatch cultivations. The low feeding at the end does not harm the productivity of the entire process, because there is an unusual high peak flow during the earlier parts.

9.5.2
Economic Optimization

For more accurate optimization, the performance index should be uniquely determined by an economic input-output balance of the process. All economic indices for a process such as quality, productivity, and efficiency are interconnected. An optimization of one of them also influences the others in a complicated manner. It is therefore preferable to take the total economic profit in dependence on the other indices as criterion for multi-objective optimization. Such a procedure was applied in [76] to an industrial scale airlift reactor with a volume of $V_R = 60\,\text{m}^3$. The total economic profit per unit of time, J, is calculated by

$$J = \frac{P_E - P_S - P_{Ae} - P_D - P_F}{T_T} \tag{9.45}$$

where

$$T_T = T_F + T_P \tag{9.46}$$

is the total duration for the completion of the process cycle, T_F the fermentation period, and T_P the time for preparation and downstream processing. The specific costs for an industrial process were about 2% for inoculum, 80% for raw materials, 9% for energy, and 9% for fixed costs (capital, staff, and so on). For the economic optimization, additional model equations are introduced to consider in a simplified form the costs and time requirements of other processing steps besides cultivation as included in Eq. (9.45): medium preparation, cleaning, inoculum propagation, aeration, operation, and downstream processing. The additional model equations are summarized in Table 9.9, where p_{Me} and p_{Sa} are the specific prizes for 1 kg of pure molasses and media components, p_{Ae} the specific cost for $1\,\text{m}^3$ of compressed air, and p_{DD} the specific costs for drying of 1 kg of yeast product. For consideration of the product quality, its exact influence on the proceeds should be known; here, an empirical factor was introduced, where K_Q is a quality index of the selling prize, and p_E the specific maximum selling prize for 1 kg of best quality dry yeast.

Based on the extended process model the total economic profit of the baker's yeast production was optimized,

$$J \xrightarrow{\;F(t),T_F,C_X^0,C_S^0,C_S^R\;} \text{max} \tag{9.47}$$

Subject to the optimization were, as constant operating parameters, the substrate concentration in the feed and inoculum, the aeration rate, the mass of pitching yeast, and the fermentation period, T_F. For the molasses feed $F(t)$ an optimum control profile is determined with an hourly change of flow rate. For technical reasons, several constraints for the optimization have to be taken into account: the maximum working volume, the maximum and minimum aeration rate, the maximum molasses flow rate, and the maximum substrate concentration in the feed that is given either by the sugar content of the molasses or the maximum viscosity that can be handled. For the given economic indices the point of optimal operation of a 60-m³ airlift reactor was obtained for a cultivation period $T_F = 10\,\text{h}$, aeration rate $F_G = 90\,\text{m}^3$ min^{-1}, and a substrate concentration of $C_S^R = 114\,\text{kg m}^{-3}$.

The time course of flow rate and fraction of budding cells resembles in general the pulsed feeding pattern of the pure quality optimization, as already shown in

Fig. 9.16. But the optimization result is strongly influenced by the relative weights of the economic indices in Eq. (9.44). For example, if the energy costs for aeration increase, lower aeration rates would be preferred, which on the other hand reduces oxygen supply to the reactor. This can be compensated for by slower feeding and an extended cultivation period. Generally, if the yield is the most important aspect of yeast production, the exponential-linear fed-profile is optimal because it avoids oxygen limitation and any ethanol production. The exponential increase at the beginning becomes more distinct with slightly higher weight on productivity. Then there is moderate overfeed of sugar, accompanied by some ethanol accumulation. In the final cultivation period, the flow rate is decreased to convert the remaining ethanol. By a proper balance of sugar feeding and ethanol uptake, the decrease in yield is very limited, because ethanol is only used for oxidative energy metabolism via TCC. If much emphasis is put on high productivity the optimization results in a different two-phase strategy: in the first phase there is a strong over-feeding with ethanol production and sugar accumulation, the growth conditions corresponding to batch operation. In the second phase the feed rate is lowered to small values so that the ethanol can be used up completely. The loss of yield can be limited to about 16%. This strategy is very similar to the combined batch/fed-batch-scheme of De-Loffre [76].

Where the product quality is considered, the optimization tends towards higher productivity with slight loss of yield since the substrate pulses for growth synchronization lead to over-feeding with subsequent ethanol uptake. But since this effect is moderate, it was concluded that the quality can be improved with low additional effort and small cut-off on the other economic indices.

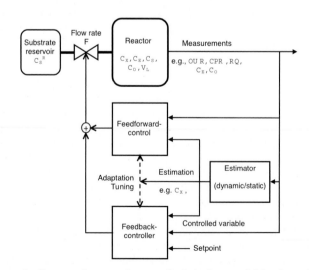

Fig. 9.17. Schematic diagram of a control system for baker's yeast fed-batch production

9.6
Automatic Control of Fed-Batch Processes

9.6.1
General Remarks

In the production of baker's yeast, the costs for substrates contribute the biggest part of the total production cost. Therefore, the main goal for process control is to obtain the maximum yield at the highest possible productivity during the cultivation. This means the growth rate has to be kept at its highest possible value that just avoids fermentative growth by either Crabtree-effect or oxygen limitation in the reactor. The simplest method from a control-point of view would be continuous cultivation where the dilution rate can be fixed at the optimum point. But by the well-known problems of operational procedures and of strain stability in continuous cultivations, fed-batch operation is preferred for industrial production, although its control is much more difficult.

In an ideal fed-batch process the specific growth rate would be fixed at the optimum value resulting in exponential growth. For a number of reasons such a simple strategy cannot be realized in practice. Usually the oxygen transfer capacity of the reactor is limited so that the growth rate has to be reduced at high cell mass concentrations when oxygen becomes limiting, by keeping the substrate flow rate constant. This results in the so-called exponential-linear feeding pattern. As discussed in the previous section this basic strategy is modified to some extent when other criteria than yield, e.g., productivity, product quality, or economic profit, are used for determination of the optimum control strategy. Further modifications have to consider the true growth kinetics of the yeast including adaptation phenomena. Therefore, the pattern of the specific growth rate is more complicated, but can in principle be translated into a suitable dosage scheme for the substrate feed rate when all operating parameters are known. For many years, fed-batch processes were carried out using such predetermined fixed feeding profiles that were at most manually adjusted during the running process. As an advantage this simple open-loop control does not require measurement equipment, but it is also difficult to guarantee reproducible cultivations.

The open-loop strategy cannot ensure that the cultivation runs near the desired optimum trajectory since it does not consider disturbances during the process, or variations in operational parameters: e.g., power supply, temperature of the cooling medium, composition of the substrate, initial biomass density, operational problems with equipment. In addition, the growth kinetics of the yeast may vary in response to such disturbances or variations in the precultures. When measuring information from the running process is available, the actual flow rate can be corrected for these deviations in the cultivation conditions by automatic feedback control, to keep the process near to the desired optimum. A number of different concepts for automatic control have been established by control engineers (see, e.g., [77, 78] for an overview), which all have their specific advantages and draw-backs when applied to biotechnical processes, and baker's yeast production in particular. Several of these concepts have also been adopted and experimentally tested for control of baker's yeast fed-batch production.

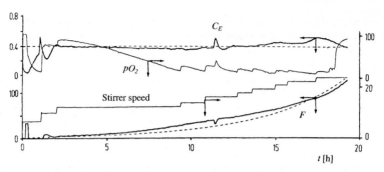

Fig. 9.18. Experimental results for a fed-batch process controlled by the ethanol concentration in the medium; ethanol concentration C_E [kg m^{-3}], percentage of dissolved oxygen pO_2 [%], stirrer speed [s^{-1}], and flow rate F [10^{-9} m^{-3} s^{-1}]. Reprinted from [87]

Another general problem for control of fed-batch processes is the increase of both volume and cell mass. Under the desired constant growth conditions the latter results in an exponentially increasing disturbance. Since all reaction rates are proportional to the cell mass due to the autocatalytic nature of growth, the process has – from the control engineering point of view – variable gain and time constants. Conventional controllers are designed and then tuned to have best performance at a certain operating point. But this is steadily changing in the fed-batch process. Therefore, the controller response may become slow or oscillatory in phases of the cultivation where the controller was not tuned for. In the worst case the control can become unstable and then drives the process far away from the desired state. A stabilization can be made only of the cost of controller performance, which then may become unsatisfactory for the entire process. The methodology of adaptive control targets this problem by changing the controller parameters according to the actual state and parameters of the process, which then, in addition to the controlled variable, must be measured, calculated from other variables, or estimated by means of a mathematical model (self-tuning control). Summarizing the above remarks, a general control-scheme can be drawn as shown in Fig. 9.17, although in particular not all blocks may be present and not all variables used. The control itself consists of the feedforward part for reducing known or measurable disturbances, e.g., by the increasing cell mass, and of the feedback controller (regulator) for fine regulation around the rough value of the flow rate determined by the feedforward block. Usually the two blocks will not use all the same measured variables. The feedback controller corrects the flow rate to make the controlled variable as close as possible to the setpoint. This may be constant or varied according to a given profile that meets a specific, possibly optimum strategy for the cultivation. There is a wide choice of controller types, ranging from simple switching control or PID control to advanced model based approaches. The third block for estimation of cell mass or other variables can be added to improve the disturbance compensation, adapt the employed control law, or control non-measured variables. The estimation method can range from simple stoichiometric calculations up to model-based dynamic observers for the state variables of the process.

In a conventional control system, the controlled variable is measured, compared to a given set point value, and the difference is looped back to the process via controller and controlling device. In baker's yeast production there are several possibilities for measuring a characteristic variable that is suitable for control. But to begin with negative examples, control of substrate concentration in the medium is not practical, not only due to lack of reliable and accurate sensors, but also because there is no unique relation of substrate concentration or substrate flux – as input of the metabolic system – to the critical point for onset of fermentative growth. As already mentioned, the latter also depends on the intracellular state of the respiratory system and the oxygen supply. The same argument can be used for control schemes that track a predefined profile of cell mass or specific growth rate, as investigated, e.g., by [81–83], or only one of the rates of gaseous metabolism, CPR and OUR. These quantities per se give no information on the metabolic state of the cells, fermentative or oxidative, and therefore cannot be used as single control variable if the yield or ethanol concentration has to be kept in a narrow range in the presence of disturbances. Nevertheless, profile control schemes can greatly reduce the variability among the production runs. This can also be an important control objective, because it promotes a constant product quality and facilitates scheduling of the entire plant in fixed time intervals.

9.6.2
Examples for Applied Control Systems

Direct information on the metabolic state of growth can be obtained from ethanol concentration, respiratory quotient, or equivalently, the difference between CPR and OUR. In 1961 the principle of automatic control of the molasses feed rate by the ethanol concentration in the gas phase was proposed by Rungeldier and Braun [79]. The ethanol content in the medium can be calculated from mass transfer equilibrium. This control was realized only years later because of the lack of cheap and reliable sensors. Nowadays, cheap in-situ membrane probes based on semiconductors are available [80]. The ethanol concentration is usually kept constant below 0.1 vol.%. This means that during fed-batch with increasing volume there is always moderate ethanol production, but strong over- or underfeeding can be avoided. Actually, some limited ethanol production during baker's yeast production is even desirable to ensure high growth activity of the cells and reduce bacterial contamination. Furthermore, in reactors with non-ideal mixing, ethanol production could only be avoided with great loss in productivity, which is not acceptable. But the loss of yield can be limited by reducing the feed rate shortly before the end of the cultivation to allow for reuse of ethanol. When applied in the liquid phase, the response time of the ethanol sensor is very short and control becomes much easier compared to gas-phase measurements. The feasibility of the control concept was tested in several papers with simple PI control up to non-linear adaptive control [82, 92, 97, 102]. Axelson et al. [87] showed that with this type of sensor during the entire cultivation a stable control can be obtained by a simple PID rule. An example is given in Fig. 9.18. The stability problems reported by other authors [84, 92], mainly during oxygen limited growth at high cell concentration, could be avoided. This relatively robust and reliable type of control is also used in industry. Nevertheless, an adaptation scheme covering the increasing cell mass can improve the controller performance and disturbance rejection. Such a system, including a Luenberger observer for the exponential disturbance by biomass growth, was presented in [103].

An observer-based approach that allowed a very accurate estimation of cell mass and specific growth rate was published by Pomerleau and Perrier [104, 105]. The algorithm included as state variables the concentration of glucose, ethanol, cell mass, and dissolved O_2 and CO_2. Glucose feed, OUR, and CPR were used as measured variables. Another means for model-based state and parameter estimation is the Kalman-filter. Although its theoretical concept – founded on a stochastic approach – is quite smooth, it can be difficult in practice to ensure stability of the estimation. This technique was also applied for baker's yeast adaptive state and parameter estimation, including cell mass and growth rate [85, 86], but seemingly there is no on-line application within a closed loop control system.

The ethanol control concept by set-point control was slightly extended by Schubert et al. [107]. A predetermined optimum profile for the ethanol concentration was followed during the fed-batch. An internal-model-control including an artificial neural network was developed to improve the performance compared to PID-control. There were also reports in the literature about fuzzy state recognition [108] and fuzzy control of baker's yeast production [110, 111]. Although this rule-based approach offers a new view on the problem of control and seems to work reasonable well on the laboratory scale, it is objected to by many control engineers because the performance and stability is theoretically difficult to evaluate. It can be seen as an advantage that backtracking of the inference rules for determination of the controlling variable's values from the measurements gives some explanation of the control action.

Another common method for automatic control of the molasses feed rate uses the respiratory quotient, RQ. Compared to control by ethanol concentration this method not only requires higher effort for measuring devices, i.e., O_2- and CO_2-analyzers, but also has inherent stability problems. As explained earlier, RQ equals one for oxidative growth on sugars with a sugar uptake rate below the critical threshold for Crabtree-effect. At fermentative growth above the critical value RQ is greater than one, and during underfeeding of sugar with ethanol uptake, RQ is less than one. When no ethanol is present in the medium, the respiratory quotient gives no information about the degree of underfeeding to allow for a proper adjustment of the flow rate by the controller. The usual way to reduce this problem is to choose a set point in the range $RQ=1.1-1.2$, which also means moderate ethanol production. A very accurate gas phase balancing is then necessary to prevent stability problems by a bias in the RQ-calculation. An over-estimation of RQ can lead to complete ethanol uptake and drive the process to relatively low growth rates in the purely oxidative region, while under-estimation gives rise to pronounced ethanol accumulation and loss of yield. Further problems in this control are the additional delay by the gas analyzers and a strong cross-coupling of the RQ-calculation to pH in the cultivation medium by dissolved carbon dioxide. An early investigation on the RQ-control was by Aiba et al. [88]. A switching control was used that increased stepwise the substrate flow rate for $RQ<1$, and decreased it for $RQ>1$. The control action occurred in intervals of about 0.5 h. Based on a mathematical model the authors discussed possible improvements of the control. For presetting the flow rate, two formulas were derived, which are still used for feed-forward compensation of the growing cell mass. The first uses the estimated cell mass,

$$F = k \frac{C_X V_L}{Y_{XS} C_S^R} \tag{9.48}$$

and the second the measured oxygen uptake rate,

$$F = k\frac{OUR \cdot V_L}{C_S^R} \tag{9.49}$$

In the first equation, k should equal a desired specific growth rate, μ_{set}, and in the second the ratio of yield coefficients for oxygen and substrate. The parameter k is in general not constant during the entire fed-batch, but can in principle be estimated by an observer as mentioned above.

Control by RQ and other variables obtained from exhaust gas analysis was studied further in a number of papers with different types of controllers including adaptive ones [89–97]. The same basic idea was worked out further by O'Connor et al. [101]. They investigated the performance and failure conditions of different control schemes for feed-forward control of the flow rate, including the calculated cell mass and oxygen uptake rate. A parallel operating inferential feedback controller, as shown in principle in Fig. 9.17, corrects the actual flow rate for disturbance and errors in the cell mass. Controlled variable is the calculated ethanol reaction rate. A satisfactory control of a lab-scale reactor was obtained by the proportional-integral control law

$$F = \overbrace{\frac{\mu_{set}C_X V_L}{Y_{XS}C_S^R}}^{Feed-forward} - \overbrace{K_P V_L R_E - K_I C_X \sum \frac{V_L R_E}{C_X}}^{PI-control} \tag{9.50}$$

The cell mass was calculated from either oxygen uptake rate, ammonium uptake rate, or by balancing that also considered the total oxygen uptake and carbon dioxide production. The ethanol reaction rate is inferred from the following linear relation (OUR, CPR in molar quantities):

$$\begin{aligned}R_E &= 1.04\,(CPR - OUR) \\ &= 1.04\,OUR\,(RQ - 1)\end{aligned} \tag{9.51}$$

According to the last line, the above control law may also be considered as RQ-control.

A quite different approach is the L/A-control proposed by Lakrori and Cheruy [98]. It can also be formulated as adaptive control. The simple structured controller uses the ratio of the controlled variable to the set-point as basic non-linear element, and its power is used as a tuning or adaptation parameter. Such a control system for baker's yeast fed-batch production was developed by Dantigny and Lakrori [99] in simulations, then modified and experimentally tested [100]. The control law used was

$$F = \mu V_L \left(\frac{C_X}{Y_{XS}C_S^R}\right)^k \tag{9.52}$$

where C_X and μ were estimated from ammonia uptake. The adaptation parameter of the control law was determined by measurements of the respiratory quotient and ethanol concentration as

$$k = \frac{1}{5}\left(\frac{C_E}{gL^{-1}} + RQ + 3.5\right) \tag{9.53}$$

In practical applications by this control, ethanol could be kept below $2\,\mathrm{g\,l^{-1}}$ and RQ close to one. The control law, Eq. (9.52), can also be interpreted as gain adjustment of some kind of feed-forward compensation, similar to Eq. (9.48), by a feedback action due to Eq. (9.53).

A control by directly using Eq. (9.48) with fine-tuning of the gain k was presented in [109]. Based on process information obtained from state observation by a mathematical model, k was stepwise up/down-adjusted to obtain high growth rate without by-product formation in baker's yeast fed-batch cultivation.

References

1. Rose AH, Harrison JS (eds) (1987) The yeasts. Academic Press, London
2. Slater ML (1981) In: Arnold WN (ed) Yeast cell envelops: biochemistry, biophysics, and ultrastructure, vol 2. CRC Press:65
3. Hartwell LH, Unger MW (1977) J Cell Biol 75:422
4. Lord PG, Wheals AE (1980) J Bacteriol 142:808
5. Thompson PW, Wheals AE (1980) J Gen Microbiol 1221:401
6. Wheals AE (1987) In: Rose AH, Harrison JS (eds) The yeasts, vol 1. Academic Press, London, p 283
7. Duboc P, Marison I, von Stockar U (1996) J Biotechnol 51:57
8. Nielsen J, Villadsen J (1992) Chem Eng Sci 47:4225
9. Steinmeyer DE, Shuler ML (1989) Chem Eng Sci 44:2017
10. Coppella SJ, Dhurjati P (1990) Biotechnol Bioeng 35:356
11. Cortassa S, Aon JC, Aon MA (1995) Biotechnol Bioeng 47:193
12. Jin S, Ye K, Shimizu K (1997) J Biotechnol 54:161
13. van Gulik WM, Heijnen JJ (1995) Biotechnol Bioeng 48:681
14. Vanrolleghem PA, de Jong-Gubbels P, van Gulik WM, Pronk JT, van Dijken JP, Heijnen S (1996) Biotechnol Prog 12:434
15. Rizzi M, Theobald U, Querfurth E, Rohrhirsch T, Baltes M, Reuss M (1996) Biotechnol Bioeng 49:316
16. Theobald U, Mailinger W, Baltes M, Rizzi M, Reuss M (1997) Biotechnol Bioeng 55:305
17. Rizzi M, Baltes M, Theobald U, Reuss M (1997) Biotechnol Bioeng 55:592
18. Cortassa S, Aon MA (1994) Enzyme Microb Technol 16:761
19. Auling G, Bellgardt KH, Diekmann G, Thoma M (1984) Appl Microbiol Biotechnol 19:3533
20. Sonnleitner B, Rothen SA, Kuriyama H (1997) Biotechnol Prog 13:8
21. Barford JP, Phillips PJ, Orlowski JH (1992) Bioproc Eng 7:297
22. Barford JP, Phillips PJ, Orlowski JH (1992) Bioproc Eng 7:303
23. Lee YS, Lee WG, Chang YK, Chang HN (1995) Biotechnol Letters 17:791
24. Gadgil CJ, Bhat PJ, Venkatesh KV (1996) Biotechnol Prog 12:744
25. Wasungu KM, Simard RE (1982) Biotechnol Bioeng 24:1125
26. Grosz R, Stephanopoulos G (1990) Biotechnol Bioeng 36:1030
27. van Urk H, Mak PR, Scheffers WA, van Dijken JP (1988) Yeast 4:283
28. Sonnleitner B, Käppeli O (1986) Biotechnol Bioeng 28:927
29. Enfors SO, Hedenberg J, Olsson K (1990) Bioproc Eng 5:191
30. Sonnleitner B, Hahnemann U (1994) J Biotechnol 38:63
31. Barford JP (1990) Biotechnol Bioeng 35:921
32. Bellgardt KH (1994) J Biotechnol 35:19
33. Bellgardt KH (1994) J Biotechnol 35:35
34. Beuse M, Kopmann A, Diekmann H, Thoma M (1999) Biotechnol Bioeng 63:410
35. Strässle C (1988) PhD thesis No 8598, ETH Zürich
36. Strässle C, Sonnleitner B, Fiechter A (1988) J Biotechnol 7:299
37. Strässle C, Sonnleitner B, Fiechter A (1989) J Biotechnol 9:191
38. Cazzador L, Mariani L, Martegani E, Alberghina L (1990) Bioproc Eng 5:175

39. Münch T, Sonleitner B, Fiechter A (1992) J Biotechnol 24:299
40. von Meyenburg K (1969) PhD thesis, ETH Zürich
41. Münch T (1992) PhD thesis No 9772, ETH Zürich
42. Beuse M, Bartling R, Kopmann A, Diekmann H, Thoma M (1998) J Biotechnol 61:15
43. Heinzle E, Furukawa K, Tanner RD, Dunn IJ (1983) In: Halme A (ed) Modelling and control of biotechnical processes. Pergamon Press, New York, p 57
44. Parulekar SJ, Semones GB, Rolf MJ, Lievense JC, Lim HC (1986) Biotechnol Bioeng 28:700
45. Bellgardt KH (1991) In: Rehm HJ, Reed G (eds) Biotechnology, vol 4, p 267
46. Dantigny P (1995) J Biotechnol 43:213
47. Deckwer WD, Yuan JQ, Bellgardt KH, Jiang WS (1991) Bioproc Eng 6:265
48. Rieger M, Käppeli O, Fiechter A (1983) J Gen Microbiol 129:653
49. Yuan JQ, Bellgardt KH, Deckwer WD, Jiang WS (1993) Bioproc Eng 9:173
50. Yuan JQ, Bellgardt KH (1994) J Biotechnol 32:261
51. George S, Larsson G, Olsson K, Enfors SO (1998) Bioproc Eng 18:316
52. Strauß G (1990) PhD thesis, Universität Hannover
53. Abel C, Hübner U, Schügerl K (1994) J Biotechnol 32:45
54. Lübbert A, Korte T, Larson B (1987) App Biochem Biotech 14:207
55. Lübbert A, Larson B (1987) Chem Eng Tech 10:27
56. Pollard DJ, Ayazi Shamlo P, Lilly MD, Ison AP (1996) Bioproc Eng 15:279
57. Deckwer WD (1985) Reaktionstechnik in Blasensäulen, Verlag Salle/Sauerländer
58. Bello RA, Robinson CW, Moo-Young M (1985) Biotechnol Bioeng 27:369
59. Hills JH (1976) Chem Eng J 12:89
60. Weiland P (1978) PhD thesis, Universität Dortmund
61. Adams J, Rothmann ED, Beran K (1981) Math Biosci 53:249
62. Hjortso M (1995) J Biotechnol 42:271
63. Hjortso M (1987) Biotechnol Bioeng 30:825
64. Hjortso MA, Nielsen J (1994) Chem Eng Sci 49:1083
65. Hjortso, MA, Bailey JE (1983) Math Biosci 63:121
66. Hjortso, MA, Bailey JE (1984) Biotechnol Bioeng 26:528
67. Porro D, Martegani E, Ranzi B, Alberghina L (1988) Biotechnol Bioeng 32:411
68. Alberghina L, Mariani L, Martegani E, Vanoni M (1983) Biotechnol Bioeng 25:1295
69. Mariani L, Martegani E, Alberghina L (1986) IEE Proceedings 133D:210
70. Hjortso MA, Nielsen J (1995) J Biotechnol 42:255
71. Takamatsu T, Shioya S, Chikatani H, Dairaku K (1985) Chem Eng Sci 40:499
72. Becker MJ, Rapoport AI (1987) Adv Biochem Eng/Biotechnol 35:127
73. Suomalainen H (1975) Eur J Appl Microbiol 1:1
74. Ginterova A, Mitterhauszerova L, Janotkova O (1966) Branntweinwirtschaft 106:57
75. Hautera P, Lovgren T (1975) Baker's Digest 49:36
76. Bellgardt KH, Yuan JQ (1991) In: Rehm HJ, Reed G (eds) Biotechnology, vol 4, p 383
77. Fisher DG (1991) Canad J Chem Eng 69:5–26
78. Johnson A (1987) Automatica 23:691
79. Rungeldier K, Braun E (1961) US Pat 3,002,894
80. Schügerl K (1998) Bioreaction engineering III. Wiley, Chichester, Amsterdam
81. Wu WT, Chen KC, Chiou HW (1985) Biotechnol Bioeng 27:756
82. Dairaku K, Izumoto E, Morikawa H, Shioya S, Takamatzu T (1982) Biotechnol Bioeng 24:2661
83. Takamatsu T, Shioya S, Okada Y, Kanda M (1985) Biotechnol Bioeng 27:1675
84. Dairaku K, Izumoto E, Morikawa H, Shioya S, Takamatzu T (1983) J Ferment Technol 61:189
85. Bellgardt KH, Kuhlmann W, Meyer HD, Schügerl K, Thoma M (1986) IEE Proceedings 133D:226
86. Shi Y, Yuan WK (1988) Chem Eng Sci 43:1915
87. Axelson JP, Mandenius CF, Holst O, Hagander P, Mattiasson B (1988) Bioproc Eng 3:1
88. Aiba S, Nagai S, Nishizawa Y (1976) Biotechnol Bioeng 18:1001
89. Cooney CL, Wang HY, Wang DIC (1977) Biotechnol Bioeng 19:55

90. Wang HY, Cooney CL, Wang DIC (1977) Biotechnol Bioeng 19:69
91. Peringer P, Blachere HT (1979) Biotechnol Bioeng Symp 9:205
92. Woehrer W, Hampel W, Roehr M (1981) Proc Ferm Symp 6:419
93. Verbruggen HB, Eelderinck G, Van den Broeck PM (1985) In: Johnson A (ed) Modelling and control of biotechnical processes. Pergamon, Oxford, p 91
94. Verbruggen HB, Eelderinck G, Van den Broeck PM (1987) BTF 4:185
95. Deckers R, Voetter M (1985) In: Johnson A (ed) Modelling and control of biotechnical processes. Pergamon, Oxford, p 73
96. Yamane T, Matsuda M, Sada E (1981) Biotechnol Bioeng 23:2509
97. Williams D, Yousefpour P, Wellington EMH (1986) Biotechnol Bioeng 28:631
98. Lakrori M, Cheruy A (1987) Bio-Sci 6:154
99. Dantigny P, Lakrori M (1992) Biotechnol Bioeng 39:246
100. Dantigny P, Lakrori M (1991) Appl Microbiol Biotechnol 36:352
101. O'Connor GM, Sanchez-Riera F, Cooney CL (1992) Biotechnol Bioeng 39:293
102. Chen L (1992) PhD thesis, Universite Catholique de Louvan
103. Axelson JP (1989) PhD thesis, Lund Institute of Technology
104. Pomerleau Y, Perrier M (1990) AIChE Journal 36:207
105. Pomerleau Y, Perrier M (1992) AIChE Journal 38:1751
106. Tyson CB, Lord PG, Wheals AE (1979) J Bacteriol 138:92
107. Schubert J, Simutis R, Dors M, Havlik I, Lübbert A (1994) J Biotechnol 35:51
108. Shimizu H, Miura K, Shioya S, Suga KI (1995) Biotechnol Bioeng 47:165
109. Ringbom K, Rothberg A, Saxén B (1996) J Biotechnol 51:73
110. Siimes T, Linko P, von Numers C, Nakajima M, Endo I (1995) Biotechnol Bioeng 45:135
111. Zhang XC, Visala A, Halme A, Linko P (1994) J Biotechnol 37:1

10 Modeling of the Beer Fermentation Process

Andreas Lübbert

10.1
Introduction

Over the last decades, huge, centrally located breweries with factory-style production and considerable brewing capacities were built. At the same time the beer consumption has been decreasing in central Europe. Thus, the breweries fight over market shares. Consequently, optimization of the benefit/cost ratio becomes decisive for breweries to keep themselves at the competitive edge. As the beer quality is not extremely different between breweries in a given area, the beer price and hence the production cost becomes an overwriting criterion. Cost must be cut wherever possible, but labor and energy costs are the primary targets of process optimization.

Central in the creation of beer is fermentation. Here we focus on this fermentation step, which can be regarded as the heart of the entire production, since the conditions of this fermentation process have a marked effect on the flavor and taste of the beer (e.g., [1, 2]). The beer fermentation process is a batch process where a *Saccharomyces* yeast (*Saccharomyces cerevisiae, Saccharomyces carlsbergensis, Saccharomyces uvarum*, etc.) is used to produce ethanol from barley malt. This anaerobic process typically takes 7 days. Here we discuss optimization procedures for the fermentation process with respect to the operational procedure and do not consider constructive aspects of the equipment. In other words, we are dealing with everyday work in breweries: optimizing a running fermentation system for the production of a particular brand of beer.

Beer fermentation is one of the large-volume industrial processes that are essentially controlled manually. There are several reasons that excluded automation from this process so far. One is that the essential process state variables are difficult to measure on-line. Another is that the fermentation process appears to be a complex system, which had not been modeled precisely enough so far. On the other hand, however, the brewmasters make excellent beers with their current technology and any automatic process supervision and control system must compete with these highly skilled human controllers. Automatic systems would only be of interest to breweries if they would help the brewmasters to produce the same quality of beer in a much less expensive way. There are some obvious entry points for further automation in beer fermentation. In beer breweries, many tasks presently being performed by men, such as manual cleaning, filling, pitching, measuring, temperature control, yeast harvesting, pumping into lager vessels, etc., could be automated [3].

A main issue in automating beer fermentation is sensing the actual state of the process and making the necessary decisions concerning eventually necessary corrections of the process based on this information. The key variable by which the brewers characterize the state of their fermentation is the extract (substrate) concentration. It is measured off-line, and thus the measurement values become available with a considerable time delay. The decisions the brewmasters make upon deviations from the data they expect are essentially based on their experience with the time profiles of the extract concentration. At least in Germany (Reinheitsgebot), they can influence a running fermentation by changing the process temperature only. Adding yeast extract or water for control purposes is seldom performed in real practice.

Automatic process supervision and control would require one to formulate the relevant knowledge about the fermentation process in such a way that it can be exploited numerically with computers. The software, representing the process in such a control system, is the way by which we implement the process model. Obviously it does not suffice to know only the profiles of a perfectly running fermentation process. Control actions are required whenever the process significantly deviates from those paths, and hence the dynamics of the process around the typical trajectories must be known as well. Thus, dynamic process models are required.

Here we discuss the problem of how to formulate and process the knowledge required to perform process optimization, supervision, and control tasks. The accompanying examples used throughout are taken from industrial-scale beer fermenters (Fig. 10.1) operated in one of the breweries of the Gilde Brauerei AG, Hannover, Germany. The brand brewed in the fermenters is based on a wort with 110 g/l extract

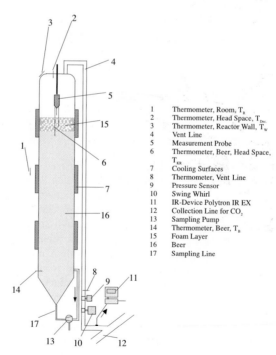

1	Thermometer, Room, T_R
2	Thermometer, Head Space, T_{Dec}
3	Thermometer, Reactor Wall, T_W
4	Vent Line
5	Measurement Probe
6	Thermometer, Beer, Head Space, T_{KR}
7	Cooling Surfaces
8	Thermometer, Vent Line
9	Pressure Sensor
10	Swing Whirl
11	IR-Device Polytron IR EX
12	Collection Line for CO_2
13	Sampling Pump
14	Thermometer, Beer, T_B
15	Foam Layer
16	Beer
17	Sampling Line

Fig. 10.1. Production-scale beer fermenter operated in one of the breweries of Gilde Brauerei AG, Hannover

(barley malt extract) concentration. The initial yeast concentration is of the order of 1 g(dry weight)/L. The fermenters are modern cylindro-conical fermenters with a total volume of $300\,m^3$ and a diameter of 4.2 m. They are cooled by means of three cooling elements each, welded as jackets onto the reactor walls around their cylindrical parts. The cooling is performed by evaporation of ammonia within the closed cooling system. All data were taken during their usual operation [4, 5a, 6, 7].

10.2
Process Optimization

10.2.1
Different Knowledge Representation Techniques

In order to optimize a process, one first of all needs sufficient knowledge about its dynamics. Beer fermentation is a classical biochemical production process, which has been performed by humans for many centuries. Hence, there is a lot of heuristic knowledge available. From the process engineering point of view, however, systematic optimization requires quantitative models that can be exploited numerically.

Since the fermentation process is basically a biochemical process, namely the conversion of the raw material barley malt (sugar) into ethanol and carbon dioxide, the overall biochemical reaction equation is the first approach to a process model. In the beer community this is referred to as the Balling formula. It is usually formulated on a mass base (component changes measured in kg) in the following form:

$$1\,kg\ Sugar \rightarrow 0.484\,kg\ EtOH + 0.463\,kg\ CO_2 + 0.053\,kg\ Biomass \qquad (10.1)$$

As the process is exothermic, the mass balance can be complemented by

$$\Delta H = 568\,kJ/kg \qquad (10.2)$$

leading to the heat generated per kg sugar consumed.

This basic balance does not take into account that a fraction of the fermentable carbohydrates, here summarized as sugar, is utilized for the production of further metabolites such as glycerol, organic acids, acetaldehyde, higher alcohols, and esters. The reason is that, although these other fermentation products are crucial from the point of view of the beer flavor, they do account for only 0.27% of the sugar utilized (e.g., [8]).

Of practical interest is to keep the biomass production rate relatively low in order to get a high beer/substrate yield. In practice this is widely achieved, and thus the first rough models of the beer fermentation process neglect biomass. That means that over a wide part of the fermentation time the net conversion is essentially sugar to ethanol and carbon dioxide. Such a reaction equation is a static relationship between the amount of sugar consumed and the amounts of products built. It tells what the decisive quantities are and how they are interrelated. However, it does not tell us about the process dynamics, i.e., the time course of the individual concentrations in the fermenter.

10.2.1.1
Classical Approach

The classical approach to dynamic process models is to start with mass balance equations for those components, which are consumed or being built to a significant extent during the conversion process. These are the components discussed in the stoichiometric model (Eq. 10.1). Since beer fermentation is operated in the batch mode, the first approach to a dynamic model is a rather simple component balance equation. When the balance is drawn over the entire mass of young beer, we obtain

$$\frac{dc}{dt} = R \tag{10.3}$$

where R is the biochemical rate of change of the components considered, the concentrations of which are combined in a vector c=[biomass; sugar; ethanol; carbondioxide].

The essential additional aspect in this balance equations is the kinetics, which is coming into play via the conversion rate vector R. The most obvious way to specify it is first to try the simple kinetic expressions formulated by Monod. However, one immediately finds that this does not work sufficiently well. It is necessary to note that in a batch fermentation the initial substrate concentration is too large for the yeast cells to work at full speed so that substrate inhibition must be taken into account. At the end of the fermentation the concentration of the product ethanol becomes inhibiting. So product inhibition must be taken into account as well.

For both effects simple extensions of the Monod expressions are known from the literature. σ, the specific substrate consumption rate, essentially determines the dynamics. It may be described as a function of the concentration S of the substrate (extract) and of the product ethanol (E) being built:

$$\sigma' = \sigma_{max} \frac{S}{K_m + S} \frac{K_{si}}{K_{si} + S} \frac{K_{ei}}{K_{ei} + E} \tag{10.4}$$

The second factor is the pure Monod expression; the third considers the substrate inhibition and the last the product inhibition. The corresponding parameters K_m, K_{si}, and K_{ei} must be determined from measurement data.

The rate vector R might be formulated by:

$$R = X[\sigma Y_{xs}; -\sigma; \sigma Y_{es}; \sigma Y_{cs}] \tag{10.5}$$

where X is the biomass concentration, and Y_{xs}, Y_{es}, and Y_{cs} the yields, the values of which can be estimated from the stoichiometric relationship at Eq. (10.1).

Since beer fermentations are controlled by the temperature only, the temperature influence on the yeast activity must be quantified. σ is affected in two ways, being increased in essentially the same way as any chemical reaction. This can be described by an Arrhenius relationship. On the other hand, at higher temperatures a decrease in the activity is observed, which can also be described by an Arrhenius-like expression. The entire temperature dependency of σ then has the form

$$\sigma = \sigma \frac{\exp(-E_{akt}/RT)}{(1 + K_d \exp(-E_{deak}/RT))} \tag{10.6}$$

The corresponding activation energies E_{akt} and E_{deak} as well as the intensity factor K_d must be determined from experimental data. R is the gas constant.

Unfortunately, these extensions of the pure Monod equation are not sufficient either. The reason is that, as known from literature, the concentration of the yeast and its activity is not readily described by simple stoichiometric or kinetic relationships.

Sufficient knowledge about the concentration of biomass in production-scale beer fermenters is usually not readily available. Values published in the literature depict much scatter. They also much depend on the particular control strategy of the brewery. Reliable data about the active part of the biomass during the beer fermentation is missing in the literature. Only some measurements performed with hemacytometers after staining the yeast cells with methylene blue to determine viability (e.g., [9]) are available.

Hence, the fermentation must be describe in an alternative way.

10.2.1.2
Heuristic Approach

Heuristically it is straightforward to divide the entire process into phases, which can separately be described in the classical way. Such a partition makes sense whenever the process behavior will be dominated by different mechanisms in the different phases. For instance, in the batch at the beginning of beer fermentation, substrate inhibition takes place which is completely unimportant at the end of the fermentation. Hence, this type of inhibition only needs to be taken into account in a model for an early phase of the fermentation.

This simplification of the models for the individual phases has the advantage that the models can be kept smaller, i.e., they contain a smaller number of free parameters. Consequently, the accuracy by which the model parameters can be identified from given data sets is much higher than in a comprehensive "world model."

The price one has to pay for this advantage is that one must decide when to switch over from one phase to the next in a practical simulation of the entire process. For that decision, the heuristic knowledge about the process accumulated by brewers can be used. Since this knowledge can often be formulated in terms of simple "if ... then ... – rules", it is straightforward to used small rule-based systems in order to make use of that knowledge. Havlik et al. [4] showed that such rule systems can easily be formulated by fuzzy rules which can be processed in computers using fuzzy logic. In this way, missing mechanistic knowledge is partly compensated for by heuristic process knowledge together with the basic mechanistic knowledge contained in the various classical models for the individual phases. The simultaneous utilization of such different representations of a priori knowledge in a process model is referred to as a hybrid model approach. It could be shown that such hybrid models can be used for process state estimation or process supervision purposes. Utilization of heuristic knowledge in addition to the classical models leads to an added value in the formulation of process knowledge and thus to an enhanced performance of model supported techniques where the models are used.

In practice one proceeds as follows. One starts with the model for the first phase of the fermentation and, with the passage of time, particularly during the time interval where a switch to the next phase is expected, the next model is also exploited. If more than a single model is considered, an appropriate weighting of the individual models becomes necessary. It is straightforward to do this by fuzzy reasoning. This allows a smooth switch over from one model to the next, thus avoiding unrealistic jumps in the signals for the key process variables.

It must be clearly stated that the reason for formulating such hybrid models is not to replace classical models, but instead to extend the latter by making available additional knowledge in order to increase the performance of the model. Where there is no further knowledge to extend the existing models on the basic mechanistic level, one must make use of the relevant heuristic knowledge, and the fuzzy rule processing provides a powerful tool to formulate these heuristics and to add this knowledge to a classical model.

10.2.1.3
Alternative Methods to Describe the Kinetics

When even additional heuristic knowledge about the part of the process under consideration is missing, one is forced to look for further information about the process dynamics. For production processes measurement data records are usually available and it is straightforward to proceed on the traditional path of engineering, namely to develop so-called engineering correlations. As is well known, such correlations are black-box models relating the desired key quantities of the process, i.e., those determining the bioreactor performance, to quantities which are adjustable independently. For instance, the Monod formula is such an expression which formally describes the dependency of the key variable specific substrate consumption rate σ from the quantity substrate concentration which can be adjusted in a fermentation.

Recently, artificial neural networks were shown to depict a much higher capacity for representing complex multidimensional nonlinear relationships between process variables. Their flexibility and accuracy in describing complex nonlinear relationships in biotechnical processes was demonstrated by many researchers, e.g., Simutis et al. [5b], who showed in several examples that the kinetics in biotechnological production systems can be accurately described by artificial neural networks. When the kinetics can be represented by an artificial neural network this can easily be used during the solution of the basic dynamic mass balance equation system just by calling the network evaluation routine as a subroutine for determining the actual rates $\mathbf{R}$.

The advantage of using this most flexible representation for kinetic expressions in hybrid models must be paid for by dispensing with the conventional network training procedures, since the relevant rates cannot usually be measured directly. For such situations, Schubert et al. [10] developed a special training technique: the sensitivity approach that uses the data usually available during the cultivation. Evolutionary algorithms can also be used to train such a hybrid model on the available process data.

When it is possible to describe the kinetics alternatively by a network or the classical Monod-like approach, there are two reasons for using both. The first is that artificial neural networks cannot be used for extrapolation, i.e., they cannot be used out of the area in the state space from which data were taken to train them. In cases where an extrapolation becomes necessary, it is believed that the classical models are much more reliable. However, one should generally be careful in using models (may they be classical ones or those based on networks) outside of the area which was experimentally explored before. Serious predictions can only be based on models that have been validated, and where there are no data no validation can be performed. The second reason is that the Monod-type kinetic models are known to have excellent global description properties in contrast to the networks which perform better when locally describing the process around the actual point in the state space. This suggests the use of both representations simultaneously just as if considering two votes given from different points of view.

Obviously, the weighting of the votes must be based on the amount of data available in the relevant area of the state space. By means of cluster analysis it is possible to evaluate the relevant area in the state space with respect to the sufficiency of data, and to express this by some kind of reliability or evidence measure E, with values between 0 and 1. This evidence measure can be used to weight the neural network component of the model, and the classical representation then gets the complementary weight (complement to 1) in the superposition of the model outputs.

Before one can make use of the models described for process optimization it must be ensured that they are reliable enough for long-term process state prediction, i.e., for predicting the course of the fermentation process over long time horizons up to the scale of the entire fermentation time. Brewers will only make decisions on the outcome of such state predictions after the underlying models have be sufficiently validated on measurement data. This can only be done by comparison with experimental data from the particular fermentation process under consideration. Of course, such a comparison must be made with independent data, i.e., data that have not been used during the development of the process model. This test procedure is referred to as the cross validation procedure in the literature.

10.2.2
State Prediction for Process Optimization

State predictions are of immediate interest to the industrial beer brewery. They allow one to estimate the end of the current fermentation runs more accurately than was possible before. Hence, scheduling of the tank capacity in the brewery can be made more precisely. This is a cost argument. On the other hand, one is interested in monitoring the quality of the young beer.

In the beer brewing industry, the characteristic buttery taste of vicinal diketones (VDK) has long been known to be a major problem. Human taste thresholds of the two vicinal diketones diacetyl and pentanedione, which may appear at significant concentrations in lager beers, are rather low: 50–100 μg/l for diacetyl and 100–500 μg/l for pentanedione [11]. Because of the considerably higher taste threshold of pentanedione, which appears in lager beers at roughly the same concentration as diacetyl, breweries mostly concentrate on keeping the diacetyl concentration low. At the end of the fermentation, the diacetyl concentration should be below the threshold value in order to avoid extended lager times or blending. Thus, it is of interest to predict the final diacetyl concentration at the earliest possible time in order to be able to reduce it by changing the fermenter control, i.e., the temperature profile, appropriately. Since diacetyl is not considered in the basic model discussed above, the latter must be extended appropriately.

There is enough experimental evidence showing that the diacetyl concentration in a normal beer fermentation rises in the first few days, crosses a maximal value [12, 13] about the third day, and then decays monotonously with time. The formation of vicinal diketones result from the oxidative decarboxylation of excess α-acetolactate hydroxyacid that leaked from the isoleucine-valine biosynthetic pathways (e.g., [14]).

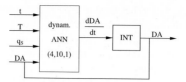

Fig. 10.2. Artificial neural network (ANN) structure used by Manikowski et al. [7] to represent the kinetics of the diacetyl (DA) degradation. The ANN is complemented by an integrator (INT) as explained in the text

When the yeast cell produces too much α-acetolactate, it exports it into its culture medium. The biochemical explanation for the degradation of this diketone is that the yeast cells take up and degrade it enzymatically.

The formation of diacetyl appears to be a very complex process with many degrees of freedom, while the degradation of diacetyl seems to be significantly more simple as it depends mainly on the metabolic activity of the yeast cells. Since both the knowledge about the diacetyl degradation and, in particular, its formation are not so clear that a mechanistic model could be built, a first approach for a quantitative representation was made by means of a black box approach based on the available experimental data only. As already mentioned, the most universal and flexible way of black-box modeling is using artificial neural networks.

Fig. 10.2 shows the network structure used by Manikowski et al. [7]. It is a simple feed-forward network with a single hidden layer containing 10 nodes. As input signals, the fermentation time t, the young beer temperature T, the substrate consumption rate q_s, and, finally, the diacetyl concentration DA, as interpolated from the measured values, were taken. The only output value is the net rate of change $d(DA)/dt$ of the diacetyl concentration. In order to get the diacetyl concentration, an integrator must follow the network output. This construction was chosen since the rate depends more sensitively on the input variables. The a priori information used in the network representation is restricted to the assumption that the temperature T is influencing the process as well as the metabolic activity of the yeast as represented by the extract degradation rate q_s. Furthermore, the heuristic knowledge about the sensitivity of the diacetyl conversion rate on the input variables is used.

In order to allow for a dynamical determination of this rate of change, the outputs were smoothed and fed back to the entrance of the network. In this way it is possible to determine the time development of the output quantity when only a single start value is provided. For the network training, however, complete data sets for the diacetyl concentration as a function of time are required.

This network must be trained with data records from several fermentations. Such a training is stopped at the minimal deviation between the network's output and the corresponding measurement data. As already mentioned, further independent data sets from additional production runs must then be used to validate the network.

The graphs in Fig. 10.3 provide an impression of the quality of the neural network representation of the diacetyl profile. In this figure the trained network is compared with the measurement data from one of the training data records. It provides a good representation of the process.

However, when it is then applied to the validation data set that was *not* known to the network before, much less agreement is observed. Figure 10.4 shows that the data

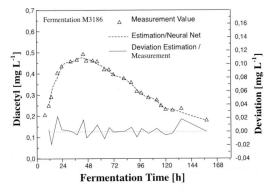

Fig. 10.3. Modeling result of the total diacetyl profile (training data) together with experimental data. The *curve in the lower part of the figure* represents the deviation of both results

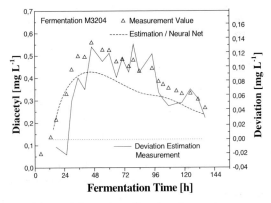

Fig. 10.4. Modeling of the total diacetyl profile using the test data

used for the training were not sufficient to train the neural network in such a way that it becomes possible to describe the diacetyl profile to a sufficient accuracy.

This is an excellent example to demonstrate that modeling without a solid validation can provide misleading results. After a closer investigation of the problem it became clear that the entire process, i.e., both diacetyl formation and degradation, cannot be modeled on the available data. However, it was found that at least the simpler diacetyl degradation phase can be modeled with the available measurement values.

The degradation process was again modeled by means of a dynamic artificial neural network of the same structure as before, where the input variables were the actual fermentation time, the substrate degradation rate q_s as determined from the extract data, the temperature of the young beer, and the measured diacetyl concentration.

Alternatively, for comparison, a simple mathematical model can be developed based on the assumption that the diacetyl degradation follows a first order reaction as suggested by Dellweg [15] and Inoue [16]. The corresponding temperature dependency of the rate was chosen according to Rice and Helbert [17] and Yamauchi et al. [18]. As with the model of García et al. [19] one can formulate as follows:

$$dC_{DA}/dt = k_1 \times q_S + k_2 \times C_{DA} \times \exp(-B/T) \tag{10.7}$$

with
$k_1 = 1.588e\text{-}3 \ [-]$
$k_2 = 4.694e\text{-}2 \ [h^{-1}]$
$B \ = 18.26 \ [^{\circ}C]$

The three parameters can be obtained by fitting this model to the experimental data. The results are depicted in Fig. 10.5

The full lines in Figs. 10.5 and 10.6 show typical results of the two diacetyl estimation techniques. The simple mathematical model (dashed curve) and the artificial neural network (full curve), both not used in MM and ANN training procedures, agree quite well with the measurement data. Such tests were repeated with several other test data records. Thus, by cross validation, it was ensured that this model can be

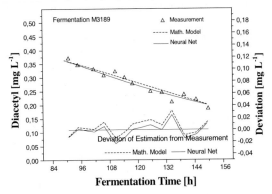

Fig. 10.5. Estimation of the diacetyl degradation in a typical fermentation supported by means of a simple mathematical model and an artificial neural network

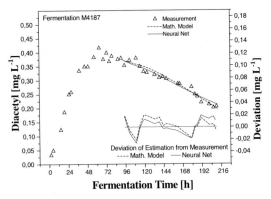

Fig. 10.6. Model-supported estimation of the diacetyl degradation in a typical fermentation using a simple mathematical kinetic model

used throughout the fermentations in the fermenters from which the data were taken for training.

10.2.3
Remarks on Hybrid Models

The beer fermentation process can be described quite accurately by means of hybrid process models that combine mechanistic information about the process in the form of general mass balance expressions with heuristic information about the various process phases and detailed local information extracted from the extended data records, usually available from production fermenters, to describe kinetic details by artificial neural networks. Since, compared to classical mathematical models, the hybrid models make use of additional knowledge and measurement information, they perform better for state prediction.

It must be stressed however, that hybrid models cannot be validated when the data base is insufficient. The important problem of diacetyl generation cannot be predicted on the data usually acquired during an industrial beer fermentation process. When its development is to be predicted as well, additional measurement information is required. On the other hand, the diacetyl degradation can be described; however, this requires a start value which expresses the amount of diacetyl to be degraded. Thus, as shown before, at least one measurement value of the diacetyl concentration must be measured at a time where the diacetyl degradation dominates over the development process to be able to predict its path for the rest of the fermentation.

One important remark which should be made here is that all the models discussed so far assume that the fermenter is a well mixed reactor. This is at least questionable, since it is well known that big fermenters often involve considerable inhomogeneity. The incorporation of fluid dynamics into these models to an accuracy comparable with the modeling of the conversion process is inhibitively difficult and time consuming at the moment. Experience showed that detailed fluid dynamics is not indispensable for this type of model. The only consequence that must be drawn from recognizing this problem is that the kinetic parameters used in the models must be taken from experiments in the fermenters of the original size, since they contain all the information about the transport problems not resolved in the models.

In this section it was described how to formulate the knowledge about the process that is quantitatively available in order to predict the behavior of the process under different operational conditions and to optimize it with respect to a predefined objective function. When this task has been performed the real process must be driven along that control path. In reality one must usually cope with distortions so that an open loop control must be augmented by a closed loop feedback control in order to counteract deviations from the desired path. In the next section we first discuss online state estimation and in a later section control. Since it is only possible to manipulate temperature, we will concentrate on temperature control.

10.3
Process Supervision

In order to supervise a running production process, information about its actual state is required. The definition of the state is problem dependent. With the state variables, those aspects must be described that are relevant to the task to be performed. Here, this task is producing a particular beer at minimum cost. What brewers found by experience over centuries is that the main state variable is the extract concentration, and if one would like to go into more details ethanol concentration and the diacetyl concentration are the next candidates. For state estimation it is indispensable to make on-line measurements.

10.3.1
On-Line Measurements are Difficult to Perform

Obviously, one first tries to measure the state variables discussed above. The most important problem appearing in this context is that none of the concentrations of the key components can easily be measured on-line. Thus, indirect measurements are required which make use of measurement data that can easily be obtained during the process and which are uniquely related to the desired state variables, i.e., the concentrations of the components in the basic reaction equation (Eq. 10.1).

As already mentioned, beer brewers traditionally take the extract concentration as their key process state variable. They assume that the extract is uniquely related to the other quantities that significantly change during the fermentation by means of the stoichiometric equation (Eq. 10.1). Since even direct off-line measurements of the extract concentration are not reasonably possible, they perform indirect off-line measurements. From every point of view, extract concentration measurements via manual density measurements proved to be the most economical technique in the past. Even these measurements require a careful pretreatment of the samples drawn from the fermenter in order to avoid distortions by CO_2-gas bubbles developing from beer as known to everybody. Hence, these measurements are time-consuming and thus cannot be performed very often in a commercial beer fermentation. The measurements required to keep a production-scale brewery under control are a significant cost factor and large breweries try to reduce the amount of measurement wherever possible. In practice, they are most often measured only once per day and fermenter.

Density measurements are not easy to perform online during a fermentation. Hence, much work has been devoted to finding alternatives. On-line measurability is not the only criterion; the measurement technique must also be robust enough to be applicable in the harsh everyday working environment in breweries at acceptable expenses. Here expenses essentially not only means investment costs; more relevant is the expense of manpower for operation and maintenance, which must be clearly below that necessary for the currently used technique.

The alternatives to extract concentration measurements can be found in the basic reaction equation (Eq. 10.1), which at the same time shows how they are related to each other. Apart from biomass, which did not prove to be a good candidate, the others – CO_2 development rate, ethanol production, and heat development rate – are possible alternatives and have been discussed as such in the literature.

CO_2 development rates can be measured on-line by volumetric or mass flow measurements in the vent line of the fermenters. The problem is that appropriate measurement devices for the relevant conditions are rare on the measuring instrument market. A more significant problem, however, is that CO_2 dissolves in the young beer to a large extent, leading to supersaturation effects. The maximum level of supersaturation of CO_2 reached during active fermentation increases with the fermentation vessel capacity. The amount of CO_2 that remains dissolved at the end of brewery fermentation is essentially constant (at a degree of supersaturation of approximately 1.5, equivalent to 11% of the total CO_2 production) and could be calculated. The profile of change in dissolved CO_2 concentration during fermentation followed the same pattern as that of the rate of CO_2 evolution. The maxima for both profiles were found to coincide [20].

Although the major portion of the dissolved carbon dioxide exists as the aqueous species $CO_{2(aqu)}$, in typical ethanol fermentations, the solubility of the CO_2 is a function of the concentration of other non-polar and ionic species, in particular ethanol. Thus, one has to consider the other molecular species that are in equilibrium with CO_2:

$$CO_2 + H_2O \leftrightarrow H_2CO_3 \leftrightarrow HCO_3^- + H^+ \tag{10.8}$$

Because $CO_{2(aqu)}$ is able to react with different components dissolved in the culture medium, e.g., with free amino groups of proteins, and H_2CO_3 to associate with positively charged groups on proteins via a dipole-protein interaction, the total amount of CO_2 in the solution can increase significantly, causing high supersaturation levels.

The gas exchange $CO_{2(aqu)}=CO_{2(gas)}$ further depends on the rate of nucleation. This rate could be a rate-limiting effect for the liquid-gas mass transfer. The resulting supersaturation leads to a non-ideal behavior of CO_2 in solution and the presence of increasing amounts of HCO_3 and $CO_{2(aqu)}$ [21]. This discussion shows that it is not simple to determine the progress of the fermentation with simple point measurements of the specific weight or the carbon dioxide evolution rate.

Since the CO_2-transfer into a bubbly gas phase is also dependent on the local temperature and the local mixing conditions, the actual gas flow rate depicts heavy fluctuations of more than 100% during the main fermentation, where the measurement values would be needed to supervise the process. From that point of view, CO_2 is not a preferred measurement variable.

The situation is much better with ethanol, since the ethanol concentration does not fluctuate very much during the fermentation. However, direct measurements of the dissolved ethanol is also difficult to obtain. An alternative is to make use of the fact that there is an equilibrium ethanol concentration in the gas phase of the reactor. Hence, some ethanol leaves the reactor via the vent line. It was shown that the ethanol concentration in the off gas nicely reflects the dissolved ethanol concentration in the young beer. The partial pressure of ethanol within the gas-phase can be measured with several devices. For on-line measurements in a brewery, infrared sensors are best. Measurement devices for supervising the ethanol concentration in room air that can be used to monitor the ethanol in the vent line are commercially available. The same applies to devices used to detect ethanol in the breath of humans. Both work in the relevant concentration interval and can be adapted to the needs in a fermenter.

Finally, heat development measurements can be performed by measuring the heat transferred into the cooling system of the beer fermenters. In new fermenters the appropriate measurement devices can easily be installed. Already producing fermentation halls, however, cannot be retrofitted easily.

The reason why biomass is not an appropriate measure for the state of the fermentation is that it is not developed in a stoichiometrically simple and stationary way from the sugar being consumed during the fermentation. Thus, the basic stoichiometric equation is changing with time. Hence, we need dynamic process models even for process supervision. This discussion shows that process supervision is not simple and must be performed indirectly, using so-called software sensors for the most important process variables.

10.3.2
Estimation of the Extract Degradation

As the state variables cannot be measured on-line, they must be measured indirectly. Hence we need a relationship between the measurement quantities and the desired state variable. It would be ideal if there were a simple correlation between the measured quantities and the extract degradation. Unfortunately things are more complex so that one needs a more detailed description. An entirely mechanistic model of the brewery fermentation is not yet possible. Furthermore, the exact description of the extract degradation suffers from the fact that many parameters of the natural raw material influencing the fermentation cannot be described quantitatively to a sufficient accuracy. Also, the parameter values fluctuate from charge to charge. Manikowski et al. [7] proposed a first approach in the form of a simple first-order mathematical model relating the extract concentration to the ethanol measurement data.

10.3.2.1
Simple Mathematical Model

The steady state ethanol concentration $C_{E,g}$ in the off gas generally depends on the ethanol concentration $C_{E,l}$ in the beer and the temperature T. Raoult's law allows one to determine the partial pressure p_E of the ethanol in the liquid phase (e.g., [22]:

$$p_E = x_E \times p_{E,0} \times g_E \tag{10.9}$$

where x_E is the relative amount of ethanol in the liquid, $p_{E,0}$ the vapor pressure of pure ethanol, and γ_E the activity coefficient of ethanol. By Dalton's law one gets

$$p_E = y_E \times p_{ges} \tag{10.10}$$

where y_E is the relative amount of ethanol in the gas phase, and p_{ges} the total pressure of the gas mixture. By combination of both formulae, we obtain the Raoult-Dalton law for non ideal solutions of ideal gases:

$$y_E \times p_{ges} = x_E \times p_{E,0} \times g_E \tag{10.11}$$

The activity coefficient γ_E is a function of the ethanol concentration in the beer. The relative amount of ethanol in the beer usually does not rise to values beyond a value of 0.016. In that region of low EtOH concentrations, the activity coefficient γ_E can be considered constant and the transformation of relative amounts to concentrations can be linearly approximated. In a fermentation performed at normal atmospheric pressure in the head space of the fermenter we then get

$$\gamma_E = \text{const.}, \qquad p_{ges.} = \text{const.}$$
$$x_E = a_1 \times C_{E,l} \qquad a_1 = \text{const.}$$

It follows:
$$k_2 = a_1 \times \gamma_E / p_{ges.}$$
$$y_E = k_2 \times C_{E,l} \times p_{E,0}$$

and, since $y_E = p_M - k_1$, the ethanol concentration $C_{E,I}$ in the beer is:

$$C_{E,I} = \frac{y_E - k_1}{k_2 \times P_{E,0}} \qquad (10.12)$$

Since the coefficients k_1 and k_2 depend on the properties of the concrete fermenter, they must be determined separately. Fits to the available data led to values
$k_1 = -222.9845 \times 10^{-6} \ [-]$
$k_2 = 13.5006 \times 10^{-4} \ [l\,g^{-1}\,bar^{-1}]$

According to Balling, the relationship between the fermentable sugar consumed and the ethanol concentration $C_{E,I}$ formed in the beer is
$S = STW - C_{E,I}/0.484$
where $S \ [g\,l^{-1}]$ is the total concentration of the fermentable sugars and STW the concentration of the original wort $[g\,l^{-1}]$. With the sugar table, the true extract E_w and by *Balling's* formula the virtual extract E_s can be determined [23].

By *Antoine's* equation, the vapor pressure $p_{E,0}$ over a pure ethanol phase dependents on the temperature T:

$$\log_{10}(p_{E,0}/p^*) = A - B/(C + T_g) \qquad (10.13)$$

The parameters A, B, and C are material dependent. They are tabulated by Gmehling et al. [24].

Since the temperature T_g at the gas-liquid interfacial area cannot be measured, Manikowski et al. [7] tried to find out which of the available temperature values can best be used to approximate its true value. Two main alternatives were tested – the gas temperature T_{KR} in the head space of the fermenter as measured with the sampling station and the beer temperature T_B. The results indicated that the head space temperature should be used rather than the beer temperature.

However, there is usually no thermometer in the head space of a production fermenter. Since it is too cost intensive to install additional thermometers in an existing production fermenter, a simple alternative was looked for. The result of this investigation showed that it is possible to obtain fairly reliable estimates for the relevant temperature from a contact-thermometer attached to the outer surface of the wall of the fermenter directly at the upper edge of the cylindrical part below the insulation layer (Fig. 10.1).

10.3.2.2
Estimation of the Extract Degradation by Artificial Neural Networks

As an alternative to the simple mathematical model described in the previous section, Manikowski et al. [7] used a black box model: An artificial neural network was taken to make direct use of the original measurement data from the plant for representing the relationship between the extract concentration and the available measurement values without referring to potentially imprecisely known physical details. Apart from the data, the only a priori information they used concerned which of the available measurement variables will influence the substrate degradation.

Different possibilities for the construction of artificial neural networks were tried. Two static feed-forward three-layer sigmoidal networks with three input nodes, ten nodes in the hidden layer, and a single output node were trained. Both use the fermentation time t and the temperature T at the reactor wall as input variables. They differ by the third input variable which, in the first network is the EtOH concentration in the off-gas line, and the corresponding EtOH value determined by the simple mathematical model described above in the other one. The networks were trained on the available process data using the conventional error backpropagation technique [25].

In order to allow a cross validation of the artificial neural networks, the available data were divided into a set of training data and a set of test data. A set of 20 fermentation records, which were obtained under the same operational conditions, were used as the validation data set. The results of the extract estimations are de-

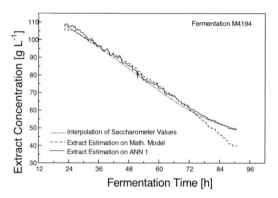

Fig. 10.7. Estimated extract values for a typical fermentation

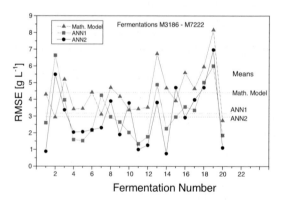

Fig. 10.8. Comparison of the rms error values of the individual fermentations

Table 1. Summary of the mean errors of the three models for the extract estimation

$[gl^{-1}]$	Mean deviation (math. model)	Mean deviation (ANN1)	Mean deviation (ANN2)	RMSE (math. model)	RMSE (ANN1)	RMSE (ANN2)
Mean	3.67	2.69	2.83	4.42	3.17	2.96
σ_E	1.28	1.33	1.50	1.37	1.53	1.70

picted in Fig. 10.7 for a typical example. Neural networks proved to deliver better results than the mathematical model approach. Both networks depict to about a similar quality. Quantitative results obtained with this method from the fermentations investigated are summarized in Table 10.1.

The mean values of the deviations were taken over all fermentations investigated. The corresponding standard deviations σ_E characterize the consistency of the extract estimations. The rms errors show that the neural network models depict a significantly lower error as compared to the simple mathematical model.

In order to provide an overview over the individual measurement data, the means and the rms errors of the predictions of all three estimations were plotted in Fig. 10.8.

10.3.2.3
Hybrid Modeling

Hybrid models are used either when there are different models available to describe the behavior of a process or when different parts of the process are described by different models. The first alternative is of advantage when the different models can be viewed as descriptions of the same process from different points of view. The second alternative is proposed for the case where the knowledge about different aspects of the process is available at different levels of sophistication. For example, the basic mass balance equations are known to a high accuracy, while our knowledge of kinetic details of the biochemical conversion and growth processes involved are at a much lower level.

The latter case led to the hybrid model, schematically depicted in Fig. 10.9. In addition to the basic mass balance, which is formulated by means of a set of ordinary differential equations, the kinetics may be described by means of an artificial neural network. This type of hybrid model (with respect to the process representation technique) can be extended by using an alternative representation of the kinetics, e.g., by classical Monod expressions.

The training of the hybrid model containing the artificial neural network and a differential equation system was performed by the method of Schubert et al. [10], since the conventional training methods cannot be used for hybrid models. The data sets used for training and the subsequent cross validation were the same as used before for the simple artificial neural networks.

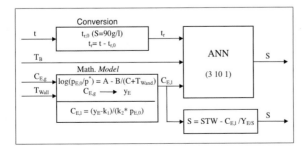

Fig. 10.9. Scheme of the hybrid model for estimating the extract degradation

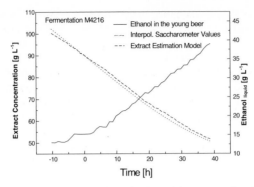

Fig. 10.10. Extract estimation with the hybrid model as example of a beer fermentation. In order to keep the graph simple, the offline measured values were depicted by a line obtained through a spline interpolation

The rms error obtained with the hybrid model approach was between $1.26\,\mathrm{g\,l^{-}}$ and $2.52\,\mathrm{g\,l^{-}}$ for the fermentations investigated. This is significantly smaller than the errors obtained with the other models. where the rms errors were between $2.64\,\mathrm{g\,l^{-}}$ and $3.17\,\mathrm{g\,l^{-}}$. Thus, the extract concentration can be estimated more consistently with a hybrid model as demonstrated in Fig. 10.10.

10.3.3
Kalman Filters, and an Advanced Method for State Estimation

The different techniques used for state estimation are distinguished by the quality and the amount of knowledge they use about the measurement data and their relationships to the extract degradation. While the model-based indirect measurement used so far only use very general assumptions about the data and assume that the relationships are essentially correct, more detailed assumptions about the measurement noise and the modeling error are taken into account in Kalman filters.

Kalman filters assume the dynamic process model to be given by a set of differential equations in the state variables **c**, which can be influenced by some control variables **u**. The state variables in the case of beer fermentation are the biochemical state variables, i.e., the concentrations of the species appearing in the basic reaction equation. As already discussed, the basic control variable in a beer fermentation is the temperature T. The parameters of the model might be combined to a parameter vector **p**. The uncertainties of the measurement values and in the model are provided by covariance matrices. The Kalman filter is a one-step-ahead prediction method, which estimates the state vector at the current time instant from the current measurement values and the value computed by the underlying process model. Since the latter must be based on the estimate made for the last step, one needs a total of three covariance matrices, since the last value estimated was obviously also not completely correct.

The Kalman filter is considered the optimal state estimator for a linear system with predetermined parameters. Unfortunately, it is not directly applicable to fermentation systems due to their nonlinearity and difficulties in determining model parameters. To cope with nonlinear models as well, Extended Kalman Filter (EKF) were developed.

Originally the Extended Kalman Filter was designed to estimate the state of a nonlinear process for which a model is already available (including its parameter values). However, Jazwinski [26], and later on many other authors, showed that it can also be applied to estimate the parameters of the process model even when these are time-varying or ill-defined model parameters are used. Simutis et al. [27] applied Extended Kalman Filters to estimate the extract degradation in production-scale beer fermenters. They extended the process model needed as one base of the Kalman filter by a heuristic component, namely the division of the entire process into several phases, which were described by different models. The different model components they handled with a small fuzzy expert system, just as shown for the modeling of the beer fermentation with fuzzy supported artificial neural networks described above.

The detailed formulae were often presented in the literature [28, 29] and will not be repeated here, particularly since the application of Kalman filters in industrial breweries currently seems to be quite unrealistic: Although conceptionally excellent, extended Kalman filters are known to be difficult to implement, the main problem being the tuning of the covariance matrices, which are the decisive components of the underlying concept. In particular, it was shown that the accuracy of process state estimation by means of Kalman filters is seriously degraded, when the uncertainties in the process model are not properly known. Thus, its use in control may be limited to companies which own control specialists that are able to deal with these advanced techniques.

10.4.
Process Control

Since the temperature is the only variable that can be manipulated during a beer fermentation process, it is of primary interest to keep it close to a predefined set point. In order to keep the temperature fluctuations safely within a limited interval around the temperature set point, it is controlled in all breweries. In practice, simple temperature controllers, often two-point controllers, are used. With this technique it is possible to keep the temperature in the production fermenters during the main fermentation phase within an interval of ± 1 K around the set point, which is usually in the range of 9–13 °C. Here we are speaking about the temperature fluctuations as measured with a single temperature sensor installed at the fermenter.

One essential point to note is that experimental (e.g., [30]) and model-based investigations [31] showed that the temperature inhomogeneity, i.e., the spatial variation of the measured temperatures, is of the same size as the temporal temperature fluctuation. It was observed that the mean temperatures over a fermenter of several hundred cubic meters differ by about 1 K only. Hence, with respect to temperature, beer production fermenters are quite well mixed reactors, a fact which, a posteriori, justifies the modeling approach discussed before.

Tightening the temperature variances by improved control, however, is a nontrivial problem, since production-scale brewery fermenter vessels are very big and thus not able to follow immediately changes in the temperature set point profiles. In order to reduce the variance of the temperature fluctuations, a model supported control is required which can make use of information about the fermenter fluid dynamics, particularly with respect to the heat transfer into the cooling system and the model for the heat generation by the biochemical reaction (Eq. 10.2).

Fluid dynamic models of beer fermenters are now coming to the stage where details about the temperature field can be calculated [31, 32]. However, such calculations currently need weeks of computing time so that an easier-to-use technique must be chosen, when dealing with control aspects.

The detailed response in the young beer temperature upon changes in the tank's cooling systems, i.e., switching on or off the coolant flow through the cooling jackets, is complicated. The stirring action of the CO_2 bubbles rising in the liquid phase is, as already mentioned, by no means continuous, since considerable eruptions characterize the carbon dioxide gas flow through the vent line which influence the heat transfer. Thus supersaturation levels cannot be predicted. The flow in the fermenter is chaotic and since it depends on the CO_2 transfer from the liquid into the gas phase it is not really stationary. The flow of CO_2 through the vent line of the fermenter and the temperature T at one point in the reactor are the measured quantities in this example.

Since most of the parameters determining the heat transfer coefficients are unknown, the cooling behavior of the tank must be determined experimentally by analyzing the temperature response to changes in the cooling control signals. In order to keep the dynamical representation simple, a linear difference equation was used by Gvazdaitis et al. [33]. In this equation, the temperature sampling values T_k, where k is the time index, are related to the registered control signals S_{ik} of the i-th cooling element:

$$
\begin{aligned}
T_k = a_0 \;&+ a_1 \, T_{k-1} + a_2 \, T_{k-2} + a_3 \, T_{k-3} + a_4 \, T_{k-4} + a_5 \, T_{k-5} \\
&+ b_1 \, S_{1,k-1} + b_2 \, S_{1,k-2} + b_3 \, S_{1,k-3} + b_4 \, S_{1,k-4} + b_5 \, S_{1,k-5} \\
&+ b_1 \, S_{2,k-1} + b_2 \, S_{2,k-2} + b_3 \, S_{2,k-3} + b_4 \, S_{2,k-4} + b_5 \, S_{2,k-5} \\
&+ b_1 \, S_{3,k-1} + b_2 \, S_{3,k-2} + b_3 \, S_{3,k-3} + b_4 \, S_{3,k-4} + b_5 \, S_{3,k-5} \\
&+ c_1 \, K_k \\
&+ c_2 \, C_k
\end{aligned}
\tag{10.14}
$$

The authors determined the parameters a_j, b_j, c_j by appropriate identification procedures. The variables K_k characterize the heat production, while the C_k are characteristic for the CO_2-off-gas mass flow. As a characteristic CO_2 mass flow through the off-gas line, a smoothed signal (typical average over 2–3 h) of the CO_2 mass flow data was taken. The characteristic heat production parameter K_k can be determined in two alternative ways, either from the long term CO_2 development, which is directly proportional to the heat production, or by directly incorporating the output of the neural net describing the kinetics of the substrate degradation, as described above. According to the general reaction(Eqs. 10.1 and 10.2), the heat production was assumed to be proportional to the substrate conversion rate. The identification of the other parameter was performed with standard optimization procedures.

In order to get an impression of the performance of this model, the temperature was predicted for periods of 2 h in advance using the available measuring data sampled up to the actual point in time. These simulated temperatures are depicted in Fig. 10.11 together with the temperature measurement values obtained 2 h later.

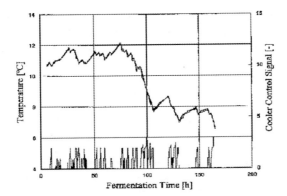

Fig. 10.11. Comparison of the simulated temperatures together with the temperatures measured 2 h later. Both curves cannot really be distinguished. Also, the control profile of the cooling system is shown in the *lower part of the graph* (label on the right). Its values 1, 2, and 3 denote that only the upper cooling section, the upper two only, and all three cooling sections are switched on

10.4.1
Controllers that Consider the Dynamics of the Fermenters

Temperature predictions over a time horizon of 2 h can be used in a model supported feedback controller. For this purpose it is straightforward to use a model predictive controller that compares the predictions with the desired temperature profiles and changes the cooling control signal in such a way that the deviations between the expected and the desired temperature profiles become minimal.

Gvazdaitis et al. [33] computed the required control vector for the three cooling elements of the real fermenter every 8 sampling time steps (12 min each) only, in order to reduce the controller actions. As already shown in the previous section, the simple model was able to predict the temperature profile over such time horizons. They used the time between two such predictions to adapt the model parameters to the process. Hence, after each 8 time steps, the next control profile segment was computed with an updated model. The controller used in this example was an adaptive controller. It is quite easy to determine the optimal temperature control vectors for the 8 time steps ahead. They can simply be chosen from all possible combinations of activating the three cooling sections in such a way that the mean squared deviation from the set point profile in that interval becomes minimal. In the case where the actual temperature is below the set point temperature, nothing is done because it is not possible to heat the fermenter by means of the cooling system. In this case, one must wait until the internal heat production enhances the temperature to a value close to the set point.

The most important parameter in the model-supported controller is the adaptation gain by which the deviations ΔT of the actual temperature value from the temperature predicted by the model influence the model parameters used in the controller. If the gain is high, i.e., the influence of the temperature deviations on the model adaptation is high, the controller might become unstable. If it is too low, the controller might not follow changes in the system quickly enough. Since the response of the reactor on the actions in the cooling system is rapidly changing during the fermentation, a high gain is desirable.

In order to minimize the risk of obtaining an unstable controller behavior it is recommended to make use of a second controller with the task of compensating for larger deviations of the actual temperature from the desired level. This can be done by a robust fuzzy expert controller based on the available knowledge on the cooling process. Gvazdaitis et al. used an additional watch-dog module in the control software in order to supervise the controller. If this adaptive controller is not able to control the temperature, the watch-dog switches over to the more robust but not as accurate fuzzy controller. Figure 10.12 depicts a schematic view of the controller structure.

An example of the improvement of the temperature signal by means of the controller is shown in Fig. 10.13. It can be seen that there is a significant improvement over the previous situation.

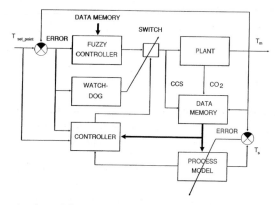

Fig. 10.12. Schematic view of the temperature controller consisting of the adaptive controller, the fuzzy controller, and the watch-dog which works as supervisor

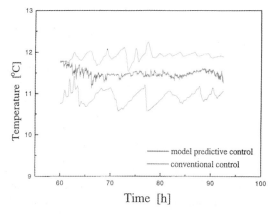

Fig. 10.13. Comparison of typical temperature signals obtained with conventional temperature control (*upper and lower signals*) and with model predictive control (*middle*)

One might argue that the controller controls the temperature at a single point only, namely at the measurement point. This might be true; however, by the model supported controller one does not change the mixing behavior of the fermenter, and thus one does not change the general behavior of the tank. The operational temperature in a production fermenter is most often measured at one point only and we know from measurements and from detailed simulations that the inhomogeneity of the mean temperature values (mean with respect to the time) is less than 1 K during the main fermentation phase. We can assume that the temperature inhomogeneity does not become larger with an improved control.

From the experience with beer fermentations we know that temperature fluctuations of $\pm 1\,°C$ around the set point do not influence the beer quality. When this variance is reduced by means of an improved control, we do not change the beer quality, although the predictability of the extract degradation and its reproducibility will be better. This might not be a strong enough argument to justify investments for an improved temperature control. The benefit of an improved controller becomes more obvious, if one considers its integration into an optimized feedback control strategy where, as well as the quality arguments, the cooling costs are taken into account.

10.4.2
Reduction of Energy Costs by Temperature Profile Optimization and Control in a Production-Scale Brewery

As already mentioned at the beginning, cost reduction is a major aim in process optimization. Besides manpower, energy consumption is a main target in this respect. Prerequisite to all changes in the process operation and control is that the product quality goals are assured. This being assured, breweries make significant efforts to reduce the cost of the energy consumption. For instance, one entry point to cost reduction is to benefit from the reduced "off-peak" electricity price during the night. Since electric energy is not conservable in sufficient amounts, some breweries, e.g., Ashai in Japan, installed so-called ice-banks, which are cooled down overnight and deliver a cooling medium at a sufficiently low temperature level during the day time (e.g., [34]). From the Guinness brewery in Ireland a similar approach was published [35].

In order to avoid the significant expenses in investment, maintenance, and ground space for such an ice-bank, one can control the fermentation process in such a way that the cooling expenses area mainly required overnight [33]. The obvious constraint to such an approach is that the temperature must not go outside the known temperature variation observed in a fermenter that is conventionally controlled, since otherwise the beer quality would be affected. The basic idea is to first improve the temperature control so that the temperature fluctuations become significantly smaller than before. Then the temperature set point can be guided through the allowed temperature interval in such a way that the temperature decreases overnight and reaches the lower bound of the interval in the morning, when the electric power becomes more expensive again. Then the temperature is allowed to rise until the upper level of the interval is reached. From there until the time instant where power becomes cheaper again the temperature is kept just below the higher bound.

Since the metabolic heat production changes over the course of the fermentation, a combination of the heat generation and the heat removal processes must be incorporated into the model. In order to find an optimal temperature profile, one first of all needs a measure of optimality.

The following cost functional J was used by Gvazdaitis et al. [33]:

$$J = k_1(S_f - S_g)^2 + \quad \text{(Deviation from extract set point)}$$
$$k_2D^2 + \quad\quad\quad \text{(Deviation from diacety set point)}$$
$$\Sigma u_k P \Delta t \quad\quad\quad \text{(Cooling cost)}$$

This is a weighted sum of three major contributions to the optimal solution. Its first component describes the influence of a deviation of the actual extract concentration from the model value. It says that the process must not deviate too much from the predefined optimal way. The coefficient k_1 is typically chosen to be 0.05. The second term describes the influence of the diacetyl concentration D. It is assumed that concentration D influences the quality loss quadratically, expressing the cost of the lager process which becomes necessary to reduce the diketone concentration below its acceptable limits. Parameter k_2 is zero as long as the concentration is below some threshold value (usually 0.2 mg/l). Above that concentration it might be chosen to be 100. The third term sums up the direct cooling cost up to the actual point of time. Δt is the time increment. u_k, with k=1, 2, 3, respectively, is a binary control signal indicating whether the individual cooling elements are switched on or off. P=P(t), a function of time t, is the price of electric power for the cooling aggregate. The minimization of the cost functional can be obtained with classical optimization techniques.

Figure 10.14 shows the structure of the feedback controller schematically. The controller contains an on-line part and an off-line component. In the off-line part, the optimal control profile is calculated every 24 h in advance. Then the on-line part controls the temperature along this profile. During this closed loop control, the de-

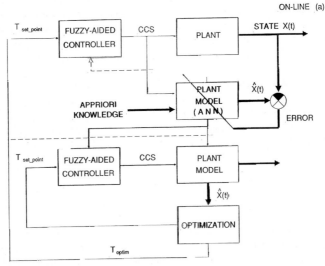

Fig. 10.14. Schematic of the optimal feedback controller used to minimize the fermentation cooling cost in a running production [33]

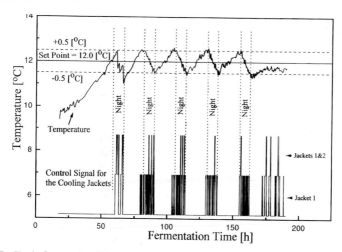

Fig. 10.15. Typical example of the optimized temperature signal in a production fermenter. By means of a feedback controller the temperature is kept close to the optimal temperature profile beforehand obtained in an optimization run [32]

viations of the temperature from the predefined profile are used to correct the parameters of the model within the controller. Every 24 h, after new extract values became available, the kinetic representation by the neural networks was improved. So, for the next day the profile can be calculated using an updated model.

Figure 10.15 presents a typical example of a temperature profile measured during such a controlled experiment. The temperature set point due to the fermentation scheme used before by the brewmaster was constant at 12 °C. It is shown together with the ±0.5 °C margins within which the temperature was hopefully kept with the simple conventional controller. The optimized set point profile is a saw-tooth-like profile where the extreme values were kept within the margins given by the classical control approach. In the figure the measured temperature is shown which was kept very close to the set point profile by model supported feedback control. The optimized controller fully exploits the temperature interval allowed during the extract degradation. As far as possible, cooling actions are performed overnight.

For a single fermenter, the cost reductions may come up to more than 20% in this way. This is in the same order of magnitude as the energy reduction claimed by Asahi and Kinoshita [34], who used an unspecified AI computer system in combination with their cold storage system to maintain an optimized economical balance.

10.5
Conclusion

There is enough knowledge and quantitative data that can be used to optimize the process operation and to supervise and control the beer fermentation process. The knowledge about this process is now at a state where the fermentation plants can be automated in order to reduce the cost of manpower and other operational costs like electric energy.

Two main problems influence the success. The first touches the question of how to solve the measurement problem. There are several possible solutions discussed in the literature, the most promising seeming to be heat removal and ethanol concentration measurements. The second main problem is how to keep the process models on the actual state. This requires an educational change in the brewers' community and an adaptation to the developments in other process industries like the pharmaceutical industry (e.g., [14]).

In this environment, hybrid modeling provides a chance to make direct use of model-supported optimization and control in beer fermentation processes, since it allows one to make use of the wide range of relevant heuristic knowledge of the process dynamics as well as the extended data records. The alternative would be to dispense with improvements possible by modern control technology and to make long term investigations into the biological and transport processes ruling this process before starting with improved optimization and automatic control. There is no doubt that the cost of beer production can be significantly reduced when fermentation control is automated.

10.5.1
Summary of the Application of the Techniques to Beer Fermentation

Modeling only makes sense when it leads to some obvious advantages. In beer fermentation such advantages would be to simplify achieving improvements of the product quality or cost reduction in terms of consumables, energy, or manpower.

The approach taken here is process improvement by optimization of the operational procedure and making sure through feedback control that the optimal procedure is maintained during the production. This requires tight process supervision.

Systematic optimization of the process operational procedure first of all requires a clear-cut criterion of optimality. The objective of the process must be defined quantitatively in all necessary detail. In beer fermentation the first issue is to produce beer with the desired quality. Second, this beer must be produced at a minimum of cost. Traditionally the state of the fermentation process is observed by means of the actual extract concentration, the operational procedure for a particular beer being determined by the initial conditions of substrate and yeast concentration as well as the temperature and the extract degradation profile.

Here we have discussed models designed to provide a sufficiently accurate quantitative description of the beer fermentation process such that they can be applied in numerical optimization procedures. They are built to represent the knowledge of the process needed for optimizing the operational procedure (initial conditions as well as temperature and substrate degradation profiles). Furthermore, models are required for improved process supervision and feedback control. In order to guarantee a benefit over trial and error based design methods, the models must meet some accuracy requirements.

Prediction of the state over a larger time horizon poses the biggest challenge to the models. Such predictions are required for process optimization in the sense of determining the optimal process control path. Sufficiently accurate predictions of the path of the beer fermentation over the entire fermentation time of roughly 5–7 days were not possible with simple classical models that are often used in biochemical engineering. Hence, extensions had to be made. With hybrid models, e.g., fuzzy supported artificial neural networks, additional knowledge or a priori information about the process was exploited. This allowed one to predict the extract degradation profile accurately enough. One essential heuristic extension was to divide the process virtually in several phases, which are separately described by individual models. This procedure reflects the fact that one speaks of different phases of a process when its dynamics is dominated by different mechanisms.

Process supervision poses a different requirement to the model. It is not so much the long term prediction capability of the model that is important, as the detailed local description. The main problem with process supervision is state estimation. Obviously supervision requires one to know the actual state of the process as accurately as necessary for the general task to be fulfilled. Such a task could be closed loop process controlled.

Process control is a measure to make sure that the process runs strictly along the predetermined control path. Closed loop or feedback controllers are designed to correct automatically for deviations of the actual process state from the desired one.

References

1. Narziß L (1990) Über einige Faktoren, die den Biergeschmack beeinflussen. Brauwelt 130:1554–1556, 1558–1562
2. Mitsui S, Shimazu T, Abe I, Kishi S (1991) Simulation and optimization of aeration in beer fermentation. MBAA Techn Quart 28:119–122
3. Dymond G (1991) The brewer in control: modern brewery automation. Brewers Guardian 120/3:17–21
4. Havlik I, Dors M, Beil S, Simutis R, Lübbert A (1992) State prediction in production scale beer fermentation using fuzzy aided Extended Kalman Filters. Dechema Conferences 5b:623–626
5a. Simutis R, Havlik I, Lübbert A (1993) Fuzzy aided neural network for real time state estimation and process prediction in a production scale beer fermentation. J Biotechnol 27:203–215
5b. Simutis R, Havlik I, Schneider F, Dors M, Lübbert A (1995) Artificial neural networks of improved reliability for industrial process supervision. In: Munack A, Schügerl K (eds) Proceedings of the 6th International Conference Computer Appl in Biotechnology. Elsevier, Oxford
6. Manikowski M (1996) Modellgestützte Überwachung und Steuerung des Gärprozesses in einer Bierbrauerei, Doctoral Dissertation, University of Hannover
7. Manikowski M, Havlik I, Simutis R, Lübbert A (1999) Model-supported estimation of the diacetyl concentration during a production-scale beer fermentation. J Inst Brew (submitted)
8. Daoud IS, Searle BA (1990) On-line monitoring of brewery fermentation by measurement of CO2 evolution rate. J Inst Brew 96:297–302
9. Austin GD, Watson RWJ, Nordstrom PA, DÁmore T (1994) An on-line capacitance biomass monitor and its correlation with viable biomass. MBAA Techn Quarterly 31:85–89
10. Schubert J, Simutis R, Dors M, Havlik I, Lübbert A (1994) Bioprocess optimization and control: Application of hybrid modelling. J Biotechn 35:51–68
11. Landaud S, Lieben P, Picque D (1998) Quantitative analysis of diacetyl, pentanedione, and their precursors during beer fermentation by an accurate GC/MS method. J Inst Brew 104:93–99
12. Nakatani K, Takahasho T, Nagami K, Kumada J (1984) Kinetic study of vicinal diketones in brewing (I): Formation of total vicinal diketones. MBAA Technical Quarterly 21:73–78
13. Nakatani K, Takahasho T, Nagami K, Kumada J (1984) Kinetic study of vicinal diketones in brewing (II): Theoretical aspect for the formation of total vicinal diketones. MBAA Technical Quarterly 21:175–183
14. Masschelein CA (1997) A realistic view on the role of research in the brewing industry today. J Inst Brew 103:103–113

15. Dellweg H (1985) Diacetyl und Bierreifung, Monatsschr Brauwiss 6:262–266
16. Inoue T (1977) Forecasting the vicinal diketone of finished beer. Rept Res Lab Kirin Brewery 20:19–24
17. Rice JF, Helbert JR (1973) The kinetics of diacetyl formation and assimilation during fermentation. Proc Annual Meeting ASBC, pp 11–17
18. Yamauchi Y, Okamoto T, Murayama H, Kajino K, Amikura T, Hiratsu H, Nagara A, Kamiya T, Inoue T (1995) Rapid maturation of beer using an immobilized yeast bioreactor. 1. Heat conversion of α-acetolactate. J Biotechnol 38:101–108
19. Garcia AI, Garcia LA, Diaz,M (1994) Modeling of diacetyl during beer fermentation. J Inst Brew 100:179–183
20. Pandiella SS, García LA, Díaz M, Daoud IS (1995) Monitoring the production of carbon dioxide during beer fermentation. MBAA Technical Quarterly 32:126–131
21. Doelle HW, Kirk L, Crittenden R, Toh H (1993) Zymomonas mobilis – science and industrial applications. Crit Rev Biotechnol 13:57–98
22. Kempe E, Schallenberger W (1983)Measuring and control of fermentation processes, part I. Proc Biochem 18:7–12
23. DeClerk J (1965) Lehrbuch der Brauerei, Bd II Analysenmethoden und Betriebskontrolle. Versuchs und Lehranstalt für Brauerei, Berlin
24. Gmehling J, Onken U, Arlt W (1977) Vapor-liquid equilibrium data collection, vol 1, pt 1a, pp 116–155
25. Werbos P (1990) Backpropagation through time: what it does and how to do it? Proc IEEE 78:1550–1560
26. Jazwinski AH (1970) Stochastic process and filtering theory. Academic Press, New York
27. Simutis R, Havlik I, Lübbert A (1992) A fuzzy-supported extended Kalman filter: a new approach to state estimation and prediction exemplified by alcohol formation in beer brewing. J Biotechnol 24:211–234
28. Stephanopoulos G, Park S (1992) Bioreactor state estimation. In: K Schügerl (ed) Biotechnology, measuring, modelling, and control, vol 4. VCH, Weinheim, pp 225–249
29. Shioya S (1992) Optimization and control in fed-batch bioreactors. Adv Biochem Eng Biotechnol 46:111–142
30. Schuch C (1996) Temperaturverteilung in zylindrokonischen Tanks, Brauwelt 13:594–597
31. Lapin A, Lübbert A (1994) Dynamic simulation of the gas-liquid flows in bubble columns, demonstrated at the example of industrial-scale beer fermenters; paper presented at the 13th International Symposium on Chemical Reaction Engineering, Baltimore
32. Lapin A, Lübbert A (1999) Fluid dynamics in industrial-scale cylindroconical beer fermenters (to be published)
33. Gvazdaitis G, Beil S, Kreibaum U, Simutis R, Havlik I, Dors M, Schneider F, Lübbert A (1994) Temperature control in fermenters: application of neural nets and feedback control in breweries. J Inst Brew UK, 100:99–104
34. Asahi K, Kinoshita M (1992) Asahi's new high-tech and energy efficient "Ibaraki" brewery, Technical Quarterly of the Master Brewer's Association of the Americas 29:48–52
35. O'Shea JA (1990) Ice-banks at Guinness Dublin. The Brewer 76:478–480

11 Lactic Acid Production

John Villadsen

11.1
Introduction

Lactic acid bacteria (LAB) comprise a diverse group of microorganisms with many common traits, but also with a great variation in phylogenetic characteristics. They are all Gram positive bacteria, and they all produce lactic acid as part of their metabolism, although different genera – even closely related LAB species – have different yields Y_{sp} of lactic acid (p=HLac) on sugar substrate (s), and widely different rates of HLac production from the same sugar. One particular species of LAB may even change its metabolic product pattern quite dramatically when the environment changes from one sugar concentration to another.

LAB are very well adapted to life in a nutrient which is rich in fermentable sugars and also rich in nitrogen-containing compounds which can be taken up directly or after proteolysis and thereafter used as building blocks for biosynthesis. The auxotrophy of LAB, i.e., the lack of machinery for de novo synthesis of a majority of the essential amino acids, is an evolutionary draw-back, and in the nutrient rich media in which they are forced to survive a great many other organisms would also thrive.

The production of HLac is the main defense action of LAB, and their ability to acidify the surroundings to a pH of 4 or below where few other microorganisms survive will give LAB a competitive advantage which outweighs their poor biosynthetic capability. LAB lack the ability to synthesize the heme group and are consequently unable to gain energy by respiration. *Escherichia coli* has a fully operational oxidative phosphorylation system with a P/O ratio of 2 and is able to grow with a maximum specific growth rate of $1.8-2\,h^{-1}$. Still LAB grow remarkably well, with specific growth rates in the range $0.7-1\,h^{-1}$ on preferred sugars. This is explained by their astounding capacity for energy production by substrate level phosphorylation of the sugar. Not only is the uptake of the preferred sugars high, but the glycolytic machinery is fast and well modulated to handle different environmental situations.

It is the ability to acidify the environment that has, since the dawn of history, made LAB such valuable fermenting agents for mankind, inhibiting the growth of many spoilage and pathogenic bacteria. We use LAB to make a wide variety of fermented dairy products from milk, to protect and to develop desirable sensory qualities of meat products such as salami and sausages, and to produce pickled vegetables and fruit (olives), soy sauce, sour dough breads, and silage. Detailed accounts of the use of LAB in the food industry are available in a large number of text books and review articles of which [1–3] are recent examples.

The present study is, however, focused on the production of lactic acid as a bulk chemical rather than on the use of lactic acid fermentation in the development of specialized food products. Here the available literature is much more sparse since – at present – the potential of lactic acid production from agricultural waste or from byproducts in the food industry has not yet been realized. There are some reports from academia – some are to be cited in Sects. 11.6 and 11.7 but almost no data from industrial scale production.

Still, the successful design of a large-scale, fermentation-based process for lactic acid is critically dependent on the choice of the right LAB species, the right nutrients, and the right operating conditions. Consequently considerable attention must be paid to these items in order to minimize both the capital costs and the cost of nutrients.

The following Sects. 11.2–11.5 will therefore give a brief outline of the genomic organization of LAB, the energy production from sugar – leading to synthesis of HLac, and the additional nutritional requirements of the bacteria. Finally the attempts to derive expressions which relate the rate of HLac production to the concentrations of key substrates will be reviewed.

11.2
Classification of Lactic Bacteria

The majority of organisms used in industry are members of the genera *Lactococcus*, *Lactobacillus*, *Streptococcus*, and *Leuconostoc*. The genera *Pediococcus* and *Carnobacterium* also supply important LAB species with specialized use in the meat industry and for the production of bacteriocins. *Lactococcus lactis* has been the most closely analyzed LAB species in the upsurge of molecular biology since the early 1980s, and the chromosome and the plasmid content of this group of cocci is well described, i.e., for the common laboratory strain *L. lactis* MG1363 and other genetic relatives of *L. lactis* subsp. *cremoris*. The plasmids which serve a wide variety of metabolic purposes [4] are normal components of *L. lactis* strains – many plasmids are conjectured to be recent additions to the genome of *L. lactis* which enables these bacteria to grow well on milk.

The genus *Lactobacillus* (rod shaped bacteria in contrast to the cocci) contains a number of important species (*Lb. delbrückii* (e.g., subsp. *bulgaricus*), *Lb. helveticus*, *Lb. plantarum* and *Lb. casei*) which are used to acidify milk. In general lactobacilli are more acid tolerant than other LAB [5] and a number of interesting strains can be chosen from this genus of LAB as potential candidates for bulk lactic acid production. The genetics of lactobacilli and their plasmids is described in a number of recent reviews of which [6] appears to be particularly complete.

11.3
Sugar Metabolism of LAB

As mentioned in the Introduction, the most noteworthy feature of LAB metabolism is the enormous capacity for degradation of carbohydrates, thereby producing ATP for biosynthetic purposes. The predominant metabolic product is lactic acid and the yield Y_{sp} of lactic acid on the sugar can be as high as 95–97%.

Most sugars are taken up by lactic acid bacteria either by a PEP (phosphoenol pyruvate) dependent phosphotransferase system (PTS) or via a permease. In the first case the sugar enters the cell in an already energized form, having been phosphorylated at the expense of a PEP molecule during the transmembrane transport process. In the latter case a kinase mediated phosphorylation takes place in the cytoplasm.

11.3.1
An Example Showing the Functioning of PTS Systems

The best described PTS system is the plasmid encoded, constitutively expressed mannose (man) -PTS which exists in a number of LAB, including *Lactococcus lactis* subsp. *cremoris* FD1 which was studied by Benthin et al. [7–10]. The following experimental results are compiled from the work of Benthin et al.

The man-PTS has a much higher affinity for glucose than for mannose or in general for any other of the sugars or sugar analogues that can be translocated via this transmembrane protein. In fact it has received its somewhat inappropriate name in order to distinguish it from a glucose (glc) -PTS which was first studied in *E. coli* [11].

Figure 11.1 shows the metabolism of a pulse of sugar ($0.87\,\mathrm{g\,l^{-1}}$ fructose+$0.87\,\mathrm{g\,l^{-1}}$ glucose, both in mutarotational equilibrium) added to a glucose limited chemostat fed with $7\,\mathrm{g\,l^{-1}}$ glucose+$10\,\mathrm{g\,l^{-1}}$ yeast extract, casein peptone (YECP) at D=$0.11\,\mathrm{h^{-1}}$. Before t=0 the glucose concentration in the chemostat was less than $5\,\mathrm{mg\,l^{-1}}$. At t>0 the flow to the chemostat was stopped and therefore the time profiles on Fig. 11.1 show batch degradation of the pulse. Throughout the experiment the only measurable metabolic product was lactic acid.

The figure contains much information concerning the functioning of the man-PTS:

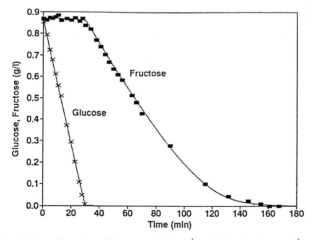

Fig. 11.1. Metabolism of a pulse of fructose ($0.87\,\mathrm{g\,l^{-1}}$) and glucose ($0.87\,\mathrm{g\,l^{-1}}$) added to a steady state glucose limited chemostat (D=$0.11\,\mathrm{h^{-1}}$) Feed: $7\,\mathrm{g\,l^{-1}}$ glucose+$10\,\mathrm{g\,l^{-1}}$ YECP before t=0. Feed was stopped at t=0. Microorganism *L. lactis* subsp. *cremoris* FD1. Simulation of the fructose profile by Eq. (11.1)

- A rapid glucose metabolism immediately starts when the glucose concentration increases above $10-20\,\mathrm{mg\,l^{-1}}$. Before $t=0$ the rate of glucose conversion is $7\times0.11=0.77\,\mathrm{g\,l^{-1}\,h^{-1}}$. For $t>0$ the rate is more than doubled.
- Assuming a constant biomass concentration of $1.4\,\mathrm{g\,l^{-1}}$ during the pulse, glucose is metabolized with an almost constant specific rate $r_s=1.24\,\mathrm{h^{-1}}$ during the pulse. The corresponding specific lactic acid production is an impressive $1.24\,\mathrm{gHLac\,g\,biomass^{-1}\,h^{-1}}$.
- Fructose must be taken up by the same transport system since no fructose is consumed before the glucose is gone. Thereafter fructose is consumed according to

$$r_{fru} = \frac{0.65\,(h^{-1})\,c_{fru}}{c_{fru} + 160\,(mg\,L^{-1})} \quad g\ fructose\ g\ biomass^{-1}\,h^{-1} \tag{11.1}$$

- The corresponding saturation constant for glucose is of the order of $1\,\mathrm{mg\,l^{-1}}$. The lines on Fig. 11.1.show simulated results corresponding to the two kinetic expressions.
- If the fructose pulse is given without the corresponding quota of glucose the fructose consumption after $t=30\,\mathrm{min}$ on Fig. 11.1 is exactly followed. This confirms that the two sugars pass through the same transport system – which is blocked for transport of fructose as long as there is a measurable glucose concentration in the medium.

In Fig. 11.2 the consumption of a fructose pulse of $0.5\,\mathrm{g\,l^{-1}}$ added to a fructose limited chemostat at $D=0.10\,\mathrm{h^{-1}}$ (feed: $7\,\mathrm{g\,l^{-1}}$ fructose, $10\,\mathrm{g\,l^{-1}}$ YECP) is shown. For a biomass concentration of $0.4\,\mathrm{g\,l^{-1}}$ one obtains approximately the same very high specific sugar consumption rate of $\sim1.4\,\mathrm{h^{-1}}$ (and saturation constant of less than $3\,\mathrm{mg\,l^{-1}}$) as obtained for glucose in Fig. 11.1. Again the only metabolic product found is lactic acid. Clearly a high capacity fructose -PTS system has been induced after prolonged fermentation of fructose without glucose present in the medium.

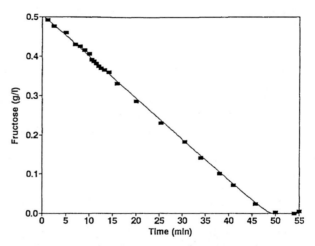

Fig. 11.2. Metabolism of a pulse of fructose ($0.50\,\mathrm{g\,l^{-1}}$) added to a steady state fructose limited chemostat ($D=0.10\,\mathrm{h^{-1}}$). Feed: $7\,\mathrm{g\,l^{-1}}$ fructose, $10\,\mathrm{g\,l^{-1}}$ YECP before $t=0$. Feed was stopped at $t=0$. Microorganism as in Fig. 11.1

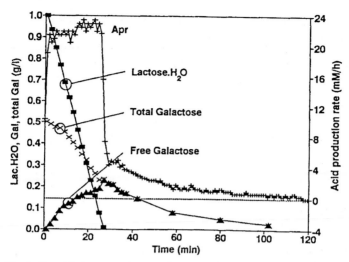

Fig. 11.3. Metabolism of a lactose pulse $(1\,g\,l^{-1})$ added to a steady state glucose limited chemostat $(D=0.11\,h^{-1})$. Feed and microorganism: as in Fig. 11.2. Apr=volumetric HLac production rate in mmol $l^{-1}\,h^{-1}$

From Figs. 11.1 and 11.2 combined we see that
- A LAB (*L. lactis* subsp. *cremoris* FD1) can have a constitutively expressed PTS with a preference order for a range of different sugars. A specific PTS with much higher affinity for a particular sugar can be induced.
- If the same pulse of fructose is added to a chemostat operating at the much higher dilution rate $D=0.46\,h^{-1}$ and with the feed of Fig. 11.2 the pulse is consumed much faster – with a constant specific rate of $2.9\,h^{-1}$. This shows that the catabolic machinery – in this case the quantity of or the activity of the fructose transporter and of the glycolysis pathway enzymes – is much more efficient when the organism has been grown at high specific growth rate $\mu=D$.

L. lactis subsp. *cremoris* also has a constitutively expressed, plasmid encoded PTS for lactose uptake. Figure 11.3 shows the metabolism of a $1\,g\,l^{-1}$ lactose pulse added to a steady state glucose limited chemostat at $D=0.11\,h^{-1}$ (feed $7\,g\,l^{-1}$ glucose, $10\,g\,l^{-1}$ YECP). The consumption of lactose starts immediately, showing that the transport protein is not repressed by glucose nor inhibited by the low (less than $5\,mg\,l^{-1}$) glucose concentration present before t=0.

The rate of lactose consumption is constant and high – in 27 min the lactose pulse is consumed. The production of lactic acid (Apr) $(20-24\,mmol\,l^{-1}\,h^{-1})$ does not quite correspond to the rate of lactose consumption $(4\times6.17=24.7\,mmol\,l^{-1}\,h^{-1})$. The reason is that a considerable fraction of the galactose moiety of the lactose is excreted to the medium and is only taken up, and very slowly metabolized long after the glucose moiety has been metabolized. The rapid drop in acid production rate followed by a long tail closely mirrors the slow drain from the medium of almost half of the original galactose formed. This is the phenomenon of a "secondary acidification" which is highly undesirable in cheese manufacture.

Physiologically the phenomenon is explained by the following uptake mechanism [10, 12, 13]:

$$(Lac)ex + (PEP)in \rightarrow (Lac - 6P)in + (PYR)in \tag{11.2}$$

$$(Lac - 6P)in \rightarrow (Gal - 6P)in + (Glc)in \tag{11.3}$$

$$(Gal - 6P)in \leftrightarrow (Gal)ex + P^* \tag{11.4}$$

$$(Gal - 6P)in \rightarrow\rightarrow (Tag - 1, 6P) \leftrightarrow (GAP)in \tag{11.5}$$

$$(GAP)in \leftrightarrow (Fru - 1, 6P)in \leftrightarrow (Fru - 6P) \tag{11.6}$$

$$(GAP)in \rightarrow HLac + byproducts + 2\,ATP \tag{11.7}$$

The glucose moiety formed in Eq. (11.3) is rapidly phosphorylated by glucokinase and enters the Glc-6P, Fru-6P pool to be used for HLac synthesis and for synthesis of biomass precursors in the Pentose Phosphate pathway. Gal-6P is metabolised in the Tagatose-6 phosphate pathway of Eq. (11.5). The resulting glyceraldehyde-3 phosphate (GAP) can either be used as a gluconeogenic substrate, entering the Fru-6P, Glc-6P pool via fructose 1,6 diphosphate, Eq. (11.6), or it can be metabolized to lactic acid. Apparently the enzymes of the Tagatose-6 phosphate pathway are inefficient in *L. lactis* subsp. *cremoris* FD1, and Gal-6P accumulates to unacceptably high levels and is secreted out of the cell, donating the activated phosphate group to glucose.

The inefficiency of the Tagatose-6 P pathway is also known for other starter cultures, especially for thermophilic LAB [14, 15], while the inability to dephosphorylate Fru 1,6 P to Fru-6 P in Eq. (11.6) (due to lack of an FDP-ase) is a specific trait of the FD1 strain used in [10].

11.3.2
Sugar Uptake by LAB in General

The experimental results shown for *L. lactis* subsp. *cremoris* FD1 can certainly not be generalized to all LAB – not even to strains which are closely related to it. But the description in some detail of the sugar uptake by one particular strain does show the great variation in uptake pattern for different sugars.

Most industrial starter cultures (at least from the *L. lactis* genus) have plasmid encoded constitutive PTS systems for glucose, lactose, mannose, and sometimes also for sucrose and galactose, although these PTS systems are likely to be inducible and severely glucose repressed. This is what makes the starter cultures such efficient catalysts for converting sugar to lactic acid.

However, an equally common way to take up lactose in LAB is by means of a permease followed by cleavage of the non-phosphorylated (Lac)$_{in}$ by β-galactosidase (β-gal as opposed to the enzyme β-gal P used to cleave Lac-6P in Eq. 11.3). Glucose and galactose can also be transported to the cell via permeases. Glucose is phosphorylated to Glc-6P, while galactose is phosphorylated to Gal-1P which is readily converted to Glc-6P in the Leloir pathway (in contrast to Gal-6 P which is formed when galactose enters via a specific PTS system or as a result of cleavage of Lac-6 P).

Some strains have both transport systems which then serve as a high affinity (the PTS) and a low affinity (permease) system. Other strains have only one transport system for a given sugar.

Thus *L. lactis* subsp. *cremoris* FD1 studied in [7–10] has no lactose permease (no activity of β-gal is detectable) and no galactose permease, but has PTS systems for both sugars. Consequently it cannot grow on galactose since it lacks an FDP-ase, but if a small amount of lactose is fed together with galactose, sufficient precursors for biosynthesis are formed from the glucose moiety of lactose to support growth at the same specific rate as if fed by lactose alone – an indication that energy generation is more critical than precursor synthesis.

In contrast *L. lactis* subsp. *cremoris* MG 1363, a laboratory strain used for research purposes by a large number of research groups, lacks a lactose PTS and therefore it grows much more slowly on lactose than the related strain FD1.

The two transport systems for fructose found in *L. lactis* subsp. *cremoris* FD1 gives rise to an interesting phenomenon [9]. The mannose-PTS has very low affinity for fructose, but translocates fructose to Fru-6P. The dedicated fructose-PTS yields Fru-1P which can be phosphorylated to Fru 1,6 diphosphate, and thereafter it enters glycolysis, but the absence of an FDP-ase makes Fru-1P useless for synthesis of biomass precursors. The flow through the fructose transporter furthermore disturbs the functioning of the mannose PTS. The combined action of the two transporters is [9] that the biomass concentration in a steady state chemostat increases with increasing dilution rate, a very atypical feature. It also gives rise to instabilities, and apparently stable oscillations with periods of 5–20 h are observed in both HLac production rate and in biomass concentration.

11.3.3
Homolactic vs Heterolactic Fermentation

LAB from the *Leuconostoc* genus and some *Lactobacilli* metabolise Glc-6P through a variant of the Pentose phosphate pathway. At the level of Xylulose-5 phosphate the enzyme phosphoketolase splits the pentose into a C3 (glyceraldehyde-3P) and a C2 (acetylphosphate) compound. The end result is generation of 1 mol each of HLac and ethanol, together with 1 ATP from one mol of glucose. If these LAB have an NADH oxidase activity they may under aerobic conditions produce acetic acid (HAc) instead of ethanol. In either case these strictly heterofermentative LAB, which own their special fermentation pattern to the lack of an FDP-aldolase to split Fru-1,6 P into two C_3 compounds, will be of no interest in the production of lactic acid since half the carbon is lost to CO_2 and to a C_2 compound.

A vast majority of LAB including Lactococci and most Lactobacilli ferment the sugars through the EMP pathway. These are the homofermentative LAB, and they generate 2 ATP and a maximum of 2 HLac per metabolized hexose molecule.

The main features of the EMP pathway from glucose to pyruvate, and further to HLac and the most common byproducts, are shown in Fig. 11.4.

The linear pathway from glucose or any of the other fermentable sugars to pyruvate is common for all the homolactic LAB and is traversed in the same fashion for low and for high sugar flux. For a high sugar flux (and this corresponds to sugar levels higher than about 20–50 mg l^{-1} in the exterior medium) the phosphoenol-pyruvate (PEP) pool is very small while the fructose 1,6 diphosphate (FDP) pool is

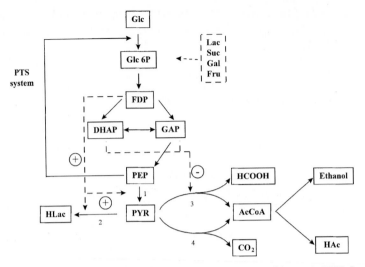

Fig. 11.4. Schematic of sugar metabolism in LAB: 1: PK=Pyruvate kinase, 2: LDH=Lactate dehydrogenase, 3: PFL=Pyruvate formate lyase, 4: PDH=Pyruvate dehydrogenase

large. The pool of triosephosphates (GAP and DHAP) is also large, but at all conditions much smaller than the FDP pool.

At low sugar flux through the pathway the PEP pool increases in size while the FDP pool, and in particular the pool of triosephosphates, diminishes. The change between "high" and "low" flux can occur quite abruptly, e.g., somewhere between $20 \, mg \, l^{-1}$ and $5 \, mg \, l^{-1}$ glucose in the medium. For sugars which are taken up by permeases rather than by the highly efficient PTS systems the condition of "low" flux is, however, obtained at much higher levels of sugar in the exterior medium.

The metabolism of pyruvate is highly dependent on whether the sugar flux through the EMP is "high" or "low." At "high" flux only lactic acid is formed; at "low" flux a considerable amount of formic acid (anaerobic conditions) or CO_2 (aerobic conditions) is formed in one of the two branches leading from pyruvate to Acetyl coenzyme A (AcCoA). When the PDH catalysed pathway is inactive (under anaerobic conditions) and the sugar flux is low, perhaps 50–60% of the sugar is converted to formic acid, ethanol, and acetic acid in the ratio 2:1:1:

$$C_6H_{12}O_6 \; \rightarrow \; 2 \, HCOOH \; + \; CH_3COOH \; + \; C_2H_5OH \; - \; H_2O \qquad (11.8)$$

When the flux through the EMP pathway increases there is an immediate relocation of fluxes at the pyruvate branchpoint. This is clearly seen in Fig. 11.5 which shows the development of the concentration of lactic acid and formic acid after the dilution rate in an anaerobic chemostat fed with $7 \, g \, l^{-1}$ glucose, $10 \, g \, l^{-1}$ YECP has been changed from $0.075 \, h^{-1}$ to $0.409 \, h^{-1}$. The strain is *L. lactis* subsp. *cremoris* FD1. At the low dilution rate the glucose concentration (measured by FIA [16]) is between $1 \, mg \, l^{-1}$ and $5 \, mg \, l^{-1}$, after the transient the steady state glucose concentration at D=$0.409 \, h^{-1}$ is $8 \, mg \, l^{-1}$. The remarkable change in product formation pattern is of course not directly a result of this small change in exterior glucose concentration, but rather of the considerable change in glucose flux through the EMP pathway which results when the dilution rate is increased.

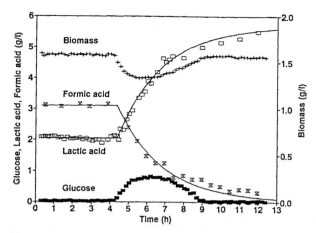

Fig. 11.5. Profiles of metabolic products after a change of dilution rate (from D=0.075 h^{-1} to D=0.409 h^{-1}). Feed and microorganism as in Fig. 11.1

Before the change of dilution rate, 2.1 g HLac l^{-1} and 3.2 g HCOOH l^{-1} is produced from the feed. Since the production of formic acid as measured by its concentration is 20% higher than what would be calculated by Eq. (11.8) from the glucose feed, 7 g l^{-1} corrected for the produced lactic acid some part of the YECP must also be catabolized, but this is not important in the present context. The key observation in Fig. 11.5 is that the formic acid production immediately stops when the flux through the EMP pathway increases, and that production of HLac immediately increases. The two lines are simulations of this situation. The increasing flux cannot be seen directly, but at the end of the transient more than five times as much glucose is metabolized by approximately the same amount of biomass (1.6 g l^{-1}) and the hump of glucose extending for about 5 h and with more than 500 mg l^{-1} glucose shows that the fermentation is in no way glucose limited until the excess glucose has been metabolized.

Figure 11.6 gives additional information related to the shift in metabolism. The specific growth rate μ is of course equal to D both before the transient and after the transient. In between, μ is first seen to be smaller than D=0.409 h^{-1} but for almost 3 h μ is larger than 0.409 h^{-1}. The first part of the μ (t) curve is explained by a higher catabolism, i.e., a larger generation of ATP than corresponding to the ATP need for anabolism. The overshoot of μ is an effect of the large enzymatic, including biosynthetic, machinery produced during the long exposure of the culture to a rich glucose medium.

Of particular interest is the jump in r_p, the specific rate of lactic acid production occurring as soon as the flux increases. This is another sign that the machinery for HLac production, i.e., the high activity of the LDH is immediately available as soon as the glucose flux through the EMP pathway increases.

Another interesting feature of Fig. 11.6 is the slow increase of the catabolic machinery which gives rise to an almost doubling of r_p over 5 h. Also shown in the figure.is the RNA content of the cell as measured by a very precise method [17]. The increase of r_p faithfully follows the increase of RNA, i.e., of cell activity in general as the cells become acclimatized to the high sugar concentration during the

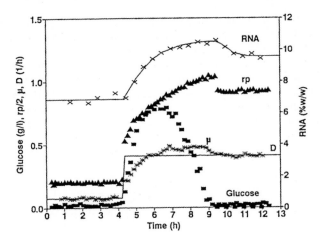

Fig. 11.6. RNA, glucose profile and specific rates of production of HLac (r_p) and biomass (μ) after the shift in dilution rate as Fig. 11.5

transient. When the sugar concentration again drops below 20–30 mg l^{-1} r_p abruptly decreases and the RNA level slowly settles on the value corresponding to D=0.409 h^{-1}. Again it is remarkable that r_p is three times as high at the end of the transient than at its start – and this in spite of the seemingly small difference in extracellular glucose level from 1–5 mg l^{-1} to 8 mg l^{-1}.

For a long time the dramatic change in product formation pattern was ascribed to the allosteric regulations shown in Fig. 11.4.

In vitro experiments demonstrated that FDP acts as an allosteric activator of LDH, and consequently it was argued that lactic acid production was enhanced by the high FDP level which was found at high glucose flux through the EMP pathway. Recently, however, it has been shown that the FDP level can have only a minor effect on the product formation pattern, since with the in vivo concentrations found by more accurate assays [18] the FDP concentration is probably always high enough to saturate the LDH.

It is more likely that the inhibition of PFL activity by the pool of triose phosphates GAP and DHAP is responsible for the mixed acid fermentation at low sugar flux through the EMP pathway since the level of this metabolite pool is shown to decrease appreciably under these conditions. With the focus on GAP/DHAP rather than on FDP, a dominating influence of the NADH/NAD$^+$ ratio has recently been postulated [19, 20] and strongly supported by measurements of the ratio [20] at different growth conditions. First, the activity of LDH is very sensitive to the NADH/NAD$^+$ ratio, and by in vitro experiments the enzyme is found to be totally inhibited when the ratio has a value lower than 0.03. Also an almost linear relation was found between the value of NADH/NAD$^+$ and the specific sugar (glucose, galactose, or lactose) consumption rate. For r_s=19 mmol g^{-1} h^{-1} the ratio was 0.08 and the LDH was fully activated. Second, a low value of NADH/NAD$^+$ will speed up the oxidation of GAP since this reaction has NAD$^+$ as cofactor. This will lead to a small pool of triosephosphates and the negative allosteric regulation of PFL is alleviated.

The details of the regulatory system which determines whether almost 100% of the sugar is fermented to HLac or whether a substantial portion is directed towards by-products may not yet have been settled, but as outlined above the main mechanisms are known. Qualitatively speaking this knowledge may be used to plan an industrial production of lactic acid by fermentation.

It is desirable to have a high sugar flux through the central metabolism of the LAB. Transport of sugars to the cell by PTS systems gives a higher flux than transport via permease systems, and one might consider importing a lac-PTS to strains (mostly of plantal origin) which normally lack this transport system, typical of starter cultures from the dairy industry. One may also consider an overexpression of genes encoding for the enzymes in the EMP pathway and for LDH. This is less likely to give a noticeable advantage due to the very tight regulatory control of the enzymes in the central metabolism.

In any case very low sugar concentrations in the medium must be avoided since this will necessarily lead to activation of the PFL or the PDH enzymes – a very useful option for the organism which will gain an extra ATP (i.e., a 50% increase in the ATP produced by metabolism of sugar) in the energy limited situation at low specific growth rate, but hardly an advantage for the production of lactic acid. In this context it should be remembered that a "low" sugar concentration in the medium is $20\,\mathrm{mg\,l^{-1}}$ or lower for sugars transported by a PTS. These transport systems are incredibly efficient, and they will furthermore spring into action as soon as the sugar concentration in the environment increases – at low flux through the glycolysis the PEP level is high due to inactivation of PK, and the uptake of sugar through the PTS can start immediately.

11.4
Nitrogen Uptake and Metabolism

In contrast to yeast and to *E. coli*, LAB lack the ability to synthesize most amino acids from ammonia or another inorganic N-source. The requirements for amino acids varies between species and it is difficult to predict which amino acids must be added. Thus some strains of *L. lactis* subsp. *lactis* are prototrophic for most amino acids whereas most *L. lactis* subsp. *cremoris* strains have requirements for about 15 amino acids, and almost all strains grow slowly on a chemically defined medium compared to growth on a complex, ill defined nitrogen source like YECP from which the proteolytic systems of the LAB can secure an ample supply of amino acids.

Even when an apparently rich N-source has been used, the LAB may experience a deficit in one or more amino acids during a batch fermentation. Incorrectly interpreted this may easily lead to the erroneous conclusion that the culture is sugar (energy) limited.

Figure 11.7 shows the progress of a batch fermentation with initially $18\,\mathrm{g\,l^{-1}}$ glucose and $20\,\mathrm{g\,l^{-1}}$ YECP inoculated with $1\,\mathrm{mg\,l^{-1}}$ *L. lactis* subsp. *cremoris* FD1. At first the culture grows with a specific growth rate of $\mu=0.82\,\mathrm{h^{-1}}$, but after 8 h and at a biomass concentration of $0.3\,\mathrm{g\,l^{-1}}$ the growth rate abruptly decreases, probably due to depletion of some component in the N-source.

In Fig. 11.8 the change in specific growth rate is seen to be closely correlated with a change in the specific acid production rate r_p. No trace of byproducts can be detected in either growth period and the lactic acid produced corresponds closely to the glucose metabolized (Fig. 11.7, accounting for the volume change due to addition of $2\,\mathrm{mol\,l^{-1}}$ NaOH to keep pH constant at 6.2). It appears that after a while both μ

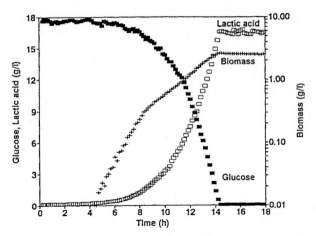

Fig. 11.7. Batch fermentation with a rich N-source ($10\,g\,l^{-1}$ glucose and $20\,g\,l^{-1}$ YECP) *L. lactis* subsp. *cremoris* FD1

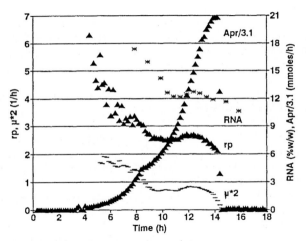

Fig. 11.8. Specific (r_p) and volumetric (Apr) production rate of HLac during the batch of Fig. 11.7. Also shown: specific growth rate μ and RNA profile

and r_p pick up again and μ reaches an approximately constant value of $0.4\,h^{-1}$ until the sugar is depleted and both r_p and μ abruptly drop to zero.

In Fig. 11.9 r_p is plotted as a function of μ. The two specific production rates are clearly linearly dependent throughout the batch fermentation:

$$r_p = 1.75 + 2.24\,\mu \quad (g\ g\,biomass^{-1}h^{-1}) \tag{11.9}$$

or

$$r_p = k_{ATP} + Y_{xATP} \cdot \mu = 19 + 25\,\mu \quad (m\,mole\ g\,biomass^{-1}h^{-1}) \tag{11.10}$$

Since in homolactic fermentation the rate of lactic acid production corresponds exactly to the rate of ATP generation (in $mmol\,h^{-1}$) the relation at Eq. (11.10) also

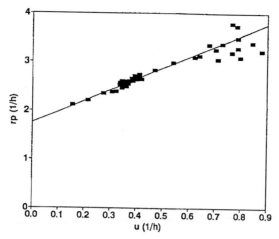

Fig. 11.9. Cross-plot of μ and r_p from Fig. 11.8

expresses the specific rate of ATP formation needed to support growth. In repeated fermentations (results not shown) with the same feed, somewhat lower values of the parameters of Eq. (11.10) are found. An average for three fermentations gives $Y_{xATP}=17$ mmol g biomass^{-1} and $k_{ATP}=21$ mmol g^{-1} h^{-1} with an estimated RSD of 6% and 11% respectively.

An ATP consumption of 17 mmol g biomass^{-1} is far below the theoretical value 37–40 mmol g biomass^{-1} for biosynthesis of bacteria [21] from a minimal medium.

Considering that LAB have a low protein content (45 wt %) compared to that of many other bacteria (60 wt %), and that some of the building blocks can be taken up directly from the yeast extract-peptone medium, one may calculate a smaller lower bound of 26 mmol g^{-1} for the theoretical ATP requirement Y_{xATP}, but this is still much smaller than 17 mmol g^{-1}.

The reason for the ATP requirement being apparently below the theoretical limit must be that the cell gains energy concomitant with the uptake of N-containing compounds from the medium.

In Figs. 11.10 and 11.11 this assumption is verified experimentally using less N-source than in the previous batch experiment. In Figs. 11.10 and 11.11 the feed composition is 20 g l^{-1} glucose and 7 g l^{-1} YECP.

For $\mu>0.3$ h^{-1} the energetic parameters $Y_{xATP}=15$ mmol g^{-1} and $k_{ATP}=18$ mmol g^{-1} h^{-1} correspond closely to the parameters determined from Fig. 11.9. For $\mu<0.3$ h^{-1}, Y_{ATP} is 50 mmol g^{-1} and the non-growth related maintenance consumption k_{ATP} is only 7 mmol g h^{-1}.

In Fig. 11.11 the net consumption of primary amino groups (AN) is shown together with the total consumed N and the specific growth rate – all as functions of the biomass concentration during the batch fermentation.

The net rate of AN consumption is the difference between the AN groups consumed for synthetic purposes (or metabolized) and the AN groups formed by hydrolysis of peptides.

In the first part of the fermentation until μ first reaches the value 0.3 h^{-1} the production and consumption of AN balance. There is a constant but small net consump-

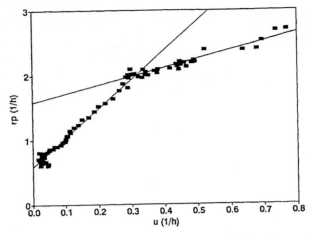

Fig. 11.10. Cross-plot of μ and r_p for a batch fermentation with less N-source than in Fig. 11.7: $20\,\mathrm{g\,l^{-1}}$ glucose and $7\,\mathrm{g\,l^{-1}}$ YECP at t=0

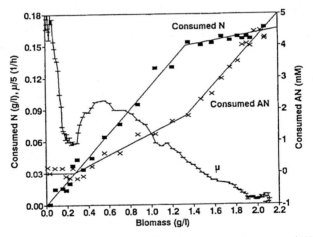

Fig. 11.11. Consumed total nitrogen (N, $\mathrm{g\,l^{-1}}$) and primary amino groups (AN, $\mathrm{mmol\,l^{-1}}$) during the batch fermentation of Fig. 11.10. A cross-plot of μ and the biomass concentration is also shown

tion of AN during the next part of the fermentation where μ increases to $0.45\,\mathrm{h^{-1}}$ and again decreases to $0.3\,\mathrm{h^{-1}}$. Throughout the first two periods there is a constant yield coefficient Y_{xN} of $0.113\,\mathrm{g\,g\,biomass^{-1}}$. In the last part of the fermentation $Y_{xN}=0.018\,\mathrm{g\,g\,biomass^{-1}}$ is so small that normal growth with cell division is severely inhibited. Rather the cells grow in mass, but with decreasing N content. The high rate of AN consumption together with the small rate of total N consumption is best explained by a considerable cell lysis which takes place at the same time as cells with a different composition are formed at an extra expense in ATP derived by glycolysis of glucose to lactic acid.

In [22] the growth energetics of *Lactococcus cremoris* FD1 is analyzed for conditions of energy-, carbon-, and N-limitation respectively. Amongst other hypotheses tested, it appears that import of C_3-C_4 peptides from the YECP medium and excretion to the medium of some of the amino acids released by hydrolysis of the peptides in the cytoplasm during the first phase of a batch cultivation is the explanation for the very low measured Y_{xATP}. Import of a peptide costs only 1 ATP while export of each of the amino acids from the hydrolyzed peptide gives rise to the production of 1 ATP.

The unknown amounts of trace compounds of possible stimulating effect and the general variability of any complex nitrogen source such as the yeast extract, casein peptone medium used in [22] makes it attractive to study the nitrogen metabolism with a feed which contains known quantities of each of the 20 "essential" amino acids. The composition of one particular "defined" medium is given in [23]. When the concentration of each of the added amino acids is followed during a batch fermentation or at different steady states in a chemostat, a very detailed picture of the uptake and metabolism of amino acids by a given LAB strain is obtained. Thus in a study of *L. lactis* subsp. *cremoris* MG 1363 it was found [24] that the organism degraded arginine almost quantitatively to ornithine, gaining ATP in the process. If the arginine present in the feed was lower (but still much higher than required for biosynthesis) the organism became heterofermentative. An increasing serine concentration (a non-essential amino acid for *L. lactis*) led to a drastic change in product distribution. With $11.6\,\mathrm{mmol\,l^{-1}}$ serine in the feed the catabolism of sugar changed from strictly homofermentative (with no serine in the feed) to almost entirely heterofermentative with $Y_{sHLac}=0.2\,\mathrm{g\,g^{-1}}$, the remainder being the 2:1:1 mixture of by-products shown in Eq. (11.8). Many of the phenomena observed up to now have not been adequately explained by reference to well defined metabolic reactions, but obviously the study of LAB fermentation with a defined medium can give much insight into the formation of a desirable flavor spectrum of the final product.

It must, however, be realized that no defined medium is likely to contain all the "growth factors," "stimulating species," etc., which in general make LAB grow better on a complex medium than on a defined medium.

11.5
Growth Kinetics and Product Formation Kinetics

The purpose of setting up a kinetic model for any chemical reaction is to predict the selectivity for different products and to find the rate of consumption of the main reactant at reaction conditions outside the experimental range used to construct the kinetic model. This type of research has proved highly successful in the petrochemical and related industries, and the rate expressions derived for typical gas phase catalytic reactions can often be used at much higher pressures and far from the temperature range of the original experiments.

Similar successes are not to be expected in fermentation studies. In contrast to the simple reaction scheme which describes, e.g., the formation of NH_3 from N_2 and H_2 with only one final product and a limited number of plausible mechanistic models for the surface reactions on the metal oxides used as catalysts, the process of converting sugar and other substrates to biomass and metabolic products is far too complex to be described by a single overall kinetic model.

For all of the industrially important microbial systems – yeast, *E. coli*, LAB, *Bacillus subtilis*, *Penicillium chrysogenum*, etc.– many attempts have been made to set up kinetic models which could be used in the design of fermentors, as the basis for process control of fermentation processes or to gain insight into the metabolism of the microorganism. Kinetic models may be of some help in a crude fermentor design, and if they are sufficiently equipped with results from different fermentation conditions they can be used in a computer run expert system for control purposes. Where they fail is in scientific studies of metabolic behavior.

The preceding sections have been illustrated with a number of experimental results of relevance for lactic acid fermentations. When considered together these results should make it abundantly clear why any overall kinetic model for LAB fermentation must fail if used to extrapolate beyond the range of the original experiments. This pessimistic forecast is not a peculiar feature of LAB which in comparison with other microorganisms have a relatively simple metabolism. The prediction of the behavior of eukaryotic microorganisms – e.g., in an attempt to calculate by a general model the production of proteins from yeast or mammalian cells at conditions somewhat different from those of the data basis used to construct the model – would probably fail even more miserably.

An overall kinetic model used to predict the specific rate of lactic acid production r_{HLac} (g g biomass^{-1} h^{-1}) or volumetric production rate q_{HLac} (g l medium^{-1} h^{-1}) would fail for the following reasons:

– Different species of LAB have different substrate requirements and different metabolic machinery. Therefore the influence of environmental parameters may be very different, both concerning the metabolic rates and, in particular concerning the product formation pattern.

– Experiments made with the same strain of LAB give different results depending on the time scale of the experiment. Chemostat experiments such as those shown in Fig. 11.6 can be used to construct relations between specific growth rate μ, the specific acid production rate r_p or the RNA content, and the concentration of the limiting substrate (glucose) in *the steady state*. These rate expressions cannot be used to predict what happens during a transient. The immediate jump in $r_p = r_{HLac}$ in Fig. 11.6 followed by a slow increase over 5 h are features which are not even mentioned in the steady state model, nor is the abrupt shift (Fig. 11.5) in the rate of formic acid production from 3.2×0.075 g l^{-1} h^{-1} to zero as soon as the dilution rate is changed to D=0.409 h^{-1}.

– Also in experiments with the same strain the influence of one sugar on the uptake of another sugar, or the combined effect of uptake of one sugar through several membrane transport systems (Figs. 11.1–11.3) is as yet difficult to predict. The same holds for the influence of the N-source on the batch kinetics, Fig. 11.7. The specific growth rate decreases by a factor of 2 at a biomass concentration which will depend not only on the ratio of carbon to nitrogen source, but also on the composition of the nitrogen-containing substrate. This composition is very dependent on the origin of this substrate (yeast extract, casein peptone, hydrolysed whey protein, etc.). Hence the yield coefficients Y_{sx} (g biomass g sugar^{-1}) and Y_{sp} (g HLac g sugar^{-1}) as well as the non-growth related maintenance requirement k_{ATP} are heavily dependent on the nitrogen source as seen in Figs. 11.9 and 11.10. They are also dependent on the sugar: in the experiment of Fig. 11.2 Y_{sx} is practically zero since the strain cannot grow on fructose admitted through the fru-PTS, whereas Y_{sx} can be high if a little glucose is used in the feed and the transport of both sugars occurs through the man-PTS.

The value of kinetic models as data fitters for a large amount of experimental data should, however, not be underrated. In this sense the mathematical model can certainly be used for design purposes when the desired operating conditions are within the parameter space spanned by the experimental data. Each detail of the physiological map of the organism can of course also be modeled quite faithfully, and mechanistic features of the metabolism can be discovered by mathematical modeling of carefully designed experiments. Thus the fructose uptake in Fig. 11.1 is very well described by Eq. (11.1) as seen from the simulation (full line on figure), the decrease of formic acid concentration on Fig. 11.5 is adequately modeled by the assumption that the formic acid production immediately stops after the shift in D, and the stepwise uptake of the N-source is understood by the simulations in Fig. 11.11. An intuitively appealing explanation for the existence of a double stranded man-PTS in contrast to the single trans-membrane strand observed for other PTS is the result of a study [7, 8] of the transport of α-d- and β-d-glucose to *L. lactis* subsp. *cremoris* FD1.

An overall model for growth and product formation should be simple and should account only for the most basic features of the growth process. In this case it can express in a compact form our empirical knowledge of the process without unwarranted claims of exactness.

Practically all LAB will in some sense be inhibited when the lactate concentration increases. In non-pH-controlled cultures an inhibiting effect of the acidification of the medium during the process is also observed.

Consequently a model of the form

$$q_x = \mu x = \frac{\mu_{\max} s}{s + K_s} \left(1 - \frac{p}{p_{\max}} \right) x \qquad (11.11)$$

for the volumetric rate q_x of biomass production has some value. s is the concentration of the limiting substrate and p is the concentration of HLac. At $p > p_{\max} \sim 50\,\mathrm{g\,l^{-1}}$ the fermentation stops and the cells lyse.

The trouble with Eq. (11.11) is that the growth limiting substrate may be different in different phases of the fermentation. Consequently the value of $\mu_{\max}$ may change during a batch fermentation as seen in Fig. 11.7. If the limiting substrate is assumed to be the sugar it is evident from the results both of pulse experiments and of steady state experiments that K_s is extremely low, of the order of $20\,\mathrm{mg\,l^{-1}}$ or less, and the first factor of Eq. (11.11) is independent of s throughout the batch experiment. The difficulty of assigning a reasonably constant value to $\mu_{\max}$ for growth on different sugars in N-rich medium has also been touched upon: if galactose or fructose is not able to enter he Fru-6P/Glc-6P pool after uptake $\mu_{\max}$ will be practically zero for these sugars due to a failure to produce biomass precursors.

The value of $p_{\max}$ varies considerably for different species of LAB which in itself makes Eq. (11.11) difficult to use in general. Even the concept of a reduced specific growth rate at higher lactic acid concentration has been questioned. Thus in [25] it is concluded that the inhibition of HLac on growth of *Lb. delbrükii* subsp. *bulgaricus* at constant pH is primarily seen as an extremely prolonged lag phase increasing from 2 h to 39 h when the concentration of HLac added at the beginning of the batch increases from $0\,\mathrm{g\,l^{-1}}$ to $25\,\mathrm{g\,l^{-1}}$ and a reduced maximum cell yield. There is virtually no effect of added HLac on the specific growth rate once the batch fermentation takes off. The effect of l-lactic acid is also much more pronounced than the effect of d-lactic acid, the isomer naturally produced by the organism. Clearly none of these effects are even suggested by the rate equation (Eq. 11.11).

In a classical study of lactic fermentation by *Lb. delbrükii*, Luedeking and Piret [26] modeled the product formation kinetics as the sum of a growth associated and a non-growth associated (maintenance) term

$$r_p = Y_{xp}\mu + m_p \qquad\qquad\qquad\qquad (11.12)$$

For any fermentation where product formation is a result of the catabolic metabolism leading to ATP formation it is expected that the product formation kinetics has the structure indicated in Eq. (11.12). In Fig. 11.9 and Eq. (11.10) we have seen that the linear relation between r_p and μ holds true throughout a batch fermentation with excess nitrogen source. Unfortunately the two parameters of Eq. (11.12) are heavily dependent on the fermentation conditions. In the original paper on batch fermentations at constant pH Luedeking and Piret [26] had already noted that Y_{xp} increases by a factor of 2 in the narrow pH range from 5.2 to 4.5 whereas the yield coefficient was constant between pH 6 and 5.4. In the same pH range the maintenance term decreases by a factor of 2.5 – one may assume that the microorganism down-regulates its ATP expenditure in futile cycles etc. when at low pH the growth becomes more energy demanding.

The results in Fig. 11.10 for a somewhat N-limited batch fermentation show that the nitrogen source also has a significant influence on the parameters of Eq. (11.12), and the great variability of the two parameters when using different sugars or mixtures of sugars [22] gives further evidence to prove that Eq. (11.12) can only be used with extreme caution for predictive purposes.

In recent years a number of structured kinetic models have been proposed. The aim of these models has been to predict on the basis of steady state experiments certain features such as the length of the lag phase in a batch fermentation, the course of a wash-out experiment, and the diauxie observed in a mixed sugar fermentation. The concept of these models is that the biomass is divided into an "active" part X_A and a "structural part" X_G which is synthesized by the "active" part and at the expense of X_A. The "limiting" substrate is consumed in one or in both reactions. It is evident that the structure of these models where the major part of the biomass X_G is synthesized from a small, active portion of the cell X_A in a consecutive fashion will make it possible to predict a lag phase. Also the slow increase of RNA coupled to a parallel increase of r_p (Fig. 11.6) and a delayed increase in the rate of total biomass production (Fig. 11.5) lend much credibility to simple structured models.

In the context of lactic acid fermentation the structured models of Nielsen et al. [27] are based on a very large database of steady state continuous fermentations. They are therefore quite robust in their predictions of batch fermentations and transients with time constants of several hours. Processes with time constants in the range of minutes, e.g., the alleviation of enzyme inhibition by metabolites in the catabolic pathways when the dilution rate is changed are not accounted for in the models of [27]. Hence the jumps in r_p in Fig. 11.6 are not envisaged.

The structure of the models in [27] is as follows.

Lactic acid formation by catabolism:

$$sugar(s) \rightarrow HLac; \quad r_1 = k_1 \frac{s}{s + K_1} X_A \qquad (11.13)$$

Formation of active compartment X_A : $\quad Y_{X_A s} s + Y_{X_A N} s_N \rightarrow X_A$

Formation of structural compartment X_G : $\quad Y_{X_G X_A} X_A \rightarrow X_G \qquad (11.14)$

One may assume that the formation of macromolecules is rate limiting in both reactions of Eq. (11.14) and that the kinetics of the two reactions is similar:

$$r_{X_A} \text{ or } r_{X_G} = k f_1(s) f_2(s_N) X_A \qquad (11.15)$$

where the rate constant as well as the parameters of f_1 and f_2 may be different for the two reactions whereas the autocatalytic nature of all three reactions in Eq. (11.13) and Eq. (11.14) is acknowledged through the factor X_A, the fraction of the total biomass ($X_A + X_G = 1$) which is "active." In [27] simple saturation type expressions were used for both f_1 and f_2 in Eq. (11.15). The two parameters of Eq. (11.13) and the six parameters of Eq. (11.15) can be determined from experiments with either sugar or nitrogen limitations.

It is not expected that a further compartmentalization of the cell will give much more generality to the model – except of course if the production of a particular enzyme is highlighted. Also the structured models of [27] must be conceived as purely empirical although more biochemical knowledge is included in order to explain phenomena which are foreign to the unstructured models of Eqs. (11.11) and (11.12). These unstructured models are incapable of explaining a change in the cell activity and consequently in the capacity for producing more biomass and more product when the environment, e.g., the substrate concentrations s or s_N changes.

Very recently a family of models [28] which are capable of predicting both short term and long term changes in cell metabolism have been advocated. Such changes result after an abrupt change in the environment of a chemostat, e.g., following addition of a substrate pulse or a step change in the dilution rate. The experimental basis for the models is built on anaerobic and aerobic yeast fermentations, but the concepts of the models can certainly be applied to explain the events following a step change in D (Figs. 11.5 and 11.6) in a continuous lactic acid fermentation.

The almost dormant cells at the low dilution rate before the transient are assumed to possess a hidden capacity for energy production in the catabolic pathways. As soon as the sugar concentration in the medium increases this capacity is activated and r_p jumps to a higher value. Biomass synthesis is considered to start at the rate corresponding to the low dilution rate, and more biosynthetic machinery (e.g., RNA, PTS proteins, proteins in the glycolysis, and finally proteins in the anabolic pathways) is produced by first order processes starting from the original level. All the features of Figs. 11.5 and 11.6 are qualitatively explained by this oral model: the peak in glucose concentration which is consumed once the active machinery (RNA) has increased with the corresponding overproduction of biomass which compensates for the initial wash out of biomass (Fig. 11.5). The accumulation of PEP at the low dilution rate could be the "hidden" capacity of the catabolic pathways which is mobilized as soon as the PTS system for glucose senses an increasing glucose concentration in the medium.

11.6
Lactic Acid Production on the Industrial Scale

Lactic acid is a commodity chemical with many different applications. The current global production is approximately 60,000 tons year^{-1} in different qualities, e.g., food grade quality 88% by weight with a sale price of 1.9–2.4 US\$ kg^{-1} (fob works), [29]. The market price has been sliding since 1990 (2.5–2.75 US\$ kg^{-1}), probably related to an increasing production volume, especially by the fermentation route.

Besides being an inhibitor of bacterial spoilage of food, lactic acid is used as an acidulant, flavoring, pH buffering agent in soft drinks, jams and jellies, candy and bakery products. Specialized dairy products and pickled vegetables such as sauerkraut and olives rely on lactic acid both for taste and for preservation. Approximately 50% of food-related lactic acid applications is found in conditioners and emulsifiers, particularly in the bakery. The emulsifying agents are esters of lactic acid with long-chain fatty acids.

Other, but in terms of volume much smaller, outlets for lactic acid are in drugs (e.g., as the antiacne preparation ethyl lactate), in cosmetics, and as biodegradable polymers for controlled release of drugs and in surgical sutures.

An exciting new and potentially enormous outlet for lactic acid has been created by the recent development of technologies for commercial production of transparent polymers for food packaging purposes. The properties of lactic copolymers approach those of petroleum derived polymers such as polystyrene and flexible PVC. The polylactide polymers have a good shelf life, degrading slowly by hydrolysis, but they are easily compostable when deposited in dumps after use. Other recent applications of lactic acid is in food safe, biodegradable esters with low molecular weight alcohols. These can be used as solvents and plasticizers. Together with the biodegradable and (from a carcinogenic point of view) absolutely safe polymer films for food packaging, lactic acid derivatives are almost certain to win a solid position as regulations and consumer preferences rapidly change our attitude towards conventional chemical products.

It is therefore not strange that many patents appear which relate to either production processes of lactic acid (of a purity and d-, l-composition suitable for polymerization purposes) or relate to new applications of lactic acid in the food industry and in production of bulk chemicals.

Chemically based production routes to lactic acid from acetaldehyde via lactonitrile or from propane via nitrolactic acid [30] are bound to become uncompetitive as larger supplies of otherwise almost useless agricultural byproducts such as corn steep liquor, a byproduct of the corn wet-milling industry, and with a lactic acid content of 20–25 wt %, become available. Another route to lactic acid is of course the fermentation of carbohydrate sources, e.g., molasses, corn syrup, or whey.

Consider production of lactic acid from whey as an example. A large, modern dairy (*Golden Cheese* in Coronado, California) has access to milk from 60–70,000 head of cattle. From its raw material, 13–1500 tons of milk per day, the factory produces approximately 100–115 tons of cheese, 1–1.5 tons of whey protein, and 1000–1100 tons of water with 45 g l^{-1} lactose together with a number of minerals. It is easy to calculate that if this lactose was dedicated to lactic acid production the factory could produce 15–18,000 tons per year of lactic acid, thus doubling the production of lactic acid in the US. The waste water with 4.5% lactose is of little value –

and it could even be considered as a liability, incurring considerable costs for biodegradation. The lactic acid could have a sales value of more than 1.5 US\$ kg^{-1} in food grade quality. It could immediately be used as a substitute for citric acid in beverages and in other food products. The mild taste of lactic acid would be preferred by many customers and the huge citric acid market (550,000 tons per year in 1990) could easily be scavenged, although the price of citric acid is "soft" (~1.9 US\$ kg^{-1} [29]) due to competition from new producers in, e.g., China. In the longer term the lactic acid might find applications for polymer production or elsewhere in the bulk chemical industry.

Considering the importance of having a cheap raw material available in sufficient quantities, the dominating role of Cargill Inc. in development of large scale integrated lactic acid fermentation processes is easily understood. This Mid-West based company has access to raw materials from the corn-belt of the US and is involved in dairies throughout the continent. The role of Du Pont and its associated companies in polymer production is also obvious. Thus both Cargill and Du Pont have a number of patents, e.g., [31] for production of high purity lactide polymers and both companies are developing their particular processes to a commercial scale, intending to manufacture biodegradable polymers [32].

The lactic acid market is, however, not an easy one in which to operate. In late 1994 the Ecochem (Ecological Chemical Company) of Wisconsin, USA dropped out of lactic acid production after 3 years of operation as a Du Pont-Con Agra partnership [33]. The reason for the failure was stated to be that production of lactic acid for the commodity market was not successful due to production difficulties, while the parallel research in pilot scale to produce polylactides was claimed to be successful. Both the Ecochem plant and a recent entrant in the market, Chronopol Inc. of Colorado, US [34] have produced lactic acid by purification of agricultural waste (corn steep liquor). The fermentation route from whey is possibly a safer one; a closing remark in [33] concerning the possible re-entry of Ecochem in the market might suggest this to be the case.

11.7
Process Technology in Lactic Acid Fermentation

Considering that the raw material for a lactic acid fermentation is always going to have a very modest price, the design of an industrial plant is critically dependent on:
- A low capital cost for the total plant, including the down-stream operation
- Low energy costs, both in the fermentation and in the down-stream processes

The low cost of the substrate does not imply that the yield of lactic acid on sugar is of little importance. On the contrary, an almost complete conversion of the sugar to lactate, and to lactate alone, is crucial for the cost of the down-stream processes, and consequently for the total process economics. The investment costs, both in the fermentation unit and in the down-stream operations, will be small if the cell density in the fermentor is high. This means that the inhibitory effect of the product on the process kinetics must be minimized.

Further design considerations are centered on the cost of the non-sugar substrates. The LAB need a rich nitrogen source to grow, and yeast-extract and other commercially available nitrogen sources may well constitute a substantial portion of the production costs.

Finally one needs to consider the alternatives of batch and continuous fermentation. Several patented lactic acid production processes are built around a batch fermentation process, but neither batch nor fed batch operation can be the preferred design solution when the product is as cheap as lactic acid and has to be produced in such large quantities as indicated by the potential scale of the *Golden Cheese* example (1000 tons per day of feed). The LAB are genetically quite stable organisms and the desired product is formed as part of the central metabolism which is not going to change over time. Hence the usual argument for batch production of chemicals or proteins from the secondary metabolic pathways will not hold for lactic acid fermentation. Finally, the process control, e.g., separation of lactic acid from the broth is much simpler in a continuous fermentation than in a batch where the level of the sugar substrate decreases and the level of the product increases during the fermentation.

Different strains of LAB have been used by different research groups working with pilot plant lactic fermentation. *Lb. helveticus* was used in [35, 36], *Lb. casei* by [37–41], *Lb. delbrückii* by [42, 43], and *Lb. plantarum* by [44].

Industry has picked one or several of the above-mentioned organisms (unpublished information). It is interesting that they are all chosen from the genera of *Lactobacillus*. Perhaps this is due to their higher acid tolerance, but it appears that several fast acid producers from the *L. lactis* genera might offer a good alternative to the *Lb.* strains since in any case the lactic acid has to be removed continuously in all industrial productions with high cell density.

Although any type of cheap sugar source can be used it appears that all academic research groups and (as far as can be conjectured) all industrial producers prefer to use whey permeate – the $45 \, \mathrm{g} \, l^{-1}$ lactose aqueous solution which results when the whey proteins have been filtered from the tail-end of the cheese production lines. Yeast extract, which can of course be used in academic research, is too expensive as a nitrogen source, and hydrolyzed whey protein is also recommended in academic research reports focused on the economics of the lactic acid process [36, 45]. When the fermentation process is located at the site of a large dairy the use of whey protein rather than any other nitrogen source must be the only rational choice. To make the nitrogen from whey protein more accessible, enzymatic degradation of the protein may be advisable as recommended in the Lactoscan process [46].

The two major considerations in the fermentation step are (1) to have a high biomass concentration and (2) to remove the lactic acid to avoid slowing down or stopping the fermentation.

Cell recirculation and continuous removal of lactic acid is included in most of the pilot-scale studies cited above. With recycling of cells one may in principle obtain an infinitely high capacity of a given reactor volume.

Let the cell density of the effluent from the plant be x_E which is lower than the biomass concentration x in the fermentor due to a separation step (filtration, cell sedimentation, hydrocyclone) between the reactor outlet and the plant-outlet. A recirculation stream of cell density x_R and flow rate Rv (R=recirculation rate and v net volumetric flow through the plant) is returned to the reactor inlet where it is mixed with fresh feed.

Define the net dilution rate D=v/V where V is the reactor volume.

Further define $f=x_E/x$ and the cell separation factor $\beta=x_R/x$.

A steady state continuous operation can now be maintained for any given value of D and at any given specific growth rate μ by selecting f such that $\mu = f \times D$. The corresponding recirculation rate R is given by

$$f = 1 - R(\beta - 1) \qquad (11.16)$$

Simultaneous removal of lactic acid and recirculation of biomass can be achieved in a membrane type reactor – ultrafiltration with a cut-off of 20 Da has been proposed. In a continuous LAB fermentation with a reasonably small dilution rate the residual sugar concentration is very low as seen in Sects. 11.2–11.4. Consequently the loss of sugar is negligible. If the lactic acid content of the retentate is considered to be a nuisance, one may run the retentate through an ion-exchanger to trap the acid before the retentate is returned to the fermentor. Occasional back flushing of the ion exchange column with NaOH adds more lactic acid to the permeate from the UF unit. Various types of ion exchange resins (e.g., Amberlite RA-400 from Rohm and Haas [41]) have been suggested.

Concentration of the lactate stream from the UF-unit is most economically done by electrodialysis [39, 45] which is easy to scale up and can be run at a low electric power requirement (1 kwh kg HLac^{-1} [30]). The process is continuous and lactate will pass through a (positively charged) anion electrodialysis membrane. It is advisable to remove calcium and magnesium ions from the solution by inserting a cation-chelating resin column before the electrodialysis unit. Ca and Mg salts of lactic acid are very sparsely soluble and they will invariably foul the electrodialysis membranes and decrease the current efficiency. After removal of divalent cations and electrodialysis the process stream is free of sugar and protein residues and the lactate is usually in the form of NH_4Lac. In a final separation step bipolar electrodialysis is recommended [46] to give a pure lactic acid which can easily be concentrated to 80 wt% lactic acid free of other organic acid residues by evaporation of water in a low pressure continuous distillation column.

Each of the down-stream purification process steps must be carefully considered and as far as possible optimized since otherwise a substantial loss of lactic acid and prohibitive energy costs may be incurred. The know-how of these process steps and the best integration of fermentor and down-stream operations (e.g., best cell concentration, optimal bleed of biomass from the reactor) is of course proprietary information in the companies engaged in commercialization of the process, but the major stages of both the fermentor design and of the design of the purification process are probably always those described above.

It is encouraging to follow the intense research effort both in academia and in industrial companies to prove the economic viability of a really large scale fermentation process with cheap raw materials and a product which, after suitable further process steps, can become quite valuable. Lactic acid fermentation is in some sense deeply rooted in the current efforts to develop "green" products from "sustainable" raw materials. The integration of all process steps is certainly required to make the product competitive with other cheap products like citric acid. But many of the secondary processes derived from the main production process – such as the production of electrical energy by anaerobic digestion of the process effluent in the Lactoscan process [46] – are indicative of the ecological awareness of entrepreneurs in the modern biotechnology based industry.

References

1. Salminen S, Wright A von (ed) (1998) Lactic acid bacteria, microbiology and functional aspects, 2nd edn. Marcel Dekker, New York
2. Roissart H de, Luquet FM (ed) (1994) Bactéries lactiques. Aspects fondamentaux et technologiques, vols 1 and 2. Lorica, Uriage (France)
3. Venema G, Huis in't Veld JHJ, Hugenholtz J (eds) (1996) Lactic acid bacteria, genetics, metabolism and applications. Kluwer Academic Publishers, Dordrecht Holland for FEMS-reprinted from Antonie van Leeuwenhoek (1996) 70:2–4
4. McKay LL (1983) Antonie van Leeuwenhoek 49:259
5. Kashket ER (1987) FEMS Microbiol Rev 46:233
6. Pouwels PH, Leer RJ (1993) Antonie van Leeuwenhoek 64:85
7. Benthin S, Nielsen J, Villadsen J (1992) Biotechnol Bioeng 40:137
8. Benthin S, Nielsen J, Villadsen J (1993) Biotechnol Bioeng 42:440
9. Benthin S, Nielsen J, Villadsen J (1993) Appl Environ Microbiol 59:3206
10. Benthin S, Nielsen J, Villadsen J (1994) Appl Environ Microbiol 60:1254
11. Postma PW, Lengeler JW, Jacobsen GR (1993) Microbiol Rev 57:543
12. Thompson J (1979) J Bacteriol 140:774
13. Thompson J (1987) FEMS Microbiol Rev 46:221
14. Hickey MW, Hillier AJ, Jago GR (1986) Appl Environ Microbiol 51:825
15. Hutkins RW, Morris HA (1987) J Food Prot 50:876
16. Benthin S, Nielsen J, Villadsen J (1991) Anal Chim Acta 247:45
17. Benthin S, Nielsen J, Villadsen J (1991) Biotechnol Tech 5:39
18. Jensen NBS, Jokumsen K, Villadsen J (1998) Biotechnol Bioeng (submitted)
19. Snoep JL, Teixeira MJ, Neijssel OM (1991) FEMS Microbiol Lett 81:63
20. Garrigues C, Loubière P, Lindley ND, Cocaign-Bousquet M (1997) J Bacteriol 179:5282
21. Nielsen J, Villadsen J (1994) Bioreaction engineering principles. Plenum Press, New York
22. Benthin S, Schulze U, Nielsen J, Villadsen J (1994) Chem Eng Sci 49:589
23. Jensen PR, Hammer K (1993) Appl Environ Microbiol 59:4363
24. Melchiorsen CR (1997) MSc thesis, Techn Univ, Denmark
25. Benthin S, Villadsen J (1995) J Appl Biotechnol 78:647
26. Luedeking R, Piret EL (1959) J Biochem Microbiol Technol Eng (Biotechnol Bioeng) 1:393
27. Nielsen J, Nikolajsen K, Villadsen J (1991) Biotechnol Bioeng 38:1, 11
28. Duboc P, Stockar U von, Villadsen J (1998) Biotechnol Bioeng 60:680–689
29. Chem Mktg Rep 251:(10) 7 March 1997
30. Kirk-Otmer (ed) (1995) Encyclopedia of chemical technology, vol 13 (4th edn). Wiley, New York
31. Gruber PR (1992) US. Pat 5,142,023 (Aug 25, 1992) (to Cargill Inc.); Bhatia KK (1989) US Pat 4,835,293 (May 30, 1989) (to EI du Pont de Nemours & Co. Inc.)+about 15 follow-up patents
32. Chem Eng News 6 (Aug 6, 1992)
33. Chem Mktg Rep 246: (22) 28 November 1994
34. Chem Mktg Rep 249:(15) April 8 1996
35. Äschlimann A, Stockar U von (1990) Appl Microbiol Biotechnol 32:398
36. Boyaval P, Goulet J (1988) Enzyme Microb Technol 10:725
37. Vaccari G, Gonzales-Vara RA, Campi AL, Dosi E, Brigidi P, Matteuzzi D (1993) Appl Microbiol Biotechnol 40:23
38. Hjörleifsdottir S, Seevaratnam S, Holst O, Mattiasson B (1990) Curr Microbiol 20:287
39. Krischke W, Schröder M, Trösch W (1991) Appl Microbiol Biotechnol 34:573
40. Tuli A, Sehthi RP, Khanna PK, Marwaha SS (1985) Enzyme Microb Technol 7:164
41. Senthuran A, Senthuran V, Mattiasson R, Kaul R (1997) Biotechnol Bioeng 55:841
42. Stenroos S-L, Linko Y-Y, Linko P (1982) Biotechnol Lett 4:159
43. Major NC, Bull AT (1989) Biotechnol Bioeng 34:592
44. Shamata TR, Sreekantiah KR (1988) J. Ind Microbiol 3:175
45. Tejayadi S, Cheryan M (1995) Appl Microbiol Biotechnol 43:242

46. Final report on lactic acid fermentation and purification (1998) (confidential report published by BIOSCAN A/S, Odense 5230 Denmark from whom further information may be obtained)

12 Control Strategies for High-Cell Density Cultivation of *Escherichia coli*

Clemens Posten, Ursula Rinas

12.1
Introduction

Growth of microorganisms under conditions of substrate saturation in batch processes is often accompanied by substrate inhibition, by-product formation, or enduring adaptation. These problems can be overcome, in principle, by fed-batch or continuous cultivation, where the concentration of one substrate (in general the carbon source) is kept at a limiting level and, furthermore, where the growth rate can be kept at a desired value. Production systems with *E. coli* in batch culture suffer mainly from the formation of acetate, which is produced in response to oxygen limitation or excess carbon. The prevention of the accumulation of toxic levels of acetic acid is the main task for the achievement of high cell *and* high product concentrations in the bioreactor. The formation of acetate can be circumvented when cell growth occurs at reduced growth rates, thus preventing carbon overflow metabolism. The critical growth rate above for which acetate formation occurs is influenced by the growth medium, temperature, and, more general, by the overall physiological status of the cells. In addition, it is important, to allow cell growth at a constant growth rate to permit the synthesis of a product (e.g., a recombinant protein) of reproducible and defined quality. In continuous cultivation, a constant and non-critical growth rate with respect to acetate formation can be achieved, but problems with plasmid stability and, therefore, decreasing productivity may occur. So in many research and industrial applications, fed-batch processes for the production of recombinant proteins are employed. With an accurate choice of the process and control parameters, very high cell concentrations in the bioreactor can be achieved. The following will give a concise mathematical description of the related microbial reactions, of the reactor dynamics, and of feasible control strategies, which finally can be called a basic model for high-cell density cultivation (HCDC).

12.2
Basic Modeling of a Fed-Batch Strategy

A desired improvement in process performance by means of feedforward or feedback control can only be achieved by employing a process model. Such a model should contain as much available knowledge as possible on the one hand but on the other should be easy to obtain, robust, and feasible under process conditions. Many process models in bioprocess engineering consist of a physiological sub-model representing the metabolic stoichiometry and kinetics and a reactor model to describe the relation between the process parameters, kinetic parameters, and material concentrations inside the reactor. Since the main goal of control strategies in HCDC is the estimation and control of the specific growth rate, the model has to relate this non-measurable physiological variable with measurable process variables, a task of physiological control.

12.2.1
The Physiological Model

For modeling and control of high-cell density cultivation (HCDC), a simple structured physiological model for a typical aerobic microorganism is sufficient. The simplified structure of the metabolic pathways of such a micoorganism, namely *Escherichia coli*, is given in Fig. 12.1.

With the assumption that the time constants of accumulation of intracellular metabolites are very small compared to reactor time constants, a set of linear equations for the process stoichiometry can be obtained by applying material balances and stoichiometric relations to the reaction network. According to network theory, a suitable choice of balance boundaries has to guarantee that:
- Each network knot lies inside at least one balance boundary
- Each pathway is cut by at least one balance boundary
- For each boundary with n pathways only n-1 linearly independant equations are possible

Details are given in the chapter Metabolic Flux Analysis.

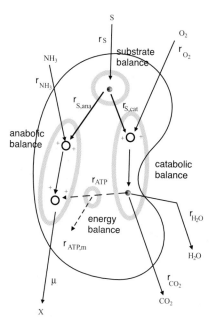

Fig. 12.1. Simplified metabolic structure of an aerobic microorganism for set-up of a partially structured process model; the *circles* represent the balance boundaries

For the indicated case, one obtains a linear set of equations

$$\mathbf{E} \cdot \mathbf{r} = 0 \tag{12.1}$$

with the vector of specific turnover rates

$$\mathbf{r} = \left[r_S, r_{NH3}, r_{O2}, r_{S,ana}, r_{S,cat}, \mu, r_{CO2}, r_{H2O}, r_{ATP}, r_{ATP,m} \right] \tag{12.2}$$

and the system matrix

$$\mathbf{E} = \begin{bmatrix} 1 & 0 & 0 & -1 & -1 & 0 & 0 & 0 & 0 & 0 \\ 0 & 1 & 0 & 1 & 0 & -1 & 0 & 0 & 0 & 0 \\ 0 & e_{N,NH_3} & 0 & 0 & 0 & -e_{N,X} & 0 & 0 & 0 & 0 \\ 0 & 0 & 1 & 0 & 1 & 0 & -1 & -1 & 0 & 0 \\ 0 & 0 & 1 & 0 & e_{O,S} & 0 & -e_{O,CO_2} & -e_{O,H2O} & 0 & 0 \\ 0 & 0 & 0 & 0 & e_{C,S} & 0 & -e_{C,CO2} & 0 & 0 & 0 \\ 0 & 0 & y_{ATP,O2} & 0 & 0 & y_{X,ATP} & 0 & 0 & 0 & -1 \end{bmatrix} \tag{12.3}$$

Here, the rows represent the mass balance of the substrate utilization, the anabolic mass balance, the anabolic nitrogen balance (assuming a constant elemental composition of the biomass), the catabolic mass balance, the catabolic oxygen balance, the catabolic carbon balance (the latter three items representing the substrate oxidation stoichiometry), and, finally, the energy (ATP) balance, which couples the anabolism and the catabolism, thus substituting the missing stoichiometry of the substrate utilization balance.

From these considerations, seven linear equations are obtained for the eight unknown metabolic fluxes. Since only the ratio between energy generation and energy consumption is required for the modeling purposes here, r_{ATP} is not calculated as an absolute value, while $r_{ATP,m}$ is taken as an unknown parameter. Finally, the system is completely described by an additional kinetic equation for the specific substrate uptake

$$r_S = r_{S,max} \cdot \frac{c_S}{k_S + c_S} \tag{12.4}$$

which is, from a modeling point of view, a link between the physiological and the reactor model. Oxygen and nitrogen limiting conditions are not considered, since they are usually avoided during HCDC. These conditions would lead to additional kinetic uptake equations for the respective medium components.

To describe the performance of an HCDC strategy, which is basically a fed-batch process strategy [1] leading to a constant specific growth rate μ, the kinetic Eq. (12.4) and a subset of Eq. (12.1), namely at least one yield equation for biomass production from nutrient uptake, giving, e.g.,

$$\mu = y_{X,S} \cdot r_S - \mu_m, \tag{12.5}$$

are sufficient. The parameters can be derived from Eq. (12.1) by linear combination of the mass balance of substrate utilization and the energy balance. However, in the representation according to Eq. (12.5), the parameters can be determined experimentally.

12.2.2
The Reactor Model

In case of a continuously stirred tank reactor (CSTR), no spatial distribution of concentrations have to be considered. The reactor equations can be obtained by concise application of the material balances for balance boundaries around the reactor.

The material balance for biomass, including accumulation and reaction term, gives

$$\frac{dm_X(t)}{dt} = \mu(t) \cdot m_X(t) \tag{12.6}$$

while the carbon substrate balance with an additional transport term for the carbon substrate pumped into the reactor can be written as

$$\frac{dm_S(t)}{dt} = f_{S,f}(t) \cdot c_{S,f} - r_S(t) \cdot m_X(t) \tag{12.7}$$

The corresponding differential equations for the concentrations can be derived from the total differentials

$$\frac{dm_X(t)}{dt} = \frac{d(c_X((t) \cdot V_R(t)))}{dt} = \frac{dc_X(t)}{dt} \cdot V(t) + \frac{dV_R(t)}{dt} \cdot c_X(t)$$
$$= \mu(t) \cdot m_X(t) \tag{12.7a}$$

and

$$\frac{dm_S(t)}{dt} = \frac{d(c_S(t) \cdot V_R(t))}{dt} = \frac{dc_S(t)}{dt} \cdot V(t) + \frac{dV_R(t)}{dt} \cdot c_S(t)$$
$$= f_S(t) \cdot c_{S,f} - r_S(t) \cdot m_X(t) \tag{12.8}$$

Relating to the reactor volume $V_R(t)$ and rearranging finally yields

$$\frac{dc_X(t)}{dt} = \mu(t) \cdot c_X(t) - \frac{f_{S,f}(t)}{V_R(t)} \cdot c_X(t) \tag{12.9}$$

and

$$\frac{dc_S(t)}{dt} = -r_S(t) \cdot c_X(t) + \frac{f_{S,f}(t)}{V_R(t)} \cdot (c_{S,f} - c_S(t)) \tag{12.10}$$

Since these equations include a dilution term $\frac{f_{S,f}(t)}{V_R(t)}$, the total biomass but not the biomass concentration will increase exponentially assuming constant μ.

The volume $V_R(t)$ of the fluid phase in the reactor itself can be regarded as

$$\frac{dV_R(t)}{dt} = f_{S,f}(t) \tag{12.11}$$

while differences of the density of the feed medium and the reactor broth are ignored as well as evaporation, sampling, and water production from the oxidation reactions of the microorganisms. A robust process strategy should not rely on a precise knowledge of the reactor volume on- or off-line. However, Eq. (12.11) can be used for simulations and to compare modeling results with measurements.

For the gas transfer into the reactor, separate balances for the fluid and for the gas phase are necessary. The oxygen balance in the fluid phase reads

$$\frac{dc_{O2}(t)}{dt} = OTR(t) - OUR(t) \qquad (12.12)$$

where the oxygen transfer rate $OTR(t)$ into the reactor follows a linear material transport law, namely Fick's first law related to the volumetric gas/fluid exchange interface a,

$$OTR(t) = k_L a \cdot (c_{O2}^*(t) - c_{O2}(t)) \qquad (12.13)$$

while the volumetric oxygen uptake rate $OUR(t)$ is given as

$$OUR(t) = r_{O2}(t) \cdot c_X(t). \qquad (12.14)$$

Quasi-steady-state conditions with respect to dissolved oxygen

$$\frac{dc_{O2}}{dt} = OTR(t) - OUR(t) \approx 0 \Rightarrow OTR(t) = OUR(t) \qquad (12.15)$$

can be assumed unless dynamics of pO_2-control or sudden substrate depletion are considered.

From the gas phase balance equations for oxygen, carbon dioxide and inert gases, mainly nitrogen, one obtains

$$x_{O2}^{out} = \frac{MW_{CO2} \cdot (OTR - Q_{G,mol}^{in} \cdot MW_{O2} \cdot x_{O2}^{in})}{OTR \cdot MW_{CO2} - MW_{O2} \cdot (Q_{G,mol}^{in} \cdot MW_{CO2} + CPR)} \qquad (12.16)$$

for the oxygen molar fraction in the off-gas, and

$$x_{CO2}^{out} = \frac{-MW_{O2} \cdot (Q_{G,mol}^{in} \cdot MW_{CO2} \cdot x_{CO2}^{in} + CPR)}{OTR \cdot MW_{CO2} - MW_{O2} \cdot (Q_{G,mol}^{in} \cdot MW_{CO2} + CPR)} \qquad (12.17)$$

for the carbon dioxide molar fraction in the off-gas. Details of the gas phase balance can be found in [2].

12.3
Growth Rate Control via Substrate Feeding

For maintaining a constant specific growth rate,

$$\mu(t) = \mu_{set} \qquad (12.18)$$

a limiting and, in case of constant k_S, a constant substrate concentration c_S is required.

Equation (12.6) can be solved directly yielding

$$m_X(t) = m_{X,0} \cdot e^{\mu_{set} \cdot t} \qquad (12.19)$$

for the biomass time profile that will be obtained.

Quasi-stationary conditions $dc_S/dt=0$ for the substrate concentration c_S can be derived from Eq. (12.10) by substitution of $r_S=(\mu+\mu_m)/y_{X,S}$ (Eq. 12.5) and rearranging to

$$\frac{dc_S(t)}{dt} = f_{S,f}(t) \cdot (c_{S,f} - c_S) - \frac{\mu_{set} + \mu_m}{y_{X,S}} \cdot m_{X,0} \cdot e^{\mu \cdot t} = 0 \qquad (12.20)$$

This is only achieved, as can be shown by comparison of the coefficients, by feeding with an exponentially increasing feeding rate

$$f_{S,f}(t) = \frac{(\mu_{set} + \mu_m) \cdot m_{X,0}}{y_{X,S} \cdot (c_{S,f} - c_S(t))} \cdot e^{\mu_{set} \cdot t} \qquad (12.21)$$

As usually $c_S \ll c_{S,f}$, the feeding rate is often referred to as

$$f_{S,f}(t) = \frac{(\mu_{set} + \mu_m) \cdot m_{X,0}}{y_{X,S} \cdot c_{S,f}} \cdot e^{\mu_{set} \cdot t} \qquad (12.22)$$

which can be interpreted as the feeding rate for the amount of substrate that the microorganisms use for maintaining the desired growth rate, while the neglected part (difference between Eqs. 12.21 and 12.22) is the amount that is necessary to keep the substrate concentration constant despite the increasing culture volume.

Simulation results for this basic strategy are given in Fig. 12.2. Values for process variables, model parameters, and starting values of the state variables are obtained from [3].

As can be seen from a comparison of the exponentially increasing feeding rate, the biomass concentration also increases in the first part of the fed-batch phase nearly exponentially as long as the additional feeding volume is low compared to the starting volume in the bioreactor. Later, the increase slows down to a theoretical limit of $c_{X,final}=c_{S,f} \times y_{X,S}$. This value can hardly be reached because of the limited volume of the reactor. A smaller starting volume would result in a higher final biomass concentration, but also causes problems such as longer cultivation time, low volumetric

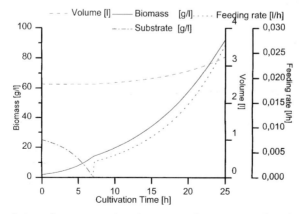

Fig. 12.2. Simulation of an HCDC using glucose as carbon source and, with the equations given above (Eqs. 12.1–12.5, 12.9–12.11, and 12.21), the parameters have been chosen from [3] as $c_{S,f}=795$ g/l, $V_{R,0}=2.5$ l, $c_{S,0,batch}=25$ g/l, $c_{X,0,batch}=0.05$ g/l, $\mu_{max}=0.45$ h^{-1}, $y_{X,S}=0.5$, $m=0.025$ g/(g×h)$\rightarrow\mu_m=0.0125$ h^{-1}, $\mu_{set}=0.12$ h^{-1} (0.14 h^{-1} in [3])

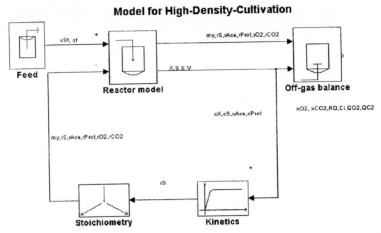

Fig. 12.3. Basic model structure of a fed-batch process, here in the graphical representation of SIMULINK, the program used for the described simulation studies

productivity, and the need for pumps working in several orders of magnitude. An optimization approach can be found in [4]. For production purposes, the initial volume is in most cases not smaller than one-third of the final volume, which means a maximum of three to four doubling times of the biomass after the start of the fed-batch phase of the HCDC. However, the feed concentration should be as high as possible and is often chosen to be at the limit of solubility. Therefore, even feed of crystalline substrate has been proposed in the literature [5].

After depletion of the initial carbon source present in the batch phase of the cultivation, the carbon substrate concentration remains of the order of magnitude of the limitation constant k_S of the *E. coli* cells as long as $\mu_{set} < \mu_{max}$. This self-stabilizing property of fed-batch processes without external feedback control can be understood by the process internal feed-back loop as shown in Fig. 12.3. An infinitesimal perturbation of the glucose concentration, e.g., an increase, will result in a higher glucose uptake rate and a higher specific growth rate. These reactions will then reduce the glucose concentration in the next time instant.

This previously described strategy of achieving a constant growth rate by exponential feeding depends strongly on the assumptions of absence of substrate accumulation and constant yield of biomass from substrate consumption $y_{X,S}$. Short distortions of feeding can be compensated by estimation of the actual biomass using the same yield equations as shown above. However, in practical applications, the substrate yield may not be known exactly or may change during the cultivation. Therefore, the question arises what the effect of an error in the assumed substrate yield with regard to the true specific growth rate can be. The fully relative sensitivity functions

$$S_{cx,yxs} = \frac{\partial\, c_X(t)}{\partial\, y_{X,S}} \cdot \frac{y_{X,S}^0}{c_X^0(t)} \tag{12.23}$$

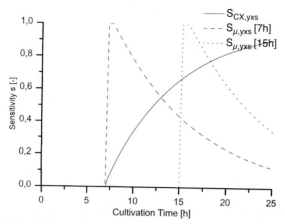

Fig. 12.4. Fully relative sensitivity functions of biomass and specific growth rate with respect to changes in the yield coefficient of the bacterial metabolism at different time instances; the nominal parameter values are the same as given in Fig. 12.2

and

$$s_{\mu,yxs} = \frac{\partial \mu(t)}{\partial y_{X,S}} \cdot \frac{y_{X,S}^0}{\mu_{set}} \tag{12.24}$$

can be evaluated. These functions give information about the time course of relative changes in biomass concentration and specific growth rate in case a relative error of the yield occurs which is not compensated in the feeding profile. Figure 12.4 shows simulation results where the sensitivity functions are computed via numerical differentiation. Here, a sudden change of $y_{X,S}$ at the beginning of the fed-batch-phase is assumed.

As expected, if the assumed yield is too low, lower biomass concentrations are obtained. The sensitivity function exhibits an asymptotic behavior with a limiting value 1. This means that an error in yield causes the same error in biomass.

A change in the yield parameter results in a corresponding change of the specific growth rate. Fortunately, this error is damped and vanishes slowly. This is due to the fact that, e.g., a higher yield results in a higher growth rate but also, after some time, in a higher biomass. The amount of substrate per cell fed to the reactor then falls slowly under the level calculated beforehand thus compensating the yield effect. Similar sensitivity functions are obtained for errors in the initial feeding rate, e.g., by an error in the measurement of the initial biomass. So the strategy of direct substrate feeding is to some extent robust to changes in yield or other parameters used for calculating the initial feeding rate (compare Eq. 12.21 for t=0).

12.4
Growth Rate Control via Oxygen Supply

As shown earlier, a constant growth rate is achieved only in case of constant substrate yield and in case of a preset growth rate lower than the critical growth rate which results in the formation of acetic acid. These parameters may also change during cultivation. So additional control strategies are advisable. From the simple linear model (Eq. 12.1) different control approaches can be deduced and have been proposed in the literature. However, all of them assume at least one constant yield coefficient, i.e., one constant element in the system matrix. Since the oxygen uptake is closely related to the energy metabolism of aerobic microorganisms and molecular oxygen is not incorporated into the biomass, some authors based their control strategy on the specific oxygen uptake rate, which is related to the specific growth rate by

$$\mu = y_{X,O2} \cdot r_{O2} - \mu_m \tag{12.25}$$

as shown above for the specific substrate uptake rate as an analogous case. Exponential growth with a constant specific growth rate is obtained by an exponential increase of the oxygen transfer rate via increasing stirrer speed, aeration rate, oxygen fraction, or pressure. This strategy, proposed and applied by Riesenberg et al. [6] dates back to Mori et al. [7] and Cutayar and Poillon [8]. It is up to this point analogous to the strategy using an exponential substrate feeding profile. However, there are two basic differences. First, an exact relation between the oxygen mass transfer and process parameters is hardly available. Second, since oxygen is not a limiting substrate, special attempts have to be made to force the cells to use the exponentially increasing oxygen supply. These two problems define the task of establishing two control loops as will be pointed out below.

For both control loops, it is necessary to estimate the oxygen uptake and the oxygen mass transfer. The oxygen transfer rate can be calculated from the off-gas analysis from application of balances such as

$$q_{O2}(t) = \frac{f_{G,in} \, MW_{O_2}}{V_{mol}} \cdot \left[\frac{x_{O2}^{in}(1 - x_{CO2}^{out}(t)) - x_{O2}^{out}(t)(1 - x_{CO2}^{in})}{1 - x_{O2}^{out}(t) - x_{CO2}^{out}(t)} \right] \tag{12.26}$$

which is a rearrangement of the Eqs. (12.16) and (12.17), here in a representation for on-line application without considering the reactor volume. Now $q_{O2}(t)$ is observable and controllable as long as there is a gradient in the oxygen partial pressure between fluid and gas phase, and can be chosen to serve for the demand of the *E. coli* cells as

$$q_{O2}(t) = r_{O2}(t) \cdot m_X(t) = \frac{\mu_{set} + \mu_m}{y_{X,O2}} \cdot m_X(t) \tag{12.27}$$

where for constant $\mu = \mu_{set}$ and an exponential increase of $m_X(t)$, an exponential increase in oxygen supply is required in analogy to the substrate feeding strategy described above. Using the approach via oxygen supply, potential overfeeding can easily be detected by a measurable increase of the dissolved oxygen concentration. In practical applications, disturbances may occur causing a remarkable deviation of the biomass profile from the precalculated value. It is also possible that a non-constant growth profile is desired. In such cases, an estimation of the actual biomass in the reactor has to be provided in order to calculate the oxygen demand according to Eq. (12.27). The biological model information for such an estimator of the biomass can be based on the assumption of constant yield coefficients. Employing again Eqs. (12.6) and (12.25) as well as $r_{O2}=q_{O2}(t)/m_X(t)$ (Eq. 12.14a), the model equation

$$\frac{d\hat{m}_X(t)}{dt} = y_{X,O2} \cdot q_{O2} - \mu_m \cdot \hat{m}_X(t) \tag{12.28}$$

can be used as an observer model. This differential equation can be either computed numerically on-line or can be solved (MAPLE V) yielding

$$\hat{m}_X(t) = y_{X,O2} \cdot e^{-\mu_m \cdot t} \int_0^t e^{\mu_m \cdot \tau} \cdot q_{O2}(\tau) \, d\tau + m_{X,0} \cdot e^{-\mu_m \cdot t} \tag{12.29}$$

where integration is only necessary during the process. Moreover, with the knowledge of the biomass, the specific growth rate can be estimated as

$$\hat{\mu} = y_{X,O2} \cdot \frac{q_{O2}}{\hat{m}_X(t)} - \mu_m \tag{12.30}$$

For negligible maintenance or constant growth rate, the integration of Eq. (12.29) can be simplified, yielding

$$\hat{\mu} = \frac{q_{O2}}{k1(y_{X,O2}, c_{X,0}) + \int_{t0}^t q_{O2}(\tau)d\tau} \tag{12.30a}$$

as given in the literature [2, 6]. So all prerequisites are given to supply the culture with an oxygen mass flow suitable for the current biomass and the chosen growth rate. Finally, the task of this first control loop is to control the estimated specific growth rate at the desired value μ_{set} via control of the oxygen supply rate q_{O2}. Therefore, according to Eq. (12.13), either k_La or c_{O2}^* can be influenced by the process parameters mentioned above as control variables.

Since oxygen is not the limiting substrate, a second control loop has to be established in order to feed enough substrate to allow the cells to utilize the offered oxygen for respiration. At first glance, the stoichiometry between glucose and oxygen consumption could be considered which would end up in a glucose feeding profile proportional to the given oxygen transfer profile. However, even small changes in the related yield coefficients could lead to substrate accumulation on one hand or loss of oxygen controllability on the other hand. At this point, another difference between the glucose and the oxygen feeding profile approach comes into action, namely the fact that, in contrast to the substrate concentration, the oxygen partial pressure in the reactor can be measured on-line. As long as $dpO_2/dt=0$, the oxygen transfer rate OTR equals the oxygen uptake rate OUR which is – via metabolic stoichiometry – a function of the substrate consumption rate. The basic idea of this second control loop is to control the oxygen partial pressure at a constant value by substrate feeding, which guarantees a substrate consumption suitable to reach finally OUR=OTR. A special point here is that no knowledge of $y_{X,S}$ or $y_{O2,S}$ is required.

Simulation results of this strategy are shown in Fig. 12.5 with the assumption of ideal controller performance. Values for process variables, model parameters, and starting values of the state variables are obtained from [6]. While the aeration rate was constant, a linear relation between the k_La-value and the stirrer speed is assumed. However, this is not essential for the process performance.

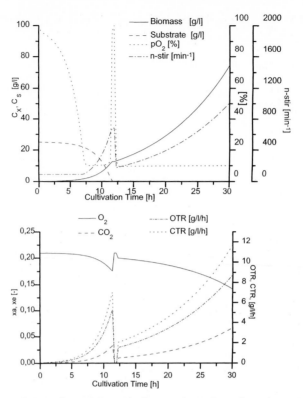

Fig. 12.5a,b. Simulation of an HCDC with the equations given above (Eqs. 12.1–12.17, 12.21, 12.26, and 12.28); the parameters have been chosen from [6] as $c_{S,f}$=700 g/l, $V_{R,0}$=37 l, $c_{S,0}$=25 g/l, $c_{X,0}$=0.1 g/l (0.07 g/l in sim.), μ_{max}=0.45 h^{-1},($r_{S,max}$=0.9 h^{-1}), $y_{X,S}$=0.5 g/g, m=0.0 g/(gh)→μ_m=0.0 h^{-1}, μ_{set}=0.11 h^{-1}, $f_{G,in}$=4440 l/h, $p_{O2,set}$=10%, k_La=f(n_{stir}) assumed: **a** measurable concentrations and process parameters in the fluid phase; **b** measurable concentrations in the off-gas and calculated volumetric mass transfer rates

Of course, the trajectories of glucose and biomass concentrations are comparable to the case of growth rate control via substrate feeding. The volumetric oxygen transfer coefficient k_La is adjusted at the beginning of the batch phase so that the partial oxygen pressure can drop to a small but not limiting value $pO_{2,set}$. Then, pO_2 is usually controlled by control of the oxygen transfer via stirrer speed. After depletion of the glucose, no more carbon source is available for respiration and, consequently, the oxygen uptake stops, while the continuing oxygen transfer leads to a sudden increase in pO_2. In some cases other peaks occur in response to glycerol and acetate depletion, which can be taken as a starting signal for glucose feeding [9]. In the further course of the cultivation, OTR and CTR increase exponentially as expected. When pO_2 is indeed constant, these quantities can be calculated from off-gas analysis.

Similar to the question of accuracy in the case of the substrate feeding strategy, the robustness of this strategy to inaccurate assumptions is investigated in the following paragraph. As already mentioned, the sensitivities to unknown substrate yield $y_{X,S}$ namely $s_{cx,yxs}$=$s_{\mu,yxs}$=0 vanish. This is indeed one of the main advantages of the oxy-

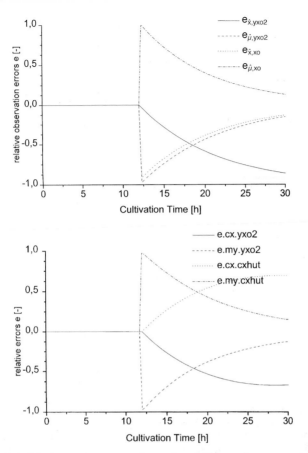

Fig. 12.6.a,b. Relative observation errors of growth and biomass in response to erroneous assumptions of biomass yield with respect to oxygen and starting values of the observer for biomass $\hat{m}_X(t=0)$ as a result of an erroneous measurement of $m_{X,0}$. **b** Relative process errors of growth and biomass in response to erroneous assumptions of biomass yield with respect to oxygen and starting values of the observer during application of the control strategy via oxygen supply

gen feeding strategy. However, the oxygen yield y_{X,O_2} may also change from one cultivation to another or during an experiment. In such a case, the observation of biomass and of the specific growth rate deviates from the real values. Simulations of this observation errors (Fig. 12.6) reveal that an error in the assumed oxygen yield causes an error in the estimated biomass and the estimated growth rate as well. These errors do not completely vanish during the entire cultivation. In a similar way, differences between the assumed and the real biomass concentration at the beginning of the fed-batch phase (caused, e.g., by an inaccurate off-line measurement) will cause an observation error which only slowly vanishes.

This situation continues considering a controlled cultivation as pointed out above. Even if it is possible to control the estimated specific growth rate at its setpoint as $\hat{\mu}=\mu_{set}$ an error in y_{X,O_2} will cause a process error, e.g., if the biomass yield with respect to oxygen is overestimated, then the real growth rate is too low. It should be noted that the biomass and the growth rate observer do not have independent measurement information which could be taken for convergence of the estimation results. However, this strategy with the measurement of pO_2 gives at least the opportunity to detect and avoid carbon substrate overfeeding. This can be considered as the second advantage of the described strategy. Altogether, the main merit of this approach is to prove that it is possible to combine the features of the self-stabilizing property of the limiting substrate and the observability of a co-substrate like oxygen or eventually a product like carbon dioxide.

12.5
Considerations for Improved Observation and Control

Altogether, the robustness of fed-batch strategies in general depends on the reliability of the considered yield coefficients. Since the oxygen yield is coupled to the dissimilatory carbon flux by stoichiometry, it is expected that a low substrate yield is accompanied by a low oxygen yield, e.g., in cases of high maintenance, decoupled growth, or other factors which affect the energy flux of the cells. While the energy generating processes may be more easy to calculate, the energy consuming processes are subjected to unknown influences [10], e.g., unbalanced metabolism during recombinant protein synthesis. From the viewpoint of process control, $y_{X,S}$ and $y_{X,O2}$ do not contribute independent information regarding the actual growth rate since both rely on a constant energy efficiency of the cells.

More reliable methods for estimation of growth processes could be based on specific uptake rates which are more closely linked to the anabolism of the cells. The uptake of ammonia could be one possible candidate, because the nitrogen demand is directly coupled to the biomass (e.g., protein, nucleic acids) synthesis reactions. While no maintenance has to be considered, $y_{X,NH3}$ depends on the cellular composition. The macromolecular composition may change during fed-batch cultivation [11] and, especially, during the production phase to some extent, but, fortunately, this does not effect the elemental composition very strongly. So for the purpose of biomass estimation, the elemental balance

$$c_X(t) = \frac{1}{e_{N,X}} \cdot \frac{MW_N}{MW_{NH3}} \cdot \int_0^t q_{NH3,f}(\tau)\, d\tau \tag{12.31}$$

can be applied with sufficient accuracy. However, while the amount of ammonia fed for pH-control can certainly be measured, it is neither the limiting substrate nor as easily measurable on-line as the dissolved oxygen concentration. So a reliable estimation based on ammonia consumption has to make sure that no acid formation leads to ammonia accumulation at constant pH. This drawback can be circumvented by employing estimators for acetic acid formation. So modern concepts of observation of HCDC should include, apart from the feeding strategies pointed out above, an additional consideration of other yield coefficients e.g. $y_{X,NH3}$ which will permit a more reliable biomass estimation which is valid at different physiological conditions [12] and, furthermore, the estimation of accumulation, e.g., of acetate [13], ammonium, or carbon substrate, which will also give the possibilities for on-line optimization of the feeding strategy.

The control task of complex fed-batch strategies has been discussed previously (e.g., [14]; for a more recent review see also [15]). In most cases linear control with constant parameters has been applied. This is in contrast to the exponentially changing system time constants and gains. Linearization of the process equations along the exponential trajectory, e.g., by MAPLE, delivers a linear system, where the linear system parameters depend on the actual biomass concentration. This offers the opportunity to use the biomass estimation for an adjusted linear controller, where the control parameters are optimally set on-line according the process state employing simple tuning rules. Furthermore, the actual $k_L a$-value, which represents the main time constant of the gas phase, can be calculated from the off-gas analysis and the pO_2, thus allowing for an adjusted linear controller as well.

12.6
A Case Study: Kinetics of Acetate Formation and Recombinant Protein Synthesis in HCDC

Even during a well-designed high-cell density-cultivation, small amounts of acetate may be produced (e.g., during the initial batch-phase or at the end of the recombinant protein production phase). Therefore, it is of some interest to understand this kind of by-product formation. The basic idea for modeling is the assumption of an overflow metabolism which has been proposed, e.g., [16] and has been supported by data from, e.g., [17].

If more glucose is taken up by the cell than can be processed in the oxidative pathway, the glucose in excess is transformed to acetate to yield additional metabolic energy, even if this is much less efficient compared to complete oxidation. This concept has been already successfully applied to understand and model the Crabtree effect in yeast [18].

The stoichiometric model (Eq. 12.1) can be modified by the inequality

$$r_{O2} < r_{O2,crit} \qquad (12.32)$$

where $r_{O2,crit}$ represents the maximum capacity of the respiratory chain. Consequently, a critical catabolic substrate flux

$$r_{S,cat,crit} < r_{S,cat} \qquad (12.32a)$$

can be introduced according to stoichio metry. The specific acetate formation rate r_{Ace} is then obtained from the catabolic mass balance as

$$r_{Ace} = y_{Ace,S} \cdot (r_S - r_{S,cat,crit}), \qquad (12.33)$$

which is valid in case of $r_S > r_{S,cat,crit}$ and therefore $r_{O2} = r_{O2,crit}$.

The yield $y_{Ace,S,ferm}$ for acetate from glucose in excess is given by the stoichiometry of acetate formation

1 mol glucose $\rightarrow$ 2 mol acetate + 2 mol CO_2 as
$$y_{Ace,S,ferm} = 2\, M_{Ace}/M_{Glu} \approx 0.666 \text{ g/g}. \qquad (12.34)$$

Accordingly, an additional term $y_{ATP,Ace} \cdot r_{Ace}$ has to be added to the energy balance. Acetate inhibition [19] and possible inhibitions by small molecules at high biomass concentrations [20] are not considered here.

From the viewpoint of downstream processing, a high intracellular protein concentration is at least as important as the achievement of high cell concentrations. After induction, the protein is produced with an initial rate which depends on the actual growth rate and, of course, on the expression system itself. In many cases, a decrease of the production rate with time is observed. This may be due to intracellular regulation or by other inhibition mechanisms leading to a decrease of gene expression with a concomitant decrease of recombinant protein synthesis. In the simulation study given below, an exponentially decreasing specific protein production rate

$$r_{Prot} = r_{Prot,ini} \cdot e^{r_p \cdot (t - t_{ind})} \tag{12.35}$$

is considered. In efficient expression systems, the amount of recombinant protein produced is so high that a major part of cell energy is channeled into the synthesis of this protein. This has been formulated in the concept of metabolic burden [21], where it has been estimated that the amount of energy needed for the production of protein is nearly twice as high as for the other cellular compounds (e.g., DNA, lipids). In the model presented here, this is accounted for by introducing another term $y_{Prot,ATP} \cdot r_{Prot}$ into the energy balance, which finally reads

$$\mu = y_{X,S} \cdot r_S + y_{X,Ace} \cdot r_{Ace} - y_{X,Prot} \cdot r_{Prot} - \mu_m \tag{12.36}$$

in the condensed parameter form. With these assumptions (Eqs. 12.32–12.36) in addition to the model shown above, simulation studies are presented in Fig. 12.7. using data from [22], where exponential carbon substrate feeding has been employed without further control and assuming constant yield and maintenance coefficients. The model parameters related to acetate and product formation were estimated directly from the data.

With these simple assumptions described above, the data are represented reasonably well (Fig. 12.7). During the batch and the first fed-batch phase (both at 30 °C),

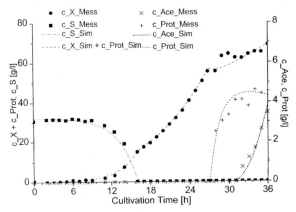

Fig. 12.7. Simulation of HCDC with temperature-induced production of an insulin B-chain fusion protein; data are from [22] with process and model parameters given as $c_{S,0}$=30 g/l, $V_{R,0}$=25 l, $c_{S,f}$=0.6 g/g, μ_{set}=0.12 h^{-1}(bef. ind.), μ_{set}=0.08 h^{-1}(after ind.); estimated parameters are $y_{X,S}$=0.48 g/g (bef. ind.), $y_{X,S}$=0.23 g/g (after ind.), $y_{X,Ace}$=0.05 g/g (assumed), $y_{X,Prot}$=1.80 g/g, $y_{Ace,S,ferm}$=0.666 g/g, $r_{S,crit}$=0.22 h^{-1}, $r_{Prot,ini}$=0.054 h^{-1}, τ_p=0.6 h^{-1}

the system behaves as expected. Less good agreement of experimental data and model prediction after induction may result from the deleterious effects associated with the overexpression of the heterologous gene. Beside the influence of the protein formation on the energy flux of the cells – as considered by the production kinetics and the energy balance – the accumulated protein itself may have an additional feed-back influence on the metabolic activity, namely reduction of growth and protein synthesis itself [23]. Also, in addition to heterologous protein production, heat shock proteins are being produced which can be considered as additional metabolic burden [24]. However, these effects cannot be considered on the level of lumped parameter models since the non-uniform distribution of recombinant protein in the population makes the concept of a uniform biomass dubious.

In the cultivation shown above, the feeding rate suitable to allow for a specific growth rate of $\mu_{set}=0.12\,h^{-1}$ under non-inducing conditions was reduced simultaneously with the induction of recombinant protein synthesis according to $\mu_{set}=0.08\,h^{-1}$ assuming no change in the biomass yield with respect to the carbon substrate [22]. However, the true growth rate and the true biomass yield were lower, therefore leading to a lower biomass profile than expected. So more and more glucose was fed per cell, and the specific substrate uptake rate increased and, finally, reached the critical value which results in progressively increasing acetate excretion. So, the experimental results here can indeed be modeled in accordance with the concept of carbon overflow metabolism. Nevertheless, the critical specific substrate uptake rate $r_{S,crit}$ during the induction phase is lower compared to those of the batch and the first fed-batch phase, where no acetate production was observed at even higher specific substrate turnover rates. Again, this should not be overinterpreted. At least it indicates that an even better process performance can be obtained by adjusting the feeding rate during the induction phase to the real energy demand of the cells. This can be done, e.g., by employing on-line state estimation to eventually occurring acetate formation. Other observations such as the increase of the oxygen uptake rate after induction of recombinant protein synthesis are not considered here. Yet it supports the assumptions of a high energy demand of the cells during the protein production period. Furthermore, a decrease of the yield coefficient $y_{X,S}$ and an increase of maintenance μ_m [25] cannot be distinguished because of the piecewise constant growth rate not allowing for a linearly independent estimation. Altogether, it has to be stated that much more work has to be carried out to understand the influence of induction and protein production with respect to energy turnover and intracellular regulation, thus allowing for an even improved coordination of gene expression and process strategy.

References

1. Yamane T, Shimizu S (1984) Fed-batch techniques in microbial processes. In: Fiechter A (ed) Adv Biochem Eng 30. Springer, Berlin Heidelberg New York
2. Luttmann R, Bitzer G, Hartkopf J (1994) Development of control strategies for high density cultivations. Mathematics and Computers in Simulation 37:153–164
3. Korz DJ, Rinas U, Hellmuth K, Sanders EA, Deckwer W-D (1995) Simple fed-batch technique for high cell density cultivation of *Escherichia coli*. J Biotechnol 39:59–65
4. Waldraff W, King R, Gilles ED (1997) Optimal feeding strategies by adaptive mesh selection for fed-batch bioprocesses. Bioprocess Eng 17:221–227
5. Matsui T, Yokota H, Sato S, Mukataka S, Takahashi J (1989) Pressurized culture of *Escherichia coli* for a high concentration. Agric Biol Chem 53:2115–2120

6. Riesenberg D, Schulz V, Knorre WA, Pohl H-D, Korz D, Sanders EA, Roß A, Deckwer W-D (1991) High density cultivation of *Escherichia coli* at controlled specific growth rate. J Biotechnol 20:17–28
7. Mori H, Yano T, Kobayashi T, Shimizu T (1979) High density cultivation of biomass in fed-batch system with DO-stat. J Chem Eng Jpn 12:313–319
8. Cutayar J, Poillon D (1989) High cell density of *E. coli* in a fed-batch system with dissolved oxygen as substrate feed indicator. Biotechnol Lett 11:155–160
9. Gollmer K, Posten C (1996) Supervision of bioprocesses using dynamic time warping algorithm. Control Eng Practice 4:1287–1295
10. Russell JB, Cook GM (1995) Energetics of bacterial growth: balance of anabolic and catabolic reactions. Microbiol Rev 59:48–62
11. Pramanik J, Keasling JD (1997) Stoichiometric model of *Escherichia coli* metabolism: incorporation of growth-rate dependent biomass composition and mechanistic energy requirements. Biotechnol Bioeng 56:398–421
12. Schmidt M, Viaplana E, Hoffmann F, Marten S, Villaverde A, Rinas U (2000) Secretion-dependent proteolysis of heterologous protein by recombinant *Escherichia coli* is connected to an increased activity of the energy-generating dissimilatory pathway. Biotechnol Bioeng (in press)
13. Åkesson M, Nordberg Karlsson E, Hagander P, Axelsson JP, Tocaj A (1999) On-line detection of acetate formation in *Escherichia coli* cultures using dissolved oxygen responses to feed transients. Biotechnol Bioeng 64:590–598
14. O'Connor GM, Sanchez-Riera F, Cooney CL (1992) Design and evaluation of control strategies for high cell density fermentations. Biotechnol Bioeng 39:293–304
15. Riesenberg D, Guthke R (1999) High-cell-density cultivation of microorganisms. Appl Microbiol Biotechnol 51:422–430
16. Majewski RA, Domach MM (1990) Simple constrained-optimization view of acetate overflow in *E. coli*. Biotechnol Bioeng 35:732–738
17. Han K, Lim HC, Hong J (1992) Acetic acid formation in *Escherichia coli* fermentation. Biotechnol Bioeng 39:663–671
18. Sonnleitner B, Käppeli D (1986) Growth of *Saccharomyces cerevisiae* is controlled by its limited respiratory capacity: formulation and verification of a hypothesis. Biotechnol Bioeng 28:927–937
19. Nakano K, Rischke M, Sato S, Märkl H (1997) Influence of acetic acid on the growth of *Escherichia coli* K12 during high-cell-density cultivation in a dialysis reactor. Appl Microb Biotechnol 48:597–601
20. Märkl H, Zenneck C, Dubach AC, Ogbonna JC (1993) Cultivation of *Escherichia coli* to high cell densities in a dialysis reactor. Appl Microb Biotechnol 39:48–52
21. Da Silva NA, Bailey JE (1986) Theoretical growth yield estimates for recombinant cells. Biotechnol Bioeng 28:741–746
22. Schmidt M, Babu KR, Khanna N, Marten S, Rinas U (1999) Temperature-induced production of recombinant human insulin in high-cell density cultures of recombinant *Escherichia coli*. J Biotechnol 68:71–83
23. Kurland CG, Dong H (1996) Bacterial growth inhibition by overproduction of protein. Mol Microbiol 21:1–4
24. Rinas U (1996) Synthesis rate of cellular proteins involved in translation and protein folding are strongly altered in response to overproduction of basic fibroblast growth factor by recombinant *Escherichia coli*. Biotechnol Prog 12:196–200
25. Bhattacharya SK, Dubey AK (1995) Metabolic burden as reflected by maintenance coefficient of recombinant *Escherichia coli* overexpression target gene. Biotechnol Lett 17:1155–1160

13 β-Lactam Antibiotics Production with *Penicillium chrysogenum* and *Acremonium chrysogenum*

Karl-Heinz Bellgardt

13.1
Introduction

The history of antibiotics production began in 1941 following the discovery of Penicillin by Fleming in 1928/29 and the first therapeutic applications in 1940 by Florey and Chain. Then the productivity and product concentration was increased by several orders of magnitudes by development of strains and improvements in cultivation media and process control. Surface culture was soon replaced by batch suspension culture in stirred tank reactors on complex media, followed by further improvement with a two phase fed-batch strategy. This extended the length of the production phase and avoided repression during high substrate levels. The industrial production capacity is in the range of several tens of thousands of tons, a market of billions of US$. In the production of secondary metabolites by filamentous fungi, such as the β-lactam antibiotics, Penicillin by *Penicillium chrysogenum*, or Cephalosporin C by *Acremonium chrysogenum* (*Cephalosporium acremonium*), many physiological, chemical, and technical factors influence the growth of the cells and product formation. Small alterations in preculture, media, and operating conditions change the productivity in a seldom predictable way. On the basis of new and improved analytical techniques, mathematical models can help to provide new insight into details, as well as an integral view of the entire process, by putting together the knowledge of the different levels from biology and cultivation to process control.

The morphology of the cells, which is a property of the strains but is also influenced by the cultivation conditions, is an important parameter for antibiotics production that is closely related to the fungal life-cycle. Antibiotics are often formed in connection with hindered growth, aging and maturation of cells, or metamorphosis. It is a general view that mainly non-growing morphological states contribute to antibiotics synthesis. Mycelial growth can range from filamentous to pelleted with any combination in between [13–16]. Important processes during the growth of filamentous fungi are shown in Figs. 13.1 and 13.2. After germination of the spores, the hyphae grow by linear extension in the apical region of the active tips, and by branching, into a complex network, the mycelial flocs. Their size is restricted by breakage or fragmentation due to mechanical shear forces. But under proper conditions dense pellets that can be rather stable may develop from hyphal filaments by increase in spatial hyphal cell density (see Fig. 13.2), up to sizes in the millimeter range. Agglomeration of spores before outgrow of filaments and attachment to solid particles supports pellet formation. The pellet size is strongly influenced by shear forces during preculture and in the production reactor [17, 18, 127]. High energy

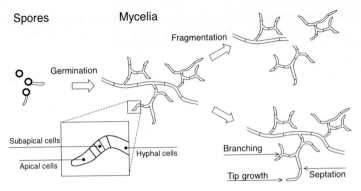

Fig. 13.1. Processes at hyphal growth and morphological development of *Penicillium chryso-genum* (septa are indicated)

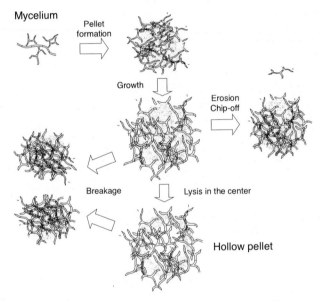

Fig. 13.2. Processes at the development of *Penicillium chrysogenum* pellets

input supports small, smooth, and compact pellets by an erosion process, while under mild shear stress the pellets can grow larger and develop a broad outer hairy region. Small pellets often have a decreasing biomass density over the radius [10, 11], but generally the biomass density can pass through a maximum [7]. Large old pellets are often hollow in the center as result of autolysis and can then easily break apart. This is caused by transport limitation of substrate and mainly oxygen into the inner zones. Measured pH and oxygen profiles by microprobes clearly support the autolysis hypothesis [137]. The kinetics of autolysis were investigated under excess and limitation of oxygen in [114, 115]. Although some autolysis was also observed in the growth phase, it was clearly induced by limitation of substrate or nitrogen. Penicillin degradation was then accelerated. Under additional oxygen limitation the onset was more sudden but hyphal disintegration was less. After restoring proper growth conditions, some regrowth of cells could be observed.

Morphology also indirectly influences production kinetics by determining the rheology of the broth. Free filaments increase viscosity and hinder gas-liquid oxygen transfer, which then limits cell density and productivity. Pellets result in low viscosity compared to free filaments and thus facilitate oxygen transfer to the liquid phase, but an additional transport resistance for substrates and oxygen from bulk liquid to the inner parts of the pellets can become important. Experimental results show that pellets up to a size of 200–400 μm are optimal and have no negative effect on the productivity, because the viscosity still remains low and there is no serious transport limitation into the pellets as is observed for larger pellets. The mechanisms of the inner-pellet transport are still under discussion. In [7, 16] it was found that with molecular diffusion, turbulent and directed convective flows have to be expected in the same order of magnitude. Diffusion seems to be the main mechanism in the center of the pellets while turbulence penetrates only the outer parts. Generally, the transport-reaction phenomena can be described by the dispersion model given in Chap. 2. The rheology of pellet suspensions was also examined in a number of papers, e.g., [19, 83, 111]. The respective correlations should be included in a complete process model to permit calculation of power supply and maximum oxygen transfer as parameters for optimization.

The production strains are far from their natural ancestors, but the reason for antibiotics synthesis might have been originally to provide some protection of the cells against competing species in critical growth situations [37]. Secondary metabolites such as many antibiotics are not part of the central metabolic pathways or of immediate use for growth. Synthesis of β-lactam-antibiotics is in a complex reaction network coupled with catabolism and anabolism and subject to various kinds of metabolic regulation. The biochemistry, metabolism and its regulation are reviewed, e.g., in [1–4, 54]. Maximum productivity is only possible in the presence of sufficient amounts of carbon source and additional precursors in the medium. Sugars such as glucose or lactate, dextrin, starch hydrolyzates, fatty oils, and other complex substrates may serve as carbon sources for the cultivation. The disadvantages of oil are increased oxygen demand and hindrance of the oxygen transfer by bubble coalescence.

Sugar substrates are catabolized via FDP- or pentose-phosphate-pathway, which are both used in a similar order of magnitude. The metabolic flux distribution was studied by Jørgensen et al. [116] during fed-batch cultivation for production of Penicillin V with a detailed stoichiometric model. The pentose-phosphate pathway was used at 20% during rapid growth but up to 40% in the production phase. By addition of the three amino acids precursors the productivity could be increased by 20%. In another investigation of chemostat cultures, the pentose-phosphate-pathway was even expected to be used up to 60% at high producing conditions [126]. The difference was explained by the higher specific production rate compared to fed-batch cultivation. The theoretical maximum Penicillin V yield was calculated as 0.43 mole per mole glucose. The antibiotics synthesis is subject to catabolite repression or inhibition during growth phases with high concentrations of easily convertible sugars, or at high rates of carbon source uptake [4, 80].

Precursors for the antibiotics come from the metabolism of amino acids (see also Figs. 13.5 and 13.10). Therefore, the kind of C- and N-sources and their amounts present in the medium have a strong influence on growth and productivity [5–7]. Soybean extract, yeast extract, peanut flour, or corn steep liquor (CSL) serve as organic nitrogen sources, beside ammonium. Addition of protein to the substrates improves the productivity by freeing resources for product formation in the amino-acid pathways.

The metabolic properties and kinetics determine the generally employed two-phased control strategy for antibiotics production: in the first phase of the process high growth rates are supported by an easily usable carbon source in batch mode or under fed-batch with high feeding rate. The repression in the following production phase can be either avoided by using carbon sources with lower maximum growth rate compared to glucose, such as lactose or oil, or by limited feed of the carbon source. In the first case, it is not necessary to establish a fed-batch strategy for the production phase. But this is preferred because it extends the possibilities for control of the process as further discussed in Sect. 13.4. Another and often the limiting factor for productivity is the oxygen supply to the culture because it sets bounds on the maximum attainable growth rate and cell density. Furthermore, the cyclization reaction of the tripeptide and the last steps of Cephalosporin C synthesis are directly oxygen-dependent [2, 8]. Even short oxygen limitations can have lasting negative effects on the product synthesis.

Summing up, the antibiotics production processes can be characterized by a diversity of influencing factors originating from medium, growth conditions, and biological properties of the strains. Thus the kinetics of production are much more complicated than for primary metabolites as was already outlined in Chap. 2. The difficulties in quantifying these effects and relating them to the morphological heterogeneity of the culture on one side, and to resulting variations in the productivity on the other side, are the cause for the so-called bad reproducibility of the process. Mathematical models can help to bring some light into the kinetics and mechanisms on the cellular level and connect them to the observed macro-kinetics. Mathematical modeling of antibiotics production covers all process levels beginning with the intracellular reaction network, growth of hyphae and pellets in space and time, over kinetics of substrate uptake and product formation, up to population dynamics for mycelial flocs and pellets. Since production is closely related to morphological changes, morphologically structured or segregated models are the proper approach for modeling of production processes for β-lactam-antibiotics. In the following, several models for *Penicillium chrysogenum* and *Acremonium chrysogenum* are presented as examples on the different modeling levels. In a first step, these models can improve our understanding of the observed macro-kinetic phenomena during the cultivations, and of the interrelation of preculture, operating conditions, growth, and production. In a second step, the models can be used to evaluate control strategies and optimize the production process. The biochemical and morphological aspects of cultivations of filamentous fungi and their modeling are discussed in several papers and reviews, e.g., [9–12, 76, 79, 93, 99, 109, 110, 112]. The aim of this chapter is to give by examples an overview of the modeling methods without looking at all details of the models. For more information the reader is referred to the more specialized reviews on biological aspects and modeling, or to the original papers.

13.2
Modeling of Penicillin Production

13.2.1
Unstructured and Simple Segregated Models

The complexity of models for Penicillin production has increased significantly over the years. Most of the early models were purely unstructured and tried only to give a formal description of the distinct growth and production phases. A simple model for growth as logistic law was chosen by Constantinides et al. [20, 70]. The production was assumed to be growth-independent but starting only after some dead-time due to induction of product synthesis. Temperature dependence of growth and production were modeled as polynomials or Arrhenius-type functions and used for optimization of batch cultivations. In a similar way, Calam and Russel [22] used a model with extremely long lag-phase to explain the delayed product formation. In some models an age dependence of the product synthesis is proposed to explain the delayed onset of production and its stagnation at the end of prolonged cultivations [33, 38].

Heijnen et al. [31] developed a model based on elemental and enthalpy balances, extended by kinetic expressions for substrate uptake, maintenance, product formation, and product hydrolysis. The slowdown of the production at the end of the cultivation is simply referred to a dilution effect due to the substrate feeding, and to product hydrolysis. In the model, a direct coupling of growth and production up to a critical growth rate of $0.01\,h^{-1}$ is proposed. Balancing methods were also employed for computer control of cultivations. For this task, correlations of penicillin production, growth, and maintenance metabolism to the CO_2-production is of special interest [34–36].

A mechanistic model based on the Contois-kinetics for carbon source and oxygen was developed by Bajpai and Reuß [28]. Yield variations are considered by a maintenance term in the balance equations. The unstructured model (see Table 13.1 for model equations and Table 13.3 for parameter values) included a substrate inhibition term in the rate of product synthesis to cover the low production during the growth phase. The model was adapted by Menezes et al. [129] to industrial Penicillin G pilot-plant fed-batch cultivations. Product synthesis was found not to be repressed at high growth rate and described by Monod-kinetics, and an autolysis-term for the cell mass was added. The uptake of components form complex substrates (CLS) was dealt with by combining them with glucose to a representative single substrate. Due to its simplicity, the model of Bajpai and Reuß served as a basis for a number of further investigations, not only to improve the validity of the model, but also to use it for extensive process optimization studies. Montague et al. [29] extended the model for carbon dioxide production and then applied it for on-line state estimation and parameter-adaptive control purposes. Further extensions and improved parameter estimation were given by Nicolai et al. [30] and van Impe et al. [125]. To overcome model inconsistencies during batch and fed-batch simulations, the authors introduced a variation in maintenance metabolism, endogenous metabolism, and yield in relation to the substrate concentration. Hegewald et al. [27] incorporated into their model not only the dependence of the growth and production kinetics on carbon source and dissolved oxygen, but also on nitrogen sources.

Table 13.1. Unstructured model of Bajpai and Reuß for Penicillin production [28]

	Description	Model equations
a	Specialized vectors, **r** considers maintenance metabolism, see Eqs. (1.12) and (2.2)	$C = \begin{pmatrix} C_X \\ C_P \\ C_S \\ C_O \end{pmatrix} \quad q = \begin{pmatrix} q_X \\ q_P \\ q_S \\ q_O \end{pmatrix} = Yr$ $Y = \begin{pmatrix} 1 & 0 & 0 \\ 0 & 1 & 0 \\ -Y_{XS}^{-1} & -Y_{PS}^{-1} & m_S \\ -Y_{XO}^{-1} & -Y_{PO}^{-1} & m_O \end{pmatrix} \quad r = \begin{pmatrix} \mu \\ r_P \\ -1 \end{pmatrix}$
b	Intrinsic rate expression based on Contois-kinetics, substrate inhibition kinetics for product	Growth $\mu = \mu_{S\,max} \dfrac{C_S}{K_S C_X + C_S} \dfrac{C_O}{K_O C_X + C_O}$ Production $r_P = r_{P\,max} \dfrac{C_S}{K_{SP} + C_S\left(1 + \frac{C_S}{K_I}\right)} \dfrac{C_O}{K_{OP} C_X + C_O}$
c	Explicit model equations for the liquid phase, the product balance considers first order decay, see Table 1.3	$\dfrac{dC_X(t)}{dt} = -D C_X(t) + \mu C_X(t)$ $\dfrac{dC_P(t)}{dt} = -D C_P(t) + r_P C_X(t) - k_D C_P(t)$ $\dfrac{dC_S(t)}{dt} = D(t)\left(C_S^R(t) - C_S(t)\right) - \left(\dfrac{\mu}{Y_{XS}} + \dfrac{r_P C_X(t)}{Y_{PS}} + m_S\right) C_X(t)$ $\dfrac{dC_O(t)}{dt} = OTR(t) - \left(\dfrac{\mu}{Y_{XO}} + \dfrac{r_P C_X(t)}{Y_{PO}} + m_O\right) C_X(t)$

The morphological development of the hyphae cannot be described by simple unstructured models. As shown in Fig. 13.1, the hyphae are only growing at their tip that forms the extending end of the growth-active apical cells. It is believed that the antibiotics are mainly synthesized in the non-growing subapical region, while the elder parts of the hyphae are more or less inactive. This heterogeneity of the biomass is more adequately described by segregated models or morphologically structured models. Simple segregated models for the antibiotics production can be taken as a first approximation to the inhomogeneities in the population, or the effects of morphological differentiation and aging of cells. For cultivations in the presence of pellets, the macrokinetics are also influenced by inhomogeneous biomass distribution in the pellets, transport limitation of substrates, and alterations in the size and age distribution of the pellet population. Growing cells are located mainly in the outer layers of the pellets, while production is assumed to take place preferably by the non-growing hyphae in the inner regions of the pellets. Also these effects can be described in a first attempt by simple segregated models.

In the morphologically structured model of Nestaas and Wang [39] the cell mass is distinguished by three types: growing tips, non-growing but producing hyphae, and degenerated non-producing inactive cells. Different sub-models were proposed for growth and production phases. Product synthesis proceeds via a postulated intermediate and is by this indirectly coupled to the specific growth rate; penicillin production starts above a minimum growth rate. Beside endogenous metabolism of the fraction of active biomass, a cell lysis mechanism is also used to explain variations in the yield. In the model, the specific growth rate, estimated from experimental data, is used as input to the balance equations instead of kinetic expressions in dependence on the substrate concentration. Although the model is, therefore, not predictive, it was used by Guthke and Knorre [40] to investigate the optimal control of repeated fed-batch cultivations. The tendency of the predicted optimal control policy was in agreement with experimental data. Kluge et al. [42] developed a segregated model as extension of the ideas of Heijnen et al. [31] that considers lactose and lysed biomass as additional substrates. For modeling of a lag-phase in product synthesis, a first-order delay system for its induction at glucose limitation is introduced. The model considers active and inactive biomass, as well as a temperature dependency of some model parameters, and thus could be used for simulation of experiments with temperature shifts or for temperature profile optimization.

Tiller et al. [47] presented a segregated model that was aimed for cultivations of a highly-producing strain of *Penicillium chrysogenum* with pronounced pellet formation. In a first batch growth phase, a mixture of pharma-medium and glucose was used as a substrate. During the fed-batch production phase, glucose was continuously added to the culture to avoid catabolite repression of the production. In the growth phase up to 90% of the total biomass were found as pellets with a diameter of >250 μm. Pronounced fragmentation and lysis of bigger pellets was observed in the later phase of the cultivation and explained by aging effects and changes in the macroscopic morphology. The model, summarized in Table 13.2, subdivides the entire biomass into two fractions, as shown in Fig. 13.3, to take this morphological change into account – growing and producing, X_1, and a non-growing but producing, X_2. The specific growth rates on glucose and pharma-medium are taken as Monod-kinetics. It is assumed that cell lysis contributes to pharma-medium with a

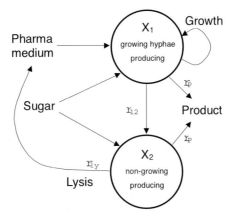

Fig. 13.3. Structure of the segregated model of Tiller et al. [47]

Table 13.2. Excerpts of the segregated model of Tiller et al. [47]

	Description	Model equations
a	Specialized model vectors (see Chap. 2.4.1 for explanation)	$C = \begin{pmatrix} C_{X1} \\ C_{X2} \\ C_S \\ C_{PM} \\ C_P \end{pmatrix}$ $\quad C_X = C_{X1} + C_{X2}$

$$\mathbf{Y}_1 = \begin{pmatrix} 1 & 1 & 0 & -1 & 0 \\ 0 & 0 & 0 & 1 & 0 \\ -Y_{XS}^{-1} & 0 & -Y_{PS}^{-1} & 0 & -1 \\ 0 & -Y_{XPM}^{-1} & 0 & 0 & 0 \\ 0 & 0 & 1 & 0 & 0 \end{pmatrix} \quad \mathbf{r}_1 = \begin{pmatrix} \mu_S \\ \mu_{PM} \\ r_P \\ r_{12} \\ m_S \end{pmatrix}$$

$$\mathbf{Y}_2 = \begin{pmatrix} 0 & 0 & 0 \\ 0 & -1 & 0 \\ -Y_{PS}^{-1} & 0 & -1 \\ 0 & 1 & 0 \\ 1 & 0 & 0 \end{pmatrix} \quad \mathbf{r}_2 = \begin{pmatrix} r_P \\ r_{ly} \\ m_S \end{pmatrix}$$

b	Intrinsic growth rate expressions	$\mu_S = \mu_{S\,max} \dfrac{C_S}{K_S + C_S} \qquad \mu_{PM} = \mu_{PM\,max} \dfrac{C_{PM}}{K_{PM} + C_{PM}}$
c	Metamorphosis reaction	$k_{12} = f_{12} A$
d	Lysis of non-growing hyphae	$k_{ly} = a_{ly} + b_{ly} A$
e	Specific rate of sugar uptake for maintenance	$m = b_m + a_m A$
f	Mean culture age	$A(t) = \dfrac{1}{C_X(t)} \displaystyle\int_0^t C_X(\tau)\,d\tau$
g	Specific product formation	$r_P = \begin{cases} \dfrac{r_{P\,max}}{\mu_{P1}} \mu & \mu \le \mu_{P1} \\ r_{P\,max} & for \quad \mu_{P1} < \mu \le \mu_{P2} \\ \dfrac{r_{P\,max}}{\mu_{P2}} (2\mu_{P1} - \mu) & \mu > \mu_{P2} \\ 0 & \mu > 2\mu_{P2} \end{cases}$

yield of one. The amino acid and protein constituents of pharma-medium are an important source of nitrogen during the cultivation. The rates of morphological differentiation, r_{12}, of lysis, r_{ly}, and of the maintenance requirements, m_S, are introduced as age-dependent. Both types of cells are assumed to produce penicillin by a specific rate r_P. The balance also considers a product decay by a first order mechanism with rate constant k_D. Product synthesis (see Table 13.2, row g) is maximum for a region of moderate growth rates, while at strict limitation or above a critical level for total catabolite inhibition, $\mu > 2\mu_{P2}$, no product is synthesized. The model was successfully applied for simulation of fed-batch cultures. Selected model parameters are given in Table 13.3. The simulation results for substrate, cell mass and product concentration are reported in Fig. 13.4. The growth phase during the experiment lasts for

Table 13.3. Selected parameters of the models of Tiller et al. [47] and Bajpai and Reuß [28]

Parameter	Tiller et al. (Table 13.2)	Bajpai and Reuß (Table 13.1)	Unit
μ_{Smax}	0.06	0.09	h^{-1}
r_{Pmax}	0.0046	0.005	h^{-1}
K_S	0.07	0.15	$g\,l^{-1}/-$
μ_{PMmax}	0.03		h^{-1}
K_{PM}	2.0		$g\,l^{-1}$
Y_{XS}	0.47	0.45	$g\,g^{-1}$
Y_{XPM}	0.51		$g\,g^{-1}$
Y_{PS}	1.2	0.9	$g\,g^{-1}$
k_D	0.0006	0.04	h^{-1}
μ_{P1}	0.003		h^{-1}
μ_{P2}	0.014		h^{-1}
k_{12}	0.00046		h^{-2}
a_m	0.001	0.014 (m_S)	h^{-1}
b_m	0.00015	0	h^{-2}
a_{ly}	−0.0008		h^{-1}
b_{ly}	$3\ 10^{-6}$		h^{-2}

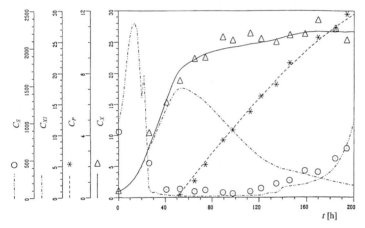

Fig. 13.4. Experimental data (*symbols*) and simulation results of a fed-batch experiment of *Penicillium chrysogenum*: total biomass (C_X), active biomass (C_{X1}), product concentration (C_P), all in $kg\,m^{-3}$; glucose concentration C_S in $g\,m^{-3}$. Reprinted from [47] "Segregated mathematical model for the fed-batch cultivation of a high producing strain of *P. chrysogenum*", Copyright (1999), with permission of Elsevier Science

about 50 h. Glucose becomes already limiting at 30 h. After that point, the growth is supported by the glucose feed. The transition to production coincides with the point, where pharma-medium becomes limiting, because then the specific growth rate falls below the upper critical value $2\mu_{P2}$. The pattern of active biomass (C_{X1}) showed a striking correspondence to the pellet sieve fraction with diameter $>250\,\mu m$, as shown in Fig. 13.8a.

13.2.2
Biosynthesis Model of Penicillin V

For optimization of the Penicillin production a detailed analysis of the biosynthesis of the antibiotics is necessary to understand its kinetics in dependence on the precursors, dissolved oxygen, and activities of the involved enzymes, or the accumulation of intermediates and by-products. A mathematical model can help to identify rate-limiting steps of the synthesis, which then may be attacked by metabolic engineering. The biosynthesis pathway is shown in Fig. 13.5. Penicillin is synthesized in several steps from a metabolite of the lysine pathway, L-α-aminoadipic acid; further amino acid precursors are valine and cysteine. The cyclization step of the tripeptide ACV (L-α-aminoadipyl-L-cysteinyl-D-valine) to Isopenicillin N is directly oxygen-dependent. The following reactions may proceed by a one-step or two-step mechanisms via 6-APA. The precursors phenylacetic acid (for Penicillin G) or phenoxyacetic acid (for Penicillin V) are incorporated into the penicillin molecule in the last synthesis step.

Jørgensen et al. [117] and Nielsen and Jørgensen [118] investigated the biosynthesis of Penicillin V. The Penicillin V biosynthetic pathway as shown in Fig. 13.5 was described by the kinetics in Table 13.4 with model parameters given in Table 13.5. Conversion of IPN to Penicillin V was assumed to take place either in the one-step reaction, where the alpha-aminoadipate side chain is exchanged directly with the precursor without release of 6-APA from the enzyme, or in a two step reaction via free 6-APA. The kinetics of carboxylation of 6-APA to 8-HPA were assumed to follow first order, since CO_2-concentration was considered as constant. They found that in a production phase of a fed-batch cultivation Isopenicillin N synthetase (IPNS) is flux limiting, and afterwards the ACV synthetase [93]. In a culture without feeding of

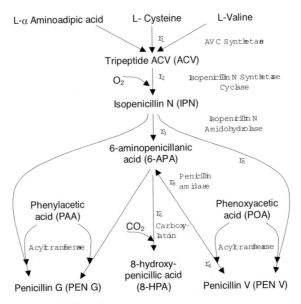

Fig. 13.5. Penicillin biosynthetic pathway

Table 13.4. Model for Penicillin V biosynthesis [117, 118]

Step	Kinetics[a]
ACV formation by ACV synthetase	$r_1 = k_1 X_{ACVS} \dfrac{1}{1 + \frac{K_{AAA}}{C_{AAA}} + \frac{K_{cys}}{C_{cys}} + \frac{K_{val}}{C_{val}}} \dfrac{1}{1 + \frac{C_{ACV}}{K_{ACV}}}$
Isopenicillin N formation by IPN synthetase	$r_2 = k_2 X_{IPNS} \dfrac{C_{ACV}}{C_{ACV} + K_O\left(1 + \frac{C_{glut}}{K_I}\right)C_O}$
Formation of 6-APA from Isopenicillin N by Isopenicillin N Amidohydrolase (IAH)	$r_3 = k_3 X_{AT} \dfrac{C_{IPN}}{C_{IPN} + K_{IPN}}$
Formation of Penicillin V from activated side-chain precursor and 6-APA by Acyl-CoA: 6-APA Acyltransferase (AAT)	$r_4 = k_4 X_{AT} \dfrac{1}{1 + \frac{K_{6APA-POA}}{C_{6APA}} + \frac{K_{POA}}{C_{POA-CoA}}}$
One-step conversion of IPN to Penicillin V	$r_5 = k_5 X_{AT} \dfrac{1}{1 + \frac{K_{IPN-POA}}{C_{IPN}} + \frac{K_{POA}}{C_{POA-CoA}}}$
Carboxylation of 6-APA to 8-HPA	$r_6 = k_6 X_{AT} C_{6APA}$
Cleaving of penicillin to 6-APA and phenoxyacetic acid by Penicillin amidase	$r_8 = k_8 X_{AT} \dfrac{C_{PenV}}{C_{PenV} + K_{PenV}}$

[a]X is the activity of the enzymes

Table 13.5. Parameters of the Penicillin V biosynthesis model

Parameter	Value	Unit
$K_{IPN-POA}$	0.023	$mmol\,l^{-1}$
K_{POA}	0.006	$mmol\,l^{-1}$
K_{IPN}	4.0	$mmol\,l^{-1}$
$K_{6APA-POA}$	0.0093	$mmol\,l^{-1}$
K_I	8.9	$mmol\,l^{-1}$
K_O	0.13	$mmol\,l^{-1}$
K_{AAA}	0.63	$mmol\,l^{-1}$
K_{val}	0.3	$mmol\,l^{-1}$
K_{cys}	0.12	$mmol\,l^{-1}$
K_{ACV}	12.5	$mmol\,l^{-1}$
K_{PenV}	2.0	$mmol\,l^{-1}$
k_1	17.77	$mmol\,l^{-1}\,h^{-1}$
k_2	16.85	$mmol\,l^{-1}\,h^{-1}$
k_3	4.03	$mmol\,l^{-1}\,h^{-1}$
k_4	1.95	$mmol\,l^{-1}\,h^{-1}$
k_5	13.74	$mmol\,l^{-1}\,h^{-1}$
k_6	0.065	$mmol\,l^{-1}\,h^{-1}$
k_8	0.4	$mmol\,l^{-1}\,h^{-1}$
$X_{ACVS}:P_{X1}$	615	$mmol\,l^{-1}\,h^{-1}$
$X_{IPNS}:P_{X2}$	330	$mmol\,l^{-1}\,h^{-1}$
$X_{AT}:P_{X3}$	420	$mmol\,l^{-1}\,h^{-1}$
$C_{POA-CoA}$	0.05	$mmol\,l^{-1}$

precursors, amplification of both enzymes by genetic engineering would be advantageous. It was also expected from simulation that oxygen concentrations above 45% of saturation could increase productivity. This effect was further evaluated in a subsequent paper [128] with an extended model that was verified with experimental data. The model indeed predicted a significant increase in specific penicillin production rate. Above 50% saturation, ACV synthetase became the only rate-limiting step.

13.2.3
Morphologically Structured Models for Growth of Hyphae

Based on the early experimental results [13, 24], Righelato [25], Bull and Trinci [26], and van Suijdam and Metz [10] proposed models for the development of mycelial flocs and average pellet populations. Hyphae are only growing in length at the free tips, as shown in Fig. 13.1. New free tips are formed by a branching mechanisms so that their number also increases during growth. Synthesis of growth metabolites mainly takes place in the apical region before the first septum. The subapical cells, separated by septa, have a cellular composition very similar to the apical cells and are considered as being metabolically active. The elder cells in the hyphal compartment often have large vacuoles and are assumed to have lower metabolic activity. The average hyphal length, the number of free tips, and the hyphal extension rate were found to be proportional to the specific growth rate. An important parameter for characterization of the filamentous morphology is the hyphal growth unit length (l_{hgu}) as the ratio of total hyphal length to number of free tips [100],

$$l_{hgu} = \frac{l_t}{n_t} \quad \begin{array}{l} \text{total hyphal length} \\ \text{total number of tips} \end{array} \tag{13.1}$$

Similar characteristic parameters can be defined by using the mass, i.e., the hyphal growth unit mass,

$$m_{hgu} = \frac{m_t}{n_t} \tag{13.2}$$

or volume of hyphae. Since their diameter and density are relatively constant, these definitions are practically equivalent. l_{hgu} and m_{hgu} are almost independent of the growth rate, and therefore the linear extension rate and the branching frequency of the hyphae are proportional to μ [10].

A morphologically structured model for an *Aspergillus* strain was presented by Megee et al. [23] in 1970. They discriminated the actively growing tips, active hyphae (subapical), and further morphological states. Growth and differentiation were described by kinetic expressions in dependence on the substrate concentration. Formation of growth and non-growth associated products was also modeled in connection with the morphological states. Using the basic ideas of Megee et al. [23], Paul and Thomas [46] presented, as extension of an earlier model [130], an interesting structured-segregated model for submerged growth of *Penicillium chrysogenum* filaments with far-reaching experimental verification by data from image-analysis. In this model the vacuole formation is considered as an important physiological process during growth and aging of hyphae. The model divides the biomass into basically three distinct regions according to the activities and structure of hyphal compartments: these are actively growing tips, non-growing Penicillin-producing regions, and degenerated or metabolic inactive regions that can be subject to fragmentation and autolysis. The non-growing region is considered to consist of cytoplasm and vacuoles. Growing tips are transformed by septation into non-growing cells that initially contain no vacuoles. Under substrate-limited conditions vacuole formation is initiated; these grow in size until they fill the entire cell volume and finally lead to the degenerated form. The vacuole size distribution in the hyphae is described by a population balance equation. Penicillin production is assumed to take place in the cytoplasm at low substrate concentrations. The differentiation phenomena and product formation are represented by kinetic equations in dependence on the substrate concentration. Unknown model parameters were estimated from experimental data for the development of cell-dry-mass equivalents of the four morphological types during fed-batch cultivations. This quantitative information was obtained by methods of digital image analysis. The model was in well agreement with the experimental data and had remarkable predictive capabilities for a number of fed-batch cultivations with varying sugar-feeding strategies. The highest product concentration was obtained by a feeding profile with gradually decreasing glucose-feed during the production phase.

A detailed hyphae model was proposed by Aynsley et al. [43] as an extension of the ideas of Prosser and Trinci [44]. The hyphae are considered as self extending tubular reactors, where substrate diffusion into the cells take place all over the hyphae. The substrate is converted into growth-precursor containing vesicles, which are transported to the tip. The flow of vesicles determines the tip extension rate, and their level the branching frequency. Fragmentation was assumed to occur mainly at substrate limitation. The model could be fitted to experimental data of total biomass and hyphal growth unit.

A morphological model was combined with a population balance model by Nielsen [45] to analyze morphological data from different batch and continuous cultivations of filamentous microorganisms in more detail, with respect to the growth mechanisms tip extension and branching, as well as fragmentation. The model considers three compartments, growing apical cells, subapical cells which still participate directly in the tip extension process, and hyphal cells that are inactive with respect to the growth process but provide material for the growth of apical cells by contributing to the stream of vesicles. Branching points are assumed to be located at the subapical region. It is an important point for the description of submerged cultures that the model discriminates between active and inactive tips since fragmentation is a frequent event under agitated conditions. If one assumes that practically every mycelia floc was subject to fragmentation, it must contain exactly two inactive tips which add to the measurable total number of tips. On average, the total number of tips and the number of active tips are related by

$$\bar{n}_t = \bar{n} + 2 \tag{13.3}$$

The morphological model is complemented by balance equations for the average number of hyphal elements and actively growing tips as well as the hyphal mass. These equations are derived from a distribution model that is summarized below. The dynamic balance for the hyphal growth unit mass was found to be

$$\frac{d\bar{m}_{hgu}}{dt} = \left(\mu - (\Phi + 2\Psi)\bar{m}_{hgu} \right)\bar{m}_{hgu} \tag{13.4}$$

where Φ is the specific branching frequency and Ψ the specific rate of fragmentation. For most species, the hyphal growth unit mass is constant when no fragmentation occurs. It can then be calculated from the average specific growth rate as

$$m_{hgu} = \frac{\mu}{\Phi} \tag{13.5}$$

and the rate of tip extension is then proportional to the average specific growth rate,

$$r_{tip} = \mu m_{hgu} \tag{13.6}$$

The model was in good agreement with experimental data of continuous cultures of *Penicillium chrysogenum*. It was also found that the specific rate of fragmentation is linearly correlated with the energy input, regardless of batch or continuous operation. The morphological model was extended by kinetics for growth and product synthesis by Zangirolami et al. [98] under the assumption of three hyphal sections (see Chap. 2, Table 2.14): growing apical tips, subapical sections with glucose inhibited product formation, and inactive parts. A complex substrate (CSL) was considered by an equivalent amount of glucose. The model was successfully compared to experimental data of fed-batch and repeated fed-batch cultivations by simulation.

The macroscopic models for hyphal growth on average can easily be compared to experimental data. But for in-depth theoretical studies and more accurate investigation of the microscopic morphology and related mechanisms, population balance models are a proper approach. Takamatsu et al. [136] developed a discrete distribution model for growth of *Aspergillus niger* in the form of mycelial flocs. The population is characterized by a number density function for flows with a certain number of cells and branches. The model included the mechanisms of branching and breakage and was used to establish a simplified segregated model of the population. Another morphological model for growth of hyphae based on the population balances of Ramkrishna (see Chap. 2, Sect. 2.4.6) was published by Nielsen [95, 113] and recently worked out further [97]. Hyphal elements are characterized by their total length l_t and number of tips n. Their number density f is given by the population balance

$$\frac{\partial f(l_t, n, t)}{\partial t} + \frac{\partial \left(q_{tip}(l_t, n, \mathbf{z}) n f(l_t, n, t) \right)}{\partial l_t} + \frac{\partial \left(q_{bra}(l_t, n, \mathbf{z}) f(l_t, n, t) \right)}{\partial n}$$
$$= h_g(l_t, n, \mathbf{z}, t) + h_{fra}(l_t, n, \mathbf{z}, t) - Df(l_t, n, t) \tag{13.7}$$

where q_{tip} is the average tip extension rate, and q_{bra} the branching frequency. Net formation of hyphal elements is by spore germination with rate h_g and fragmentation with rate h_{fra}, both also depending on the conditions in the bioreactor denoted by the general vector $\mathbf{z}$. It is assumed that by spore germination newly formed hyphal elements have fixed properties n_g and l_{tg}, and fragmentation results in binary fission, then

$$h_g(l_t, n, \mathbf{z}, t) = g(\mathbf{z}, t)\delta(l_t - l_{tg})\delta(n - n_g) \tag{13.8}$$

$$h_f(l_t, n, \mathbf{z}, t) = 2\int_n^\infty \int_{l_t}^\infty q_{fra}(l_t', n', \mathbf{z}) p(l_t', n', l_t, n) f(l_t', n', t) dn dl_t - q_{fra}(l_t, n, \mathbf{z}) f(l_t, n, t)$$

where q_{fra} is the breakage function, p the partitioning function, and g spore germination frequency. The latter was fitted to a B-distribution to estimate the spore viability and germination time interval from experimental data. From the above model balance equations for macroscopic population parameters, the total number of hyphal elements, e, average number of tips, n_{av}, and average hyphal length, l_{av}, as defined by

$$
e = \int_{n_g}^{\infty} \int_{l_{tg}}^{\infty} l_t f(l_t, n, t) \, dl_t \, dn
$$

$$
l_{tav} = \frac{1}{e} \int_{n_g}^{\infty} \int_{l_{tg}}^{\infty} l_t f(l_t, n, t) \, dl_t \, dn \qquad (13.9)
$$

$$
n_{av} = \frac{1}{e} \int_{n_g}^{\infty} \int_{l_{tg}}^{\infty} n f(l_t, n, t) \, dl_t \, dn
$$

were derived for the situation when all spores have germinated as follows:

$$
\frac{de}{dt} = \left(q_{fra}(l_{tav}, \mathbf{z}) - D \right) e
$$

$$
\frac{dl_{tav}}{dt} = n_{av} q_{tip}(l_{tav}, \mathbf{z}) - q_{fra}(l_{tav}, \mathbf{z}) l_{tav} \qquad (13.10)
$$

$$
\frac{dn_{av}}{dt} = q_{bra}(l_{tav}, \mathbf{z}) - q_{fra}(l_{tav}, \mathbf{z}) n_{av}
$$

The following kinetic expressions were in well agreement with experimental data:

$$
q_{tip}(l_{tav}, \mathbf{z}) = k_{tip\,max} \frac{C_S}{C_S + K_S} \frac{l_{tav}}{l_{tav} + K_l} \qquad (13.11)
$$

$$
q_{bra}(l_{tav}, \mathbf{z}) = \begin{cases} 0 & for \quad l_{tav} < 44 \ \mu m \\ k_{bra}(\mathbf{z}) l_{tav} & for \quad l_{tav} \geq 44 \ \mu m \quad with \ \mu = \sqrt{k_{tip} k_{bra}} \end{cases}
$$

$$
q_{fra}(l_{tav}, \mathbf{z}) = \begin{cases} 0 & for \quad l_{eav} < l_{eeq} \\ k_{fra}(\mathbf{z})(l_{eav}^2 - l_{eeq}^2) & for \quad l_{eav} \geq l_{eeq} \end{cases} \quad with \ l_{eav} = \frac{n_{av}+1}{2 n_{av}}(l_{tav} - l_{tip})
$$

where k_{fra} was correlated to the specific power input of the reactor and geometric parameters. The average effective length of hyphal elements is l_{eav}. Growth and fragmentation of single hyphae were described by a modified population model that was solved for steady states by Monte Carlo simulation [96]. It was shown that it is possible, from measurements of the number of tips and total hyphal length, to discriminate between four sets of models consisting of two different fragmentation kinetics and two partitioning functions.

Turbulent breakage of hyphae in mechanically stirred bioreactors was also investigated by Shamlou et al. [132] and found to be approximately of first order, i.e., the mean main hyphal length could be described by

$$
\frac{dl_{tav}}{dt} = k(l_{eq} - l_{tav}) \qquad (13.12)
$$

where l_{eq} is the equilibrium length. The rate constant k was found to depend on the product of mean energy dissipation rate and reciprocal of the impeller circulation time.

13.2.4
Models for Growth of Fungal Pellets

Similar to the models for growth of hyphae, the early models for pellet populations considered the development of pellets only on average assuming uniform size and density. The growth of a pellet in mass can in a very simplified way be described by the cube-root law under assumption of constant radius growth:

$$m_p(t) = \left(m_p^{\frac{1}{3}}(0) + kt \right)^3 \tag{13.13}$$

The kinetics correspond to the following volumetric and specific rate equations for growth of cell mass:

$$Q_X = 3kC_X^{\frac{2}{3}} \quad \mu = 3kC_X^{-\frac{1}{3}} \tag{13.14}$$

The growth of fungal pellets on average with consideration of oxygen transport limitation can be described by simple unstructured models in connection with the diffusion equation, Eq. (2.9) in Chap. 2 [10, 12, 14, 41, 131], assuming constant biomass density within the pellet. This allows one to estimate the penetration depth of oxygen, which determines the active portion of a pellet, and its total oxygen and substrate turnover. For very dense pellets, the dispersion model has to consider the maximum possible cell density that is limited by two mechanisms: available space for hyphae and reduced free outer-cell volume for transport of substrates. Both effects can be modeled in analogy to the dusty-gas model by a cell mass dependent diffusion coefficient, Eq. (2.13) in Chap. 2.

Pellet breakage and disruption due to mechanical sheer stress as a consequence of high energy input for agitation of the cultivation medium was studied by van Suijdam and Metz [12] in stirred tank reactors. They derived the following theoretical correlation for the mean pellet diameter under shear stress that was in good agreement with experimental data:

$$d_p = kN^{-1.2}d_{Stir}^{-0.8} \tag{13.15}$$

where d_p is the pellet diameter, N the stirrer speed, and d_{Stir} the stirrer diameter.

In order to proceed toward a better understanding of influencing factors for growth and product formation of *Penicillium chrysogenum* under the presence of pellets, Meyerhoff et al. [49] developed a detailed model for the growth of single hyphae up to mycelia flocs and pellets. The model adopted the basic concepts of Yang et al. [50, 51] for the growth of *Streptomyces tendae*. This is a combined mechanistic-stochastic approach for the description of the three-dimensional development of mycelia by linear extension of hyphae through apical growth, septation, and branching. But in addition, kinetics for growth limitation by space, substrate or oxygen, cell inactivation, or lysis under starvation condition, e.g., in the center of pellets), and mechanisms for erosion of hyphae at the surface of pellets were introduced. Growth of hyphae is determined by a key-compound that is synthesized from substrate transported into the cells through the outer wall along the branches of the hyphae. The event of septation and location of the septum are also controlled by the concentration of the key-compound. The linear extension rates of the tips are determined by diffusion of a key-compound along the branches to the tips, where it is used to build new cell material. The diffusion equation can be solved analytically and gives for the quasi-stationary case the maximum linear extension rate of a compartment with one tip:

$$\alpha_{1,max} = \alpha_{max} \frac{\beta e^{2\kappa L_1} + \gamma}{\beta e^{2\kappa L_1} - \gamma} \tag{13.16}$$

with known auxiliary variables β, γ, and κ that depend on the geometry of the branches [49]. The events of branching and the spatial direction for growth of new tip-sectors are given by probability functions in the stochastic part of the model. Transport processes of substrate and oxygen from the bulk medium to the center of the pellet are described by the diffusion equation, Eq. (2.9) in Chap. 2, with cell mass dependent effective diffusion coefficient. The growth limitation mechanism for the linear extension rates of individual tips,

$$\alpha_i = \alpha_{i,max} \frac{C_O}{K_O + C_O} \frac{C_S}{K_S + C_S} \tag{13.17}$$

is assumed to follow Monod-kinetics for each limiting compound, oxygen, and substrate. Cell inactivation with subsequent lysis is assumed to take place when the substrate concentration falls below a critical value. Shear forces are an important factor for the development of the pellets that limits the growth in diameter by the erosion process. It was assumed that only hyphae growing out from the denser parts of the pellet are subject to breakage according to the probability function

$$P_{bre} = 100 \left(1 - e^{-\lambda_{Shear} \frac{r_{tip} - r_{thr}}{r_{thr}}} \right) \tag{13.18}$$

where $r_{thr}(\rho_{thr})$ is the critical radius above which breakage can occur, r_{tip} is the radial position of the tip, and λ_{Shear} is the characteristic parameter of the probability distribution that can be varied to adjust the model to experimental conditions. The critical radius is determined by the local hyphal density ρ_{thr}. Simulations of a single pellet have been carried out with the model that were quite realistic by visual inspection. Under mild shear conditions, simulated pellets developed a broad and more hairy outer region than the compact and dense pellets resulting under enforced shear stress; these can have a very smooth surface. In the simulation studies, the cell density within the pellets approaches a maximum volume fraction of about 0.4. After the cell density has reached this maximum, the active growth zone keeps a thickness of about 200 μm since no oxygen reaches the inner parts due to transport limitation, and the density profile then moves more or less only outside without altering the shape or maximum value.

The detailed model above is not well suited for more far reaching simulation studies of bigger pellets or pellet ensembles, because the demand for computer resources is tremendous. Therefore, the model was simplified by neglecting individual hyphae and looking at average morphological quantities, while keeping the morphological basis and information on the microscopic structure of the pellets [52]. In the modified model the pellet is considered as having spherical symmetry and averaging is done in small radial layers. The simulation proceeds in discrete time steps. The simplified model considers only the total length of hyphae within a layer as representative variable. The hyphal growth unit length is then reconstructed using a correlation derived from the detailed model as a function of the radius by

$$l_{hgu} = 70\mu m + \frac{360\mu m - 2r_R}{\sqrt[3]{\sum_R n_{tR}}} \tag{13.19}$$

where n_{tR} is the total number of tips including inactive ones. The number of tips n_R in the radial layer is then evaluated from Eq. (13.1). Similarly, the linear extension is only calculated on the average. The total hyphal length increment in the simulated pellet layer by growth is

$$\frac{\Delta l_{tR}^+}{\Delta t} = \begin{cases} \alpha_{max} n_R \tilde{r}(C_O) \tilde{r}(C_S) & \rho < 0.6, C_S > C_{S,Crit}, C_O > C_{O,Crit} \\ \alpha_{max} n_R \tilde{r}(C_O) \tilde{r}(C_S)(0.79 - \rho) & \text{where} \quad \rho \geq 0.6, C_S > C_{S,Crit}, C_O > C_{O,Crit} \\ -\alpha_{ly} n_R & C_S \leq C_{S,Crit} \text{ or } C_O \leq C_{O,Crit} \end{cases} \tag{13.20}$$

The first condition is for weakly limited growth while the second also considers spatial limitation; the most dense packing of cylinders of equal diameter is 0.79. The third condition describes reduction of the active hyphal length by inactivation at growth limitation. The inactive parts could then be subject to lysis. Length reduction by pellet erosion is given by

$$\frac{\Delta l_{tR}^-}{\Delta t} = -l_{Shear} n_{ShearR} \tag{13.21}$$

where the number of broken tips, n_{ShearR} is calculated as in the detailed model, and the average mycelia length loss per break event, l_{Shear} is a model parameter.

Simulation results of the simplified model were compared to microprobe measurements for glucose and oxygen profiles as described in [137], and to cell density profiles obtained by digital image analysis of microtome preparations of a pellet. Figure 13.6 depicts the simulated concentration profile of a small pellet by using the cell

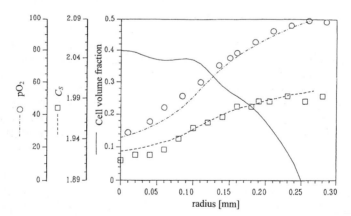

Fig. 13.6. Comparison of radial oxygen (pO2 in %) and glucose profiles (C_S in kg m^{-3}) within a small pellet as obtained by microprobe measurements (*symbols*) to simulation results (*lines*) by using the measured cell volume fraction, 216h after germination; parameters: α_{max}=5 μm h^{-1}, l_{Shear}=200 μm, α_{ly}=50 μm h^{-1}, C_{Ocrit}=8 10^{-5} g l^{-1}. Reprinted from [52]

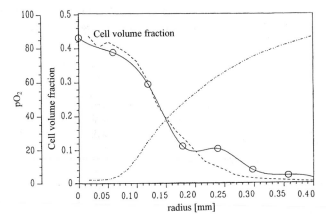

Fig. 13.7. Simulation results (*dashed lines*) for radial profiles of dissolved oxygen (pO2) in %) and cell volume fraction, the latter in comparison to experimental data (*symbols with interpolating line*) as obtained from digital image processing; de novo simulated pellet after 115 h of shake flask cultivation; parameters: $\alpha_{max}=18\,\mu m\,h^{-1}$, $l_{Shear}=214\,\mu m$, $\alpha_{ly}=50\,\mu m\,h^{-1}$, $C_{Ocrit}=10^{-5}\,g\,l^{-1}$. Reprinted from [52]

volume fraction from image analysis. The simulation can reproduce the measured profiles quite well. In a second step it was tested whether the model yields realistic cell density profiles. Shake flasks were inoculated with spores and cultivated up to pellets under mild shear forces. Pellets were taken out at different time and the cell volume fraction determined. Simulations over an identical time interval yielded very similar cell density profiles, as shown in Fig. 13.7 by one example.

13.2.5
Models for Growth of Pellet Populations

In a bioprocess the pellet population is never homogeneous but shows a wide distribution in shape, size, density, and growth activity that all influence the macrokinetics of the process. This is due to the stochastic nature of the important processes during the pellet development as shown schematically in Fig. 13.2: germination, erosion, and breakage events follow probability functions instead of being exactly determined in every single occurrence. The unequal development can be further pronounced by transport limitation of substrates into the inner regions of the differently developing pellets. The inhomogeneity can be described by population models with continuous variation in characteristic variables of a pellet.

A purely theoretical model for growth of pellet populations was developed by Edelstein and Hadar [134], later modified by Tough et al. [135] in the growth equation, and translated into a finite elements description that generated more stable solutions in the pellet size distribution. Pellet fragments after breakage were assumed to have all the same size; chip-off of hyphae at the outer region of the pellet was not considered. Nielsen [95] presented a population balance model in analogy to the model of hyphal development given in Sect. 13.2.3. The pellets are characterized by their diameter d_p. The number density of pellets f_p with this diameter is described by the population balance

$$\frac{\partial f_p(d_p, t)}{\partial t} + \frac{\partial \big((q_{gro}(d_p, \mathbf{z}) - q_{pfra}(d_p, \mathbf{z}))f_p(d_p, t)\big)}{\partial d_p}$$

$$= h_{for}(d_p, \mathbf{z}, t) - h_{bre}(d_p, \mathbf{z}, t) - Df_p(d_p, t)$$

(13.22)

where q_{gro} is the rate of increase of pellet diameter, q_{pfra} the rate of hyphal fragmentation (erosion) at the pellet surface, h_{for} the net formation rate of pellets, and h_{bre} the loss rate due to breakage. Functional expressions were not suggested. In analogy to hyphal growth, it was proposed to describe the growth in average pellet diameter as

$$\frac{d(d_{pav})}{dt} = q_{pgro}(d_{pav}, \mathbf{z}) - q_{pfra}(d_{pav}, \mathbf{z})$$

(13.23)

with the kinetic expressions

$$q_{pbre}(l_{tav}, \mathbf{z}) = k_{bre}(\mathbf{z})d_{pav}^{3.2}N^{6.65}d_{Stir}^{8.72}$$

$$q_{pfra}(l_{tav}, \mathbf{z}) = \begin{cases} 0 & \text{for} \quad d_{pav} < d_{peq} \\ k_{pfra}(\mathbf{z})(d_{pav}^2 - d_{peq}^2) & \text{for} \quad d_{pav} \geq d_{peq} \end{cases}$$

(13.24)

where k_{pfra} depends on the specific power input of the reactor and geometric parameters. The equilibrium diameter, d_{peq}, was correlated with the specific power input [12],

$$d_{pav} = d_{frag}\left(\frac{P}{V}\right)^{-k}$$

(13.25)

The agitation induced fragmentation of pellets was successfully described by Jüsten et al. [94] by a population balance model founded on detailed experimental data from image analysis [133]. Correlations to agitation parameters were also developed on this basis. The projected area A_p of a pellet was used as characteristic variable in the population balance for the number density function $f(A_p, t)$,

$$\frac{\partial f(A_P, t)}{\partial t} = -\frac{\partial \big(f(A_P, t)(-k)(A_P - A'_P)\big)}{\partial A_P}$$

(13.26)

The breakage mechanism is of first order in some driving force with fragmentation rate constant k. Driving force is the size difference in the actual and the minimum projected area A_p' of a pellet, below which no breakage occurs. The rate constant was found to depend only on the agitation conditions and could be correlated to impeller tip speed, specific power input, and the "energy dissipation/circulation" function [132]. No direct significant effect of fermentation conditions was found. The simulated distributions were in very nice agreement with the experimental data.

In an investigation on pellet morphology with *Aspergillus awamori*, Cui et al. [138] found that pellet breakage is not important. The effect of agitation influenced mainly the rate of hyphal chip-off in the outer hairy zone of the pellet. This mechanism restricts the size of pellets and increases the density of pellets as well as the mass of free hyphae. Nevertheless, when hollow pellets are formed due to transport limitation, pellet breakage seems to be more likely.

The simplified pellet model of Meyerhoff and Bellgardt [52] presented in Sect. 13.2.4 also provides a tool for an integrated, morphology-based simulation of fed-batch processes in the presence of pellets with non-uniform property distribution. In [53] the distribution function was approximated by simulation of up to 100 different pellets that represent all of the pellets in the cultivation. The simulation started from the very beginning of the process, i.e., inoculation of spores in the preculture. This opens the additional chance of studying the influence of variations in the preculture conditions on the main cultivation. For this task, the model of [52] was extended by a description of the germination time distribution of spores. A modified truncated Gauss-distribution

$$\Phi(\ln x) = \frac{1}{\sqrt{2\pi}} \int\limits_{0}^{\ln x} e^{-\frac{\ln x^2}{2}} d\ln x \tag{13.27}$$

was found to be in agreement with experimental data, where the actual germination time t_g is

$$t = (\beta \ln x + \gamma)t_{g\,max}, \ x > 0, t_g \geq 0 \tag{13.28}$$

where β and γ are model parameters. Since the observed pellet size distribution could only be explained by assuming breakage events, this mechanism was also included in the model. Breakage events are influenced by the actual state of the pellet with respect to cell density profile, size, and age. Although the model provides this information, a simple correlation for the breakage probability of a pellet, P_{bre}, in a time interval Δt was introduced for the beginning:

$$P_{bre} = k_{bre}\lambda_{Shear}C_X \frac{\Delta t}{0.2h} \tag{13.29}$$

The model was applied to simulations of pellet size distributions in shake-flask cultivations in the presence of glass beads under varying conditions and for interpretation of the data. Also complete fed-batch cultivations were simulated. For the process shown in Fig. 13.4, the simulations of the usual process variables based on this pellet population model were quite close to the segregated model. The simulated pellet size distribution is compared to experimental data in Fig. 13.8. At the beginning there is an exponential growth phase with significant formation of pellets with diameter >250 μm. During the production phase with only slightly increasing total biomass, the formation of bigger pellets slows down at first. Later on, after about 100 h, the pellets begin to disappear due to breakage and autolysis. After 140 h, there remain only filaments and pellet fragments with small diameter. The simulation showed that the pellets are most susceptible to oxygen limitation in the early phases of the cultivation although the dissolved oxygen concentration remains high. The reasons for this effect are the high growth rate and large mean pellet diameter at the beginning of the cultivation. Afterwards oxygen transport limitation into the pellets plays no role because of the reduced growth rates and decreasing mean diameter.

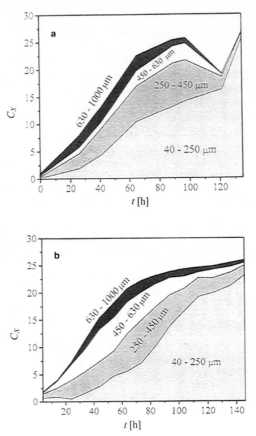

Fig. 13.8.a Experimental data for pellet sieve fractions. **b** Simulations results of the pellet size distribution for the cultivation in Fig. 13.4. Parameters: α_{max}=20 μm h^{-1}, l_{Shear}=214 μm, α_{ly}= 20 μm h^{-1}, λ_{Shear}=0.1. Reprinted from [53] "A morphology-based model for fed-batch cultivations of *P. chrysogenum* growing in pellet form" Copyright (1999), with permission of Elsevier Science

13.3
Modeling of Cephalosporin C Production

The morphology and metabolism of *Acremonium chrysogenum* deviates somewhat from *Penicillium chrysogenum* which has consequences for the modeling of the process. In submerged cultures *A. chrysogenum* can be found in three main morphological types as shown in Fig. 13.9: slim thin-walled filamentous hyphae (mycelia), swollen thick-walled hyphae or hyphal fragments, and arthrospores. The organism has no clear tendency to pellet formation. At high growth rate mainly slim hyphae and a low percentage of swollen hyphae are formed. Gradual limitation of the carbon source accelerates the transition from the slim filamentous to the second thicker form. Under growth limitation this develops septa and can subsequently break down to arthrospores. Cephalosporin C (CPC) production is usually maximum in process

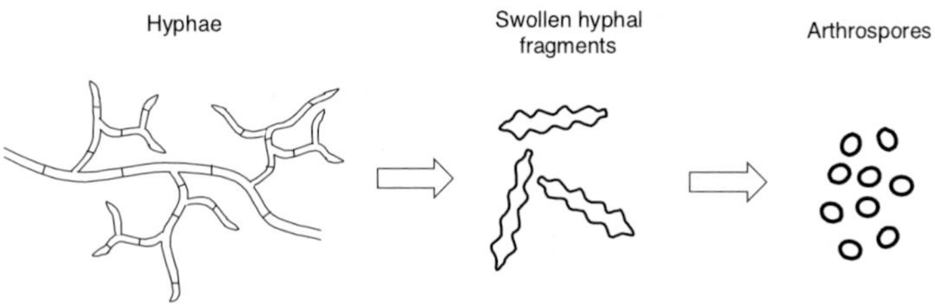

Fig. 13.9. Morphological development of *Acremonium chrysogenum*

phases with a high rate of differentiation from swollen hyphae to arthrospores [2, 56–58, 77]. Thus it correlates with a high number of swollen hyphae as well. The arthrospores can germinate and grow out to new hyphae, but seem to be non-producing.

Similar to Penicillin production, high productivity can be obtained by a two-phase control strategy with clearly distinguishable growth and production phase. Oxygen limitation has to be avoided and product formation must be supported by sufficient feed of substrate. The CPC-production is stimulated by addition of methionine during the growth phase, but inhibitory high concentrations or overfeed is contra-productive [57, 68]. Methionine is converted to cysteine [1] and can supply sulfur to the CPC-pathway, inducing its enzymes as well as stimulating morphological differentiation from thin hyphae to produce swollen hyphae. During later phases of the cultivation, the production slows down and it must be stopped before the product degradation surmounts production. In all, 25–40% of the product can undergo such hydrolysis, which follows first order kinetics [2]. Another problem of the cultivation is the accumulation of undesired precursors of the Cephalosporin C synthesis, e.g., deacetylcephalosporin C, which is rather stable.

13.3.1
Biosynthesis of Cephalosporin

The synthesis steps of Penicillin and Cephalosporin C up to Isopenicillin N are equivalent, as is shown in Fig. 13.10. This is then converted to Penicillin N without precursor, and in a further oxygen-dependent reaction sequence to Cephalosporin C. Under oxygen concentrations below 20% of saturation, these further steps are inhibited and Penicillin N is accumulated [106]. Glucose and other rapidly utilized carbon sources repress the ring-expansion enzyme deacetoxycephalosporin-C-synthetase and inhibit the first enzyme of the CPC-pathway, the ACV synthetase. The ring expansion enzyme is not inhibited after induction [55], but may be rate limiting [107, 108].

Malmberg and Hu [59] presented a detailed kinetic model for analysis of rate limiting steps in the CPC biosynthetic pathway under conditions of constant dissolved oxygen tension. The model, given in Table 13.6, includes the six reaction steps to Cephalosporin C, starting with the amino acids L-α-aminoadipic acid, L-cysteine, and L-valine. Cell growth was simulated according to a formal balance with fixed specific growth rate, and loss of Penicillin N into the medium was considered. In vivo data of enzyme activities were converted to the intracellular conditions required for the model simulations. From the simulation results the authors conclude that the

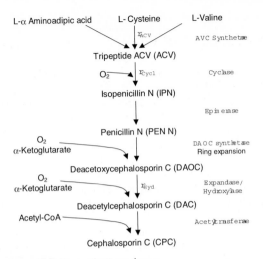

Fig. 13.10. Cephalosporin C biosynthetic pathway

Table 13.6. The Cephalosporin biosynthesis model of Malmberg and Hu [59]

Description	Model equation
Intrinsic tripeptide ACV	$\dfrac{dc_{ACV}}{dt} = k_{ACV}\dfrac{c_{AAA}}{K_{AAA}+c_{AAA}}\dfrac{c_{cys}}{K_{cys}+c_{cys}}\dfrac{c_{val}}{K_{val}+c_{val}}$ $-k_{Cycl}\dfrac{c_{ACV}}{K_{ACV}+c_{ACV}}-\mu c_{ACV}$
Intrinsic Penicillin N	$\dfrac{dc_{PenN}}{dt}=\dfrac{1}{2}\left(k_{Cycl}\dfrac{c_{ACV}}{K_{ACV}+c_{ACV}}-k_{Exp}\dfrac{c_{PenN}}{K_{PenN}+c_{PenN}}\right.$ $\left.-k_{SE}(c_{PenN}-kC_{PenN})\right)-\mu c_{PenN}$
Intrinsic deacetoxy-cephalosporin C (DAOC)	$\dfrac{dc_{DAOC}}{dt}=k_{Exp}\dfrac{c_{PenN}}{K_{PenN}+c_{PenN}}-k_{Hyd}\dfrac{c_{DAOC}}{K_{DAOC}+c_{DAOC}}-\mu c_{DAOC}$
Cephalosporin C	$\dfrac{dC_{CPC}}{dt}=\rho k_{Hyd}\dfrac{c_{DAOC}}{K_{DAOC}+c_{DAOC}}C_X$
External Penicillin N	$\dfrac{dC_{PenN}}{dt}=\rho k_{SE}(c_{PenN}-kC_{PenN})C_X$

ACV-tripeptide formation catalyzed by the ACV-synthetase is the rate limiting step of CPC-synthesis. Increase of the intracellular level by genetic engineering methods should enhance the production.

13.3.2
Simple Cybernetic Model for Growth and Production on Sugar and Soy-Oil

An expert system for modeling and control design of biotechnical processes [104] was used for modeling of Cephalosporin C production on sugar syrup (main component maltose) and soy-oil [105]. The process is divided into a growth phase on sugar with parallel uptake of soy-oil and moderate production, and a second production phase on soy-oil only with continuous feeding. According to the data analysis and knowledge acquisition by the expert system, the cybernetic metabolic regulator model (see Chap. 2.3.4) was established as summarized in Table 13.7.

The initial lag-phase on sugar and the diauxic lag-phase for adaptation from sugar substrate (C_S) to soy-oil (C_{Oil}) are modeled by metabolic regulators, Eq. (2.85) in Chap. 2, for $r_{1max}(t)$ and $r_{2max}(t)$; see Table 13.7, row d). Limitation terms for soy-oil or oxygen were not considered because both concentrations were kept above the critical values. The production of Cephalosporin C with specific rate r_P was assumed to proceed more or less with constant specific rate during growth on soy-oil, but slight inhibited by sugar and product. The product inhibition terms in Table 13.7, rows d and e, were suggested by the expert system to describe the reduced product formation and growth at the end of the cultivation, because the data did not allow for clear discrimination of other possible mechanisms. The effect could also be caused by degradation of the product, variation in the synthetic activity of the cells, or in the active portion of the cell mass. Simulation results for a fed-batch cultivation are given in Fig. 13.11 with the model parameters in Table 13.8. The structured unsegregated model is in good agreement with the experimental data when considering the measurements errors. In

Table 13.7. Cybernetic model for Cephalosporin C production in a fed-batch process [102, 103]

	Description	Model equation
a	Stoichiometric model	$\mathbf{q} = \mathbf{Y}\mathbf{r} + \mathbf{m}$ $$\begin{pmatrix} q_X \\ q_S \\ q_{oil} \\ q_P \\ q_{CO2} \end{pmatrix} = \begin{pmatrix} 1 & 0 & 0 & 0 \\ 0 & 1 & 0 & 0 \\ 0 & Y_{S1}^{-1} & 1 & Y_{P1}^{-1} \\ 0 & 0 & 0 & 1 \\ 0 & Y_{C1}^{-1} & Y_{C2}^{-1} & Y_{CP}^{-1} \end{pmatrix} \cdot \begin{pmatrix} \mu \\ r_{S1} \\ r_{S2} \\ r_P \end{pmatrix} + \begin{pmatrix} 0 \\ 0 \\ 0 \\ 0 \\ m_C \end{pmatrix}$$
b	Specific growth rate	$\mu = Y_{XS1} r_{S1} + Y_{XS2} r_{S2}$
c	Metabolic coordination	$J = \mu(r_{S1}, r_{S2}) \rightarrow \max$
d	Constraints for metabolic coordination	$0 \le r_{S1} \le \min\left(r_{1\,max}(t), r_{S1\,max} \dfrac{C_S}{K_{S1} + C_S}\right)$ $0 \le r_{S2} \le \min\left(r_{2\,max}(t), r_{S2\,max} \dfrac{K_{IS}}{K_{IS} + C_S}\left(1 - \dfrac{C_P}{K_{P1}}\right)\right)$
e	Specific rate of Cephalosporin C synthesis	$r_P = r_{P\,max} \dfrac{K_{IP}}{K_{IP} + C_S}\left(1 - \dfrac{C_P}{K_{P2}}\right)$

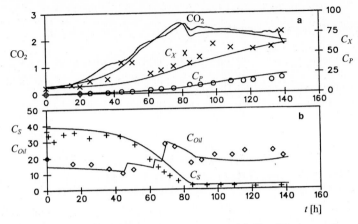

Fig. 13.11a,b. Simulation results with the cybernetic model (*lines*) in comparison to experimental data (*symbols*) for a fed-batch cultivation of *Acremonium chrysogenum* [102]: **a** concentrations of cell mass (C_X), Cephalosporin C (C_P), and mole fraction of CO_2 in the exhaust gas; **b** concentrations of sugar (C_S) and soy oil (C_{Oil}); all liquid phase concentrations are in $kg\,m^{-3}$

Table 13.8. Parameters in the metabolic regulator model, Fig. 13.11 and Table 13.7

Parameter	Value	Unit
r_{S1max}	0.04	h^{-1}
r_{S2max}	0.024	h^{-1}
r_{Pmax}	0.013	h^{-1}
Y_{XS1}	0.75	–
Y_{XS2}	1.0	–
Y_{S1}	5.5	–
Y_{P1}	0.9	–
Y_{C1}	0.9	–
Y_{C2}	0.7	–
Y_{CP}	2.0	–
K_{S1}	1.5	$kg\,m^{-3}$
K_{IS}	0.5	$kg\,m^{-3}$
K_{IP}	22.0	$kg\,m^{-3}$
K_{P1}	35.0	$kg\,m^{-3}$
K_{P2}	35.0	$kg\,m^{-3}$
m_C	0.008	h^{-1}

the simulation, the initial and diauxic lag-phase can be seen on the exhaust gas CO_2-concentration.

13.3.3
Segregated Models Describing Morphological Differentiation

Matsumura et al. [56] presented a structured-segregated model describing the morphological differentiation and regulation of the Cephalosporin C synthesis by the carbon source and methionine. They discriminate the three cell types slim hyphae (X_1), swollen hyphae (X_2), and arthrospores (X_3) as shown in Fig. 13.9. The fundamental equations of the segregated model that emphasizes the importance of an intracellular methionine pool are given in Chap. 2, Table 2.13. It is assumed that methionine stimulates product synthesis and metamorphosis of slim hyphae to swollen, thick-walled hyphae. The intrinsic methionine concentrations in slim hyphae, c_{S21}, swollen hyphae, c_{S22}, and arthrospores, c_{S23}, are calculated by

$$c_{S21} = \frac{C_{S21}}{C_{X1}} \quad c_{S22} = \frac{C_{S22}}{C_{X2}} \quad c_{S23} = \frac{C_{S23}}{C_{X3}} \tag{13.30}$$

and the totally measured intrinsic methionine concentration is then

$$c_{S2} = \frac{c_{S21}C_{X1} + c_{S22}C_{X2} + c_{S23}C_{X3}}{C_X} \tag{13.31}$$

The intrinsic balance of methionine in slim hyphae considers uptake from the medium with rate q_{S21}, endogenous de novo synthesis from inorganic sulfate with rate r_{NS21}, utilization for protein synthesis with rate r_{GS21}, first order decay, and dilution by growth. Since metamorphosis of hyphal cells to thick-walled cells does not change the intrinsic concentration of hyphal cells, the balance becomes

$$\frac{dc_{S21}}{dt} = \overbrace{q_{S21}}^{\text{Uptake}} + \overbrace{r_{NS21}}^{\text{Synthesis}} - \overbrace{k_{D1}c_{S21}}^{\text{Degradation}} - \overbrace{r_{GS21}}^{\text{Utilization}} - \overbrace{\mu_{12}c_{S21}}^{\text{Dilution by growth}} \tag{13.32}$$

The following methionine balance in swollen hyphae has to consider by the last summand the inflow of methionine from the newly formed cells by metamorphosis,

$$\frac{dc_{S22}}{dt} = \overbrace{q_{S22}}^{\text{Uptake}} + \overbrace{r_{NS22}}^{\text{Synthesis}} - \overbrace{k_{D2}c_{S22}}^{\text{Degradation}} - \overbrace{r_{GS22}}^{\text{Utilization}} + \overbrace{\mu_{12}\frac{C_{X1}}{C_{X2}}(c_{S21} - c_{S22})}^{\text{Inflow by metamorphosis}} \tag{13.33}$$

In arthrospores there is no growth activity. Therefore, the balance simplifies to

$$\frac{dc_{S23}}{dt} = - \overbrace{k_{D3}c_{S23}}^{\text{Degradations}} + \overbrace{\mu_{23}\frac{C_{X2}}{C_{X3}}(c_{S22} - c_{S23})}^{\text{Inflow by morphogenesis}} \tag{13.34}$$

A fictive rate limiting enzyme, e.g., the ring-expansion enzyme, in the swollen hyphae that is repressed by glucose is assumed to be responsible for the CPC synthesis. Formation of this enzyme in the swollen hyphae is assumed to follow a delayed activation by methionine with lag-time t_1 and immediate repression by glucose, and is described by the intrinsic balance

$$\frac{dc_E(t)}{dt} = \frac{r_{\max E}c_{S22}(t - t_1)}{K_E + c_{S22}(t - t_1)} \cdot \frac{1 + K_{I1}C_{S1}(t)}{1 + K_{I2}C_{S1}(t)} \cdot \frac{C_{X2}(t - t_1)}{C_{X2}(t)}$$
$$- \left(k_{DE} + \mu_{12}\frac{C_{X1}(t)}{C_{X2}(t)} \right) c_E(t) \tag{13.35}$$

The dilution by newly formed X_2-cells being deficient of enzyme is considered in the last term. The above balance is actually inconsistent in regard to structured models, because the intrinsic concentration c_{S22} – being averaged over the population – is used in the non-linear Michaelis-Menten kinetics. The averaged intrinsic rate can only be calculated from averaged intrinsic concentrations when the model is linear, i.e., the first factor must be either constant or proportional to c_{S22}. Please refer to Chap. 2 for further discussion of structured models. The model was used to simulate qualitatively a process with linear-increasing substrate feed-rate over time. By the feeding of glucose and methionine the process duration could be extended and the product concentration was increased.

A model considering only thin hyphae (X_1) and swollen hyphae (X_1), which are formed at impaired growth conditions, was developed by Chu and Constantinides [55] for batch cultivations on glucose and sucrose. The basic segregated model is described in Table 2.12, Chap. 2. Both morphological forms catabolize substrate, can grow, and form new slim hyphae. In the model, production of CPC, formed by swollen hyphae, is repressed by glucose. The diauxic lag-phase for the switch to growing on sucrose is modeled by a formal delay function in a non-mechanistic way,

$$\mu_{S2} = \begin{cases} \mu_{S2\,max}\tilde{r}(C_{S2})\left(\frac{1}{\pi}\arctan\left(k_1\left(\frac{t-t_1}{t_{lag}}-1\right)\right)+\frac{1}{2}\right) & t > t_1 \\ 0 & t \le t_1 \end{cases} \tag{13.36}$$

where t_1 is the time of glucose limitation. Functional expressions for t_{lag} in dependence on the total cell mass and sucrose concentration were afterwards proposed by Basak et al. [101]. Similar to the previous model, delayed synthesis of a fictive rate limiting enzyme is assumed to be responsible for a lag-phase before start of the CPC-production after depletion of glucose. Product synthesis is proportional to the concentration of the enzyme in the thick-walled cells that is synthesized during metamorphosis. The intrinsic balance for the rate-limiting enzyme given in the paper violates the conservation law of mass and is not repeated here. A modified expression with a synthesis rate $\mu_{12}kC_{X1}$ and a decay constant k_D could read

$$\frac{dc_E}{dt} = \mu_{12}\frac{C_{X1}}{C_{X2}}(k - c_E) - k_D c_E \tag{13.37}$$

This means that when the percentage of thick-walled cells is high, the rate of change in enzyme concentration by the inflowing enzyme produced during metamorphosis is slowed down, and vice versa. The concentration will not change if the newly formed cells have the same state as the existing ones, i.e., $k=c_E$. The model was supplemented by introducing a dependency of kinetic parameters on temperature and pH. The extended model was then used to generate optimal control profiles for these control variables, which increased the productivity by more than 50%.

A segregated-structured model aimed at industrial fed-batch cultivations on complex substrate mixtures was developed for Cephalosporin C production by Meyerhoff [60]. In the experiments, dry glucose syrup, which contains mainly maltose and oligomers of this sugar, and soy-oil were used as carbon source [61]. There is no clear separation of the growth phases on sugars and oil, both substrates being used in parallel. Nevertheless, after limitation of the sugars a lag-phase can be observed, where the cells adapt further to growth on oil as sole carbon source. The high-producing strain used in the experiments also showed no clear separation of growth and production phase. Production starts very early when still significant amounts of sugars are present in the medium. In this model, no precursors were considered, since during the experiments limitation of methionine, ammonia, or phosphate were avoided.

Table 13.9. Segregated model for a high-producing strain of *Acremonium chrysogenum* [60]

	Description	Model equation
a	Growth of active hyphae	$\mu = \mu_S + \mu_{Oil}$
		$\mu_S = \mu_{S\,max}\dfrac{C_S}{K_S + C_S}$
		$\mu_{Oil} = \mu_{Oil\,max}\dfrac{C_{Oil}}{K_{Oil} + C_{Oil}}c_E + \mu_{Oil\,maxS}\dfrac{C_S}{K_S + C_S}$
b	Total biomass	$C_X = C_{X1} + C_{X2} + C_{X3} + C_{X4}$
c	Metamorphosis reactions of inactive cells to active hyphae, active hyphae to swollen hyphae, and swollen hyphae to arthrospores	$\mu_{12} = k_{12}$ $\mu_{23} = k_{23}$ $\mu_{34} = k_{34}\dfrac{K_1}{K_1 + K_S}$
d	Product formation by swollen hyphae	$q_P = k_P\dfrac{C_{Oil}}{K_{Oil}^+ + C_{Oil}}$
e	Sugar uptake	$Q_S = -\dfrac{\mu_S}{Y_{XS}}C_{X2} - \dfrac{q_P}{Y_{PS}}C_{X3}$
f	Soy oil uptake	$Q_{Oil} = \dfrac{\mu_{Oil}}{Y_{XOil}}C_{X2} + \left(m_{Oil}(C_{X2} + C_{X3}) + \dfrac{q_P C_{X3}}{Y_{POil}}\right)\dfrac{C_{Oil}}{K_{Oil}^* + C_{Oil}}$
g	Intrinsic balance of oil catabolizing enzyme	$\dfrac{dc_E}{dt} = \mu_{Oil\,max}\dfrac{C_{Oil}}{K_{Oil} + C_{Oil}}c_E + k_{max}\dfrac{K_I}{K_I + C_S}(1 - c_E) - \mu c_E$

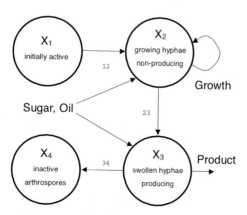

Fig. 13.12. Structure of the segregated model of Meyerhoff [60]

The model given in Table 13.9 considers initially inactive (X_1), growing but non-producing (X_2), and swollen producing hyphae (X_3), as well as inactive arthrospores (X_4) as shown in Fig. 13.12. By the inactive form an initial lag phase after inoculation of the reactor is modeled. These cells are transformed into the active growing form, which then can differentiate to swollen hyphae, and, after limitation of the sugars, to arthrospores. Growing hyphae use the sugar and oil substrates with specific growth rates μ_S and μ_{Oil}. Arthrospores are formed from swollen hyphae at low sugar concentrations. The sugar uptake, mainly maltose, is determined by growth and product formation in the respective morphological states. Since in the measured sugar concentration some sucrose is contained, which cannot be metabolized by the fungus, the total sugar concentration is modeled as the sum of maltose and a constant value of sucrose. The balance equation for soy oil considers the effects of growth, maintenance, production, and feeding. The experimental results suggest that the oil uptake in parallel to sugar follows a different mechanism than for oil as a sole

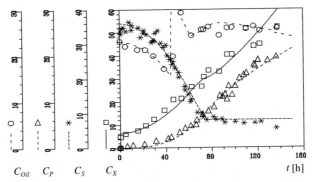

Fig. 13.13. Experimental data (*symbols*) and simulation results (*lines*) by the segregated model [60] for a fed-batch cultivation of *Acremonium chrysogenum*. Concentrations of total biomass (C_X), total sugar (C_S), Cephalosporin (C_P), and soy oil (C_{Oil}), all in kg m^{-3}

Table 13.10. Model parameters for simulation of the fed-batch cultivation of *Acremonium chrysogenum* in Fig. 13.13 by the segregated model in Table 13.9

Parameter	Value	Unit
μ_{Smax}	0.028	h^{-1}
μ_{Oilmax}	0.032	h^{-1}
k_{max}	0.115	h^{-1}
Y_{XS}	0.53	–
Y_{XOil}	0.8	–
Y_{PS}	1.5	–
Y_{POil}	1.2	–
K_S	3.3	g l^{-1}
K_I	0.3	g l^{-1}
k_{12}, k_{23}	0.027	h^{-1}
k_{34}	0.032	h^{-1}
k_P	0.023	h^{-1}
m_{Oil}	0.0027	h^{-1}

carbon source. The first one is assumed to be controlled by the sugar uptake rate, and the second one by Monod-kinetics dependent on the oil concentration. The total specific growth rate on soy oil then includes both terms as a sum. The key enzyme for pure growth on oil is inducible and described by an intrinsic balance, Table 13.9, row g).

The model was able to describe experimental data for several cultivations with only small variations in the parameters. The simulation results of one cultivation are given in Fig. 13.13 with the model parameters in Table 13.10. Maltose is the preferred substrate for this organism. Therefore, the oil uptake rate in the first growth phase is much lower than after the sugar limitation and adaptation of the oil catabolizing system. For this process, the unique discrimination in growth and production phase is difficult. The presence of sugar has no strong repression effect on the product synthesis. Therefore, Cephalosporin C is synthesized just from the beginning. The lag-phase in growth and oil uptake after 70 h, when maltose became limiting, can be deduced from the temporary increasing oil concentration, and more clearly from exhaust gas CO_2 that is not shown.

13.4
Process Control and Optimization

13.4.1
Problems and Possibilities

The field for process control can be divided into four areas: determination of the optimum mode of operation of the bioreactor, including optimum conditions for inoculation under consideration of the preculture; dynamic optimization of process variables; feedback control of operating parameters and process variables to constant setpoints or optimum dynamic profiles; and finally optimum scheduling of the entire plant. Process variables of interest for control are temperature, pH, substrate feed, oxygen supply, precursor feed, and nitrogen supply.

On the plant level, the optimization can be done by dynamic scheduling of the process steps and determination of a suitable harvest time for the cultivation. Using the on-line estimated profit and the natural variance in the productivity, Yuan et al. [75] suggest a strategy for deciding when a cultivation should be stopped. In comparison to a set of historic data, the actual cultivation run is classified as bad, normal, or good. By stopping bad runs earlier and prolonging the cultivations classified as good ones, the profit can be increased without changing the control scheme for the cultivations. The advantage of this method is that it does not directly rely on a mathematical model. This is only used for estimation of the product concentration when direct measurements are not available. Further improvements can be expected by combining the dynamic scheduling with optimal dynamic control of each run.

The frame for process control is determined by the physiology of the fungi. The catabolite repression of antibiotics synthesis requires roughly a two-phase process control strategy with a first phase of optimum growth and a second phase with low growth rate but maximum production. This procedure was extensively studied experimentally and is considered as optimal. It can be realized in two ways. In a batch process, a mixture is provided of a rapidly catabolizable first substrate, e.g., glucose, and a slowly degradable second substrate, e.g., lactose. When the first substrate is used up, the process switches from growth to production phase by itself. But the growth rate is then fully determined by the choice of the substrate and cannot be further influenced directly, so it may deviate from optimum conditions for production. Therefore, additional precautions such as nitrogen limitation must then be employed. These disadvantages are avoided by fed-batch cultivation, where the growth rate in the production phase can in principle be controlled to any value below the maximum rate. In the first growth phase the process is usually still run as batch cultivation. The two-phase strategy may be supported by a proper control of temperature and pH levels, because these variables also have different optimum values for growth and production; pH strongly influences morphological differentiation and degradation of the product. The biomass should grow as much as possible in the first growth phase to ensure high productivity afterwards, but oxygen limitation must be avoided [19, 62]. The optimal initial amount of substrate and the maximum attainable biomass depends on the oxygen transfer capacity of the reactor [22, 27, 28, 48]. The production phase is usually continued into the stationary phase without net growth. But the process has to be stopped, e.g., by cooling below 10 °C, before the onset of autolysis induces a strong increase in pH that leads to degradation of the product. Other stopping criteria with respect to downstream processing are the rheology or filtration properties of the broth.

The optimum mode of operation of a general bioprocess, where substrate concentration determines the growth and production kinetics, was investigated by Modak and Lim [124]: batch, fed-batch, repeated batch/fed-batch, or continuous. According to their kinetics, the processes were classified into four types: for maximization of yield, batch operation is optimal when the intrinsic yield (the ratio of specific product formation to substrate uptake) increases with substrate concentration; continuous operation is optimal when the specific yield decreases, fed-batch when it goes through a maximum, and repeated operation for constant specific yield. For maximization of productivity, continuous operation is always optimal from the theoretical point of view, but in practice this is avoided by reasons of operational and strain stability. The process can be prolonged and the productivity also increased by repeated batch/fed-batch operation mode, which reduces the required time for reactor preparation and growth phases. This could be shown generally in theoretical investigations for production of antibiotics and non-growth associated products [32, 85–87]. Hasegawa et al. [84] used a model for batch penicillin production [70] to investigate repeated batch strategies for enhancing productivity. The improvement is due to increased production rate, prolonged cultivation, and higher final concentration. Guthke and Knorre [40] investigated the optimal control of repeated fed-batch cultivations by the model of Nestaas and Wang [39], and found an optimal cycle period of 48 h and a draw-off ratio of about 80%. This is lower than the result in [32] but still unusually high. The time for addition of fresh media can be very critical, and the production may even stop under adverse conditions [33]. The maximum number of repeated cycles is bounded by the capacity of the plant for substrate preparation, accumulation of toxic or inhibitory substances, or degeneration of the strain. Nevertheless, repeated fed-batch cultivation is of great interest for industrial production.

Several authors have dealt with optimal control of fed-batch processes in general, and antibiotics production in particular [21, 63–68]. Generally unstructured kinetic models are used and comparison to experimental data is rare. Mostly the substrate feed or concentration is optimized. The optimum control profile consists of a sequence of maximum feeding, batch, singular feeding during production phase, and a short final batch phase. Depending on the boundary conditions for the optimization, not all phases of the sequence may be present. The optimum initial substrate concentration was found to be higher when productivity is maximized instead of yield. Numerical algorithms for practical computation of the control were given in Lim et al. [78] for penicillin production. The results were later improved with Iterative Dynamic Programming by Luus [120]. A penalty function was used to handle constraints of the state variables and the global optimum was supposed to be found. Diener and Goldschmidt [122] published a suboptimal control strategy for maximizing product concentration, yield, or productivity in fed-batch cultivations using phase partitioning. In this scheme, the global optimization problem defines the sequence of phases being computed off line, while during each phase only local criteria have to be fulfilled which can easily be translated into feedback control laws. The method was applied for penicillin production using the model of Bajpai and Reuß [28]. The results of the suboptimal strategy were quite close to the global optimum that requires a much higher effort for its determination. Guthke and Knorre [69] calculated optimum concentration profiles for substrate during antibiotics production by using a general model where the specific product formation is maximum below the maximum growth rate. Suboptimal strategies were proposed that can be implemented more easily. Bajpai and Reuß [48] evaluated different feeding strategies on the basis of simulations with their simple unstructured model. The effect of oxygen limitation is discussed. It was found that the differences in productivity among the feeding strategies are less than expected, so the robustness in practical application should be decisive for their selection. San and Stephanopoulos [121] investigated the optimal control of the substrate inlet concentration at fixed flow rate under constraints in maximum cell mass and substrate concentration by the model of Bajpai and Reuß [28]. Rodrigues and Filho [123] optimized the productivity numerically by the above model using a simplex method. Temperature and pH optimization of Cephalosporin production was investigated by Chu and Constantinides [55]. Suitable temperature profiles were determined by Constantinides et al. [20, 70] for batch penicillin production by means of a simple unstructured model and application of the continuous maximum principle.

Automatic feedback control of the feeding rate of carbon sources for antibiotics production was not so extensively studied as for other biotechnical processes, e.g., baker's yeast production. This has a number of reasons. The setpoint – or better time profile – of operation for maximum productivity is not so clearly defined due to the still somewhat limited knowledge of the process dynamics, although the mathematical models may delude a clearer image. Furthermore, there is no direct sensitive and unique indication for the metabolic state of the culture on easily accessible measurements, such as RQ for baker's yeast. Due to the morphological heterogeneity of the culture the optimum rate of substrate supply depends on the biomass fraction in the producing morphological state. So even for identical total biomass the optimum substrate feed is different when the morphological composition of the population changes, and up to now this is difficult to determine on-line. Another problem is the use of ill-defined substrates from natural resources that may only be catabolized partially during the process. Application of a single-loop control for substrate flow seems to be generally problematic in antibiotics production. When there is indeed a maximum for the specific production over growth rate as assumed by a number of models, it is impossible to decide whether the substrate supply should be reduced or increased without additional knowledge. So further information from other process variables has to be taken into account. Therefore, practical feedback control schemes are mostly realized as tracking control according to a predetermined profile that was found experimentally or with assistance of a model. The preset profile can be for cell mass; but because this is difficult to measure, estimation schemes must be established by balancing or mathematical models. These can use, for example, exhaust gas measurements of O_2 and CO_2. Variants of this control take the specific growth rate as controlled variable instead of cell mass. Nevertheless, these control schemes are at most suboptimal since they do not consider the true actual morphological or physiological state of the culture. But often the productivity and reproducibility can be improved even by simple control, because critical situations during the process can be avoided or reduced.

Mou and Cooney studied the computer control of penicillin production by means of balancing principles [35] with the aim of increasing reproducibility. Carbon dioxide production was used to estimate cell mass and specific growth rate, and to control both according to predefined schemes. The concept was extended for cultivations containing corn-steep-liquor [119] as complex substrate by using an approximate carbon balance. Fed-batch control by different measured variables was the subject of a number of papers: pO_2 [62, 88]; pH, [89, 90]; carbon dioxide production [91]; specific growth rate [22, 34, 36, 92]. Montague et al. [29] used an extended Bajpai-Reuß-model for estimation of cell mass by a Kalman-filter by means of the carbon dioxide production rate. By adaptive control of substrate feed, the cell mass was kept on a predetermined trajectory. Carbon dioxide production and RQ were used by Nelligan and Calam [36] for estimation of μ and desired oxygen uptake rate. On the basis of different predetermined strategies the process was then controlled by both variables while avoiding oxygen limitation. Nestaas and Wang [39] used the morphologically structured model for open-loop and feedback control of fed-batch penicillin processes according to the desired cell mass and specific growth rate. Feedback control was achieved by measuring the hyphal density with a filtration probe. Suboptimal control strategies were investigated by Van Impe et al. [125]. A heuristic control law for the feed-rate during production phase by the substrate concentration as control variable gave good performance in simulations and was quite close to the global optimum. Adaptive extensions, either based on an a simple observer for the substrate concentration or on the carbon dioxide production rate, were also proposed.

13.4.2
Example for Dynamic Optimal Control of Fed-Batch Antibiotics Production

In the work of Meyerhoff [60], summarized in [99], the model given in Sect. 13.3.3, Table 13.9, was applied for dynamic optimization of the sugar and oil feeding for cultivations of *Acremonium chrysogenum* by using the Iterative Dynamic Programming (IDP) algorithm [71–74]. This is easy to use and has good convergence properties compared to methods based on Pontryagin's Maximum Principle. Several performance criteria were tested: total mass of product, product yield per supplied substrate, and economic profit. During the simulation studies, a maximum filling volume of the reactor and a maximum oil concentration of $8\,g\,l^{-1}$ were introduced as additional restrictions. Since the oil promotes coalescence of the air bubbles, the sense of the latter restriction is to maintain a sufficient oxygen supply to the reactor.

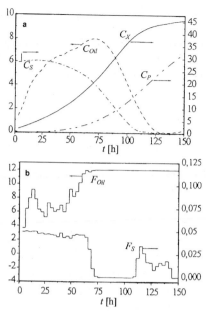

Fig. 13.14a,b. Simulation of the optimal control of sugar and soy-oil feeding for a fed-batch cultivation of *Acremonium chrysogenum* by a segregated model [60]: **a** concentrations of total biomass (C_X), sugar (C_S), Cephalosporin C (C_P), and soy oil (C_{Oil}), all in kg m^{-3}; **b** flow rates of sugar (F_S in l h^{-1}), and soy oil (F_{Oil} in g h^{-1}). Reprinted from [99]

For the optimization, the total process duration (t_f =150 h) and maximum feeding rates were fixed ($F_S <0.051$ l h^{-1}, $F_{Oil}<12$ g h^{-1}).

Figure 13.14 shows the optimization results for both of the control variables, sugar (F_S) and soy oil feed (F_{Oil}), for a performance index of final total mass of the product. Compared to the original process, the growth phase is prolonged to about 105 h by a higher sugar feed under near-batch conditions. The sugar flow rate also supports sufficient growth in the production phase when the oil concentration decreases. Soy oil is used up to the maximum amount that can be afforded under the given restrictions on states and feeding rate. From the simulation, the optimized process shows a clearly increased product concentration and productivity by enforced growth. This is due to the seemingly weak catabolite repression of product synthesis that allows for relative high product formation even under growth favoring conditions.

13.4.3
Economic Optimization for Mycelia Fed-Batch Cultivation

Optimization of a bioprocess is mostly a multi-objective problem tackling several criteria such as productivity, yield, and quality of the product. These are usually not independent of each other, and the manipulating variables influence them simultaneously. A more general way is to optimize the economic profit directly. The profit function is an objective description of a process under the aspect of economy. Global economic optimization of a general antibiotics plant was investigated by Okabe and Aiba [66] based on data of an industrial process. They considered the sections cultivation, filtration, extraction, and final treatment of the product, e.g., by crystallization. A major part of the costs is for cultivation, i.e., raw material and energy, and therefore its optimization is of greatest interest. The influence of operational procedures, variation in down-stream processing steps, disposal of solid waste, and selection of cultivation medium could be successfully evaluated with respect to the economic effectiveness. Genon and Saracco [32] extended the Bajpai-Reuß model [28] by economic balances to evaluate the profitability of fed-batch, repeated fed-batch, and continuous cultivations in one- and two-stage reactor systems. It was found that with this simple unstructured model the last system performed best when neglecting the known operational difficulties for continuous cultivations. Repeated batch with a cycle period of 96 h and a draw-off ratio of >95% of the reactor operating volume was the second choice.

The on-line profit estimation and optimization of industrial multi-reactor plants for mycelia fed-batch cultivations is further discussed by Yuan et al. [75]. For a single batch, the gross profit is the revenue from product sales minus the total input for raw materials, energy, capital, personnel, and other costs, which were expended for winning the product. The goal of optimization is to get the highest gross profit within a minimal operation period. The profit function is then given by

$$J(t_f) = \frac{\text{revenue} - \text{total costs}}{t_f + t_{pr}} \tag{13.38}$$

where t_f is the process duration and t_{pr} the time interval between two successive batches in the same reactor in which emptying, cleaning, batch medium preparing, sterilization, and inoculation are done. The total costs as summarized in Table 13.11 were divided into two parts – the direct production-related costs for raw material and energy consumption from pretreatment to product package, and the remaining indirect costs. The material consumption in downstream processing is product concentration-dependent. So the related terms, rows e–g in Table 13.11, are multiplied by a correction factor, φ, being the ratio of actual product concentration to desired product concentration under design conditions. The operating factor, representing the fraction of time where the plant is in full-scale production, was assumed as 90% in row h). The profit function is then in detail

$$J(t_f) = \frac{V_L(t_f)C_P(t_f)0.91P_{Sale} - P_{ini} - P_{in} - P_{opt}t_f - V_L(t_f)\left(P_{extr} + P_{cryst} + P_{dry}\right)}{t_f + t_{pr}}$$

$$- P_{indir}V_R \tag{13.39}$$

where P_{Sale} is the sales price of product, and the separation yield was estimated to 0.91.

The profit function can be used either for off line optimization similar to the examples above, or for real-time control and scheduling of the process. In this case its development in time is of great interest. Figure 13.15 shows the typical time course of the profit function of a batch. At the beginning, the profit is negative because of the cost for inoculum and initial batch medium preparation. During the following cultivation, the profit increases in connection with the amount of product. The beginning of the profitable operation is at time t_1; the stationary phase of the cultiva-

Table 13.11. Costs per m^3 of Penicillin production for a 50-m^3 production reactor [75, 81, 82]

Cost	Calculation formula
Direct	

a   Seed inoculum, approximated as direct production costs:
$P_{INO}=0.23P_{Imp}+ 1730P_{Air}+1.0P_W+251P_{CW15}+1.01P_{Ele}+0.012P_{Steam}+30P_{Glu}+10P_{Lac}+0.5P_{KH2P}+ 5P_{CaCO3} +2P_{NH4S}+15P_{Starch}$

b   Initial input, includes expenses for sterilization, medium preparation and inoculum:
$P_{ini}=33P_W+6P_{Steam}+6P_{INO}+1.8P_{Desmo}+108P_{Glu}+45P_{Starch}+14P_{KH2P}+180P_{CaCO3}+180P_{NH4S}+90P_{K2SO4}+360P_{Pharma}$

c   Feed solution: nitrogen source, carbon source and precursor, the factor of 1.15 considers the costs for pretreatment:

$$P_{in} = 1.15 \int_0^{T_f} \left[F_S(t)C_S^R(t)P_{Glu} + F_{PAA}(t)P_{PAA} + F_{NH3}(t)P_{NH3} \right.$$
$$\left. + F_{NH4S}(t)C_{NH4S}^R(t)P_{NH4S}\right] dt$$

d   Energy for aeration, agitation and cooling, including electric power and cooling water: $P_{opt} =28P_{CW4}+132P_{Ele}+4000P_{Air}$

e   Product extraction from broth: consumption of extract solvent, cooling water and electric power: $P_{extr}=2.24P_{BuAc}\,\varphi +0.2P_{Net}\,\varphi +0.08P_{H2SO4}+P_{CW15}+2P_{CW4}+97P_{Ele}$

f   Product crystallization: salts, cooling water and electric power:
$P_{cryst}=10.6P_{KAc}\,\varphi +0.6P_{CW4}+50P_{Ele}$

g   Drying: consumption of steam, nitrogen, wash solution, and so on:
$P_{dry}=2P_{Isop}\,\varphi +146P_{N2}\,\varphi +1.5P_{CW4}+5P_{Ele}+0.01P_{Steam}$

Indirect

h   Total indirect costs: credit repayment, labor expense, maintenance and repairs, research and development, insurance, storage and transportation, sales expenses:
$P_{indir}=$Total indirect costs/($360\times24\times0.9\times$Number of reactors$\times V_R$)

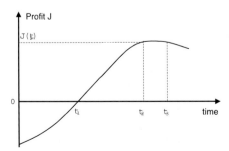

Fig. 13.15. Typical time course of a profit function during fed-batch antibiotics production

Table 13.12. Prices of raw materials and product for penicillin production [75]

Price	Symbol	Value	Unit
compressed air	P_{Air}	0.004	DM m^{-3}
butyl acetate	P_{BuAc}	2.37	DM kg^{-1}
CaCO$_3$	P_{CaCO3}	0.45	DM kg^{-1}
cooling water (4 °C)	P_{CW15}	0.15	DM m^{-3}
cooling water (15 °C)	P_{CW4}	0.3	DM m^{-3}
anti-foam solution	P_{Desmo}	5.0	DM l^{-1}
electricity	P_{Ele}	0.15	DM (kWh)$^{-1}$
glucose	P_{Glu}	0.53	DM kg^{-1}
H$_2$SO$_4$ (98%)	P_{H2SO4}	0.46	DM kg^{-1}
inoculum liquid with a volume of 100 ml	P_{Imp}	10.0	DM per flask
approximate indirect costs	P_{indir}	1.86	DM h^{-1}m^{-3}
isopropanol	P_{Isop}	1.0	DM kg^{-1}
K$_2$SO$_4$	P_{K2SO4}	0.99	DM kg^{-1}
KH$_2$PO$_4$	P_{KH2P}	4.09	DM kg^{-1}
lactose	P_{Lac}	20.0	DM kg^{-1}
nitrogen	P_{N2}	0.05	DM m^{-3}
net promote	P_{Net}	2.0	DM kg^{-1}
ammonia solution (16.5%)	P_{NH3}	0.15	DM kg^{-1}
(NH$_4$)$_2$SO$_4$	P_{NH4S}	1.18	DM kg^{-1}
phenylacetic acid	P_{PAA}	3.25	DM l^{-1}
pharma-medium	P_{Pharma}	0.25	DM kg^{-1}
sales price of penicillin G	P_{Sale}	80.0	DM kg^{-1}
corn starch	P_{Starch}	0.85	DM kg^{-1}
steam	P_{Steam}	0.028	DM kg^{-1}
process water	P_W	1.6	DM m^{-3}

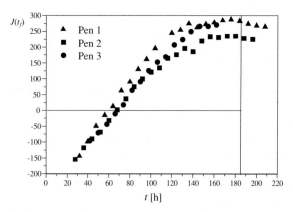

Fig. 13.16. Profit functions of three industrial Penicillin G fed-batch cultivations. Reprinted from [75] "Profit optimization for mycelia fed-batch fermentation" Copyright (1999), with permission of Elsevier Science

tion is more or less reached at time t_2, where the profit approaches the maximum value; later on the profit function may drastically decrease, which can be caused by autolysis of mycelia, hydrolysis of product, or contamination. The cultivation must be stopped and downstream processing started before this point at t_3. The estimated profit function, converted to equivalents of a 50-m^3 reactor, for three industrial Penicillin production runs is presented in Fig. 13.16. In Table 13.12 approximate prices of raw materials and product for the profit function estimation of Penicillin G production are summarized. In the stationary phase, the difference of the profit function between runs Pen 1 and Pen 2 is about 50 DM h^{-1}. When these two batches are stopped at 186 h and t_{pr} is set to 14 h, the gross profit difference is $50 \times (186+14) = 10,000$ DM. From Fig. 13.16 the potential of dynamic scheduling by using the natural variance in the profit function is also obvious. Compared to fixed harvesting time, the profit can be increased by stopping those cultivations in advance that have a lower profit or reach their maximum profit earlier, while processes with high maximum profit may be prolonged even when the increase in profit function is delayed. The possible gain in profit by such procedure was estimated at about 5% without changing the control scheme of the cultivation.

References

1. Vandamme EJ (ed) (1984) Biotechnology of industrial antibiotics. Dekker, New York
2. Moo-Young M (ed) (1984) Comprehensive biotechnology, vol 3: the practice of biotechnology, current commodity products. Pergamon, Oxford
3. Queener S, Neuss N (1982) The biosynthesis of β-lactam antibiotics. In: The chemistry and biology of β-lactam antibiotics, vol 3. Academic Press, New York, p 141
4. Martin JF, Revilla G, Zanca DM, Lopez-Nieto MJ (1983) Carbon catabolite repression of penicillin and cephalosporin biosynthesis. In: Trends in antibiotics research. Antibiotics Research Association, Japan, p 258
5. Court JR, Pirt SJ (1981) J Chem Technol Biotechnol 31:235
6. Niehoff J (1987) PhD thesis, Universität Hannover
7. Wittler R, Schügerl K (1985) Appl Microbiol Biotechnol 21:834
8. Hilgendorf P, Heiser V, Diekmann H, Thoma M (1987) Appl Microbiol Biotechnol 27:247
9. Nielsen J (1992) Adv Biochem Eng Biotechnol 46:187
10. van Suijdam JC, Metz B (1981) Biotechnol Bioeng 23:111
11. Metz B, de Bruijn EW, van Suijdam JC (1981) Biotechnol Bioeng 23:149
12. van Suijdam JC, Metz B (1981) J Ferment Technol 59:329
13. Trinci APJ (1970) Arch Microbiol 73:353
14. Metz B, Kossen NWF (1977) Biotechnol Bioeng 19:781
15. Whitaker A (1987) Int Ind Biotechnol 2:85
16. Wittler R, Baumgartl H, Lübbers DW, Schügerl K (1986) Biotechnol Bioeng 28:1036
17. Hotop S, Möller J, Niehoff J, Schügerl K (1993) Proc Biochem 28:99
18. Edwards N, Beeton S, Bull AT, Merchuck JC (1989) Appl Microbiol Biotechnol 30:190
19. Wittler R (1983) PhD thesis, Universität Hannover
20. Constantinides A, Spencer JL, Gaden EL (1970) Biotechnol Bioeng 12:803
21. Fishman VM, Biryukov VV (1974) Biotechnol Bioeng Symp 4. Wiley Interscience, New York, p 647
22. Calam CT, Russel DW (1973) J Appl Chem Biotechnol 23:225
23. Megee RD, Kinoshita S, Fredrickson AG, Tsuchiya HM (1970) Biotechnol Bioeng 12:771
24. Trinci APJ (1974) J Gen Microbiol 81:225
25. Righelato RC (1979) The kinetics of mycelial growth: fungal walls and hyphal growth. Cambridge University Press, London, p 385
26. Bull AT, Trinci APJ (1977) Adv Microbiol Physiol 15:1

27. Hegewald E, Wolleschensky B, Guthke R, Neubert M, Knorre WA (1981) Biotechnol Bioeng 23:1133
28. Bajpai RK, Reuß M (1980) J Chem Technol Biotechnol 30:332
29. Montague A, Morris AJ, Wright M, Aynsley M, Ward A (1986) Can J Chem Eng 64:567
30. Nicolai BM, Van Impe JF, Vanrolleghem PA, Vandewalle J (1991) Biotechnol Letters 13:489
31. Heijnen JJ, Roels JA, Stouthamer AH (1979) Biotechnol Bioeng 21:2175
32. Genon G, Saracco G (1992) Chem Biochem Eng Q 6:75
33. Calam CT, Ismail BAK (1980) J Chem Technol Biotechnol 30:249
34. Cooney CL, Mou DG (1982) Application of computer monitoring and control to the penicillin fermentation. In: 3rd International Conference on Computer Applications in Fermentation Technology, Manchester, England, 1981, p 217
35. Mou DG, Cooney CL (1983) Biotechnol Bioeng 25:225
36. Nelligan I, Calam CT (1983) Biotechnol Letters 5:561
37. Calam CT (1979) Folia Microbiol 24:276
38. Germain B, Mele M, Kremser M (1980) Biotechnol Bioeng 22:255
39. Nestaas E, Wang DIC (1983) Biotechnol Bioeng 25:781
40. Guthke R, Knorre WA (1987) Bioproc Eng 2:169
41. van Suijdam JC, Hols H, Kossen NWF (1982) Biotechnol Bioeng 24:177
42. Kluge M, Siegmund D, Diekmann H, Thoma M (1992) Appl Microbiol Biotechnol 36:446
43. Aynsley M, Ward AC, Wright AR (1990) Biotechnol Bioeng 35:820
44. Prosser JI, Trinci APJ (1979) J Gen Microbiol 111:153
45. Nielsen J (1993) Biotechnol Bioeng 41:715
46. Paul GC, Thomas CR (1996) Biotechnol Bioeng 51:558
47. Tiller V, Meyerhoff J, Sziele D, Schügerl K, Bellgardt KH (1994) J Biotechnol 34:119
48. Bajpai RK, Reuß M (1981) Biotechnol Bioeng 23:717
49. Meyerhoff J, Tiller V, Bellgardt KH (1995) Bioproc Eng 12:305
50. Yang H, Reichl U, King R, Gilles ED (1991) Biotechnol Bioeng 39:44
51. Yang H, King R, Reichl U, Gilles ED (1991) Biotechnol Bioeng 39:49
52. Meyerhoff J, Bellgardt KH (1995) Bioproc Eng 12:315
53. Meyerhoff J, Bellgardt KH (1995) J Biotechnol 38:201
54. Demain AL, Ahronowitz Y, Martin JF (1983) Metabolic control of secondary biosynthetic pathways. In: Vining LC (ed) Biochemistry and genetic regulation of commercially important antibiotics. Addison-Wesley, Reading, MA
55. Chu WBZ, Constantinides A (1988) Biotechnol Bioeng 32:277
56. Matsumura M, Imanaka T, Yoshida T, Tagichi H (1981) J Ferment Technol 59:115
57. Matsumura M, Imanaka T, Yoshida T, Tagichi H (1980) J Ferment Technol 58:197
58. Matsumura M, Imanaka T, Yoshida T, Tagichi H (1980) J Ferment Technol 58:205
59. Malmberg LH, Hu WS (1992) Appl Microbiol Biotechnol 38:122
60. Meyerhoff J (1995) PhD thesis, Universität Hannover
61. Tollnik C (1993) PhD thesis, Universität Hannover
62. Giona AR, De Santis R, Marelli L, Toro L (1976) Biotechnol Bioeng 18:493
63. Modak JM, Lim HC (1987) Biotechnol Bioeng 30:528
64. Modak JM, Lim HC, Tayeb YJ (1986) Biotechnol Bioeng 28:1396
65. Modak JM, Lim HC (1989) Biotechnol Bioeng 33:11
66. Okabe M, Aiba S (1975) J Ferment Technol 53:730
67. Hong J (1986) Biotechnol Bioeng 28:1421
68. Vicik SM, Fedor AJ, Swartz RW (1990) Biotechnol Prog 6:333
69. Guthke R, Knorre WA (1981) Biotechnol Bioeng 23:2771
70. Constantinides A, Spencer JL, Gaden EL (1970) Biotechnol Bioeng 12:1081
71. Luus R (1989) Hung J Ind Chem 17:523
72. Luus R (1990) Int J Control 52:239
73. Bojkov B, Luus R (1994) Chem Eng Res Des 72:72
74. Bojkov B, Luus R (1992) Ind Eng Chem Res 31:1308
75. Yuan JQ, Guo SR, Schügerl K, Bellgardt KH (1997) J Biotechnol 54:175
76. Paul GC, Kent CA, Thomas CR (1993) Biotechnol Bioeng 42:11

77. Bartoshevic YE, van den Heuvel JC (1990) Chem Eng J 44:B1
78. Lim HC, Tayeb YJ, Modak JM, Bonte P (1986) Biotechnol Bioeng 28:1408
79. Prosser JI, Tough AJ (1991) Crit Rev Biotech 10:253
80. Keim J, Shen YQ, Wolfe S, Demain AI (1984) Europ J Appl Microbiol Biotechnol 19:232
81. Grant EL, Ireson WG, Leavenworth RS (1982) Principles of engineering economy II, 7th edn. Wiley, New York
82. Ulrich GD (1984) A guide to chemical engineering process design and economics. Wiley, New York
83. Wittler R, Matthes R, Schügerl K (1983) Eur J Appl Microbiol Biotechnol 18:17
84. Hasegawa S, Shimizu K, Kobayashi T, Matsubara M (1985) J Chem Tech Biotechnol 35B:33
85. Shimizu K, Kobayashi T, Nagara A, Matsubara M (1985) Biotechnol Bioeng 27:743
86. Abulesz EM, Lyberatos G (1987) Biotechnol Bioeng 24:1059
87. Guthke R, Knorre WA (1982) Biotechnol Bioeng 24:2129
88. Giona AR, Marelli L, Toro L, De Santis R (1976) Biotechnol Bioeng 18:473
89. Pan CH, Hepler L, Elander RP (1972) Dev Ind Microbiol 13:103
90. Jensen FG, Nielsen R, Emborg C (1981) Eur J Appl Microbiol Biotechnol 13:29
91. Itsygin SB, Biryukov VV, Lurie LM, Berezovskaya AI (1977) Pharm Chem J 10:119
92. Ryu DDY, Hospodka J (1980) Biotechnol Bioeng 22:289
93. Schügerl K, Seidel G (1998) Chem Ing Tech 70:1596
94. Jüsten P, Paul GC, Nienow AW, Thomas CR (1998) Bioproc Eng 18:7
95. Nielsen J (1995) Physiological engineering aspects of *Penicillium chrysogenum*. Polyteknisk Forlag, Lyngby
96. Krabben P, Nielsen J, Michelsen ML (1997) Chem Eng Sci 52:2641
97. Krabben P, Nielsen J (1998) Adv Biochem Eng Biotechnol 60:125
98. Zangirolami TC, Johansen CL, Nielsen J, Jorgensen SB (1997) Biotechnol Bioeng 56:593
99. Bellgardt KH (1998) Adv Biochem Eng Biotechnol 60:153
100. Caldwell IY, Trinci APJ (1973) Arch Microbiol 88:1
101. Basak S, Velayudhan A, Ladisch MR (1995) Biotechnol Prog 11:626
102. Ludewig D (1999) Expertensystem zur Entwicklung von Prozeßmodellen für biotechnische Prozesse, Fortschritt-Berichte 20 Nr 289, VDI Verlag, Düsseldorf
103. Ludewig D, Bellgardt KH (1997) BioTec 4:31
104. Gomersall R, Grumann S, Ludwig D, Seeger M, Hitzmann B, Bellgardt KH, Munack A (1998) Automatisierungstechnik 46:368
105. Tollnick C (1996) PhD thesis, Universität Hannover
106. Zhou W, Holzhauer-Rieger K, Dors M, Schügerl K (1992) Enzyme Microb Technol 14:848
107. Weil J, Miramonty J, Ladisch MR (1995) Enzyme Microb Technol 17:88
108. Weil J, Miramonty J, Ladisch MR (1995) Enzyme Microb Technol 17:85
109. Martin JF, Liras P (1989) Adv Biochem Eng/Biotechnol 39:153
110. Vining LC, Sharpiro S, Madduri K, Stuttard C (1990) Biotech Adv 8:159
111. Pedersen AG, Bungaard-Nielsen M, Nielsen J, Villadsen J, Hassager O (1993) Biotechnol Bioeng 41:162
112. Packer HL, Thomas CR (1990) Biotechnol Bioeng 35:870
113. Nielsen J, Krabben P (1995) Biotechnol Bioeng 46:588
114. McNeil B, Berry DR, Harvey LM, Grant A, White (1998) Biotechnol Bioeng 57:297
115. Harvey LM, McNeil B, Berry DR, White S (1998) Enzyme Microb Technol 22:446
116. Jørgensen H, Nielsen J, Villadsen J, Møllgardt H (1995) Biotechnol Bioeng 46:117
117. Jørgensen H, Nielsen J, Villadsen J, Møllgardt H (1995) Appl Microbiol Biotechnol 43:123
118. Nielsen J, Jørgensen H (1995) Biotechnol Prog 11:299
119. Mou DG, Cooney CL (1983) Biotechnol Bioeng 25:257
120. Luus R (1993) Biotechnol Bioeng 41:599
121. San KY, Stephanopoulos G (1989) Biotechnol Bioeng 34:72
122. Diener A, Goldschmidt B (1994) J Biotechnol 33:71
123. Rodrigues JAD, Filho RM (1996) Chem Eng Sci 51:2859

124. Modak JM, Lim HC (1992) Chem Eng Sci 47:3869
125. Van Impe JF, Nicolai BM, De Moor B, Vandewalle J (1993) Chem Biochem Eng Q 7:13
126. Henriksen CM, Christensen LH, Nielsen J, Villadsen J (1996) J Biotechnol 45:149
127. Möller J, Niehoff J, Hotop S, Dors M, Schügerl K (1992) Appl Microbiol Biotechnol 37:157
128. de Noronha Pissara P, Nielsen J, Bazin MJ (1996) Biotechnol Bioeng 51:168
129. Menezes JC, Alves SS, Lemos JM, de Azevedo SF (1994) J Chem Tech Biotechnol 61:123
130. Thomas CR, Paul GC, Kent CA (1994) Alberghina L, Frontali L, Sensi P (eds) Proceedings of the 6th European Congress on Biotechnology, p 849
131. Michel FC, Grulke EA, Reddy CA (1992) AIChE J 38:1449
132. Shamlou PA, Makagiansar HY, Ison AP, Lilly MD, Thomas CR (1994) Chem Eng Sci 49:2621
133. Jüsten P, Paul GC, Nienow AW, Thomas CR (1996) Biotechnol Bioeng 52:672
134. Edelstein L, Hadar Y (1983) J Theor Biol 105:427
135. Tough AJ, Pulham J, Prosser JI (1995) Biotechnol Bioeng 46:561
136. Takamatsu T, Shioya S, Furuya T (1981) J Chem Tech Biotechnol 31:697
137. Cronenberg CCH, Ottengraf SPP, van den Heuvel JC, Pottel F, Sziele D, Schügerl K, Bellgardt KH (1994) Bioproc Eng 10:209
138. Cui YQ, van der Lans RGJM, Luyben KCAM (1997) Biotechnol Bioeng 55:715

Part D
Metabolite Flux Analysis, Metabolic Design

14 Quantitative Analysis of Metabolic and Signaling Pathways in *Saccharomyces cerevisiae*

Klaus Mauch, Sam Vaseghi, Matthias Reuss

14.1
Introduction

The yeast *S. cerevisiae* has been used in the oldest biotechnical workshops and industries, baking, brewing, and wine making, from the earliest days of recorded histories. In our times it is in particular the application as bakers yeast which has resulted in production processes with very impressive scales of operation. With the advent of recombinant DNA technology, the yeast has also proved to be an excellent host for the production of a number of different proteins with exciting commercial applications. With increasing information on metabolic processes and physiology it can be expected that this organism becomes a very suitable cell factory for the production of high-added-value products.

In 1996, the complete genome sequence of *S. cerevisiae* was deposited in the public data libraries. However, very soon after this important event, attention has shifted to the so-called functional analysis which is characterized by different approaches grouped in the domains genome, transcriptome, proteome, and metabolome. The purpose of this concerted action is to derive a better knowledge and to enable us to understand, predict, and eventually control the metabolism of the yeast to mold it to the purpose of old and new production processes. It is very clear that the rather recent field of metabolic engineering, defined as the purposeful and rational design of metabolic networks, will play an important role in this field. The challenge for the new discipline in this difficult and ambitious endeavor is to bridge the gap between the fingerprints of molecular assemblies and the physiological behavior of the system. This requires an integrative biological systems analysis as the quantitative description at the hierarchical levels of molecular, cellular, and phenotypic functions including their interaction with the environment is extremely complex. Indeed, there is a growing awareness that this complex interplay between the genome and physiological functions of the cell needs a new holistic and fully integrative view by placing the organism, rather than the genome, at the center of investigations.

This ambitious goal, first of all, points to a pivotal problem: more specific requirements for the design of experiments necessary to elucidate the functions of unknown genes as well as the complex interplay of metabolic and signaling networks. The demand of better definition of the physiological state of the cells and the need of new methods for sampling to analyze gene products (transcripts or proteins) as well as metabolites has been recognized in the scientific community engaged in the functional analysis of the yeast genome [1]. Experimental methods described here address this issue of investigations at defined physiological conditions. Also measurements and analysis in ways that accurately reflect the status within the living cell are discussed. A systematic and consequent application of these experimental strategies for quantitative analysis of the transcriptome (DNA arrays) and proteome (two-dimensional-gel electrophoresis) should enable a deeper understanding at these levels of protein biosynthesis and its regulation.

The concepts regarding the quantitative analysis of the metabolome are less clear. Indeed, in context with the holistic and integrative view of cell metabolism there is growing recognition that the quantitative description of metabolic and signaling networks is strongly limited by the insufficiency of the experimental methods as well as quantitative analysis of the data. The implications of these missing links are quite striking in context with attempts to predict structural and functional aspects of metabolic networks from the genome [2–6]. It must be stressed that mathematical analysis of reconstructed metabolic systems to predict control mechanisms of functional blocks or even dynamic behavior of the systems are based on information about metabolic models available in data banks. The practical feasibility of these approaches including the application of the tool boxes of Metabolic Control Analysis (MCA) obviously depend on the quality of the data available for the kinetics of intracellular reactions.

The biochemical research in the last few decades has been concerned with purifying the individual enzymes and studying in isolation the chemistry of the reactions they bring about. Most of the present knowledge about the kinetics has been derived from these investigations in which the enzymes are isolated and thus free from intracellular interference. There are two pivotal questions regarding the application of these kinetic for studying metabolic networks: (1) to what extent does the multitude of interacting processes inside the cell lead to kinetic behavior which differs from in vitro conditions and (2) what is the influence of the functioning of the entire ensemble; in other words – can we simply sum up every enzyme reaction to understand the system quantitatively?

The answers to these questions depend on the possibilities of our determining the enzyme kinetics under in vivo conditions of the living and growing cell and analyzing these kinetics by taking into account the open character of the system. The essential tools to tackle these ambitious tasks are steady state metabolic flux analysis, measurement of intracellular metabolites at dynamic conditions, and strategies for deriving the structure of in vivo enzyme kinetics and their parameters from these data.

14.2
Metabolic Flux Analysis

Accurate quantification of intracellular fluxes under stationary conditions (Metabolic Flux Analysis) has become an indispensable tool in metabolic engineering [7, 8]. Basic requirement for metabolic flux analysis is the experimental determination of an adequate number of metabolic fluxes as well as sufficient knowledge of the metabolic reaction network. Particularly with regard to the validation of stoichiometric models, labeling experiments provide a major contribution [9, 10]. The flux distribution observed is always an immediate consequence of regulation phenomena on the genetic and metabolic level. Hence, metabolic flux distributions in no way own a predictive character.

Spatial separation of enzymes in cellular compartments is a characteristic property of eukaryotic organisms to which the yeast *S. cerevisiae* belongs. For estimating metabolic fluxes in eukaryotes, compartmentation into cytosol and mitochondria is highly relevant because in those compartments the predominant part of the reactions within the central metabolic pathways proceed. With regard to the control of metabolism, the observation of transport fluxes between cell compartments is of high interest. Moreover, by using compartmented metabolic networks, the diversity in the usage of metabolites in individual compartments can be represented.

Most of the work published on flux distributions in *S. cerevisiae* [11–14], however, is based upon non-compartmented or partly compartmented stoichiometric models. This seems mainly due to the incomplete knowledge on transport systems, but technical difficulties in treating the usually large compartment models may have circumvented a broader application as well.

In this section, metabolic flux distributions in *S. cerevisiae* CBS 7336 are observed at three physiological states by means of a completely compartmented stoichiometric model. The main emphasis is placed on the estimation of fluxes crossing the inner mitochondrial membrane as well as on energetic aspects.

14.2.1
Metabolite Balancing in Compartmented Systems

Under stationary conditions, a compartmented metabolic network can be described by the superposition of reaction and transport according to

$$\Gamma r + Tt = 0 \tag{14.1}$$

where matrix $\Gamma(m \times n)$ contains the stoichiometric coefficients $\gamma_{i,j}$ of the n biochemical reactions. m denotes the sum of metabolites present in each compartment. The membership of metabolites with respect to a compartment is indicated by a superscript. Matrix T ($m \times t$) comprises the stoichiometric coefficients $t_{i,j}$ of t transport equations. r denotes the vector of reaction rates, t the vector of transport rates. A transport equation can formally be treated similar to a biochemical equation. If, for example, metabolite A is transported from compartment α to compartment β by facilitated diffusion, the transport equation is written as

$$A^\alpha = A^\beta \tag{14.2}$$

whereas, when A is transported by the symport of one proton H, the transport equation is formulated according to

$$A^\alpha + H^\alpha = A^\beta + H^\beta \tag{14.3}$$

The stoichiometric coefficient $t_{i,j}$ of transport matrix T is defined to be negative when metabolite m_i is consumed and positive at the metabolites production. When r and t in Eq. (14.1) are used on a volumetric basis, the coefficient $t_{i,j}$ has to be weighted by the ratio of volumes V^α, V^β in between which metabolite m_i is exchanged. The transport equations, Eqs. (14.2) and (14.3), for instance, are then re-formulated according to Eq. (14.4) and Eq (14.5), respectively:

$$\frac{V^\alpha}{V^\beta} A^\alpha = A^\beta \tag{14.4}$$

$$\frac{V^\alpha}{V^\beta} A^\alpha + \frac{V^\alpha}{V^\beta} H^\alpha = A^\beta + H^\beta \tag{14.5}$$

For a metabolic network located solely in a single compartment, multiplication of transport matrix $\mathbf{T}$ with the transport vector $\mathbf{t}$ immediately leads to vector $\mathbf{Q}$ comprising the systems net-conversion rates. Hence, Eq. (14.1) can be simplified to

$$\Gamma \mathbf{r} = \mathbf{Q} \tag{14.6}$$

Matrices Γ and $\mathbf{T}$ can be combined to matrix $\mathbf{N}$ according to

$$[\Gamma \mid \mathbf{T}]\begin{bmatrix}\mathbf{r}\\\mathbf{t}\end{bmatrix} = \mathbf{N}\mathbf{v} = 0 \tag{14.7}$$

where vector $\mathbf{v}$ now contains both the reaction vector $\mathbf{r}$ and the transport vector $\mathbf{t}$. Methods for estimating metabolic fluxes [8, 12], data reconciliation, and error diagnosis [16–18] can directly be applied to Eq. (14.7).

14.2.2
Stoichiometric Model

In the following, characteristic properties of a compartmented stoichiometric model for the yeast *S. cerevisiae* CBS 7336 are described. All biochemical reaction equations, abbreviations for metabolite names, and the C-mol composition of macromolecules are listed in the appendix. The organism for which the model is proposed, has been grown in continuous culture with glucose as sole carbon and energy source and ammonia as nitrogen source at dilution rates of $D=0.10\,\mathrm{h}^{-1}$, $D=0.24\,\mathrm{h}^{-1}$, and $D=0.33\,\mathrm{h}^{-1}$. Details on fermentation conditions, media, sampling methods and methods for the experimental determination of extracellular metabolites can be extracted from Theobald et al. [19].

The stoichiometric model is partitioned into the compartments cytosol and mitochondria. EMP pathway, PPP, synthesis of nucleotides, amino acids, polysaccharides, and reactions of the C_1-transfer are all located in the cytosol. Pyruvatecarboxylase is assumed to be present in the cytosol as well, just as the pathways for the formation of ethanol, acetate and glycerol. Cytosolic acetyl CoA synthetase produces acetyl CoA which is further used as precursor for the synthesis of fatty acids and amino acids in the cytosol. Malate dehydrogenase and citrate synthase are present in both compartments [20]. All other enzymes of the citric acid cycle and pyruvate dehydrogenase are located in mitochondria.

No transport systems are formulated for NADP, NADPH, NAD, NADH [21], acetyl CoA, and oxaloacetate. As a consequence, those metabolites are blocked by the inner mitochondrial membrane. For the transfer of inorganic phosphate [22], pyruvate [23], and succinate, transport equations for proton symport are formulated. For the remaining metabolites simultaneously occurring in the cytosol and mitochondria, equations for facilitated diffusion/passive transport enable transport across the inner mitochondrial membrane.

Biochemical reactions of the respiratory chain as well as ATP regeneration by ATP synthase are located in the mitochondria [20]. It is assumed that cytosolic NADH is oxidized by ubiquinone and not coupled to ATP synthesis [20]. Moreover, shuttle systems for cytosolic NADH (for an overview see [24]) are assumed to be inactive. The efficiency of oxidative phosphorylation η is introduced to take into account the unknown numbers of protons pumped from mitochondria to the outside and in turn used to regenerate ATP. The symbolic stoichiometric coefficient η is normalized such that for $\eta=1$ a P/O-ratio of 2 results for the oxidation of cytosolic NADH and mitochondrial $FADH_2$ whereas a P/O-ratio of 3 results for the oxidation of mito-chondrial NADH. η can be considered as a measure of the degree at which the re-spiratory chain is coupled with ATP-regeneration driven by ATP synthase. Thus, for $\eta=0$ reduction equivalents are oxidized, but no ATP is produced, indicating a com-. plete uncoupling.

A reaction for ATP hydrolysis is included in the stoichiometric model to make allowance for ATP consumption through growth dependent and growth independent maintenance (e.g., turnover of macromolecules, "futile cycles," repair pro-cesses).

Within the present stoichiometric model, the PPP is assumed to be the sole source of NADPH production. In turn, NADPH serves as reduction equivalent in anabolic reactions. Activity of transhydrogenases is assumed to be negligible. Data on the amino acids composition of cellular protein was taken from [13], whereas the nucleic composition of RNA has been adopted from [25]. Lipid composition (saturated and non-saturated fatty acids, glycerol) has been analyzed by Maser [26] and was found to be virtually constant at different dilution rates. Therefore, the stoichiometry of lipid formation is kept at a constant value. Similarly, for the composition of protein with respect to amino acids as well as the composition of RNA with respect to nucleic acids, almost constant values are reported for differ-ent growth rates in the literature [12, 25]. For each prolongation step, a demand of four ATP for translation and a demand of one ATP for transcription is as-sumed, and one ATP is used for the extension of a polysaccharide with one car-bohydrate unit [27].

As the fraction of macromolecules constituting biomass is a distinct function of the specific growth rate, the weight fraction of protein, lipids, RNA, and polysacchar-ides has been determined experimentally for *S. cerevisiae* CBS 7336 [28]. Results are depicted in Fig. 14.1. The data is in good agreement with data published on other strains of *S. cerevisiae* [29, 30].

Most remarkable, by increasing the dilution rate, the protein fraction increases at the expense of polysaccharides. For the derivation of the stoichiometry for biomass formation, an ash content of 3.8% [31] has been considered for each physiological state.

14.2.3
Computational Aspects

Weight fractions of monomers within macromolecules as well as the weight fractions of macromolecules within biomass are converted to integers and fractions of integers and subsequently translated into biochemical equations for polymerization by the programming system METAFLUX [32]. Based upon biochemical reactions edited textually in a form similar to equations found in textbooks, the matrices Γ and $\mathbf{T}$

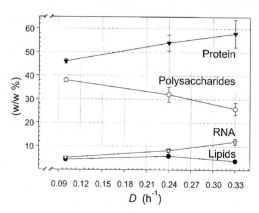

Fig. 14.1. Weight fraction of protein, lipids, RNA and polysaccharides in *S. cerevisiae* CBS 7336 as a function of the dilution rate

are generated automatically. To prove the consistency of the biochemical equations, matrix E comprising the elemental composition and the charge of the metabolites included in the stoichiometric model is created from a metabolite composition database. Since matrix Γ multiplied by matrix E equals the zero matrix, a comprehensive topological analysis (for an overview see [33]) is carried out revealing, for instance, degrees of freedom, optimal and suboptimal yields, conserved moieties, unbranched subsystems, and dead ends of the metabolic network. By analyzing its topological structure, unexpected or incorrect properties of the metabolic network can be detected in an early phase of the modeling process. Application of matrix multiplication and matrix inversion by using integers and fractions of integers allows a precise computation of flux distributions independent of the network's size. Hence, computing the condition number of matrix N becomes dispensable. Due to the built in symbolic computation facilities of MAPLE [34], METAFLUX allows the treatment of unknown stoichiometric factors. Those unknowns (i.e., the P/O-ratio) may be incorporated in the stoichiometric matrix symbolically.

14.2.4
Results

The degree of freedom of the stoichiometric model proposed in this contribution is found to be five. As seven substrate uptake and product production rates have been determined experimentally at every physiological state, the degree of redundancy is two. This redundancy has been used to reconcile the data according to a method published by van der Heijden et al. [16, 17]. Absolute values per liter of reactor volume for the measured and reconciled net conversion rates are given in Table 14.1.

The error criteria ε computed from the differences in measured and reconciled data [16] has been compared to the χ^2-distribution by using a confidence level of 95%. For each data set the ratio $\chi^2_{95} \eta^{-1}$ is below 0.03. From there, a significant error resulting, for example, from excreted compounds omitted in the measurement vector cannot be diagnosed. According to Table 14.1, metabolic flux distributions within the central metabolic pathways are summarized in Fig. 14.2. All numbers shown are normalized to the glucose uptake rate ("100").

Table 14.1. Measured (1st column) and reconciled (2nd column) net conversion rates

D (h^{-1})	$Q_{Glucose}$ (mmol l^{-1}h^{-1})		$Q_{Biomass}$ (C-mmol l^{-1} h^{-1})		$Q_{Ethanol}$ (C-mmol l^{-1} h^{-1})		$Q_{Glycerol}$ (C-mmol l^{-1} h^{-1})		$Q_{Acetate}$ (C-mmol l^{-1} h^{-1})		Q_{CO_2} (mmol l^{-1}h^{-1})		Q_{O_2} (mmol l^{-1}h^{-1})	
0.10	16.7	16.4	64.8	65.4	0.15	0.15	0.03	0.03	<0.1	0	32	32.6	30	29.6
0.24	39.9	39.9	150.7	150.5	1.6	1.6	0.13	0.13	<0.1	0	83	85.4	79	76.9
0.33	53.1	51.9	71.6	72.4	71.6	70.5	0.18	0.18	0.9	0.9	93	93.6	19	19

At a dilution rate of $D=0.10\,h^{-1}$ a normalized flux of 50 is branched off into the PPP whereas a flow of 23 is used for the formation of polysaccharides. The remaining flow of 27 at the G6P branch point is funneled into the EMP pathway. The flux into the PPP appears high compared to values obtained by radioactive labeling experiments [35]. However, the flux into the PPP largely depends upon the linkage of GDH either to NADH or NADPH which varies according to the nitrogen source [20]. A linkage of GDH exclusively to NADH results in a reduced flux of 20 into the PPP (instead of 50 for an NADPH linked GDH).

Around a half of the pyruvate produced in the cytosol is consumed for the formation of amino acids (22), acetaldehyde (21), and cytosolic oxaloacetate (23). The other half of pyruvate formed in the cytosol is directly transported to mitochondria by means of proton symport. The flux to acetaldehyde is further transferred to cytosolic acetyl CoA where the flux is split into the flux forming fatty acids (13) and amino acids (8). The bigger part of oxaloacetate produced in the cytosol (15) is used for the production of aspartate whilst a flux of 8 leads to the formation of cytosolic malate. The latter is then transported to the mitochondria where it is used to replenish the citric acid cycle. Fumarate also serves as anaplerotic flux even though to a lower extent (4).

As malate and fumarate are transported into mitochondria, α-ketoglutarate crosses the inner mitochondrial membrane from mitochondria to the cytosol (12) where it is transformed at the high rate of 58 to glutamate. The relatively low net flux of α-ketoglutarate transferred to the cytosol (12) is mainly due the high rate at which α-ketoglutarate is recovered from amino acids synthesis (46) in the cytosol.

Pyruvate dehydrogenase transforms mitochondrial pyruvate to acetyl CoA (48), which further fuels the citric acid cycle. Carbon dioxide formed at a rate of 131 by pyruvate dehydrogenase (48), malate dehydrogenase (48), and citrate synthase (48) in the mitochondria clearly outweighs the carbon dioxide produced in the cytosol (67). Likewise, the amount of reduction equivalents produced in mitochondria (NADH 181; FADH$_2$ 36) is large compared to NADH (142) formed in the cytosol.

Between mitochondria and cytosol a flow of 328 ATP is observed. ATP transported to the cytosol supplies the anabolic ATP demand which exceeds glycolytic ATP production. At the same rate (328), ADP and inorganic phosphate is transported from the cytosol to mitochondria. The high value of the ATP/ADP counter-exchange is noteworthy as enzymes transforming metabolites at a high rate usually underlie tight control. Otherwise pool concentrations of metabolites formed or consumed quickly overshoot or deplete. The consequences for cell metabolism through enzymes operating at a high rate are all the more distinct when globally potent metabolites (i.e., ATP/ADP, inorganic phosphate) are involved.

Although the absolute value of the glucose influx significantly increases (240%), only minor changes in the distribution of metabolic fluxes are observed when shifting

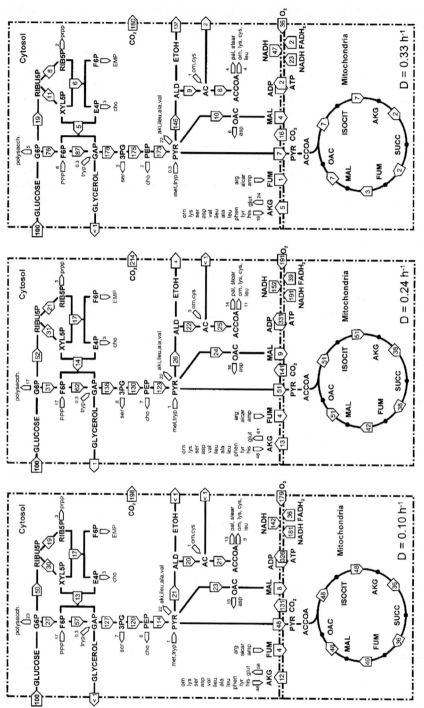

Fig. 14.2. Metabolic flux distributions in *S. cerevisiae* CBS 7336 as a function of the dilution rate D

from a dilution rate of $D=0.10\,h^{-1}$ to $D=0.24\,h^{-1}$. Due to an increased proportion of protein in biomass, the flux branched into the PPP and the ATP/ADP counter-exchange is slightly higher. Furthermore, a rising efflux of ethanol (4) can be noticed.

Compared to flux distributions discussed previously, the situation is completely different at a dilution rate of $D=0.33\,h^{-1}$. The drastic increase of an ethanol efflux coincides with a decrease of the biomass yield on glucose from $Y_{S,X}=0.51$ ($D=0.24\,h^{-1}$) to $Y_{S,X}=0.17$ ($D=0.33\,h^{-1}$). In the literature, this phenomenon has been described as long-term Crabtree effect [15]. Consequently, an increased flux through the EMP is observed. Associated with the lower biomass production rate, the flux into the PPP decreases from 50 ($D=0.24\,h^{-1}$) to 20 ($D=0.33\,h^{-1}$). At the cytosolic pyruvate branch point, the flow towards acetaldehyde increases drastically (137). Solely a flux of 7 reaches mitochondria as pyruvate. Accordingly, the rate at which reduction equivalents are produced in mitochondria is significantly reduced. Moreover, at a dilution rate of $D=0.33\,h^{-1}$ a disproportionate efflux of acetate (2) is observed.

Most remarkable, however, is the collapse of the ATP/ADP counter-exchange. In contrast to a value of 331 estimated at a dilution rate of $D=0.24\,h^{-1}$, a value of 2 is observed for the counter-exchange at a dilution rate of $D=0.33\,h^{-1}$. At the corresponding physiological state, the need of ATP in anabolic reactions is almost completely covered by substrate chain phosphorylation in the cytosol. Despite the low net flow of ATP to the cytosol, a significant amount of reduction equivalents is still oxidized in the respiratory chain. From this finding it can be concluded that either (a) the respiratory chain is uncoupled from oxidative ATP regeneration to a high degree or (b) the use of ATP for maintenance increases drastically. To elucidate this result more closely, the linear relation between the consumption of ATP by maintenance $r_{m,ATP}$ and the efficiency of oxidative phosphorylation η have been derived symbolically. By weighting η with the proportion at which mitochondrial $FADH_2$, NADH, and cytosolic NADH are oxidized in the respiratory chain, the effective overall P/O- ratio is obtained and we write

$$r_{ATP,m} \sim \eta \ \text{or} \ r_{ATP,m} \sim P/O \tag{14.8}$$

By relating the amount of ATP consumed for maintenance to the biomass produced (in C-mol), we obtain the maintenance coefficient k introduced by Verduyn et al. [36]. Hence, at the three physiological states under investigation, the energetic relations can be expressed as

$$D = 0.10\,h^{-1}: \eta = 0.44\,k + 0.74 \ \text{or} \ P/O = 1.04\,k + 1.73 \tag{14.9}$$

$$D = 0.24\,h^{-1}: \eta = 0.44\,k + 0.74 \ \text{or} \ P/O = 0.91\,k + 1.64 \tag{14.10}$$

$$D = 0.33\,h^{-1}: \eta = 0.44\,k + 0.74 \ \text{or} \ P/O = 1.77\,k + 0.03 \tag{14.11}$$

From Eqs. (14.10) and (14.11), minimal values for η and P/O are easily obtained for $k=0$. The minimal values are given in Table 14.2.

Minimal P/O values of 0.74 and 0.70 at the dilution rates $D=0.10\,h^{-1}$ and $D=0.24\,h^{-1}$, respectively, differ only to a minor extent. Importantly, the minimal P/O values reported here are already well above the P/O ratios of 1.09 [13] and 1.2 [14] reported in the literature for S. cerevisiae at a dilution rate $D=0.10\,h^{-1}$. The reason can be specified by a detailed implementation of active transport across the inner mitochondrial membrane and the oxidative ATP-regeneration in this contribution. Both

Table 14.2. Minimal values for η and P/O

$D \, h^{-1}$	η (–)	P/O$_{min}$ (–)
0.10	0.74	1.73
0.24	0.70	1.64
0.33	0.01	0.03

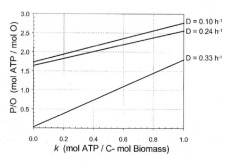

Fig. 14.3. P/O-ratio as a function of the maintenance coefficient k

processes are driven by the same proton gradient. Since ATP-generation and active transport essentially have to be lumped together when a non-compartmented approach is used, the effective P/O ratio is apparently lower. Or, in other words, the apparent loss in the efficiency of oxidative phosphorylation is partly caused by active transport.

While η is a measure of the degree by which ATP regeneration is coupled to the respiratory chain, $1-\eta$ can be regarded as a measure for the degree of the uncoupling. Accordingly, the maximal degree of uncoupling at the lower dilution rates $D=0.10 \, h^{-1}$ and $D=0.24 \, h^{-1}$ is comparatively small whereas at dilution rate of $D=0.33 \, h^{-1}$ the degree of uncoupling can almost be as high as 100%.

Figure 14.3 shows the linear relations between the P/O-ratio and the maintenance coefficient k. A constant P/O-ratio of 1.73 for all three dilution rates would result in $k=0$ for $D=0.10 \, h^{-1}$, $k=0.12$ for $D=0.24 \, h^{-1}$, and $k=0.94$ for $D=0.33 \, h^{-1}$. In contrast, assuming a constant k of 0.6 reported for at an average for aerobic [13, 14] and anaerobic [30] fermentations, a vanishing degree of uncoupling results for $D=0.10 \, h^{-1}$, 6% for $D=0.24 \, h^{-1}$, and 49% for $D=0.33 \, h^{-1}$.

The quest for the actual values for k and P/O cannot be resolved from the experiments analyzed in this section. However, the breakdown of transport fluxes across the inner mitochondrial membrane observed at a dilution rate of $D=0.33 \, h^{-1}$ supports the thesis of a distinct uncoupling of the respiratory chain with oxidative phosphorylation during the long-term Crabtree effect as both findings are strongly related to the properties of the mitochondrial membrane.

14.3
Measurements of Intracellular Compounds

Beside measurement of fluxes [9, 10], it is the quantitative determination of various biotic state variables which occupies a central place in metabolic engineering. The ambitious goals in the quantitative description of the interplay between the hierarchical levels of the genome, transcriptome, proteome, and metabolome require in vivo measurements which can be grouped into:
- Macromolecular pools (specific compounds and integral values)
- Enzyme activities
- Metabolites and signals

For a quantitative functional analysis it is an essential prerequisite to define the physiological state of the cells used for these measurements. Of course, this imperative requires experimental conditions and related process operations which are defined and reproducible. We also need to devise methods for sampling, quenching, and extraction that ensure that the results of the subsequent analysis accurately reflect the status of the living cell. The concrete design of the appropriate tools and operations in the sequence:
- Process operation (steady state, fed batch, transient)
- Sampling (time span after disturbances and frequency)
- Quenching
- Extraction
- Analysis

depend on (a) the biotic variables to be measured and (b) the information to be derived from these observations. Attributes related to (a) include chemical and biological stability of the compound, turn-over times, analytical methods applied, etc. The focus of the second point is the purpose for these measurements. The interest in such measurements may be related to dynamic responses of metabolite pools to extracellular disturbances for identification of in vivo kinetics including modulation at the metabolic level. Another focal point of these measurements could be regulation phenomena involved in transcription, translation, or posttranslational processing.

In the following, examples for this sequence of operations are provided from work conducted in the laboratory authors'. It is important to emphasize, that this is by no means a complete review of the activities in this area. In particular, the interesting field of measurements of macromolecular pools including the more specific information about the transcriptome and proteome are left out.

14.3.1
Measurements of Intracellular Metabolites and Signals – General Tools

Now as far as we have formulated the general framework for the measurements of intracellular compounds, let us address some of the more specific problems associated with the observations of concentrations of intracellular metabolites. In the preceding section it was mentioned that design of the experimental conditions is an essential prerequisite. This can be pursued by an analogy between measurement and modeling. Bailey [37] correctly complained that sometimes modelers have not clearly defined the purpose of modeling and then referred to Casti [38, 39] who asserts an inextricable coupling between a model and its intended application. Because measurements are an integral part of the modeling cycle, the same statement holds for the experimental design. It is impossible to evaluate the success or failure of experimental observations without specification of the demand or use for the data.

With the measurements of intracellular metabolites presented in the following, the goal is clear. This data should be used for identification of in vivo enzyme kinetics which afterwards could be applied for the purpose of Metabolic Control Analysis (MCA).

The strategy for this in vivo diagnosis of intracellular reactions uses experimental observations of intracellular metabolites under transient conditions. For this purpose, a continuous culture (or fed-batch process at a controlled specific growth rate) is disturbed by a pulse of glucose and changes in metabolite concentrations which are measured within milliseconds, seconds, or minutes after the disturbance. There are two reasons for choosing this time scale:
– Regulation at the metabolic level occurs within seconds or even faster
– Within this time scale, changes are caused only by metabolic regulation. Biosynthetic reactions can be regarded as staying in a "frozen" state.

As will be shown later, there are, however, some special reactions related to fast enzyme modifications (such as phosphorylation) which need special consideration. Precise measuring of intracellular concentrations in the time window of seconds requires appropriate techniques for rapid sampling, inactivation of metabolic enzymes (quenching), and extraction of metabolites, taking into account the high metabolic turnover rates of the compounds of interest.

Figure 14.4 illustrates two different techniques developed for the aforementioned rapid sampling and quenching experiments. Both sampling devices are connected with a stirred tank bioreactor operating in a continuous mode. In the first approach

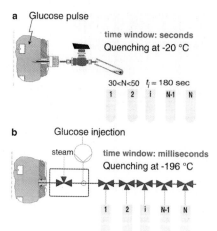

Fig. 14.4a,b. Two different sampling techniques for measurement of intracellular metabolites at transient conditions: **a** manual sampling after a pulse of glucose in continuous culture; **b** stopped flow technique with glucose shift with the sampling valve [40]

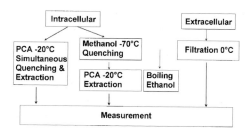

Fig. 14.5. Overview on quenching and extraction procedure [40, 44]

a pulse of glucose is injected into the bioreactor with a syringe to give an initial glucose concentration of, for example, $1\,g\,l^{-1}$ (steady state concentration of glucose before the pulse is ca. $20\,mg\,l^{-1}$). Samples are then rapidly taken aseptically with vacuum-sealed, precooled glass tubes through a special sampling device [19, 40]. The frequency of sampling is indicated in Fig. 14.5. The sample tubes contain an appropriate quenching fluid depending on the metabolite to be measured (perchloric acid solution: $-20\,°C$; Methanol: $-70\,°C$, liquid nitrogen: $-196\,°C$). Systematic investigations have indicated that the most important quenching effect is due to the low temperature [43].

This sampling technique can easily be automated to increase the frequency of sampling [41]. However, as far as the very fast and initial response of intracellular metabolites in the millisecond range is concerned, this method shows an inherent limitation. The time span for the first sample after the disturbance is determined by the mixing time of the glucose pushed into the bioreactor. Even in small laboratory reactors mixing times are of the order of 2–3 s.

Figure 14.4b shows a new sampling technique designed to overcome this problem. It is based on the well-known stopped-flow-method used for fast measurements of enzymatic reactions. As shown in Fig. 14.4b in its application to sampling from bioreactors, a continuous stream of biosuspension leaving the bioreactor is being mixed with a glucose solution in a turbulent mixing chamber (mixing time a few milliseconds). The position of the valves in the cascade illustrated in the figure then determines the residence time of the biosuspension before being quenched in the corresponding sampling tube.

The main features of this sampling device may be characterized as follows. (1) The culture remains at steady state because the organisms are stimulated by the glucose in the mixing chamber within the valve. (2) The sampling time and reaction time are decoupled. The volume of the individual sample can be chosen independently. (3) The time span between glucose stimulus and first sample can be less than 100 ms. The only limitation of the technique is the problem of oxygen limitation at aerobic growth. Thus, the longest reaction time is determined by the oxygen consumption rate in the sampling tube, which then depends on the organism, growth rate, and cell concentration in the reaction. For studying the complete response we therefore recommend use of both sampling devices, the stopped-flow-technique for the first seconds and the pulse technique with manual (or automated) sampling for longer time periods.

Figure 14.5 summarizes some aspects of the various techniques used for the simultaneous or sequential operations of quenching and extraction for extra- and intra-

cellular concentrations of metabolites. In utilizing any of these methods, attention should be paid to ensure that:
- Metabolic activity is rapidly and irreversibly inactivated
- All the metabolites to be measured are completely extracted
- The quenching/extraction solutions used do not interfere with the assay used for the analysis

The left hand side of Fig. 14.5 shows the well-known and often applied method of a simultaneous quenching and extraction with the aid of perchloric acid solutions [42]. There are some critical points with this technique which should draw attention. Often it is claimed that the sample is quenched at –20 °C if 25% solution of PCA is used as quenching/extraction fluid. However, careful measurements with thermoelements [43] indicate that, depending on the volume ratio, the temperature of the mixture (sample+quenching/extraction fluid) might be as high as 10 °C for several seconds immediately after sampling. Thus, in general, attention should be paid to the choice of the volume ratio. Because of the additional neutralization, PCA-extractions always result in high salt concentrations, which may create problems in HPLC measurements. For optimal extraction results with PCA solutions, it is important to perform at least three freeze-thaw cycles [43].

As an alternative to the quenching with methanol – which is very efficient as far as the time course of the temperature during sampling is concerned [43] – and additional extraction with PCA, it is also possible to combine this quenching method with other extraction procedures. An interesting procedure has been suggested and systematically applied by Gonzales et al. [81]. After quenching in methanol at –70 °C and centrifugation at –20 °C the yeast pellet is incubated in boiling ethanol (80 °C). After evaporation of ethanol, resuspension in water, and centrifugation at 4 °C, the metabolites are measured in the supernatant.

For measurements of metabolites, a variety of analytical techniques are available. In the laboratory authors' mainly enzymatic methods and HPLC have been used. For more detailed information about these analytical techniques the reader is referred to the original papers of Theobald et al. [40, 44] and Mailinger et al. [80].

14.3.2
Dynamic Response of Metabolite Pools of Glycolysis

Figure 14.6a–c show illustrative examples of some metabolite concentrations during steady state and after a glucose pulse introduced into the fermentor. The application of two different sampling techniques allows a resolution either in the range of milliseconds or seconds.

The extracellular measurements show the consumption of substrate glucose after the pulse (marked by the arrows) and excretion of glycerol, ethanol, and acetic acid. The changes in intracellular metabolites show the feedback regulation of glucose uptake. Immediately after the pulse glucose appears in the cell as G6P due to the high activity of hexokinase and a low activity of the next enzyme, PFK1. PFK1 must be activated by the decreasing ATP and the increasing F6P, AMP, and ADP concentrations. Next, glucose uptake stops the decrease in ATP level. One can imagine that a further decrease in ATP could limit PFK1. After 30–40 s, however, the PFK1 is activated to work at a level higher than the glucose uptake and the G6P-concentration

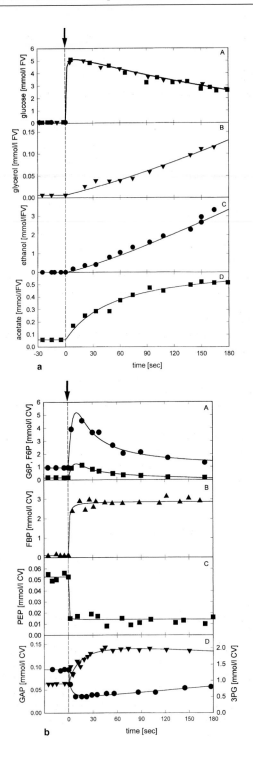

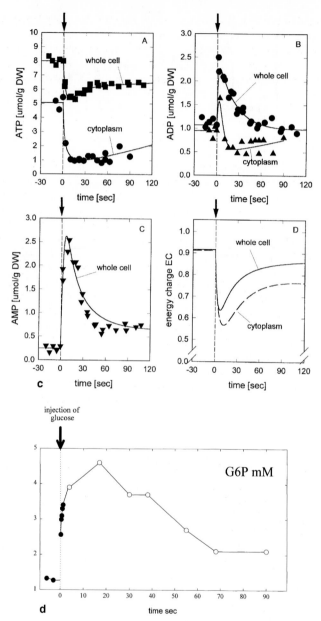

Fig. 14.6.a. Changes in the levels of glucose (A), glycerol (B), ethanol (C), and acetic acid (D). **b** Changes in the levels of G6P, F6P (A), F1,6bP (B), PEP (C), and GAP/3PG (D) during steady state conditions and after a glucose pulse. **c** Changes in the levels of ATP (A), ADP (B), AMP (C), and energy charge in cytoplasm and the whole cell (D) during steady state conditions and after a glucose pulse. **d** Time course of G6P: Comparison of injection- and pulse-experiments (ll) and injection experiment (λλ) long term pulse-experiment

decreases. Note that phosphoglucose-isomerase maintains near-equilibrium between G6P and F6P even under dynamic conditions.

Whereas the first two intracellular metabolites of the glycolysis have a concentration maximum shortly after the glucose pulse, F1,6bP, and GAP remain at a high level for a long time. This behavior may result from the interact on between glycolysis and the pentose-phosphate-cycle. The initial increase of G6P concentration is accompanied by a delayed increase of 6PG, the first metabolite in the PPP. This will cause an increasing flux through the PPP, which most likely will serve to keep F1,6bP, and GAP concentrations at higher levels. The most striking result, however, concerns the opposite trend of the next metabolites 3-GP and PEP. This dynamic behavior may be explained by the well known activation of the pyruvate kinase by F1,6bP. Thus, the increasing F1,6bP-level activates the conversion of PEP on pyruvate and leads to a sharp decrease of the PEP-concentration. If we assume that the reactions catalyzed by enolase and mutase are very close to equilibrium, then 3-PG may show the same trend. The forward activation of PK also helps to keep the 1,3-DPG concentration low. At first glance it seems that GAP-DH would act as a bottleneck because of the increased level of cytoplasmatic NADH. Alcohol dehydrogenase cannot use this NADH because it is still substrate limited (pyruvate decarboxylase shows a co-operative effect with respect to its substrate, pyruvate). However, three effects counteract the increase of NADH level: (1) the cytoplasmic *ortho*-phosphate level which is also a substrate of GAP-DH increases, (2) the activated PK withdraws the product (1,3-DPG) of the reversible reaction of GAP-DH, and (3) glycerol excretion reduces the level of cytoplasmic NADH. Moreover, direct oxidation of cytosolic NADH by ubiquinone [20] or shuttle systems [24] capable of transferring cytosolic NADH to mitochondria may also influence strongly the dynamics of the NADH concentration in the cytosol. The intracellular adenine nucleotides respond extremely rapidly. Only 2 s after the glucose pulse, the cytoplasmic ATP decreases to 42% of the steady state value. The first steps

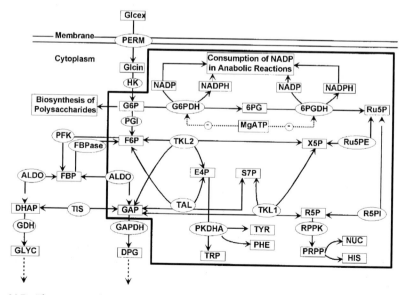

Fig. 14.7. The pentose-phosphate pathway

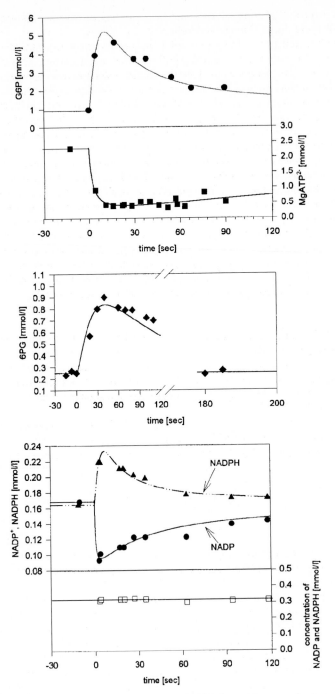

Fig. 14.8. Time course of G6P, NADP, NADPH and ATP after a glucose pulse in a glucose limited continuous culture at $D=0.1\,h^{-1}$ [46]

of glycolysis (HK, PFK1) use a large amount of ATP before the subsequent steps can create ATP. In addition, the ATP level remains at this lower level due to further ATP-consuming reactions stimulated by the increasing level of glycolytic intermediates. This finding may be superimposed by both a reduced efficiency of oxidative phosphorylation and an increased ATP demand for maintenance. However, the total amount of adenosine nucleotides has been found to decrease. A potential explanation for this observation is an enhanced synthesis of RNA in consequence of stress triggered by the glucose pulse. One can exclude the production of glycogen as a reason for additional ATP consumption. Previous work showed that the glycogen level remains constant although G6P increases. A possible explanation is the inhibition of glycogen synthetase by increasing *ortho*-phosphate concentration. Figure 14.6d illustrates an example for application of the stopped-flow sampling techniques (data points depicted with ● indicate reaction times of 150, 300, 450, 600, and 800 ms respectively; the open symbols o show the same measurements illustrated in Fig. 14.6b).

The data [40, 44] have been applied to identify the kinetic rate expressions and parameters of the intracellular enzymes involved. The detailed results and the complete dynamic model have been described in the paper of Rizzi et. al. [45]. In the following, more simple examples will serve to demonstrate some of the important tools for identification of the structure of rate expressions as well as estimation for the kinetic parameter for intracellular reactions at in vivo conditions.

14.3.3
Dynamics of the Pentose Phosphate Pathway – an Example for in vivo Diagnosis of Intracellular Enzyme Kinetics

Transient experiments as described above were performed to study the in vivo dynamics of the PPP in *S. cerevisiae* [46]. Figure 14.7 summarizes the metabolites and enzymes of interest.

Vaseghi et al. [46] have derived a complete model for the quantitative description of the dynamics of the PPP. The first step towards quantitative analysis of the shunt involves the assumption, that the first two reactions catalyzed by G6P-DH and 6PG-DH are irreversible whereas the other reactions are assumed to be reversible and in near-equilibrium.

The balance equation for 6PG is given by

$$\frac{dC_{6PG}}{dt} = r_{G6P-DH} - r_{6PG-DH} - \mu \, C_{6PG} \tag{14.12}$$

The last term in Eq. (14.12) represents the dilution caused by the growth of yeast. The measurements required for the identification of the kinetics of the two enzymes are summarized in Fig. 14.8.

In addition to the substrates G6P and 6PG we need to know the dynamic response of the concentrations of the co-substrate NADP. Because NADPH is known as a product inhibitor for both reactions, this co-metabolite must also be measured. A careful inspection of the measured data illustrates that the level of G6P at steady state already results in a substrate saturation of G6P-DH. Therefore, the increase of NADPH and 6PG cannot be attributed to the influence of increased G6P concentrations. A careful consideration of the literature and in vitro measurements with the isolated G6P-DH [46] provided an interesting solution to the problem. ATP turned out to be a strong inhibitor of both reactions. The increased flux through the PP-shunt can therefore be easily explained by the drop of ATP after the pulse of glucose (Fig. 14.8). Summarizing the aforementioned characteristics of the two reactions, the following rate expressions are suggested:

$$\frac{dC_{6PG}(t)}{dt} = r_{G6P-DH}^{max} \frac{C_{NADP}(t)}{\left(C_{NADP}(t) + K_{NADP,1}\left(1 + \frac{C_{NADPH}(t)}{K_{i,NADPH,1}}\right)\right)\left(1 + \frac{C_{MgATP}(t)}{K_{i,MgATP,1}}\right)}$$

$$- r_{G6P-DH}^{max} \frac{C_{NADP}(t)}{\left(C_{NADP}(t) + K_{NADP,2}\left(1 + \frac{C_{NADPH}(t)}{K_{i,NADPH,2}}\right)\right)\left(1 + \frac{C_{MgATP}(t)}{K_{i,MgATP,2}}\right)} - \mu C_{6PG}(t) \quad (14.13)$$

Notice, that the concentrations at the right hand-side of this equation are time-dependent. A comprehensive solution would require an extension of the balance equations for NADP, NADPH, and ATP. To reduce the complexity of the problem, measured data for the co-metabolites are approximated with the aid of approximate analytical functions. The following functions have been used to fit the observed changes of concentrations after the glucose pulse.

$$\hat{C}_{G6P}(t) = 0.9 + \frac{44.1\,t}{48.0 + t + 0.45\,t^2} \quad (14.14)$$

$$\hat{C}_{NADP}(t) = 0.17 - \frac{1.48\,t}{9.71 + 16.1\,t + 0.48\,t^2} \quad (14.15)$$

$$\hat{C}_{NADPH}(t) = 0.16 - \frac{0.516\,t}{25.39 + 0.37\,t + 0.5\,t^2} \quad (14.16)$$

$$\hat{C}_{ATP}(t) = 2.3 - \frac{29.83\,t}{29.77 + 13.42\,t + 0.05\,t^2} \quad (14.17)$$

These functions are simply used to fit the data and do not have any mechanistic background.

The next step towards estimation of the parameters is the calculation of the maximal rates $r^{max}{}_j$. Estimates from measured enzyme activities under in vitro conditions are questionable, because of the possible removal of effectors during cell disruption, shear sensitivity, effects of ion strength, protein-membrane, and protein-protein interactions, etc. An alternative estimate is based on the rate equation and

Table 14.3. Results of the parameter identification (in vivo diagnosis) for the enzymes involved in the pentose-phosphate pathway in *S. cerevisiae*

Parameter	Value mmol/l
$K_{NADP,1}$	0.116
$K_{i,NADPH,1}$	1.702
$K_{i,MgATP,1}$	0.33
$K_{NADP,2}$	1.848
$K_{i,NADPH,2}$	0.055
$K_{i,MgATP,2}$	0.109

measured intracellular concentration under steady-state conditions. Let us split the rate equation into two terms:

$$\tilde{r}_j = r_j^{max} f_j(\tilde{C}_j, P_j) \tag{14.18}$$

with concentration vector C containing all compounds that influence the activity and P parameter vector of the reaction j. Let us next assume that r_j has been estimated from Metabolic Flux Analysis, C has been measured, and P is available as a first estimate. The unknown maximal rates are given as

$$r_j^{max} = \frac{\tilde{r}_j}{f_j(\tilde{C}_j, P_j)} \tag{14.19}$$

This approach assumes that during the transient enzyme concentrations remain at their steady state value.

Combining Eqs. (14.13)–(14.19) leads to

$$r_{G6P-DH} = r_{G6P-DH}^{max} f_{G6P-DH}\left(\hat{C}_{MgATP}(t), \hat{C}_{NADP}(t), \hat{C}_{NADPH}(t), P_{G6PDH}\right) \tag{14.20}$$

and

$$r_{6PG-DH} = r_{6PG-DH}^{max} f_{6PG-DH}\left(\hat{C}_{MgATP}(t), \hat{C}_{NADP}(t), \hat{C}_{NADPH}(t), P_{6PGDH}\right) \tag{14.21}$$

which may be written as

$$r_{G6P-DH} = r_{G6P-DH}^{max} f_{G6P-DH}(t, P_{G6P-DH}) \tag{14.22}$$

$$r_{6PG-DH} = r_{6PG-DH}^{max} f_{6PG-DH}(t, P_{6PG-DH}) \tag{14.23}$$

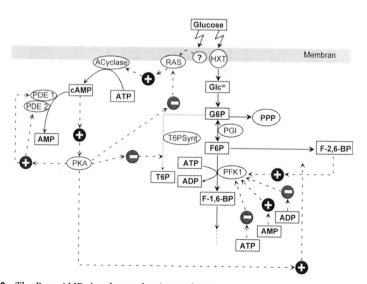

Fig. 14.9. The Ras-cAMP signal transduction pathway

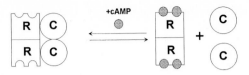

Fig. 14.10. Mechanism of activation of PKA

Thus the balance equation for 6PG (Eq. 14.13) takes the form

$$\frac{d\hat{C}_{6PG}}{dt} = -r^{max}_{6PG-DH}\, f_{6PG-DH}(t, P_{6PGDH})$$
$$+ r^{max}_{G6P-DH}\, f_{G6P-DH}(t, P_{G6PDH}) - \mu\, \hat{C}_{6PG}(t) \qquad (14.24)$$

with f_{G6P-DH} and f_{6PG-DH} defined in Eqs. (14.20) and (14.21) respectively.

The software packages ACSL (Mitchel and Gauthier, Concord USA, integration procedure Gear algorithm) and OptdesX (Design Synthesis Inc., Orem, UT, optimization strategy simulated annealing) have been applied for the estimation of the parameter vectors P_{G6P-DH} and P_{6PG-DH}. The results of the parameter identification procedure are shown in Table 14.3 The relative error square sum ε_{rel} for the time course of the different metabolites and co-metabolites concentrations were calculated in order to assess the quality of the identification procedure. The functional used for parameter estimation is

$$\varepsilon_{rel,6PG} = \sum_k \left(\frac{C_i(t_k) - \hat{C}_i(t_k)}{C_i(t_k)} \right)^2 \qquad (14.25)$$

The comparison between measurements and model predictions for 6PG is shown in Fig. 14.8.

This example not only illustrates the strategy for the in vivo identification of intracellular kinetics. – a more detailed description of this approach is presented in the papers of Rizzi et. al. [45] and Vaseghi et. al. [46] – it also indicates an interesting result regarding the modular structure, which is important for complex network analysis. According to the result of the identification procedure, the flux through the PPP is independent from the concentration of the substrate G6P and only depends on the ATP pool as well as the NADP/NADPH ratio. It therefore seems that the flux is adjusted to the energy state and balanced to the demand for biosynthesis. The PP-shunt thus acts as a functional unit, which is modulated by the energy state of the cell and the demand in biosynthesis.

14.4
Quantitative Analysis of Glucose Induced Signal Transduction

Signal transduction is rapidly evolving into one of the major topics in biology [47]. Quantitative analysis based on in vivo measurements and mathematical modeling will be crucial for fundamental understanding of regulation phenomena as well as its integration into Metabolic Control Analysis (MCA). Much prior research rests on in vitro analysis or in vivo measurements with mutants at physiological conditions, which are not well defined. Actual focus of experimental investigations is the unra-

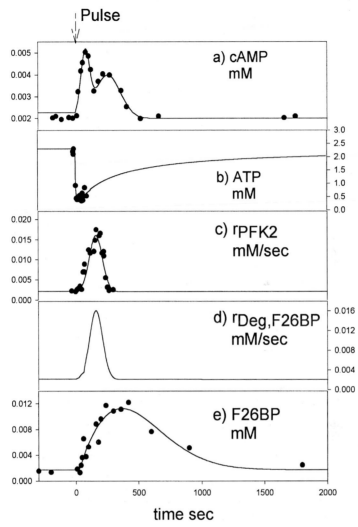

Fig. 14.11a–e. Time course of cAMP, ATP, PFK2 activity and the rate of F2,6bP degradation after a glucose pulse in a glucose limited continuous culture at $D=0.1\,h^{-1}$ [43,56]

veling of the genes involved and much less emphasis is put on the kinetic analysis of transduction cascades.

The glucose induced RAS-adenylate cyclase signaling pathway is one of the best known and most intensively studied regulatory pathways in *S. cerevisiae* [47]. An overview of the signaling pathway in response to the glucose-sensing system and its interaction with the glycolysis is schematically illustrated in Fig. 14.9.

When glucose is added to derepressed cells of the yeast *S. cerevisiae*, a cAMP signal is induced which triggers a protein phosphorylation cascade. The mechanism by which glucose initially activates the adenylate cyclase is still the subject of research.

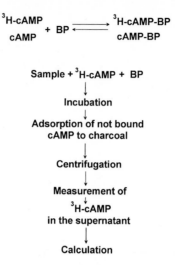

Fig. 14.12. Assay for the measurement of cAMP; BP: binding protein

It is known, that both glucose addition and acidification trigger an increase of the cAMP level.

There is increasing evidence that only acidification causes an increase in the GTP/GDP ratio on the Ras proteins which in turn leads to the activation of adenylate cyclase [48]. The stimulation by glucose seems to depend upon the presence of an additional Gα-protein Gpa2. These different phenomena are important for the kind of experiments described above, because the glucose pulse results in a drop of intracellular pH. We therefore anticipate a superposition of the two effects, acidification and glucose.

The increased cAMP leads to an activation of the PKA. The mechanism of this activation is illustrated in Fig. 14.10. The inactivated form of PKA is a tetramer consisting of two subunits, the catalytic C-PKA and the regulatory R-PKA [50–55]. Two cAMP molecules are reversibly bounded to R-PKA and are responsible for the dissociation of the two catalytic subunits. These catalytic subunits are representing the active form of PKA which in turn is catalyzing the phosphorylation of the target proteins in the glycolysis as well as the cell cycle [47].

As far as the aforementioned signal cascade is concerned, the most important enzymes in the glycolysis are the PFK2, the F6P-synthetase, and the trehalose-6-phosphate-synthetase. The dynamics of the activation of PFK2 may serve as an example for the quantitative analysis of such activation modules for enzymes. For in vivo diagnosis of intracellular reactions this is a special challenge because the maximal rate r_{max} in the rate expression changes during the transient and, thus, the enzyme does not stay any more in a "frozen" state.

PFK2 is activated by PKA causing an increase of fructose-2,6-bisphosphate. This "metabolic messenger" stimulates PFK1 and strongly inhibits FBPase1. T6P-synthetase, also shown in Fig. 14.9, is additionally modulated by PKA. The increase of T6P caused by the stimulation of the enzyme results in an inhibition of HK I and II [78]. Quantitative analysis of these regulation phenomena requires the in vivo measurement of the dynamics of signal compounds such as cAMP and F2,6bP, as well as

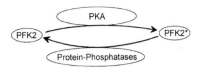

Fig. 14.13. Phosphorylation of PFK2 through PKA

transient behavior of an enzyme activity (PFK2) as a rapid response to the glucose stimulus. Both problems represent new challenges for the measurement of intracellular compounds because of very low concentrations and application of modified quenching and extraction methods to maintain enzyme activities, respectively.

14.4.1
Measurement of Intracellular cAMP

The results of the observed dynamic response illustrated in Fig. 14.11a [56] were obtained by quenching samples in methanol at –70 °C and afterwards extracting with boiling ethanol [55]. The measurements of the cAMP concentrations were performed with a commercialy available competitive protein binding assay (Amershan International TRK 432). The method is schematically illustrated in Fig. 14.12. The time course of the signal (Fig. 14.11a) indicate a rapid response to the glucose concentration which is in qualitative agreement with the observations of Beullens et al. [57].

Unravelling of the action mechanism of this signal is complicated because most of the experiments reported so far in the literature have been performed with glucose-starved cells. Alternatively, glucose has been added to cells grown on a non fermentable energy and carbon source. As outlined by Thevelein [47], the effects observed during these experiments might be masked by the superposition of the time course of ATP level. This is particularly important if observations are extended to the dynamic behavior of protein kinases which are obviously affected by the ATP pool. As such, the experiment reported here presents for the first time a quantitative analysis of the dynamics of the cAMP signal at physiologically defined growth conditions. This simultaneous observation leads to interesting clues regarding the influence of intracellular pH and ATP upon the cAMP cascade. Due to the sharp drop in intracellular ATP pool (Fig. 14.11b), pH will inevitably also drop. According to the observations of Colombo et al. [49] we must conclude that both activation effects upon

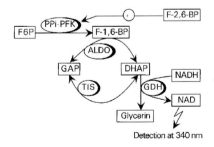

Detection at 340 nm

Fig. 14.14. Assay for the determination of F2,6bP [61]

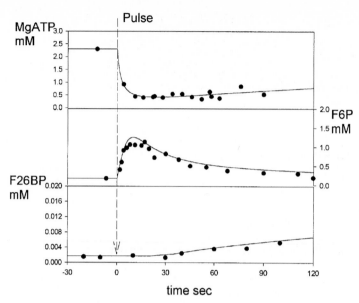

Fig. 14.15. Time course of ATP, F6P and F2,6bP after a glucose pulse in a glucose limited continuous culture at $D=0.1\,h^{-1}$ [43]

adenylate cyclase (acidification via Ras and glucose via Gpa2 protein) act simultaneously. However, we are also able to confirm quantitatively the conjectures of Colombo et al. [49], that ATP drops to limiting levels for the adenylate cyclase. Careful inspection of the time course of ATP (Fig. 14.11) clearly shows a prolonged period of ATP levels which are far below the K_M-value [49]. The increase of adenylate cyclase activity may therefore be counterbalanced by the substrate limitation of ATP immediately after the glucose pulse. The result is a delayed increase of the cAMP signal.

14.4.2
Measurement of the PFK2 Activity

A further step for elucidation of the interaction between the signaling pathway and glycolysis is the observation of the dynamics of enzyme activation. In what follows PFK2 will serve as a typical example. According to the scheme in Fig. 14.13, PFK2 is phosphorylated by PKA which in turn is activated by cAMP. This example also serves to introduce the experimental tools for rapid measurements of intracellular enzyme activities. The quenching and extraction methods applied to measure enzyme activities are those suggested by Francois et al. [58]. Samples are taken after the glucose pulse and quenched in methanol at –70 °C. After centrifugation at –9 °C and resuspension in a buffer the cells are mechanically disrupted by intensive shaking with glass beads. To prevent temperature increase, the disruption procedure is interrupted four times to cool the sample down to 0 °C. After centrifugation, the extract is incubated with the substrate F6P. Product concentration F26BP is determined with the assay described below. To measure the time course of product for-

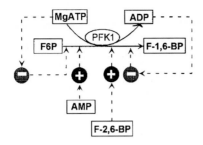

Fig. 14.16. The reactions around PFK1

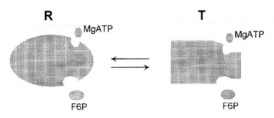

Fig. 14.17. The mechanism of the M-model [65]

mation, the reaction is quenched at different times by adding NaOH and incubated 10 min at 80 °C.

The results of the measurements are shown in Fig. 14.11c [43]. The time course of activity is similar to the cAMP signal in Fig. 14.11a. Attempts to correlate the two signals with a simple Michaelis Menten type of kinetic (graded response of activated protein to its stimulus) as well as taking into account a Hill type cooperativity (switch-like response) [59] failed. However, recently Vaseghi could show [43] that plotting the activity of PFK2 against cAMP resulted in a hysteresis, which points to a "bi-stable true switch" [43, 59]. Further measurements of PKA activities are required to elucidate the source of the positive feed-back, essential for this special type of kinetics.

14.4.3
Measurements of F2,6bP

The measurement of F2,6bP concentrations rests on a special assay developed by van Schaftingen et al. [60]. This assay uses the activation of pyrophosphate dependent PFK1 from potato tubers through F26bP. The formation of F26bP is measured with the aid of the enzyme cascade illustrated in Fig. 14.14. Results of the measurements after pulsing glucose to the steady state continuous culture are summarized in Fig. 14.11c [43]. It must be emphasized that only part of the time course is directly correlated with the activity of the PFK2. The decline of the signal is also determined by the degradation of F2,6bP, which has been studied with permeabilized cells using a special in situ assay suggested by Vaseghi [43].

From the comparison of the time courses of ATP, F6P, and F2,6bP (Fig. 14.15) it can be concluded that the rate of formation of F2,6bP is not influenced by the two

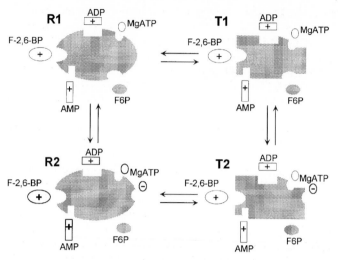

Fig. 14.18. The mechanism of the V-model [43]

substrates of the PFK2, ATP and F6P. We may therefore assert that the formation of F2,6bP is determined by the activity of PFK2, which in turn is regulated via the cAMP cascade [43].

Recently Vaseghi [43] has shown, that the quantitative information, which can be extracted from these measurements could be significantly expanded by considering the degradation of F2,6bP. This is an important issue for the dynamic transition to the previous steady state of the culture because PFK1 is strongly activated by this signal.

The balance equation for F2,6bP reads:

$$\frac{dC_{F2,6bP}}{dt} = r_{PFK2} - r_{Deg,F2,6bP} - \mu C_{F2,6bP} \tag{14.26}$$

The rate of degradation $r_{Deg,F2,6bP}$ summarizes the action of the two specific phosphatases with different affinity and one unspecific phosphatase [61–64]. Using suitable approximation functions for $C_{F2,6bP}(t)$ and $r_{PFK2}(t)$, it is possible to estimate the time course of the degradation rate. The results of these calculations are depicted in Fig. 14.11d [43]. The important observation is that the dilution of the F2,6bP pool due to growth considerably contributes to the decrease of the signal [43]. This again indicates the importance of the physiologically defined growth conditions for the quantitative investigations of the dynamics of signal transduction processes.

14.5
Comparison Between in vitro and in vivo Kinetics –
Illustrated for the Enzyme PFK1 (Phosphofructokinase 1)

Unfortunately, most of the published work in the application of Metabolic Control Analysis (MCA) – the important framework for rational design of cell factories – is based on in vitro kinetics. Provided that intracellular pool concentrations are known, we may ask the important question: to what extend does the situation inside the cell leads to kinetic behavior, which differs from the in vitro condition in which the enzyme has been taken out of the intracellular milieu? Presumably the answer to this question depends on the enzyme under study. Because of its complex structure and function, the enzyme PFK1 is one of the most challenging examples.

The kinetic behavior of this enzyme has attracted a lot of attention because of its important role in the regulation of the glycolysis. The allosteric enzyme catalyses the phosphorylation of F6P to F1,6bP and is modulated by several effectors as schematically illustrated in Fig. 14.16.

The most important attempts to model the allosteric behavior of this enzyme goes back to the work of Monod et al. [65]. The basic structure of this model is illustrated in Fig. 14.17.

Two conformations are considered: conformation T has a low- and confirmation R a high affinity to the substrate F6P. In the case of PFK1, each enzyme molecule consists of 8 subunits. The basic model of Monod et al.[65] only considers the allosteric behavior with respect to the substrate F6P and a non-allosteric effect of ATP. Freyer [66] additionally considered the modulation through AMP, ADP, and ATP and its influence upon the allosteric effects of F6P. This model is also known as Reuter-model [67]. In a similar way to that illustrated in the more sophisticated V-model (Fig. 14.18), this model assumes that:
- Each protomer of the octameric protein appears in two basic conformations R1 and R2 as well as two sub-conformations R2 and T2
- Substrate F6P is bonded to R1 and T1 with different affinities
- ATP binds to all different conformations with the same affinity
- ATP binds as inhibitor to R1 and T1
- AMP and ADP bind as activators to R1/T1 or R2/T2, respectively

A further approach to the kinetics of this enzyme has been suggested by Hofmann and Kopperschläger [68]. These authors reduced the complexity of the model by assuming that the low affinity of the enzyme in the T1 conformation can be neglected.

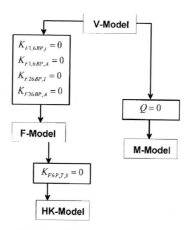

Fig. 14.19. Strategy for the reduction of the general V-model [43]

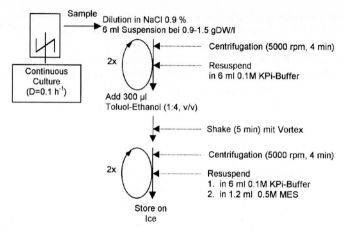

Fig. 14.20. Procedure for permeabilization of yeast cells [76]

None of these models considered the strong activation through F2,6bP [69]. Also the competitive inhibition between F2,6bP and F1,6bP reported by several research groups [70, 71] should be taken into account.

Vaseghi [43] has derived a complete model for PFK1 considering the concerted action of all the aforementioned regulation phenomena. The model is schematically illustrated in Fig. 14.18 and can be represented by the following set of equations:

$$
\begin{cases}
\alpha_{F2,6bP,I} = C_{F2,6bP}\, K_{F2,6bP,I} \\
\alpha_{F2,6bP,A} = C_{F2,6bP}\, K_{F2,6bP,A} \\
\alpha_{F1,6bP,I} = C_{F2,6bP}\, K_{F1,6bP,I} \\
\alpha_{F2,6bP,A} = C_{F2,6bP}\, K_{F2,6bP,A}
\end{cases}
\tag{14.27}
$$

$$
\begin{cases}
\alpha_{ATP,S} = C_{ATP}\, K_{ATP,S} \\
\alpha_{ATP,I} = C_{ATP}\, K_{ATP,I} \\
\alpha_{ATP,A} = C_{ATP}\, K_{ATP,A} \\
\alpha_{ADP,I} = C_{ADP}\, K_{ADP,I} \\
\alpha_{ADP,A} = C_{ADP}\, K_{ADP,A} \\
\alpha_{ADP,C} = C_{ADP}\, K_{ADP,C} \\
\alpha_{AMP,I} = C_{AMP}\, K_{AMP,I} \\
\alpha_{AMP,A} = C_{AMP}\, K_{AMP,A} \\
\alpha_{F6P,T} = C_{F6P}\, K_{F6P,T} \\
\alpha_{F6P,R} = C_{F6P}\, K_{F6P,R}
\end{cases}
\tag{14.28}
$$

$$
Z = \left(1.0 + \alpha_{ATP,I}\right)\left(1.0 + \alpha_{ADP,I}\right)\left(1.0 + \alpha_{AMP,I}\right) \\
\left(1.0 + \alpha_{F1,6bP,A} + \alpha_{F2,6bP,A}\right)
\tag{14.29}
$$

$$
N = \left(1.0 + \alpha_{ATP,A}\right)\left(1.0 + \alpha_{ADP,A}\right)\left(1.0 + \alpha_{AMP,A}\right) \\
\left(1.0 + \alpha_{F1,6bP,A} + \alpha_{F2,6bP,A}\right)
\tag{14.30}
$$

Table 14.4. Comparison of the parameters of the V-Model for in situ and in vitro conditions. Except for L_0, all other parameters are expressed in mM^{-1}

Parameter	in vitro	in situ
L_0	40.57	3.4e-9
$K_{ATP,S}$	27.15	6.55
$K_{ATP,I}$	0.4	0.11
$K_{ATP,A}$	6.48	1.1
$K_{F6P,R0}$	4.48	9.61
$K_{F6P,T0}$	0.92	200.18
$K_{ADP,I}$	1.22	2.15
$K_{ADP,A}$	0.52	1.13
$K_{ADP,C}$	3.07	1.7e-8
$K_{AMP,I}$	NI[a]	NI[a]
$K_{AMP,A}$	NI[a]	NI[a]
$K_{F26BP,I}$	NI[a]	NI[a]
$K_{F26BP,A}$	NI[a]	NI[a]
$K_{F16BP,I}$	28.43	0.0
$K_{F16BP,A}$	273.2	1.74e+4

[a]NI: Not identifiable

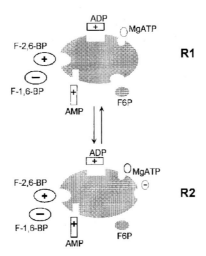

Fig. 14.21. The reduced V-model for in situ conditions according to Vaseghi [43]

$$Q = \frac{Z}{N} \tag{14.31}$$

$$K_{F6P,R} = K_{F6P,R,0} \, Q \tag{14.32}$$

$$K_{F6P,T} = K_{F6P,T,0} \, Q \tag{14.33}$$

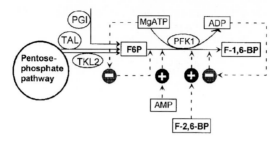

Fig. 14.22. Reactions involved in the balance of F6P

$$L = \frac{T}{R} = L_0 \left(\frac{1 + \alpha_{F6P,T}}{1 + \alpha_{F6P,R}}\right)^8 \tag{14.34}$$

$$M_{F6P,R} = \frac{\alpha_{F6P}}{\alpha_{F6P} + 1} \tag{14.35}$$

$$M_{F6P,T} = \frac{c\alpha_{F6P}}{c\alpha_{F6P} + 1} \tag{14.36}$$

$$N_{F6P} = M_{F6P,R}\, R + M_{F6P,T}\, T \tag{14.37}$$

$$N_{ATP} = \frac{\alpha_{ATP,S}}{1.0 + \alpha_{ATP,S} + \alpha_{ADP,C}} \tag{14.38}$$

$$\frac{r_{PFK}}{r_{PFK}^{max}} = N_{F6P}\, N_{ATP} \tag{14.39}$$

Figure 14.19 shows that this general model can easily be reduced to the existing models of Monod et al. [65] (M-Model), Freyer [66] (F-Model), and Hofman and Kopperschläger [68] (HK-Model).

For a more rational interpretation of the differences between in vitro and in vivo conditions, it was decided to complement the experimental investigations with some in situ experiments performed with permeabilized yeast cells. The procedure to prepare permeabilized cells, which may offer an interesting tool for studying intracellular enzymatic rate expressions, is depicted in Fig. 14.20 according to Serrano et al. [76].

The assay for determination of in situ activities for PFK1 has been described by Gancedo and Gancedo [72]. It consists of an enzymatic coupling of the PFK1 reaction with a glycerol-3-phosphate-dehydrogenase and a measurement of NADH.

Table 14.4 summarizes the results of the identification of the parameters of the kinetic model using the in vitro data of Hofman and Kopperschläger [68] as well as the in situ measurements of Vaseghi [43]. The differences in the values of the parameters are pronounced. According to Vaseghi [43] there are two striking phenomena, which attract attention:

– Compared to the in vitro situation, the modulation strength of all the effectors are less pronounced under in situ conditions

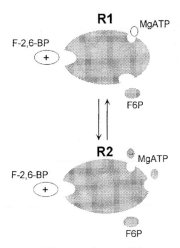

Fig. 14.23. The reduced V-model for in vivo conditions according to Vaseghi [43]

– The parameter value of L_0 tends to zero. This result leads to the interesting conclusion that the enzyme only acts in the R conformation (high affinity)

On the basis of these observations, the model can be reduced to a much more simple structure illustrated in Fig. 14.21 and represented by the following expressions [43]:

$$Z = \left(1.0 + \alpha_{ATP,I}\right) \left(1.0 + \alpha_{ADP,I}\right) \left(1.0 + \alpha_{AMP,I}\right)$$
$$\left(1.0 + \alpha_{F1,6bP,A} + \alpha_{F2,6bP,A}\right) \tag{14.40}$$

$$N = \left(1.0 + \alpha_{ATP,A}\right) \left(1.0 + \alpha_{ADP,A}\right) \left(1.0 + \alpha_{AMP,A}\right)$$
$$\left(1.0 + \alpha_{F1,6bP,A} + \alpha_{F2,6bP,A}\right) \tag{14.41}$$

$$Q = \frac{Z}{N} \tag{14.42}$$

$$K_{F6P,R} = K_{F6P,R,0}\, Q \tag{14.43}$$

$$M_{F6P,R} = \frac{\alpha_{F6P}}{\alpha_{F6P} + 1} \tag{14.44}$$

$$N_{F6P} = M_{F6P,R}\, R + M_{F6P,T}\, T \tag{14.45}$$

$$N_{ATP} = \frac{\alpha_{ATP,S}}{1.0 + \alpha_{ATP,S} + \alpha_{ADP,C}} \tag{14.46}$$

$$\frac{r_{PFK}}{r_{PFK}^{max}} = N_{F6P}\, N_{ATP} \tag{14.47}$$

Next, the identification of the kinetic parameters of the original V-model (Fig. 14.18) is performed with the aid of the in vivo diagnosis [45], which has been exemplary explained with the PPP in Sect. 14.3. For this purpose the measured time courses of F6P, F1,6bP, F2,6bP, MgATP, ADP, and AMP (experimental results summarized in Fig. 14.6b,c) are approximated by suitable analytical functions f_i [43]. The balance equation for F6P (see Fig. 14.22) reads:

$$\frac{dC_{F6P}}{dt} = -r_{PFK1}^{max} f_{PFK1}(t, C_{F6P}, \mathbf{P}_{PFK1}) + r_{PGI,G6P}^{max} f_{PGI}(t, C_{F6P}, \mathbf{P}_{PGI}) + r_{TAL}(t)$$
$$+ r_{TKL2}(t) - \mu\, C_{F6P}(t) \tag{14.48}$$

The contribution from the reactions Transaldolase (TAL) and Transketolase (TKL2) are computed from the model of the pentose phosphate shunt [46] (Sect. 14.3). The other rate expressions are those described in the original paper of Rizzi et al. [45]. The parameter were then estimated by minimizing the functional

$$\varepsilon_{rel,F6P} = \sum_k \left(\frac{C_{F6P}(t_k) - \hat{C}_{F6P}(t_k)}{C_{F6P}(t_k)}\right)^2 \tag{14.49}$$

The results of this estimation procedure further support the conclusions from the in situ experiments [43]. Again, the value of L_0 tends to zero and indicates, that the enzyme only acts in the R configuration. In case of the in vivo diagnosis, the sensitivity of the individual parameters were further investigated by comparing the time-dependent elasticity coefficient [43]. Accordingly, flux control coefficients are given by [79]:

$$\frac{dr_{PFK1}}{dP_j} = \frac{\partial\, r_{PFK1}}{\partial\, P_j} + \left[\sum_i \frac{\partial\, r_{PFK1}}{\partial\, C_i} \frac{\partial\, C_i}{\partial\, P_j}\right] \tag{14.50}$$

By quantitative analysis of the hierarchy of these coefficients, Vaseghi [43] clearly showed, that modulation of ADP, AMP, and F1,6bP is less important under in vivo conditions. F6P, ATP, and F2,6bP are the metabolites and effectors with strongest influence on the activity of the enzyme. Together with the result $L_0 = 0$ and

$$L = \frac{T}{R} = L_0 \left(\frac{1 + \alpha_{F6P,T}}{1 + \alpha_{F6P,R}}\right)^8 = 0 \tag{14.51}$$

we conclude that cooperativity between the eight subunits of the enzyme may exist but cannot be identified under the chosen conditions. The model therefore finally reduces to the simple structure shown in Fig. 14.23 and rate expression:

$$Z = \left(1.0 + \alpha_{ATP,I}\right)\left(1.0 + \alpha_{F2,6bP,A}\right) \tag{14.52}$$

$$N = \left(1.0 + \alpha_{ATP,A}\right)\left(1.0 + \alpha_{F2,6bP,A}\right) \tag{14.53}$$

$$Q = \frac{Z}{N} \tag{14.54}$$

$$K_{F6P,R} = K_{F6P,R,0}\, Q \tag{14.55}$$

$$M_{F6P,R} = \frac{\alpha_{F6P}}{\alpha_{F6P} + 1} \tag{14.56}$$

$$N_{F6P} = M_{F6P,R}\, R + M_{F6P,T}\, T \tag{14.57}$$

$$N_{ATP} = \frac{\alpha_{ATP,S}}{1.0 + \alpha_{ATP,S} + \alpha_{ADP,C}} \tag{14.58}$$

$$\frac{r_{PFK}}{r_{PFK}^{max}} = N_{F6P}\, N_{ATP} \tag{14.59}$$

To extrapolate the results obtained from the kinetic analysis of PFK1 to other enzymes, it is worthwhile to speculate about possible clues for these pronounced differences between in vivo and in vitro conditions. A reasonable explanation needs to consider the well known effects of homologous and heterologous protein interaction inside the cell. The first effect is related to the concentration of the enzyme. Srere [73] indicated that concentrations of glycolytic enzymes inside the cells are usually 1000 times higher than those normally applied to in vitro assays. According to Aragón and Sanchez [74] the concentration of PFK1 in *S. cerevisiae* is of the order of 190–550 μg/ml. The concentration of in vitro assays is 1–10 μg/ml. The influence of high intracellular concentrations of the enzyme on the regulation of PFK1 has been shown by Aragón and Sanchez [74]. The second effect – heterologous intereactions – stands for associations between enzymes and structure proteins of the cytoskeleton [77]. Particularly for PFK1 in *S. cerevisiae*, Kopperschläger [75] has shown for the first time an organized association between the enzyme molecule and the microtubules under in vivo conditions. The impacts of such associations on the kinetic behavior of the enzymes is not known yet. However, because of the strong relationship between structure and function, these are strong candidates for the explanation of the aforementioned differences between in vitro and in vivo conditions. Because of the closer agreement between in situ and in vivo observations, the in situ experiments could be viewed as interesting and alternative tools for the kinetic analysis of intracellular reactions.

Appendix

Biochemical Reactions in the Cytosol

glc + atp = g6p + adp + h
g6p = f6p
f6p + atp = fdp + adp + h
fdp = gap + dhap
dhap = gap
gap + nad + p = 3 pg + nadh + h
3 pg + adp = g3p + atp
g3p = pep + h2o
pep + adp + h = pyr + atp

glut + nh4 + atp = glum + adp + p + h

glut + atp + 2 nadph + 2 h = pro + adp + p + h2o + 2 nadp

nh4 + akg + nadph + h = glut + nadp + h2o

2 atp + nh4 + co2 + h2o = carp + 2 adp + 3 h + p

orn + carp + asp + atp + h2o = arg + fum + amp + 3 p + 2 h

g3p + glut + nad + h2o = ser + akg + p + h + nadh

ser + thf = gly + methf + h2o

so4 + 2 atp + 4 nadph + 2 h = sulfide + 4 nadp + amp + adp + 3 p + h2o + h

oac + glut = asp + akg

asp + glum + atp + 2 h2o = asn + 2 h + amp + 2 p + glut

asp + atp + 2 nadph + 2 h = hom + adp + p + 2 nadp

hom + atp + h2o = thr + adp + p + h

2 pyr + glut + nadph + 2 h = val + akg + h2o + nadp + co2

thr + pyr + nadph + glut + 2 h = ileu + nh4 + nadp + h2o + co2 + akg

pyr + glut = ala + akg

2 pyr + nadph + 2 h = aki + co2 + nadp + h2o

2 pep + e4p + nadph + atp = cho + adp + 4 p + nadp

cho + glut + h = phen + akg + co2 + h2o

cho + glut + nad = tyr + akg + co2 + nadh

2 glut + accoA + atp + nadph + h2o = orn + akg + coA + adp + p + nadp + ac + h

2 glut + accoA + atp + 2 nadph + 2 nad + 2 h2o = lys + co2 + akg + coA + amp + 2 nadp + 2 p + h + 2 nadh

cho + glum + prpp + ser = tryp + 2 p + co2 + gap + glut + h + pyr + h2o

rib5p + atp = prpp + amp + h

prpp + glum + atp + 2 nad + 5 h2o = his + aicar + akg + 2 nadh + 7 h + 5 p

hom + succoA + cys + mythf + 2 h2o = met + coA + suc + pyr + nh4 + h + thf + h2o

ser + accoA + sulfide + h = cys + coA + ac

aki + glut + accoA + h2o + nad = leu + akg + coA + nadh + h + co2

pyr + atp + h2o + co2 = oac + adp + p + 2 h

pyr + h = ald + co2

ald + nadh + h = etoh + nad

ald + nad + h2o = ac + nadh + 2 h

dhap + nadh + h = glycerol3p + nad

glycerol3p + h2o = glycerol + p

prpp + 2 glum + gly + 5 atp + asp + fthf + 4 h2o + co2 = aicar + 5 adp + 7 p + 2 glut + thf + fum + 9 h

aicar + fthf = thf + imp + h2o

imp + asp + atp = amp + adp + p + fum + 2 h

imp + nad + 2 atp + glum + 3 h2o = gmp + 2 adp + 2 p + glut + nadh + 4 h

carp + asp + nad + prpp = ump + nadh + co2 + 3 p + h

ump + 2 atp = utp + 2 adp

utp + glum + atp + h2o = ctp + adp + p + 2 h + glut

ctp + 2 adp = cmp + 2 atp

amp + atp = 2 adp

g6p + 2 nadp + h2o = ribu5p + 2 nadph + co2 + 2 h

ribu5p = rib5p

ribu5p = xyl5p

rib5p + xyl5p = sed7p + gap

sed7p + gap = f6p + e4p

xyl5p + e4p = f6p + gap

ac + coA + atp + h2o = accoA + amp + 2 p + h

8 accoA + 7 atp + 14 nadph + 6 h + h2o = pal + 8 coA + 7 adp + 14 nadp + 7 p

pal + 2 nadph + h + accoA + atp = stear + adp + p + 2 nadp + coA

pal + nadh + o2 + h = palen + nad + 2 h2o

stear + nadh + h + o2 = ol + nad + 2 h2o

ol + nadh + h + o2 = linol + nad + 2 h2o

mal + nad = oac + nadh + h

oac + accoA + h2o = isocit + coA + h

thf + atp + nadh + co2 = fthf + adp + p + nad

thf + co2 + 3 nadh + 3 h = mythf + 3 nad + 2 h2o

methf + 2 nad + 2 h2o = thf + co2 + 2 nadh + 2 h

Synthesis of Macromolecules and Biomass

11 g6p + 10 atp + 10 h2o = polysacch + 20 p + 10 adp + 10 h

91872/317671 pal + 196680/317671 ol + 330660/317671 palen + 23661/317671 stear + 600/317671 linol + 309540/317671 glycerol = lipids + 2 h2o

193920/23593 glut + 68175/23593 glum + 106050/23593 pro + 103020/23593 arg + 184830/23593 lys + 119685/23593 ser + 187860/23593 gly + 4444/23593 cys + 190890/23593 asp + 65650/23593 asn + 124230/23593 thr + 31815/23593 met + 124230/23593 ileu + 293910/23593 ala + 172710/23593 val + 199980/23593 leu + 84840/23593 phen + 66660/23593 tyr + 17574/23593 tryp + 42420/23593 his + 400 atp + 300 h2o = -protein + 400 adp + 400 h + 400 p

2770277/1189 amp + 8320832/3567 gmp + 3640364/1189 ump + 280028/123 cmp + 10000 atp = rna + 10000 adp + 10000 p + 10000 h

5763419200/228923467 lipids + 12475694000/228923467 polysacch + 3202100/228923467 rna + 4650031400/228923467 protein = 100 bio

Biochemical Reactions in the Mitochondria

atp + h2o = adp + p + h

pyr + nad + coA = accoA + nadh + co2

oac + accoA + h2o = isocit + coA + h

isocit + nad = akg + nadh + co2

akg + coA + nad = succoA + nadh + co2

succoA + adp + p = suc + atp + coA

suc + fad = fum + fadh2

fum + h2o = mal

mal + nad = oac + nadh + h

Oxidative Phosphorylation

Komplex I: $nadh^{mit} + h^{mit} + UQ + 3\,h^{mit} = UQH2 + nad^{mit} + 3\,h^{cyt}$

Komplex II: $fadh2^{mit} + UQ = UQH2 + fad^{mit}$

$nadh^{cyt} + h^{cyt} + UQ = UQH2 + nad^{cyt}$

Komplex III: $UQH2 + CYTc + 3\,h^{mit} = CYTcH2 + UQ + 3\,h^{cyt}$

Komplex IV: $CYTcH2 + 1/2\,o2^{mit} + 3\,h^{mit} = CYTc + h2o^{mit} + 3\,h^{cyt}$

ATP-Synthase: $h^{cyt} + 1/3\,\eta\,adp^{mit} + 1/3\,\eta\,p^{mit} + 1/3\,\eta\,h^{mit} = h^{mit} + 1/3\,\eta\,atp^{mit} + 1/3\,\eta\,h2o^{mit}$

Transport Mitochondria-Cytosol

$adp^{mit} = adp^{cyt}$

$akg^{mit} = akg^{cyt}$

$atp^{mit} = atp^{cyt}$

$co2^{mit} = co2^{cyt}$

$coA^{mit} = coA^{cyt}$

$fum^{mit} = fum^{cyt}$

$h^{mit} = h^{cyt}$

$h2o^{mit} = h2o^{cyt}$

$isocit^{mit} = isocit^{cyt}$

$mal^{mit} = mal^{cyt}$

$o2^{mit} = o2^{cyt}$

$pyr^{mit} + h^{mit} + adp^{mit} + p^{mit} = pyr^{cyt} + atp^{mit} + h2o^{mit}$

$2\,p^{mit} + h^{mit} + adp^{mit} = p^{cyt} + atp^{mit} + h2o^{mit}$

$suc^{mit} + 2\,h^{mit} + 2\,adp^{mit} + 2\,p^{mit} = suc^{cyt} + 2\,atp^{mit} + 2\,h2o^{mit}$

Transport Cytosol-Extracellular

$ac^{cyt} = ac^{ext}$

$bio^{cyt} = bio^{ext}$

$co2^{cyt} = co2^{ext}$

$etoh^{cyt} = etoh^{ext}$

$glc^{cyt} = glc^{ext}$

$glycerol^{cyt} = glycerol^{ext}$

$h^{cyt} = h^{ext}$

$h2o^{cyt} = h2o^{ext}$

$o2^{cyt} = o2^{ext}$

$nh4^{cyt} + h^{cyt} + adp^{cyt} + p^{cyt} = nh4^{ext} + atp^{cyt} + h2o^{cyt}$

$2\,p^{cyt} + h^{cyt} + adp^{cyt} = p^{ext} + atp^{cyt} + h2o^{cyt}$

$so4^{cyt} + h^{cyt} + adp^{cyt} + p^{cyt} = so4^{ext} + atp^{cyt} + h2o^{cyt}$

Metabolites: Cytosol

accoA   acetyl-CoA

ac	acetate
adp	adenosine diphosphate
aicar	5-amino-4-imidazolecarboxamide ribotide
akg	alpha-ketoglutarate
aki	alpha-ketoisovalerate
ala	L-alanine
ald	formaldehyde
amp	adenosine monophosphate
arg	arginine
asn	asparagine
asp	aspartate
atp	adenosine triphosphate
bio	biomass
carp	carbamyl phosphate
cho	chorismate
cmp	cytidin-5-monophosphate
co2	carbon dioxide
coA	coenzyme A
ctp	cytidin-5-triphosphate
cys	cystein E
dhap	dihydroxyacetone phosphate
e4p	erythrose 4-phosphate
etoh	ethanol
f6p	fructose-6-phosphate
fdp	fructose-1,6-phosphate
fthf	formyltetrahydrofolate
fum	fumarate
g6p	glucose-6-phosphate
gap	glyceraldehyde-3-phosphate
glc	glucose
glum	glutamine
glut	L-glutamate
gly	glycine
glycerol	glycerol
glycerol3p	glycerol-3-phosphate
3 pg	3-phospho-d-glyceroyl-phosphate
g3p	glycerol-3-phosphate
gmp	guanosin-5-monophosphate
h	proton
h2o	water
his	histidine
hom	homoserine
ileu	isoleucine
imp	inosin-5-monophosphate
isocit	iso-citrate
leu	leucine
linol	*cis*-9,12-octadecenoate (linoleic acid)
lipids	lipids
lys	lysine
mal	malate
met	methionine
methf	methylentetrahydrofolate
mythf	methyltetrahydrofolate
nad	diphosphopyridindinucleotide, oxidized
nadh	diphosphopyridindinucleotide, reduced
nadp	diphosphopyridindinucleotide-phosphate, oxidized

nadph	diphosphopyridindinucleotide-phosphate, reduced
nh4	ammonium ion
o2	oxygen
oac	oxalacetate
ol	*cis*-octadecenoate (oleic acid)
orn	ornithine
p	inorganic phosphate
pal	hexadecanoate (palmitic acid)
palen	*cis*-hexadecenoate (palmitinoleic acid)
pep	phosphoenolpyruvate
phen	phenylalanine
polysacch	polysaccharides
pro	prolin
protein	protein
prpp	phosphoribosylpyrophosphate
pyr	pyruvate
rib5p	ribose 5-phosphate
ribu5p	ribulose 5-phosphate
rna	ribonucleic acid
sed7p	sedheptulose 7-phosphate
ser	serine
so4	sulfate ion
stear	octadecenoate (stearic acid)
suc	succinate
succoA	succinyl CoA
sulfide	sulfide
thf	tetrahydofolate
thr	threonine
tryp	tryptophan
tyr	tyrosine
ump	uridin-5-monophophate
utp	uridin-5-triphophate
val	valine
xyl5p	xylulose 5-phosphate

Metabolites: Mitochondria

accoA	acetyl-CoA
adp	adenosine diphosphate
akg	alpha-ketoglutarate
atp	adenosine triphosphate
co2	carbon dioxide
coA	coenzyme A
CYTc	cytochrome c, oxidized
CYTcH2	cytochrome c, reduced
fad	flavin-adenine-dinucleotide, oxidized
fadh2	flavin-adenine-dinucleotide, reduced
fum	fumarate
h	proton
h2o	water
isocit	iso-citrate
mal	malate
nad	diphosphopyridindinucleotide, oxidized
nadh	diphosphopyridindinucleotide, reduced
oac	oxalacetate
p	inorganic phosphate

pyr   pyruvate
suc   succinate
succoA   succinyl CoA
UQ   ubiquinone, oxidized
UQH2   ubiquinone, reduced

Cmol-Composition, Charge and Molecular Weight (MW) of Macromolecules and Biomass

Protein	$C_1 H_{1.58} N_{0.271} O_{0.315} S_{0.0032}$	charge: -0.00843	MW 22.5 g $Cmol^{-1}$
RNA	$C_1 H_{1.024} N_{0.383} O_{0.747} P_{0.1056}$	charge: -0.2113	MW 33.6 g $Cmol^{-1}$
Lipids	$C_1 H_{1.798} O_{0.134}$	charge: -0.5516	MW 16.0 g $Cmol^{-1}$
Polysaccharides	$C_1 H_{1.682} O_{0.894} P_{0.0152}$	charge: -0.0303	MW 28.5 g $Cmol^{-1}$
Biomass D=0.33 h^{-1}	$C_1 H_{1.61} N_{0.188} O_{0.45} P_{0.0047} S_{0.00214}$	charge: -0.0187	MW $23{,}6$ g $Cmol^{-1}$
D=0.24 h^{-1}	$C_1 H_{1.59} N_{0.181} O_{0.58} P_{0.0103} S_{0.00185}$	charge: -0.0302	MW 24.2 g $Cmol^{-1}$
D=0.10 h^{-1}	$C_1 H_{1.61} N_{0.161} O_{0.521} P_{0.0091} S_{0.00171}$	charge: -0.0267	MW 24.6 g $Cmol^{-1}$

References

1. Oliver SG (1997) Curr Opin Genet Dev 7:405
2. Selkov E, Basmanova S, Gaasterland T, Goryanin I, Gretchkin Y, Maltsev N, Nenashev V, Overbeek R, Panyushkina E, Pronevitch L, Selkov EJR, Yunus I (1996) Nucl Acids Res 24:26
3. Selkov E, Maltsev N, Olsen GJ, Overbeek R, Whitman WB (1997) Gene 197:GC11
4. Overbeek R, Panyushkina E, Pronevitch L, Selkov EJR, Yunus I (1996) Nucl Acids Res 24:26
5. Overbeek R, Larsen N, Smith W, Maltsev N, Selkov E (1997) Gene 191:GC1
6. Bork P, Koonin EV (1998) Nature Genet 13:313
7. Bailey JE (1991) Toward a science of metabolic engineering. Science 252:1668–1675
8. Stephanopoulos GN, Aristidou AA, Nielsen J (1998) Metabolic engineering. Academic Press, New York
9. Marx A, de Graaf AA, Wiechert W, Eggeling L, Sahm H (1996) Biotechnol Bioeng 49:111
10. Zupke C, Stephanopoulos G (1995) Biotechnol Bioeng 45:229
11. Cortassa S, Aon JC, Aon M (1995) Biotechnol Bioeng 47:193
12. Nissen T, Schulze U, Nielsen J, Viladsen J (1997) Microbiology 143:203
13. Vanrolleghem PA, de Jong-Gubbels P, van Gulik WM, Pronk JT, van Dijken JP, Heijnen JJ (1996) Biotechnol Prog 12:434
14. van Gulik WM, Heijnen JJ (1995) Biotechnol Bioeng 48:681
15. Nielsen JVJ (1994) Bioreaction Engineering Principles. Plenum-Press, New York
16. van der Heijden RTJM, Heijnen JJ, Hellinga C, Romein B, Luyben KCHAM (1994) Biotechnol Bioeng 43:11
17. van der Heijden RTJM, Heijnen JJ, Hellinga C, Romein B, Luyben KCHAM (1994) Biotechnol Bioeng 43:3
18. van der Heijden RRJM, Romein B, Heijnen JJ, Hellinga L, Lyben KCHAM (1994) Biotechnol Bioeng 44:781

19. Theobald U, Mailinger W, Baltes M, Rizzi M, Reuss M (1997) Biotechnol Bioeng 55:305
20. Gancedo CSR (1989) Energy- yielding metabolism In: Rose AHH (ed) The yeasts, vol 3. Academic Press, London, p 205
21. Krämer R, Palmierei F (1992) Metabolite carriers in mitochondria. In: Ernster L (ed) Molecular mechanisms in bioenergetics. Elsevier Amsterdam, p359
22. Cockburn M, Earnshaw P, Eddy AA (1975) J Biochem 146:705
23. Bünger R, Mallet RT (1993) Biochim Biophys Acta 1151:223
24. de Vries S, Marres CAM (1987) Biochim Biophys Acta 895:205
25. de Robichon-Szulmajster H, Surdin-Kerjan H (1971) Nucleic acid and protein synthesis in yeast:regulation of synthesis and activity. In: Rose AHH (ed) The yeasts, vol 2. Academic Press, London, p 335
26. Maser F (1997) Experimentelle und modellgestützte Untersuchungen zum Lipid-und Proteingehalt in *S. cerevisiae* unter definierten physiologischen Bedingungen. Institut für Bioverfahrenstechnik, Universität Stuttgart
27. Stryer L (1988) Biochemistry. Freeman, New-York
28. Fuchs H (1997) Makromolekulare Zusammensetzung von *S. cerevisiae* bei verschiedenen Verdünnungsraten im aeroben, glucoselimitierten Chemostaten. Institut für Bioverfahrenstechnik, Universität Stuttgart
29. Oura E (1972) Biotechnol Bioeng 28:415
30. Verduyn C, Postma E, Scheffers WA, van Dijken JP (1990) J Gen Microbiol 136:395
31. Spieth A (1997) Experimentelle Untersuchungen zur Bestimmung des Gehalts an Polysacchariden, RNA und Asche in *S. cerevisiae* unter definierten physiologischen Bedingungen. Institut für Bioverfahrenstechnik, Universität Stuttgart
32. Mauch K, Arnold S, Posten C, Reuss M (1997) Computer algebra systems in model building and model analysis for bioprocesses. 15th IMACS World Congress
33. Heinrich R, Schuster S (1996) The regulation of cellular systems. Chapman and Hall, New-York
34. Char BW, Geddes KO, Keith O, Gonnet GH, Leong BL, Monagan MB, Watt SM (1991). Maple V reference manual. Springer, Berlin Heidelberg New York
35. Gancedo JM, Lagunas R (1973) Plant Science Letters 1:193
36. Verduyn C, Stouthamer AH, Scheffers WA, van Dijken JP (1991) Antonie van Leeuwenhoek 59:49
37. Bailey JE (1998) Biotechnol Prog 14:8
38. Casti JL (1992) Reality rules: I. Picturing the world in mathematics, the fundamentals. Wiley, New York
39. Casti JL (1992) Reality rules: II. Picturing the world in mathematics, the frontier. Wiley, New York
40. Theobald U, Mailinger W, Reuss M, Rizzi M (1993) Anal Biochem 214:31
41. Schäfer U, Weuster-Botz D, Wandrey C (1998) Dechema Jahrestagungen 98, 26–28. Mai 1998, 16. Jahrestagung der Biotechnologen, Band 1:353
42. Kopperschläger G, Augustin HW (1967) Fehlermöglichkeiten bei der Bestimmung von Metabolitgehalten in Hefezellen. Experientia 23:623
43. Vaseghi S (1999) Modellgestützte Analyse des PFK1/PFK2 Systems in *S. cerevisiae*. PhD Thesis, Universität Stuttgart, Stuttgart, Germany
44. Theobald U (1995) Untersuchungen zur Dynamik des Crabtree-Effekts. PhD Thesis, Universität Stuttgart, Stuttgart, Germany und VDI Verlag, Düsseldorf, Germany
45. Rizzi M, Baltes M, Theobald U, Reuss M (1997) Biotechnol Boeng 55:592
46. Vaseghi S, Baumeister A, Rizzi M, Reuss M (1999) Metabol Eng 1:128
47. Thevelein JM (1994) Yeast 10:1753
48. Thevelein JM, Beulens M, Honshoven F, Hoebeeck G, Detremerie K, Griewel B, den Hollander JA, Jans AWH (1987) Gen Microbiol 133:2197
49. Colombo S, Ma P, Cauwenberg L, Winderickx J, Crauwels M, Teunissen A, Nauwelaers D, de Winde JH, Growa M-W, Colavizza D, Thevelein JM (1998) EMBO J 17:3326
50. Hixson CS, Krebs EG (1980) J Biol Chem 255:2137
51. Matsumoto K, Uno I, Oshima Y, Ishikawa T (1982) Proc Natl Acad Sci USA 79:2355
52. Wingender-Drissen R (1983) FEBS Lett 163:28

53. Cannon JF, Tatchell K (1987) Mol Cell Biol 7:2653
54. Toda A, Cameron S, Sass P, Zoller M, Wigler M (1987) Cell 50:277
55. Toda A, Cameron S, Sass P, Zoller M, Scott JD, McBullen B, Hurwitz M, Krebs EG, Wigler M (1987) Mol Cell Biol 7:1371
56. Macherhammer F, Vaseghi S, Rizzi M, Reuss M (1998) Modelling of glycolysis in yeast: consideration of signal transduction dynamics. Poster presentation, Metabolic Engineering II, Elmau, Germany
57. Beullens M, Mbonyi K, Geerts L, Gladins D, Detremerie K, Jans AWH, Thevelein JM (1988) Eur J Biochem 172:227
58. François J, van Schaftingen E, Hers H-G (1984) Eur J Biochem 145:187
59. Ferrel JE (1998) TIBS 23:461
60. van Schaftingen E, Lederer B, Bartons R, Hers H-G (1982) Eur J Biochem 129:191
61. François J, van Schaftingen E, Hers H-G (1988) Eur J Biochem 171:599
62. Kretschmer M, Schellenberger W, Hofmann E (1987) J Theoret Biol 127:181
63. Kessler R, Gärtner G, Schellenberger W, Hofmann E (1991) Biomed Biochim Acta 50:851
64. Purwin C, Laux M, Holzer H (1987) Eur J Biochem 164:27
65. Monod J, Wyman J, Changeux, J-P (1965) J Mol Biol 12:88
66. Freyer R (1977) Phosphofructokinase aus Hefe: Kinetisches Verhalten verschiedener Enzymformen und Diskussion eines mathematischen Modells. PhD Thesis, Karl-Marx-Universität Leipzig, Leipzig
67. Reuter R, Eschrich K, Schellenberger W, Hoffmann E (1979) Acta Biol et Med Germ 38:1067
68. Hofmann E, Kopperschläger G (1982) Meth Enzymol 90:49
69. Kessler R, Schellenberger W, Nissler K, Hofmann E (1988) Biomed Biochim Acta 47:221
70. Lederer B, Vissers S, van Schaftingen E, Hers H-G (1981) Biochem Biophys Res Commun 103:1281
71. Przybylski F, Otto A, Nissler K, Schellenberger W, Hofmann E (1985) Biochim Biophys Acta 831:350
72. Gancedo JM, Gancedo C (1971) Arch Microbiol 109:132
73. Srere PA (1967) Science 158:936
74. Aragón JJ, Sánchez V (1985) Biochem Biophys Res Commun 131:849
75. Kopperschläger G (1999) personal communication
76. Serrano R, Gancedo JM, Gancedo C (1973) Eur J Biochem 34:479
77. Ovadi J (1995) Cell architecture and metabolism. Springer, Berlin, Heidelberg New York
78. Entian KD, Zimmerman FK (1997) Yeast sugar metabolism: biochemistry, genetics, biotechnology and application. Technomic Publishing, Lancaster, Basel
79. Mauch K, Arnold S, Reuss M (1997) Dynamic sensitivity analysis for metabolic systems. Chem Eng Sci 52:2589
80. Mailinger W, Baumeister A, Reuss M, Rizzi M (1998) J Biotechnol 63:155
81. Gonzales B, Frqancois J, Renaud M (1997) A rapid and reliable method for metabolite extraction in yeast using boiling buffered ethanol. Yeast 13(14):1347–1355

15 Metabolic Analysis of *Zymomonas mobilis*

Albert A. de Graaf

15.1
Introduction

15.1.1
Zymomonas mobilis

Zymomonas mobilis is an anaerobic, Gram-negative bacterium producing ethanol from glucose via the Entner-Doudoroff (2-keto-3-deoxy-6-phosphogluconate, KDPG) pathway in conjunction with the enzymes pyruvate decarboxylase and alcohol dehydrogenase. The organism was originally isolated from fermenting sugar-rich plant saps, e.g., in the mexican pulque drink made from agave sap, or in palm wines of tropical Africa [1]. The carbohydrate metabolic pathways of *Z. mobilis* have been reviewed recently [2] and are shown in Fig. 15.1.

Being an anaerobic bacterium, *Z. mobilis* still has retained functions of the electron transport chain and grows in microaerobic environments [3], illustrating its closeness to aerobic relatives *Gluconobacter* and *Acetobacter* [4]. Indeed, the behavior of this anaerobic bacterium under aerobic conditions is currently receiving increased attention in connection with its membrane-associated ATPase [5, 6], respiratory chain-linked NADH oxidase system [7], and a d-lactate oxidase component of its respiratory chain [8].

The ethanol yield from glucose fermented by this remarkable organism can be as high as 97–98% of the theoretical value [2, 9]. Only insignificant amounts of by-products are produced by *Z. mobilis* growing on glucose [10, 11]. However, although the pathway for glucose and fructose degradation is almost identical (it differs only in the first two steps), biomass and product yields on fructose are lower than on glucose and more by-products are generated. Especially at high fructose concentrations, growth is slower, biomass yields are only half of those on glucose, and ethanol yields are only 90% of the theoretical value [12–14]. This is due to an increased formation of side products, especially glycerol and dihydroxyacetone [15–18]. While these compounds were shown to derive from glyceraldehyde-3-phosphate via dihydroxyacetone phosphate and glycerol-3-phosphate [19], the metabolic regulations leading to the increased by-product formation are not yet understood.

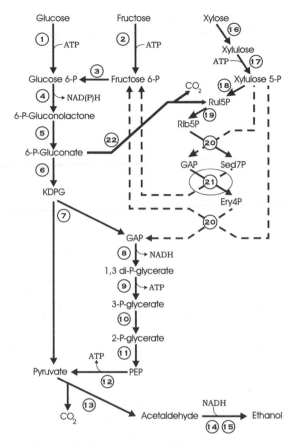

Fig. 15.1. Scheme of the catabolic pathways in wild-type and xylose-degrading recombinant *Zymomonas mobilis*. Abbreviations: KDPG=2-keto-3-deoxy-6-phosphogluconate, GAP=glyceraldehyde-3-phosphate, PEP=phosphoenolpyruvate, Rul5P=Ribulose 5-phosphate, Rib5P=Ribose 5-phosphate, Sed7P=sedoheptulose 7-phosphate, Ery4P=erythrose 4-phosphate. Enzyme activities are as follows: 1, glucokinase; 2, fructokinase; 3, phosphoglucose isomerase; 4, glucose 6-P dehydrogenase; 5, 6-P-gluconolactonase; 6, 6-P-gluconate dehydratase; 7, KDPG aldolase; 8, glyceraldehyde-P dehydrogenase; 9, phosphoglycerate kinase; 10, phosphoglycerate mutase; 11, enolase; 12, pyruvate kinase; 13, pyruvate decarboxylase; 14,15, alcohol dehydrogenases (two isoenzymes); 16, xylose isomerase; 17, xylulokinase; 18, ribulose 5-P epimerase; 19, ribose 5-phosphate isomerase; 20, transketolase; 21, transaldolase; 22, 6-phosphogluconate dehydrogenase. Adapted from [20, 159]

15.1.2
Substrate Spectrum Engineering

While *Z. mobilis* is superior to yeasts with respect to ethanol productivity (3–5 times higher), sugar tolerance (up to $400\,g\,l^{-1}$), and resistance to high ethanol concentrations (12–13%), it has not become a serious competitor because of its narrow substrate spectrum. The wild-type organism only grows on glucose, fructose, and sucrose. Several attempts to engineer *Z. mobilis* into an efficient ethanol producer from abundant and renewable lignocellulosic biomass (wood, straw, milk whey, agricultural residues) have been carried out. Studies until 1993, reviewed in [20] met with limited success: only a partial and very slow conversion of lactose (which constitutes up to 75% of the dry weight of whey [21]) and cellobiose (a main product of cellulase-catalyzed breakdown of cellulose) could be achieved, mainly due to limitations in the substrate uptake. However, in recent years several breakthroughs were achieved. Zhang et al. [22] reported the successful construction of a *Z. mobilis* strain that was able to convert xylose with an 86% of theoretical yield to ethanol. To this end, *Escherichia coli* genes encoding xylose isomerase, xylulokinase, transaldolase, and transketolase had been introduced and expressed in *Z. mobilis*. Growth and productivity on xylose, however, were about five times slower than on glucose. More recently, the same research group succeeded in engineering an efficient arabinose-fermenting *Z. mobilis* strain [23]. While xylose is the predominant pentose sugar derived from the hemicellulose of most hardwood feedstocks, arabinose is an important constituent of various agricultural residues and other herbaceous crops. The arabinose-fermenting strain was constructed by introducing the *E. coli* genes for L-arabinose isomerase, L-ribulokinase, L-ribulose-5-phosphate-4-epimerase, transaldolase, and transketolase in *Z. mobilis*. Growth and ethanol productivity on the new sugar were only 2–3 times slower than on glucose and the ethanol yield based on consumed substrate was as high as 98% of the theoretical yield. Weisser et al. [24] constructed a strain that used mannose as a novel substrate after introduction of the phosphomannose-isomerase gene from *E. coli*. Uptake of this substrate was shown to be due to the glucose facilitator of *Z. mobilis*, while phosphorylation was due to a side activity of the resident fructokinase. After plasmid overexpression of fructokinase activity, growth rates of $0.25\,h^{-1}$ were obtained and the growth yield was better than on fructose. The construction of an alternative xylose-degrading strain using the genes encoding xylose isomerase and xylulokinase from *Klebsiella pneumoniae* in addition to the *E. coli* genes for transaldolase and transketolase has also recently been completed (G. Sprenger et al., manuscript in preparation).

15.1.3
Purpose

The recent progress in the engineering of a broader substrate spectrum for *Z. mobilis* shows that the organism in principle is able to ferment different carbon sources to ethanol with yields generally above 85–90% of theoretical yields. However, growth rates, growth yields, and by-product formation vary widely between the reported strains. This suggests that, although the glycolytic enzymes are always able to direct the main carbon flux to ethanol, metabolic bottlenecks and/or a considerably disturbed secondary metabolism occur in different strains or with different substrates. Therefore, to a certain extent the intracellular metabolism of *Z. mobilis* may still be regarded as black box machinery where we are able to exchange or modify some parts, but of which the details of its functioning are poorly understood. It is the purpose of this contribution first to give an overview of state-of-the-art methods for metabolic analysis and, second, to give an overview of those studies that have been devoted to the characterization by such techniques of the intracellular metabolism of *Z. mobilis*, with the aim of providing a more detailed insight into the metabolic processes and their regulation as they function in vivo in this intriguing microorganism. The availability of such knowledge is a prerequisite for a continuing improvement of biotechnologically relevant strains by purposeful metabolic engineering.

15.2
Methods for Metabolic Analysis

15.2.1
Introduction

The process of characterizing the turnover rates of intracellular metabolism can be referred to as Metabolic Flux Analysis. For the purpose of this overview, only regulation of metabolic rates on the level of enzyme activity is considered. Thus, we will be dealing with metabolic events that are either in a steady state or represent a short-term transient reaction to some change of environmental parameters. Studies focusing on metabolic regulation at the molecular level and its long-term effects on gene and enzyme expression are not covered. The relevant parameters to describe metabolism then are, first, the metabolic reaction rates themselves, second, the enzyme levels, and third, the concentrations of the total of all effectors influencing the activities of the enzymes involved. It is important to note here that enzyme activities alone, especially as determined from cell-free extracts, can at best give a qualitative impression of which pathways are active in vivo and do not allow one to draw conclusions about the relative activities of parallel pathways. Enzyme activity determination assays are more or less standard and will not be discussed here. It is nevertheless worth pointing out that determination of certain enzymes in Z. mobilis is far from trivial and requires extensive treatment of the cell extract preparations to remove competing overwhelming activities of the glycolytic enzymes [25, 26]. The remaining parameters for a description of metabolism are metabolite concentrations (i.e., pool sizes) and metabolic reaction rates (i.e., fluxes). Several approaches that can be used to quantify these can be distinguished and will be reviewed now.

15.2.2
Metabolite Pool Determination

15.2.2.1
Invasive Approaches

The usual approach to measure the intracellular pool sizes is to take a sample of cells from the reactor, stop the metabolism by some procedure, then extract the metabolites of interest and finally determine their concentrations by standard analytical methods like gas chromatography, capillary electrophoresis, HPLC, mass spectrometry, NMR, or via enzymatic methods. Superior sensitivity is obtained in recent approaches that employ enzymatic cycling [27–29]. By far the least sensitive method is NMR spectroscopy, which is however applied because of its high dynamic range and its ability to distinguish many different compounds in a single sample, thus avoiding the need for partial purification of samples to select only the wanted class of metabolites. For instance, a large number of different sugar phosphates can be identified from a single ^{31}P NMR spectrum at suitable pH [30, 31] and ^{13}C NMR is very specific in its identification of, e.g., sugars [32] and amino acids [33].

The application of invasive sampling techniques to measure intracellular metabolite pool sizes requires [34] that:

1. The sample can be taken without disturbing the metabolic state of the cells
2. The inactivation of the metabolism is complete and fast as compared to the intracellular metabolic reaction rates
3. The extraction of the wanted metabolites and the denaturation and separation of the intracellular enzymes is complete
4. The total procedure of sampling, inactivation and extraction does not affect the stability of the metabolites
5. The dilution by inactivation and extraction is controlled, reproducible and minimal

Z. mobilis is a very critical organism with respect to the second requirement (i.e., that the inactivation be fast as compared to the intracellular metabolic rates) because it has an extremely high rate of glycolysis. Glucose uptake rates as high as $900\ \mu\text{mol min}^{-1}$ (g dry wt)$^{-1}$ [35] have been observed with this organism. Assuming an average intracellular volume of $2\ \mu\text{l}$ (mg dry wt)$^{-1}$ [36], an intracellular sugar phosphate pool of $0.1\ \text{mmol l}^{-1}$ would completely turn over once every 16 ms under these conditions. Likewise, since two reactions of the ED pathway produce ATP, a $2\ \text{mmol l}^{-1}$ ATP pool would turnover once every 0.16 s. This illustrates the point: the quenching of metabolism should be faster than these turnover times as a mere prerequisite to ensure a reliable preservation of the indicated pools. Therefore, the classical method for inactivation within a few ms has become the direct quenching of the sample in liquid nitrogen or liquid CO_2, where the heat transfer rate is maximized by the high temperature difference and by a large sample-liquid exchange area obtained by spraying of the sample [37]. Recently, spraying the sample in 60% methanol at $-40\,^{\circ}\text{C}$ has become popular [38, 39] (inactivation within 100 ms) as well as the injection in precooled perchloric acid [29, 40] which results in inactivation times of 200–500 ms [34].

15.2.2.2
In vivo Techniques

Since the relative errors associated with all consecutive steps involved in the metabolite pool measurement procedures described in Sect. 15.2.2.1 are cumulative, methods that avoid some or most of these steps seem desirable. Thus, approaches for in situ, in vivo pool measurements have attracted considerable interest. Two important representatives of this category are in vivo Nuclear Magnetic Resonance spectroscopy [41] and in vivo fluorescence spectroscopy [42]. In vivo NMR has a long and steadily increasing record of application to the characterization of biochemical processes in living cells. A number of early in vivo NMR studies on microorganisms were conducted in the Shulman group and have become classical examples [43–45]. Since then, many more applications to microorganisms and also to cell culture work, studies on animals and human subjects, have been reported [46]. Thus ^{31}P NMR allows the intracellular measurement of pH, sugar phosphates, inorganic phosphate, polyphosphate, nucleoside di- and tri-phosphates, UDP-sugars, PEP, nicotine-adenine dinucleotides, cell wall phospholipids, and phosphorylated sugar polymers (Fig. 15.2). Currently, with the best systems a metabolite concentration (related to the total sample volume) of approximately $250\ \mu\text{mol l}^{-1}$ is needed to produce

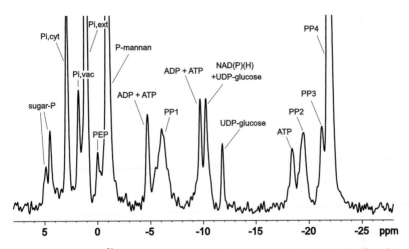

Fig. 15.2. Typical in vivo ^{31}P NMR spectrum of *Saccharomyces cerevisiae* aerobically cultivated at a density of 75 (g dry wt) ml^{-1} in a hydrocyclone bioreactor system specifically designed for NMR studies [47]. Assignments: sugar-P, sugar phosphates; Pi,cyt, cytoplasmic inorganic phosphate; Pi,vac, vacuolar inorganic phosphate; Pi,ext, extracellular inorganic phosphate; PEP, phospho*enol*pyruvate; P-mannan, phosphomannan. Peaks labeled PP1–PP4 stem from polyphosphate resonances [176]. The differences in the positions of the inorganic phosphate resonances arise from differences in pH between the compartments [30]. Recording time was 2.5 min (Gonzalez B, de Graaf AA (1998) unpublished results)

a useful in vivo ^{31}P signal (S/N=5) in 1 min [47]. Since the measurement time varies inversely to the squared concentration, a 10-fold dilution of the cell suspension would require a 100-fold longer measurement time for intracellular pools. Therefore, in vivo NMR studies usually employ rather high cell densities (typically in excess of 100 mg dry weight per ml). This has triggered the development of in situ NMR/bioreactor systems that are capable of adequately maintaining highly concentrated cell suspensions in a well-defined steady state for a prolonged period of time [47–49]. (For a recent review, see [34]). Signals in the in vivo ^{31}P NMR spectrum are generally very broad, leading to strong overlap and a very limited capability to resolve individual sugar phosphate signals (Fig. 15.2). ^{15}N NMR spectroscopy in principle has considerable potential for the in vivo monitoring of, e.g., amino acids and amidated sugars (e.g., glucosamine) [50, 51] but its application is impeded by the very low sensitivity and natural abundance of the ^{15}N nucleus. The ^{13}C nucleus has a large chemical shift dispersion (approx. 200 ppm) and the spectra are generally well-resolved and easy to interpret (Fig. 15.3). Thus, in vivo ^{13}C NMR has a great potential for the measurement of many different intracellular metabolites (sugars, amino acids, sugar phosphates, compatible solutes, etc.). However, since the intrinsic NMR sensitivity of ^{13}C is four times lower than that of ^{31}P and the natural abundance of ^{13}C is only 1.1%, these advantages can only be exploited in tracer studies using ^{13}C-enriched precursors (see also Sect. 15.2.3.3).

In vivo fluorimetry for the purpose of monitoring intracellular NAD(P)H levels at 450 nm was introduced in 1957 [52]. The method has been applied for the on-line monitoring of biomass concentration [53, 54], sugar [55], and phenol concentration [56] during fermentation processes. It has also been used to characterize intracellu-

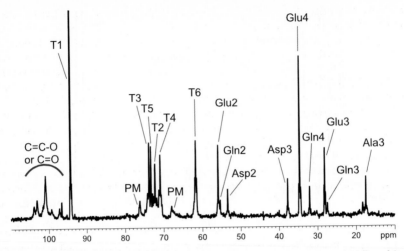

Fig. 15.3. Typical in vivo ^{13}C NMR spectrum of *Saccharomyces cerevisiae* aerobically cultivated at a density of 75 (g dry wt) ml^{-1} in a hydrocyclone bioreactor system specifically designed for NMR studies [47], taken 1 h after the start of infusion of [1-^{13}C] labelled glucose. Assignments: T, trehalose; PM, phosphomannan; Glu, glutamate; Gln, glutamine; Asp, aspartate; Ala, alanine. The numbers refer to the respective carbon atoms. Recording time was 5 min (Gonzalez B, de Graaf AA (1998) unpublished results)

lar pH [57, 58] in yeast. The tryptophan fluorescence was reported to correlate better with cell mass than that of NAD(P)H [59]. Fluorescence peaks are very broad in general, making it difficult to identify more than a single compound when employing a limited measuring range. The use of a wide spectral range of excitation and emission wavelengths, allowing one to detect a series of important compounds, e.g., riboflavin, FAD, FMN, NAD(P)H, pyridoxinic compounds, tryptophan, tyrosine, and phenylalanine [60] was integrated in a new a technique for two-dimensional fluorescence spectroscopy [61]. Thus, an increase of NAD(P)H and a decrease of FAD/FMN could be monitored in *Saccharomyces cerevisiae* during the aerobic-anaerobic transition and the formation of biomass and ergot alkaloids in a culture of *Claviceps purpurea* could be quantified [61]. The time resolution of the method is about 1 min. While this technique certainly has significant potential, a wider and truly quantitative application of in vivo fluorimetry is still awaited.

15.2.2.3
Rapid Sampling

The methods for metabolite pool determination as described above reveal very little about the dynamic properties of the metabolism. Worse, any subsequent modeling of the data has to rely very heavily on enzyme kinetic data as determined from in vitro measurements. Dynamic methods with a time resolution in the subsecond range, applied to a metabolic system that is perturbed from the steady state by, e.g., a sudden substrate pulse, may provide a wealth of information on the dynamic properties of the metabolism because the system passes through a whole range of different states during its reaction on the perturbation. Clearly, invasive techniques with very short inactivation times (Sect. 15.2.2.1) must be applied in order to achieve a subsecond time resolution. As early as 1964, incubations of cells were performed with a freeze-quench apparatus with a four-jet tangential mixer [37] and the changes of glycolytic intermediates after a pulse of glucose to the carbon-starved culture were measured one by one by varying the incubation time (0.015–5 s). However, this method did not allow for a defined metabolic state of the microorganisms, whence later developments concentrated on connecting rapid sampling devices to a controlled bioreactor where the organisms are in a balanced steady state. The first such device employed a miniature valve coupled with an HPLC capillary in a hypodermic needle which was connected to the mixing zone of the fermentor [40]. Samples were taken manually every 5 s with vacuum sealed, precooled glass tubes containing the inactivation/extraction solution. The achieved sampling rate of 0.2 per second did not however allow one to monitor fast dynamic changes. This system was adapted later to generate several additional samplings at a rate of 5 per second during the first phase of the dynamic response (M. Reuss, personal communication). In a different approach, continuous sampling of microorganisms from a controlled bioreactor with rapid inactivation of metabolism and extraction of metabolites using precooled –40 °C perchloric acid solution (35%) was achieved with a sampling tube [29]. This method, which used a total sampling time of 200 s, resulted in the fixing of fast dynamic reactions at a certain position in the tube. Immediately after stopping the sampling, the tube was frozen in liquid nitrogen. The tube was then divided into identical parts and the metabolites were analyzed enzymatically. After deconvolution for the axial dispersion in the tube, a time resolution of approximately 70 ms was achieved. Very recently, a completely automated rapid sampling device for monitoring intracellular metabolite dynamics was described that employed sample flasks fixed in transport magazines that were moved by a step engine such that subsequent flasks were filled at a rate of 4 per second for a total time of about 40 s [62]. The flasks contained 60% methanol of –50 °C for quenching of the metabolism.

Analysis of the large number of samples resulting from these type of experiments (typically 100) has to be performed by laboratory robots. Extensive macrokinetic modeling must be applied to extract information on fast metabolic regulation from the data (see Sect. 15.2.4).

15.2.3
Metabolic Flux Analysis

15.2.3.1
Basic Carbon Balancing

Besides pool sizes, a second important parameter to describe metabolism is constituted by the metabolic fluxes. The first step to quantitate these is the establishment of elementary balances for, e.g., phosphorus, nitrogen, oxygen, and especially carbon. The measurement of the amounts of carbon source consumed, of products excreted, as well as of biomass synthesized, offers a first important quantitative insight into the metabolic transitions that must occur inside the cell. In the industrial environment, careful and exhaustive analysis of the composition of the fermentation medium before and after a fermentative production run may give important clues as to which medium components may be limiting productivity, or which components are present in excess and can be saved on. Thus, although the cell at this stage is still largely considered as a black box, the fluxes in its metabolic network can be confined within certain boundaries by this basic input-output analysis.

15.2.3.2
Metabolite Balancing

A more detailed flux analysis requires that the cell wall boundaries be crossed and a quantitative analysis based on biochemical reaction network stoichiometries be performed. This has culminated in the past decade in the appearance of numerous methods and their computer implementations for so-called metabolic flux balancing. The physical basis of this method is that it assumes a metabolic steady state, whence metabolic fluxes into and out of every metabolite pool must exactly match in order to comply with the conservation of mass. This principle is combined with detailed quantitative knowledge about the cellular biomass composition and the available biochemical knowledge of the pathways involved in biosynthesis. A detailed quantitative description of *Escherichia coli* biomass composition and biosynthetic pathways can, e.g., be found in [63, 64]. Thus, from the biomass growth rate, the biomass protein content, and its amino acid composition, the fluxes through all amino acid synthesis pathways can be calculated. Likewise, from the contents and composition of DNA/RNA and cell (phospho)lipids, the fluxes through the biosynthetic pathways of nucleotides and fatty acids, respectively, can be evaluated. All these biosynthetic building blocks are synthesized from a rather small set of 11 elementary precursor metabolites generated by the central metabolism, namely glucose-6-phosphate, fructose-6-phosphate, ribose-5-phosphate, erythrose-4-phosphate, glyceraldehyde-3-phosphate, 3-phosphoglycerate, phospho*enol*pyruvate (PEP), pyruvate, acetyl-coenzyme A, oxaloacetate, and α-ketoglutarate [64]. Therefore, these calculations of the drain-offs from the central metabolism result in the end in a set of linear constraints for the distribution of the metabolic fluxes over glycolysis, pentose phosphate pathway, citric acid cycle, glyoxylate pathway, and the anaplerotic C3-carboxylating reactions in the cell. Together with the measured values of substrate uptake rate and (by-)product excretion rates, the equation system for the complete metabolic flux network usually is determined to within 1–4 degrees of freedom. This indeterminacy can then be resolved either by assuming additional reaction stoichiometries (e.g., P/O ratio for oxidative phosphorylation or an exact balance of NADPH regeneration), by assuming inactivity of a redundant pathway based on enzyme measurements (e.g., the glyoxylate cycle during growth on glucose), or by acquiring additional data of a different class (e.g., isotopic tracer data, see Sect. 15.2.3.3).

The metabolite balancing approach, first formulated in the form outlined above by Holms [63], was made a focus of the rapidly evolving field of metabolic engineering by publications of Vallino and co-workers [65–67] that concentrated on the use of intracellular fluxes for the elucidation of metabolic control. Since then, an ever increasing number of applications of the method has been reported [68–70].

15.2.3.3
Stable Isotope Labeling

As mentioned above, the metabolite balancing method usually still leaves several degrees of freedom for the flux network equation system. The application of additional assumptions on the metabolism to render the system (over)determined may not be feasible. Moreover, even if the system is fully determined, the metabolite balancing approach does not allow one to draw any conclusions about the reversibility of certain reaction steps in vivo, since such reversibility is simply not reflected in the net mass balances. Tracer data represent an alternative source of information. For a considerable period, the radioactive ^{14}C isotope was used for metabolic flux quantitation (see, e.g., [71] for a state-of-the-art review)). While the ^{14}C isotope offers a very high sensitivity, experimental procedures using this isotope were extremely tedious since metabolites had to be extracted and chemically degraded to obtain the specific enrichment values for the single carbon atom positions [71–73] that are necessary for a comprehensive analysis. A much more elegant, though less sensitive, approach evolved that uses stable isotopes in combination with NMR spectroscopy. Undoubtedly the most important of these is the ^{13}C isotope, but it should also be kept in mind that ^{2}H [74, 75] and ^{15}N [76, 77] have considerable potential for flux determination. While the history of ^{13}C NMR metabolic flux analysis dates back to the early 1970s [78], with classical examples already from the late 1970s [43–45] until about a decade ago [79–82], a significant improvement was achieved by the integration of metabolite balancing and ^{13}C labeling techniques to analyze complete metabolic networks, with pioneering work presented in [83–85]. The latter approach [85] used a truly comprehensive isotopic labeling data set that was obtained using NMR analysis of a variety of proteinogenic amino acids isolated from cells exclusively grown on [1-^{13}C]glucose as the sole carbon source. Those integrated approaches finally enabled to achieve an overdetermined system of equations that simultaneously describe the metabolic fluxes in vivo and the ensuing ^{13}C labeling, thus solving the flux analysis problem for the stationary case [86]. Subsequent analyses showed that ^{13}C labeling data also allow one to quantify the degree of reversibility of various intracellular bidirectional reaction steps [87–89]. While this adversely implies that the number of unknowns in the flux determination procedures increases considerably, this may be more than compensated for [90, 91] by the increase of information content of the ^{13}C labeling experiment that will result from the use of multiply-labeled ^{13}C substrates in combination with refined measurement techniques for isotopomer analysis such as two-dimensional NMR [92, 93], heteronuclear spin-echo NMR [94, 95] ,and mass spectrometry [96, 97]. While the stationary flux analysis problem may thus be considered solved, it only represents a single point analysis which using current approaches is still rather time-consuming. New methodological developments therefore will concentrate on more efficient analytical techniques that will enable one to perform dynamic flux analysis, such that, e.g., the evolution of the metabolic flux distribution during a fed-batch process can be monitored.

15.2.3.4
NMR Magnetization Transfer

While the techniques described so far are more or less indirect in that the metabolic fluxes are inferred either from measured changes of metabolite pool sizes or from the isotopic labeling of metabolites, the NMR magnetization transfer method enables the direct determination of unidirectional rate constants of metabolic reactions at steady state. It does so by monitoring the transfer of magnetization from one molecular species to another in real time. The accepting species may be another metabolite or the same compound in a different compartment, where it has a different NMR resonance frequency. Several examples of in vivo determinations of especially glycolytic fluxes [98, 99] and membrane transport rates (see [100] for a review) have been published. The method, however, requires that the turnover rates of the involved metabolic pools be from the same order of magnitude as the longitudinal relaxation rates of their nuclear spins. This in practice limits the application of NMR spin transfer experiments to reactions that involve rather high pool turnover rates of 0.1 to 10 per second. Moreover, the adequate supply of substrate and oxygen to the dense bacterial suspensions needed for NMR often presents a major problem [101].

15.2.4
Metabolic Modeling

The experimental data resulting from the methods described above cannot be used just as such to extract the relevant metabolic information. The structure, dynamics and regulatory properties of metabolic networks are extremely complex. Mathematical models are the only way that net consequences of simultaneous, coupled, and often counteracting processes can be evaluated consistently and quantitatively [102]. Therefore, the importance of mathematical modeling of metabolic phenomena has grown strongly in recent years. Within the scope of this review, only a few key references to the various methods can be given.

The most frequently used modeling approach is mechanistic modeling, i.e., a quantitative description of the system's dynamics based on enzyme kinetic properties. In this type of approach, each individual enzyme-catalyzed reaction or transport step is represented by a kinetic equation for the reaction velocity as a function of the concentrations of the involved substrates and effectors [103–105]. This results in a set of balance equations for each intracellular metabolic pool that, depending on the presentation, manifest themselves as stationary equations describing a stationary state, or as differential equations describing transient phenomena (e.g., as used in combination with rapid sampling procedures [106]). Several approaches based on analytical methods from the field of nonlinear system dynamics have been developed for the theoretical analysis of such equation systems. They all require a comprehensive mechanistic model of the metabolic pathways considered as well as the accurate and precise values of the reaction kinetic parameters and result in general model structures of the type presented in [107, 108]. This general structure allows one to calculate stationary states as well as to carry out stability analyses and sensitivity analyses [104, 107, 109, 110]. The well-known Metabolic Control Theory [111–115] can be considered as a special development of this technique. This theory is based on a linearization of the system equations on a logarithmic scale. The ensuing state sensitivity matrices are built up by control coefficients that allow a quantitative characterization of the regulatory influences of the kinetic model parameters. Since, however, detailed knowledge of the kinetic parameters is often at best fragmentary, alternative approaches were developed. One of these is the important universal *S*-Systems modeling framework [116–119] where each intracellular material balance is approximated by an exponential expression involving all metabolite concentrations. Positive effectors of the reaction velocity enter with a positive exponent, negative effectors with a negative exponent. Non-participating metabolites enter with an exponent equal to zero. This is equivalent to a linearization of the mechanistic model on a logarithmic concentration scale and as such it is very similar to Metabolic Control Theory. The approach, which may be used independent of a mechanistic model, is able to generate at least an approximate description of the system's behavior. It has served as a basis for the recent formulation of a production-oriented metabolic optimization framework [119, 120]. Of a still more semi-quantitative nature are the order-of-magnitude-calculation [121, 122] modeling and the approach to incorporate thermodynamic data as constraints for the intracellular reaction steps [123, 124]. The latter approach can be integrated with Metabolic Control Theory [125].

A special case is formed by the stationary balancing approaches involved in metabolic flux analysis (see Sect. 15.2.3.2). They employ rather straightforward matrix formalisms to handle the linear equation systems [65, 68, 69] that describe the material balances for all intracellular metabolite pools considered. The information obtained from isotopic labeling experiments (see Sect. 15.2.3.3) may either be added as a set of additional constraints to such equation systems (as was done, e.g., in [83, 84, 126], or it may be incorporated in a special isotopic labeling model. Classical examples of the latter can be found in [79, 81, 97]. These and subsequent models considered only flux distributions over a limited number of pathways at a few specific branchpoints in metabolism. Only recently, a truly comprehensive modeling framework for stationary flux analysis based on isotope labeling was established [85, 87, 88, 127] which in fact represents nothing else than the formulation of the material balances for all individual carbon atoms of the metabolic network in a single matrix. The metabolite pool balances emerge as a special case from these. Basic elements of this integrated isotopic labeling model, i.e., the so-called atom mapping matrices, were described in [128]. The modeling of ^{13}C isotopomer data, a current focus of metabolic flux analysis, also has some history. Classical examples can be found in [80, 82, 129, 130]. These models also suffer from a limited applicability. Recent developments therefore exploit matrix formulation to achieve a more general isotopomer modeling framework [131–133]. Very recently, Wiechert et al. [90] and Möllney et al. [91] succeeded in establishing a completely general modeling framework for isotopomer experiments based on a very fast, non-iterative solution strategy for the isotopomer balance equations. The previously developed comprehensive model for stationary flux analysis based on isotope labeling [85, 87, 88] emerges as a special case within this new framework.

The coming decade will yield an enormous increase of information not only from the familiar data sources like dynamic metabolite concentration measurements and metabolic flux analysis but also from very different sources, i.e., genome sequencing, gene expression studies using DNA microarrays and protein two-dimensional gel electrophoresis. Therefore, methods of bioinformatics that combine these different types of often imprecise and fragmentary data are expected to play an increasingly important role in metabolic engineering. Several approaches for holistic modeling now under development are already worth pointing out [134–136].

15.3
Metabolic Analysis of *Zymomonas mobilis*

15.3.1
Introduction

A specialty of *Z. mobilis* metabolism is that only 1 mol of ATP is formed per mol glucose consumed. Accordingly, only 2% of the substrate carbon typically ends up in the biomass. Nevertheless, this organism can grow at growth rates as high as

$0.25\,h^{-1}$. It succeeds in doing so by maintaining an extremely high glycolytic rate (specific glucose uptake rates of 0.9–$1.0\,U$ (mg dry wt)$^{-1}$ have been reported [137]). A number of metabolic studies concentrated on the question of how this flux is regulated and where eventually the rate-limiting step in the glycolytic pathway is located. These included the substrate uptake (which proceeds via a glucose facilitator) and product excretion (i.e., rapid membrane diffusion) steps.

Z. mobilis exhibits decreased growth and ethanol yields as well as increased by-product formation during growth on fructose. Several studies therefore concentrated on the differences between glucose and fructose metabolism. Adaptation of *Z. mobilis* to highly osmotic media has been a topic of interest because the natural environment of the organism features high glucose and fructose concentrations. As the principal product ethanol is known to become a limiting factor for *Z. mobilis* growth at high concentrations (above $80\,g\,l^{-1}$), several studies concentrated on effects of this compound on the glycolytic enzymes. Since the Entner-Doudoroff glycolytic pathway of *Z. mobilis* is a linear pathway and the wild-type organism does not possess a complete pentosephosphate pathway nor a closed citric acid cycle, the central metabolism of this organism contains no important branching points. This certainly is a reason why very few isotopic labeling studies for flux analysis have been carried out with *Z. mobilis*. Only the recent construction of pentose-fermenting recombinant strains has motivated research in this area. In the following review, establishment of a more or less coherent view on *Z. mobilis* metabolism based on the information provided by studies on the various topics indicated above will be attempted. The discussion will be focused on ethanol production on media containing a single monomeric sugar as carbon source. Thus, the biotechnologically relevant conversion of glucose plus fructose to gluconolactone and sorbitol by the action of a glucose-fructose-oxidoreductase in *Z. mobilis* [138] as well as the bioconversion of sucrose ([139] and references therein) are not covered in this contribution.

15.3.2
Enzymatic Studies

Specific activities of the enzymes involved in *Z. mobilis* glycolytic and other pathways have been reported in several studies [140–142], giving rather complete compilations of data. However, data in [142] were obtained at $23\,°C$ and therefore cannot be used in a predictive manner for metabolic rates at $30\,°C$. Typical values are compiled in Table 15.1.

While a considerably reduced specific growth rate and cell yield were observed during growth on fructose as compared to glucose, approximately equal specific consumption rates for the two substrates of 10–$11\,g\,g^{-1}\,h^{-1}$ (equivalent to 0.9–$1.0\,U$ (mg dry wt)$^{-1}$ were found [137]. Scopes et al. [145] reported that glucose-6-phosphate dehydrogenase was inhibited by its product, glucose-6-phosphate and glucose 6-phosphate dehydrogenase by ATP, but only with rather high inhibition constants K_I of $15\,mmol\,l^{-1}$ for glucose 6-phosphate and $1.4\,mmol\,l^{-1}$ for ATP, respectively. No effectors for the other five glycolytic enzymes glyceraldehyde 3-phosphate dehydrogenase, phosphoglycerate kinase, phosphoglycerate mutase, enolase, and pyruvate kinase, were identified [146].

From these studies and the data as shown in Table 15.1, it was originally inferred that the phosphorylation of glucose by glucokinase (or of fructose by fructokinase, respectively) and the oxidation of glucose-6-phosphate by glucose 6-phosphate dehy-

Table 15.1. Average reported activities at 30 °C of the enzymes involved in *Z. mobilis* sugar catabolism determined from cell-free extracts. In converting values per g protein to values per g dry mass, it was assumed that protein constitutes 60% of *Z. mobilis* dry mass [1]

Enzyme	Typical activity $(U \ (mg \ dry \ wt)^{-1})$	Source
Glucose facilitator	1.8[a]	[143]
Glucokinase	1.0	[141]
Fructokinase	0.6[b]	[141]
Phosphoglucose isomerase	0.8[b]	[141]
Glucose 6-phosphate dehydrogenase	1.4	[141]
6-phosphogluconolactonase	2.5	[141]
6-phosphogluconate dehydratase	1.8	[141]
KDPG aldolase	3.0	[141]
Glyceraldehyde 3-phosphate dehydrogenase	3.0	[141]
Phosphoglycerate kinase	6.0	[141]
Phosphoglycerate mutase	12.0	[141]
Enolase	1.5	[141]
Pyruvate kinase	4.5	[141]
Pyruvate decarboxylase	2.8	[141]
Alcohol dehydrogenase-1	3.0	[141]
Alcohol dehydrogenase-2	10.0	[141]

[a] Value for glucose. V_{max} for fructose and xylose were reported to be 23% and 121% higher, respectively [24]

[b] Reported to be three times higher during growth on fructose [144]

drogenase are the major flux-controlling steps in the absence of high ethanol concentrations. This conclusion was recently confirmed with substantial evidence: Snoep et al. [147] reported that glucose 6-phosphate dehydrogenase exerts a very large flux control over the glycolytic flux in *Z. mobilis*, and two studies [148, 149] demonstrated that the enzyme is indeed allosterically controlled by phosphoenolpyruvate (PEP), presumably in order to maintain a balance of glycolytic throughput and ATP consumption in *Z. mobilis*.

A discussion of metabolic activity based on metabolite pool measurements (see next section) must take into account the kinetic constants of the pathway enzymes. A number of glycolytic enzymes from *Z. mobilis* have been kinetically characterized. A compilation of kinetic constants reported in the literature is given in Table 15.2. From this table, it can be seen that with the exception of pyruvate decarboxylase, all glycolytic enzymes will reach substrate-saturating conditions at submillimolar concentrations of the substrates.

15.3.3
Metabolite Pool Measurements

15.3.3.1
Overview

A number of studies have been devoted to the study of intracellular glycolytic metabolite pools in *Z. mobilis*. Lazdunski and Belaich [156] measured the static ATP con-

Table 15.2. Literature data on kinetic constants for various glycolytic enzymes of *Z. mobilis*[a]. Abbreviations: G6P, glucose 6-phosphate; PEP, phospho*enol*pyruvate; 6PG, 6-phosphogluconate; GP, glycerophosphate; Pi, inorganic phosphate; 3-PGA, 3-phosphoglycerate; 2-PGA, 2-phosphoglycerate

Enzyme	K_m-value(s)(mmol l^{-1})	Source
Glucose facilitator	3.2 (glucose)	[143]
	39 (fructose)	[150]
	40 (xylose)	[24]
Glucokinase	0.1 (glucose)	[151]
	0.2 (ATP)	
	0.05 (phosphate)	
	15 (G6P)[b]	[145]
Fructokinase	0.7 (fructose)	[145]
	0.45 (ATP)	
Glucose 6-phosphate dehydrogenase	0.2 (G6P)	[149]
	0.022 (NADP)	
	0.5–1.1 (NAD)[d]	
	0.05 (PEP)[c]	
	1.4 (ATP)[b]	[145]
6-phosphogluconate dehydratase	0.04 (6PG)	[152]
	0.3 (D-α-GP)[b]	
	2.5 (Pi)[b]	
	2.0 (3-PGA)[b]	
KDPG aldolase	0.25 (KDPG)	[153]
Phosphoglycerate kinase	1.5 (3-PGA)	[146]
	1.1 (ATP)	
Phosphoglycerate mutase	1.1 (3-PGA)	[146]
Enolase	0.08 (2-PGA)	[146]
Pyruvate kinase	0.08 (PEP)	[146]
	0.17 (ADP)	
Pyruvate decarboxylase	4.4	[154]
Alcohol dehydrogenase-1	0.086 (acetaldehyde)	[155]
	0.027 (NADH)	
	4.8 (ethanol)	
	0.073 (NAD)	
Alcohol dehydrogenase-2	1.3 (acetaldehyde)	[155]
	0.012 (NADH)	
	27.0 (ethanol)	
	0.11 (NAD)	

[a] No kinetic data for phosphoglucoseisomerase, 6-phosphogluconolactonase and glyceraldehyde 3-phosphate dehydrogenase were found
[b] Inhibitory constant, i.e., K_i
[c] PEP is an allosteric inhibitor; Hill coefficient varies from 1.35 at [PEP]=0 μmol l^{-1} to 2.0 at [PEP]>100μmol l^{-1}
[d] Depends both on [PEP] and on [G6P]

tent to describe the energetic cellular state at different uncoupling conditions of growth. Barrow et al. [157] studied the levels of nucleoside triphosphates, sugar phosphates, UDP-sugars, and P_i with a time resolution of 1 min in intact fermenting cells as well as in perchlorate extracts using ^{31}P NMR spectroscopy. Algar and Scopes [140] studied metabolite concentrations in cell-free extracts of *Z. mobilis* to which glucose was added continuously, in order to investigate the controls on the glycolytic enzymes. Osman et al. [142] determined intracellular levels of glycolytic intermediates and nucleotides in several phases during a batch glucose fermentation using ^{31}P NMR. Strohhäcker et al. [158] investigated ethanol inhibition of glucose catabolism in *Z. mobilis* using ^{31}P NMR spectroscopy in vivo and of perchloric acid extracts from cell suspensions incubated with various concentrations of ethanol. Weuster-Botz [29] used a sampling tube device with rapid inactivation of metabolism for continuous sampling of *Z. mobilis* cells from a controlled bioreactor after application of a glucose pulse, enabling a dynamic investigation of *Z. mobils* glycolytic intermediates with a time resolution of only 0.64 s. De Graaf et al. [159] compared the intracellular levels of nucleoside di- and triphosphates, sugar phosphates, UDP-sugars, and P_i in glucose- and fructose continuous cultures of *Z. mobilis* as measured by ^{31}P NMR using a special membrane-cyclone NMR bioreactor system.

Intracellular sugar concentrations have also been studied. Struch et al. [160] determined intracellular glucose and xylose concentrations in an investigation of the physiological basis of the exceptionally high sugar tolerance of *Z. mobilis*. In a study of pentose metabolism in wild-type and recombinant *Z. mobilis* strains, Feldmann et al. [25] determined intracellular concentrations of various pentoses and their phosphates. Schoberth and de Graaf [161] used in vivo ^{13}C Nuclear Magnetic Resonance spectroscopy to follow xylose uptake in *Zymomonas mobilis*. In a study using in vivo NMR spin transfer of ethanol transmembrane diffusion in *Z. mobilis*, Schoberth et al. [36] monitored intra- and extracellular glucose and ethanol concentrations simultaneously in intact fermenting cells in a single experiment.

In the next sections an attempt will be made to put together the fragmentary information presented in the studies indicated above in order to get a more complete view on *Z. mobilis'* metabolic properties.

15.3.3.2
Glycolytic Intermediates

The concentration time profiles of *Z. mobilis* glycolytic intermediates after a glucose substrate pulse as described in [140, 157, 158] reveal several common features that are sketched in Fig. 15.4. The NTP content rises rapidly to a plateau value of 3–5 mmol l^{-1} reached already after 1–2 min. Glucose 6-phosphate and glyceraldehyde 3-phosphate both rise to a maximum value within 30 s. Hereafter, they decline again, such that glucose 6-phosphate reaches a steady-state value at t=1 min while glyceraldehyde 3-phosphate continues to decrease to a level of 0.3–0.2 mmol l^{-1}, reached at t=4 min. 6-Phosphogluconate also rises within 1 min to a plateau value of about 0.5 mmol l^{-1}. 3-Phosphoglycerate rises to a steady-state value of about 2 mmol l^{-1} in 2–3 min, i.e., slower than glucose 6-phosphate, 6-phosphogluconate, and glyceraldehyde 3-phosphate. In accordance with the NMR studies, Weuster-Botz [29] also reported a rapid increase and subsequent decline of glucose 6-phosphate, glyceraldehyde 3-phosphate, and 3-phosphoglycerate after application of a glucose pulse. However, due to the fact that this study applied extremely rapid sampling in a well-con-

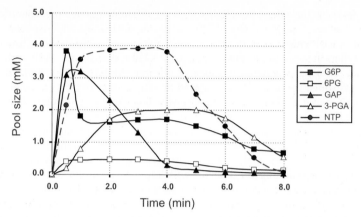

Fig. 15.4. Time courses for selected glycolytic intermediates and NTP (ATP plus UTP) in *Z. mobilis* after application of a glucose pulse of several hundred mmol l^{-1} to very dense cell suspensions (100–200 mg dry wt. ml^{-1}). The traces represent smoothed averages of concentrations reported in [157, 158] determined by ^{31}P NMR. More or less similar patterns, although on a roughly tenfold longer time scale due to crude extract dilution, are reported in [140]

trolled bioreactor with a mixing time of about 500 ms it was possible to demonstrate that glucose 6-phosphate and 3-phosphoglycerate reach their peak values essentially within 1 s, where in the NMR studies this occurred only after 30 s. This probably reflects a much longer mixing time of the glucose pulse in the dense cell suspensions used in these studies. In the rapid sampling experiment, glyceraldehyde 3-phosphate reacted slower and reached a peak value after 25 s [29]. Unfortunately, no data on other glycolytic intermediates and ATP are given. In metabolic studies on cell-free extracts of *Z. mobilis,* Algar and Scopes [140] demonstrated that the levels of the glycolytic intermediates depended strongly on the ATPase level added to the extracts, with limiting ATPase activity causing accumulation of metabolites from the Entner-Doudoroff pathway, especially glucose 6-phosphate. ATP levels were very low both at limiting ATPase levels and at excess ATPase levels, and high at intermediate levels. This partly paralleled earlier results [156] showing that ATP levels in rapidly growing *Z. mobilis* (corresponding to high ATPase activity) were significantly lower than in cells that were growth-limited by panthotenate.

A first macrokinetic model of *Z. mobilis* catabolism, based on the enzyme characteristics given in Tables 15.1 and 15.2 (except for the allosteric inhibition of glucose 6-phosphate dehydrogenase by PEP), was reported by Wulf et al. [103] and used for dynamic simulations of glucose and fructose metabolism. This model qualitatively reproduced the accumulation of glucose 6-phosphate found initially after application of a substrate pulse.

The increased by-product formation by *Z. mobilis* during growth on fructose [13] has long remained a puzzling fact since both sugars share essentially the same glycolytic pathway (Fig. 15.1). Only very recently was a clue to a possible explanation reported [159]. In vivo ^{31}P NMR experiments on *Z. mobilis* growing in a controlled bioreactor showed that inorganic phosphate, NDP, NTP, and UDP-sugar levels were two-fold lower during fructose metabolism, and that the total sugar phosphate pool was almost five times higher than during glucose metabolism. Subsequent ^{31}P NMR measurements on cell-free chloroform extracts demonstrated elevated levels of fructose 6-phosphate, ribose 5-phosphate, sedoheptulose 7-phosphate, 3-phosphoglycerate, and dihydroxyacetonephosphate. It was hypothesized [159] that this global alteration of the levels of intracellular phosphorylated metabolites is primarily caused by an elevated concentration of intracellular fructose-6-phosphate during growth on fructose. This hypothesis is supported by the finding that overexpression of fructokinase results in a severe growth retardation of *Z. mobilis* on fructose media [24].

15.3.3.3
Sugars

From the enzyme activities as given in Table 15.1, it is to be expected that the uptake of glucose, fructose, and xylose is faster than the maximum glycolytic rate. Thus, it is to be anticipated that the substrate accumulates intracellularly, even during high fermentative activity. Several authors indeed reported this observation. Belaich et al. [143] already concluded from microcalorimetric measurements that an excess of glucose entry, compensated by an outflow, most probably occurs in *Z. mobilis* exponentially growing cells. Struch et al. [160] were the first to demonstrate elevated intracellular levels of glucose in metabolizing *Z. mobilis*. These authors found that the intracellular glucose concentration in cultures with a specific glucose uptake rate of $4.3\,g\,g^{-1}\,h^{-1}$ (i.e., about half-maximal) was always about $200\,mmol\,l^{-1}$ lower than the extracellular concentration. This gradient constituted the driving force for the net glucose uptake. The equilibration of intra- and extracellular glucose was shown to provide an almost complete osmotic balance between internal and external space [160], which explains the exceptionally high sugar tolerance of *Z. mobilis* (up to 40% glucose [1]. Using an in vivo NMR method developed by Schoberth and de Graaf [161], Schoberth et al. [36] were able to monitor intracellular levels of both the α and the β anomers of glucose in 3-ml cultures of intact *Z. mobilis* converting $780\,\mu mol$ of glucose to ethanol within 20 min. The total intracellular glucose concentration was as high as $160\,mmol\,l^{-1}$ when the extracellular level was $400\,mmol\,l^{-1}$.

15.3.3.4
Ethanol

The studies of Schoberth et al. [36] not only measured intracellular sugar concentrations but also generated unique data on intracellular ethanol concentrations and unidirectional rate constants of ethanol transmembrane diffusion. From the data given [36] it can be calculated that ethanol efflux would match ethanol production in cells fermenting glucose at the highest reported specific uptake rates $(0.9–1.0\,U\,(mg\,dry\,wt)^{-1}$ [137]) already when the intracellular ethanol concentration is only $2\,mmol\,l^{-1}$ higher than the extracellular concentration. Thus, this study ruled out any possibility that a limitation in ethanol efflux leading to strong intracellular accumulation would block glycolysis in *Z. mobilis*.

Nevertheless, ethanol at high concentrations obviously affects metabolism adversely. Accumulation of ethanol during fermentation was reported to cause a decrease of the growth rate and the specific ethanol production rate [9]. In the presence of $100\,g\,l^{-1}$ ethanol, glycolysis is severely slowed down and 3-phosphoglycerate accumulates [158]. Data from [140] obtained with cell-free extracts suggest that glyceraldehyde 3-phosphate, KDPG, acetic aldehyde, and pyruvate levels also accumulate and that the NAD concentration increases with increasing ethanol concentration. Under the conditions tested (i.e., cells grown on glucose concentrations of 10–12%, showing enzyme activities as given in Table 15.1), inhibition of enolase by ethanol might be the primary responsible cause of these effects [158]. However, in fermentations involving higher ethanol concentrations (i.e., 70–110 g l^{-1}) the effects of ethanol on *Z. mobilis* metabolism very likely are much more diverse. Relevant studies indicate that ethanol primarily exerts inhibitory effects on cellular growth [162, 163]. Responsible mechanisms include an increase of maintenance energy requirement [164], irreversible deactivation of anabolic enzymes [26], and a loss of cofactors and metabolites [165]. Hermans [166] showed that the energy charge and the ATP concentration remained high in *Z. mobilis* continuously cultivated at 100 g l^{-1} ethanol, but that the ability to maintain pH homeostasis decreased due to inhibition of the membrane-bound H$^+$-ATPase. Moreover, it was observed that cell division as well as synthesis of DNA and fatty acids in *Z. mobilis* were completely inhibited at ethanol concentrations above 80 g l^{-1} while biomass still increased due to protein synthesis. At ethanol concentrations above 110 g l^{-1}, severe leakage of metabolites into the medium indicates that the cells are in fact extracted [166].

15.3.4
Flux Analyses

15.3.4.1
Overview

As pointed out in Sect. 15.2.3.1, the first step in the quantitation of metabolic reaction rates is the establishment of elementary balances, especially for carbon. Several of the physiological studies of *Z. mobilis* referred to in the preceding sections include a basic level of this type of flux analysis, in the form of fermentation balances. Since these typically indicate a conversion of 96–98% of the substrate into equimolar amounts of ethanol and carbon dioxide [1, 18], there has been little interest in a characterization involving detailed metabolic balancing procedures (see Sect. 15.2.3.2) of the remaining 2–4% of the carbon. Likewise, the largely unbranched structure of the Entner Doudoroff pathway did not make *Z. mobilis* a promising object for investigation by stable isotope labeling methods and NMR, since the potential of these methods is only fully exploited with metabolic networks involving multiple branching points and reversible reaction steps (Sect. 15.2.3.3). Hence, only a rather limited number of flux analysis studies dealing with *Z. mobilis* have been reported.

15.3.4.2
Metabolite Balancing

Variable data on the macromolecular biomasss composition [1, 167] and the protein composition [167, 168] of *Z. mobilis* have been reported. After combination of these with those on another Gram-negative organism, *E. coli* [64], the following overall composition of *Z. mobilis* dry mass can be proposed: protein 60.5%; RNA 19.5%; DNA 2.7%; lipid 8.5%; peptidoglycan 2.5%; glycogen 2.5%; polyamines 0.3%; metabolites and ions, 3.5% [159]. From these data, following the approach of Neidhardt et al. [64], a monomeric composition showing the detailed precursor requirement for

Table 15.3. Precursor requirements for biomass synthesis of *Z. mobilis* calculated from data in [1, 64, 167, 168]. Adapted from [159]. Abbreviations: G6P, glucose 6-P; F6P, fructose 6-P; RI5P, ribose 5-P; E4P, erythrose 4-P; GAP, glyceraldehyde-3-P; PGA, phosphoglycerate; PEP, phospho*enol*pyruvate; PYR, pyruvate; AcCoA, acetyl-CoA; OAA, oxaloacetate; AKG, 2-oxoglutarate

Amino acid	Amount (μmol g^{-1})	G6P	F6P	RI5P	E4P	GAP	PGA	PEP	PYR	AcCoA	OAA	AKG	CO$_2$
Ala	1088								1				1
Arg	181											1	1
Asx	478										1		
Cys	20						1						
Glx	343											1	
Gly	920						1						
His	82			1									
Ile	369								1		1		-1
Leu	369								2	1			-2
Lys	249								1		1		-1
Met	81										1		
Phe	11				1			2					-1
Pro	210											1	
Ser	202						1						
Thr	224										1		
Trp	54			1	1			1					-1
Tyr	70				1			2					-1
Val	569								2				-1
polymer		G6P	F6P	RI5P	E4P	GAP	PGA	PEP	PYR	AcCoA	OAA	AKG	CO2
Protein		0	0	136	135	0	1142	216	3582	369	1401	734	–
RNA				600			350				250		600
DNA				87			44				44		87
Lipids						120	120			1976			-1976
Peptidoglycan			55					28	83	55	28	28	-55
Glycogen		154											
C1-Units							49						
Polyamines												59	
total		154	55	823	135	120	1705	244	3665	2400	1723	821	–

biomass synthesis of *Z. mobilis* as given in Table 15.3 can be calculated [159]. These data provide a basis for detailed metabolite balancing studies of *Z. mobilis*.

15.3.4.3
NMR and Stable Isotope Labeling

^{13}C NMR has been used in *Z. mobilis* research for the identification and quantitation of the major lipid components [169], for the structure elucidation of a novel poly-fructoside (-α-fructofuranosyl-(2-1)-β-fructofuranosyl-(2-6)-) [170], and for the identification of sorbitol in sucrose and fructose plus glucose fermentations [171]. However, these studies used the natural abundance ^{13}C present in these metabolites. In fact, true stable isotope labeling has only been applied in very few cases to eluci-date specific metabolic characteristics of *Z. mobilis*. Barrow et al. [172] studied the deuterium-labeling patterns and stereochemistry of the ethanols produced from the metabolism of [1-^{2}H]glucose and unlabelled glucose in ^{2}H$_2$O by a combination of ^{2}H, ^{13}C, and ^{1}H NMR. The labeling patterns were explained in terms of enzyme me-chanisms and stereospecificity of the dehydrogenases involved in the conversion of sugars to ethanol in *Z. mobilis*, and metabolite enolization. Rohmer et al. [173] stu-died the biosynthesis of the side-chain of bacteriohopanetetrol and of a carbocyclic pseudopentose from ^{13}C-labeled glucose as well as the early steps in isoprenoid bio-synthesis [174] in *Zymomonas mobilis* using ^{13}C NMR.

As pointed out already in Sect. 15.1.1, the intracellular differences between glucose and fructose metabolism leading to increased by-product formation in *Z. mobilis* during growth on fructose have largely remained unexplained. While a hypothetical biosynthetic pathway for the main by-products glycerol and dihydroxyacetone was already proposed in 1986 [13], the formation of these compounds has only recently been elucidated [19] using a combination of enzymatic methods and ^{31}P- as well as ^{13}C NMR spectroscopy. The latter was used to study the incorporation of label from [2-^{13}C]fructose into glycerol, dihydroxyacetone, lactate, and acetoin. These studies identified a novel pathway for the formation of glycerol 3-phosphate which branches off the Entner-Doudoroff pathway at the intermediate glyceraldehyde 3-phosphate and proceeds via dihydroxyacetone phosphate, dihydroxyacetone, glycerol to glycer-ol 3-phosphate [19].

To date, only a single comprehensive metabolic flux analysis study combining me-tabolite balancing and ^{13}C NMR of chemostat-grown *Z. mobilis* has been carried out [159]. This study involved stable isotope-aided flux analysis of *Z. mobilis* wild type growing on glucose and fructose, and of a xylose-degrading recombinant strain that was constructed in a similar approach to [22] but using the genes for xylose isomer-ase and xylulokinase from *Klebsiella pneumoniae* in addition to the genes for trans-ketolase and transaldolase from *E. coli* [175]. The recombinant strain was able to grow on xylose and to produce ethanol at 86% of the theoretical yield but specific growth and production rates were 4–5 times lower than those typically observed on glucose. The flux analysis of [2-^{13}C]glucose- and [2-^{13}C]fructose-grown *Z. mobilis* shows clearly that 95% of the sugar is rapidly metabolized over the ED pathway with very little drain-offs to by-products and cell mass synthesis. While this is evident from the carbon balances alone, the analyses also give detailed insight in the meta-bolic activities of the non-oxidative pentose phosphate pathway enzymes in vivo. Only negligible net fluxes are carried by the transketolase, 6-phosphogluconate, ri-bulose-5-phosphate epimerase, and phosphoribose isomerase reactions, but the flux

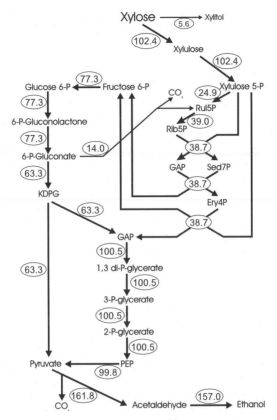

Fig. 15.5. Intracellular flux distribution during [1-^{13}C]xylose-fed chemostat fermentation (D=0.018 h^{-1}) of a recombinant strain of *Zymomonas mobilis* in which heterologous activities of xylose isomerase, xylulokinase, transketolase and transaldolase were expressed. All fluxes are given as μmol per gram dry weight and minute. Data taken from [159]. Fluxes were determined by non-linear least squares fitting of a flux model to joint NMR and fermentation data. Fluxes towards minor by-products as well as precursor fluxes for biomass synthesis, included in the original analysis [159], are not shown

analysis predicts a high degree of reversibility of the transketolase, ribulose-5-phosphate epimerase, and phosphoribose isomerase reactions as well as of the phosphoglucose isomerase reaction. Because the principal ^{13}C label in ribose 5-phosphate was found in C-2, it was concluded that this precursor is synthesized primarily via the non-oxidative pentose phosphate pathway in *Z. mobilis* [159], in accordance with an earlier study [173]. The flux analysis of the [1-^{13}C]xylose-grown recombinant strain gives a detailed view of the heterologous catabolic activities of the pentose phosphate pathway enzymes (Fig. 15.5). Moreover, this study identified heterologous xylulokinase as the rate-limiting enzyme of the strain and hypothesizes that stronger overexpression of this enzyme would probably completely prevent accumulation of the by-products xylitol and xylulose [159]. This would improve the ethanol yield to 96%, equal to that on glucose.

15.3.5
Summary

Clearly, no consistent single set of pool measurement data covering all glycolytic intermediates as well as pyruvate, acetaldehyde, ethanol, and the nucleoside mono-, di- and triphosphates measured under defined conditions in *Z. mobilis* is available. However, the material discussed in Sects. 15.3.2. and 15.3.3 allows one to establish a coherent picture of *Z. mobilis* glycolytic metabolism.

The capacity of the uptake of the monomeric sugars glucose and fructose is (far) greater than the capacity of the glycolytic system for sugar concentrations above the K_m of the facilitated diffusion uptake system (3.2 mmol l^{-1} and 39 mmol l^{-1} for glucose and fructose, respectively (Table 15.2.)). This strong overcapacity enables *Z. mobilis* to maintain effortlessly an osmotic balance in its high-sugar natural environments.

The maximum for the glycolytic rate is ultimately set by the glucokinase V_{max} (Table 15.1). However, the consistent presence of glucose 6-phosphate in concentrations well above K_m for glucose 6-phosphate dehydrogenase (0.2 mmol l^{-1}, Table 15.2) [29, 157, 158] together with the finding that glucose 6-phosphate dehydrogenase is allosterically inhibited by PEP [148, 149] identifies this enzyme as the principal rate-controlling step of glycolysis in non-optimal conditions, as was also concluded from metabolic control analyses [147]. Apparently, the overall regulatory control structure of *Z. mobilis* catabolism established by the reported characteristics of its glucose 6-phosphate dehydrogenase is that the pace of glycolysis is set by the total ATPase activity resulting from, e.g., cellular growth processes and maintenance of transmembrane gradients of pH and ion concentrations. The ATP breakdown by this ATPase activity must stoichiometrically match the overall ATP production of 1 mol ATP per mol glucose consumed in glycolysis [140, 156]. When ATPase activity is less than the glucokinase potential, e.g., due to limiting concentrations of certain growth factors in the medium, the level of regenerated ADP will fall, thereby limiting pyruvate kinase activity which will lead to an accumulation of PEP. The increased PEP level, together with the increased ATP concentration, downregulates glucose 6-phosphate dehydrogenase activity [141, 145] (Table 15.2). The resulting accumulation of glucose 6-phosphate will then slow down glucokinase activity (Table 15.2) to the appropriate level, even in the presence of saturating ATP concentrations for this enzyme.

At elevated ethanol concentrations, no simple view of the regulation of metabolism of *Z. mobilis* can be established since a variety of effects occur simultaneously (see Sect. 15.3.3.4). However, the inhibition of the membrane-bound H$^+$-ATPase as well as of DNA- and fatty acid synthesis, together with the observation that ATP concentrations seem little affected by ethanol [166], suggest that an inhibition of the effective total ATPase activity of the cell may be an important causal factor in the progressive limitation of the glycolytic rate in *Z. mobilis* with increasing ethanol concentrations.

The flux analyses performed thus far with *Z. mobilis* show that the non-oxidative pentose phosphate pathway plays an important role in the biosynthesis of ribose 5-phosphate. Furthermore they indicate that properly engineered recombinant strains of *Z. mobilis* will always be able to direct about 95% of the carbon to the product ethanol, irrespective of the type of monomeric sugar chosen as substrate.

15.4
Concluding Remarks

The many physiological studies conducted with *Zymomonas mobilis* in the past 15 years have yielded a basic understanding of the short-term regulation of the catabolic fluxes in this organism. Interestingly, several very important metabolic characteristics (i.e., the high flux control by and allosteric regulation of glucose 6-phosphate dehydrogenase as well as the overall alteration of phosphorylated pools during growth on fructose) of *Z. mobilis* have only very recently been discovered.

    Z. mobilis typically converts sugars at extremely high rates with over 95% yield into ethanol. While worldwide attempts to transform this organism into an efficient ethanol producer from abundant and renewable carbon sources have long failed, the recent successful engineering of xylose-, arabinose-, and maltose-fermenting strains now brings this perspective into closer view. First flux analyses reveal the remarkable fact that *Zymomonas mobilis* demonstrates the same high yield and capacity of ethanol formation on these sugars as it does on glucose and fructose. These characteristics guarantee an ongoing interest in and use of this organism and its genes for metabolic engineering purposes.

Acknowledgements. The author wishes to thank S. Bringer and G. Sprenger for helpful comments, and Prof. H. Sahm for continuous support.

References

1. Swings J, de Ley J (1977) Bacteriol Rev 41:1
2. Sprenger GA (1996) FEMS Microbiol Lett 145:301
3. Kalnenieks U, de Graaf AA, Bringer-Meyer S, Sahm H (1993) Arch Microbiol 160:74
4. Sprenger G (1992) The genus *Zymomonas*. In: Balows A, Trüper HG, Dworkin M, Harder W, Schleifer KH (eds) The prokaryotes, vol III, 2nd edn. Springer, Berlin Heidelberg New York, p 2287
5. Reyes L, Scopes RK (1991) Biochim Biophys Acta 1068:174
6. Toh H, Doelle H (1997) Arch Microbiol 168:46
7. Kim YJ, Song KB, Rhee SK (1995) J Bacteriol 177:5176
8. Kalnenieks U, Galinina N, Bringer-Meyer S, Poole RK (1998) FEMS Microbiol Lett 168:91
9. Rogers PL, Lee KJ, Scotnicki ML, Tribe DE (1982) Adv Biochem Eng 23:37
10. Viikari L (1988) CRC Critical Rev Biotechnol 7:237
11. Amin G, Van den Eynde E, Verachtert H (1983) Eur J Appl Microbiol Biotechnol 18:1
12. Rogers PL, Lee KJ, Scotnicki ML, Tribe DE (1982) Adv Biochem Eng 23:37
13. Viikari L, Korhola M (1986) Appl Microbiol Biotechnol 24:471
14. Toran-Diaz I, Jain VK, Baratti JC (1984) Biotechnol Lett 6:389
15. Dawes EA, Ribbons DW, Large PJ (1966) Biochem J 98:795
16. Johns MR, Greenfield PF, Doelle HW (1991) Adv Biochem Eng 44:97
17. Viikari L, Korhola M (1986) Appl Microbiol Biotechnol 24:471
18. Viikari L (1988) CRC Critical Rev Biotechnol 7:237
19. Horbach S, Strohhäcker J, Welle R, de Graaf AA, Sahm H (1994) FEMS Microbiol Lett 120:37
20. Sprenger GA (1993) J Biotechnol 27:225
21. Reiser J, Käppeli O, Fiechter A (1988) Bio/Technol 6:1335
22. Zhang M, Eddy C, Deanda K, Finkelstein M, Picataggio S (1995) Science 267:240
23. Deanda K, Zhang M, Eddy C, Picataggio S (1996) Appl Environ Microbiol 62:4465
24. Weisser P, Krämer R, Sprenger GA (1996) Appl Environ Microbiol 62:4155
25. Feldmann SD, Sahm H, Sprenger GA (1992) Appl Microbiol Biotechnol 38:354

26. Bringer-Meyer S, Sahm H (1989) Appl Microbiol Biotechnol 31:529
27. De Koning W, van Dam K (1992) Anal Biochem 204:118
28. Bergmeyer HU (1985) Methods of enzymatic analysis, vols VI and VII. VCH, Deerfield Beach, Florida
29. Weuster-Botz D (1997) Anal Biochem 246:225
30. Moon RB, Richards JH (1973) J Biol Chem 248:7276
31. Salhany JM, Yamane T, Shulman RG, Ogawa S (1975) Proc Natl Acad Sci USA 72:4966
32. Breitmaier E, Voelter W (1989) Carbon-13 NMR spectroscopy, 3rd edn. VCH, Weinheim New York
33. Barrett GC, Davies JS (1985) Nuclear Magnetic Resonance spectra of amino acids and their derivatives. In: Barrett GC (ed) Chemistry and biochemistry of the amino acids. Chapman and Hall, London New York, p 525
34. Weuster-Botz D, de Graaf AA (1996) Reaction engineering methods to study intracellular metabolite concentrations. In: Scheper T (ed) Advances in biochemical engineering/biotechnology, vol 54: Metabolic engineering. Springer, Berlin Heidelberg New York, p 76
35. Lee KJ, Skotnicki ML, Tribe DE, Rogers PL (1980) Biotechnol Lett 2:339
36. Schoberth SM, Chapman BE, Kuchel PW, Wittig RM, Grotendorst J, Jansen P, de Graaf AA (1996) J Bacteriol 178:1756
37. Chance B, Eisenhardt RH, Gibson QH, Louberg-Holm KK (eds) (1964) Rapid mixing and sampling techniques in biochemistry. Academic Press, New York
38. De Koning W, van Dam K (1992) Anal Biochem 204:118
39. Seiler M, Sauer U, Bailey JE (1994) Intracellular metabolite determination in *Escherichia coli*. Institute of Biotechnology, ETH Zürich
40. Theobald U, Mailinger W, Reuss M, Rizzi M (1993) Anal Biochem 214:31
41. Gadian DG (1995) NMR and its applications to living systems, 2nd edn. Oxford University Press, Oxford
42. Srivastava AK, Volesky B (1991) Appl Microbiol Biotechnol 34:450
43. Shulman RG, Brown TG, Ugurbil K, Ogawa S, Cohen SM, Den Hollander JA (1979) Science 205:160
44. Den Hollander JA, Brown TG, Ugurbil K, Shulman RG (1979) Proc Natl Acad Sci USA 76:6096
45. Ugurbil K, Rottenberg H, Glynn P, Shulman RG (1978) Proc Natl Acad Sci USA 75:2244
46. Gillies R J (1994) NMR in physiology and biomedicine. Academic Press, London New York
47. Hartbrich A, Schmitz G, Weuster-Botz D, de Graaf AA, Wandrey C (1996) Biotechnol Bioeng 51:624
48. de Graaf AA, Wittig RM, Probst U, Strohhäcker J, Schoberth SM, Sahm H (1992) J Magn Reson 98:654
49. Gillies RJ, Galons J-P, McGovern KA, Scherer PG, Lien Y-H, Job C, Ratcliff R, Chapa F, Cerdan S, Dale BE (1993) NMR Biomed 6:95
50. Haran N, Kahana ZE, Lapidot A (1983) J Biol Chem 258:12,929
51. Altenburger R, Abarzua S, Callies R, Grimme LH, Mayer A, Leibfritz D (1991) Arch Microbiol 156:471
52. Duysens LNM, Amesz J (1957) Biochim Biophys Acta 24:19
53. Zabriskie DW, Humphrey AE (1978) Appl Eur Microbiol 35:337
54. Luong JHT, Carrier DJ (1986) Appl Microbiol Biotechnol 24:65
55. Einsele A, Ristroph DL, Humphrey AE (1978) Biotechnol Bioeng 20:1487
56. Boyer PM, Humphrey AE (1988) Biotechnol Lett 2:193
57. Slavik J, Kotyk A (1984) Biochim Biophys Acta 766:697
58. Pena A, Ramirez J, Rosas G, Calahorra M (1995) J Bacteriol 177:1017
59. Horvath JJ, Glazier SA, Spangler CJ (1993) Biotechnol Prog 9:666
60. Schulmann SG (1985) Molecular luminescence spectroscopy, methods and applications. In: Chemical analysis, vol 77. Wiley, New York, p 170
61. Marose S, Lindemann C, Scheper T (1998) Biotechnol Prog 14:63
62. Schäfer U, Boos W, Takors R, Weuster-Botz D (1999) Anal Biochem 270:88

63. Holms WH (1986) Curr Top Cell Reg 28:69
64. Neidhardt FC, Ingraham JL, Schaechter M (1990) Physiology of the bacterial cell: a molecular approach. Sinauer Associates, Sunderland, Massachusetts
65. Vallino JJ (1991) PhD Thesis, Massachusetts Institute of Technology, Cambridge, MA
66. Vallino JJ, Stephanopoulos G (1994) Biotechnol Progr 10:320
67. Vallino JJ, Stephanopoulos G (1994) Biotechnol Progr 10:327
68. van der Heijden RTJM, Heijnen JJ, Hellinga C, Romein B, Luyben KChAM (1994) Biotechnol Bioeng 43:3
69. van der Heijden RTJM, Heijnen JJ, Hellinga C, Romein B, Luyben KChAM (1994) Biotechnol Bioeng 43:11
70. van Gulik WM, Heijnen JJ (1995) Biotechnol Bioeng 48:681
71. Blum JJ, Stein RB (1982) On the analysis of metabolic networks. In: Goldberger R (ed) Biological regulation and development, vol 3A. Plenum Press, New York, p 99
72. Di Donato L, Des Rosiers C, Montgomery JA, David F, Garneau M, Brunengraber H (1993) J Biol Chem 268:4170
73. Cohen SM, Rognstad R, Shulman RG, Katz J (1981) J Biol Chem 256:3428
74. London RE (1992) In: Berliner LJ, Reuben J (ed) Biological magnetic resonance, vol 11. Plenum Press, New York, p 277
75. Ross BD, Kingsley PB, Ben-Yoseph O (1994) Biochem J 302:31
76. Kanamori K, Weiss RL, Roberts D (1988) J Biol Chem 263:2871
77. Tesch M, de Graaf AA, Sahm H (1999) Appl Environ Microbiol 65:1099
78. Eakin RT, Morgan LO, Gregg CT, Matwiyoff NA (1972) FEBS Lett 28:259
79. Walker TE, Han CH, Kollman VH, London RE, Matwiyoff NA (1982) J Biol Chem 257:1189
80. Chance EM, Seeholzer SH, Kobayashi K, Williamson JR (1983) J Biol Chem 258:13,785
81. Walsh K, Koshland DE (1984) J Biol Chem 259:9646
82. Malloy CR, Sherry AD, Jeffrey FMH (1988) J Biol Chem 263:6964
83. Sharfstein ST, Tucker SN, Mancuso A, Blanch HW, Clark DS (1994) Biotechnol Bioeng 43:1059
84. Zupke C, Stephanopoulos G (1995) Biotechnol Bioeng 45:292
85. Marx A, de Graaf AA, Wiechert W, Eggeling L, Sahm H (1996) Biotechnol Bioeng 49:111
86. Wiechert W, de Graaf AA (1996) In vivo stationary flux analysis by ^{13}C labeling experiments. In: Scheper T (ed) Advances in biochemical engineering/biotechnology, vol 54: Metabolic engineering. Springer, Berlin Heidelberg New York, p 111
87. Wiechert W, de Graaf AA (1997) Biotechnol Bioeng 55:101
88. Wiechert W, Siefke C, de Graaf AA, Marx A (1997) Biotechnol Bioeng 55:118
89. Follstad BD, Stephanopoulos G (1998) Eur J Biochem 252:360
90. Wiechert W, Möllney M, Isermann N, Wurzel M, de Graaf AA (1999) Biotechnol Bioeng 66:69
91. Möllney M, Wiechert W, Kownatzki D, de Graaf AA (1999) Biotechnol Bioeng 66:86
92. Szyperski T (1995) Eur J Biochem 232:433
93. Szyperski T, Bailey JE, Wüthrich K (1996) Trends Biotechnol 14:453
94. Wendisch VF, de Graaf AA, Sahm H (1997) Anal Biochem 245:196
95. de Graaf AA, Mahle M, Möllney M, Wiechert W, Stahmann P, Sahm H (2000) J Biotechnol 77:25
96. Park SM, Shaw-Reid C, Sinskey AJ, Stephanopoulos G (1997) Appl Microbiol Biotechnol 47:430
97. Inbar L, Lapidot A (1987) Eur J Biochem 162:621
98. Brown TR, Ugurbil K, Shulman RG (1977) Proc Natl Acad Sci USA 74:5551
99. Alger JR, Shulman RG (1984) Q Rev Biophys 17:83
100. Kirk K (1990) NMR Biomed 3:1
101. de Graaf AA, Wittig RM, Probst U, Strohhäcker J, Schoberth SM, Sahm H (1992) J Magn Reson 98:654
102. Hatzimanikatis V, Emmerling M, Sauer U, Bailey JE (1998) Biotechnol Bioeng 58:154
103. Wulf G, Wiechert W, de Graaf AA, Krämer R, Posten C, Sprenger G, Weuster-Botz D, Munack A (1997) Modelling, simulation and measurement of metabolic pathways in

Zymomonas mobilis. In: Gertler JJ, Cruz JB, Peshkin M (eds) Proceedings of the 13th World Congress, International Federation of Automatic Control, on CD-ROM. Elsevier Science, Amsterdam

104. Torres NV (1994) Biotechnol Bioeng 44:104
105. Wright BE, Butler MH, Albe KR (1992) J Biol Chem 267:3101
106. Rizzi M, Baltes M, Theobald U, Reuss M (1997) Biotechnol Bioeng 55:592
107. Reder C (1988) J Theor Biol 135:175
108. Liao JC, Lightfoot EN (1988) Biotechnol Bioeng 31:847
109. Shiraishi F, Savageau MA (1992) J Biol Chem 267:22,926
110. Torres NV (1994) Biotechnol Bioeng 44:112
111. Heinrich R, Rapoport TA (1974) Eur J Biochem 42:89
112. Heinrich R, Rapoport TA (1974) Eur J Biochem 42:97
113. Kell DB, Westerhoff HV (1986) FEMS Microbiol Rev 39:305
114. Kacser H (1988) Regulation and control of metabolic pathways. In: Bazin MJ, Prosser JI (eds) Physiological models in microbiology. CRC series in mathematical models in microbiology, vol 2. CRC Press, Boca Raton, FL, USA
115. Fell DA (1998) Biotechnol Bioeng 58:121
116. Voit EO (ed) (1991) Canonical nonlinear modeling. *S*-system approach to understanding complexity. Van Nostrand Reinhold, New York
117. Shiraishi F, Savageau MA (1992) J Biol Chem 267:22,912
118. Torres NV, Voit EO, Gonzalez-Alcon C (1996) Biotechnol Bioeng 49:247
119. Hatzimanikatis V, Floudas CA, Bailey JE (1996) Biotechnol Bioeng 52:485
120. Hatzimanikatis V, Floudas CA, Bailey JE (1996) AIChE Journal 42:1277
121. Mavrovouniotis ML, Stephanopoulos G (1988) Comp Chem Eng 12:867
122. Mavrovouniotis ML, Stephanopoulos G (1989) Biotechnol Bioeng 34:196
123. Mavrovouniotis ML (1993) Identification of localized and distributed bottlenecks in metabolic pathways. In: Hunter L, Searls D, Shavlik J (eds) Proceedings of the First International Conference on Intelligent Systems for Molecular Biology. AAAI Press, Menlo Park, CA, USA
124. Mavrovouniotis ML (1993) Identification of qualitatively feasible metabolic pathways. In Hunter L (ed) Artificial intelligence and molecular biology. AAAI Press, Menlo Park, CA, USA
125. Nielsen J (1998) Biotechnol Bioeng 58:125
126. Sauer U, Hatzimanikatis V, Bailey JE, Hochuli M, Szyperski T, Wüthrich K (1997) Nat Biotechnol 15:448
127. Marx A, Striegel K, de Graaf AA, Sahm H, Eggeling L (1997) Biotechnol Bioeng 56:168
128. Zupke C, Stephanopoulos G (1994) Biotechnol Progr 10:489
129. Katz J, Wals P, Lee W-NP (1993) J Biol Chem 268:25,509
130. Lee W-NP (1993) J Biol Chem 268:25,522
131. Schmidt K, Carlsen M, Nielsen J, Villadsen J (1997) Biotechnol Bioeng 55:831
132. Zupke C, Tompkins R, Yarmush D, Yarmush M (1997) Anal Biochem 247:287
133. Schmidt K, Nielsen J, Villadsen J (1999) J Biotechnol 71:175
134. Yen J, Lee B, Liao JC (1996) In: Proceedings of the 5th IEEE International Conference on Fuzyy Systems, vol 1. IEEE Press, Piscataway, NJ, USA, p220
135. Edwards JS, Palsson BO (1998) Biotechnol Bioeng 58:162
136. Breuel G, Gilles ED, Kremling A (1995) 6th Conference on Computer applications in Biotechnology, DECHEMA, p 199
137. Lee KJ, Scotnicki ML, Tribe DE, Rogers PL (1981) Biotechnol Lett 3:207
138. Hardman MJ, Scopes R (1988) Eur J Biochem 173:203
139. Doelle MB, Greeenfield PF, Doelle HW (1990) Appl Microbiol Biotechnol 34:160
140. Algar EM, Scopes RK (1985) J Biotechnol 2:275
141. Scopes RK (1987) Austral J Biotechnol 1:58
142. Osman YA, Conway T, Bonetti SJ, Ingram LO (1987) J Bacteriol 169:3726
143. Belaich JP, Senez JC, Murgier M (1968) J Bacteriol 95:1750
144. Hesman TL, Barnell WO, Conway T (1991) J Bacteriol 173:3215
145. Scopes RK, Testolin V, Stoter A, Griffiths-Smith K, Algar EM (1985) Biochem J 228:627

146. Pawluk A, Scopes RK, Griffiths-Smith K (1986) Biochem J 238:275
147. Snoep JL, Arfman N, Yomano LP, Westerhoff HV, Conway T, Ingram LO (1996) Biotechnol Bioeng 51:190
148. Anderson AJ, Dawes EA (1985) FEMS Microbiol Lett 27:23
149. Scopes RK (1997) Biochem J 326:731
150. Weisser P, Krämer R, Sahm H, Sprenger GA (1995) J Bacteriol 177:3351
151. Scopes RK, Bannon DR (1995) Biochim Biophys Acta 1249:173
152. Scopes RK, Griffiths-Smith K (1984) Anal Biochem 136:530
153. Scopes RK (1984) Anal Biochem 136:525
154. Hoppner TC, Doelle H (1983) Eur J Appl Microbiol Biotechnol 17:152
155. Kinoshita S, Kakizono T, Kadota K, Das K, Taguchi H (1985) Appl Microbiol Biotechnol 22:249
156. Lazdunski A, Belaich JP (1972) J Gen Microbiol 70:187
157. Barrow KD, Collins JG, Norton RS, Rogers PL, Smith GM (1984) J Biol Chem 259:5711
158. Strohhäcker J, de Graaf AA, Schoberth SM, Wittig RM, Sahm H (1993) Arch Microbiol 159:484
159. de Graaf AA, Striegel K, Wittig RM, Laufer B, Schmitz G, Wiechert W, Sprenger GA, Sahm H (1999) Arch Microbiol 171:371
160. Struch T, Neuss B, Bringer-Meyer S, Sahm H (1991) Appl Microbiol Biotechnol 34:518
161. Schoberth SM, de Graaf AA (1993) Anal Biochem 210:123
162. Lee KJ, Rogers PL (1983) Chem Eng J 27:B31
163. Joebses IML, Roels JA (1986) Biotechnol Bioeng 28:554
164. Joebses IML, Hiemstra HCH, Roels JA (1987) Biotechnol Bioeng 29:502
165. Ingram, LO (1986) Trends Biotechnol 4:40
166. Hermans M (1992) PhD Thesis, University of Düsseldorf
167. Low KS, Rogers PL (1984) Appl Microbiol Biotechnol 19:75
168. Sutter B (1991) PhD Thesis, Université de Haute-Alsace, Mulhouse, France
169. Barrow KD, Collins JG, Rogers PL, Smith GM (1983) Biochim Biophys Acta 753:324
170. Barrow KD, Collins JG, Rogers PL, Smith GM (1983) Eur J Biochem 145:173
171. Barrow KD, Collins JG, Leigh DA, Rogers PL, Warr RG (1984) Appl Microbiol Biotechnol 20:225
172. Barrow KD, Rogers PL, Smith GM (1986) Eur J Biochem 157:195
173. Rohmer M, Sutter B, Sahm H (1989) J Chem Soc Lond, Chem Commun 19:1471
174. Rohmer M, Knani M, Simonin P, Sutter B, Sahm H (1993) Biochem J 295:517
175. Laufer B (1998) PhD Thesis, University of Düsseldorf, Germany
176. Navon G, Shulman RG, Yamane T, Eccleshall R, Lam KB, Baronsfski J, Marimur J (1979) Biochem 18:4487

16 Metabolic Flux Analysis of *Corynebacterium glutamicum*

Albert A. de Graaf

16.1
Introduction

Corynebacterium glutamicum is an aerobic gram-positive bacterium intensively used in the industrial production of a variety of amino acids. L-Glutamate and L-lysine are produced in annual quantities of several hundreds of thousands of tons [1] by this organism. In addition, processes for producing L-tryptophan, L-tyrosine, L-phenyla-lanine [2], L-valine [3], and histidine [4] have been patented. The organism also has a high potential for the production of L-threonine [5, 6] and L-isoleucine [7], and probably for other amino acids as well. Extensive process development and genetic engineering has been carried out over the years with *C. glutamicum* and relevant studies are well-documented in several recent review articles, e.g., specifically on glutamate production [8], L-isoleucine biosynthesis [7], L-lysine biosynthesis [9], and aspartate-family amino acid synthesis [10, 11], or more generally on recombinant DNA technology [12] and metabolic engineering [13, 14].

The physiology and mechanism of secretion of amino acids has been intensively studied [15–19] and several transport genes have been identified [20], including that of the recently found lysine export carrier [21] which structurally represents a new type of translocator.

The central carbon metabolic pathways in *Corynebacterium glutamicum* are shown in Fig. 16.1; they include the Embden-Meyerhoff-Parnas (EMP) pathway, the pentose phosphate pathway (PPP), the tricarboxylic acid (TCA) cycle, as well as several enzymes for acetate catabolism.

Of the glycolytic pathway, the gene cluster encoding glyceraldehyde-3-phosphate dehydrogenase, 3-phosphoglycerate kinase, triosephosphate isomerase, and phosphoenolpyruvate carboxylase has been transcriptionally analyzed [22] and the genes for fructose-1,6-bisphosphate aldolase [23], the three enzymes glyceraldehyde-3-phosphate dehydrogenase, 3-phosphoglycerate kinase, and triosephosphate isomerase [24], and pyruvate kinase [25] have been cloned.

Very recently, the *C. glutamicum* transaldolase gene was cloned [26] and the enzymes glucose-6-phosphate dehydrogenase and 6-phosphogluconate dehydrogenase from the oxidative branch of the PPP were kinetically characterized [27].

The TCA cycle genes for citrate synthase [28], isocitrate dehydrogenase [29], and 2-oxoglutarate dehydrogenase [30] have been cloned. During growth on acetate, activities of phosphotransacetylase and acetate kinase as well as of the glyoxylate cycle enzymes isocitrate lyase and malate synthase are strongly elevated. The respective genes have all recently been cloned [31–33].

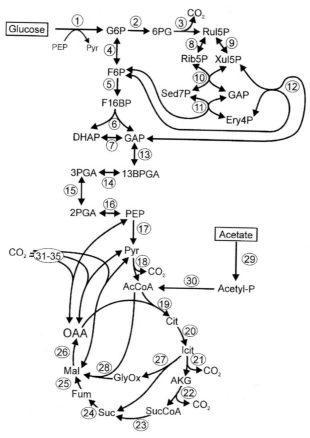

Fig. 16.1. Scheme of the pathways of central metabolism in *Corynebacterium glutamicum*. Enzymes involved: (1) phosphotransferase system; (2) glucose-6-phosphate dehydrogenase; (3) 6-phosphogluconate dehydrogenase; (4) phosphoglucose isomerase; (5) phosphofructokinase; (6) fructose-1,6-bisphosphate aldolase; (7) triose phosphate isomerase; (8) phosphoribose isomerase; (9) ribulose-5-phosphate epimerase; (10) transketolase; (11) transaldolase; (12) transketolase; (13) glyceraldehyde-3-phosphate dehydrogenase; (14) phosphoglycerate kinase; (15) phosphoglycerate mutase; (16) enolase; (17) pyruvate kinase; (18) pyruvate dehydrogenase complex; (19) citrate synthase; (20) aconitase; (21) isocitrate dehydrogenase; (22) 2-oxoglutarate dehydrogenase; (23) succinate thiokinase; (24) succinate dehydrogenase; (25) fumarase; (26) malate dehydrogenase and/or malate:quinone oxidoreductase [114]; (27) isocitrate lyase; (28) malate synthase; (29) acetate kinase; (30) phosphotransacetylase; (31)–(35) anaplerotic enzymes (see Fig. 16.2). Abbreviations: G6P, glucose 6-phosphate; 6PG, 6-phosphogluconate; Rul5P, ribulose-5-phosphate; Rib5P, ribose-5-phosphate; Xul5P, xylulose-5-phosphate; Sed7P, sedoheptulose-7-phosphate; GAP, glyceraldehyde-3-phosphate; Ery4P, erythrose-4-phosphate; F6P, fructose 6-phosphate; F16BP, fructose-1,6-bisphosphate; DHAP, dihydroxyacetone phosphate; 13BPGA, 1,3-bisphosphoglycerate; 3PGA, 3-phosphoglycerate; 2PGA, 2-phosphoglycerate; PEP, phospho*enol*pyruvate; Pyr, pyruvate; AcCoA, acetyl-CoenzymeA; Cit, citrate; Icit, isocitrate; AKG, 2-oxoglutarate; SucCoA, succinyl-CoenzymeA; Suc, succinate; Fum, fumarate; Mal, malate; OAA, oxaloacetate; acetyl-P, acetyl phosphate. *Double-pointed arrows* indicate reactions that are likely to operate in a reversible manner in vivo

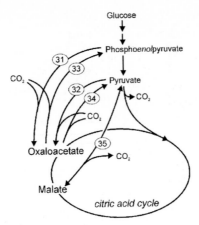

Fig. 16.2. Scheme of the pathways involved in the anaplerosis of *Corynebacterium glutamicum*. Enzymes involved: (31) PEPcarboxylase; (32) pyruvate carboxylase; (33) PEPcarboxykinase; (34) oxaloacetate decarboxylase; (35) malic enzyme

Regarding the anaplerotic reactions, *Corynebacterium glutamicum* possesses a rather complete spectrum of enzymes that interconvert C3- and C4-metabolites (Fig. 16.2). The anaplerotic enzymes phospho*enol*pyruvate(PEP) carboxylase [34], pyruvate carboxylase [35], and malic enzyme [36] were all shown to be present. Moreover, the organism possesses PEP carboxykinase [37], oxaloacetate decarboxylase [38], and probably also PEP synthetase [39]. Of these, the genes for PEP carboxylase [40, 41] and pyruvate carboxylase [42, 43] have been cloned.

Efficient nitrogen assimilation is of key importance for amino acid production. When *Corynebacterium glutamicum* is grown with a sufficient nitrogen supply, the important nitrogen sources ammonium and urea cross the cytoplasmic membrane by passive diffusion. Under conditions of nitrogen starvation, energy-dependent uptake systems for urea and ammonium are synthesized [44, 45]; the (methyl)ammonium carrier has been cloned [44]. *C. glutamicum* was shown to possess glutamate dehydrogenase [46, 47] and glutamine synthetase [48] as the principal ammonium-assimilating enzymes; genes for these enzymes were cloned [47, 48]. Under conditions of ammonium limitation or in glutamate dehydrogenase-deficient mutants, the glutamine:2-oxoglutarate aminotransferase system is expressed as alternative glutamate-producing enzyme [49, 50]. Figure 16.3 shows the principal ammonium-assimilating enzyme system of *C. glutamicum*.

The biosynthesis of the amino acids from the aspartate family in *C. glutamicum* (Fig. 16.4) is one of the best studied pathway complexes in micro-organisms. Nevertheless, despite the detailed knowledge of the genetics, physiology, and regulation of the pathways for lysine [9] and isoleucine [7], reported selectivities for these products are still significantly lower than theoretical maxima. In the case of isoleucine, the transport step may be limiting [51]. In the case of lysine, several authors have claimed that selectivities are restricted by a non-optimal coordination of central metabolism and amino acid biosynthesis, resulting in a limitation due to, e.g., inadequate precursor supply [52] or energy excess [53]. These suppositions have greatly stimulated the detailed investigation of the interplay between central metabolism

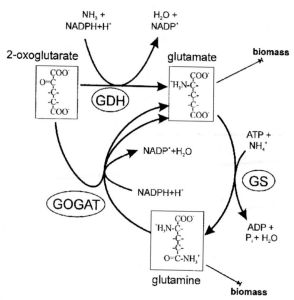

Fig. 16.3. Scheme of the principal ammonium-assimilating pathways in *Corynebacterium glutamicum*. Abbreviations: GDH, glutamate dehydrogenase; GS, glutamine synthetase; GOGAT, glutamine 2-oxoglutarate aminotransferase

and amino acid biosynthesis in *C. glutamicum*. This contribution presents a state-of-the art overview and analysis of studies that employed intracellular metabolic flux analysis by means of metabolite balancing and/or isotopic labeling combined with NMR spectroscopy in order to correlate activities of the central metabolic pathways in vivo with amino acid overproduction.

16.2
Fundamentals of Intracellular Metabolic Flux Analysis in *Corynebacterium glutamicum*

16.2.1
Metabolite Balancing

This part presents the elementary data and flux model necessary to perform metabolite balancing with *C. glutamicum*.

16.2.1.1
Biomass Composition

During non-limited aerobic growth of *C. glutamicum*, up to 60% of the carbon may be converted to biomass. Therefore, the first step in metabolic flux analysis is the detailed calculation of the precursor requirements for anabolic purposes. For *C. glutamicum*, this has been done, e.g., in [54–56]. These analyses were all based on the

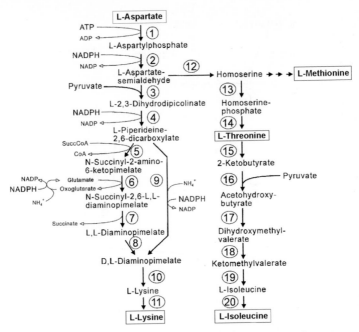

Fig. 16.4. Scheme of the biosynthetic pathways of amino acids from the aspartate family in *Corynebacterium glutamicum*. Enzymes and genes involved: (1) aspartate kinase (*lysC*); (2) aspartate semialdehyde dehydrogenase (*asd*); (3) dihydrodipicolinate synthase (*dapA*); (4) dihydrodipicolinate reductase (*dapB*); (5) tetrahydrodipicolinate succinylase (*dapD*); (6) N-succinyl aminoketopimelate transaminase; (7) N-succinyl diaminopimelate desuccinylase (*dapE*); (8) diaminopimelate epimerase; (9) diaminopimelate dehydrogenase (*ddh*); (10) diaminopimelate decarboxylase (*lysA*); (11) permease (*lysE*); (12) homoserine dehydrogenase (*hom*); (13) homoserine kinase (*thrB*); (14) threonine synthase (*thrC*); (15) threonine dehydratase (*ilvA*); (16) acetohydroxyacid synthase (*ilvB, ilvN*); (17) isomeroreductase (*ilvC*); (18) dihydroxyacid dehydratase (*ilvD*); (19) transaminase; (20) permease. Cofactors are shown only for the lysine biosynthetic pathway

approach and data of Neidhardt et al. [57] for *E. coli*, a Gram-negative organism. Since *C. glutamicum* is a Gram-positive organism, the data have to be corrected for the different cell wall composition, i.e., the higher peptidoglycan content. This can be done on the basis of the diaminopimelate content of *C. glutamicum*, as given in [56]. The calculated precursor requirements are given in Table 16.1.

This table can also be used to calculate that *C. glutamicum* requires 9816 μmol/g dry wt of nitrogen for its elementary composition, of which 1773 μmol/g dry wt are accounted for by glutamine amidotransferase reactions in the synthesis of arginine, histidine, tryptophan, N-acetylglucosamine, and the pyrimidine and purine nucleotides of RNA and DNA. The remaining 8043 μmol/g dry wt are accounted for by glutamate dehydrogenase (7137 μmol/g dry wt) as well as a number of directly aminating enzymes (906 μmol/g dry wt).

Table 16.1. Precursor as well as carbon dioxide requirements in μmol/g of dry weight for biomass synthesis of *Corynebacterium glutamicum*. The amino acid composition was determined from *C. glutamicum* strain MH20-22B *leuA* grown in continuous culture [56]. The peptidoglycan content was inferred from the diaminopimelate (Dap) content. The high intracellular glutamate and glutamine pools of *C. glutamicum* [50] were explicitly taken into account. The distinction between glutamate and glutamine in the cell protein was made based on *C. glutamicum* codon usage data (Eikmanns B, Tesch M, de Graaf AA, unpublished results). Relative amounts of the (deoxy)nucleotides were estimated taking into account that the relative GC content of *C. glutamicum* is 56%. The data on other cellular constituents were taken from the literature [57]. NADPH stoichiometries were inferred from data presented in [113] Abbreviations: PEP, phosphoenolpyruvate; AcCoA, acetyl-coenzymeA

Amino acid	Amount (μmol/g dry weight)	G6P	F6P	Ri5P	E4P	GAP	PGA	PEP	Pyr	AcCoA	OAA	AKG	CO2	NADPH
Ala	606								1				1	1
Arg	189											1	1	4
Asx	399										1			1
Cys	87						1							5
Glu	360											1		1
Gln	147											1		1
Glu_pool	250											1		1
Gln_pool	49											1		
Gly	361						1							1
His	71			1										1
Ile	202								1		1		-1	5
Leu	440								2	1			-2	2
Lys	202								1		1		-1	4
Met	146								1		1		-1	8
Phe	133				1			2						2
Pro	170											1	-1	3
Ser	225						1							1
Thr	275										1			3
Trp	54			1	1			1					-1	3
Tyr	81				1			2					-1	2
Val	284								2				-1	2
Dap	146								1		1		-1	4

Precursor stoichiometry (mol/mol amino acid)

Table 16.1. Continued

Polymer		G6P	F6P	Ri5P	E4P	GAP	PGA	PEP	Pyr	AcCoA	OAA	AKG	CO2	NADPH
		Precursor amount (μmol/g dry weight)												
Protein	total	0	0	125	268	0	673	482	2604	440	1370	1165	-1647	10548
RNA	ATP			152			152						152	456
	GTP			218			218						218	436
	UTP			125							125		125	125
	CTP			135							135		135	135
DNA	dATP			22			22						22	88
	dGTP			28			28						28	84
	dCTP			28							28		28	56
	dTTP			22							22		22	44
Lipids		51				129	129			2116			-2116	3612
LPS			16	24			24	24		329			-329	470
Peptido-glycan			292					146	0	292	0	0	-292	146
Glycogen		154					49							0
C1-Units												59		49
Poly-amines														180
total		205	308	879	268	129	1295	652	2604	3177	1680	1224	-3654	16429

16.2.1.2
Condensed Bioreaction Network

The condensed bioreaction network of C. *glutamicum* can be obtained from Figs. 16.1, 16.2, and 16.4 by lumping all reactions that occur in a linear sequence without branching points as well as all sets of reactions that cannot be mutually discriminated because they have identical overall reaction equations. This results in the following set of equations describing central metabolism (water is omitted, **upt** denotes glucose uptake via the PTS system, **ace** denotes acetate uptake and activation, **ppp** denotes pentose phosphate pathway, **emp** denotes Embden-Meyerhof-Parnas pathway, **ana** denotes anaplerosis, **tcc** denotes citric acid cycle, **gs** denotes glyoxylate shunt):

upt:	$Glucose+PEP \rightarrow G6P+Pyr$
ace:	$Ace+ATP+CoA \rightarrow AcCoA+ADP$
ppp1:	$G6P+2\,NADP \rightarrow Rul5P+CO_2+2\,NADPH$
ppp2:	$Rul5P \rightarrow Xul5P$
ppp3:	$Rul5P \rightarrow Rib5P$
ppp4:	$Xul5P+Ery4P \rightarrow F6P+GAP$
ppp5:	$Xul5P+Rib5P \rightarrow Sed7P+GAP$
ppp6:	$Sed7P+GAP \rightarrow Ery4P+F6P$
emp1:	$G6P \rightarrow F6P$
emp2:	$F6P+ATP \rightarrow F16BP+ADP$
emp3:	$F16BP \rightarrow 2\,GAP$
emp4:	$GAP+ADP+NAD \rightarrow PGA+ATP+NADH$
emp5:	$PGA \rightarrow PEP$
emp6:	$PEP+ADP \rightarrow Pyr+ATP$
ana1:	$PEP+CO_2 \rightarrow OAA$
ana2:	$Pyr+CO_2+NADPH \rightarrow Mal+NADP$
ana3:	$Pyr+CO_2+ATP \rightarrow OAA+ADP$
tcc1:	$Pyr+CoA+NAD \rightarrow AcCoA+CO_2+NADH$
tcc2:	$AcCoA+OAA \rightarrow Icit+CoA$
tcc3:	$Icit+NADP \rightarrow AKG+CO_2+NADPH$
tcc4:	$AKG+NAD+ADP \rightarrow Suc+CO_2+NADH+ATP$
tcc5:	$Suc+FAD \rightarrow Mal+FADH$
tcc6:	$Mal+NAD \rightarrow OAA+NADH$
gs1:	$Icit \rightarrow Suc+GlyOx$
gs2:	$GlyOx+AcCoA \rightarrow Mal+CoA$

The sequence **emp6–ana3** is completely identical to the single reaction **ana1**. Therefore, **ana1** and **ana3** must also be lumped, i.e., **ana3** is omitted from the set of equations. The sequences **emp6–ana2** and **ana1–tcc6** (reversed) are distinguishable in principle because of their different cofactor requirements.

The two different pathways for lysine synthesis in C. *glutamicum* [58] are lumped into:

lys: $OAA+Pyr+ATP+4\,NADPH \rightarrow Lys+CO_2+ADP+4\,NADP$

The NADPH is used directly as shown in Fig. 16.4, or indirectly in the regeneration of glutamate from 2-oxoglutarate. Glutamate is nitrogen donor in the transamination of oxaloacetate to aspartate, a precursor of lysine.

NAD, FAD, and ATP are regenerated by oxidative phosphorylation with different P/O stoichiometries S1 and S2 (water is again omitted):

oxp1:   NADH+1/2 O2+S1 ADP→NAD+S1 ATP

oxp2:   FADH+1/2 O2+S2 ADP→FAD+S2 ATP

The reactions that describe the withdrawal of precursor metabolites (G6P, F6P, Rib5P, Ery4P, GAP, PGA, PEP, Pyr, AcCoA, OAA, AKG) and the concomitant use of NADPH as well as the fixation and release of CO_2 can be written using the stoichiometries given in Table 16.1.

16.2.1.3
Approaches to Resolve Network Underdeterminacy

Rather than by using matrices, the extent to which the condensed bioreaction network can be determined from extracellular (i.e., involving net import into or output from the cells) measurements can be judged rather conveniently from Fig. 16.1, as described in the following. Once the biomass yield on the substrate is measured, all reactions removing precursors for anabolic purposes are known from the yield value and the stoichiometries given in Table 16.1. Now consider as a typical network node glucose 6-phosphate (G6P) (Fig. 16.5). At metabolic steady state, the G6P pool does not change, whence the total flux into the pool must exactly match the total efflux from the pool. Or,

$$upt = ppp1 + emp1 + bs\_G6P \qquad (16.1)$$

where the latter denotes the anabolic precursor requirement of G6P. Since the specific glucose uptake rate is also measured, **upt** and **bs_G6P** are both known. Therefore, only the sum **ppp1+emp1** is determined by the measurements. Two conclusions can be drawn at this stage. First, the anabolic precursor requirement fluxes do not have to be considered in the judgment of the network determinacy, since all are known (i.e., omitting **bs_G6P** from Eq. (16.1) still leaves only the sum **ppp1+emp1** determined by the measurements). Second, a value for **ppp1** must be chosen in order to determine **emp1**. Having done this and applying the first conclusion, it can easily be verified from Fig. 16.1 that **ppp2, ppp3, ppp4, ppp5, ppp6, emp2, emp3, emp4,** and **emp5** are all determined by the measured specific glucose uptake rate, the measured biomass yield, and our choice of **ppp1** in combination with the reaction stoichiometries. For the next branching point, phosphoenolpyruvate (PEP), we then have:

$$emp5 = anal + upt + emp6 \ (+bs\_PEP), \qquad (16.2)$$

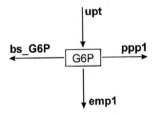

Fig. 16.5. Metabolite fluxes at the glucose-6-phosphate node. The flux names are as defined in the text

forcing us to choose a value for **ana1** in order that **emp6** be determined. At the next node, pyruvate (Pyr), a value for **ana2** must be chosen in order that **tcc1** be determined. This then suffices to determine all fluxes since the total of oxaloacetate and malate synthesized via **ana1**, **ana2**, and **gs2** must equal the precursor amounts of oxaloacetate plus 2-oxoglutarate, i.e., **gs2** is determined by our choice of **ana1** and **ana2**. With **tcc1**, **gs2**, **ana1**, and **ana2** known, **tcc2**, **gs1**, **tcc3**, **tcc4**, **tcc5**, and **tcc6** all follow consecutively.

Summarising, the metabolic network at this stage still has three degrees of freedom: **ppp1**, **ana1**, and **ana2** (using different choices at the branch points G6P, PEP, and Pyr, one could consider as completely equivalent **emp1**, **emp6** and **tcc1** as the free fluxes in the model). Unless further action is taken, the flux analysis problem cannot be solved. If no further measurements can be performed, basically three options remain:

1. Further lumping of reactions
2. Exclusion of reactions based on biochemical considerations
3. Inclusion of other balances than carbon, e.g., redox or energy balances

In the case of C. glutamicum, these three approaches can all, respectively, be used: **ana1** and **ana2** may be lumped into a single reaction originating from a combined PEP/pyruvate pool; **gs2** can be assumed zero during growth on glucose based on enzyme measurements; and **ppp1** can be determined by assuming that no net synthesis or consumption of NADPH occurs in the network defined by the reactions presented in the previous paragraph. The latter in fact creates one additional balance equation:

$$2 \times \mathbf{ppp1} + \mathbf{tcc3} = \mathbf{ana2} + 4 \times \mathbf{lys} + \mathbf{bs\_NADPH} \qquad (16.3)$$

where **bs_NADPH** denotes the total anabolic precursor requirement of NADPH. The combined application of all three measures results in a completely determined equation system. As an alternative to the lumping of **ana1** and **ana2**, one of these two reactions may be assumed absent, enabling one to retain separate pools of PEP and pyruvate. One might also include **oxp1** and **oxp2** in order to create an additional balance equation based on NADH conservation. Application of this procedure requires the measurement of the oxygen consumption rate in addition to knowledge of the stoichiometries S1 and S2 (see preceding paragraph).

16.2.1.4
Theoretical Lysine Selectivity

An important benefit of the metabolite balancing procedure is that it allows the theoretical consideration of optimal product yields as well as of the flux configurations necessary to produce these. Different maximum product yields will be arrived at, depending on the constraints applied. Lysine synthesis is a good example. For the purpose of this illustration, let it be assumed that we are dealing with a population of non-growing C. glutamicum.

If one considers only the carbon balance and disrespects all biochemistry, an overall reaction stoichiometry of

1 glucose→1 lysine

would be allowed, or a molar yield of 100%. Considering that 4 moles of NADPH are consumed per mole of lysine formed, and assuming that NADPH can only be regen-

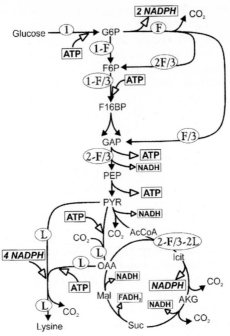

Fig. 16.6. Reaction network with cofactor stoichiometries for non-growing, l-lysine-producing *Corynebacterium glutamicum. Symbols in ovals* denote molar fluxes

erated in the oxidative pentose phosphate pathway with concomitant loss of CO_2, one arrives at the following reaction stoichiometry:

    4 glucose→3 lysine+6 CO_2,

or a molar yield of 75%.

Now consider a somewhat more realistic calculation that takes into account ATP, NAD, and NADP. Figure 16.6 shows the reaction network in the absence of growth. It is assumed that pyruvate carboxylase is the only active anaplerotic reaction. The glucose uptake rate is normalized to 1. It is easily verified that the fluxes as given in Fig. 16.6 yield a closed carbon balance. The need for a closed NADPH balance yields the following requirement for the activity F of the oxidative pentose phosphate pathway as a function of the lysine production rate L:

$$F = (18 \times L - 6)/5 \tag{16.4}$$

Inspection of the flux distribution (cf. Fig. 16.6) shows that even at the imaginary yield of 100% (i.e., L=1), no fructose-1,6-bisphosphatase activity is required. This result differs from that of [53].

Setting up the balance for NADH and substituting Eq. (16.4) yields a net synthesis of NADH:

$$R_{NADH} = 8 - 4 \times F/3 - 6 \times L = (48 - 54 \times L)/5 \tag{16.5}$$

For FADH we have analogously

$$R_{FADH} = 2 - F/3 - 2 \times L = (12 - 16 \times L)/5 \tag{16.6}$$

The net requirement for ATP is given by

$$R_{ATP} = 2 \times L \tag{16.7}$$

This ATP has to be generated by oxidative phosphorylation (reactions **oxp1** and **oxp2**). Assuming a P/O ratio of 2 [59], i.e., P/O stoichiometry S1=2 in **oxp1** and S2=1 in **oxp2**, the oxidation of NADH and FADH yields an amount O_{ATP} of ATP:

$$O_{ATP} = 2 \times R_{NADH} + R_{FADH} = (108 - 124 \times L)/5 \tag{16.8}$$

Requiring that $O_{ATP}-R_{ATP}>0$ and substituting Eqs. (16.7) and (16.8) yields an upper value of 108/134=0.806 for L based on the ATP balance. Inspection of Fig. 16.6 reveals, however, that this would imply a reductive operation of the TCA cycle and a negative pyruvate dehydrogenase flux. Therefore, a more or less realistic uppermost lysine yield results when all pyruvate is channeled to lysine. From Fig. 16.6, it is easily seen that this implies

$$2 \times L = 2 - F/3 \tag{16.9}$$

which after substitution of Eq. (16.4) yields a value of 0.75 for L and a net synthesis of 1.5 mol ATP per mol glucose. The translocation of protons for lysine transport [60] requires only half that amount of ATP if an effective stoichiometry of 1 mol ATP per mol of lysine transported is adopted (it is likely to be less). The required pentose phosphate pathway activity at the yield L=0.75 according to Eq. (16.4) is equal to 1.5. No transhydrogenase activity would be necessary.

16.2.1.5
Limitations

The fundamental limitation of the metabolite balancing approach is of course that the assumptions necessary to arrive at a fully determined equation system may not be valid. Membrane-associated additional NADPH oxidase activity not accounted for in the analysis would for instance affect the NADPH balancing and result in an underestimation of the pentose phosphate pathway activity. Likewise, a residual activity of the glyoxylate pathway not accounted for would result in an overestimation of the anaplerotic carboxylating reaction rates.

A second, also very important limitation of the technique, is that it is not able to resolve a lumped pathway as well as the degree of reversibility of reversible enzyme reactions. Thus, metabolite balancing can for instance give no clues as to which anaplerotic enzyme is the predominant supplier of oxaloacetate for lysine biosynthesis in vivo, a very important issue in the chemical engineering of *C. glutamicum* [61]. Also, metabolic cycles cannot be allowed for by metabolite flux balancing since the technique considers only net fluxes. However, metabolic cycles may play an important regulatory role, as well as having an impact on the energy balance of the cell. To answer questions related to such topics, stable isotope labeling approaches are required.

16.2.2
Isotopic Labeling Combined with NMR Spectroscopy

The use of stable isotopes allows the metabolic fate of single atoms to be traced through the metabolic network. The analysis of timecourses from dynamic labeling experiments allows one to calculate directly absolute fluxes for the pathways involved. The measurement of label at isotopic steady state allows the determination of the relative fluxes, since the fractional enrichment of a metabolite is determined by the ratio of the fluxes coming from different source metabolites and by the fractional enrichment of these source molecules. Thereby, the limitations of the metabolite balancing technique mentioned above can be overcome. NMR spectroscopy appears to be particularly suitable for the analysis of isotopic labeling experiments since it does not require chemical derivatization of the labeled compounds.

   This part introduces the application of ^{13}C and ^{15}N labeling to the study of metabolic rates in *C. glutamicum*. It will be shown that ^{13}C labeling data from pyruvate, oxaloacetate, and lysine (or from their successors, e.g., alanine, aspartate) allow one to resolve the flux distribution at the glucose-6-phosphate and PEP/pyruvate branchpoints, in addition to the flux distribution over the parallel diaminopimelate pathways in lysine biosynthesis. ^{15}N labeling allows one to determine the flux distribution over the primary ammonium-assimilating enzymes of *C. glutamicum*.

16.2.2.1
Isotopic Atom Balancing

In order to calculate metabolic fluxes from measured isotopic enrichments, mathematical expressions relating both must be formulated. The basic procedure is to establish specific isotope mass balances around single atoms. This is exemplified here for the case of ^{13}C. Thus, for an arbitrary atom Q of an arbitrary metabolite pool that receives inputs F_k from several other metabolite carbon atoms with fractional ^{13}C enrichments P_k while Q itself is supplying several other pools with rates E_j (Fig. 16.7), we can write for the rate of change of the isotopic content:

$$d(^{13}C\_Q)/dt = F_1 \times P_1 + F_2 \times P_2 + ... + F_N \times P_N \\ - (E_1 + E_2 + ... + E_M) \times (^{13}C\_Q)/[Q] \tag{16.10}$$

where ^{13}C_Q denotes the absolute ^{13}C content of pool Q, and [Q] denotes the pool size (concentration) of Q. Equations of this type can be used for dynamic simulations. At isotopic steady state, $d(^{13}C\_Q)/dt=0$ and Eq. (16.10) transforms to

$$(^{13}C\_Q)/[Q] = (F_1 \times P_1 + F_2 \times P_2 + ... + F_N \times P_N)/(E_1 + E_2 + ... + E_M) \tag{16.11}$$

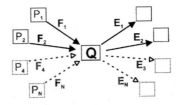

Fig. 16.7. Principle of carbon atom balancing. See text

The formulation of these simple-type isotopic balance equations allows one to gain important basic insights in flux analysis by isotope labeling as will be shown in the next sections.

16.2.2.2
Resolving Glycolysis and Pentose Phosphate Pathway

As a first example, the flux partition over glycolysis and the pentose phosphate pathway at the branchpoint glucose 6-phosphate will be analyzed under the simplifying condition that all fluxes are unidirectional. It will be shown how the ^{13}C label in C-3 of glyceraldehyde 3-phosphate (precursor of pyruvate) under isotopic steady state conditions reflects the pentose phosphate pathway activity. In the following equations, the ^{13}C fractional enrichment of the i-th carbon atom of metabolite X is denoted by X_i.

The relevant part of the metabolic network, along with the definition of the fluxes, is shown in Fig. 16.8. The substrate glucose is labeled only in C-1, with enrichment L_0. Since the fructose-1,6-bisphosphate pool is supplied only from fructose 6-phosphate, its ^{13}C-labeling at steady state is identical to that of the latter. The pentose phosphates are assumed to be in rapid equilibrium. GAP_3 results from three different sources for which equations must be formulated: F16BP_1 (equal to F6P_1), F16BP_6 (equal to F6P_6), and Xul5P_5 (in the two transketolase reactions, see Fig. 16.8).

For glucose 6-phosphate, Eq. (16.11) translates to

$$G6P\_1 = f_0 \times L_0/(f_1 + f_2) = L_0$$

$$G6P\_6 = f_0 \times 0/(f_1 + f_2) = 0$$

Since the pentose phosphates are only synthesized from glucose 6-phosphate, they will be unlabelled in all positions. Sedoheptulose 7-phosphate is only synthesized from the pentose phosphates and will therefore also be unlabelled. Erythrose 4-phosphate, which is only synthesized from sedoheptulose 7-phosphate, will also be unlabeled. We then have for fructose 6-phosphate (Eq. 16.11, Fig. 16.8):

$$F6P\_1 = (f_1 \times G6P\_1 + f_3 \times Sed7P\_1 + f_3 \times Xul5P\_1)/(f_1 + 2f_3) = f_1 \times L_0/(f_1 + 2f_3),$$

$$F6P\_6 = (f_1 \times G6P\_6 + f_3 \times GAP\_3 + f_3 \times Ery4P\_4)/(f_1 + 2f_3) = f_3 \times GAP\_3/(f_1 + 2f_3),$$

and for glyceraldehyde-3-phosphate:

$$GAP\_3 = \{2f_3 \times Xul5P\_5 + (f_1 + 2f_3) \times (F16BP\_1 + F16BP\_6)\}/(2f_1 + 6f_3)$$
$$= \{(f_1 + 2f_3) \times (F6P\_1 + F6P\_6)\}/(2f_1 + 6f_3)$$
$$= \{f_1 \times L_0 + f_3 \times GAP\_3\}/(2f_1 + 6f_3).$$

Rearranging terms and substituting $f_3=f_2/3$ yields

$$GAP\_3 = L_0 \times f_1/(2f_1 + 5f_3) = L_0 \times (f_0 - f_2)/(2f_0 - f_2/3) \qquad (16.12)$$

Figure 16.9 shows GAP_3 as a function of f_2. According to this graph, NMR measurement of the ^{13}C enrichment in C-3 of glyceraldehyde-3-phosphate (or one of its successors such as pyruvate) enables the sensitive and unequivocal determination of the pentose phosphate pathway activity.

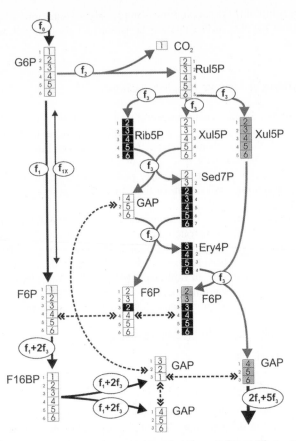

Fig. 16.8. Carbon atom transitions in the pentose phosphate pathway. The *rectangular boxes* represent individual carbon atoms, with numbers according to their position in the original glucose molecule. *Small numbers next to the boxes* indicate the carbon numbering in the respective metabolites. *Double-pointed arrows* symbolize complete equilibration of different labeled species within one metabolite pool. The fluxes f_0–f_3 and f_{1X} are used for modeling purposes (see text). The bidirectional flux of the phosphoglucose isomerase reaction f_{1X} is only used in Eq. (16.17) and following. Abbreviations: G6P, glucose 6-phosphate; F6P, fructose 6-phosphate; F16BP, fructose-1,6-bisphosphate; GAP, glyceraldehyde 3-phosphate; Rul5P, ribulose 5-phosphate; Rib5P, ribose 5-phosphate; Xul5P, xylulose 5-phosphate; Sed7P, sedoheptulose 7-phosphate; Ery4P, erythrose 4-phosphate

16.2.2.3
Resolving the Parallel Lysine Biosynthetic Pathways

The parallel lysine biosynthetic pathways (see Fig. 16.4), not accessible to metabolite balancing, are readily resolved using ^{13}C labeling and NMR. Figure 16.10 shows the relevant labeling scheme (adapted from [62]). Operation of the succinylase variant of the diaminopimelate pathway results in label scrambling since a symmetric molecule is involved. For the labeling of lysine we have (Eq. 16.11, Fig. 16.10):

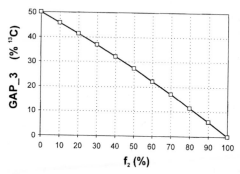

Fig. 16.9. Fractional ^{13}C enrichment GAP of C-3 of glyceraldehyde 3-phosphate as a function of the activity f_2 of the oxidative pentose phosphate pathway (expressed as % of the glucose uptake rate), calculated using Eq. (16.12)

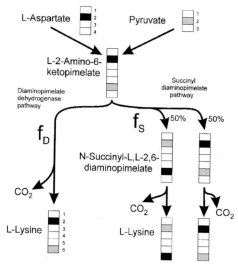

Fig. 16.10. Carbon atom transfer scheme for the biosynthesis of lysine from its precursors aspartate and pyruvate. *Black and shaded boxes* indicate the routes of aspartate C-2 and pyruvate C-2, respectively. The intermediate N-succinyl-l,l-2,6-diaminopimelate is symmetric, which leads to label scrambling in the succinyl diaminopimelate pathway

$$Lys\_2 = (f_D \times Asp\_2 + 1/2 \times f_S \times Pyr\_2 + 1/2 \times f_S \times Asp\_2)/(f_D + f_S)$$

$$Lys\_6 = (f_D \times Pyr\_2 + 1/2 \times f_S \times Asp\_2 + 1/2 \times f_S \times Pyr\_2)/(f_D + f_S)$$

Addition of the two equations yields

$$Lys\_2 + Lys\_6 = Asp\_2 + Pyr\_2$$

or

$$Asp\_2 = Lys\_2 + Lys\_6 - Pyr\_2$$

Subtracting the two equations yields

$$Lys\_2 - Lys\_6 = f_D \times (Asp\_2 - Pyr\_2)/(f_D + f_S)$$

Substitution of Asp_2 and rearrangement of terms yields for the flux partition of the dehydrogenase variant of the diaminopimelate pathway:

$$f_D/(f_D + f_S) = (Lys\_2 - Lys\_6)/(Lys\_2 + Lys\_6 - 2 \times Pyr\_2) \qquad (16.12A)$$

Thus, NMR measurement of the ^{13}C-label in the relevant carbons of pyruvate-derived alanine as well as of lysine resolves the fluxes.

16.2.2.4
Resolving Anaplerosis, Citric Acid Cycle, and the Glyoxylate Shunt

The analysis of the fluxes over anaplerosis (f_A), citric acid cycle (f_T), and glyoxylate pathway (f_G) using simple isotopic atom balances is somewhat more demanding. The relevant labeling scheme along with the fluxes are shown in Fig. 16.11. It is assumed that pyruvate carboxylase is the only active anaplerotic carboxylating enzyme. The total of the anaplerotic and glyoxylate pathway fluxes $f_A + f_G$ must equal the sum of the precursor fluxes of 2-oxoglutarate plus oxaloacetate, bs_AKG+bs_OAA. It is important to note that the reaction from succinate to oxaloacetate involves complete scrambling of the ^{13}C label between C-1 and C-4, and between C-2 and C-3.

The strategy is to express the labeling of the oxaloacetate carbons as a function of the labeling of pyruvate. For our purpose, it suffices to consider only C-2 and C-3 of oxaloacetate, denoted as O_2 and O_3, and C-2 and C-3 of pyruvate, denoted as P_2 and P_3, respectively.

The basic equations are (Eq. 16.11, Fig. 16.11):

$$O\_2 = \{f_A \times P\_2 + 1/2 \times f_T \times (O\_2 + P\_3) + \\ 1/2 \times f_G \times (O\_2 + P\_3) + f_G \times O\_3\}/(f_A + f_T + 2f_G) \qquad (16.13)$$

$$O\_3 = \{f_A \times P\_3 + 1/2 \times f_T \times (O\_2 + P\_3) + \\ 1/2 \times f_G \times (O\_2 + P\_3) + f_G \times P\_3\}/(f_A + f_T + 2f_G). \qquad (16.14)$$

For ease of description it is assumed that pyruvate is labeled only in C-3, i.e., P_2=0. Subtracting Eq. (16.14) from Eq. (16.13) gives

$$O\_2 - O\_3 = \{f_G \times O\_3 - (f_A + f_G) \times P\_3\}/(f_A + f_T + 2f_G)$$

or

$$O\_3 = \{(f_A + f_T + 2f_G) \times O\_2 + (f_A + f_G) \times P\_3\}/(f_A + f_T + 3f_G) \qquad (16.15)$$

Substitution of Eq. (16.15) in Eq. (16.13) yields the desired function

$$O\_2 = (R/S) \times P\_3 \qquad (16.16)$$

with

$$R = f_T + f_G + 2f_G \times (f_A + f_G)/(f_A + f_T + 3f_G)$$

and

$$S = 2f_A + f_T + 3f_G - 2f_G \times (f_A + f_T + 2f_G)/(f_A + f_T + 3f_G)$$

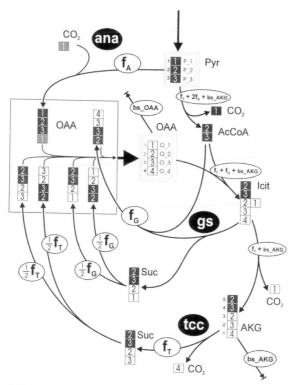

Fig. 16.11. Model of carbon atom transfer in the anaplerotic carboxylation reactions (ana; flux: f_A), the glyoxylate shunt (gs; flux: f_G) and the tricarboxylic acid cycle (tcc; flux: f_T). *The rectangular boxes* represent individual carbon atoms, with numbers according to their original position in pyruvate (*dark-shaded boxes*) or oxaloacetate (*shaded boxes*). *Small numbers next to boxes* indicate the carbon numbering in the molecule. Abbreviations: Pyr, pyruvate; AcCoA, acetyl-coenzymeA; Icit, isocitrate; AKG, 2-oxoglutarate; Suc, succinate; OAA, oxaloacetate. The precursor requirement fluxes of OAA and AKG for anabolic reactions are represented by bs_OAA and bs_AKG, respectively

Figure 16.12 shows a set of curves obtained using Eq. (16.16) for realistic values of P_3 (i.e., 30%) and the total anaplerotic flux (f_A+f_G) (i.e., 20% of the glucose uptake rate). Considering that the typical label measurement inaccuracy is 0.5% ^{13}C, Fig. 16.12 illustrates that the isotopic labeling approach enables a good resolution of the fluxes over anaplerotic carboxylation and the glyoxylate pathway, which is impossible by the metabolite balancing approach.

16.2.2.5
Resolving the Principal Ammonium-Assimilatory Pathways

By monitoring the preferred site of incorporation of ^{15}N label upon incubation with ^{15}N-labeled ammonium, researchers have been able to identify the principal ammonium-assimilating enzymes in several microorganisms [63–65]. For our purpose, the

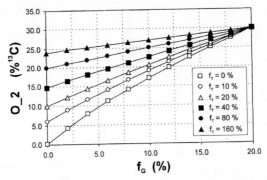

Fig. 16.12. Fractional ^{13}C enrichment O_2 of C-2 of oxaloacetate as a function of the activity f_G of the glyoxylate shunt (expressed as % of the glucose uptake rate) at various citric acid cycle fluxes f_T as indicated, calculated using Eq. (16.16) under the assumption that C-3 of pyruvate is 30% ^{13}C-enriched, and that the total anaplerotic flux is 20% of the glucose uptake rate (molar units)

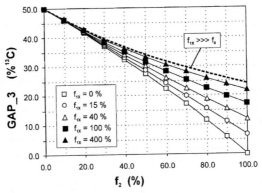

Fig. 16.13. Fractional ^{13}C enrichment GAP_3 of C-3 of glyceraldehyde-3-phosphate as a function of the activity f_2 of the oxidative pentose phosphate pathway at various bidirectional fluxes f_{1X} of the phosphoglucoseisomerase reaction (see Fig. 16.8) as indicated, calculated using Eq. (16.23). *The broken line* represents the curve that results when glucose 6-phosphate and fructose 6-phosphate are in extremely fast equilibrium. All fluxes are expressed as % of the glucose uptake rate

dynamic behavior of the ^{15}N labeling of the glutamate and glutamine nitrogen atoms is modeled with a differential equation approach analogous to Eq. (16.10). Three principal ammonium-assimilating pathways exist in *C. glutamicum* ([46–50] and references therein): the glutamate dehydrogenase (GDH) pathway, the glutamine synthetase (GS) pathway, and the glutamine 2-oxoglutarate-aminotransferase (GO-GAT) (see Fig. 16.3). Denoting the corresponding fluxes with F_{GDH}, F_{GS}, and F_{GOGAT}, respectively, and the precursor fluxes of glutamate and glutamine for biomass biosynthesis as B_{Glu} and B_{Gln}, respectively, the following differential equations describe the simplified system for the case that only net fluxes are considered.

For the N_α pool of glutamate:

$$[Glu] \times d^{15}N_\alpha Glu/dt = F_{GDH} \times {}^{15}NH_4 + F_{GOGAT} \times \{{}^{15}N_\alpha Gln + {}^{15}N_\delta Gln\} - \{F_{GS} + B_{Glu}\} \times {}^{15}N_\alpha Glu$$

For the N_α and N_δ pools of glutamine:

$$[Gln] \times d^{15}N_\alpha Gln/dt = F_{GS} \times {}^{15}N_\alpha Glu - \{F_{GOGAT} + B_{Gln}\} \times {}^{15}N_\alpha Gln$$

$$[Gln] \times d^{15}N_\delta Gln/dt = F_{GS} \times {}^{15}NH_4 - \{F_{GOGAT} + B_{Gln}\} \times {}^{15}N_\delta Gln$$

As initial condition, the ^{15}N content of all intracellular nitrogen pools is assumed to be zero, and the labeling of ammonium, i.e., $^{15}NH_4$, is set to the constant nitrogen-15 substrate enrichment. Substituting this in the above set of equations, we obtain for the first point on the time courses (i.e., the first measuring point):

$$[Glu] \times {}^{15}N_\alpha Glu \cong F_{GDH} \times {}^{15}NH_4 \cdot \Delta t = M_1$$

$$[Gln] \times {}^{15}N_\alpha Gln \cong 0$$

$$[Gln] \times {}^{15}N_\delta Gln \cong F_{GS} \times {}^{15}NH_4 \times \Delta t = M_2,$$

where M_1 and M_2 denote the measured ^{15}N contents (in $mmol\,l^{-1}$) after a short time Δt of incubation with the labeled ammonium.

Thus, provided the time resolution is fast enough and the ^{15}N quantitation is sufficiently accurate, the ratio M_1/M_2 of the relative rates of ^{15}N accumulation in the amino nitrogen of glutamate and the amido nitrogen of glutamine, respectively, directly reflects the ratio of the GDH and GS fluxes.

For the second iteration step, the values for $^{15}N_\alpha Glu$, $^{15}N_\alpha Gln$, and $^{15}N_\delta Gln$ derived from M_1 and M_2 at the first time point are substituted in the differential equations for the N_α and N_δ pools of glutamine. This, after some rearrangement, yields for the ratio M_3/M_4 of the *increments* of the glutamine N_δ and N_α signals in the second time interval:

$$M_3/M_4 = [Glu]^{15}NH_4/M_1 - ([Glu]/[Gln]) \times (M_2/M_1) \times (F_{GOGAT} + B_{Gln})/F_{GS},$$

showing that now F_{GOGAT} is also reflected in the time courses of ^{15}N incorporation. The fluxes F_{GDH}, F_{GS}, and F_{GOGAT} are then unequivocally determined from the measured ^{15}N incorporation time courses together with the measured specific ammonium uptake rate and values for B_{Glu} and B_{Gln} inferred from Table 16.1.

16.2.2.6
Influence of Reaction Reversibility

The presence of reversible reaction steps may have a pronounced effect on isotopic labeling depending on conditions. This on the one hand complicates the analysis since the simplified pictures drawn up above are no longer valid. On the other hand, it offers the possibility of identification of such reversible reactions, which may have significant informative value for metabolic engineering purposes. As an illustrative example the analysis of the pentose phosphate pathway will be redone, now including a variable reversibility of the phosphoglucose isomerase reaction. Figure 16.8 again applies, however now with the additional bi-directional flux f_{1x} interconverting glucose 6-phosphate and fructose 6-phosphate included. For glucose 6-phosphate we then have

$$G6P\_1 = (f_0 \times L_0 + f_{1X} \times F6P\_1)/(f_1 + f_{1X} + f_2) \tag{16.17}$$

$$G6P\_6 = f_{1X} \times F6P\_6/(f_1 + f_{1X} + f_2), \tag{16.18}$$

i.e., the label on C-6 of glucose 6-phosphate is no longer 0.

Since the pentose phosphates are only synthesized from glucose 6-phosphate, they will only be labeled in C-5. Sedoheptulose 7-phosphate is only synthesized from the pentose phosphates and will only be labeled in C-7. Erythrose 4-phosphate, which is only synthesized from Sedoheptulose 7-phosphate, will be labeled only in C-4. We then have for fructose 6-phosphate (Fig. 16.8):

$$F6P\_1 = (f_1 + f_{1X}) \times G6P\_1/(f_1 + f_{1X} + 2f_3) \tag{16.19}$$

$$F6P\_6 = \{(f_1 + f_{1X}) \times G6P\_6 + f_3 \times GAP\_3 + f_3 \times Ery4P\_4\}/(f_1 + f_{1X} + 2f_3) \tag{16.20}$$

Substitution of Eq. (16.17) in Eq. (16.19) yields

$$F6P\_1 = f_0 \times L_0 \times (f_1 + f_{1X})/\{(f_1 + f_{1X} + f_2) \times (f_1 + f_{1X} + 2f_3) - (f_1 + f_{1X}) \times f_{1X}\} \tag{16.21}$$

Since Ery4P_4=Sed7P_7=Rib5P_5=G6P_6 we can rearrange Eq. (16.20) after substitution of Eq. (16.18) into

$$F6P\_6 = Q \times GAP\_3 \tag{16.22}$$

where

$$Q = f_3 \times (f_1 + f_{1X} + f_2)/\{(f_1 + f_{1X} + 2f_3) \times (f_1 + f_{1X} + f_2) - (f_1 + f_{1X}) \times f_{1X} - f_3 \times f_{1X}\}$$

For glyceraldehyde-3-phosphate we again have

$$\begin{aligned} GAP\_3 &= \{2f_3 \times Xul5P\_5 + (f_1 + 2f_3) \times (F16BP\_1 + F16BP\_6)\}/(2f_1 + 6f_3) \\ &= \{2f_3 \times G6P\_6 + (f_1 + 2f_3) \times (F6P\_1 + F6P\_6)\}/(2f_1 + 6f_3) \end{aligned}$$

Substitution of Eqs. (16.18), (16.21), and (16.22) after rearranging of terms yields

$$GAP\_3 = L_0 \times R/\{(2f_1 + 6f_3) - Q \times S\} \tag{16.23}$$

with

$$R = f_0 \times (f_1 + f_{1X}) \times (f_1 + 2f_3)/\{(f_1 + f_{1X} + 2f_3) \times (f_1 + f_{1X} + f_2) - (f_1 + f_{1X}) \times f_{1X}\}$$

$$S = \{2f_3 \times f_{1X} + (f_1 + 2f_3) \times (f_1 + f_{1X} + f_2)\}/(f_1 + f_{1X} + f_2)$$

and Q as defined in Eq. (16.22). As the final step, $f_3 = f_2/3$ is to be substituted in these expressions. It can be verified that Eq. (16.23) transforms to Eq. (16.12) in the case when $f_{1X}=0$.

Figure 16.13 shows a set of curves of GAP_3 as a function of f_2 for various values of f_{1X}, obtained using Eq. (16.23). Obviously, the degree of reversibility of the phosphoglucose isomerase reaction strongly influences the estimation of the pentose phosphate pathway activity from the ^{13}C label in glyceraldehyde 3-phosphate, or pyruvate.

16.2.2.7
Isotopomers

The fact that the degree of reversibility of enzyme reactions may be quantitated from its effects on the ^{13}C label distribution adversely implies that the number of unknowns in the flux determination procedure increases considerably. Therefore, it is of crucial importance to increase the information content of the labeling experiment as much as possible. Positional isotopic enrichments clearly are not the optimal choice, since a molecule containing N carbon atoms in this case can provide at most N independent pieces of information. In contrast, a molecule with N carbon atoms can exist as 2^N different isotope isomers, or isotopomers, since each carbon may be present as ^{12}C or as ^{13}C. Thus, a small molecule like pyruvate can provide up to 8 independent pieces of information from its ^{13}C isotopomer distribution. This number may be extended even further by combining ^{13}C labeling with heteronuclear stable isotope labeling, e.g., ^{15}N, ^{2}H, or ^{17}O. Modern, refined measurement techniques such as two-dimensional NMR [66, 67], heteronuclear spin-echo NMR [68, 69], or mass spectrometry [70, 71], have helped isotopomer analysis to evolve into an extremely powerful tool for metabolic flux analysis. As a qualitative example, consider again the labeling scheme in Fig. 16.11. It may be helpful to think of isotopomer analysis as a method to measure the relative abundance of intact carbon backbone fragments (Fig. 16.14). It is now assumed that pyruvate exists in two isotope isomeric forms: [^{12}C$_3$]pyruvate (90%) and [^{13}C$_3$]pyruvate (10%). [^{13}C$_3$]Pyruvate entering the citric acid cycle via pyruvate dehydrogenase is decarboxylated, leaving an intact ^{13}C2-fragment that is subsequently transferred to oxaloacetate via one turn of the citric acid cycle. It has a probability of only 10% to form a C-C bond with another ^{13}C-labeled carbon. This probability can easily be corrected for [66] and is not considered here. In contrast, [^{13}C$_3$]pyruvate entering the citric acid cycle via pyruvate carboxylase will appear as an intact ^{13}C3 fragment in oxaloacetate. Thus, to a first order approximation, the ratio of the pyruvate carboxylase flux to the total of anaplerotic and citric acid cycle fluxes into the oxaloacetate pool is simply proportional to the relative isotopomeric abundance of intact 1,2,3-^{13}C3 fragments within the ox-

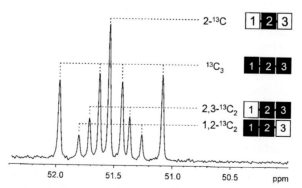

Fig. 16.14. Part of the one-dimensional 100 MHz ^{13}C NMR spectrum of alanine isolated from hydrolyzed protein of *Corynebacterium glutamicum* grown on ^{13}C-labeled glucose. The resonance of C-2 is shown, with different multiplet contributions from the various isotopomer species as indicated

aloacetate pool (or within an oxaloacetate-derived amino acid such as aspartate). The proof of the possible presence of back reactions of oxaloacetate to the symmetric intermediate fumarate also seems almost trivial, since this would give rise to unique intact 2,3,4-^{13}C3 fragments within the oxaloacetate pool. The presence of the respective ^{13}C3 fragments can unequivocally be inferred from multiplet structures in the ^{13}C NMR spectrum (Fig. 16.14).

These rather qualitative arguments already clearly illustrate the potential of isotopomer measurements for metabolic flux analysis. The mathematical analysis, however, is rather involved.

16.2.2.8
Sources of Isotopic Measurement Data

Collecting the biogenic material from which the labeling data are to be measured presents an important experimental problem in isotopic labeling studies. The simplest approach is to use the culture supernatant and to analyze the labeling patterns in metabolites excreted by the cells. However, under optimal conditions for exponential growth, aerobic *C. glutamicum* may produce only carbon dioxide, from which (at least at isotopic steady state) little or no information can be derived. Or, under optimal production conditions, only a single or at best a few metabolites are excreted that can be used to obtain labeling data from. Moreover, the labeling of excreted metabolites may be diluted by the presence of unlabeled material accumulated before addition of the labeled substrate, or resulting from the metabolization of unlabelled storage compounds. Thus, analysis solely of excreted compounds generally results in a very small isotopic labeling data set with a poor and possibly biased information content. A better, but more elaborate, procedure is to perform a cytoplasmic extraction of the cells. This gives access to the labeling of a larger number of compounds with high information content, such as, e.g., alanine, aspartate, and glutamate. These amino acids are derived from their precursors in a single enzymatic step. At isotopic steady state, the labeling state of the carbon backbones of these amino acids therefore is identical to that of their respective precursors pyruvate, oxaloacetate, and 2-oxoglutarate. When using cytoplasmic extracts, one has to take care that the metabolism is quenched quickly enough in order to prevent significant alterations of the metabolic pools and/or their labeling state during the extraction procedure. Also, it must be assured that a true isotopic steady state of all cytoplasmic metabolites has been established. This may take several hours in *C. glutamicum* due to the presence especially of a very high cytoplasmic glutamate pool (250 mmol l^{-1} in the wildtype ATCC13032 [50]). Since, however, a very large number of different metabolites with very different concentrations are present in the cytoplasmic extract, the determination of ^{13}C fractional enrichments by NMR often requires chemical isolation and purification of compounds, since otherwise the NMR spectra are largely unsuited for quantitative analysis. In practice, the use of cytoplasmic extracts for labeling determinations is also rather limited because many important metabolites are present in too low a concentration to be accessible by NMR. The analysis typically requires 1 μmol of the labeled compound. This implicates for a compound with a cytoplasmic concentration of x mmol l^{-1} that a cytoplasmic volume of 1/x ml must be extracted with 100% efficiency, i.e., at least approximately 0.5/x g dry weight of cells.

Because of the limited availability of labeled biogenic material even in cytoplasmic extracts, current most efficient methods [66, 56] use the macromolecular cell constituents to retrieve the labeling information, since these function as storage devices for precursor metabolites. Hydrolysis of the cellular protein for example gives access to the complete range of amino acids in which the carbon backbones of the respective precursor metabolites (Table 16.1) are preserved in a defined and well-known way. Thus, erythrose 4-phosphate, ribose 5-phosphate, 3-phosphoglycerate, PEP, pyruvate, acetyl-CoA, oxaloacetate, and 2-oxoglutarate can all be accessed using this procedure. Glyceraldehyde-3-phosphate, 3-phosphoglycerate, and acetyl-CoA can be accessed via the cell lipids, and the cellular RNA constitutes an additional source of especially ribose 5-phosphate labeling data. An added advantage of using the macromolecular components is that very large amounts of precursor metabolites are stored in them; i.e., hydrolysis of only 10 mg of dry biomass yields sufficient material of a precursor with a very modest content of $100\,\mu$mol/g dry weight for NMR analysis. Sole requirement is that the macromolecular constituents have been synthesized entirely from precursors that were in an isotopic steady state, and that this state did not vary significantly during the incubation period. These conditions are best satisfied when using cells incubated with the labeled substrate for at least three dilution times in chemostat culture, cells exponentially grown in very thinly-inoculated batch cultures, or excreted proteins.

16.2.2.9
A Comprehensive Modeling Framework

The metabolite balancing approach essentially amounts to the solution of a set of linear equations (typically 20–30). It is therefore readily appreciated that this can be done most conveniently by standard matrix calculation procedures [54, 72]. The isotopic labeling experiment for which only positional enrichments at steady state are considered can be regarded as a metabolite balancing procedure for single carbon atoms (cf. the examples given above). As such, it also gives rise to a set of linear equations (typically 70–120). Again in this case, one would expect that the most efficient approach to handle these equation systems would be the application of matrix calculation methods. This, however, was not realized until fairly recently, when the first procedures involving so-called atom mapping matrices appeared in the literature [73]. While these formulations did not immediately reveal the analogy with the metabolite balancing solution approaches, this was clearly the case in the simultaneously developed formalism of Wiechert and co-workers [56, 74, 75]. This procedure enables a fully integrated analysis of mixed data resulting from measurements of fluxes as well as positional ^{13}C enrichments, and includes efficient numerical procedures for the estimation of statistical variances of the calculated fluxes. Other approaches also developed. A recent overview that describes analytical methods for ^{13}C-labeled metabolites, ^{13}C labeling strategies, metabolite balancing approaches, as well as their synergetic integration can be found in [76].

The modeling and analysis of isotopomer data is much more complicated since bimolecular reactions give rise to bilinear terms in the isotopomer balance equations [77]. Only in special cases will these equations transform to linear equations, namely when no bimolecular reactions are present in the network, or when for each bimolecular reaction the labeling state of one of the inputs is given. This for instance applies to a model that involves the citric acid cycle, the glyoxylate cycle and an anaplerotic carboxylation reaction in the case that the labeling state of the common substrate of all bimolecular reactions involved, i.e., pyruvate, is given. The reason is, that this isotopomer distribution can be considered as a constant which results in linear equations relating the remaining unknown isotopomers and fluxes. This subsequently allows one to formulate the appropriate convenient analytical expressions [78, 79]. It has long been thought that only iterative numerical strategies allow one to calculate the isotopomer distributions in the general case. Several generalized modeling approaches involving so-called atom mapping and isotopomer mapping matrices [80, 81] were devised in order to facilitate the modeling. However, Wiechert and co-workers [82, 83] very recently demonstrated that the isotopomer balance equations can indeed also be solved for the general case using standard matrix approaches and following a recursive procedure after suitable coordinate transformation. This enabled to construct a completely general isotopomer modeling framework where only the basic carbon atom transitions in the various reaction steps have to be specified by the user. The strongly enhanced computational efficiency allows one to carry out optimal design studies of labeling experiments by searching for minimal simulated standard deviations of calculated fluxes resulting from the use of different labeled substrates *in silico*.

16.3
Metabolite Balancing Studies

16.3.1
Overview

After their key article [52] introducing the concepts of flux balancing and network rigidity in relation with metabolic engineering, Vallino and Stephanopoulos wrote an important paper [59] about metabolic flux distributions in *C. glutamicum* during growth and lysine overproduction. Together with their first paper on the subject [54], these publications have been highly instrumental in the establishment of metabolic flux analysis as the invaluable tool for metabolic engineering purposes it is has become today. In order to apply a network rigidity analysis as proposed in [52], these authors applied several metabolic perturbations that enabled one to characterize the flexibility of the pyruvate and glucose-6-phosphate branch points in lysine-producing *C. glutamicum* ATCC 21253 [61, 84].

A series of flux analyses at different growth rates in lactate-limited as well as in glucose-limited chemostat cultures of *C. glutamicum* ATCC 17965 were reported by Cocaign-Bousquet and co-workers [55, 85]. These authors used reported and measured enzyme activities to decide which pathways are active in vivo, so as to yield a determined system of equations for metabolic balancing. This group also reported metabolite balancing studies performed in order to characterize the metabolic changes associated with a transient period of oxygen-limited growth of batch cultures *of C. glutamicum* ATCC 17965 [36]. However, the complexity of the observed changes together with the enormous variation of the calculated carbon flux distribution within the central metabolic pathways does not permit a clear analysis of the results.

In the following sections, an illustrative selection of the most important results of the reported metabolite balancing studies of *C. glutamicum* are given and discussed.

16.3.2
Comparison of Fluxes During Growth and Lysine Production

Vallino and Stephanopoulos [59] studied a control lysine fermentation of *C. glutamicum* ATCC 21253 by analyzing the culture broth at several time points during the initial growth phase and during several following different lysine production phases, while applying the metabolite balancing approach. This strain produces lysine under threonine limitation. Two illustrative flux distributions are shown in Fig. 16.15. During the growth phase (Fig. 16.15a), the main products from glucose are biomass, carbon dioxide, and trehalose, while only trace amounts of lactate, acetate, and lysine are observed. For the analysis it was assumed that the glyoxylate pathway is inactive, that no other NADPH-regenerating enzymes than isocitrate dehydrogenase, glucose-6-phosphate dehydrogenase, and 6-phosphogluconate dehydrogenase are active, and that PEPcarboxylase is the only active anaplerotic carboxylation enzyme. The two different lysine biosynthetic pathways (Fig. 16.4) were lumped. This strategy results in a fully determined equation system as explained above in Sect. 16.2.1.3. Since measurements of the oxygen and ammonium uptake rates as well as the carbon dioxide evolution rate were also included, the network was in fact overdetermined. This analysis indicates a significant activity of the oxidative pentose phosphate path-

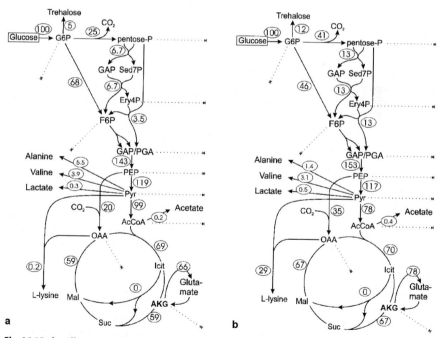

Fig. 16.15a,b. Illustrative flux distributions in *Corynebacterium glutamicum* ATCC 21253 during: a the growth; b early lysine production phase, as determined by metabolite balancing. Data were taken from [59]. Fluxes are normalized by the glucose uptake rate. *Dotted arrows* indicate precursor fluxes for biomass synthesis. While these fluxes were not explicitly stated in [59], 60% of total carbon was used for biomass synthesis in A and 15% in B

way (PPP) already under conditions of balanced growth: 25% of glucose 6-phosphate are metabolized over the PPP.

As soon as threonine in the medium was depleted, lysine production started at a high rate while growth continued but gradually slowed down. The metabolic flux distribution determined towards the end of this phase is given in Fig. 16.15b. The main products are now biomass, lysine, and carbon dioxide. Trehalose synthesis is increased and there is still a minor excretion of acetate and lactate. Lysine is produced at 29% molar yield. A number of very characteristic changes in the flux distribution are apparent when comparing Fig. 16.15a and 16.15b: an almost twofold increase of the activity of the PPP and the anaplarotic carboxylation and a significant increase of the glutamate dehydrogenase activity. The citric acid cycle rate is little affected [59].

16.3.3
The Search for Yield-Limiting Flux Control Architectures

16.3.3.1
The Pyruvate Branch Point

Two experiments to assess the control structure around the pyruvate branch point were carried out [61]. In the first experiment, a permanent potential increase of the pyruvate availability for lysine synthesis was created by selecting a mutant with 98% attenuated pyruvate dehydrogenase activity. The metabolic flux distribution map obtained from this experiment is shown in Fig. 16.16. Since it very closely resembles the flux distribution obtained with the control fermentation (Fig. 16.15b), it was concluded that the availability of pyruvate is not a limiting factor for lysine synthesis. However, the attenuation of the pyruvate dehydrogenase activity resulted in an overall flux attenuation in the network of 65–75% while metabolite excretion patterns were unchanged. This was considered evidence for the presence of a rigid branch point somewhere else in the metabolic network [61].

In the second experiment, an instantaneous increase of the pyruvate availability for lysine synthesis was created by the addition of fluoropyruvate at the onset of lysine production. Fluoropyruvate is a strong inhibitor of the pyruvate dehydrogenase complex. The flux distribution resulting from the analysis reveals a pattern that is largely identical to the one obtained with the control fermentation at the same production phase, with the exception that approximately 40% of the pyruvate previously metabolized in the citric acid cycle is diverted from that pathway and excreted [61]. The fact that the lysine production is virtually unaffected again demonstrates that the lysine yield is not limited by pyruvate availability.

16.3.3.2
The Glucose-6-Phosphate Branch Point

Two experiments to assess the control structure around the glucose-6-phosphate branch point were carried out [84]. In the first experiment, a permanent potential increase of the glucose-6-phosphate availability for metabolism over the oxidative PPP was created by selecting a mutant with 90% attenuated glucose-6-phosphate isomerase activity. The results of the analysis indicated an attenuation of fluxes throughout the metabolic network, although without a significant alteration of flux

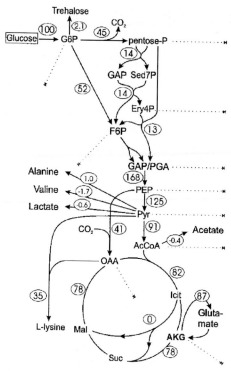

Fig. 16.16. Flux distribution in *Corynebacterium glutamicum* FPS009 during the early lysine production phase, as determined by metabolite balancing. In this strain, pyruvate dehydrogenase activity was only 2% of the activity present in the parent strain ATCC 21253. Data were taken from [61]. Fluxes are normalized by the glucose uptake rate. *Dotted arrows* indicate precursor fluxes for biomass synthesis. While these fluxes were not explicitly stated in [61], 20.5% of total carbon was used for biomass synthesis. Compare with Fig. 16.15b

partitioning. This again indicated the presence of a rigid branchpoint elsewhere in the metabolic network [84]. In the second experiment, the glucose-6-phosphate branch point was essentially bypassed by using gluconate as the primary carbon source. Gluconate enters the PPP after phosphorylation to 6-phosphogluconate. The backreaction of 6-phosphogluconate to glucose 6-phosphate must be considered kinetically unfeasible for thermodynamic reasons. However, it appeared that without the inclusion of an additional NADPH-oxidizing reaction in the network this constraint could not be fulfilled due to an overproduction of NADPH resulting from the forced metabolization of 100% of the substrate over the oxidative PPP. The flux distribution resulting after this network modification is given in Fig. 16.17. The gluconate consumption was almost as rapid as the glucose consumption in the control fermentation and the flux distribution map other than the PPP was again very similar to the control fermentation (Fig. 16.15b) except that the catabolism of gluconate produces a substantial excess of NADPH. This excess, however, does not result in an increase in lysine yield, thus ruling out a limitation by NADPH availability.

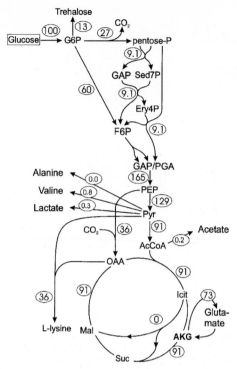

Fig. 16.17. Flux distribution in *Corynebacterium glutamicum* NFG068 during the early lysine production phase, as determined by metabolite balancing. In this strain, phosphoglucose isomerase activity was only 7% of the activity present in the parent strain ATCC 21253. Data were taken from [84]. Fluxes are normalized by the glucose uptake rate. No biomass synthesis took place; in fact, 1.1% of the carbon was derived from biomass hydrolysis [84]. Compare with Fig. 16.16

16.3.4
Growth Rate-Dependent Modulation of the Central Metabolic Fluxes

16.3.4.1
Growth on Lactate

Figure 16.18 shows two representative flux distributions for growth on lactate at low and high growth rates [55]. At the low growth yield, NADPH for anabolic purposes is supplied by the pentose phosphate pathway (PPP) and the isocitrate dehydrogenase reaction, while the anaplerotic flux was found to be distributed over PEPcarboxylase and the glyoxylate pathway. At the high growth rate, the anaplerotic flux is supplied by a pyruvate-carboxylating enzyme while a relatively large flux catalysed by malic enzyme generates additional NADPH. Moreover, a significant overflow of pyruvate was observed (Fig. 16.18). The rationale behind the seemingly arbitrary choice of pathways acting in vivo is as follows. Enzymatic determinations in cell-free extracts of *Brevibacterium flavum*, a close relative of *C. glutamicum*, had indicated

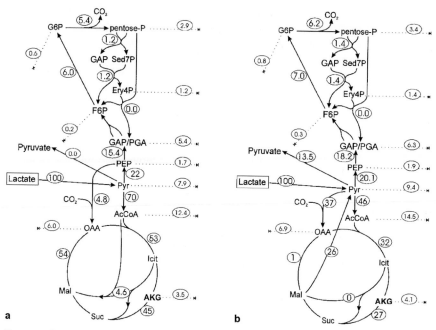

Fig. 16.18a,b. Illustrative flux distributions in *Corynebacterium glutamicum* ATCC 17965 at dilution rates of: **a** 0.17 h^{-1}; **b** 0.28 h^{-1} in chemostat cultures during growth on lactate, as determined by metabolite balancing. Data were taken from [55]. Fluxes are normalized by the lactate uptake rate. *Dotted arrows* indicate precursor fluxes for biomass synthesis. Note the proposed changes in the operation of the anaplerotic enzymes. At the high growth rate, malic enzyme covers the increased anabolic demand for NADPH

that the key enzyme of the oxidative PPP, glucose-6-phosphate dehydrogenase, was expressed approximately 7–20-fold lower during growth on organic acids (lactate, acetate, glutamate) than during growth on glucose [86]. Therefore, the authors assumed that at all growth rates, the oxidative PPP supplies precursors for ribose 5-phosphate and erythrose 4-phosphate only to the exact amounts that are needed for biomass synthesis. This activity results in a basic supply of NADPH (i.e., 2 mol per mol of glucose 6-phosphate metabolized over the oxidative PPP).

At the low growth rate, the additional NADPH required for biomass synthesis was assumed to be provided only by the isocitrate dehydrogenase reaction, in the exact amount needed. By this assumption, the flux through the citric acid cycle is fixed and the carbon balance must be closed by adjustment of the fluxes over the following two alternative routes with different overall stoichiometries: an anaplerotic carboxylation reaction (overall stoichiometry Pyr+CO_2→OAA) and the glyoxylate cycle (with overall stoichiometry 2 Pyr→OAA+2 CO_2). This adjustment is completely determined since, first, the total molar amount of oxaloacetate formed in these two reactions must exactly equal the sum of the molar amounts of the precursor metabolites oxaloacetate and 2-oxoglutarate required for biomass synthesis, and secondly, the total amount of available pyruvate is fixed. The choice of PEPcarboxylase as the anaplerotic reaction is in fact arbitrary.

At the high growth rate (Fig. 16.18b), the demands for additional NADPH and anaplerotic precursors are much larger than can be supplied by the isocitrate dehydrogenase and PEPcarboxylase reactions, respectively. From their enzyme measurements, which indicated a decrease of PEPcarboxylase activity along with a strong increase of malic enzyme and oxaloacetate decarboxylase activity when going to higher growth rates [55], the authors assumed that PEPcarboxylase does not contribute to the anaplerotic flux at the high growth rate. Instead, it was concluded that malic enzyme provides the additional NADPH needed for biomass synthesis, and that the measured oxaloacetate decarboxylase activity in reality represents a pyruvate carboxylase activity. The latter provides for the recycling of pyruvate formed by malic enzyme and fulfills the anaplerotic precursor requirement. It was also assumed that the glyoxylate cycle is inactive, since an increased pyruvate concentration had been shown to repress the synthesis of the glyoxylate bypass enzymes.

The resulting pathway configuration then yielded the fully determined flux estimation shown in Fig. 16.18b. The rather constant ratio of pyruvate dehydrogenase activity vs the calculated flux through this enzyme, as well as the pyruvate overflow observed at high growth rates, were taken as evidence that this enzyme was substrate-saturated under all conditions [55].

16.3.4.2
Growth on Glucose

Figure 16.19 shows a representative flux distribution for growth on glucose at a very high growth rate of $0.59\,h^{-1}$ [85]. NADPH for anabolic purposes is supplied to the exact amount required, but only by the pentose phosphate pathway (PPP) and the isocitrate dehydrogenase reaction. Pyruvate carboxylase was assumed to be the only active anaplerotic enzyme at this growth rate. With these choices, the flux estimation is completely determined as explained above under in Sect. 16.2.1.3. A flux of almost 60% of available glucose 6-phosphate into the PPP was calculated. This study [85] did not supply other explicit complete flux distributions, but these may easily be calculated from the data given, except for the precise flux distributions over the alternative anaplerotic enzymes PEPcarboxylase, malic enzyme, and oxaloacetate decarboxylase (thought to mask pyruvate carboxylase activity). These enzymes are shown to be present in significant amounts, but with relative activities that vary considerably with the growth rate. None of them may be considered inactive at any growth rate and one can only speculate about the actual activities of the enzymes in vivo. Here, clearly a limitation of flux analysis by the metabolic balancing method is encountered.

16.3.5
Summary

The metabolite balancing studies referred to above have yielded valuable insights in the interdependency of the metabolic fluxes in the central metabolism of *C. glutamicum*. While there may remain some doubt as to whether PEPcarboxylase is really the principal anaplerotic enzyme in *C. glutamicum* as assumed in [54, 59, 61], the experimental evidence gathered in the perturbation studies of the pyruvate branch point [61] leaves little or no alternative than to support the conclusion that pyruvate availability does not limit lysine yield. Likewise, irrespective of whether an addi-

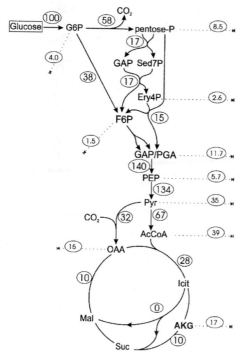

Fig. 16.19. Flux distribution in *Corynebacterium glutamicum* ATCC 17965 at a dilution rate of 0.59 h^{-1} in chemostat culture during growth on glucose, as determined by metabolite balancing. Data were taken from [84]. Fluxes are normalized by the glucose uptake rate. *Dotted arrows* indicate precursor fluxes for biomass synthesis. The 1.44 mol NADPH per mol glucose required for biomass synthesis is supplied by the pentose phosphate pathway and the isocitrate dehydrogenase reaction. Malic enzyme was concluded to be inactive [84]

tional NADPH oxidase activity is present in *C. glutamicum* or not, the experimental evidence presented in [84] unequivocally proves that lysine yield is not limited by NADPH supply. In contrast, the results of the analyses of the growth rate-dependent modulation of the central metabolic fluxes in *C. glutamicum* leave considerably more questions that ask for a definitive answer. Many choices regarding the operationality of pathways in vivo made in these studies were based on enzyme measurements. However, activities of the key enzymes of the oxidative pentose phosphate pathway and the glyoxylate pathway were not determined, whence important information necessary to support the assumptions is missing. But even if this information would be available, any conclusion about the relative activity of, e.g., anaplerotic pathways in vivo based on enzyme measurements alone must be treated with great care. Thus, additional data that give access to the true metabolic activities in vivo are required before any definitive answers can be given to such exciting questions as to whether an NADPH-generating malic enzyme-pyruvate carboxylase metabolic cycle exists in *C. glutamicum* [55]. In the next section it will be demonstrated that such data can in fact be provided by isotopic labeling experiments in combination with NMR or mass spectrometry (MS).

16.4
Studies Based on Isotopic Labeling and NMR

16.4.1
Overview

Glutamate-excreting bacteria, which are today all known by the species name *C. glutamicum*, were isolated starting in 1957 [87]. From relatively early on, isotopic labeling studies of *C. glutamicum* and its close relatives *Brevibacterium flavum* and *B. ammoniagenes* have been performed. While at first radiolabel methods were employed [88–90], studies using stable isotope labeling in combination with NMR and MS started to emerge in the 1980s. They were initiated by a ^{15}N NMR study of *Brevibacterium lactofermentum* [91] and a ^{13}C NMR study of the biosynthesis by *Microbacterium ammoniaphilum* of glutamate selectively enriched with carbon-13 [92]. This study already used mathematical expressions relating ^{13}C NMR multiplet ratios to activities of PEPcarboxylase, the citric acid cycle, and the glyoxylate cycle, and as such presents at once the first isotopomer study of a *C. glutamicum*-related bacterium. Ishino et al. subsequently used ^{13}C labeling and NMR to establish the presence of the dual pathway for lysine synthesis (Fig. 16.4) in vivo [93] and to characterize a histidine-fermenting *C. glutamicum* mutant [94]. First attempts to model and quantify the principal fluxes of the central metabolism during lysine production by ^{13}C labeling were reported in [95] for *C. glutamicum* using ^{13}C NMR and in [96] for *B. flavum* using a combination of ^{1}H NMR, ^{13}C NMR, and MS. Ishino et al. [97] performed a comparative study of the flux distribution over the Embden-Meyerhof pathway and the pentose phosphate pathway during lysine and glutamate fermentation by *C. glutamicum* using ^{13}C-labeled glucose and ^{13}C NMR. Sonntag et al. [62] presented a detailed ^{13}C labeling study of the flux distribution over the diaminopimelate pathways in lysine biosynthesis of wildtype and lysine-producing *C. glutamicum* in dependence on culture conditions. Using a model adapted from [92, 96], the principal fluxes in the central metabolism of *C. glutamicum* ATCC17965 during exponential growth on ^{13}C-labeled glucose were estimated [98]. Marx et al. [56] presented a first fully integrated metabolite balancing/^{13}C labeling flux analysis of lysine-producing *C. glutamicum* MH20–22B in chemostat culture with [1-^{13}C]glucose as the sole carbon source. In order to obtain an isotopic enrichment data set as large as possible, this method used a comprehensive set of amino acids isolated from the ^{13}C-labeled cell protein. This approach was subsequently refined and applied to exponentially growing, glutamate-producing, and lysine-producing *C. glutamicum* in batch cultures [99], to exponentially growing, and glutamate-producing isogenic strains derived from *C. glutamicum* MH20–22B in continuous cultures [100], and to lysine-producing strains with different glutamate dehydrogenase cofactor dependency derived from *C. glutamicum* MH20–22B in continuous cultures [101]. Using different ^{13}C-labeled substrates and modeling procedures, two independent studies of the anaplerotic reactions in PEPcarboxylase-deficient *C. glutamicum* were reported [70, 102]. Using a membrane cyclone bioreactor devised for in vivo NMR spectroscopy with high microbial cell densities, intracellular fluxes in lysine-producing *C. glutamicum* were studied with the help of ^{13}C-labeled glucose and in vivo ^{13}C NMR [103]. Finally, this same bioreactor system was employed to quantify the principal ammonium-assimilating fluxes in *C. glutamicum* ATCC 13032 and a glutamate dehydrogenase mutant with the help of ^{15}N-labeled ammonium and in vivo ^{15}N NMR spectroscopy [50].

In the next sections, the most important results of the reported studies using stable isotope labeling and NMR are presented and discussed.

16.4.2
The Dual Pathways of Lysine Biosynthesis

The dual pathways of lysine biosynthesis in *C. glutamicum* (see (Fig. 16.4) have received considerable interest. At the tetrahydrodipicolinate branchpoint of the lysine biosynthetic pathway of *C. glutamicum*, carbon can be shunted either through the one-step *meso*-diaminopimelate (Dap) dehydrogenase pathway, or through the four-step succinnylase pathway to form the cell wall component and immediate precursor to lysine, *meso*-Dap [58, 93]. Studies using mutants either of the one-step dehydrogenase pathway [58] or of the four-step succinylase pathway [104, 105] indicated that each pathway is in principle dispensable for growth of *C. glutamicum*, as long as the other remains unaffected. The dehydrogenase pathway was shown to be essential for high lysine productivity [58], whereas the succinylase pathway was very recently shown to be dispensable for lysine production [105]. However, an initial analysis of lysine-producing *C. glutamicum* ATCC 21543 using ^{13}C stable isotope labeling had shown that the flux over the dehydrogenase pathway was approximately only half as large as that over the succinylase pathway [95]. Consequently, the correlation of the flux distribution with lysine productivity and culture parameters has been investigated in more detail [62].

16.4.2.1
Correlation with Lysine Production

In the specialized study [62], four *C. glutamicum* strains were compared: ATCC 13032 (wildtype), DG 52–5 (a moderate lysine producer), AS72 (a mutant of strain DG 52–5 missing the one-step pathway), and MH20–22B (a good lysine producer). The cells were incubated in batch cultures with [6-^{13}C]glucose, and harvested at the end of the lysine production phase. Lysine and alanine were isolated from the culture supernatant, or from the cytoplasm in case of the non-excreting wildtype strain. It was checked that the ^{13}C labeling of pyruvate-derived alanine could be used to represent the labeling of pyruvate. The ^{13}C fractional enrichments in C-2 of alanine and C-2 as well as C-6 of lysine were determined by proton NMR spectroscopy of the isolated amino acids. The model using simple carbon atom balances presented above (Eq. 16.12A) was then used to analyze the ^{13}C enrichments. Table 16.2 shows the results. The experiment with strain AS72 served to check the model validity, i.e., that the lysine fractional enrichments of C-2 and C-6 as well as of C-3 and C-5 were mutually identical (within experimental error) upon the sole operation of the succinylase pathway.

The results clearly showed that, contrary to expectations, the lysine productivity did not correlate with the participation of the dehydrogenase pathway.

Table 16.2. Fractional ^{13}C enrichments P2 of alanine C-2, and L2 and L6 of lysine C-2 and C-6, respectively, as well as calculated contribution f of the one-step pathway to the total lysine synthesized with different strains of *C. glutamicum*. Data were taken from [62]. nd=not determined

Strain	Accumulated L-lysine (mmol l^{-1})	P2 (%)	L2 (%)	L6 (%)	f (%)
ATCC 13032	0.0	4.2	10.3	8.0	24 ± 2
AS72	19.7	nd	7.6	7.9	0
DG 52–5	29.2	1.7	5.4	3.6	32 ± 7
MH 20–22B	210^a	2.7	7.2	4.9	33 ± 3

^{a}MH20–22B was incubated with 525 mmol l^{-1} glucose, all other strains with 200 mmol l^{-1} glucose

16.4.2.2
Correlation with Culture Parameters

Sonntag et al. [62] subsequently studied the variation of the relative contributions of the dehydrogenase and succinylase pathways to lysine synthesis during a batch fermentation of *C. glutamicum* MH20–22B and found that the instantaneous flux distribution over the dehydrogenase pathway changed from an initial 72% to a final 0% towards the end of the fermentation, resulting in an overall contribution to the total synthesized lysine of 33% as found before (Table 16.2). Interestingly, in strain DG 52–5 cultivated with glutamate as the sole nitrogen source, the dehydrogenase pathway did not contribute at all to lysine synthesis. Analysis of lysine produced in short-term fermentations with strain MH20–22B upon incubation with different ammonium concentrations (Table 16.3) then unequivocally revealed that the flux over the dehydrogenase pathway was controlled by the ammonium concentration. Since no regulatory phenomena for the involved enzymes had been found [106, 107], it was concluded that this is a purely kinetic effect of the ammonium availability, and that the rather low percental participation of the one-step pathway reflects the low affinity of *meso*-diaminopimelate dehydrogenase for NH$_4^+$ (typical K$_M$ 36 mmol l^{-1}) [108].

Table 16.3. Relative contribution f of the dehydrogenase pathway to lysine synthesis by *C. glutamicum* MH20–22B at different ammonium concentrations in the culture medium. Data were taken from [62]

Ammonium concentration (mmol l^{-1})	Lysine produced (mmol l^{-1})	f (%)
23	1.9	0
38	3.7	6 ± 3
150	5.6	24 ± 4
600	4.3	51 ± 3

16.4.3
Distinct Metabolic Modes: Growth, Glutamate Production, and Lysine Production

Several isotope labeling studies with *C. glutamicum* or related strains had indicated that the contribution of the pentose phosphate pathway is high (44–69%) during lysine synthesis [95, 97], rather low (11–40%) during glutamate synthesis [88–90, 92, 97], and intermediate (45%) during growth [98]. Since the reported values vary considerably, and since the employed modeling approaches were in a need of improvement as seen in the light of the appearing new metabolite- and carbon balancing methods, it was decided to perform a series of studies employing the newest techniques of integrated metabolite balancing/^{13}C isotope labeling [56, 74, 75, 77] to characterize the metabolic flux distributions in *C. glutamicum* during growth, glutamate production, and lysine production [56, 99, 100].

16.4.3.1
Comparing Isogenic Strains in Continuous Cultures

Metabolic flux analysis in continuous culture offers the advantage that the organism is cultivated in a reproducible stationary state, and that the establishment of closed elementary balances such as is necessary for metabolic balancing is a relatively straightforward standard procedure. Moreover, once the cells have been incubated for at least three dilution times with a labeled substrate, the macromolecular cell components can be isolated and hydrolyzed to yield an abundance of amino acids or nucleotides for ^{13}C isotope labeling analysis by NMR. Thus, metabolic flux analysis was performed with two isogenic strains: lysine-producing *C. glutamicum* MH20–22B, *leuCD lysC*Ser381Phe [107] featuring a feed-back resistant aspartate kinase [56], and *C. glutamicum* LE4, *leuCD lysC+* featuring a feedback-sensitive aspartate kinase [100]. The latter strain was grown both in glucose-limited chemostat and in combined glucose- and biotin-limited chemostat cultures. Biotin limitation is one of several possibilities to trigger glutamate excretion in any *C. glutamicum* strain [109].

The carbon balances of the cultures are shown in Table 16.4. During growth, 45% of the carbon is converted into biomass, and 52% into carbon dioxide. Glutamate production results in a significantly increased CO_2 yield, and an almost twofold reduction in biomass yield. Growth-associated lysine production results in a reduction of both CO_2 and biomass yields, but not as severe (Table 16.4).

After harvesting the cells, separate hydrolyses and chemical preparations of the cellular protein and RNA fractions, respectively, were carried out. Amino acids and nucleotides were finally separated by cation exchange chromatography in order to obtain fractions of the single compounds [56, 62, 100] that were sufficiently pure to allow for highly accurate quantitative NMR analysis. A representative set of ^{13}C labeling data obtained with non-excreting strain LE4 is reproduced in Table 16.5. These enrichments were determined by ^{1}H NMR, exploiting the carbon-13 satellite signals in ^{1}H NMR spectra [56, 62, 68, 100].

Flux estimates for the three cultures, obtained by non-linear least squares fitting to the flux and labeling measurement data [56, 74, 75, 77], are summarized in Fig. 16.20. In the flux model, the PEP and pyruvate pools were merged since no difference in labeling between these pools could be detected. The results first of all

Table 16.4. Carbon balances of *C. glutamicum* MH20–22B and *C. glutamicum* LE4 in glucose-limited chemostat cultures (Data taken from [100])

Carbon[a]	Strain/conditions		
	LE4 Non-excretion	LE4 Biotin limitation L-Glutamate-production	MH20–22B L-Lysine-production
Glucose	100	100	100
Biomass	45	26	39
Product	3	18	19
CO_2	52	59	43
Sum	100	103	101
$Y_{X/S}$ ($g_X \times g_S^{-1}$)	0.46	0.27	0.38
$Y_{P/S}$ ($mol_P \times mol_S^{-1}$)	<0.01	0.22	0.19

$Y_{X/S}$ ($Y_{P/S}$) represent biomass (X) and product (P) yield on glucose (S), respectively
[a] Expressed as percentage of the consumed glucose carbon

Table 16.5. Representative data set of measured ^{13}C enrichments in precursor metabolites from non-excreting chemostat-grown *C. glutamicum* LE4. The data were obtained from proteinogenic amino acids and RNA-derived nucleotides as follows: Rib5p (ribose 5-phosphate) from guanosine, Ery4p (erythrose 4-phosphate) from phenylalanine, GAP (glyceraldehyde-3-phosphate and 3-phosphoglycerate) from glycine and serine, Pyr (pyruvate) from alanine and valine, AKG (2-oxoglutarate) from glutamate, and OAA (oxaloacetate) from aspartate and threonine. The values in italics are the corresponding enrichments as predicted by the solution of the mathematical model for flux analysis (Data taken from [100]). nd=not determined

Metabolite	^{13}C enrichment (%) at carbon				
	C-1	C-2	C-3	C-4	C-5
Rib5p	19.8 ± 0.8	2.0 ± 1.0	2.0 ± 1.0	2.0 ± 1.0	19.7 ± 0.6
Rib5p	*19.3*	*1.9*	*3.1*	*2.3*	*19.8*
Ery4p	1.5 ± 1.0	1.5 ± 1.0	1.5 ± 1.0	18.7 ± 2.0	
Ery4p	*2.6*	*3.0*	*2.2*	*18.1*	
GAP	nd	3.6 ± 0.5	28.4 ± 0.5		
GAP	*4.0*	*2.8*	*27.7*		
Pyr	3.9 ± 1.0	4.4 ± 0.5	27.0 ± 0.5		
Pyr	*5.0*	*4.6*	*26.8*		
AKG	14.1 ± 1.0	22.6 ± 0.2	12.7 ± 0.2	26.5 ± 0.2	3.0 ± 2.0
AKG	*14.5*	*22.5*	*12.5*	*26.8*	*4.6*
OAA	9.6 ± 0.5	12.1 ± 0.3	22.5 ± 0.8	14.1 ± 0.7	
OAA	*9.4*	*12.5*	*22.5*	*14.5*	

illustrate that a very detailed flux quantification is possible using the integrated metabolite balancing/^{13}C labeling approach. In addition to the net fluxes over glycolysis, the pentose phosphate pathway, the anaplerotic carboxylation reactions, the glyoxylate pathway, the citric acid cycle, and the reactions withdrawing glucose 6-phosphate, fructose 6-phosphate, ribose 5-phosphate, erythrose 4-phosphate, glyceraldehyde-3-phosphate, 3-phosphoglycerate, PEP, pyruvate, acetyl-CoA, oxaloacetate, and 2-oxoglutarate shown in Fig. 16.20, also bidirectional fluxes over phosphoglucose isomerase, transketolase, transaldolase, phosphoribose isomerase, ribulose-

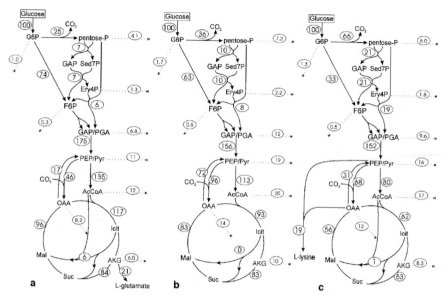

Fig. 16.20a–c. Flux distributions in *Corynebacterium glutamicum* during: **a** glutamate production with strain LE4 under biotin limitation (μ=0.05 h^{-1}); **b** growth with strain LE4 (μ=0.1 h^{-1}); **c** l-lysine production with strain MH20–22B (μ=0.1 h^{-1}) in glucose-fed chemostat cultures, as determined by integrated ^{13}C NMR/metabolite balancing analysis. Data were taken from [56, 100]. Note the presence of simultaneous forward and reverse fluxes in the PEP/pyruvate and oxaloacetate-interconverting pathways, with significantly different values in the three cases. No NADPH constraints were applied in the analyses

5-phosphate epimerase, as well as bidirectional fluxes interconverting glyceraldehyde-3-phosphate and PEP/pyruvate, oxaloacetate/malate, and PEP/pyruvate, and oxaloacetate/malate and fumarate could be quantified from the data. Typical values for the degree of reversibility (defined here as the ratio of the backward and forward fluxes) of various reactions as observed in the experiments [56, 100, 110] are given in Table 16.6. There is obviously considerable variability in these estimates. This reflects, of course, the differences in metabolic states considered, but also the sometimes limited precision of the estimates. Estimates of the reversibility of the anaplerotic reactions are generally precise, indicating that the variability reflects differences during growth, lysine production, and glutamate production. However, especially the bidirectional fluxes in glycolysis show rather large 90% confidence intervals [110], indicating that their estimate is imprecise.

The flux distributions (Fig. 16.20) reveal an impressive flexibility of *C. glutamicum* central metabolic pathways even in isogenic strains. This is illustrated in more detail for the glucose 6-phosphate and pyruvate branchpoints in Fig. 16.21. Most striking, the contribution of the PPP to glucose 6-phosphate conversion ranges from 25% during glutamate production to 66%, i.e., almost three times higher, during lysine production (Fig. 16.21). While the total flux entering PEP/pyruvate varies only by 10% (Fig. 16.20), the relative flux over PEP/pyruvate anaplerotic carboxylating reactions (PEPcarboxylase, pyruvate carboxylase, malic enzyme) ranges more than double from 26% (glutamate production) to 62% (growth), that via pyruvate dehydro-

Table 16.6. Typical degree of reversibility (i.e., 100×backward flux/forward flux) observed for bidirectional reactions in *C. glutamicum* MH20–22B and LE4 [56, 100, 110]

Reaction(s)	Enzyme/metabolites involved	Degree of reversibility (%)
emp1	phosphoglucose isomerase	10–90
ppp2	ribulose-5-phosphate epimerase	>95
ppp3	phosphoribose isomerase	>95
ppp4	transketolase	80–90
ppp5	transketolase	30–50
ppp6	transaldolase	20–70
emp4-emp5	glyceraldehyde-3-phosphatePEP/pyruvate	10–50
ana1, ana2, ana3	PEP/pyruvateoxaloacetate/malate	40–75
tcc5-tcc6	succinate/fumarateoxaloacetate/malate	1–30

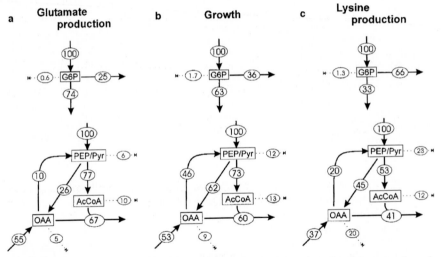

Fig. 16.21a–c. Detailed flux partitionings at the glucose-6-phosphate and PEP/pyruvate branch points for the three flux distributions of Fig. 16.20. Fluxes are normalized to the glucose uptake rate and the net glycolytic flux, respectively, for these two branch points. *Dotted arrows* at the pyruvate and oxaloacetate nodes include lysine biosynthesis, and the *dotted arrow* at the acetyl-coenzymeA node includes the glyoxylate cycle activity. These flux distributions demonstrate the high flexibility of the central metabolism of *Corynebacterium glutamicum*

genase 1.5-fold from 53% (lysine production) to 77% (glutamate production). The back flux in the anaplerotic reactions is increased more than fourfold during growth as compared to glutamate production. The glyoxylate pathway was not or only very weakly active in all cases, in accordance with the fact that the activities of the responsible enzymes are reduced 50-fold during growth on glucose as compared to acetate as the sole carbon source [31, 32], and in accordance with [98].

16.4.3.2
Comparing Different Strains in Batch Cultures

Metabolic flux analysis in *C. glutamicum* during growth and glutamate production (strain ATCC 13032) as well as lysine production (strain MH20–22B) in batch cultures was performed using a slightly modified procedure [99]. Only cytoplasmic glutamate and alanine were available for NMR analysis of ^{13}C fractional enrichments. Therefore, in order to obtain a reliable estimate of the pentose phosphate pathway flux, each experiment was performed in duplicate: first, with [1-^{13}C]glucose, and second, with [6-^{13}C]glucose as a substrate. The fraction of glucose 6-phosphate metabolized over the PPP relative to that over glycolysis was calculated as (1–A/B) [92], where A and B represent the ^{13}C labeling in C-3 of pyruvate obtained from the experiments with [1-^{13}C]glucose and [6-^{13}C]glucose, respectively. Since pyruvate could not be isolated from the cell extracts, the ^{13}C labeling of C-3 of alanine and/or C-4 of glutamate (which are both directly derived from C-3 of pyruvate) were used instead. The flux ratio of the PPP vs glycolysis was then supplied as an additional measurement value to the flux analysis software, in addition to the measured glucose uptake rates, product excretion rates, and precursor fluxes for biomass synthesis [99]. The results obtained are summarized in Fig. 16.22 as flux distributions at the glucose 6-phosphate and PEP/pyruvate branching points. While absolute values differ, several tendencies observed in the experiments with chemostat cultures are also apparent from the batch cultures. PPP activity during lysine production is again increased

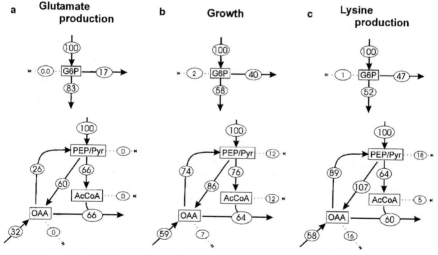

Fig. 16.22a–c. Detailed flux partitionings at the glucose-6-phosphate and PEP/pyruvate branch points for flux distributions determined from batch cultures of: **a** glutamate-producing *Corynebacterium glutamicum* ATCC 13032; **b** exponentially growing *Corynebacterium glutamicum* ATCC 13032; **c** lysine-producing strain MH20–22B. The analyses involved metabolite balancing as well as NMR analysis of cytoplasmic alanine and glutamate upon incubation with [1-^{13}C]glucose and [6-^{13}C]glucose in separate experiments. Data were taken from [99]. Fluxes are normalized to the glucose uptake rate and the net glycolytic flux, respectively, for the two branch points. *Dotted arrows* at the pyruvate and oxaloacetate nodes include lysine biosynthesis

almost threefold as compared to glutamate production, and the anaplerotic carbox-ylation during glutamate production is almost twice as low as during lysine produc-tion. The back reaction from oxaloacetate/malate to PEP/pyruvate is again lowest during glutamate production. However, forward- as well as reverse fluxes connecting PEP/pyruvate and oxaloacetate/malate are now highest during lysine production, whereas this was not the case with the continuous cultures. Also, the flux from suc-cinate to oxaloacetate during glutamate production is almost twice as low as during growth and lysine production, whereas in the chemostat cultures it was higher. While these differences may partly arise due to significant imprecision of the flux estimates for the anaplerotic reactions (not assessed in [99]), they also certainly re-flect metabolic differences between non-carbon-limited batch and glucose-limited chemostat cultures. That such differences must exist can be deduced from the sig-nificantly altered growth yields $Y_{X/S}$ during lysine and glutamate production, which are strongly reduced to 0.15 and 0.0 $g_X \times g_S^{-1}$, respectively. As with the continuous cultures, the glyoxylate pathway was not active in all cases. Comparing all experi-ments, the situation with glutamate production is obviously rather a special one, since activity of the PPP as well as of anaplerotic carboxylation reactions is consis-tently low. The low PPP activity can easily be explained by the low demand for NADPH during glutamate production (see also next section). A possible explanation for the low anaplerotic carboxylation activity would be that the activity of biotin-dependent pyruvate carboxylase is strongly reduced during biotin-limitation-in-duced glutamate production, which would lower the overall anaplerotic rate and re-duce the need to recycle excess oxaloacetate to PEP/pyruvate.

16.4.4
Perturbations of the Redox Metabolism

The flux analyses described in the previous two sections, that are based on indepen-dent ^{13}C labeling data, allow one to perform an independent analysis of the NADP(H) redox balance for each experiment. To this end, the NADPH stoichiome-tries as presented in Table 16.1 can be used. These differ slightly from those used in [56, 100, 101] due to newly incorporated (deoxy)nucleotide and peptidoglycan data. Thus, an estimated 16,385 μmol of NADPH are required in the biosynthesis of 1 g *C. glutamicum* biomass from glucose.

Furthermore, lysine biosynthesis requires effectively 4 mol NADPH per mol lysine, and glutamate 1 mol NADPH per mol glutamate. In *C. glutamicum*, NADPH is regen-erated from NADP only in the oxidative PPP (2 mol NADPH per mol glucose 6-phos-phate converted) and in the isocitrate dehydrogenase reaction (disregarding malic enzyme). Taking this into account, the relative fluxes of oxidation and regeneration of NADPH as given in Table 16.7 are arrived at. From Table 16.7, there seems to be a positive correlation between the total amount of NADPH oxidized and the amount regenerated in the PPP, and a negative correlation between the total amount of NADPH oxidized and the amount regenerated in the isocitrate dehydrogenase reac-tion. This prompted study of the effect of a defined modification of the redox meta-bolism of *C. glutamicum* [101]. In this study, glutamate dehydrogenase was chosen as the target because this single enzyme accounts for approximately 50% of the total NADPH oxidized in the synthesis of biomass. A change of the cofactor dependency of this enzyme therefore is expected to result in a significantly altered NADPH-re-lated redox metabolism. Thus, a glutamate dehydrogenase-deficient mutant of the

Table 16.7. Relative fluxes of oxidation and regeneration of NADPH (mol per mol of glucose consumed) calculated from the flux distributions presented in [56, 99, 100, 110] and the NADPH stoichiometries given in Table 16.1

Condition	NADPH (mol/mol)							
	Oxidized				Regenerated			
	Biomass synthesis	Gluta- mate synthesis	Lysine synthesis	Total	PPP	Isocitrate dehydro- genase	Total	Net regene- rated
Glutamate production (batch)	0	0.66	0	0.66	0.34	1.28	1.62	0.96
Growth (batch)	1.09	0	0	1.09	0.80	1.08	1.88	0.79
Lysine production (batch)	0.44	0	0.96	1.40	0.94	1.06	2.00	0.60
Glutamate production (chemostat)	0.80	0.21	0	1.01	0.50	1.11	1.61	0.60
Growth (chemostat)	1.36	0	0	1.36	0.72	0.93	1.65	0.29
Lysine production (chemostat)	1.09	0	0.75	1.84	1.32	0.61	1.93	0.09

lysine-producing strain MH20–22B was complemented in one case with the homologous NADP-dependent plasmid-encoded glutamate dehydrogenase, and in another case with an heterologous, NAD-dependent glutamate dehydrogenase from *Peptostreptococcus asaccharolyticus*. The two strains are henceforth termed homologous mutant and heterologous mutant, respectively.

Continuous fermentations with [1-^{13}C]-labeled glucose as the sole carbon source were carried out, and integrated metabolite balancing/^{13}C stable isotope labeling flux analyses were performed [101]. The resulting flux distributions at the glucose 6-phosphate and PEP/pyruvate branching points are given in Fig. 16.23. Obviously, the defined alteration of the redox metabolism leads to remarkable changes in the flux distribution. The effect is most striking for the PPP, of which the activity is reduced threefold in the heterologous mutant. The flux distribution at the PEP/pyruvate and oxaloacetate/malate branching points is less affected, showing a 36% increase in citric acid cycle flux and also a significant increase in oxaloacetate/malate decarboxylating flux in the heterologous mutant [101]. With the fluxes thus determined, balances of NADPH-oxidizing and -regenerating fluxes can again be established, on the assumption that the heterologous glutamate dehydrogenase is strictly NAD-dependent (in fact, enzyme measurements indicated a 10% NADP-dependent side activity). In setting up the balances, special care must be taken to correct the NADP stoichiometries (Table 16.1) of the various biosynthetic reactions for the altered glutamate dehydrogenase cofactor dependency. For example, the 1-step and 4-step pathways in lysine biosynthesis (Fig. 16.4) will now have different NADPH stoichiometries: while synthesis of lysine from pyruvate and oxaloacetate in the homologous mutant requires 4 NADPH for either pathway, synthesis via the dehydrogenase (1-step) and succinylase (4-step) pathway requires 3 NADPH and 2 NADPH, respectively, in the heterologous mutant. Since the flux distribution over these pathways was also determined in the integrated analysis, a rigorous book-keeping of the NADPH-oxidizing and -regenerating fluxes can be performed. The results are shown in Table 16.8. A very strong and flexible response of *C. glutamicum* central metabolism is the answer to the twofold decrease in requirement for NADPH regen-

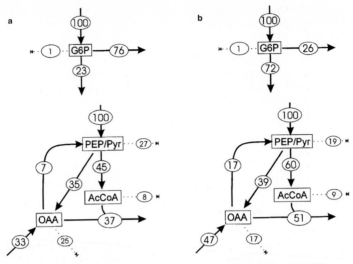

Fig. 16.23a,b. Detailed flux partitionings at the glucose-6-phosphate and PEP/pyruvate branch points for flux distributions determined from chemostat cultures of glutamate dehydrogenase mutants of lysine-producing *Corynebacterium glutamicum* MH20–22B equipped with plasmid-encoded: a homologous (i.e., NADP-dependent); b heterologous (i.e., NAD-dependent) glutamate dehydrogenases. The analyses involved integrated ^{13}C NMR/metabolite procedures. Data were taken from [101]. Fluxes are normalized to the glucose uptake rate and the net glycolytic flux, respectively, for the two branch points. *Dotted arrows* at the pyruvate and oxaloacetate nodes include lysine biosynthesis. The flux partitioning at the PEP/pyruvate node is much less affected than that at the glucose-6-phosphate node

Table 16.8. Relative fluxes of oxidation and regeneration of NADPH (mol per mol of glucose consumed) during lysine production with glutamate-dehydrogenase mutants of *C. glutamicum* MH20–22B, calculated from the flux distributions presented in [101] and the NADPH stoichiometries given in Table 16.1 after adaptation for the NAD-dependent glutamate dehydrogenase in the heterologous mutant

Strain	NADPH (mol/mol) Oxidized				Regenerated			
	Biomass synthesis	Gluta-mate synthesis	Lysine synthesis	Total	PPP	Isocitrate dehydro-genase	Total	Net re-generated
Heterologous mutant	0.55	0	0.44	0.99	0.53	0.86	1.39	0.40
Homologous mutant	0.83	0	1.20	2.03	1.53	0.58	2.11	0.07

eration. Very interestingly, the PPP adapts quantitatively to the change. However, the excess NADPH regenerated in the isocitrate dehydrogenase reaction results in a net over-regeneration of NADPH in the heterologous mutant, whereas (within experimental error) no excess regeneration of NADPH is present in the homologous mutant. The last column of Tables 16.7 and 16.8 show data regarding the NADPH over-regeneration for all cases analyzed. A clear tendency in the rate of over-regeneration of NADPH can be seen, namely that it shows a negative correlation with the rate of

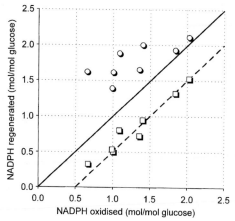

Fig. 16.24. Calculated fluxes (mol per mol of glucose consumed) of regenerated NADPH by the oxidative pentose phosphate pathway plus isocitrate dehydrogenase (*circles*) and by the oxidative pentose phosphate pathway alone (*squares*), as a function of the calculated total flux of NADPH oxidation (mol per mol of glucose consumed) for flux distributions presented in Tables 16.7 and 16.8. *The continuous line* represents points where oxidizing and regenerating fluxes match exactly. *The broken line* suggests a trend in the data points with *square symbols*. An excess of NADPH regeneration is apparent that seems to increase with diminishing NADPH oxidation

NADPH oxidation, i.e., the higher the rate of oxidation, the less NADPH is over-regenerated. This is more evident from the graphical representation of selected data from Tables 16.7 and 16.8 in Fig. 16.24. Apparently, the PPP consistently regenerates NADPH to a rate 0.5 mol/mol less than the rate of oxidation. The additional NADPH regeneration by the isocitrate dehydrogenase reaction matches the required 0.5 mol/mol at high NADPH oxidation rates (i.e., 1.8–2.0 mol/mol), resulting in a good match of total oxidation and regeneration rates, but it is increasingly in excess of this amount the less NADPH is oxidized. Several speculative explanations are possible. One is that at lower oxidation rates, additional oxidative processes are active that are not accounted for in the calculations. These processes may serve as a kind of reducing power sink, which the organism possibly needs in order to maintain a basic level of NADPH-turnover. While the cycle pyruvate-malate-oxaloacetate-pyruvate involving malic enzyme might theoretically act as a transhydrogenase system oxidizing NADPH, a functioning of malic enzyme in the anaplerotic sense in *C. glutamicum* to date could not be demonstrated. Growth-related processes are also unlikely candidates, since no correlation with biomass synthesis is apparent from the combined Tables 16.7 and 16.8. The only positively identified candidate to date is a recently discovered superoxide-generating NADPH oxidase system in the respiratory chain of *C. glutamicum* [111]. Another theoretical possibility would be that isocitrate dehydrogenase, contrary to reports, is not strictly NADP-dependent, or that an NAD-dependent isoenzyme exists in *C. glutamicum*. However, no experimental data supporting either one of these hypotheses is currently available.

16.4.5
The Ammonium-Assimilating Fluxes

A glutamate dehydrogenase mutant of *C. glutamicum* ATCC 13032 did not show any phenotype [46]. This seemed to be incompatible with knowledge of the regulation of the principal enzymes for ammonium assimilation, i.e., glutamate dehydrogenase (GDH), glutamine synthetase (GS), and glutamine 2-oxoglutarate-aminotransferase (GOGAT), which indicated that the alternative GS/GOGAT pathway was not expressed during ammonium abundance. Since it was unclear whether alternative routes are operative in the GDH-mutant or the regulation of GS/GOGAT is altered, the principal ammonium-assimilatory fluxes in *C. glutamicum* were quantitated using in vivo ^{15}N NMR in a dynamic labeling study [50]. *C. glutamicum* ATCC 13032 (wildtype) and its GDH-mutant were cultivated and analyzed with on-line NMR using a specially developed NMR membrane-cyclone bioreactor system [103, 112] which is suited for the continuous aerobic cultivation of *C. glutamicum* at cell densities above 50 g dry wt/l. At metabolic steady state, the culture ammonium concentration was increased within 2 min from 40 mmol l^{-1} to 150 mmol l^{-1} by the addition of ^{15}N-labeled ammonium sulfate. The time courses of the ensuing incorporation of the label in the N_α nitrogen of glutamate and the N_α and N_δ nitrogens of glutamine were monitored with a time resolution of 8–15 min. After proper calibration of the NMR data, the parameters of a nitrogen flux model were then fitted to the data. The differential equation model included the following reactions: GDH (reversible), GS, GOGAT (reversible), glutamate aminotransferases, glutamine amidotransferases, and incorporation of glutamate and glutamine in protein and cytoplasmic pools. Tests on the GDH mutant with a GOGAT inhibitor showed that no other primary ammonium-assimilating pathways are operative in *C. glutamicum*. The pools of glutamate and glutamine were found to be strongly altered in the GDH mutant:

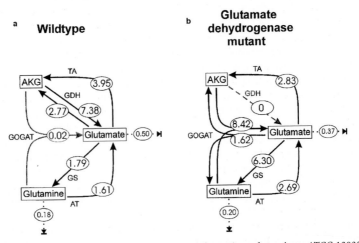

Fig. 16.25a,b. Nitrogen flux distributions in: **a** *Corynebacterium glutamicum* ATCC 13032; **b** its glutamate dehydrogenase mutant, determined in a dynamic labelling study [50] using in vivo ^{15}N NMR. TA, transaminases; GDH, glutamate dehydrogenase; GOGAT, glutamine:2-oxoglutarate aminotransferase; GS, glutamine synthetase; AT, glutamine amidotransferases

82 μmol/g dry wt and 93 μmol/g dry wt, respectively, as compared to 250 μmol/g dry wt and 49 μmol/g dry wt in the wildtype. The flux distributions resulting from the analyses [50] are represented in Fig. 16.25. They indeed showed a marked difference between the two strains; whereas in the wildtype, 72% of the ammonium was assimilated via GDH and 28% via GS, 100% were assimilated via GS in the GDH mutant. In the wildtype, glutamate resulted from operation of GDH (74%) and glutamine transaminases (26%); GOGAT was inactive. In the GDH mutant, the task of GDH was completely taken over by GOGAT, via which 72% of the glutamate was produced. The validity of the analysis was confirmed by the fact that it predicted zero activity of GDH for the mutant. The absence of GOGAT flux predicted for the wildtype was completely in accordance with enzyme determinations. Thus, nitrogen fluxes can also be quantified independently in great detail by ^{15}N labeling and NMR.

16.4.6
Summary

Flux analysis studies based on isotopic labeling and NMR detection have provided a much refined insight in the metabolic processes as they occur in vivo in a number of different strains of *C. glutamicum* and under a variety of conditions. These studies have given undoubted independent proof that the pentose phosphate pathway is very important for an adequate supply of reducing equivalents during amino acid production, and that this pathway reacts extremely flexibly to the actual demand for NADPH in any given situation [56, 99–101]. The experiments also gave independent proof that the glyoxylate pathway in *C. glutamicum* is not or only very weakly active when glucose is the sole carbon source [56, 99, 100]. Furthermore, an important insight in the regulation of the flux distribution at the tetrahydrodipicolinate branchpoint of the lysine biosynthetic pathway of *C. glutamicum* was provided [62], showing that ammonium is a key factor in the activation of the one-step dehydrogenase pathway which is essential for high lysine productivity. Several reversible reaction steps were identified from the ^{13}C labeling data. Typical degrees of reversibility for the transaldolase, transketolase, phosphoribose isomerase, ribulose-5-phosphate epimerase, phosphoglucose isomerase, as well as for the reactions interconverting glyceraldehyde-3-phosphate and PEP/pyruvate, and oxaloacetate/malate and fumarate, could be calculated from the data.

The integrated metabolite balancing/isotopic labeling analysis revealed a first clue to the elucidation of the mechanisms that contribute to the rigidity of the flux distribution observed around the PEP/pyruvate/oxaloacetate branching points in *C. glutamicum* metabolism: the presence of a very significant reverse flux of oxaloacetate/malate to PEP/pyruvate that varies considerably depending on conditions [56, 100, 101]. The available data suggest an important role of these backreactions in the regulation of the relative levels of the pools of PEP, pyruvate, and oxaloacetate, which are chief amino acid precursors. The ^{13}C labeling approach has also been highly instrumental in the identification of pyruvate carboxylase as important anaplerotic enzyme in *C. glutamicum* [70, 102].

The in vivo ^{15}N NMR dynamic labeling flux analysis study led to the discovery that glutamine, together with glutamate, plays a key role in the ammonium assimilatory metabolism of *C. glutamicum*.

As a whole, the results from the studies based on isotopic labeling and NMR demonstrate that the central metabolism of *C. glutamicum* is extremely flexible in adapting to situations with strongly different precursor demands, and that both glutamine and glutamate are key metabolites in *C. glutamicum* nitrogen metabolism. Furthermore, the complex set of PEP/pyruvate carboxylating and oxaloacetate-decarboxylating reactions is identified as a promising object for further analysis.

16.5
Concluding Remarks

While extensive genetic knowledge about the central metabolic and amino acid biosynthetic pathways in *C. glutamicum* has accumulated already over many years, metabolic flux analysis studies have now begun to yield deeper insights into the interplay of amino acid biosynthesis and the central metabolism which generates the precursors, the energy, and the reducing power necessary to produce these compounds. Several targets for genetic engineering have been identified; some of these have already been probed while others will be studied in the near future. Metabolic flux analysis has thus developed into a very powerful measurement tool. The strategies employing integrated metabolite balancing and stable isotope-labeling combined with NMR-detection described in this contribution have significantly contributed to this advancement. Present developments which aim to maximize the information content of the stable isotope labeling experiment by using ^{13}C-isotopomer analysis, and which seek to integrate the flux information with cytoplasmic pool measurements, are likely to strengthen further the role of metabolic flux analysis within the rapidly evolving field of metabolic engineering in the near future.

Acknowledgements. The author wishes to thank W. Wiechert, A. Marx, L. Eggeling, M. Tesch, K. Striegel, K. Sonntag, V. Wendisch, and A. Hartbrich for their excellent contributions to the NMR and flux analysis studies covered in this contribution, and Prof. H. Sahm for continuous support and stimulation.

References

1. Eggeling L, Sahm H, de Graaf AA (1996) Quantifying and directing metabolite flux: application to amino acid overproduction. In: Scheper T (ed) Advances in biochemical engineering/biotechnology, vol 54: Metabolic engineering. Springer, Berlin Heidelberg New York, p 2
2. Katsumata R, Ikeda M (1997) Process for producing l-tryptophan, l-tyrosine or l-phenylalanine. US Pat 5,605,818
3. Katsumata R, Hashimoto S (1996) Process for producing l-valine. US Pat 5,521,074
4. Mizukami T, Katsumata R, Oka T (1990) Process for producing histidine. US Pat 4,927,758
5. Malumbres M, Martin JF (1996) FEMS Microbiol Lett 143:103
6. Reinscheid DJ, Kronemeyer W, Eggeling L, Eikmanns BJ, Sahm H (1994) Appl Environ Microbiol 60:126
7. Eggeling L, Morbach S, Sahm H (1997) J Biotechnol 56:167
8. Kimura E, Kawahara Y, Nakamatsu T (1997) Tanpakushitsu Kakusan Koso (Japan) 42:2633
9. Eggeling L (1994) Amino Acids 6:261
10. Sahm H, Eggeling L, Eikmanns BJ, Krämer R (1996) Ann N Y Acad Sci 782:25
11. Eikmanns BJ, Eggeling L, Sahm H (1993) Antonie van Leeuwenhoek 64:145
12. Jetten MSM, Sinskey AJ (1995) Crit Rev Biotechnol 15:733
13. Sahm H, Eggeling L, Eikmanns BJ, Krämer R (1995) FEMS Microbiol Rev 16:243
14. Jetten MSM, Follettie MT, Sinskey AJ (1994) Ann N Y Acad Sci 721:12

15. Krämer R (1994) FEMS Microbiol Rev 13:75
16. Krämer R (1996) Analysis and modeling of substrate uptake and product release by prokaryotic and eukaryotic cells. In: Scheper T (ed) Advances in biochemical engineering/biotechnology, vol 54: Metabolic engineering. Springer, Berlin Heidelberg New York, p 31
17. Palmieri L, Berns D, Krämer R, Eikmanns BJ (1996) Arch Microbiol 165:48
18. Hermann T, Krämer R (1996) Appl Environ Microbiol 62:3238
19. Zittrich S, Krämer R (1994) J Bacteriol 176:6892
20. Eggeling L, Krämer R, Vrljic M, Kronemeyer W, Sahm H (1996) Ann N Y Acad Sci 782:191
21. Vrljic M, Sahm H, Eggeling L (1996) Mol Microbiol 22:815
22. Schwinde JW, Thum-Schmitz N, Eikmanns BJ, Sahm H (1993) J Bacteriol 175:3905
23. von der Osten CH, Barbas CF, Wong CH, Sinskey AJ (1989) Mol Microbiol 3:1625
24. Eikmanns BJ (1992) J Bacteriol 174:6076
25. Gubler M, Jetten MSM, Lee SH, Sinskey AJ (1994) Appl Environ Microbiol 60:2494
26. Ikeda M, Okamoto K, Katsumata R (1999) Appl Microbiol Biotechnol 51:201
27. Moritz B, Striegel K, de Graaf AA, Sahm H (2000) Eur J Biochem (submitted)
28. Eikmanns BJ, Thum-Schmitz N, Eggeling L, Ludtke KU, Sahm H (1994) Microbiol 140:1817
29. Eikmanns BJ, Rittmann D, Sahm H (1995) J Bacteriol 177:774
30. Usuda Y, Tujimoto N, Abe C, Asakura Y, Kimura E (1996) Microbiol 142:3347
31. Reinscheid DJ, Eikmanns BJ, Sahm H (1994) J Bacteriol 176:3474
32. Reinscheid DJ, Eikmanns BJ, Sahm H (1994) Microbiol 140:3099
33. Reinscheid DJ, Schnicke S, Rittmann D, Zahnow U, Sahm H, Eikmanns BJ (1999) Microbiol 145:503
34. Kinoshita S (1985) Glutamic acid bacteria. In: Demain A, Solomon N (eds) Biology of industrial microorganisms. Benjamin/Cummings Publishing, London, p 115
35 Peters-Wendisch PG, Wendisch VF, Paul S, Eikmanns BJ, Sahm H (1997) Microbiol 143:1095
36. Dominguez H, Nezondet C, Lindley ND, Cocaign M (1993) Biotechnol Lett 15:449
37. Jetten MSM, Sinskey AJ (1993) FEMS Microbiol Lett 111:183
38. Jetten MSM, Sinskey AJ (1995 Antonie Van Leeuwenhoek 67:221
39. Jetten MSM, Pitoc GA, Follettie MT, Sinskey AJ (1994) Appl Microbiol Biotechnol 41:47
40. O'Regan M, Thierbach G, Bachmann B, Villeval D, Lepage P, Viret JF, Lemoine Y (1989) Gene 77:237
41. Eikmanns BJ, Follettie MT, Griot MU, Sinskey AJ (1989) Mol Gen Genet 218:330
42. Peters-Wendisch PG, Kreutzer C, Kalinowski J, Patek M, Sahm H, Eikmanns BJ (1998) Microbiol 144:915
43. Koffas MA, Ramamoorthi R, Pine WA, Sinskey AJ, Stephanopoulos G (1998) Appl Microbiol Biotechnol 50:346
44 Siewe RM, Weil B, Burkovski A, Eikmanns BJ, Eikmanns M, Krämer R (1996) J Biol Chem 271:5398
45. Siewe RM, Weil B, Burkovski A, Eggeling L, Krämer R, Jahns T (1998) Arch Microbiol 169:411
46. Börmann-El Kholy ER, Eikmanns BJ, Gutmann M, Sahm H (1993) Appl Environ Microbiol 59:2329
47. Bormann ER, Eikmanns BJ, Sahm H (1992) Mol Microbiol 6:317
48. Jakoby M, Tesch M, Sahm H, Krämer R, Burkovski A (1997) FEMS Microbiol Lett 154:81
49. Tesch M, Eikmanns BJ, deGraaf AA, Sahm H (1998) Biotechnol Lett 20:953
50. Tesch M, de Graaf AA, Sahm H (1999) Appl Environ Microbiol 65:1099
51. Eggeling L, Morbach S, Sahm H (1996) Appl Environ Microbiol 62:4345
52. Stephanopoulos G; Vallino JJ (1991) Science 252:1675
53. De Hollander JA (1994) Appl Microbiol Biotechnol 42:508
54. Vallino JJ, Stephanopoulos G (1990) In: Sikdar SK, Bier M, Todd P (eds) Frontiers in bioprocessing. CRC Press, Boca Raton, p 205
55. Cocaign-Bousquet M, Lindley ND (1995) Enz Micr Technol 17:260

56. Marx A, de Graaf AA, Wiechert W, Eggeling L, Sahm H (1996) Biotechnol Bioeng 49:111
57. Neidhardt FC, Ingraham JL, Schaechter M (1990) Physiology of the bacterial cell: a molecular approach. Sinauer Associates, Sunderland, Massachusetts
58. Schrumpf B, Schwarzer A, Kalinowski J, Pühler A, Eggeling L, Sahm H (1991) J Bacteriol 173:4510
59. Vallino JJ, Stephanopoulos G (1993) Biotechnol Bioeng 41:633
60. Broer S, Krämer R (1991) Eur J Biochem 202:136
61. Vallino JJ, Stephanopoulos G (1994) Biotechnol Progr 10:320
62. Sonntag K, Eggeling L, de Graaf AA, Sahm H (1993) Eur J Biochem 213:1325
63. Kanamori K, Weiss RH, Roberts JD (1987) J Biol Chem 262:11038
64. Kanamori K, Weiss RL, Roberts JD (1987) J Bacteriol 169:4692
65. Kanamori K, Weiss RL, Roberts JD (1988) J Biol Chem 263:2817
66. Szyperski T (1995) Eur J Biochem 232:433
67. Szyperski T, Bailey JE, Wüthrich K (1996) Trends Biotechnol 14:453
68. Wendisch VF, de Graaf AA, Sahm H (1997) Anal Biochem 245:196
69. de Graaf AA, Mahle M, Möllney M, Wiechert W, Stahmann P, Sahm H (2000) J Biotechnol 77:25
70. Park SM, Shaw-Reid, Sinskey AJ, Stephanopoulos G (1997) Appl Microbiol Biotechnol 47:430
71. Inbar L, Lapidot A (1987) Eur J Biochem 162:621
72. van Gulik WM, Heijnen JJ (1995) Biotechnol Bioeng 48:681
73. Zupke C, Stephanopoulos G (1994) Biotechnol Progr 10:489
74. Wiechert W, de Graaf AA (1997) Biotechnol Bioeng 55:101
75. Wiechert W, Siefke C, de Graaf AA, Marx A (1997) Biotechnol Bioeng 55:118
76. Szyperski T (1998) Q Rev Biophys 31:41
77. Wiechert W, de Graaf AA (1996) In vivo stationary flux analysis by ^{13}C labelling experiments. In: Scheper T (ed) Advances in biochemical engineering/biotechnology, vol 54: Metabolic engineering. Springer, Berlin Heidelberg New York, p 111
78. Klapa MI, Park SM, Sinskey AJ, Stephanopoulos G (1999) Biotechnol Bioeng 62:375
79. Park SM, Klapa MI, Sinskey AJ, Stephanopoulos G (1999) Biotechnol Bioeng 62:392
80. Schmidt K, Carlsen M, Nielsen J (1997) Biotechnol Bioeng 55:831
81. Zupke C, Tompkins R, Yarmush D, Yarmush M (1997) Anal Biochem 247:287
82. Wiechert W, Möllney M, Isermann N, Wurzel M, de Graaf AA (1999) Biotechnol Bioeng 66:69
83. Möllney M, Wiechert W, Kownatzki D, de Graaf AA (1999) Biotechnol Bioeng 66:86
84. Vallino JJ, Stephanopoulos G (1994) Biotechnol Prog 10:327
85. Lindley ND, Cocaign-Bousquet M, Guyonvarch A (1996) Appl Environ Microbiol 62:429
86. Sugimoto SI, Shiio I (1987) Agric Biol Chem 51:101
87. Kinoshita S, Udaka S, Shimono M (1957) J Gen Appl Microbiol (Tokyo) 3:193
88. Shiio I, Otsuka S-I, Tsunoda T (1960) J Biochem 47:414
89. Oishi K, Aida H (1965) Agric Biol Chem 29:83
90. Otsuka S-I, Miyajima R, Shiio I (1965) J Gen Appl Microbiol 11:285
91. Haran N, Kahana ZE, Lapidot A (1983) J Biol Chem 258:12,929
92. Walker TE, Han CE, Kollman VH, London RE, Matwiyoff NA (1982) J Biol Chem 257:1189
93. Ishino S, Yamaguchi K, Shirahata K, Araki K (1984) Agric Biol Chem 48:2557
94. Ishino S, Kuga T, Yamaguchi K, Shirahata K, Araki K (1986) Agric Biol Chem 50:307
95. Yamaguchi K, Ishino S, Araki K, Shirahata K (1986) Agric Biol Chem 50:2453
96. Inbar L, Lapidot A (1987) Eur J Biochem 162:621
97. Ishino S, Shimomura J, Yamaguchi K, Shirahata K, Araki K (1991) J Gen Appl Microbiol 37:157
98. Rollin C, Morgant V, Guyonvarch A, Guerquin-Kern J-L (1995) Eur J Biochem 227:488
99. Sonntag K, Schwinde J, de Graaf AA, Marx A, Eikmanns BJ, Wiechert W, Sahm H (1995) Appl Microbiol Biotechnol 44:489
100. Marx A, Striegel K, de Graaf AA, Sahm H, Eggeling L (1997) Biotechnol Bioeng 56:168
101. Marx A, Eikmanns BJ, Sahm H, de Graaf AA, Eggeling L (1999) Metab Eng 1:35

102. Peters-Wendisch PG, Wendisch VF, de Graaf AA, Eikmanns BJ, Sahm H (1996) Arch Microbiol 165:387
103. Hartbrich A, Schmitz G, Weuster-Botz D, de Craaf AA, Wandrey C (1996) Biotechnol Bioeng 51:624
104. Wehrmann A, Phillipp B, Sahm H, Eggeling L (1998) J Bacteriol 180:3159
105. Shaw-Reid CA, McCormick MM, Sinskey AJ, Stephanopoulos G (1999) Appl Microbiol Biotechnol 51:325
106. Cremer J, Treptow C, Eggeling L, Sahm H (1988) J Gen Microbiol 134:3221
107. Schrumpf B, Eggeling L, Sahm H (1992) Appl Microbiol Biotechnol 37:566
108. Misono M, Soda K (1980) J Biol Chem 255:10,599
109. Shiio I, Otsuka S-I, Takahashi M (1962) J Biochem (Tokyo) 52:108
110. Marx A (1997) PhD Thesis, University of Bonn, Germany
111. Matsushita K, Yamamoto T, Toyama H, Adachi O (1998) Biosc Biotechnol Biochem 62:1968
112. Weuster-Botz D, de Graaf AA (1996) Reaction engineering methods to study intracellular metabolite concentrations. In: Scheper T (ed) Advances in biochemical engineering/biotechnology, vol 54: Metabolic engineering. Springer, Berlin Heidelberg New York, p.75
113. Gottschalk G (1986) Bacterial metabolism, 2nd edn. Springer, Berlin Heidelberg New York, chap 3
114. Molenaar D, van der Rest ME, Petrovic S (1998) Eur J Biochem 254:395

17 Analysis of Metabolic Fluxes in Mammalian Cells

Neil S. Forbes, Douglas S. Clark, Harvey W. Blanch

17.1
Applications of Metabolic Flux Analysis in Mammalian Cells

The analysis of metabolic fluxes provides a unique perspective on the functioning of mammalian cells. Metabolic flux analysis is the use of a mathematical description of central metabolism to quantify the flow of carbon through the cell. Compared to other biochemical techniques, which reduce cellular function to specific enzymes and pathways, metabolic flux analysis views the cell as a whole. Concurrently measuring the fluxes through multiple biochemical pathways can identify interactions in the regulatory machinery of the cell. When pathways are observed independently, their interactions must be inferred. However, associations between pathways can only be seen when their respective fluxes are compared. Understanding the interactions between pathways enables the prediction of cellular responses to changes in nutrient composition, the response to the genetic events involved in transformation, and the response to external stimuli and signals.

This chapter reviews the modeling methods used to quantify metabolic fluxes in mammalian cells. We describe the different systems and cell types to which metabolic flux analysis has been applied. The examples considered all employ a biochemical pathway model of the whole cell to determine metabolic fluxes. We describe the common assumptions used to create viable models and the corresponding mathematical descriptions to deconvolute typical experimental data into fluxes. Model building and mathematical manipulation is used because many fluxes can be quantified without direct measurements. Compared to measuring each flux independently, the use of a pathway model requires much less experimental effort. Studies of carbon fluxes through single pathways, independent of central metabolism, will not be considered here.

Metabolic flux analysis is not new. Because of numerous technological advances, many previously unanswered questions and new issues can now be addressed. Two experimental tools, ^{13}C- and ^{14}C-tracer studies, have traditionally been employed to study metabolic pathways. Most of the biochemical pathways of central metabolism were first elucidated using ^{14}C-labeling in the 1920s and 1930s [1]. With the advent of more powerful superconducting magnets in the 1970s, nuclear magnetic resonance (NMR) spectroscopy using ^{13}C-enriched substrates provided a new route to study biological systems [2]. For example, in 1983 Chance et al. [3] incorporated isotopomer[1] data derived from acid extraction of perfused rat hearts into a relatively

---

1 An isotopomer, as defined later in the review, is a metabolite with a specific pattern of labeled carbons. There are 2^n isotopomers in a molecule with n carbon atoms.

simple model to quantify the flux of the TCA cycle. At that time, even the modest complexity of their model required the computing power of a Cray mainframe computer for its solution. In contrast, Chatham et al. [4] reused Chance's model in 1994, with added pathways, and solved the system of equations on a laptop 486 personal computer. The striking development of more powerful computers in the last decade has allowed a re-examination of whole-cell flux analysis. The experimental techniques have not changed very much, but now more information can be derived from isotope distributions (and other measurements) using more complex models of metabolic pathways.

To date, there have been three major applications for flux analysis of whole cells: optimization of therapeutic protein production, understanding metabolic regulation in cancerous cells, and understanding metabolic regulation in normal cells. These applications will now be discussed in turn.

17.1.1
Optimization of Protein Production

Large-scale mammalian cell culture is used almost exclusively to produce therapeutic proteins. Typically, a desired protein comprises only about 10–20% of the total cellular protein [5]. Clearer understanding of the interactions between substrate consumption, product formation, and energy management could enhance productivity and reduce the costs of culturing mammalian cells. Based on this information, growth media could be rationally designed, and bioreactor design and control could be improved [6]. A more complete picture of metabolism would include limiting pathways, inter-relationships between pathways, and environmental effects on nutrient requirements [5]. As an example of the power of flux analysis to generate production-increasing strategies, Fitzpatrick et al. [7] measured metabolic fluxes in hybridomas and examined how the interaction between the consumption of glucose and glutamine affected energy production. The results allowed these authors to hypothesize that limited ATP formation might limit antibody production. With incomplete glucose oxidation much of the energy contained in the glucose is lost. These authors proposed that to increase ATP production and produce more protein would involve inhibiting lactate formation and inducing pyruvate dehydrogenase (PDH).

These applications mirror ongoing efforts to optimize bacterial fermentations. An even greater challenge than identifying the optimal environment for cell growth is the genetic manipulation of a cell line to shift its metabolism towards higher production rates. Efforts with bacteria have shown this to be a difficult task, because of complex inter-regulation of the central metabolic enzymes [8]. Increasing the flows through a given metabolic pathway often does not result in an overall increase in production of a desired end product, due to unexpected limitations and controls elsewhere in the cell. Similar efforts are compounded in mammalian cells by the difficulty of introducing stable genetic alterations.

17.1.2
Metabolic Regulation in Transformed Cells

Since primary metabolism provides the energy and cellular components necessary for proliferation, quantifying metabolic fluxes may shed light on the altered mechanisms of growth in cancerous cells. It has been known for some time that transformed cells have high rates of aerobic glycolysis [9], a characteristic typical of rapidly pro-liferating cells. Warburg postulated in 1956 that a cure for cancer would be found if the causes of the high aerobic glycolysis rates were discovered [10]. Since then, many of the genetic mutations leading to the onset of cancer have been uncovered, invali-dating much of Warburg's conjecture. However, the causes of aerobic glycolysis in proliferating cells have yet to be determined even though it remains a defining char-acteristic of the transformed state. High aerobic glycolytic rates have been observed in all of the cancerous cell studies included in this review [5–7, 11–15]. One possible explanation for these high glycolytic rates is an altered composition of glycolytic isozymes. The alteration of isozyme composition is so closely associated with trans-formation that the isozyme composition is used as a clinical marker of progression to the transformed state [16]. Changes in isozyme expression imply that enhanced glycolytic rates involve genetic modifications. Another explanation for the high gly-colytic flux is that it is induced by changes in other metabolic pathways. Portais et al. [14, 17] and Merle et al. [13] studied how activity of the TCA cycle relates to glyco-lytic rates by using NMR to quantify the metabolic fate of 1-^{13}C glucose in C6 glioma cells. These authors used the C6 cells because they have "representative tumoral characteristics," implying that fluxes measured in these cells are useful as a model of metabolic regulation in all transformed cells. The usefulness of flux quantification to the study of cancer (here specifically breast cancer) was described by Singer et al. [15]:

> Although considerable effort has been devoted to determining the genetic events in breast cell carcinogenesis, the relationship of the changes to the biochemical phenotype of the cancer cell is essentially unknown. A more complete understand-ing of the biochemical phenotype of normal cells versus cancer cells may allow for a more focused study of important genetic changes leading to these biochemical changes.

Unfortunately, achieving such a goal relies on an accurate description of normally functioning cells, and this is difficult to obtain.

17.1.3
Metabolic Regulation in Non-Transformed Cells

Unlike cancerous cell lines, which are immortal, normal mammalian cells only sur-vive a few passages in culture, making them difficult to study. Conditions in experi-ments on immortalized or transformed cell lines can be carefully controlled and reproduced with cells derived from a single source. On the other hand, normal cells must be studied as short-lived primary cultures, which are more problematic to con-trol and reproduce. Because it is difficult to duplicate physiological conditions accu-rately ex vivo, normal cells are often studied as intact tissues or in their natural environment within animal subjects.

Various rationales were proposed by the authors reviewed here for quantifying metabolic fluxes in non-transformed mammalian cells. Hyder et al. [18] quantified glycolytic and TCA cycle flux rates in rat brains to test the hypothesis that "addi-tional energy required for brain activation is provided through non-oxidative glyco-lysis." Malloy et al. [19] determined the fluxes of the TCA cycle and the activity of the anaplerotic enzyme, pyruvate carboxylase. Anaplerotic activity is relevant because it increases the concentrations of TCA cycle metabolites, and changes in these concentrations are associated with many "clinically relevant states," e.g., diabetes and ischemia. Martin et al. [20] characterized the metabolic consequences of consumed glucose and glutamine in rat cerebellum astrocytes and granule cells. These authors sought to understand how the proposed two compartmentalized TCA cycles in the brain are dependent on intercellular regulations. Chance et al. [3], Chatham et al. [4], and Schrader et al. [21] only quantified the metabolic flux though glycolysis and the TCA cycle in rat hearts [3, 4] and human erythrocytes [21] without defining a purpose beyond metabolic characterization.

17.2
Experimental Techniques

This section describes the different experimental measurements from which fluxes are calculated. They include the direct measurement of external metabolite concentrations, the determination of ^{13}C and ^{14}C distribution patterns by NMR spectroscopy and scintillation counting, respectively, and measurement of enzyme activity following extraction. Most researchers have employed more than one technique, since single techniques alone are not generally sufficient to quantify cellular fluxes uniquely [22]. The data generated by different techniques are translated into metabolic fluxes by the mathematical schemes and models described in Sects 17.3 and 17.4, respectively.

The different types of cultures and environments in which metabolic fluxes in mammalian cells have been measured are outlined in Table 17.1.

The mammalian cell lines used commercially to produce proteins include baby hamster kidney (BHK) cells, Chinese hamster ovary (CHO) cells, and murine hybridomas. All of these cell lines have been adapted to grow in suspension culture and industrial practice is to employ batch systems, although continuous and perfused systems are also used.

Tumor cell lines used for cancer research are anchorage dependent. These cells will only grow when attached to a solid substrate, typically flasks, bottles, perfusion reactors, and polymer beads. Anchorage-dependent cells must be adapted to growth in suspension. This adaptation generates cells with de-differentiated genetics and metabolism, which can undermine the purpose of the research. Tissues, excised

Table 17.1. Different cell culture types and their uses

Culture Type	Use	Refs.
Suspension (both batch and continuous)	protein production	5–7, 12, 16
Culture flask and roller bottle (batch)	protein production and metabolic research	13, 14, 17, 20
Perfusion reactor	protein production and metabolic research	23, 24
Perfused tissue	metabolic research	3, 4, 19, 25
Animal subject	metabolic research	18

from animal subjects, are usually investigated by perfusing them with specified buffers, e.g., Krebs-Henseleit buffer for the perfused rat heart [3, 4, 19]. The measurement of external fluxes in perfused tissues is done in a similar manner to perfused cell line cultures. An animal subject can also be considered as a perfused culture, although the measurement of nutrient usage is more complex [18].

17.2.1
Direct Measurement of Extracellular Production and Consumption Rates

Direct measurement of extracellular metabolites is the simplest route for flux quantification, because many extracellular metabolites exist in large amounts relative to intracellular metabolites. All that is required to determine a production or consumption flux (in addition to cell concentration measurement) is the measurement of the bulk concentration of an extracellular metabolite as a function of time. Historically, observing changes in these fluxes provided the first insights into metabolic regulation by distinguishing between altered physiologies induced by external stimuli. Further characterization of metabolic regulation, by determining the flux through internal pathways, cannot be achieved by measurement of external metabolites alone. The branched nature of intracellular pathways requires that additional measurements be made, e.g., ^{14}C- or ^{13}C-tracer measurements, or assays of extracted enzyme activities.

In the context of metabolic flux analysis, "external flux" refers to transportation of a metabolite across the membrane of individual cells. However, metabolic flux analysis is not limited to characterizing homogeneous populations of individual cells. Metabolic flux analysis has been applied to tissues perfused with medium [3, 4, 19, 25] and specific cells functioning within animal subjects [18]. In these cases, "external flux" describes the rate of exchange of a metabolite in the cytoplasm of a group of specific cells with the surroundings.

The external metabolites typically measured in cell culture are glucose, lactate, amino acids, ammonia, carbon dioxide, oxygen, fatty acids, sterols, protein products, fructose, pyruvate, and acetate. The nutrient composition of culture medium is typically designed to resemble mammalian blood plasma. Therefore, the consumption rates of cells in animal subjects are found by measuring similar external metabolites whenever possible. Because measured external fluxes only describe transmembrane uptake or excretion and not necessarily metabolic utilization [5, 7, 12], the effects of nutrient storage, e.g., glycogen production, must be considered for some cell lines.

External fluxes are calculated in a different manner for each of three different types of cultures: continuous suspension culture, perfusion culture, and batch culture. In both continuous and perfusion cultures at steady state, nutrients are steadily supplied and wastes are removed; thus concentrations in the bulk solution do not vary with time. The rates of production and consumption are dependent on the flow rates for delivery and removal. In batch cultures there is no replenishment of nutrients, so they are depleted over time, and wastes accumulate. Cells within animal subjects are treated like perfused cultures; it is assumed that tissues are supplied with a constant supply of nutrients by the blood.

17.2.1.1
Continuous Suspension Culture

Metabolic studies of suspension cultures are best performed using a chemostat [26]. At steady state, the specific flux of any extracellular metabolite (e.g., μmol glucose$\,hr^{-1}$ $(10^9\,cell)^{-1}$) can be determined from the following mass balance:

$$f_{i,ex} = \left| \frac{F(C_i - C_{i,o})}{V \cdot X} \right| \tag{17.1}$$

where F=volumetric flow rate through the chemostat, V=liquid volume of the chemostat, X=cell concentration (cells/ml), C_i=steady state concentration of metabolite i, (e.g., $mmol\,l^{-1}$), and $C_{i,o}$=feed concentration of metabolite i. In the context of metabolic flux analysis, a flux is defined as the rate of appearance of a reaction product and is therefore always positive. (Although a nutrient is consumed, and its flux is traditionally defined as negative, the corresponding transmembrane flux is always positive.) Hence the inclusion of the absolute value in Eq. (17.1).

17.2.1.2
Perfused Culture

Fluxes for perfused cultures are calculated in a similar manner as for suspension continuous culture [24]:

$$f_{i,ex} = \left| \frac{F(C_i - C_{i,o})}{N} \right| \tag{17.2}$$

where N is the cell number within the perfusion device/reactor. For some devices this value can be quite difficult to determine, and much effort has been made to calculate it accurately. For example, Mancuso et al. [27] described an NMR technique to determine the cell number within a hollow fiber bioreactor (HFBR) by measuring the number of sodium atoms excluded by the cells in a sodium rich medium.

An important consideration for perfused cultures is the effect of cell growth on cell number, N, which increases as a function of time when the system is not in true steady state. This issue can be avoided by allowing the cells to reach confluence, at which point cell growth essentially stops. Perfused tissues and animal subjects usually also have negligible cell growth. However, for non-confluent cell lines, it is generally assumed that during the long doubling time much of the nutrients are consumed for maintenance. The doubling time of many mammalian cell lines is often of the order of 1–2 days, much greater than the 20–60 min typical of bacterial systems. From kinetic tracer studies, most metabolic processes, with the exception of biosynthesis of large macromolecules, reach steady state after approximately 30 min. This two order-of-magnitude difference in time scales between growth rate and metabolic rate implies that, even in the presence of growth, the cell number can be considered constant with regard to metabolic reactions.

17.2.1.3
Batch Culture

Mammalian cells are most typically grown in batch. Nutrient concentrations in typical batch medium are high enough to maintain saturation in the rate-limiting steps of cellular transport and utilization (assuming that transport and the initial reactions of substrate utilization are regulated by enzymes that exhibit saturation kinetics, e.g., a glucose transport protein or hexokinase). Saturation can be confirmed by observing a linear concentration vs time profile of excreted and consumed metabolites in batch culture. Linear profiles lasting over 12 h have been observed for some cultures [28]. Specific fluxes can be determined from these linear profiles. Least-squares fitting is one method for determining the flux at a given point in time:

$$f_{i,ex} = \left| \frac{1}{N} \frac{dC_i}{dt} \right| \tag{17.3}$$

17.2.2
Detection of Isotope Distribution by ^{13}C-NMR

There are two levels of ^{13}C isotope identification in intracellular metabolites: the *fractional enrichment* of specific carbons and the *isotopomer distribution* of the entire metabolite. Fractional enrichment is the ratio of metabolite labeled at a specified carbon atom to the total metabolite concentration. A molecule with n carbon atoms has n fractional enrichments. An isotopomer is a metabolite with a specific labeling pattern. There are 2^n possible isotopomers in the molecule with n carbons[2]. Isotopomer analysis considers the position of label in multiply-labeled molecules, information that would be lost by fractional enrichment analysis. For example, a molecule labeled at the first and third carbons is a different isotopomer than one labeled at the first carbon alone. An experimentally useful property of fractional enrichments and isotopomer distributions is that for metabolites homogeneously compartmentalized within the cell, any portion of the metabolite will have the same enrichment or isotopomer pattern.

17.2.2.1
Measuring Fractional Enrichments

Portais et al. [14] described three different techniques to measure fractional enrichments, briefly described here in turn.

Calculation from Absolute Labeling and Metabolite Concentrations

One key advantage of NMR for determining label distributions is its ability to detect separate chemical species. In addition to the ability to distinguish different molecular species, NMR can distinguish the individual carbons on a given molecular species. Many carbons of metabolic molecules have distinct and identifiable chemical

---

2 As defined here, an isotopomer is an actual chemical species. A model based on isotopomer information will in fact use an *isotopomer fraction*, which is the ratio of isotopomer concentration to total metabolite concentration.

shifts in typical NMR spectra. Measuring the amount of label at a given carbon on a molecule distinguishes that molecule from among the ensemble of differently labeled molecules. Care must be taken when determining concentrations from NMR spectra, however, and various phenomena must be accounted for, including irradiation bandwidth [29], incomplete relaxation [14], and irregular NOE [14] or DEPT [29] enhancement.

The absolute label concentration is calculated by referencing to a standard. Total metabolite concentrations are determined by off-line analysis (enzymatic assay, HPLC, etc.). When the total metabolite concentration is not known (e.g., in in vivo experiments) only relative fractional enrichments can be measured. A relative fractional enrichment is the ratio between the concentrations of a molecule labeled at two different positions. However, relative fraction enrichments are not useful for metabolites with only one labeled and observable carbon. For multiply labeled species, one less relative fractional enrichment is available than the number of labeled and observable carbons. Much potentially useful information is obviously lost when metabolite concentrations are not known.

Determination by ^{1}H-NMR Spectroscopy

Purification of the metabolites by chromatography is usually necessary to discern the peaks of interest in the ^{1}H-NMR spectrum. Proton peaks will be split by nearest neighbor ^{13}C atoms. The ratio of the satellite peaks to the total proton peak is the fractional enrichment.

Mass Spectrometry

Portais et al. [14] also describe determination of the fractional enrichment of an aspartate carbon not detectable by NMR from the enrichment of specific ionization fragments.

17.2.2.2
Measuring Isotopomer Fractions

Determination of isotopomer fractions relies on another property of NMR spectroscopy. In a similar manner to proton splitting due to nearest ^{13}C neighbors, ^{13}C peaks will split in proportion to the amount of neighboring ^{13}C. Martin et al. [20] describe the deconvolution of three glutamate peaks into isotopomer distributions. An atom with one neighbor will appear as a triplet – the sum of a singlet and a doublet. The singlet represents label at the home carbon alone and the doublet corresponds to label at both the home and neighboring carbon. An atom with two neighbors will appear as a quintuplet, provided that the coupling constants for the two pairs are different. Most of the isotopomers can be determined by quantifying these peaks and satellites. However, carbons separated by more than one bond will not affect each other. Therefore, there are isotopomers not discernible by ^{13}C-NMR. This ambiguity must be accounted for when using isotopomer data in models of metabolism.

17.2.2.3
In Vivo NMR

In vivo and extraction NMR are two methods by which intracellular isotope distributions are determined. We define in vivo NMR as a technique whereby an isotope distributed throughout the metabolites of central metabolism is detected in living cells. In order for the cells to function normally, an experimental apparatus is needed to provide the cells with a controlled environment. Two common approaches are growth on microspheres [30, 31] and growth in hollow fiber bioreactors (HFBRs) [23, 24]. The microsphere method involves growing cells on agarose beads or some other suitable matrix within standard NMR tubes supplied with oxygenated growth media. The HFBR method similarly perfuses cells with oxygenated growth medium through the lumina of numerous (greater than 1000) porous fibers of a tubular reactor. The cells are grown in the extracapillary space of the reactor; solutes diffuse across the fiber. Reactors are scaled to fit within specially modified NMR probes.

An important advantage of in vivo NMR is that it can detect metabolites within living cells. Because cells are metabolically active throughout the analysis, label concentrations are derived directly from the actual concentrations within the cell. In addition, changes in the intracellular environment can be observed as they occur. In vivo NMR is one of only a few techniques able to monitor molecular events in living cells in real time. However, in addition to the complex apparatus required, it is necessary to increase cellular density and overcome the limitations of low intracellular metabolite concentrations. Another drawback is that only relative label fractions can be detected, because pool sizes are difficult to measure by NMR. The inability to detect absolute fractional enrichments limits the amount of data that can be provided with this approach.

17.2.2.4
Extraction NMR

On the other hand, extraction NMR, defined as the separation of specific cellular components from non-viable cells, is a much simpler procedure. It is possible to extract metabolites from cells grown on numerous types of growth supports, as well as from tissues excised from organisms [3]. The extraction method typically involves lysing cells with an acid solution (often perchloric acid) that precipitates nucleic acids, proteins, and triglycerides. The lysate is neutralized, lyophilized to remove cellular water, and reconstituted in D_2O for NMR analysis. Greater sensitivity can be achieved in extraction samples because it is easier to use more cells, and because metabolites contained within extracts can be concentrated by separation. A major drawback of extraction, however, is the possibility that the extraction process will affect the label distribution. This will, for example, occur if metabolites are contained in different cellular compartments. Likewise, isotope distributions between cellular compartments cannot be detected by in vivo NMR unless imaging techniques are applied. An additional disadvantage of extraction NMR is that measurements require the sacrificing of cells and thus only represent a single test condition.

17.2.3
Radio-Isotope Tracer Studies and Enzyme Activity Assays

Radio isotope tracer studies and enzyme activity assays are both classic methods of investigating metabolic pathways. Both of these methods have been recently combined in attempts to quantify whole cell metabolism [7, 12].

Radio isotope enriched substrates can be used to determine many cellular rates including the rate of glucose membrane transport, the rate of glycolysis, the rate of glucose carbon entering the TCA cycle, the rate of the pentose phosphate pathway, the rate of pyruvate carbon entering the TCA cycle, and the rate of glutamine oxidation [7, 12]. These assays are all based on time-course measurements of either $^{14}CO_2$ or ^3H_2O production. For each flux of interest, a substrate labeled at a specific carbon with ^{14}C is used. For example, the flux of glucose carbon entering the TCA cycle is determined by using $6\text{-}^{14}C$-glucose.

Neerman and Wagner [12] and Fitzpatrick et al. [7] described a method to determine the maximum activities of metabolic enzymes. This method first involves lysing cells with specific extraction buffers to maintain select enzyme activities. The activity can then be determined spectrophotometrically, at physiological conditions, and in the presence of excess substrates. For this method, an extensive list of enzymes must be investigated, since it is through comparison of the activities and comparison to the carbon flow that an understanding of cellular regulation can be obtained.

By combing the results from both of these methods, enzymes with activities closest to the measured pathway flux are designated as rate controlling. It is proposed that the flux through that pathway should be similar to the activity of the rate-limiting step. Enzymes with activities higher than the measured flux are termed "open" and have minimal control over the flow. However, making these designations assumes that only one enzyme is rate-controlling. Experiments with recombinant bacteria, investigated by metabolic control analysis [8], have shown that the concept of a single rate controlling enzyme in primary metabolism is not tenable.

For these types of analyses, it is assumed that CO_2 is mostly produced and consumed in a few reactions; the evolution rate of $^{14}CO_2$ provides fluxes directly without mathematical manipulation. If ^{14}C results were incorporated into a flux model, this assumption would not be needed, as more sites of CO_2 production could be considered. The *coupling* of ^{14}C-tracer studies and enzyme extraction to other methods of flux determination would no doubt give a more complete picture of metabolism and regulation.

17.3
Mathematical Descriptions to Quantify Fluxes in Metabolic Models

The central metabolism of most cell types contains branched pathways that recombine internally (e.g., the anaplerotic reactions, the pentose phosphate pathway, etc.), which makes measuring the uptake and production rates of external metabolites inadequate to characterize internal metabolic fluxes. Two techniques described herein to calculate metabolic fluxes, cometabolite measurement and ^{13}C-NMR, overcome this limitation. A metabolic model is then used to calculate internal fluxes from these additional measurements. By using metabolic models, neither of these methods requires the measurement of each internal flux explicitly, which is necessary with the $^{14}CO_2$-evolution technique.

Measuring the production of cometabolites, i.e., metabolites produced or consumed concurrently with other metabolites, provides additional constraints on the set of internal fluxes. The number of cometabolite measurements needed to solve for the fluxes in a given model can be found by assessing the degrees of freedom (DOF) within the system:

$$\text{degrees of freedom (DOF)} = n_{\text{unknown fluxes}} - n_{\text{metabolite balance equations}}$$
$$- n_{\text{measured fluxes}} \tag{17.4}$$

where $n_{\text{unknown fluxes}}$ is the number of fluxes (variables) in the model, $n_{\text{metabolite balance equations}}$ is the number of metabolites (equations) in the model, and $n_{\text{measured fluxes}}$ is the number of measured external fluxes (constraints), including those of cometabolites. To obtain a unique set of fluxes, the degrees of freedom must be greater than or equal to zero. Since $n_{\text{unknown fluxes}}$ and $n_{\text{metabolite balance equations}}$ are both defined by the system, the minimum number of measurements that have to be made, $n_{\text{measured fluxes}}$, is found when DOF=0 in Eq. (17.4). When there are no degrees of freedom (i.e., DOF=0) the system of equations is said to be "exactly determined." If no branches existed, then the number of unknown fluxes would equal the number of balances, and no cometabolite measurements would be necessary. Note that any measured flux could be considered another equation and combined with metabolite balances to form a group of "defining equations." This distinction is made to draw attention to the fact that one set of equations is inherent in the structure of the presumed pathway model and the other requires experimental measurement.

The ^{13}C-NMR technique provides additional information in the form of fractional enrichments. However, using fractional enrichments requires the inclusion of isotope or isotopomer balances, which adds unknown fractional enrichment variables to the calculation of the system's degrees of freedom:

$$DOF = n_{\text{unknown fluxes}} + n_{\text{unknown fractional enrichments}} - n_{\text{metabolite balances}}$$
$$- n_{\text{isotope balances}} - n_{\text{measured fluxes}} - n_{\text{measured fractional enrichments}} \tag{17.5}$$

Models based on tracer experiments inherently introduce non-linear equations in the form of isotope label balances, whereas cometabolite measurements do not. Thus, there is a clear association between linear models and cometabolite measurement, and non-linear models and tracer studies.

As an example, consider the sample pathway model in Fig. 17.1. It contains 25 unknown fluxes, 11 metabolite balances, and 7 measurable external fluxes, leaving 7 degrees of freedom. This implies that a minimum of seven fractional enrichments is necessary to close the system and render it determinable.

17.3.1
Determining Fluxes Using Cometabolite Measurements

Cometabolites are metabolites produced or consumed concurrently with other metabolites within cyclic reactions. How they can be used to determine intracellular fluxes is illustrated by the following example of Bonarius et al. [5] (Fig. 17.2). If only the external consumption and production rates of S, A, and B are measured (example I), internal fluxes f_{2-4} cannot be uniquely determined. However, if the rate of

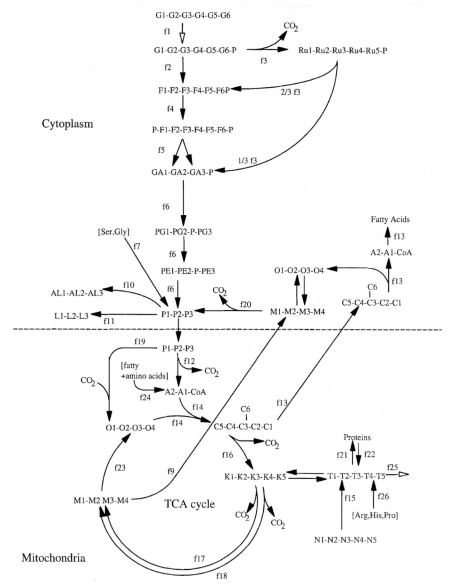

Fig. 17.1. Example pathway model. Abbreviations: A, acetyl-CoA; AL, alanine; C, citrate; F, fructose; G, glucose; GA, glyceraldehyde; K, α-ketoglutarate; L, lactate; M, malate; N, glutamine; O, oxaloacetate; P, pyruvate; PE, phosphoenolpyruvate; PG, phosphoglycerate; Ru, ribulose; T, glutamate

appearance of cometabolite C is also measured (example II), a unique solution for the internal fluxes can be found.

The cometabolites measured by Bonarius et al. [5] and Zupke and Stephanopoulos [6] were NADH, NADPH, O_2, CO_2, and NH_3. Due to the unmeasurable activity of

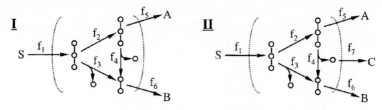

Fig. 17.2. Cometabolite measurement can determine intracellular fluxes

transhydrogenase, Bonarius et al. [5] lumped NADH and NADPH into one balance. Alternately, Zupke and Stephanopoulos [6] considered them as separate pools. Neither NADPH nor NADH is easily measured directly, but since all of the major pathways utilizing them are known, their balances are closed by the oxygen uptake rate (OUR).

Cometabolites are metabolites that participate in reaction pathways with multiple products or substrates. For our purposes, the cometabolites most useful to measure are those produced or consumed within cyclic pathways, that have accurately measurable fluxes, and that have closeable mass balances. For a cometabolite to have a closeable balance, all of the major fluxes that produce or consume the cometabolite must be included in the pathway model. Because Bonarius et al. [5] believed that there are too many biochemical reactions utilizing ATP to close its balance, these authors concluded that an ATP balance could not be used. In contrast, Zupke and Stephanopoulos [6] lumped all generic ATP utilization reactions into the biomass reaction, thus closing the balance.

17.3.1.1
Solution of the Stoichiometric Matrix

The linear model used to evaluate cometabolite measurements is made up of the three components in Eq. (17.4): unknown fluxes, metabolite balances, and measured external fluxes. The metabolite balances relate the unknown fluxes to the external fluxes [5]:

$$r_A = \sum_j \alpha_{j,A} f_j \qquad\qquad (17.6)$$

where r_A is the rate of change of metabolite A. The right hand side is the sum of the contributions of each pathway flux, f_j, to the flow through metabolite A. The reaction coefficients, $\alpha_{j,A}$, for pathway j and metabolite A have negative values if A is a substrate and positive values if A is a product. Typically, they have values of 1 or –1 although some reactions introduce fractional and integer values. Rates of change for excreted and absorbed metabolites are synonymous with $f_{i,ex}$ in Eqs. (17.1)–(17.3). For internal metabolites, the rates of change are assumed to be zero because of the pseudo-steady-state approximation (PSSA) [6].

Zupke and Stephanopoulos [6] showed that the PPSA is valid if the rate of change of a given metabolite's concentration is small relative to the flux *through* the metabolite. To illustrate this concept, consider internal metabolite M, which has flows in and out, f_{in} and f_{out}:

$$\xrightarrow{f_{in}} M \xrightarrow{f_{out}}$$

The differential form of the metabolite balance (Eq. 17.6) around M is given by

$$r_M = V(dC_M/dt) = f_{in} - f_{out} \qquad (17.7)$$

where V is the intracellular volume (assumed constant) and C_M is the intracellular concentration of M. An arbitrary tolerance, δ, can be defined and used along with the following criterion to determine whether the PSSA is valid:

$$V(dC_M/dt) < 1/2\delta(f_{in} + f_{out}) \qquad (17.8)$$

The right hand side of Eq. (17.8) is the average flow through M multiplied by the tolerance, δ. Zupke and Stephanopoulos [6] estimated the cell volume (V) and average metabolite flux (f_{in} and f_{out}) for CRL-1606 hybridomas to be 9×10^{-13} l/cell and $\sim 10^{-10}$ mmol/cell/h, respectively. These values indicate that dC_M/dt must be less than 10 mmol/l/h to satisfy the PSSA, if a tolerance, δ, of 10% is assumed. Most cellular concentrations are in the range 0–10 mmol/l with an average being approximately 1 mmol/l, which implies that the PSSA is generally true. If a metabolite has a concentration of 10 mmol/l and a rate of change of 10 mmol/l/h then that pool is changing from completely "empty" to "full" in the course of an hour. It is difficult to envisage a metabolite pool that a cell would completely cycle in an hour in the absence of an environmental stimulus. Given this, $V(dC_M/dt) \approx 0$, and Eq. (17.7) is reduced to $f_{in} = f_{out}$.

Equation (17.6) can be written in matrix form to express the entire system of equations. Solution involves the inversion of the sparse coefficient matrix $\underline{\underline{A}}$ [5, 6]:

$$\underline{\underline{A}}\underline{f} = \underline{r} \qquad (17.9)$$

$$\underline{f} = \underline{\underline{A}}^{-1}\underline{r} \qquad (17.10)$$

where $\underline{\underline{A}}$ is the stoichiometric matrix containing the reaction coefficients, $\alpha_{j,i}$. Each row in $\underline{\underline{A}}$ represents one metabolite balance; there is a column for each flux, and $\underline{f}$ and $\underline{r}$ are both column vectors containing all of the variable fluxes and the measured extracellular production/consumption rates, respectively. An exactly determinable system will have a square stoichiometric matrix, $\underline{\underline{A}}$.

If $\underline{\underline{A}}$ can be inverted, then the vector $\underline{f}$ can be obtained exactly. However, the inversion of large sparse matrices has been the focus of much study and is not necessarily straightforward. Therefore, the method of least-squares fitting is often employed to estimate $\underline{f}$. Various computational methods [32], including steepest descent, Gauss-Newton, Marquardt's method, etc., can be used to find the values of $\underline{f}$ that minimize the weighted sum of squares, F:

$$F = \sum_i \sum_j \frac{(\alpha_{j,i}f_j - r_i)^2}{\sigma_i^2} \qquad (17.11)$$

The squares of the deviations are weighted by the variance, σ_i^2, of the measurements, r_i.

17.3.1.2
The Objective Function

If fewer measurements are made than the minimum required to solve the system of equations for the assumed biochemistry, the system is underdetermined. To solve such an underdetermined system, Bonarius et al. [5, 33] applied an objective function [34, 35] to their stoichiometric matrix. Examples of objective functions include maximization of ATP production [36], maximization of NAD(P)H production [36], maximization of biosynthesis [36], and maximization of total cellular flux [33].

The appropriateness of such objective functions depends on the subsequent use of the resultant fluxes. If they are used for hypothesis testing of cells in differing states, care must be taken not to incorporate inadvertently the expected results into the objective function. For example, if the effect of different glutamine feed concentrations on the transport of NADH is being tested, it does not make sense to minimize or maximize the production of NADH. For cancerous cell lines, it may not be reasonable to claim that the cells are performing optimally for any criterion. If cell functions were optimal, then the cells would either not be cancerous or they would behave like bacteria. Clearly, mammalian cells reproduce more slowly and have slower metabolic rates than bacteria because of certain advantages this might provide the organism. The inclusion of any teleological argument potentially biases the outcome of any analysis or investigation. Is it possible to know how the *purpose* of a cancerous cell's existence differs from the *purpose* of a normal cell? Ideally, a measurement should report the state of a system without such assumptions.

17.3.2
General Principles of Isotope Balancing

17.3.2.1
Steady State Flux Analysis

In many ways ^{13}C-NMR is the most powerful of the techniques reviewed here. Steady state flux analysis is often unjustly criticized by claims that fluxes cannot be computed from non-transient fractional enrichments, since the relationship between steady state isotope distribution and pathway flux is not directly obvious. The simple illustration taken from Wiechert and de Graaf [22] in Fig. 17.3 demonstrates this relationship.

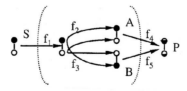

Fig. 17.3. Use of fractional enrichment data for flux determination. S and P are a consumed nutrient and an excreted product respectively; A and B are intracellular metabolites. The *circles* represent carbons; *filled ones* are isotopically labeled. The carbons of product P are *partially labeled*

External measurement alone can only determine the consumption flux f_1, which is equal to the production flux f_4 at steady state. Fluxes f_2 and f_3 remain unknown. However, reactions f_2 and f_3 scramble the carbons of S when producing the internal metabolites A and B. If the label distribution of P were measured, then fluxes f_2 and f_3 could be found.

The ratio of label in P directly relates to the ratio of the internal fluxes:

$$P1/P2 = f_2/f_3 \tag{17.12}$$

where P1 and P2 are the fractional enrichments of P on the first and second carbon, respectively. Notice that only *relative* flux information can be directly determined from fractional enrichment data. To find the absolute fluxes f_2 and f_3 it is necessary to measure also f_1 (re-emphasizing the statement that the combination of techniques is necessary for complete analysis). In this example the fractional enrichments, P1 and P2, sum to one. Combining this relationship with the defining material balance around S, $f_1 = f_2 + f_3$, and Eq. (17.12) produces:

$$f_2 = P1\, f_1 \tag{17.13}$$

$$f_3 = P2\, f_1 \tag{17.14}$$

This illustrates how the numerous branched pathways of central metabolism necessitate the use of techniques beyond direct excretion measurements. Because they recombine internally, the relative fluxes through many branches (such as f_2 and f_3 in Fig. 17.3) are not discernible by measured external fluxes. Evaluation of the label distribution following labeled nutrient infusion, as demonstrated, has the ability to determine these relative fluxes. While writing Eq. (17.13) from Fig. 17.3 is intuitive, a more formalized approach is needed for any realistic and thus more complex model.

17.3.2.2
The Isotope Balance

The construction of a *useful* mathematical description requires that the purpose of the model and its assumptions be well defined. Clearly, the goal is quantification of fluxes in a whole cell metabolic model. Here the initial information is a set of external fluxes and enrichment fractions. The external fluxes relate directly to the internal fluxes as seen in the discussion of cometabolite utilization and Fig. 17.3. The isotope balance is used to interpret the enrichment data.

Figure 17.4 shows two pathways that both produce metabolite C from metabolites A and B through reactions 1 and 2. Metabolite C is consumed by reaction 3.

In this example, the first carbons of A and C are both partially labeled and the first carbon of B is not. Intuitively, because of the dilution contributed by reaction 2, the first carbon of C is less enriched than the first carbon of A. The fractional enrichment of A on the first carbon, M_{A1}, is defined as

$$M_{A1} \equiv \frac{\text{conc. A labeled at position 1}}{\text{total conc. A}} \tag{17.15}$$

which can be rearranged as

$$M_{A1}\, C_A = L_{A1} \tag{17.16}$$

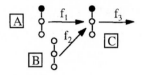

Fig. 17.4. An example to demonstrate fractional enrichment, label flux, and the label balance. Each *circle* represents carbon atoms in one of three molecules: A, B and C. The *darkened circles* are partially enriched by ^{13}C. Three-carbon molecules, of which there are only a few in central metabolism, were used to simplify the notation. The three carbon atoms of each molecule are numbered 1–3 starting at the top. This particular set of reactions represents the recombination of two pathways. Many of the metabolites in Fig. 17.1 are sites where pathways unite. Two examples of such sets of reactions are the combining of pyruvate and malate carbons in oxaloacetate through pyruvate carboxylase and malate dehydrogenase, and the combining of glutamate and citrate carbons in α-ketoglutarate through isocitrate dehydrogenase and glutamate dehydrogenase

where C_A and L_{A1} are the total intracellular concentration of A and the intracellular concentration of A labeled at position 1, respectively. The absolute labeling, L_{A1}, is measured by ^{13}C-NMR. The fractional enrichment, M_{A1}, is detected as split peaks in ^{1}H NMR following ^{13}C infusion (see Sect. 17.2.2.1). Taking the derivative of both sides with respect to time and dividing by the cell number shows how fractional enrichments relate metabolite fluxes, $f_{A \rightarrow C}$, to label fluxes, $\lambda_{A1 \rightarrow C1}$. Note that fluxes are by definition directional; hence, the designation of starting, A1, and ending, C1, positions:

$$M_{A1} f_{A \rightarrow C} = \lambda_{A1 \rightarrow C1} \qquad (17.17)$$

The initial and final metabolite notation can be dropped and substituted with the numeric flux notation of Fig. 17.4 with the following consideration: The flow of isotope through a unidirectional pathway is relative to the fractional enrichment of the initial and not the final metabolite, because of the effect of diluting flows into the product. For this reason it is also necessary to define the flux, $f_{A \rightarrow C}$, relative to the rate of change of the initial metabolite concentration. The numbered fluxes are normally defined as the specific rate of metabolite production, thus requiring the isotope balance to be scaled by the stoichiometric coefficient, α_{1A}:

$$M_{A1} \frac{f_1}{\alpha_{1A}} = \lambda_{A1 \rightarrow C1} \qquad (17.18)$$

Here α_{1A} is defined similarly to Eq. (17.6); it is the ratio of the moles of products formed to the moles of reactant consumed. There is only one flux, f_5, in Fig. 17.1 for which α_{1A} does not equal 1: two glyceraldehyde-3Ps are produced from each fructose-1,6biP.

Equations (17.16) and (17.17) demonstrate, albeit simply, how fractional enrichments relate both metabolite fluxes to label fluxes and metabolite concentrations to label distributions. This relationship permits the development of additional equations to reduce the degrees of freedom of the system (Eqs. 17.4 and 17.5) when fractional enrichment data are available.

Just as the mass of metabolites and carbon must be conserved, so must the mass of label. In Fig. 17.4 the total flow of label into C must equal the flow out:

$$\lambda_{A1 \to C1} + \lambda_{B1 \to C1} = \lambda_{C1 \to} \qquad (17.19)$$

(It is not necessary to define an end-point for the out flow of label. The in flows must balance *all* of the out flows.) Using Eq. (17.18), Eq. (17.19) can be written in the standard form:

$$A1 \cdot f_1 + B1 \cdot f_2 = C1 \cdot f_3 \qquad (17.20)$$

Here we write M_{A1} as A1 to simplify the subsequent equations. In the example in Fig. 17.4, two more balances can be written, one for each of the three carbon atoms in C.

There are 44 label balances in the model in Fig. 17.1 if only atoms through which label can pass after $1\text{-}^{13}C$ glucose infusion are considered. Note that these equations must contain the manner by which metabolic enzymes transfer carbons from metabolite to metabolite. Historical isotope (mostly ^{14}C) experiments determined the carbon transition patterns for all of the reactions in Fig. 17.1.

17.3.3
Least Squares Fitting of the Algebraic Form

Portais et al. [14] fitted their data to the complete set of label balances. The pathway model used by these authors was simpler than the model in Fig. 17.1; they reduced the number of label balances from 44 to 24. Many of the metabolites in the complete set of label balances are present at low levels within the cell, rendering their fractional enrichments difficult to measure. Therefore, these unknown fractional enrichments do not reduce the degrees of freedom in the system. In the approach of these authors, unmeasured enrichment fractions were left as variables, and the fitting algorithm was used to determine their values.

The procedure used by Portais [14] to solve for pathway fluxes can be summarized by the following steps:
1. Generate "differential equations describing time dependent variation" [14] of metabolites and ^{13}C labeled species. These equations are the same as the metabolite balances in the linear model description. Initially the accumulation term was retained so that transient data could potentially be incorporated
2. Apply the PSS assumption (Sect. 17.3.1.1), which sets all time derivative (accumulation) terms to zero. All of the differential equations generated in step 1 are now reduced to algebraic equations and are identical to metabolite and label balances
3. Use least squares fitting to "fit" the fluxes to the label fraction data
4. Calculate errors for the fluxes by the method of support planes as described in Sect. 17.3.7.2.

17.3.4
Atom Mapping/Transition Matrices

Zupke and Stephanopoulos [37] provide an alternative approach. The first step in this method is to arrange the enrichments of all atoms of each metabolite into vector form. The vectors have the same number of elements as each metabolite has carbons. In vector form a label balance is written for each *metabolite* rather than each *atom*. Any changes to the model can be made by rearranging the enrichment vectors in the matrix label balance equations. The use of vector notation reduces the number of

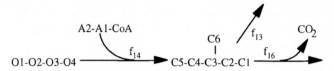

Fig. 17.5. Example of atom mapping matrices: balance around citrate extracted from Fig. 17.1

equations needed to describe the label flow in the model in Fig. 17.2 from 44 to 16. The information that is lost in reducing the number of equations is *how* the atoms move from one metabolite to another via enzyme activity. Individual label balances necessarily contain transition information. The form proposed by Zupke and Stephanopoulos compensates for this by the inclusion of atom mapping matrices (AMMs) [37].

AMMs describe the transfer of carbons from reactants to products and are generated using knowledge of specific biochemical pathways and enzyme reactions. AMMs are constructed such that the multiplication of the AMM by the reactant enrichment vector produces the product enrichment vector. For any reaction there is an AMM for each reactant-product pair. An example is provided by the balance around citrate in Fig. 17.1 as shown in Fig. 17.5.

For this example, there are three enrichment vectors:

$$\mathbf{OAA} = \begin{bmatrix} O1 \\ O2 \\ O3 \\ O4 \end{bmatrix} \quad \mathbf{AcCoA} = \begin{bmatrix} A1 \\ A2 \end{bmatrix} \quad \mathbf{Cit} = \begin{bmatrix} C1 \\ C2 \\ C3 \\ C4 \\ C5 \\ C6 \end{bmatrix} \tag{17.21}$$

Since there are two reactants and one product, there are two AMMs for this reaction:

$$[\mathbf{OAA} > \mathbf{Cit}]_{14} = \begin{bmatrix} 0 & 0 & 0 & 0 \\ 0 & 0 & 0 & 0 \\ 0 & 1 & 0 & 0 \\ 0 & 0 & 1 & 0 \\ 0 & 0 & 0 & 1 \\ 1 & 0 & 0 & 0 \end{bmatrix} \quad [\mathbf{AcCoA} > \mathbf{Cit}]_{14} = \begin{bmatrix} 1 & 0 \\ 0 & 1 \\ 0 & 0 \\ 0 & 0 \\ 0 & 0 \\ 0 & 0 \end{bmatrix} \tag{17.22}$$

Here the subscript 14 refers to flux f_{14}, the reaction of interest. Steady state isotope balances equations are formulated as the "sum of the products of the mapping matrices and the reactant (enrichment) vectors, weighted by the corresponding reaction flux" [37]. In the example in Fig. 17.5 the label balance around citrate is

$$[f_{13} + f_{16}] \mathbf{Cit} = f_{14} [\mathbf{OAA} > \mathbf{Cit}]_{14} \mathbf{OAA} + f_{14} [\mathbf{AcCoA} > \mathbf{Cit}]_{14} \mathbf{AcCoA} \tag{17.23}$$

An equation in the form of Eq. (17.23) can be written for each metabolite in the pathway.

Zupke and Stephanopoulos [6] applied this method to data from a hybridoma culture. They formulated a considerably simplified model pathway which contained only two branches. Two flux fractions were defined to describe the flow through each of the branches from the branch points. The hypothetical measurements were two relative fractional enrichments of lactate (L1/L2 and L3/L1). The isotope distributions for all possible flux fractions (from 0 to 1) were solved iteratively using the Gauss-Seidel method.

This is a reverse solution strategy; fluxes are provided as input data and the isotope distribution calculated. A graphical method was described to determine the fluxes from isotope data. The solution space (both relative enrichment fractions) was represented as overlaid contours on a plot with both flux fractions as axes. The intersection of the label ratio contour lines (determined experimentally) fixed both flux fractions. For the previous example there were two parameters and two solutions (the relative isotope distributions), so it was possible to analyze graphically. However, visualizing of more than three-dimensional space is difficult; this is the major drawback of using a reverse solution and graphical analysis.

It is not possible to use the AMM notation and determine fluxes from a label distribution without incorporating additional equations. Within each enrichment vector there can be both measured fractional enrichments and unknowns. In the approach of Portais et al., unmeasured fractional enrichments remained variables, and measured enrichments were used to compute the sum of errors squared. Any enrichment vector that is entirely measured or unmeasured can be treated in this fashion. However, any vector composed of both would have to be separated into its scalar elements. This effectively eliminates the elegance of the AMM approach.

17.3.5
Isotopomer Mapping Matrices

Given the ability to experimentally distinguish metabolite isotopomers, the inclusion of isotopomer distribution information into the mathematical description can more accurately characterize the fluxes than fractional enrichments alone. However, numerical manipulation of isotopomer information is cumbersome because of the exponential increase in the number of system variables; there are 2^n isotopomer states per metabolite containing n carbon atoms as opposed to n fractional enrichment states.

In contrast to fractional enrichment analysis, isotopomer distribution analysis must be viewed within an entire-metabolite framework. Balances cannot be written around each atom, since the state of neighboring atoms must be considered. And as with label balances, any set of balance equations would contain many more isotopomers than can be detected. A conservative estimate yields about fifteen detectable isotopomers in a mammalian cell system (it would be much higher for any bacterial system excreting detectable product) whereas there are over 600 total isotopomers in Fig. 17.1. In order to handle this computationally daunting problem a method similar to the atom mapping matrices has been employed [38]. The problem would be greatly simplified if measured isotopomers were deconvolute from unknown isotopomers to remove excess isotopomers from the problem.

Schmidt et al. [38] introduce a notation to describe the relation between isotopomers. The distribution of isotopomers in a metabolite is contained in *isotopomer distribution vectors* (IDVs) that, "in contrast to (enrichment) vectors, do not contain fractional enrichments at individual carbon atom positions but rather mole fractions of metabolite *molecules* that are labeled in a specific pattern [38]." Each isotopomer of a metabolite is represented by a binary number of length equal to its number of carbons. The labeled and unlabeled atoms are represented by ones and zeros respectively. When converted to a decimal number, each of the 2^n possible isotopomer is represented. Equation (17.24) is Schmidt's notation applied to acetyl-CoA (aca) in Fig. 17.5, which has two carbons and four possible isotopomers, is

$$\mathbf{I_{aca}} = \begin{pmatrix} I_{aca}(0) \\ I_{aca}(1) \\ I_{aca}(2) \\ I_{aca}(3) \end{pmatrix} = \begin{pmatrix} I_{aca}(00_{bin}) \\ I_{aca}(01_{bin}) \\ I_{aca}(10_{bin}) \\ I_{aca}(11_{bin}) \end{pmatrix} = \begin{pmatrix} \circ\!\!-\!\!\circ \\ \circ\!\!-\!\!\bullet \\ \bullet\!\!-\!\!\circ \\ \bullet\!\!-\!\!\bullet \end{pmatrix} \qquad (17.24)$$

The "bin" notation signifies that the preceding number is a binary number. Each element of an IDV is an isotopomer mole fraction, all of which sum to one:

$$\sum_{i=0}^{3} \mathbf{I_{aca}}(i) = 1 \qquad (17.25)$$

As with enrichment vectors, a formulation is necessary that describes the carbon transitions between the IDVs, in order to incorporate them into the metabolite model. For this purpose, Schmidt generated *isotopomer mapping matrices* (IMMs). Given a specific labeling pattern of reactants and knowledge of how carbons transition from the reactants to the products for the given reaction, the distribution of label in the products is determinable. Contrary to fractional enrichment transitions, it is possible for multiple substrate isotopomers to result in one single product. Schmidt pointed out that isotopomer mole fractions could be viewed as probabilities. In a reaction, the probability of one isotopomer form of a reactant combining with one isotopomer form of another reactant is equal to the product of their mole fractions. In Fig. 17.5 unlabeled acetyl-CoA and OAA would produce unlabeled citrate:

$$\mathbf{I_{cit}}(0) = \mathbf{I_{aca}}(0) \cdot \mathbf{I_{OAA}}(0) \qquad (17.26)$$

Acetyl-CoA labeled at the second position and OAA label at the second and fourth would produce citrate labeled at the second, third, and fifth carbon:

$$\mathbf{I_{cit}}(011010_{bin}) = \mathbf{I_{aca}}(01_{bin}) \cdot \mathbf{I_{OAA}}(0101_{bin}) \qquad (17.27)$$

The large number of possible isotope transition equations can be reduced by expressing them in matrix form. As with AMMs, columns and rows represent substrate and product labeling patterns, respectively. For reactions with multiple reactants and products, an IMM is needed for each substrate-product pair. An IMM for a reaction can be calculated if the AMM is known by systematically generating each possible reactant and determining the products. The IMMs are sparse matrices usually containing ones and zeros.

Reactions with single substrates are expressed mathematically in a manner similar to label balances written using AMMs. Figure 17.6 is pyruvate dehydrogenase [f_{12}] from Fig. 17.1.

The IDV for acetyl-CoA derived from pyruvate is given by

$$\mathbf{I_{aca}} = \mathbf{IMM_{pyr > aca}} \, \mathbf{I_{pyr}} \qquad (17.28)$$

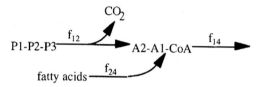

Fig. 17.6. IMM example: pyruvate dehydrogenase example from Fig. 17.1

where

$$IMM_{pyr>aca} = \begin{bmatrix} 1 & 0 & 0 & 0 & 1 & 0 & 0 & 0 \\ 0 & 1 & 0 & 0 & 0 & 1 & 0 & 0 \\ 0 & 0 & 1 & 0 & 0 & 0 & 1 & 0 \\ 0 & 0 & 0 & 1 & 0 & 0 & 0 & 1 \end{bmatrix}$$ (17.29)

The complete isotopomer balance around acetyl-CoA is therefore

$$f_{14}I_{aca} = f_{12}IMM_{pyr>aca}I_{pyr} + f_{24}\varnothing$$ (17.30)

where $\varnothing$ is a null vector with four elements; it assumes no label is present in fatty acids.

For multiple substrate reactions, the operator $\otimes$ must be introduced:

$$\otimes : \mathfrak{R}^n \otimes \mathfrak{R}^n \rightarrow \mathfrak{R}^n$$ (17.31)

where Schmidt et al. define the $\otimes$ operation as the "elementwise multiplication of two equally long column vectors [38]." The IMM-IDV product for a single substrate-product pair does not contain information about any other substrate. The elementwise multiplication of the IMM-IDV products scales the probability for each pattern of the product appropriately. The isotopomer balance around citrate in Fig. 17.5 is

$$[f_{67} + f_{16}]I_{cit} = f_{14}(IMM_{aca>cit}I_{aca}) \otimes (IMM_{OAA>cit}I_{OAA})$$ (17.32)

An isotopomer balance, like Eqs. (17.30) and (17.32), can be written around each metabolite and can be used in a manner similar to the balances written using AMMs. Schmidt's implementation used fluxes as input data and solved for isotopomer distributions in a manner similar to Zupke and Stephanopoulos's [6] implementation of AMMs. When NMR is used to measure the isotopomer distribution, the problem is reversed: metabolite fluxes must be calculated from the isotopomer balances. The IDVs of metabolites with greater than three carbons cannot be completely determined by NMR, since only nearest neighbors can be detected, e.g., a three carbon metabolite labeled on the first and third carbon can not be distinguished from those labeled on the first and third alone. Since no IDV can be completely determined, additional equations relating the known scalar components are necessary. This is a similar limitation to the use of AMM notation. However, in the isotopomer case where a large number of individual balances would be needed to describe the system, the use of matrix notation is a great simplification. While the addition of equations relating individual vector elements diminishes the elegance of a system entirely represented by matrices, their addition outweighs the cumbersome nature and potential inaccuracy of using individual isotope balances (e.g., the equations of Malloy et al. [19]).

17.3.6
Transient NMR Measurement

A classic approach to flux measurement using isotopic labeling is the detection of the transient appearance of the isotope throughout metabolic pathways. To obtain meaningful results from the transient response necessitates information be provided about the kinetics of the enzymes and/or the pool sizes of the metabolites involved in the reactions. Both of these quantities can be elusive. The kinetics of enzymes may change once removed from the cell; physiological levels of product inhibition and allosteric effectors can be difficult to simulate. Pool sizes are difficult to measure because metabolites are present in small concentrations and are distributed among various cellular compartments. Many methods have been employed to overcome these limitations, including direct measurement of pool sizes [3, 18, 21], and compilation of literature values for both pool sizes [4, 18] and kinetic constants [18].

17.3.7
Errors in the Determination of Fluxes

The process of flux determination discussed in the preceding section needs to be complemented by analysis of experimental errors. Knowledge of the error distribution for the resultant fluxes has four uses:

1. The accuracy of the flux results can be evaluated.
2. The value of alternative pathway models can be discerned based on predetermined physiological boundaries. Many models produce physiologically unreasonable results. Given that a physiological range of a flux can be defined (for most fluxes it is simply the requirement that they be positive), the errors can be directly calculated. From the set of probabilities for all the fluxes in a model, the probability that the model produces physiological meaningful results can be assessed.
3. The sensitivities of each flux to each measurement can be established. Sensitivities are useful in experiment design. The measurements with the greatest influence on the results must be performed the most accurately, or repeated to increase confidence.
4. A statistical significance can be assigned to observed changes in flux results between different test conditions. One means of calculating significance is the students-t test.

17.3.7.1
Errors in Linear Models

Random errors in experimental technique or gross errors introduced by incorrectly formulated pathways can only be detected in linear models if redundant measurements are made [6]. A redundant system is over-determined, having more measurements than the minimum required for the assumed biochemical pathways.

Within any organism there are more than 500 possible fluxes to consider. While it is imperative that this number be reduced to the number of measurements to find a solution, additional fluxes must be eliminated to allow for redundant measurements. In doing so, the advantages of error determination are offset by the disadvantage that fluxes contributing to the carbon flow are eliminated, making calculations less accurate. For the purposes of error analysis, Zupke and Stephanopoulos [6, 37] eliminated both the PPP and the anaplerotic reactions. Both of these pathways have been shown to be active in most cell lines and are critical to metabolic regulation. However, the flow of carbon through both is typically small.

With more system-defining equations than unknown fluxes, the stoichiometric matrix $\underline{\underline{A}}$ is not square; it has more rows than columns. Gaussian elimination of $\underline{\underline{A}}$ produces the permutation matrix $\underline{\underline{P}}$ which is defined such that

$$\underline{\underline{P}}\,\underline{\underline{A}} = \begin{bmatrix} \mathbf{I} \\ \varnothing \end{bmatrix} \tag{17.33}$$

where $\underline{\underline{I}}$ is the identity and $\underline{\underline{\varnothing}}$ is the null matrix. The flux vector $\underline{f}$ can then be determined in the presence of redundant measurements in a manner similar to the determination of $\underline{f}$ by Eqs. (17.9) and (17.10). Both sides of Eq. (17.9) are multiplied by $\underline{\underline{P}}$ to generate Eq. (17.34):

$$\underline{\underline{P}}\,\underline{\underline{A}}\,\underline{f} = \begin{bmatrix} \mathbf{I} \\ \varnothing \end{bmatrix} \underline{f} = \underline{\underline{P}}\,\underline{r} = \begin{bmatrix} \mathbf{S} \\ \mathbf{E} \end{bmatrix} \underline{r} \tag{17.34}$$

Equation (17.34) can be partitioned into two matrix equations; the first is used to determine $\underline{f}$ with the square upper portion ($\underline{\underline{S}}$) of $\underline{\underline{P}}$:

$$\underline{f} = \underline{\underline{S}}\,\underline{r} \tag{17.35}$$

Matrix $\underline{\underline{S}}$ can be considered equivalent to $\underline{\underline{A}}^{-1}$ in Eq. (17.10). The second portion of Eq. (17.34) defines the redundancy matrix $\underline{\underline{E}}$:

$$\underline{\underline{E}}\,\underline{r} = 0 \tag{17.36}$$

The number of rows in $\underline{\underline{E}}$ is equal to the number of redundant measurements (also the difference between the number of rows and columns in $\underline{\underline{A}}$). The product $\underline{\underline{E}}\,\underline{r}$ represents a series of equations that relate how various measurements, r_i, scaled by the coefficients of $\underline{\underline{E}}$, should sum to zero. Experimental errors and incorrectly formulated models shift the results of each redundancy equation in the product $\underline{\underline{E}}\,\underline{r}$ from this optimal result.

17.3.7.2
Errors in Non-linear Models

Either Monte Carlo simulations or the method of support planes can be used to determine the error in non-linear models. Both of these techniques are described by Chandler et al. [32] and Duggleby [39]. The Monte Carlo method involves the following steps:
1. The set of best-fit fluxes ($\hat{f}$) are calculated from the input data by minimization of the sum of squares.
2. Random data is generated from the measured values and their experimental errors. If the measured data set is assumed to be representative of all possible data sets, a random data set would be contained within a Gaussian or other distribution about the measured set.
3. The flux parameters are fit to the random data, using the best-fit fluxes ($\hat{f}$) as initial guesses.

4. Steps 2 and 3 are repeated M times. The more times the simulation is repeated the more accurate the calculated errors.

5. The standard error (E_i) of a particular flux parameter (f_i) is the root mean square of its deviations from best-fit:

$$E_i = \sqrt{\sum_{j=1}^{M} \left(f_{ij} - \hat{f}_i\right)^2 \Big/ M} \tag{17.37}$$

When the parameter space is flat around the minimum sum-of-squares in the direction of one parameter, the error in that parameter is large. Support planes, as described by Duggleby [39], are the positive and negative values of a given parameter (flux) which cause the minimum sum-of-squares value to increase by a factor (1+1/DOF), where DOF is the number of degrees of freedom in the metabolite and label balance equations (an exactly determined model has zero degrees of freedom, so support planes cannot be used). For linear systems, the (1+1/DOF) factor exactly determines the standard error. For non-linear systems (especially those that are linear in a parameter near the minimum) it is a reasonable approximation.

The method used to find the support planes is an iterative process. The chosen parameter is constrained to a guessed value. The remaining parameters are optimized to find the new minimum sum-of-squares. New guesses are made, and the process is repeated until the sum-of-squares equals (1+1/DOF) times the original minimum sum-of-squares.

17.4
Biochemical Pathway Model Formulation and Reduction

A comprehensive model of primary and secondary metabolism of mammalian cells is depicted in Fig. 17.7. All of the pathways considered in this review are included. Also, many of the pathways connected to common metabolites were included based on known biochemistry [1, 40]. It is necessary to eliminate many of these pathways to produce models useful for flux analysis.

Two assumptions were made to develop the model in Fig. 17.7:

1. All of the major pathways of central metabolism are known. This implies that the pathways have been identified in the literature or detected by direct measurement. Exclusion of major pathways significantly producing or consuming a given metabolite would invalidate the metabolite balance.

2. The pseudo steady-state approximation applies for each metabolite. This implies that the flow of total carbon (or any element) in is equal to the flow of carbon (or other element) out. Zupke and Stephanopoulos [6] make a very reasonable argument substantiating the PSSA, discussed in Sect. 17.3.1.1.

17.4.1
Reduction of Comprehensive Models

The "comprehensive" model of Fig. 17.7 cannot be used to convert ^{13}C-tracer and co-metabolite data into intracellular fluxes; it contains 98 fluxes and 45 metabolites, only 39 of which have closed mass balances. As such the system is under-determined and would require 59 additional measurements to solve for all intracellular unknowns. The deter-

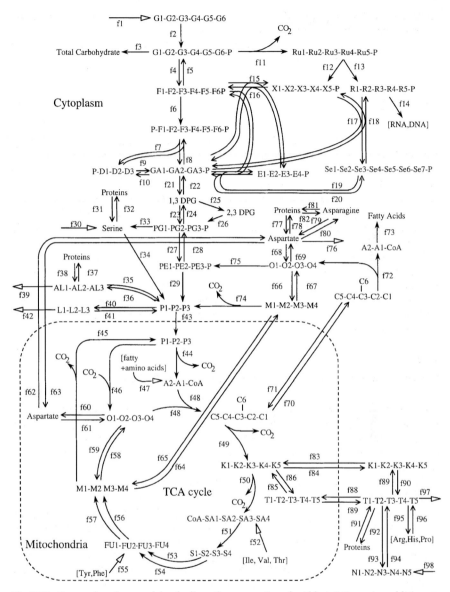

Fig. 17.7. Comprehensive model of all pathways reviewed. Abbreviations in addition to Fig. 17.1: D, dihydroxyacetone phosphate; DPG, diphosphglycerate; E, erythose; FU, fumarate; R, ribose; S, succinate; Se, sedoheptulose; X, xylulose

mination of cometabolite production rates, fractional enrichments, and isotopomers can only provide 18 additional measurements (the number is dependent on the extensiveness of the experiment). The intractability of this system makes it necessary to reduce the number of fluxes in the model. This is done by deciding which pathways are active in the particular cell line, and eliminating those with only small contributions.

The validity of various assumptions can be tested. Particular sets of data can be analyzed using a number of reduced models. For a given cell line and culture conditions, "physiologically reasonable" ranges for intracellular fluxes can be defined (e.g., flux through an irreversible reaction cannot have a negative value; the rate of the TCA cycle cannot be orders of magnitude larger than glycolysis, etc.). Simulations can be used to determine the probability that a given set of data and a given model produce a realistic physiological result.

The creation of a pathway model is not usually discussed from the standpoint of reducing a comprehensive model. A model containing all of the pathways known to exist in a mammalian cell would be quite daunting. Models are constructed in degrees of increasing complexity. The usual reasons for including pathways include:

1. Pathways are included that are obviously affected by a given measurement, e.g., the fractional enrichment of 4-^{13}C-glutamate following 1-^{13}C-glucose infusion is strongly related to pyruvate flow into the TCA cycle.
2. Pathways are included that are important for a hypothesis, e.g., pyruvate carboxylase (PC) is active. If the model containing the PC flux more accurately fits the data and has a plausible flux, then it is most likely active.
3. In ^{13}C-NMR experiments, pathways are often included that dilute the flow of isotope through the system, even if they do not affect the overall carbon flows, e.g., the exchange of amino acids into total cell protein. These pathways must be included because they affect the accuracy of other fitted fluxes.

17.4.2
Pathway Inclusion and Reduction Assumptions

Table 17.2 contains each of the pathways in Fig. 17.7 as well as past studies including and eliminating those pathways. Following the table are reasons for inclusion and arguments for elimination of pathways in various cell types and conditions. These arguments were used to reduce the comprehensive model in Fig. 17.7 to the simplified model in Fig. 17.1.

Several pathways, for the most part, are conserved in all of the models. They are glycolysis (or the Embden, Meyerhof, Parnas pathway), which is defined as the conversion of glucose to pyruvate starting with hexokinase (HK) and ending with pyruvate kinase (PK); pyruvate dehydrogenase (PDH); lactate dehydrogenase (LDH); and the TCA cycle, which includes the reactions from citrate synthase (CS) to malate dehydrogenase (MDH).

Some models do not include these pathways, usually because of special experimental conditions. For example, Malloy et al. [19] and Chance et al. [3] examined rat hearts perfused with ^{13}C-labeled acetate and pyruvate. In these cases no glycolysis or LDH activity would be detected and their inclusion in the model would not be relevant. Likewise, Schrader et al. [21] studied human erythrocytes, which do not have functioning TCA cycles when mature; PDH would not be active either.

Pyruvate carboxylase (PC) is the major anaplerotic enzyme. Its major function is the replenishment of TCA cycle intermediates. For steady or pseudo-steady state systems, PC would only be active if there was a depletion of TCA cycle metabolites. There are two sources of such depletions [1]: biosynthesis of fatty acids [f_{73}] and non-essential amino acids [$f_{78,81,92}$], and oxidation following conversion to pyruvate [$f_{74,75}$].

Table 17.2. Pathways included and eliminated in the literature

Pathway	Corresponding fluxes in Fig. 17.7.	Studies including pathway	Studies actively eliminating pathway
Pyruvate Carboxylase (PC)	f_{46}	5,12,14,19,23	4, 6, 18
Malic Enzyme (ME)	$f_{45,74}$	4,5,12,14,23,37	3, 18, 19
Aminotransferase	$f_{89,90}$; $f_{35,36}$; $f_{68,69}$ and $f_{85,86}$; $f_{60,61}$	3–5,12,14,18, 19,23,37	
Malate/aspartate shuttle	f_{58-69}	4	
Glutaminase	$f_{93,94}$	5,12,18,23	
Pentose phosphate pathway	f_{11-20}	5,12,14,21,23	37
Macromolecular synthesis	$f_{3,14,73}$	5	
Non-essential amino acids	$f_{30,39,76,97}$; $f_{33,34,79,80}$; $f_{31,32,37,38,77,78,81,82,91,92}$	5,14,23,37	
Fatty acid and cholesterol synthesis	$f_{72,73}$	5,14,23	
DNA and RNA synthesis	f_{14}	5	
Macromolecular carbo-hydrate	$f_{1,3}$	5,12,14,18,21	
Additional catabolic pathways	f_{47}; $f_{52,55,96}$	3–5,14,18,23	
Enzymatic channeling of TCA cycle intermediates	$f_{53,54,56,57}$	14	3,4,23,37
Phosphoenolpyruvate carboxykinase	f_{75}	12	
2,3 diphosphoglycerate bypass	f_{21-26}	21	

Hyder and co-workers [18] justified the elimination of PC because the rat brains they studied were incapable of growth and therefore had low requirements for biosynthesis. Chatham et al. [4] argued that the PC reaction, whose purpose is to increase TCA cycle intermediates, should not be active in their steady-state experiments. They claimed that an isotope distribution would not be affected by a small PC activity because of the rapid forward and backward rates of fumarase. The flow through PC is typically small, so it does not significantly affect carbon flows. This is not necessarily true in rapidly growing cells, however, where PC is more active [16] in order to supply biosynthetic pathways with more TCA cycle intermediates. The only other appreciable source of TCA cycle intermediates is the consumption of glutamine.

Malic enzyme (ME) has the opposite function of PC; it is the critical step in the oxidation of TCA cycle intermediates, converting malate into pyruvate. Anaplerotic reactions, including PC and catabolism of some of the amino acids (i.e., [$f_{52,55,61,94,96}$]), must be balanced by active ME or biosynthetic fluxes. Because malic enzyme is the only enzyme able to produce pyruvate from TCA cycle intermediates, it is included in almost every model.

An important function of ME has been described as the production of the NADPH that participates in de novo lipogenesis [41]. The association between malic enzyme and de novo lipogenesis has also been made because de novo lipogenesis relies on citrate transport out of the mitochondria [$f_{70,71}$], resulting in cytosolic OAA which may only be oxidized after conversion to pyruvate by ME [f_{74}].

In the works of Malloy et al. [19], Hyder et al. [18], and Chance et al. [3], no ME was included. Because Malloy et al. [19] included a PC flux, it is most likely that the overall mass balance did not close. In normal, non-cancerous cells, e.g., rat brains [18] and hearts [3], the consumption of glutamine is low and PC is inactive, so ME need not be considered.

There is some controversy about the location of ME, i.e., whether or not it is present within the mitochondria. Most models do not consider metabolite partitioning across the mitochondrial membrane [5, 12, 14, 37]. When these include ME, its location is unimportant. Portais et al. [14] include a generic flux described as "efflux from the citric acid cycle," which is most likely flux through malic enzyme.

The presence of mitochondrial compartmentalization is one of the two major differences between the metabolism of prokaryotic and mammalian cells (the other is the absence of the glycoxylate shunt). Chatham et al. [4] modeled ME as occurring inside the mitochondria [f_{45}] for rat hearts and Sharfstein et al. [23] modeled it outside [f_{74}] for hybridomas. It has been shown that the location of ME is tissue dependent [41, 42]. Newsholme and Leech [42] describe kidney metabolism where "malate [is] transported out of the mitochondria and is converted...to pyruvate via 'malic enzyme'," and malate is converted to pyruvate in the intestine by a "mitochondrial reaction catalyzed by a decarboxylating malate dehydrogenase (ME)." Biochemistry texts [1, 40] generally assume that malic enzyme is active only in the cytosol. An extramitochondrial ME, in contrast to an intramitochondrial ME, would have different effects on the flows through the malate/aspartate shuttle; the out flow of aspartate would be greater than the in flow of malate.

Aminotransferase (AT) reactions are all linked by the nitrogen mass balance [5]. Glutamate (glu) can be converted to α-ketoglutarate (α-kg) by three reactions: alanine aminotransferase (alaAT), aspartate aminotransferase (aspAT), and glutamate dehydrogenase (GDH). Glutamate dehydrogenase releases the glutamate amine as NH_3; alaAT and aspAT transfer it to pyruvate and OAA, respectively. When ammonia production is measured, the rate of GDH can be deconvoluted and a net flux through the transferases determined [5, 6]. In the absence of ammonia measurement, it is often assumed that the forward and back reactions are fast, allowing the amino and α-keto acids to be treated as a single pool [4, 14, 18, 19]. Because of the equilibrium between malate and OAA, a single pool of malate, OAA, and aspartate has been suggested [5].

The aminotransferase reactions are important for ^{13}C-NMR experiments. They provide data on the labeling of α-keto acids, all of which are centrally located metabolites and whose labeling data is required to solve most models. However, the concentrations of the α-keto acids are often too small to be detectable. If the aminotransferase reactions are assumed to be in rapid equilibrium, then the labeling pattern of the more abundant amino acids will be the same as that of their corresponding α-keto acids [19].

It is not necessary for fluxes f_{85} and f_{86} to be rapid for the relative labeling of glutamate and α-ketoglutarate to be identical. If the source of label is "upstream" of isocitrate dehydrogenase [f_{49}] then α-ketoglutarate is the only source of label for glutamate. If the system is at metabolic steady state, then any label transferred to α-ketoglutarate would also be transferred to glutamate. A slow transferase reaction between glutamate and α-kg is one possible way to explain the "dilution" of label in the TCA cycle. If the difference between the forward and back reactions [f_{85} and f_{86}] is large compared to the mean of the two values, then a reduction or "dilution" in the absolute label in glutamate would be observed.

The model of Chatham et al. [4] included the two sets of aspartate aminotransferases present in mammalian cells, an extramitochondrial [$f_{68,69}$; $f_{89,90}$] and an intramitochondrial [$f_{60,61}$; $f_{85,86}$]. Both of these aspATs are needed for a model to contain a complete malate/aspartate shuttle.

The *malate/aspartate shuttle* transports reducing equivalents (NADH) produced in the cytosol into the mitochondria. It was included in the model of Chatham et al. [4] because it is "the only mechanism for labeling the cytosolic glutamate pool from a very small mitochondrial α-ketoglutarate pool." A major effect of an active malate/ aspartate shuttle is that it provides a means by which TCA intermediates can reach the cytosol. Alternately, citrate can be transported by the tricarboxylate transport system [$f_{70,71}$] or malate can be transported by the malate/aspartate shuttle operating in reverse [f_{64}]. This has been called the malate shunt [23]. It has been proposed that cancerous cells have limited ability to transport reducing equivalents [43]. The malate/aspartate shuttle accounts for the oxidation of NADH produced under aerobic glycolysis, a function also performed by the glycerol phosphate shuttle [4].

Glutaminase is more active in transformed cells, which consume large amounts of glutamine. The reverse reaction [f_{93}] is glutamine synthetase. Hyder et al. [18] accounted for loss of glutamate label in the rat brain by an equilibrium between glutaminase and glutamine synthetase. Deamination of unlabeled glutamine would dilute label in glutamate.

The *pentose phosphate pathway (PPP)* is active in most cell types. However, there is some discrepancy in the literature about how much it contributes to the flux from glucose to pyruvate [44]. The magnitude of the flux has been shown to be dependent on the method of measurement used. Schmidt et al. [44] demonstrated that cometabolite measurement has a tendency to overestimate this flux, to the extent of not being physiologically feasible.

Many of the models treat the PPP as a single irreversible flux [5, 14, 23]. This treatment is plausible if it is assumed that the intermediates in the PPP are not involved in any other reactions of primary metabolism. Then the system of fluxes can be lumped together as a single flux as in the model of Fig. 17.1. When the DNA/RNA biosynthetic rate [f_{14}] is measured, this drain on ribose can be added [5]. Schrader et al. [21] extensively studied this system of pathways in human erythrocytes using kinetic NMR and various labeled glucose species. After treatment with 1-^{13}C glucose, these authors detected 1-^{13}C pentose 5-phosphate (P5P), which could only have been produced from 1-^{13}C fructose 6-phosphate (F6P). This implies that transketolase [$f_{16\&17}$] and transaldolase [f_{19}] are both reversible [$f_{15,18,20}$]. As with most reversible fluxes, treating them as net fluxes does not affect the general metabolite balance, but does affect the label distribution. Additionally, the rate through these reverse reactions is most likely very small compared to the glycolytic flux (about 10% of the total PPP flux [21]).

Zupke and Stephanopoulos [37] did not include the PPP. From literature values these authors expected it to be less than 5% of the glycolytic flux. Even in the presence of high PPP flux, the effect on carbon flow to pyruvate is small; equal fluxes through glycolysis and the PPP only produce 1/6 less pyruvate. The most significant effect is on the labeling distribution following 1-^{13}C glucose infusion, in which all of the isotope is lost as $^{13}CO_2$ in the phosphogluconate dehydrogenase reaction of the PPP. The PPP activity Schrader et al. [21] determined for human erythrocytes was similar to the assumption of Zupke and Stephanopoulos: 6.5–17%.

Macromolecular synthesis was shown by Bonarius et al. [5] to be necessary for models of fast growing cells. These authors measured cell composition changes and incorporated biosynthetic fluxes into their model. These measurements have been seen previously as "corrections" to the energy metabolism fluxes, since their values are often assumed to be small. However, this was not the case with the hybridoma line employed by Bonarius et al. in continuous culture [5]. Labeling experiments [14, 18, 23], which do not independently measure macromolecular synthesis, often include "generic" synthesis fluxes to explain losses in label.

Non-essential amino acids have two fates once synthesized: excretion and incorporation into total cellular protein. It is generally believed that some amino acids are excreted (usually alanine) by transformed cells to reduce the internal concentration of NH_3. Incorporation of the amino acids that are in exchange with the α-keto acids of central metabolism (alanine, aspartate, and glutamate) into proteins $[f_{37,38,77,78,91,92}]$ is an explanation for the loss of label in the TCA cycle often observed in ^{13}C-NMR experiments [14, 23]. It is generally assumed that the rate of incorporation (translation) is equal and opposite to the rate of production (protein degradation). This is a reasonable assumption in non-growing cells [14, 23] These fluxes cause label dilution because labeled amino acids are incorporated into proteins during the course of the experiment, and are replaced by unlabeled amino acids incorporated into proteins before isotope infusion.

From many ^{13}C-NMR experiments [14, 23], label appeared to be diluted in the TCA cycle in both an "overall" and a "per turn" manner. Overall dilution results from the loss of the total amount of label in the TCA cycle compared to the amount that should have entered via pyruvate dehydrogenase (PDH). Determination of the flux through PDH and pyruvate label enrichment show a dilution in the label enrichment of glutamate. This apparent "overall" dilution of label in the TCA cycle may result from the loss of glucose's first carbon in the PPP, slow aminotransferase reactions, uptake of unlabeled lactate, and catalysis of fatty acids.

The early model of Malloy et al. [19] did not contain protein exchanges and fatty acid catabolism. This model predicts that glutamate would remain equally labeled in the second and third positions. In practice, this is not observed. For label enrichments to be different, dilution fluxes from unlabeled metabolites and fluxes to undetectable macromolecules must enter and leave the TCA cycle as the label progresses from the third to the second position of glutamate on each turn. Hence the concept "per turn." There are two mechanisms that could diminish the "per turn" label in the TCA cycle: protein exchange of aspartate $[f_{77,78}]$ or of glutamate $[f_{91,92}]$. These two are experimentally indistinguishable without additional measurements. The values of some TCA cycle fluxes $[f_{48,49}]$ are slightly affected by the choice of location for this dilution [14].

Fatty acid and cholesterol synthesis are dependent on the tricarboxylate transport system $[f_{70,71}]$ which is the means by which acetyl-CoA is transported out of the mitochondria. The precursor for both fatty acid and cholesterol synthesis is cytosolic acetyl-CoA. It is somewhat misleading to think of citrate transport as directly exchanging mitochondrial for cytosolic acetyl-CoA because of the depleting effect it has on TCA cycle intermediates if the carbons return through ME $[f_{74}]$ and PDH $[f_{44}]$. Citrate efflux must be coupled with an anaplerotic reaction (either PC or glutamine catabolism) to maintain steady state, or the carbons must return via malate transport $[f_{65}]$. This synthetic flux is negligibly active in slow growing cells.

DNA and RNA synthesis both require ribose produced in the pentose phosphate pathway. This flux is often assumed to be small, even in rapidly doubling cultures. *Macromolecular carbohydrate* synthesis and degradation are responsible for most observed differences between glucose uptake and glycolytic flux. The most prominent macromolecular carbohydrate is glycogen, whose production can be measured [5] or used as a fitting flux in label experiments [14]. Membrane transport of glucose can be estimated from the literature [18], or measured using radio-labeled substrates [7, 12]. Transformed cells have such high rates of glycolysis that they are unlikely to store glycogen.

Additional catabolic pathways make small contributions to the metabolite balances, but they introduce unlabeled carbon that can dilute label enrichments. This second effect is more important because calculated fluxes are sensitive to relatively small changes in label enrichments. For this reason, many authors have included the breakdown of fatty acids and cholesterol into acetyl-CoA [f_{47}] [3, 4, 14, 18].

Measuring the amino acid composition in culture medium is a straightforward process. Catabolism of amino acids can be determined by direct measurement, in a manner similar to extracellular metabolites (see Sect. 17.2.1). Protein synthesis can be considered as an accumulation of amino acids. Caution must thus be exercised when attributing cellular uptake of essential amino acids to requirements for energy generation. In the presence of preferred energy sources (e.g., glucose or glutamine) alternate catalytic pathways are often suppressed (e.g., LDH). Therefore, it is likely that essential amino acid uptake contributes primarily to protein synthesis.

Enzymatic channeling of TCA cycle intermediates was invoked by Portais et al. [14] to explain observed label distributions. Srere [45] postulated that the enzymes of the TCA cycle are contained within a metabolon, which is a supramolecular complex bound to the inner surface of the mitochondrial membrane. Metabolites would be transferred directly from enzyme to enzyme, without being allowed to diffuse and rotate freely in the mitochondrial fluid. This is important for reactions involving succinate and fumarate, which are both symmetric molecules. With free rotation, label on the second and third positions of succinate would be equivalent as substrates for succinate dehydrogenase [$f_{53,54}$]. Sumegi et al. [46, 47] cite considerable evidence to support the metabolon concept and performed experiments demonstrating this "orientation-conserved transfer" [46]. These authors fed mammalian cells labeled substrates and observed the labeling pattern in specific metabolites. Two such experiments were 3-^{13}C propionate to alanine [47], and 5-^{13}C glutamate to aspartate [46]. In both cases, a preferred orientation of the product was detected (2-^{13}C succinyl-CoA resulted in only 2-^{13}C malate. If free rotation were possible 3-^{13}C malate would have been detected). Despite the evidence supporting this theory, many of the models still contain freely rotating succinate and fumarate [3, 4, 23, 37]. In Zupke's model "both succinate and fumarate are symmetrical, so that when fumarate is converted to malate the 1 and 4 carbons will have the same labeling [37]."

Phosphoenolpyruvate carboxykinase (PEPCK) is the transcriptionally controlled enzyme that regulates gluconeogenesis. Pyruvate kinase [f_{29}] is considered irreversible, which implies that flux through PEPCK is the only route for carbons to enter glycolysis and produce glucose. It is assumed to be inactive in most cells, except those derived from the liver, gluconeogenesis being one of the liver's major functions. In most transformed cell lines, high rates of glycolysis indicate that gluconeogenesis is not operational.

The *2,3 diphosphglycerate (2,3DPG) bypass* was included in the model of Schrader et al. [21] because the rate of label appearance in 2,3DPG determined the PPP flux. Being a cycle with only a single input and output, flux through the 2,3DPG bypass does not affect total carbon flow; nor does it affect label distribution.

The *determination of bi-directional reaction rates* is one obvious advantage of isotope labeling experiments coupled with non-linear analysis [22]. However, inclusion of both fluxes (forward and back) invariably requires that additional measurements be made, since no further metabolite balances are added. Reversible reactions can be represented by net fluxes without loss of accuracy. However, considerable information is contained in many of the exchange fluxes. As demonstrated in the discussion on aminotransferases, a positive net flux from glu to α-ketoglutarate [f_{85}] without any consideration of the reverse flux [f_{86}] precludes any label from reaching glu from α-ketoglutarate. If the exchange is assumed to be very rapid then the fractional enrichment of the two can be made equal. Alternately, the reverse flux could be included with no assumptions needed.

17.5
Observed Metabolic Flux Patterns in Mammalian Cells

It has long been known that cancerous cells have high rates of aerobic glycolysis, a characteristic typical of rapidly proliferating cells [9]. Aerobic glycolysis is inefficient, wasting much of glucose's energy potential. The magnitude of aerobic glycolysis is characterized by the molar yield of lactate from glucose ($Y_{lac/gluc}$). Neerman and Wagner [12] found that $Y_{lac/gluc}$ ranged from 1.31 to 1.56 in BHK and CHO cells. They also found that the amount of either pyruvate or glucose carbons entering the TCA cycle via PDH in BHK cells was below the detection limit of the $^{14}CO_2$-evolution assay. This observation was confirmed by the low activity detected for extracted PDH. In T47D breast cancer cells, $Y_{lac/gluc}$ was approximately 1.4 [28].

There are numerous hypotheses explaining why cancer cells survive and flourish with such apparently inefficient metabolic characteristics. In 1956 Warburg [10] postulated that the high cancerous rates of aerobic glycolysis were due to damaged respiration. Both Singer et al. [15] and Neerman and Wagner [12] found an elevated $NADH/NAD^+$ ratio in primary breast cancer cells and CHO/BHK cells compared to normal breast epithelial cells and insect cells, respectively. It was postulated that a high $NADH/NAD^+$ ratio drives the high production rate of lactate. Mazurek et al. [11] and Eigenbrodt et al. [43] postulated that the transport of reducing equivalents into the mitochondrion to drive oxidative phosphorylation is hampered. Mazurek also postulated that an accelerated glycolytic flux would increase the concentration of phosphometabolites upstream of pyruvate kinase; many of these metabolites are precursors for biosynthesis, and their increase would induce proliferation. All of these hypotheses can be further examined by flux analysis experiments.

17.5.1
Linkage of Glycolysis to the Tricarboxylic Acid Cycle

Regardless of which hypothesis most effectively explains aerobic glycolysis in cancerous cells, all of them involve a relationship between cytosolic gylcolysis and the mitochondrial TCA cycle. This relationship includes the transport of both reducing equivalents and carbons across the mitochondrial membrane.

Lia and Behar [48] postulate that in brain cells (neurons, astrocytes, and oligodendrocytes) the transfer of reducing equivalents into the mitochondria is the mechanism that mediates the coupling of glycolysis to the TCA cycle. These authors showed that in the presence of an inhibitor of aspAT (β-methylene-DL-aspartate), cortical slices have reduced oxygen uptake and increased lactate production [49]. This result implies that impairing the malate/aspartate shuttle induces aerobic glycolysis. By measuring the NADH/NAD$^+$ ratio, this hypothesis could be further confirmed.

In BHK and CHOs, Neerman and Wagner [12] did not detect any activity in the mitochondrial "linking" enzymes PEPCK, PC, and PDH. Without PDH activity, glucose cannot be oxidized, and without PEPCK and PC activity, carbon cannot be exchanged between glycolysis and the TCA cycle. In contrast, Portais et al. [14] found by ^{13}C-NMR that the oxidation of glucose produced 30% of the ATP in C6 glioma cells and claimed that oxidative phosphorylation constituted the major energy source for these cells.

17.5.2
Reducing Equivalents

Chatham et al. [4] included two reducing equivalent (NADH) transporters in their model of rat heart metabolism: the malate/asp shuttle and the glycerophosphate shuttle. As opposed to the malate/aspartate shuttle, which can be accounted for in a carbon flow model, the glycerophosphate shuttle only involves the cycling of dihydroxyacetone phosphate (DHAP) and 3-phosphoglycerol [1]. Therefore its activity cannot be detected by a carbon flow, and it can be considered a "generic" transporter, whose only observable effect is the direct transfer of cytosolic NADH to the mitochondrial inner membrane electron transport chain.

Chatham et al. [4] measured both isotopomer distributions and oxygen consumption; together these allowed the formulation of an NADH balance. When the glycerophosphate shuttle was not considered, the rate of the malate/asp shuttle was too high to explain the rate of label appearance in glutamate and it was too slow to account for the NADH produced by glycolysis. These authors concluded that the glycerophosphate shuttle had to be included in order for the NADH balance to close.

Bonarius et al. [5] detected active transhydrogenase in hybridomas by comparison of both the NADH and NADPH balances. They found that it operated in the direction producing NADPH and NAD$^+$ from NADP$^+$ and NADH. Transhydrogenase activity reduces the NADH/NAD$^+$ ratio, inducing glycolytic flux, and increases the concentration of NADPH, which is necessary for many biosynthetic reactions. In the literature there is some controversy about the activity of transhydrogenase; most authors consider it inactive. In mammalian cells further investigation of transhydrogenase is needed because of its key role in pathway regulation.

17.5.3
Glutaminolysis

Glutaminolysis is defined as the production of pyruvate from consumed glutamine. It is sometimes implied that the pyruvate is excreted as lactate, but there are many fates possible for pyruvate depending on its location within the cell. Cell lines that do not appreciably oxidize glucose are dependent on the oxidation of glutamine for energy. When glutamine carbons are consumed, they are first converted to the TCA cycle intermediate α-ketoglutarate by glutaminase and GDH. They are then oxidized to OAA by a truncated TCA cycle, producing reducing equivalents (2 NADH, and FADH$_2$) within the mitochondrion. Glutamine consumption thus has an anaplerotic nature, contributing carbons to the pool of TCA cycle intermediates.

There are three general fates for glutamine-derived OAA carbons:

1. They can be passed from the mitochondrion as aspartate where cytosolic malic enzyme can convert them to pyruvate for excretion as lactate or alanine. Pyruvate can also be transported to the mitochondria and oxidized, a route that requires PDH activity.

2. Alternatively, a mitochondrial ME could directly convert the OAA (after MDH) into pyruvate, which is then oxidized. As stated earlier, there is disagreement in the literature about the existence of a mitochondrial ME. Chatham et al. [4] did not observe any ^{13}C label in either alanine or lactate. They speculated that, because label was not present outside the mitochondrion, mitochondrial ME was active in rat hearts. It has also been shown that the level of mitochondrial ME activity correlates with the degree of de-differentiation in cancer cells [16]. Portais et al. [14] did not see appreciable levels of ^{13}C-2 lactate in C6 glioma cells and concluded that ME was not active. These authors speculated that this low ME activity was due to "a lack of enzyme or a low level of extracellular malate [14]."

3. Another consequence of the glutamine carbons is the fueling of the biosynthesis of cellular components derived from TCA cycle intermediates (non-essential amino acids or fatty acids).

Studies with U-^{14}C glutamine showed that BHKs [12] and hybridomas [7] oxidize 18% and 36% of consumed glutamine, respectively. This range of glutamine consumption is typical for most cultured cell lines. Both Neerman and Wagner [12] and Fitzpatrick et al. [7] reported little PDH activity, implying that a mitochondrial ME is active to oxidize the glutamine carbons. Extracted citrate synthase was observed to be 2- to 3-fold less active than aspAT, supporting the concept of a truncated TCA cycle producing aspartate from glutamine carbons. Most of the energy generated in the TCA cycle would be formed by the conversion of glutamine to aspartate. The truncated cycle would produce nine ATPs, whereas a full turn of the cycle produces 12. Fitzpatrick et al. [7] reported a consumption of glutamine in hybridomas half that of the glucose uptake rate. That translates to 55% of the cellular energy being derived from glutamine and 45% from glucose. Literature reports energy production from glutamine in cultured cell lines ranging from 30% to 98% [7] of that obtained from glucose.

17.5.4
Pyruvate Carboxylase

Baggetto [16] describes how glycolytic cancer cells preferentially excrete citrate from the mitochondrion. This citrate produces cytosolic acetyl-CoA through the action of ATP-citrate lyase, which fuels lipid and cholesterol synthesis. For cells to excrete carbons from the TCA cycle, an anaplerotic counter-balance is needed at steady state. This can arise through the action of either PC or glutaminolysis. Both Martin et al. [20] and Merle et al. [13] found that PC was twice as active in astrocytes as in granule cells. They used this result to explain the greater dependence on glutamine in granule cells.

17.5.5
Pentose Phosphate Pathway

Martin et al. [20] determined the flux through the PPP in astrocytes and granule cells to be 11% and 29% of the glucose flux, respectively. Bonarius et al. [5] observed much higher levels in hybridomas, ranging from 76% to 92% of the total glucose uptake. In the cells with 92% of the incoming glucose flowing through the PPP, a negative flux from F6P to G6P was calculated. It was speculated that these high values of the PPP flux were required to supply NADPH to the rapidly growing continuous hybridoma culture. NADPH is consumed primarily in biosynthetic reactions, and is therefore essential for rapidly proliferating cells. Prior to the work of Bonarius et al. [5], the range of the PPP flux relative to glucose uptake was given as 0.7% to 11% in hybridomas [5, 23].

Schmidt et al. [44] questioned the validity of this high PPP flux based on metabolite balancing alone. These authors used bacteria cultured on two different nitrogen sources as a model to demonstrate their skepticism. In a single culture, both the cometabolite measurements necessary for metabolite balancing and isotope tracers were employed. They reported the shift in PPP flux when the nitrogen source was changed from NH_3 to NO_3 to be 34% to 120% of the glucose uptake by metabolite balancing, and 53% to 60% by ^{13}C-NMR measurements. They attribute the discrepancy between these results to a dependence of metabolite balancing on an NADH/NADPH balance. It was assumed by Bonarius [5] that all of the reactions either producing or consuming NADH and NADPH were included in the model in order to close both balances. Transhydrogenase was also assumed to be activity. Neither of these assumptions is necessary in tracer experiments. Bonarius et al. [33] defend metabolite balancing, claiming that it is far simpler and less expensive for industrial on-line testing. Zupke and Stephanopoulos [6] found little difference between the fluxes determined by metabolite balancing and ^{13}C-NMR.

Schrader et al. [21] found a reverse flux of both TA and TK in the presence of a positive PPP flux in human erythrocytes. Using 1-^{13}C glucose, they detected 1-^{13}C lactate, which could only have been formed by rearrangement in the PPP. Of the total labeled lactate, 6% was labeled at the first position; this ratio approximates the relation of the reverse TK and TA flux to the total glycolytic flux.

17.5.6
Tumors as Nitrogen Sinks

Baggetto [16] claimed that GDH is not active in tumor cells, glutamate being preferentially converted to α-ketoglutarate by aminotransferases. GDH is strongly inhibited by GTP, as well as its enzymatic product NH_3, both of which are abnormally high in cancerous cells. The dependence of cancer cells on glutamine oxidation results in a high concentration of nitrogen inside the tumor, where it remains trapped, starving the host organism. Bonarius et al. [5] saw similar events in hybridoma culture. The GDH flux measured by these authors ranged from 0% to 1% of the glucose uptake rate. In their system, glutamine carbons were supplied to TCA cycle α-kg through aspAT and alaAT. The high concentration of intracellular NH_3 forced GDH to produce glutamate. Glutamate production has also been observed in T47D breast cancer cells [28]. Bonarius et al. [5] also saw a large proline synthesis flux. They attribute this to accumulation of glutamate within the cell and propose it as another means of reducing the intracellular NH_3 concentration. It is postulated that proline synthesis in hybridomas is a vestige of the intercellular proline cycle. In the presence of glutamate, NADPH, and ATP, proline is produced by blood cells, from which hybridomas are originally derived.

17.5.7
Oxidative Glycolysis in the Rat Brain

Hyder et al. [18] concluded that oxidative as opposed to non-oxidative glycolysis is the major energy source in the brain following cortical stimulation. This was dramatically demonstrated by the more rapid increase of 4-^{13}C glutamate in stimulated as opposed to resting rats. It was found that upon stimulation the rate of oxidative glucose utilization and rate of oxygen uptake increased 195% and 201%, respectively. These identical shifts in rates allowed the authors to conclude that the non-oxidative rate did not contribute to the increase in glucose usage.

17.6
Specific Uses of Flux Pattern Information

Within most discussions of metabolic flux analysis, a common question often arises: how is this knowledge about carbon fluxes through internal pathways of mammalian cells useful? A common accusation is that metabolic flux analysis is a solution without a problem. In other words, what do these lists of numbers representing flows of carbon through mammalian cells actually mean? Regardless of whether this criticism was valid in the past, it is now outmoded. Metabolic flux analysis exemplifies the recent change in how biological systems are analyzed: from a reductionist to a holistic approach. Many of the processes that control the functions of living cells can only be fully understood when viewed in the context of other cellular processes. While metabolic flux analysis by no means considers all cellular processes, it is one of the first techniques to model whole cells. As experimental and modeling techniques advance, models will soon account for more cellular processes, thereby increasing the accuracy and significance of the results.

Recent advances in metabolic flux analysis have mostly been generated by so-called metabolic engineers [50, 51] seeking explanations for the unexpected results of planned genetic manipulations. For example, it has often been observed that singular increases in expression of what were believed to be rate-limiting enzymes did not increase production of a desired metabolite. From metabolic control analysis [8, 52, 53] it was realized that many enzymes, not just one alone, control the flux of a given pathway. This discovery of the inaccuracy of the rate-limiting step concept was the driving force behind the development of the more holistic metabolic flux analysis. Because understanding and modifying an isolated pathway proved to be insufficient to increase product synthesis, it became necessary to understand how pathways interact. Metabolic flux analysis provides the means by which such interactions will be illuminated.

Metabolic flux analysis will continue to play an important role in discovering regulation mechanisms and new functions within mammalian cells. Coupled with the forthcoming automation and miniaturization of biological assays, flux analysis will enable re-evaluation of old biological questions in a new and holistic manner. The simultaneous comparison of multiple pathways will clarify intracellular regulation mechanisms. Initially, metabolic flux analysis will be critical to defining cellular phenotypes and interpreting newly discovered genetic information. The ability to quantify internal pathway fluxes enables flux analysis to define phenotypes in a way no other method can. Palsson [54] termed this new field phenomics.

In addition, there is new interest in understanding the modifications of transformed cells connected to their uncontrolled proliferation. While the concept that an altered metabolism is the cause of proliferation [9] is outdated, there is obviously a strong link between metabolism and growth rate. It is becoming more apparent from genomics research that metabolism comprises a large portion of cellular activity. As Edwards and Palsson point out, "a majority of the assigned open reading frames relate to metabolic functions" [55]. Causality notwithstanding, significant changes in the expression and function of the enzymes of central metabolism must surely be necessary for cancer cells to sustain increased and stable proliferation. Determining changes in metabolic fluxes will point to corresponding changes in gene expression.

Clearly, there are numerous potential benefits to obtaining a more detailed picture of metabolic regulation, for both industrial and medical applications. Industrially, knowing the factors that induce product formation would invariably improve process efficiency. And medically, altered isozyme expression in tumor cells, detected as changes in metabolic fluxes, may be novel drug targets. Baggetto [16] remarked about observed flux differences between normal and cancerous cells: "such metabolic particularities of cancer cells should stimulate researchers in finding specific drugs to reach and kill cancer cells." The exciting future for metabolic flux analysis is that it will uncover regulatory mechanisms within mammalian cells, which in turn will be used to develop novel drug treatments and enhance protein production.

References

1. Voet D, Voet JG (1990) Biochemistry. Wiley, New York
2. Matwiyoff NA, London RE, Hutson JY (1982) The study of the metabolism of ^{13}C labeled substrates by the ^{13}C-NMR spectroscopy of intact cells, tissues, and organs. In: Levy GC (ed) NMR spectroscopy: new methods and applications. American Chemical Society, Washington DC, chap 8
3. Chance EM, Seeholzer SH, Kobayashi K, Williamson JR (1983) J Biol Chem 258:13,785
4. Chatham JC, Forder JR, Glickson JD, Chance EM (1995) J Biol Chem 270:7999
5. Bonarius HPJ, Hatzimanikatis V, Meesters KPH, de Gooijer CD, Schmid G, Tramper J (1996) Biotechnol Bioeng 50:299
6. Zupke C, Stephanopoulos G (1995) Biotechnol Bioeng 45:292
7. Fitzpatrick L, Jenkins HA, Butler M (1993) Appl Biochem Biotech 43:93
8. Fell DA (1998) Biotechnol Bioeng 58:121
9. Warburg O (1929) Biochem Z 204:482
10. Warburg O (1956) Science 123:309
11. Mazurek S, Michel A, Eigenbrodt E (1997) J Biol Chem 272:1491
12. Neerman J, Wagner R (1996) J Cell Physiol 166:152
13. Merle M, Martin M, Portais JC, Schuster R, Canioni P (1995) J Chim Phys 92:1783
14. Portais JC, Schuster R, Merle M, Canioni P (1993) Eur J Biochem 217:457

15. Singer S, Souza A, Thilly WG (1995) Cancer Res 55:5140
16. Baggetto LG (1992) Biochimie 74:959
17. Portais JC, Merle M, Labouesse J, Canioni P (1992) J Chim Phys 89:209
18. Hyder F, Chase JR, Behar KL, Mason GF, Siddeek M, Rothman DL, Shulman RG (1996) Proc Natl Acad Sci 93:7612
19. Malloy CR, Sherry AD, Jeffrey FMH (1988) J Biol Chem 263:6964
20. Martin M, Portais JC, Labouesse J, Canioni P, Merle M (1993) Eur J Biochem 217:617
21. Schrader MC, Eskey CJ, Simplaceanu V, Ho Chien (1993) Biochim Biophys Acta 1182:162
22. Wiechert W, de Graaf AA (1996) In vivo stationary flux analysis by ^{13}C labeling experiments. In: Scheper T (ed) Advances in biochemical engineering. Springer, Berlin Heidelberg New York, p 109
23. Sharfstein ST, Tucker SN, Mancuso A, Blanch HW, Clark DS (1994) Biotechnol Bioeng 43:1059
24. Mancuso A, Sharfstein ST, Tucker SN, Clark DS, Blanch HW (1994) Biotechnol Bioeng 44:563
25. Fernandez CA, Rosiers CD (1995) J Biol Chem 270:10,037
26. Blanch HW, Clark DS (1996) Biochemical engineering. Marcel Dekker, New York
27. Mancuso A, Fernandez EJ, Blanch HW, Clark DS (1990) Bio/Technology 8:1282
28. Forbes NS, Clark DS, Blanch HW (1998) University of California, Berkeley (unpublished data)
29. Mancuso A (1993) PhD Thesis, University of California, Berkeley
30. Furman E, Rushkin E, Margalit R, Bendel P, Degani H (1992) J Steroid Biochem Molec Biol 43:189
31. Neeman M, Degani H (1989) Cancer Res 49:589
32. Chandler JP, Hill DE, Spivey HO (1972) Comput Biomed Res 5:515
33. Bonarius HPJ, Timmerarends B, de Gooijer CD, Tramper J (1998) Biotechnol Bioeng 58:258
34. Savinell JM, Palsson BO (1992) J Theor Biol 154:421
35. Savinell JM, Palsson BO (1992) J Theor Biol 155:421
36. Pramanik J, Keasling JD (1997) Biotechnol Bioeng 56:398
37. Zupke C, Stephanopoulos G (1994) Biotechnol Prog 10:489
38. Schmidt K, Carlsen M, Nielsen J, Villadsen J (1997) Biotechnol Bioeng 55:831
39. Duggleby RG (1980) Eur J Biochem 109:93
40. Stryer L (1988) Biochemistry, 3rd edn. WH Freeman, New York
41. Brdiczka D, Pette D (1971) Eur J Biochem 19:546
42. Newsholme EA, Leech AR (1983) Biochemistry for the medical sciences. Wiley, Chichester
43. Eigenbrodt E, Fister P, Reinacher M (1985) New perspectives on carbohydrate metabolism in tumor cells. In: Bietner R (ed) Regulation of carbohydrate metabolism. CRC Press, Boca Raton, chap 6
44. Schmidt K, Marx A, de Graaf AA, Wiechert W, Sahm H, Nielsen J, Villadsen J (1998) Biotechnol Bioeng 58:254
45. Srere PA (1990) Trends Biochem Sci 15:411
46. Sumegi B, Sherry AD, Malloy CR, Srere PA (1993) Biochemistry 32:12,725
47. Sumegi B, Sherry AD, Malloy CR (1990) Biochemistry 29:9106
48. Lia JCK, Behar KL (1993) Dev Neurosci 15:181
49. Fitzpatrick SM, Cooper AJL, Duffy TE (1983) J Neurochem 41:1370
50. Bailey JE (1991) Science 252:1668
51. Stephanopoulos G, Sinskey AJ (1993) Tibtech 7:392
52. Heinrich R, Rapoprt TA (1974) Eur J Biochem 42:89
53. Kacser H, Burns JA (1973) Symp Soc Exp Biol 27:65
54. Palsson BO (1998) Metabolic engineering II. Elmau, Germany
55. Edwards JS, Palsson BO (1998) Biotechnol Bioeng 58:162

Subject Index

A

abiotic phase 74, 86
accuracy 125
acetate balance 186
acetate formation 387
acetate limitation constant 191
acetate production 389
acetate uptake rate 191
acetic acid, formation 382
acetoin 241
acid titration rate 185
acid tolerance 370
activation 46
active part 366
activity, fermentative 279, 283
adaptive linearizing control 155, 156, 164
aeration 173
aerobic 61
aerobic batch 291
aerobic conditions 278
aerobic process 16, 51
age distribution model 101, 299, 308
μ/agitation controll 197
air sparging 247
airlift 293, 311
   reactor 10, 247
airlift tower loop (ATL) 34
   reactor 21
algebraic slip model 212, 227
allosteric activator 358
amino acid 359, 506
   non-essential 586
aminotransferase 584
ammonia 359
   balance 187
ammonium carrier 508
ammonium concentration 540
ammonium limitation 508
ammonium-assimilating enzymes 523
ammonium-assimilating pathway 509
AMMs (atom mapping matrices) 574
anabolism 45
anaerobic 58, 71
anaerobic conditions 278
anaerobic digestion 157, 160

anaerobic processes 17
anaplerosis 508, 522
anchorage-dependent cells 559
animal cell culture 158, 163, 164
arginine 363
Arrhenius-relation 72
Arrhenius-type function 395
arthrospores 413, 417, 420
artificial neural network 326, 335
aspartate familiy 510
aspartate kinase, feed-back resistant 541
asymptotic observer 154, 161
ATL (airlift tower loop) 34
ATL reactor 34, 36
   pilot plant 37-39
atom mapping matrices 114, 489, 529, 574
ATP consumption 361
ATP-citrate lyase 590
automation 167, 321
axial dispersion 261
   coefficient 36

B

Bacillus subtilis 241
back flushing 371
bacterial spoilage 368
Baker's yeast 37, 65
balance boundary 375
balance equation, isotopic 519
balance for the buffer 187
Balling formula 323
base titration rate 185
batch 25, 35, 80, 92, 283, 418, 422, 426, 560
batch cultivation 13, 63
batch culture 545, 562
batch kinetics 364
batch process 47
BC (bubble column) 21, 30, 34, 247
beer fermentation 321, 323
bimolecular reaction 529
biochemical engineering model 172
biodegradable ester 368
biofilm 54, 55, 90
bioinformatics 489
biological behaviour 173

Printing (Computer to Film): Saladruck, Berlin
Binding: H. Stürtz AG, Würzburg